Amino Acids, Peptides and Proteins

Volume 31

A Specialist Periodical Report

Amino Acids, Peptides and Proteins
Volume 31

A Review of the Literature Published
during 1998

Senior Reporter
J.S. Davies, *University of Wales, Swansea, UK*

Reporters
G.C. Barrett, *Oxford, UK*
A. Dutta, *AstraZeneca, Macclesfield, UK*
D.T. Elmore, *University of Oxford, UK*
E. Farkas, *Lajos Kossuth University, Debrecen, Hungary*
J.A. Littlechild, *University of Exeter, UK*
I. Sóvágó, *Lajos Kossuth University, Debrecen, Hungary*

ROYAL SOCIETY OF CHEMISTRY

ISBN 0-85404-227-X
ISSN 1361-5904

Published by The Royal Society of Chemistry
Thomas Graham House, Science Park, Milton Road, Cambridge CB4 0WF, UK

For further information see our web site at www.rsc.org

Typeset by Computape (Pickering) Ltd, Pickering, North Yorkshire, UK
Printed by Athenaeum Press Ltd, Gatehead, Tyne and Wear, UK

Preface

The task set for the authors in this volume was to survey the world's peptide/ protein literature to the end of 1998. The emphasis, as always, in this series has been to review as comprehensively as possible the specialist papers produced in each area, but the expansion in the volume of literature usually means that individual Reporters have to highlight the most significant work, at the expense of the routine. Regular readers of these annual volumes often have the chance of monitoring the ebb and flow of various specialities, and occasionally an area which has been the subjected of many annual progress reports experiences a quantum leap into fruition. The year under review experienced such a 'happening', in the culmination of the total syntheses for the antibiotic, vancomycin, by the groups of Evans and Nicolaou (see Chapter 4 for details). This is an important step for the design of new anti-bacterials, which will be vital for overcoming the growing resistance of the 'superbugs'.

As a discipline matures, early attempts to set out guidelines and standards of nomenclature and abbreviations, inevitably produce evolutionary modifications, and sometimes suffer slippage in standards. Editors and reporters are in a key position to observe and hopefully respond to such developments, so it is timely to refer to a recent publication by a past Editor of this series, Dr. John Jones, who has published a short guide (*J. Peptide Science*, 1999, **5**, 465) to peptide nomenclature.

Reviewing the core work for Chapters 1–4, has again been in the hands of four Reporters (Graham Barrett, Donald Elmore, Anand Dutta and John Davies), who have 'done it all before'. We also welcome back our two authors from Hungary (Etelka Farkas and Imre Sóvágó) who have contributed their biennial Chapter on 'Metal Complexes of Amino Acids and Peptides'. We missed Jennifer Littlechild's 'Current Trends in Protein Research' last year, but she is back in the fold with coverage of the literature for 1997/98. My grateful thanks are very much due to all these stalwarts of this series of Specialist Reports, as well as to the RSC publications staff who take our varied manuscripts and transform them into this final form. I hope all readers will make good use of the wealth of work between these covers.

John S. Davies
University of Wales, Swansea

Contents

**Chapter 3   Analogue and Conformational Studies on Peptides, Hormones
and Other Biologically Active Peptides   174**
Anand S. Dutta

Abbreviations

The abbreviations for amino acids and their use in the formation of derivatives follow in general the 1983 Recommendations of the IUB-IUPAC Joint Commission, which were reprinted as an Appendix in Volume 16 of this series. These are also published in:

Eur. J. Biochem., 1984, **138**, 9–37: *Int. J. Pept. Protein Res.*, 1984, **24**, after p.84; and *J. Biol. Chem.*, 1985, **260**, 1442.

A complete listing of the single-letter codes for amino acids appeared in the Abbreviations section of Volume 24 of these Reports, together with structures for the closely related BOP family of coupling reagents.

Chapter authors have been encouraged annually to include new abbreviations in their texts. With the ever increasing diversification in structures, lists of unusual abbreviations are periodically compiled. Some examples are listed below.

Abo	2-azabicyclo[2.2.2]octane-3-carboxylic acid
Abu	α-aminobutyric acid
A₂bu	2,4-diaminobutyric acid
ACCA	4-aminocyclohexanecarboxylic acid
εAhx	6-aminohexanoic acid
Aib	α-aminioisobutyric acid
Aic	2-aminoindane-2-carboxylic acid
A₂pr	2,3-diaminopropionic acid
Atc	2-aminotetralin-2-carboxylic acid
Ava	5-aminopentanoic acid
Aze	azetidine-2-carboxylic acid
Cha	3-cyclohexylalanine
Cpg	α-cyclopentylglycine
Cpp	1-mercaptocyclohexaneacetic acid, *or* β-mercapto-β,β-cyclopentamethylenepropionic acid, *or* Pmp (*below*)
cPzACAla	*cis*-3-(4-pyrazinylcarbonylaminocyclohexyl)alanine
Dab	2,4-diaminobutyric acid
Dap	2,3-diaminopropionic acid
Dbf	3-(2-dibenzofuranyl)alanine
Dip	3,3-diphenylalanine
Dph	α,α-diphenylglycine
Dpr	2,3-diaminopropionic acid
Gly(Ph)	phenylglycine
Har	homoarginine
Hib	α-hydroxyisobutyric acid

Hyp	*trans*-4-hydroxyproline
Iva	isovaline
Mpt	*trans*-4-mercaptoproline
1-Nal	3-(1-naphthyl)alanine
2-Nal	3-(2-naphthyl)alanine
Nap	β-(1′-naphthyl)alanine
Oic	octahydroindolecarboxylic acid
Opt	*O*-phenyltyrosine
3-Pal	3-(pyridyl)alanine
Pen	penicillamine
Phg	phenylglycine
Pip	pipecolic acid
Pmp	β,β-pentamethylene-β-mercaptopropionic acid, *or* Cpp (*above*)
Qal	3-(3-quinolyl)alanine
Qua	quinoline-2-carboxamide
Sar	sarcosine
Thi	β-thienylalanine
Tic	1,2,3,4-tetrahydroisoquinoline-3-carboxylic acid

1
Amino Acids

BY GRAHAM C. BARRETT

1 Introduction

The literature of 1998 relating to the amino acids is covered in this Chapter, which is based on the chemistry literature mainly, and on related biological studies. The format used in all preceding Volumes of this Specialist Periodical Report is adopted. Some economies, introduced last year to save space, are continued. These do not affect the depth of coverage, but there are fewer sub-headings so that some topics, previously grouped on their own, have been combined more economically with other material.

Literature coverage is based on information from *Chemical Abstracts* (Issue 10 of Volume 128 to Issue 9 of Volume 130 inclusive), and scanning the major Journals. The literature continues to expand. The accusation[1] that much of the primary scientific literature is of declining quality could be recast in less provocative language so as to become more acceptable (in the context of the present Chapter: the accusation could be 'too many fragmented reports, too many papers describing predictable outcomes of well-known reactions and obvious properties of amino acids'); but nevertheless, most of the new experimental detail published on amino acids is actually needed by researchers.

2 Textbooks and Reviews

A recent monograph deals with the synthesis of derivatives of amino acids.[2] Reviews covering synthesis include: uses of L-amino acids in synthesis,[3] uses of L-pyroglutamic acid for the synthesis of alkaloids and other natural products.[4] Further reviews covering synthesis, reactions and properties of amino acids are located in appropriate sub-sections of this Chapter.

Mycosporins[5] and excitatory amino acids,[6] occurrence and physiological role of D-amino acids,[7] and L-cysteine metabolism and toxicity[8] have been reviewed.

The nomenclature of amino acids and their approved abbreviations have been surveyed.[9]

Amino Acids, Peptides and Proteins, Volume 31

3 Naturally Occurring Amino Acids

3.1 Occurrence of Known Amino Acids – Among the non-routine topics that this Section covers, is the location of amino acids in extra-terrestrial samples, bones and fossils. The significance of amino acids as biomarkers indicative of life early in the Earth's history has been reviewed,[10] and the uncertainty in the use of ^{14}C data for dating bone samples through their γ-carboxyglutamic acid and α-carboxyglycine (*alias* aminomalonate) content has been emphasised.[11]

The other main area under this heading, the identification of known amino acids in live organisms, is similarly restricted here to non-routine examples: the high levels of L-DOPA in seeds of *Stizolobium aterrima*[12] and the presence of *cis*-3-hydroxy-N-methyl-L-proline in the South Australian marine sponge *Dendrilla*.[13]

Unusual known amino acids condensed with other compounds to give representatives of the usual families of natural products include D-proline-containing dioxopiperazines in the sponge *Calyx CF podatypa* for which a correction has been published.[14] N[ε]-[(R)-(1-Carboxyethyl)]-L-lysine in the form of its N[α]-(D-glucuronoyl) derivative in *Providencia alcalifaciens 023*,[15] γ-glutamyl-S-ethenylcysteine in seeds of the Narbon bean (*Vicia narbonensis*; breakdown products of this isopeptide are responsible for the repulsive odour associated with its germination),[16] and 1-aminocyclopropanecarboxylic acid in the *Streptomyces* sp. metabolite cytotrienin A,[17] provide further examples. Unusual modifications of common amino acids in peptides include D-tryptophan as a constituent of contryphans present in the venom of fish-hunting cone-snails *Conus radiatus*,[18] and 6-chloro-N-methyl-L-tryptophan and N-methyl-L-tryptophan as constituents of keramamides K and L respectively, from *Theonella* sp.[19] The novel indole alkaloid martefragin A (1) present in the red alga *Martensia fragilis*, is the result of an unusual *in vivo* elaboration process applied to N[α]-L-isoleucyl-L-tryptophan.[20]

3.2 New Naturally Occurring Amino Acids – Previously-known (2RS)-2-amino-4,5-hexadienoic acid and (2S)-2-amino-4-hexynoic acid accompany new natural products (2R)-2-amino-6-hydroxy-4-hexynoic acid and (2S)-2-amino-5-chloro-5-hexenoic acid in fruit bodies of *Amanita miculifera*.[21] A novel lysine

Three-dimensional features of molecules are depicted throughout this Chapter as follows: horizontally-ranged atoms and bonds and ring atoms are to be understood as being in the plane of the paper; substituent atoms and groups attached to these are to be understood to be ABOVE the page if ranged LEFTWARDS and BELOW the page if ranged RIGHTWARDS relative (2) has been identified in the Micronesian marine sponge *Axinyssa terpnis*.[22] Ascaulitoxin is an unusual phytotoxic bis(amino acid) N-glucoside (3) that has been isolated from *Ascochyta caulina*.[23]

3.3 New Amino Acids from Hydrolysates – As usual, this section collects papers that describe reports of new amino acids condensed with other compounds: into dioxopiperazines, *e.g.* mollenines A and B (4) from the sclerotoid ascostromata of *Eupenicillium molle*,[24] and into peptides: the potent trypsin inhibitor, dehydroradiosumin (5) from the freshwater cyanobacterium *Anabaena cylindrica*,[25] eurypamide A and three related cyclic tripeptides in the Palauan sponge *Microciona eurypa*, containing iodotyrosine and (2S,3S,4R)-3,4-dihydroxyarginine,[26] cyclolinopeptide CLX from linseed that contains N-methyl-4-aminoproline,[27] hibispeptin A (6) from the root bark of *Hibiscus syriacus*, that contains a novel homophenylalanine derivative (stereochemistry not yet defined),[28] and 3-amino-6-hydroxystearic acid in nostofungicidine from the terrestrial blue-green alga *Nostoc commune*.[29]

(4) Mollenine A   (5)   (6)

(2S,3R)-3-Hydroxy-3-methylproline is a new natural product, a component together with other unusual α-amino acids of the carcinoma apoptosis-inducing polyoxypeptin.[30]

A novel protein crosslink found in bovine dentin, together with dihydroxy-lysinonorleucine and hydroxylysylpyridinoline, consists of a pyrroleninone carrying three amino groups and three carboxy groups.[31]

4   Chemical Synthesis and Resolution of Amino Acids

This Section is subdivided so as to collect current papers describing new examples of applications of each of the major general amino acid synthesis approaches. These methods are mostly well-established, though many improvements are to be found in the small print of these papers. Some newly-introduced synthesis methods are described.

As in last year's Volume, syntheses of isotopically-labelled coded amino acids are not collected in a separate subsection, but are spread throughout the Chapter: [2]H, Refs. 98, 370, 599, 945, 946, 1034, 1097; ^{3}H, Ref. 1105; ^{11}C,

Ref. 161; ^{13}C, Refs. 122, 124, 126, 131, 599, 1098; ^{14}C, Refs. 123, 1105; ^{15}N, Refs. 122, 124, 131, 160, 235; ^{18}O, Ref. 599; ^{18}F, Refs. 1011, 1038; [128]I, Ref. 1035.

General reviews of synthesis methods applied to α-amino acids have appeared: asymmetric synthesis,[32] general strategies of synthesis of α-amino acids and α-methyl-α-amino acids,[33] use of sulfinimines in asymmetric synthesis,[34] 2-(α-aminoalkyl)thiazoles as masked α-aminoaldehydes,[35] and use of the Mitsunobu reaction.[36]

4.1   General Methods for the Synthesis of α-Amino Acids, Including Enantioselective Synthesis – *4.1.1 Amination of Alkanoic Acid Derivatives by Amines and Amine-related Reagents* – Descriptions of simple syntheses of common amino acids are to be found in the recent literature, and some of these emerge from the continuing fascination of prebiotic amino acid synthesis (Section 4.5). α-Phenylglycine, $H_3N^+CHPhCO_2^-$, is formed from the reaction of phenylacetic acid with Br_2 and NH_3,[37] and 3,5-dihydroxy-4-methoxybenzaldehyde leads on to the correspondingly-substituted phenylglycine through reaction with ammonia and toluene-p-sulfonyl cyanide (Strecker synthesis, see also Section 4.1.6).[38]

Palladium(0)-catalysed azidation of (2S)-1-ethoxycarbonylmethylidene-2-methylcyclopropanes is a notable feature of a route to (−)-(1R,2S)-norcoronamic acid (Scheme 1).[39] Similar treatment of the α-chloroester formed by m-chloroperbenzoic acid oxidation of sugar-derived dichloroalkenes (Scheme 2) makes use of a remarkably simple $C=O \rightarrow >CHCO_2Me$ procedure.[40]

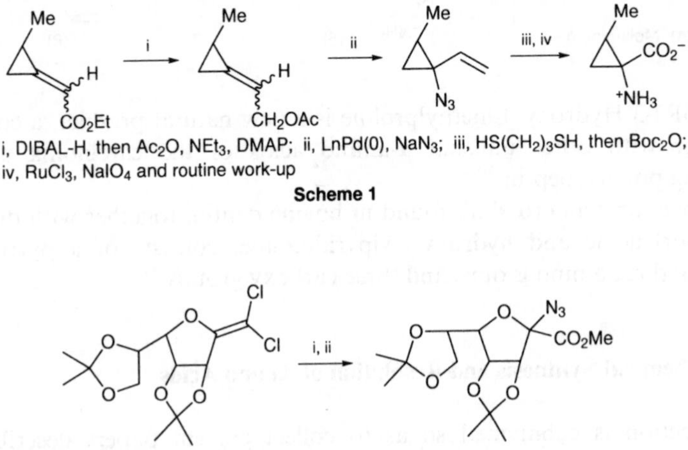

Reagents: i, DIBAL-H, then Ac_2O, NEt_3, DMAP; ii, LnPd(0), NaN_3; iii, $HS(CH_2)_3SH$, then Boc_2O; iv, $RuCl_3$, $NaIO_4$ and routine work-up

Scheme 1

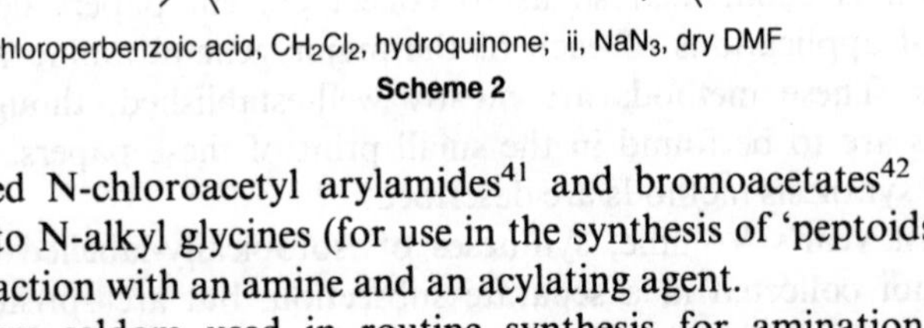

Reagents: i, m-chloroperbenzoic acid, CH_2Cl_2, hydroquinone; ii, NaN_3, dry DMF

Scheme 2

Resin-tethered N-chloroacetyl arylamides[41] and bromoacetates[42] have been converted into N-alkyl glycines (for use in the synthesis of 'peptoids') through successive reaction with an amine and an acylating agent.

Nitrenes are seldom used in routine synthesis for amination, potential hazards possibly being a deterrent, but low yields are another factor, seen in the $FeCl_2$-catalysed amination of ketene acetals by Boc-azide.[43]

Amination of aldehydes to give imines has long been a reliable method of introducing a nitrogen function as a substituent on a carbon chain, and in the special case of glyoxylic acid some useful amino acid syntheses have been established. One-pot processes starting with this step can be developed in a number of ways depending on other reagents; the formation of a ternary iminium salt (Scheme 3) from the intermediate aminal is a key step in the synthesis of syn,anti- or anti,syn-δ-hydroxy-α-amino acids.[44] Some inspired reasoning in mechanistic organic chemistry is needed to explain how *cis*-2-butenedial held at 435 K for 6 h in aqueous ammonia is converted into aspartic acid.[45]

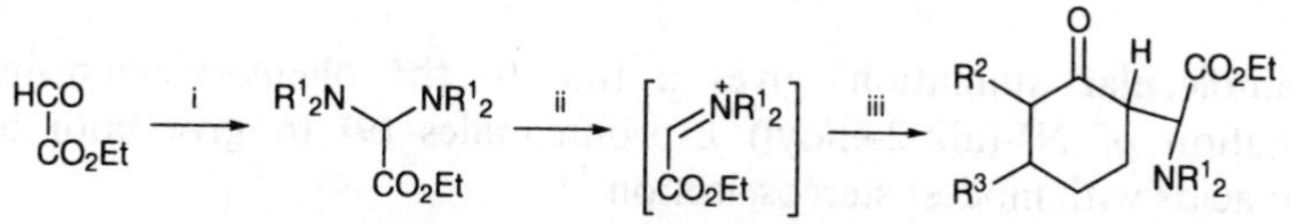

Reagents: i, R^1_2NH; ii, AcCl; iii, substituted cyclohexanone, reflux 3 h

Scheme 3

Homochiral bis(sulfinyl)oxiranes (7) are masked forms of ketones that can be aminated to give α-amino acid amides, announced as a new asymmetric synthesis.[46]

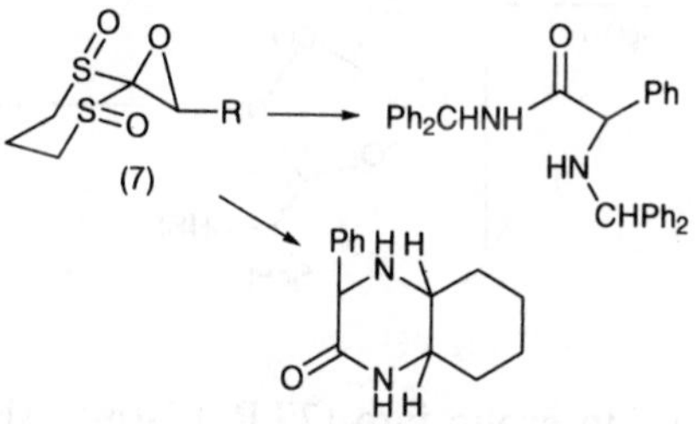

Amination of 3-nosyloxy-2-ketoesters with methyl carbamate and reduction of the resulting 4-alkoxycarbonyloxazolin-2-ones gives α-amino acid esters, but hydrogenation is sluggish.[47] Reductive amination of α-ketoesters represents a one-pot introduction of an alkylamino group without isolation of the intermediate imine, and treatment of simple pyruvates with $Na(OAc)_3BH$ and phenylethylamine,[48] and similar treatment of structurally complex homologues (protected α-D-galactohexodialdo-1,5-pyranoses → 8) has been described.[49] The trifluoroalanine analogue $CF_3CH(NHR^1)P(O)(OR^2)_2$ has been prepared in this way.[50] (p-Methoxybenzoyl)acrylic acid $PhCH_2COCH=CHCO_2H$ adds (S)-phenylethylamine to give the L-homophenylalanine derivative in poor diastereoisomeric excess (ca 10%), but gives enantiomerically pure product through equilibration, presumably involving dynamic resolution.[51]

β-Keto-esters undergoing α-oximation and reduction to the α-amino-β-keto-

(8)

ester stage, and then asymmetric hydrogenation, have been converted into syn- and or anti-β-hydroxy α-amino acids, *e.g.* Pht$(CH_2)_3$CH(OH)CH(NH_2)-CO_2H.[52] Exploitation of dynamic kinetic resolution in the last step led to a quantitative yield of the syn-isomer, (2S,3R)-3-hydroxylysine, after deprotection.

'Intramolecular amination' gives a title to the phenylselenium-induced lactamization of N^α-(alk-2-enoyl) L-prolinamides (9) to give both α- and β-amino acids with modest stereoselection.[53]

Introduction of the azido group into (2S,R_S)-1-(p-tolylsulfinyl)-butan-2-ol (NaN$_3$, PPh$_3$, CBr$_4$) and the equivalent Mitsunobu reaction with diethyl azodicarboxylate (see also Ref. 911), gives enantiomerically pure β-amino-alkanols after routine functional group changes.[54] Serine C-glucoside analogues that have an α-azido group masquerading as a protected amino group have been prepared from 1-vinyl-D-glucosides by stereoselective [2,3]Wittig rearrangement (Vol. 30, p. 12).[55]

The Evans approach to amino acids *via* amination procedures applied to N-acyloxazolidinones is illustrated in a synthesis of synthetically-useful ω-bromo-(2S)-azido acids (Scheme 4),[56] and a further example of the same approach, used many times by Hruby's group (Vol. 30, p. 10), giving four 2-amino-3,3-diarylpropanoic acids.[57] C-Linked isosteres of α- and β-glycocon-jugates carrying the serine moiety have been prepared by electrophilic amina-tion of the enolate of the Evans chiral oxazolidinone.[58] A study of the conversion of the azido function into the Boc-amino group using the Stau-

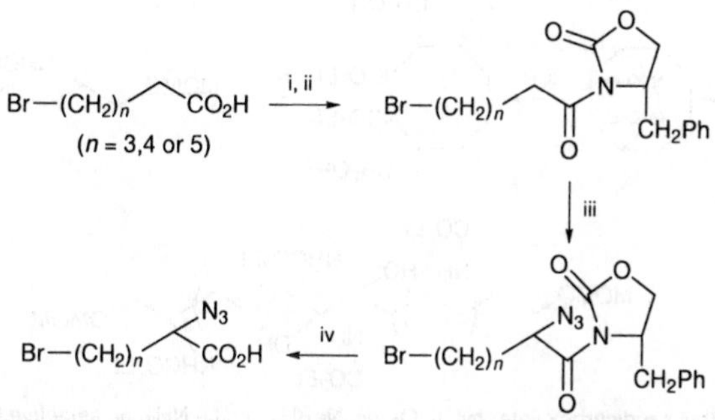

Reagents: i, Et₃N, Bu^{t}COCl; ii, Bu^{n}Li, (*S*)-(–)-4-benzyloxazolidin-2-one; iii, KHDMS, trisyl azide in AcOH; iv, LiOH, aq H_2O_2

Scheme 4

dinger reaction (PBu_3 in the presence of Boc_2O) has concentrated particularly on the side-reaction leading to ureas.[59] The work of Evans and Nelson (1997, Vol. 30, p. 6) based on magnesium bis(sulfonamide) complexes as catalysts for merged enolization and amination of N-acyloxazolidinones has been surveyed,[60,61] and a new cleavage method has been introduced that exchanges the chiral auxiliary for an alcohol moiety using lanthanum(III) iodide in an alcohol at room temperature.[62] New data on dynamic kinetic resolution of α-haloacyl imidazolidinones accompanying amination have been collected.[63]

Chiral 4-substituted-5,5-diaryloxazolidin-2-ones[64] and benzosultams[65] are readily N-acylated and can be used in the Evans way, azidation occurring with better than 95% diastereoselectivity and, no doubt, capable of improvement. For further applications of Evans methodology in the amino acids context see Refs. 142, 206, 335, 461, and 819.

A protected cyanohydrin, $ROCH_2CN$, has appeared to be an attractive starting point for αα-disubstituted glycine synthesis for many years, through double nucleophilic addition to the triple bond, and this reaction with Grignard reagents has now been found to be promoted by titanium isopropoxide.[66]

The enantioselective aminohydroxylation procedure improved recently by Sharpless has been shown to be a valuable stage on a route from alkenes to α-amino acids, *e.g.* conversion of styrenes into (R)- and (S)-N-benzyloxycarbonyl- or tert-butoxycarbonyl-arylglycinols,[67] and the use of other alkyl carbamates in this way,[68] and oxidation of the products to the corresponding arylglycine derivatives. Regioselection in aminohydroxylation of cinnamate esters can be reversed to give phenylserines if *Cinchona* alkaloid ligands with an anthraquinone core are used, or to give isoserines if the usual phthalazine ligands are used;[69] adenine N-chloro-N-sodio salts are also suitable.[70] The classic alternative route for aminohydroxylation of alkenes, *via* epoxides, is illustrated with aluminium azide as reagent.[71] Oxidation of homochiral α-amino-β-hydroxyalkanes with CrO_3 does not cause racemization.[72] Oxazolines, readily prepared from α-amino-β-hydroxyalkanes, have uses in enantioselective α-amino acid synthesis, and these uses have been reviewed.[73] Alkyl carbamates have been employed for aminohydroxylation of styrenes. Conju-

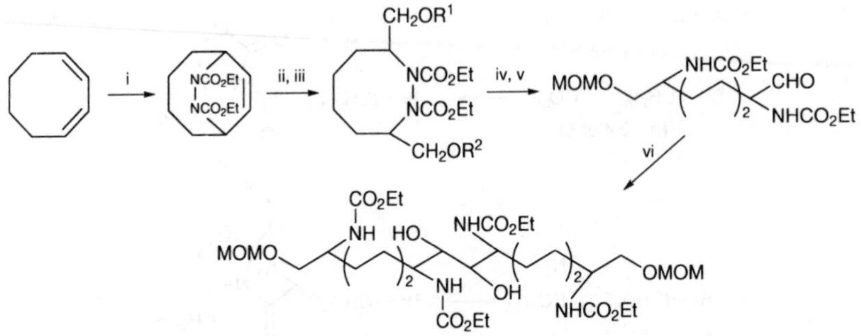

Reagents: i, Diethyl azodicarboxylate, *hv*; ii, O_3; iii, $NaBH_4$; iv, Li–NH_3; v, selective O-protection,
oxalyl chloride–DMSO; vi, pinacol formation using $[V_2Cl_3(thf)_6]_2$ $[ZnCl_6]$

Scheme 5

gated dienes can be aminated through cycloaddition to diethyl azodicarboxylate, a particular example studied this year being 1,3-cyclo-octadiene;[74] much
interest in this study lies in the development of the resulting adduct into
α-amino aldehydes and other useful products (Scheme 5). The bis(α-amino
acid)s formed through pinacol formation are particularly notable.

Nitroalkenes are also valuable sources of α-amino acids through nucleophilic addition routes, rendered enantioselective when attached to a homochiral grouping; a (+)-camphorsulfonamide gives (10) through addition to a
nitronate, and this is easily converted into the α-amino acid thiolester by
ozonolysis.[75] Further work has been reported, on addition of the potassium
salt of (R)- or (S)-4-phenyl-2-oxazolidinone to nitroalkenes, followed by
oxidative conversion into nitromethyl into carboxyl and oxazolidinone cleavage, giving D- and L-amino acids respectively.[76]

α-Ethoxycarbonylaziridines are conveniently prepared from αβ-unsaturated
esters, *e.g.* using $PhI=NSO_2Ar$,[77] and syntheses of α-amino acid develop from
these intermediates through a variety of ways; for example through nucleophilic ring opening as in the preparation of D-α-(3-phenylpropyl)glycine[78] and
D-homophenylalanine[79] using the aziridine (11) prepared from D-mannitol;
preparation of α-hydroxymethylserine derivatives (12) and (13);[80] and
D-phenylalaninol by hydrogenolysis.[81]

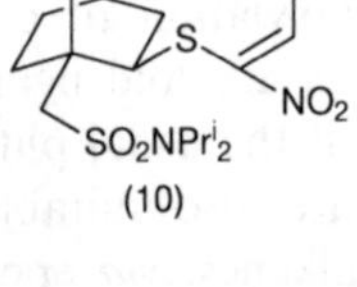

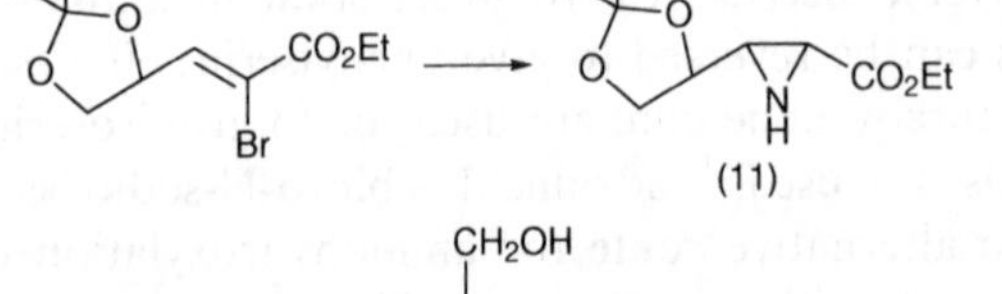

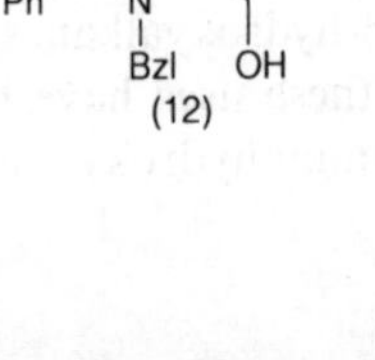

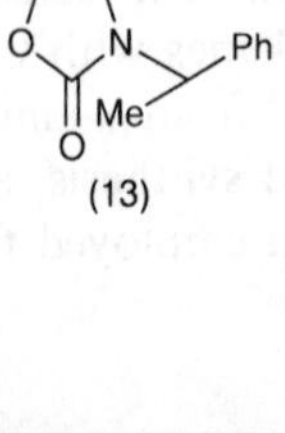

4.1.2 Carboxylation of Alkylamines and Imines, and Related Methods – Processing of N-TMS-N-aryl enamines (14; Scheme 6) gives αβ-dehydroamino acids and α-carboxyimines.[82] The insertion of imines into acyl-palladium bonds was achieved (Scheme 7), but the intended outcome, a new synthesis of amino acid amides, was not realized.[83]

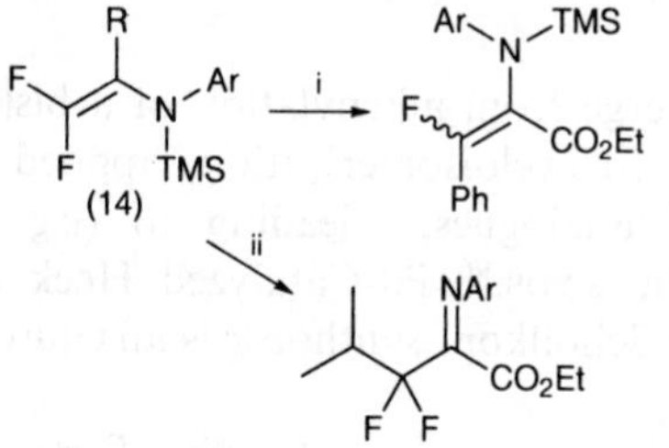

Reagents: i, PhLi (1.5 equiv.); ii, Pr^{i}I (10 equiv.), BEt$_3$

Scheme 6

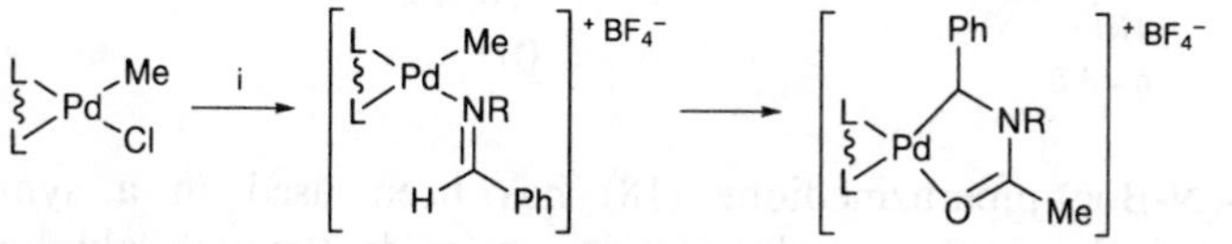

Reagents: i, CO (3,4 bar)

Scheme 7

α-Amino acid thiolesters R^{1}NHCHR^{2}COSPh can be obtained by trifluoro-acetic anhydride-induced Pummerer rearrangement of 3,N-disubstituted 4-phenylsulfinyl-β-sultams.[84] Homochiral toluene-p-sulfinylaziridines (15) have been carboxylated with ethyl chloroformate, the products being further developed, *e.g.* into N-phenyl S-α-methylphenylalaninate.[85]

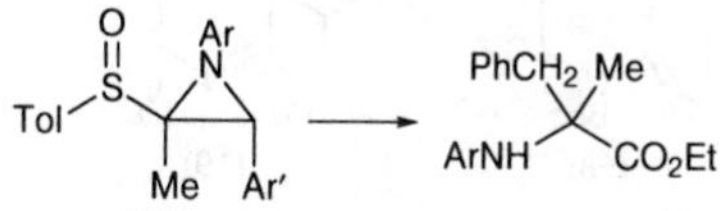

4.1.3 Use of Chiral Synthons in Amino Acid Synthesis – The use of the Schollkopf approach has continued on a broader basis, due to the exploration of modified homochiral dioxopiperazines for the purpose. Standard applications of the Schollkopf method include syntheses of 'αα'-bis(glycine)s', *i.e.* R^{1}O$_2$CCR^2(NHR^3)-R-CR^2(NHR^3)CO$_2$R^1 where -R- is an alkyl chain [R^2 = Me, R = H],[86] leading to (2R,5R)-2,5-diaminohexane-1,6-dioic acid,[87] (2S,7S)-2,7-diamino-octane-1,6-dioic acid, the dicarba-analogue of L-cystine,[88] and a di(hydroxylysine) analogue (16),[89] and syntheses of R- and S-p-(4-hydroxybenzoyl)phenylalanines,[90] 3',5'-dimethoxy-2-mercaptophenylalanines and heteroarylphenylalanines synthesized in support of an imbricatine synthesis,[91] 2,6-dibromo-L-tyrosine,[92] and 2-isopropenyl-L-tryptophan.[93] 1-Amino-

(16)

cycloalkenoic acids emerge from alkenylation of a bis(lactim) ether followed by ruthenium(I)-catalysed cycloisomerization, applied to 3-hydroxymethyl[94] and 2-alkenylcarbonyl analogues,[95] leading to (*e.g.* 17), and 1-amino-4-hydroxycyclohex-3-enoic acids.[96] Pd-Catalyzed Heck cyclization of bromodienes derived from the Schollkopf synthon[97] is an equivalent process.

The bis(N-Boc)-piperazinedione (18) has been used in a synthesis of [2,3-^{2}H$_2$]-labelled erythro-β-hydroxy-α-amino acids through aldol addition, elimination and catalysed deuteriation;[98] use of the standard Schollkopf methodology (no Boc protection on nitrogen) gives the corresponding threo isomers. Mono-N-Boc-piperazinone (19) is a new homochiral alanine template that has been advocated for the synthesis of α-methyl-α-amino acids, giving better than 94% diastereoisomeric excess.[99]

A use for ethyl (6S)-6-isopropyl-2,5-diethoxy-dihydropyrazine-3-carboxylate (20) in syntheses of α-alkylserines through alkylation with an alkyl halide[100] or with an aldehyde[101] provides a valuable new variation. These are mostly based on bis-lactim ethers derived from 3-isopropylpiperazine-2,5-dione (for which a reliable 45 g scale synthesis has been worked out, also a synthesis of the monomethyl ether for the first time[102]). Alkylation of a related mono-ethyl ether, N-[(S)- or (R)-α-phenylethyl]oxazinone (21) offers a satisfactory enantioselective route to α-methyl-α-alkylglycines.[103] A related derivative, (S)-3-isopropyl-N,N-di-(p-methoxybenzyl)piperazine-2,5-dione (22) has been shown to allow oxidative removal of the N-substituents, and to favour trans-alkylation leading to better than 90% diastereoisomeric excess;[104] similar use of this synthon has been illustrated in syntheses of both enantiomers of phenylalanine,[105] and in a broad range of similar applications.[106] N-(L-Prolyl)amino-

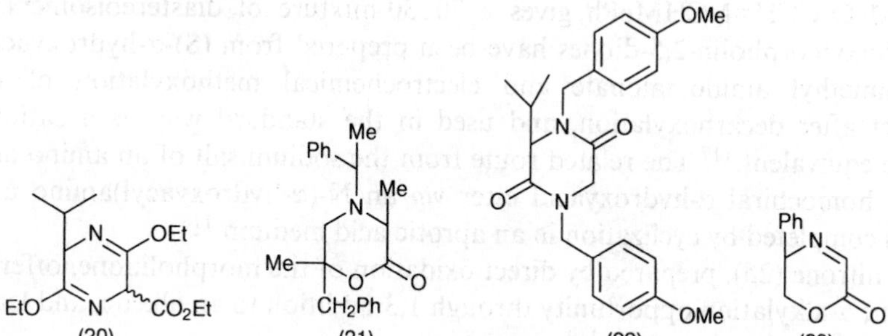

(structures 20, 21, 22, 23 shown above with labels (20), (21), (22), (23))

malonate is the source of methoxycarbonylpiperazin-2,5-dione which gives 3-methoxypiperazin-2,5-dione by electrochemical methoxylative decarboxylation; this is a useful new synthon whose potential has been demonstrated in a synthesis of D-allylglycine.[107]

Structurally-related homochiral morpholinones (e.g. 23) can be synthesized carrying a range of substituents, due to the easy availability of β-amino-alkanols used for their preparation. Standard methodology is illustrated in Lewis acid-catalysed alkylation by Grignard reagents,[108] and in synthesis of α-methyl-α-amino acids from the 3-methyl-5-isopropyl analogue.[109] A broader use of (5S)-phenylmorpholin-2-ones lies in prior conversion into iminium salts (24) formed with aldehydes (Scheme 8).[110] In the latter case, excess aldehyde

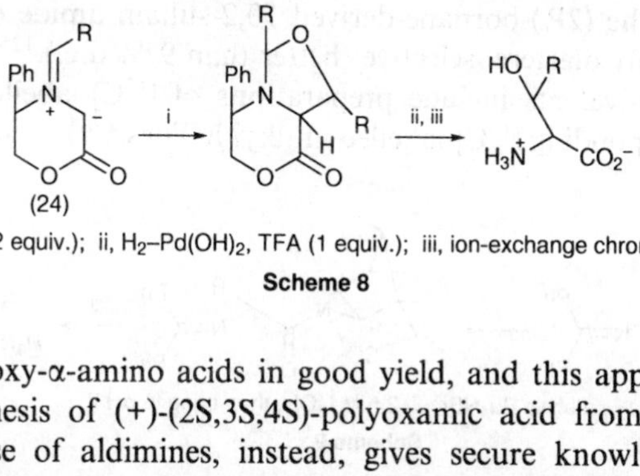

(reaction scheme with structure (24) shown)

Reagents: i, RCHO (2 equiv.); ii, H₂–Pd(OH)₂, TFA (1 equiv.); iii, ion-exchange chromatography

Scheme 8

leads to β-hydroxy-α-amino acids in good yield, and this approach has been used in a synthesis of (+)-(2S,3S,4S)-polyoxamic acid from (S)-glyceraldehyde.[111] The use of aldimines, instead, gives secure knowledge of stereochemistry for the product of cycloaddition; the cleavage product of an adduct formed in this way is a threo-(2S,3R)-3-aryl-2,3-diaminoalkanoic acid.[112]

(5S,6R)-N-Benzyloxycarbonyl-5,6-diphenylmorpholin-3-one (see also Ref. 911) has been applied to a preparation of (S)-α-methylasparagine through alkylation using a glycine enolate (see Section 4.1.7) as reactant,[113] and through another standard approach, aldolization, to give diastereoisomeric β-hydroxpipecolic acids.[114] Standard practice with this synthon has also been illustrated in a synthesis of p-(carboxydifluoromethyl)-L-tyrosine.[115] A related 5,6-disubstituted morpholinone has been used in an interesting study of alkylation via alkyl radicals at −40 °C, a process that shows complete stereocontrol; in contrast, the cycloaddition of this synthon to the homochiral imine

(E,S)-MeO$_2$CCH=NCHMePh gives a 70:30-mixture of diastereoisomers.[116] 3-Methoxymorpholin-2,5-diones have been prepared from (S)-α-hydroxyacids and dimethyl aminomalonate and electrochemical methoxylation of the product after decarboxylation, and used in the standard way as a cationic glycine equivalent.[117] The related route from the sodium salt of an amino acid with a homochiral α-hydroxyacid ester *via* an N-(α-hydroxyacyl)amino acid ester is completed by cyclization in an aprotic acid medium.[118]

The nitrone (25), prepared by direct oxidation of the morpholinone, offers a different α-alkylation opportunity through 1,3-addition to an alkene, and leads to γ-oxygenated amino acids.[119]

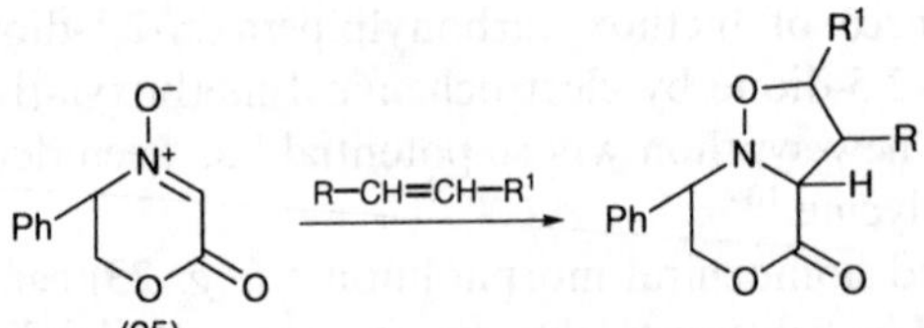

(25)

The Oppolzer camphorsultam has many supporters, new and old, who have applied it for enantioselective syntheses of N-Boc-N-methyl-(p-benzoyl)-L-phenylalanine,[120] highly-fluorinated L-(S)-2-amino-3-(7-methoxycoumarin-4-yl)propanoic acids,[121] [1-¹³C]-, [2-¹³C]-, and [¹⁵N]-labelled L-amino acids,[122] D-[3-¹⁴C]phenylalanine,[123] and L-[¹³C, ¹⁵N]prolines.[124] Phase-transfer catalysis of alkylation of the (2R)-bornane-derived 10,2-sultam amide (Scheme 9) has proved to be highly diastereoselective (better than 97% d.e.).[125] Other uses for chiral glycine equivalents include preparations of [¹³C]-labelled L-β-arylalanines from corresponding [¹³C]-labelled aralkyl halides.[126]

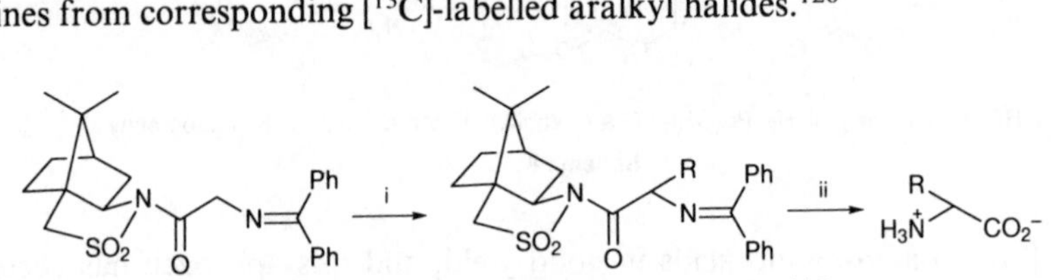

Reagents: i, RBr, K$_2$CO$_3$, catalytic Bu$_4$NBr; ii, 2.5 M LiOH, then bring to pH 2

Scheme 9

4.1.4 Use of Rearrangements Generating a Carbon–Nitrogen Bond – Trichloroacetimidate esters are the essential intermediates in the [3,3]-sigmatropic rearrangement route, *e.g.* to *cis*-5-phenylproline from (R)-mandelic acid methyl ester protected as the MOM ether.[127]

Beckmann rearrangement of tosylated oximes of homochiral acyclic β-ketoesters is advocated for a synthesis of αα-disubstituted glycines, although it is already well-documented in this context.[128] The Curtius rearrangement appears several times in this Chapter (Refs. 268, 289, 290, 434), and has been used in the conversion of cyclopent-3-ene-1,1-dicarboxylic acid into 1-aminocyclopent-3-ene 1-carboxylic acid *en route* to 1-aminocyclopentane-1,3-dicarboxylic acid,[129] and epoxysilyl ethers *via* isomeric aldehydes into S-α-methylphenylglycine.[130]

Na[13]CN and NaC[15]N have been used to prepare substrates for Neber rearrangements leading to [1-^{13}C]- and [1-^{15}N]-DL-homophenylalanine, employing conventional methodology.[131]

4.1.5 Other Rearrangements – Claisen rearrangements of allyl glycinates have been covered in this Section in previous Volumes of this Specialist Periodical Report, and current papers (Refs. 215-217) develop the same theme.

4.1.6 Amidocarbonylation and Related Multicomponent Processes – Catalysts for the synthesis of N-acetylamino acids from an aldehyde, acetamide, and carbon monoxide[132] have been investigated methodically, and the PdBr$_2$–LiBr–H$_2$SO$_4$ cocktail has proved to be the best.[133] Of course, the classic Bucherer–Bergs synthesis employing an aldehyde, ammonium carbonate, and sodium cyanide is a three-component synthesis, and has been applied to give a high yield of phenylglycine from benzaldehyde,[134] and to give ββ-disubstituted tryptophans, *via* hydantoins.[135] Combinatorial synthesis of hydantoins continues to be attractive for the preparation of libraries containing potentially useful lead compounds,[136] with resin-bound ketimines being used in a versatile approach.[137] [For other Strecker syntheses see Refs. 138, 141, and 142.] Since scandium triflate-catalysed Strecker-type reactions with aldehydes, amines and SnBu$_3$CN give α-amino nitriles in both organic and aqueous media, they have been described (stretching commonly-understood definitions somewhat) as an environmentally-friendly route to amino acids.[138] An aldehyde, together with (S)-α-methylbenzylamine, and TMSCN in LPDE gives predominantly S-α-aminonitriles.[139] Development of earlier work on asymmetric Strecker synthesis covers catalysis by the homochiral dioxopiperazine, cyclo(D-His-D-Phe) (Vol. 29, pp. 12, 82);[140] the use of a chiral sulfinimine ArCH=NSOAr for the Strecker synthesis of a vancomycin component has been reported.[141] (S)-Phenylglycinol is the component that makes a Strecker synthesis of N-Boc-D-[3,5-bis(isopropyloxy)-4-methoxyphenyl]glycine diastereoselective.[142] Double asymmetric synthesis is illustrated in the use of homochiral amine and (2S)-methylcyclopropane methyl hemiacetal in a synthesis of (1R,2S)-(+)-allonorcoronamic acid.[143] Better than 80% e.e. has been achieved for the first time using this approach with N-allylbenzamidine.[144]

Other 'multicomponent' amino acid syntheses include the Ugi four-component condensation (4CC), which is undergoing something of a revitalization. The mechanistic basis of this method has been reviewed,[145] and an application leading to a 2,6-dioxopiperazine (26) has been reported, employing an amino acid as the amine component leading to the derivative of a different amino acid.[146] A 4CC approach to 2-oxopiperazines uses an N-(2-aminoethyl) N-Boc-N-alkylamine with the usual three other reactants.[147] Synthesis of α-methyl-α-amino acids through the 4CC route in solution and in solid-phase (combinatorial synthesis) modes has been described,[148] including a polymeric amine on which isocyanoacetamide and a carboxylic acid were condensed.[149]

Rapid access to metalloproteinase inhibitors (27) is possible through the 4CC synthesis.[150] A variant of this has been described as a five-component condensation, the 4CC synthesis being followed in the same pot by hydroxy-laminolysis.[151] Protected D-glucosyl isonitrile syntheses do not inject diastereo-selectivity into the process when used in the Ugi amino acid synthesis.[152]

(26)   (27)

Further results have been reported for the formation of anti-β-amino-alkanols from alkenyl or arylboronic acids by condensation with a primary amine and an α-hydroxyaldehyde or α-hydroxydioxolane (Vol. 30, p. 8), and ruthenium(IV) oxidation into amino acids.[153]

4.1.7 From Glycine Derivatives and from Imines of Glyoxylic Acid Derivatives –

These two families of compounds, $R^1NHCH_2CO_2R^2$ and $R^1N=CHCO_2R^2$ respectively, are so closely related in their applicability to α-amino acid synthesis that there is some logic in collecting their uses under the same heading.

The classic amino acid synthesis based on C-alkylation of diethyl acetamido-malonate[154] has been used for the preparation of N-acetyl-β-(2-naphthyl)-alanine ethyl ester,[155] and perdeuterioglycine.[156] Diethyl formamidomalonate has been used to prepare substituted β-aminoalanines $R^1R^2NHCH_2CH(NH_2)$-CO_2H, the acid hydrolysis step needed to release the amino acid product needing to be conducted below 30 °C with this substrate.[157] Diethyl 2-phthali-midomalonate and diethyl 2-nitromalonate were 2-formylethylated success-fully.[158] *erythro*-3-Phenylglutamic acid diethyl ester emerges from NaOEt-catalysed Michael addition of diethyl benzylidenemalonate to diethyl acetami-domalonate.[159] Alkylation of an analogous [^{15}N]glycine synthon has been employed in a synthesis of [α-^{15}N]-DL-tryptophan,[160] and an analogous route to [β-^{11}C]-L-DOPA is based on alkylation by [α-^{11}C]-3,4-dimethoxybenzyl bromide.[161]

Aldolization of glycine by 4-acetamidobenzaldehyde using a pyridoxal-mediated abzyme system is a development of knowledge of the mode of action of L-threonine aldolase,[162] which catalyses the addition of α-adeninyl- and α-guaninylacetaldehyde to glycine.[163] D-Threonine aldolase catalyses the aldolization of glycine by the protected glyceraldehyde, giving the (2R,3R,4R)-product (28; 0.2% solutions in organic solvents form gels), mixed with its epimer (which does not share this property).[164]

C-Alkylation of alkoxycarbonylmethyl imines $Ph_2C=NCH_2CO_2Bu^t$ [165,166] is a well-established synthesis approach to α-amino acids, and has been used for preparations of 2-amino-4-fluoroprop-4-enoic acid[167] (the alanine homo-logue has been used to prepare γ-fluoroalkyl-α-methyl-α-amino acids[168]), *cis*-

(28)

2,3-dialkylpyroglutamic acid esters,[169] and, in resin-bound form, Michael addition to give libraries of glutamic acid analogues.[170] Alkylation using a neutral non-ionic phosphazene base and chiral quaternary ammonium salt derived from a *Cinchona* alkaloid is highly enantioselective.[171] Allylation followed by tandem Michael-Dieckmann cyclization gives a 4-ethoxycarbonyl pyrrolidin-3-one *en route* to substituted quinolines.[172] High enantioselectivity accompanies cinchonidinium salt-catalysed alkylation.[173] The advantages of this glycine derivative, compared with ethyl nitroacetate, for Pd-mediated coupling with 2,6-dibromo-1,6-dienes, cycloisomerization of the resulting 1,6-enynes giving bicyclic and tricyclic $\alpha\alpha$-disubstituted glycines, have been established;[174] nevertheless, nitroacetates are versatile glycine synthons, suitable for dehydroamino acid synthesis and open to alkylation with organozinc reagents.[175] A version of this glycine imine carrying a novel chiral auxiliary (29) is advocated for the synthesis of D-α-amino acids (from 29) or L-α-amino acids (from the epimer of 29).[176] Through an uneventful application of

(29)

standard reactions, the analogous Schiff base of (+)-(R,R,R)-hydroxypinanone has led to (2R,3R)-m-chloro-3-hydroxytyrosine (a vancomycin and aridicin A component)[177] and to R-2′-methyltryptophan,[178] and underwent diastereoselective aldolization to start a synthesis of sulfobacin (independently prepared from cysteine, see Ref. 1011).[179]

The long-established Schiff base complex formed between an o-substituted benzophenone and L-prolylglycine and a nickel(II) salt has been used in newer forms, *e.g.* the L-alanine homologue to prepare α-methyl-α-amino acids [L- and D-α-(3,4-dichlorophenyl)-α-methylalanine;[180] α-methyl-D-phenylalanine, D-β-allylglycine and α-allyl-D-alanine;[181] other homologues through phase transfer-catalysed alkylation employing (4R,5R)- or (4S,5S)-2,2-dimethyl-tetraphenyl-1,3-dioxolane-4,5-dimethanol, *alias* TADDOL[182]]. L,L-2,4-Diaminoglutaric acid has been prepared through Michael addition to dehydroalanine.[183] Use of homochiral sodium alkoxides to catalyse the Michael addition of methyl methacrylate with an achiral Schiff base of this family, formed from glycine and N-(2-pyridylcarbonyl)-o-aminobenzophenone, gives 4-methylglutamic acid with moderate enantioselectivity.[184] Use of this synthon continues to attract other research groups, and C-alkenylation using allyl or propargyl bromides, followed by hydroboration of the resulting ω-unsaturated α-amino acid derivatives has led to ω-boronoalkyl-α-amino acids.[185]

Isocyanoacetates are occasionally used for α-amino acid synthesis, bis(α-alkylation) with 2,3-di(2-iodoethyl)buta-1,3-diene giving 3,4-dimethylenecyclopentane-1-carboxylic acid after the usual isocyanide-to-amino group conversion.[186]

The first enantioselective imino-ene reaction shown in Scheme 10 (see also Ref. 190) is capable of providing amino acids of high optical purity.[187] The (−)-8-phenylmenthylglyoxylate has been used similarly.[188] The N-toluene-p-sulfo-

$$EtO_2C-CH \begin{matrix} CH_2=C \\ \\ N \\ Ts \end{matrix} \begin{matrix} R \\ \\ CHR^1 \\ H \end{matrix} \xrightarrow{i} EtO_2CCH \begin{matrix} CH_2-C \\ \\ NH \\ Ts \end{matrix} \begin{matrix} R \\ \\ CHR^1 \end{matrix}$$

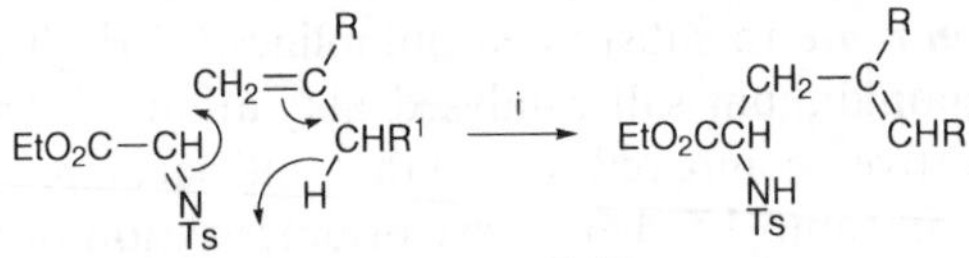

Reagents: i, (*R*)-Tol-BINAP, Cu-bis(MeCN) perchlorate, PhCF$_3$, room temp.

Scheme 10

nylimine TsN=CHCO$_2$Et can be C-alkylated diastereoselectively and enantioselectively by an enol silane when catalysis by (S)-BINAP-CuClO$_4$(MeCN)$_2$ is employed;[189] a preference for (R)-Tol-BINAP has been expressed for alkylation of this imine by an alkene R^{1}C(=CH$_2$)CH$_2$R^2 to give a γδ-unsaturated (S)-α-amino acid TsNHCH(CH$_2$CR^1=CHR^2)CO$_2$H.[190] The imine from (R)-phenylglycinol and glyoxal carrying an N-(6-trimethylsilyl-hex-4-enyl) side-chain moiety yields a transient iminium ion that undergoes intramolecular addition to give methyl *cis*- or *trans*-3-vinyl-(R)-pipecolinate.[191] Imines formed from glyoxylate esters HCOCO$_2$R + RNH$_2$ → RN=CHCO$_2$R can be alkylated using enol silyl ethers Me$_3$SiOCR=CH$_2$ and lead to γ-oxoalkylglycines in better than 90% e.e. when a chiral palladium(II) species is employed as catalyst,[192] and undergo Diels-Alder addition (R = MeOCH=CH-) to give N-aryltetrahydropyridin-4-one 3-carboxylates in high enentiomeric excess when catalysed with a homochiral Lewis acid.[193] Chiral glyoxylate ester imines undergo Staudinger β-lactam synthesis (to give 3-substituted azetidin-2-one-4-carboxylic acids).[194]

The dianion of ethyl Z-glycinate has been used in a synthesis of 3-alkoxycarbonylisoquinolin-2-one through condensation with an alkyl o-formylbenzoate.[195] Ethyl 2-phthalimidoacetate undergoes alkylation to give the corresponding α-amino acid derivative with no better than 59.5% enantioselectivity when catalysed by chiral reverse micelles,[196] but wide variation (20–98%) in the outcome with phenylalanine synthesized through the alkylation of hippuric acid (1R,2S)-2-arylcyclohexyl esters, depending on the nature of the aryl group, has been reported.[197] An unusual intramolecular cyclization, N-Boc-N-(o-bromobenzyl)-glycine N'-allylamide to pyrrolidin-2-one by Bu$_3$SnH–1,1'-azobis(cyclohexanecarbonitrile), relies on formation of a glycinyl radical intermediate.[198]

Hippuric acid coupled to (1S,2R)-1-amino-indan-2-ol provides a chiral glycine equivalent that gives D-amino acids in 90–99% d.e. through alkylation of its enolate.[199]

Oxazolidinones and dihydroimidazoles have been resolved over chiral stationary phases using the moving bed technique,[200] and the newly-prepared dihydroimidazole (30) has been resolved by chromatography over a polysaccharide stationary phase.[201] The new chiral glycine derivative (30) undergoes diastereoselective alkylation with an alkyl halide after lithiation;[202] a valuable feature is the mild release of the amino acid product after the alkylation stage, so that acid-sensitive functional groups are not affected. Alkylation of a differently-conceived chiral glycine derivative, the oxazolidine (31), prepared from (S)-phenylglycinol, is described in a preliminary study.[203] Its use in aldol reactions after lithiation and reaction of the enolate with an aldehyde is notable in giving β-hydroxy-α-amino acids with anti-selectivity.[204]

(30)   (31)

The dehydroalanine synthon (32) continues to be a useful substrate for addition reactions that lead to L-α-amino acids with extended side-chains (Vol. 30, p. 18; see also Refs. 295, 319), and an extraordinary use has been reported: photoinduced radical addition of alkanols and ethers leading to homoserine derivatives.[205] A difficulty arises when this synthon is used for these targets, since epimerization accompanies cleavage when the product is hydrolysed to the amino acid.

(32)

The ureide (33) formed from methyl glycinate and the Evans oxazolidinone undergoes allylation under radical conditions (allyltributyltin, AIBN, $-50\,^{\circ}\mathrm{C}$) with excellent stereocontrol exerted by the presence of $ZnCl_2$-Et_2O to give D-allylglycine with 93:7-diastereoselectivity.[206] The glycine synthon (34) that incorporates the analogous imidazolidinone as an amide function is also well suited for L-α-amino acid synthesis.[207] A review has appeared covering the uses of chiral imidazolidinones in asymmetric synthesis.[208]

(33)   (34)

α-Heteroatom-substituted glycinates $RNHCH[P(O)(OMe)_2]CO_2Me$ have been regularly used for dehydro-amino acid synthesis (*cf.* Section 4.9) through Wittig olefination with a chiral aldehyde, as for the preparation of novel cyclobutanes (35),[209] and with acetaldehyde (developed *via* the β-bromoalkyl amino acid into a substituted dehydrotryptophan).[210] A 3-pyrrolylalanine synthesis also uses this route.[211] Synthesis of the cyclopropylglycine (36)

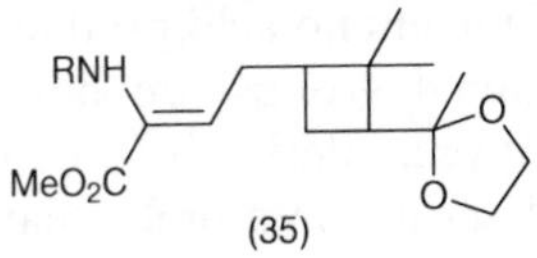

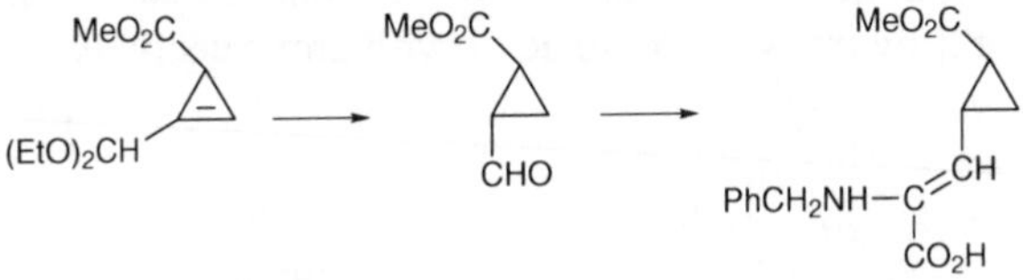

illustrates the enantioselectivity that accompanies hydrogenation of the cyclo-propene, controlled by the 2R-methyl group.[212] An extended route (Scheme 11) to C-glucopyranosyl-D- and L-serines demonstrates the construction of an extended aliphatic side-chain through the Wittig synthesis strategy.[213] The α-glycinylzinc chloride $Ph_2C = NCH(CO_2Et)ZnCl$ undergoes palladium(0)-cat-alysed cross-coupling with substituted bicyclopropylidenes and bromomethyl-enecyclopropane to give α-(alkadienyl)glycines (37).[214]

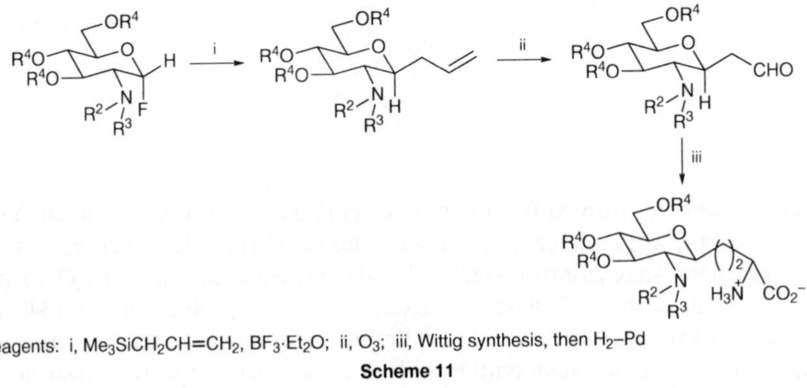

Reagents: i, $Me_3SiCH_2CH=CH_2$, $BF_3 \cdot Et_2O$; ii, O_3; iii, Wittig synthesis, then H_2–Pd

Scheme 11

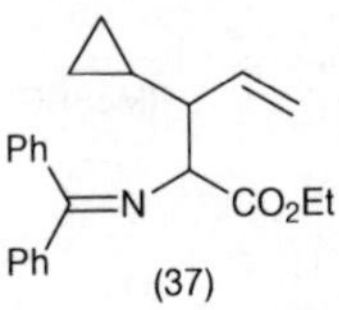

Allyl glycinates can undergo [3,3]-sigmatropic rearrangement, delivering the appropriate unsaturated side-chain onto the glycine methylene group. The topic has had tediously fragmented exposure in the recent literature (Vol. 30, p. 29) but the process provides unsaturated amino acids (see also Section 4.9) with predictable stereochemistry.[215,216] Its scope has been broadened to encompass the rearrangement of allyl ethers into γ-keto-α-amino acids $[XNHCH_2CO_2CHRC(OMEM)=CF_2 \rightarrow XNHCH(CF_2COCH_2R)CO_2H]$.[217] 2-Trichloromethyl-imidazole has been prepared from glycinal dimethyl acetal $H_2NCH_2CH(OMe)_2$.[218]

4.1.8 From Dehydro-amino Acid Derivatives – In addition to examples mentioned elsewhere in this Chapter, a number of non-routine studies cover enantioselective synthesis of α-amino acids from achiral α-aminoacrylates. These studies focus mainly on chiral phosphines as ligands for rhodium(I)-catalysed hydrogenation, such as CARAPHOS, which is less effective in this context[219] than other well-known diphosphines[220] such as DIPAMP,[221] and PROPRAPHOS.[222] Good stereoselectivity is seen for (S,S)-1,2-bis(alkyl-methylphosphino)ethanes,[223] (R,R)-ethyl-DuPHOS,[224] and hydroxy-DIOP.[225] Other less familiar ligands include homochiral phosphaferrocenes[226] and the air-stable FERRIPHOS series,[227] and an aminophosphine phosphinite prepared from (1S,2R)-1,2-diphenyl-2-aminoethanol.[228] An alternative approach using a phosphinite, $[Rh(COD)_2]BF_4$, has the benefit of high stereoselectivity;[229] this system has been used also for corresponding synthesis of β-aminoalkanols from $MOMOCH = CRNHAc$.[230]

The rhodium(I)-(S,S)-PrTRAP complex has been used to catalyse the hydrogenation of the unfamiliar substrate, ethyl (E)-$\alpha\beta$-bis(N-acylamino)-acrylate, to provide the corresponding (2S,3R)-2,3-diamino-alkanoic acid derivative.[231]

This topic is well-served with reviews.[232]

4.2 Synthesis of Protein Amino Acids and Other Naturally Occurring α-Amino Acids – Reports of enzyme-catalysed syntheses leading to familiar L-α-amino acids provide a large part of this Section, and, as usual, the literature coverage is not exhaustive on biotechnological aspects, but describes selected non-routine studies. This literature on biosynthesis studies also extends to bacterial and other whole-cell processes, but studies involving plants and higher organisms are not covered in this Chapter.

Alanine dehydrogenase (conversion of pyruvate into L-alanine) is usefully combined with glucose dehydrogenase so that NADH regeneration can accompany continuous production.[233] The classic route to L-DOPA using mushroom tyrosinase-catalysed oxidation of tyrosine can be improved by immobilizing the enzyme on chemically-modified nylon[6.6] membranes.[234] Immobilized aspartase features in a synthesis of [^{15}N]aspartic acid from a dialkyl fumarate and $^{15}NH_4OH$,[235] and this route and a whole-cell alternative based on L-aspartate-β-decarboxylase has also been developed.[236] A whole-cell approach to L-DOPA production by use of *Stizolobium hassjoo* cell culture has been described,[237] and the use of *Corynebacterium glutamicum* has been shown to be appropriate for L-lysine production.[238]

Reviews cover uses of transaminases for large-scale production of non-natural α-amino acids,[239] enzyme-catalysed manufacture of L-phenyl-alanine,[240] production of D-α-amino acids from hydantoins using N-car-bamyl-D-amino acid aminohydrolase,[241] processes leading to L-α-amino acids base on Coryneform bacteria,[242] and plant sources as the basis for amino acid production.[243]

Routine literature covering the laboratory synthesis of common naturally-occurring α-amino acids is dealt with in other sections of this Chapter. New syntheses of more complex examples have been reported for sphingofungins B and F [tin(II)-catalysed asymmetric aldol addition],[244] an alternative route to sphingofungin F through asymmetric alkylation of an azlactone catalysed by a recently-introduced homochiral palladium complex (Vol. 30, p. 23),[245] and for lactacystin (see also Refs. 479, 480; and Vol. 30, p. 12). A route[246] to the last-mentioned cysteine derivative is notable for a doubly-diastereoselective Mu-kaiyama aldol coupling of a chiral tertiary α-aminoaldehyde with an achiral silyl enol ether to give the required antiselectivity, and a route to the precursor aminoalkanol for lactacystin (Scheme 12) employs a trichloracetimidate ester for the amination step.[247]

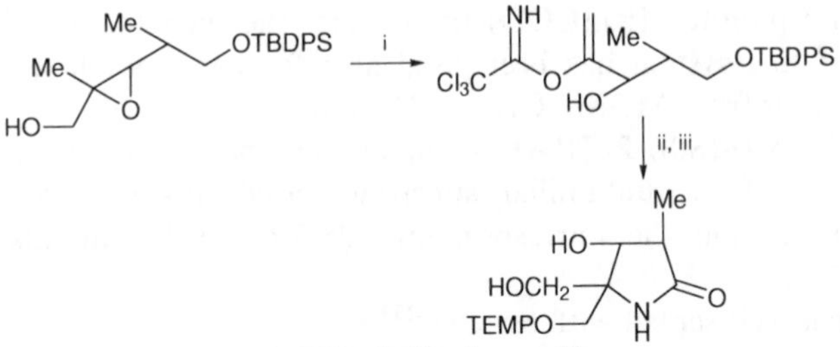

Reagents: i, LDA/thf, then Cl₃CCN; ii, Hg(OTf)₂, K₂CO₃, then aq. KBr;
iii, functional group adjustment followed by conc. HCl/Zn

Scheme 12

Polyoxamic acid is available through a relatively short synthesis based on manipulation of N-Boc-R-phenylglycinol (Scheme 13; see also Ref. 931).[248] Improved routes to an intermediate suitable for uracil polyoxin C synthesis,[249] and to carbocyclic analogues (38),[250] include carefully worked-out stages calling for mild conditions. Also noteworthy are new synthesis approaches to nicotianamine and 2′-deoxymugineic acid,[251] (2S,4S)-γ-hydroxynorvaline (a constituent of the Boletus toxin),[252] (2S,4R,5S)-piperidine-2,4,5-tricarboxylic acid from the poisonous mushroom *Clitocybe acromelalga*,[253] Actinomycin Z₁ constituents (2R,3R,5R)- and (2S,3S,5S)-3-hydroxy-5-methylpiperidine-2-car-boxylic acids,[254] and the N-terminal amino acid residue of nikkomycin B.[255]

Synthesis studies for kainic acid, and kainoids generally, have featured in the literature over several years. New work deals with ketyl radical cyclization onto an enecarbamate derived from L-pyroglutamic acid to introduce the C-4

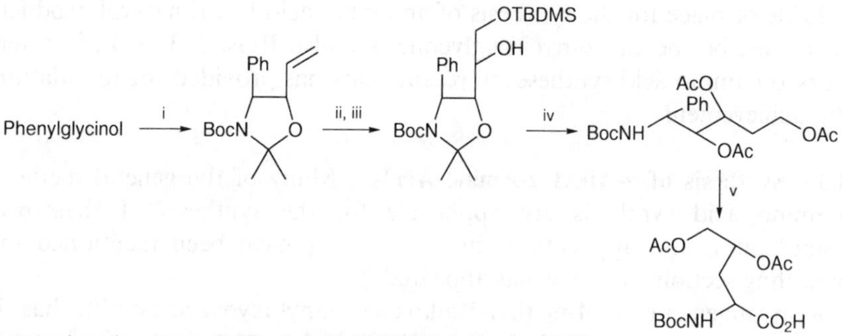

Reagents: i, routine stages; ii, AD-mix-β; iii, TBDMSCl, imidazole, DMAP;
iv, Bu₄NF/thf. then Ac₂O/DMAP; v, RuCl₃–NaIO₄

Scheme 13

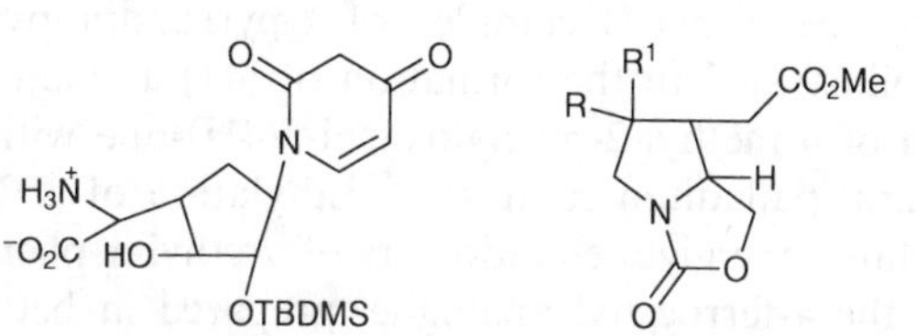

(38)   (39)

substituent on the pyrrolidine framework,[256] and an alternative approach to
racemic 4-substituted kainic acids (Scheme 14)[257] with variants that lead to
(−)-kainic acid[258] that also start from L-pyroglutamic acid. A route to an
established key intermediate (39) for a (+)-α-allokainic acid synthesis has been
established.[259] The synthesis strategy for (−)-α-kainic acid used by Bachi's
group (Vol. 30, p. 21) has been reviewed.[260] Full reports of acromelic acid A
and alloacromelic acid A syntheses, and an approach to acromelic acid B
through palladium(0)-catalysed cross-coupling to C-4-catechol precursor and
pyridones, have appeared;[261] a biomimetic approach was followed similar to
that leading to stizolobic acid (Vol. 30, p. 70) that was used earlier by this
research group.

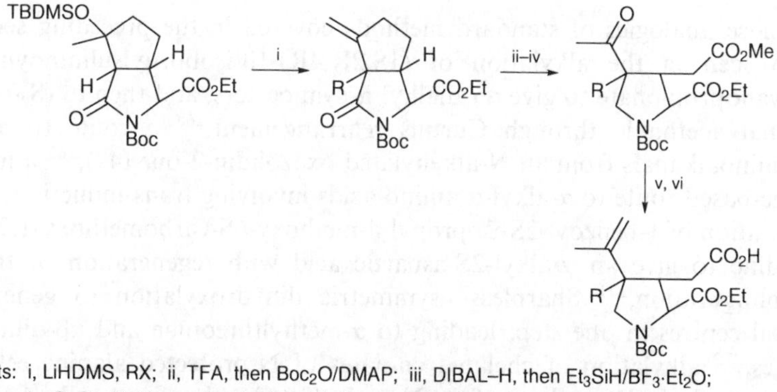

Reagents: i, LiHDMS, RX; ii, TFA, then Boc₂O/DMAP; iii, DIBAL–H, then Et₃SiH/BF₃·Et₂O;
iv, OsO₄, NaIO₄; v, Ph₃P=CH₂; vi, Jones oxidation followed by deprotection

Scheme 14

Pride of place for the synthesis of an amino acid-based natural product this year must be the vancomycin aglycone (see also Refs. 373, 374);[262] extensive work on amino acid syntheses in recent years has provided the foundation for this achievement.

4.3   Synthesis of α-Alkyl α-Amino Acids – Many of the general methods of α-amino acid synthesis are applicable for the synthesis of their α-alkyl homologues, and applications in this context have been mentioned in the preceding section. A review has appeared.[263]

Rearrangement of 4-methyl-5-alkoxycarbonyloxyoxazoles (40) has been known for many years and seems ideally suited for the synthesis of α-methyl-α-amino acids and homologues, given the extensive experience of the use of oxazol-5(4H)-ones in amino acid synthesis and the effective catalysis of the rearrangement by the iron(III) complex of 4-pyrrolidinopyrindene.[264] Standard practice is illustrated in the formation of (41) as major product (8.7:1) through reaction of 4-methyl-2-phenyloxazol-5(4H)-one with 3-acetoxycyclohexene and a chiral palladium catalyst.[265] Ethylation of (2R,4S)- or (2S,4S)-diphenyloxazolidin-5-one gives enantiomers of α-ethyl-α-phenylglycine,[266] and a similar use of the 4-ferrocenyl analogue, prepared in better than 98% e.e. from ferrocenecarbaldehyde and sodium L-alaninate followed by reaction with pivaloyl chloride, gives (R)-α-methyl-α-amino acids.[267]

Close analogies of standard methods covered in the preceding section are also seen in the alkylation of (1S,2R,4R)-10-isobornylsulfamoylisobornyl 2-cyanopropionate to give αα-dialkyl-α-cyanoesters, and then to (S)-α-methyl-serinal acetonide through Curtius rearrangement,[268] a route to α-methyl-α-aminoalkanals from an N-alkenylated oxazolidin-2-one (42),[269] a morpholinone-based route to α-alkyl-α-amino acids involving trans-induction,[270] trans-alkylation of 1-benzoyl-2S-isopropyl-4-methoxy-6S-carbomethoxy-1,2,5,6-pyrimidine to give an α-alkyl-2S-aspartic acid with regeneration of the initial S-configuration,[271] Sharpless asymmetric dihydroxylation to generate two chiral centres in one step, leading to α-methylthreonine and αβ-dimethylcysteines,[272] alkylation of chelated enolates of N-protected alanine esters [best results with tin(II) enolates of the N-arylsulfonylated amino acid],[273] alkylation of the homochiral N-toluene-p-sulfinylimine of ethyl trifluoromethylpyruvate

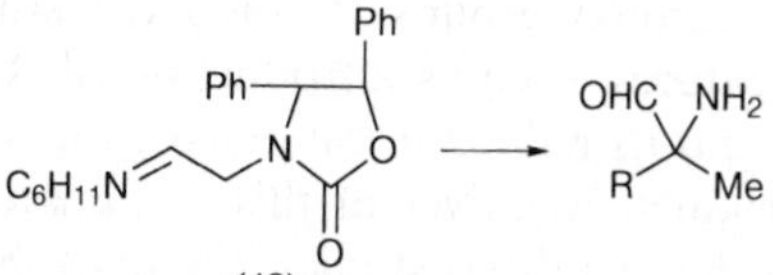

(42)

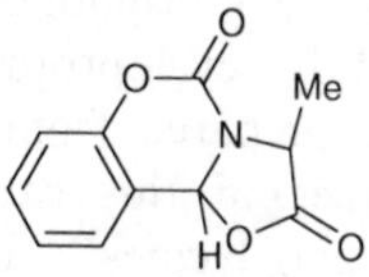

(43)

giving α-trifluoromethyl-α-amino acids with high enantioselectivity,[274] and alkylation of the tricyclic oxazolidinone (43) formed from the N-salicylalde-hyde Schiff base of L-alanine, to give R-α-methyltryptophan.[275] α-Methyl- and α-benzyl-quisqualic acids have been synthesized.[276]

Individual tailoring of a route to suit a particular synthetic target is illustrated in highly diastereoselective addition of an alkyllithium to the chiral oxime ether derived from erythrulose to give (R)-(−)-2-methylserine and its (S)-enantiomer, (R)-(+)-2-phenylserine and (R)-(−)-2-butylserine.[277] Sequential nucleophilic additions to the tartaric acid-derived nitrile (44) give imines *en route* to αα-dialkylglycines.[278]

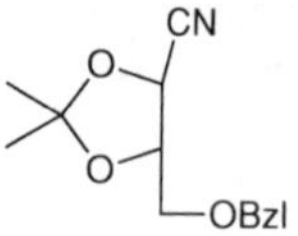

(44)

4.4 Synthesis of α-Amino Acids Carrying Alkyl Side-chains, and Cyclic Analogues – This section is in essence little different from the preceding two sections, because it covers standard methods of synthesis of α-amino acids; but the complexity of the synthesis target often constrains the choice of synthesis route and determines strategy in certain cases. Major subdivisions of this section are falling into place through the increasing interest in pharmacological properties of conformationally-constrained versions of the common amino acids; (a) α-cyclopropylglycines; (b) 1-aminocycloalkanecarboxylic acids and homologues, and (c) saturated nitrogen heterocycles carrying one or more carboxy-groups.

4.4.1 α-Cyclopropylglycines and Higher Homologues – Synthesis of stereo-isomers of α-(2-carboxycyclopropyl)glycine (45) through previously developed methods provides materials for NMR determination of conformational prefer-ences for comparison with MO calculations (see also Section 5.7);[279] another route involving cyclopropanation of trans-1,3-di(2-furyl)propenone uses the

furyl moieties as latent carboxy groups.[280] The latter study has several other
noteworthy synthesis features, such as introduction of the amine function by
way of an oxime ether and its reduction catalysed by a chiral oxazaborolidine
to give the D- or L-product. Relatives of this glutamate analogue have also
featured in numerous other synthesis studies: the phosphonoglycine analogue
(46) prepared by alkylation of the phosphonoglycine Schiff base (47) with
ethyl bromocrotonate followed by cyclopropanation;[281] and 2-(2-carboxy-
3',3'-difluorocyclopropyl)glycine prepared from (R)-2,3-O-isopropylidenegly-
ceraldehyde.[282] αα-Dialkylglycines in this category, such as α-alkyl-α-(2-
carboxy-3'-substituted cyclopropyl)glycines[283] include the extraordinary tritiated derivative (48).[284] Extraordinary in another way is (S)-2-(4'-carboxy-
cubyl)glycine (49), a new selective metabotropic glutamate receptor
antagonist.[285] (2S,1'S,2'S)- and (2S,1'S,2'R)-2-(2-carboxycyclobutyl)glycines
have been prepared through [2+2]photocycloaddition applied to a homochiral
αβ-unsaturated-γ-lactam.[286]

(45)   (46)   (47)

(48)   (49)

More flexibility is offered by the α-(substituted α-cyclohexyl)glycine (50),
prepared through [4 + 2]cycloaddition to the widely-useful chiral oxazolidine
synthon (51).[287] Relatively rigid analogues of Nα-(1-iminoethyl)-L-ornithine,
eight in number (including 52 and 53), have been prepared.[288]

(50)   (51)   (52)   (53)

4.4.2 1-Aminocyclopropanecarboxylic Acids – '2,3-Methano-amino acids', *alias*
2-substituted 1-aminocyclopropane-1-carboxylic acids, constitute a further
class of conformationally-constrained amino acid that includes the naturally-
occurring 1-aminocyclopropane-1-carboxylic acid (1-ACA) and coronamic

Reagents: i, $CH_2(CO_2Me)_2$, NaH; ii, mild hydrolysis; iii, resolution as (−)-brucine salt;
iv, Curtius rearrangement

Scheme 15

acids. New syntheses have been reported based on Curtius rearrangements; with cyclopropane-1,1-dicarboxylic acid[289] for 1-ACA, enantiomerically-pure 1,2-propanediols as starting point for the preparation of norcoronamic acids from 2-methyl-1-methoxycarbonylcyclopropanecarboxylic acid (Scheme 15),[290] and L-malic acid as the starting point for 2,3-methanoglutamic acid, 2,3-methanoarginine and trans-N-Boc-N-Z-2,3-methano-L-lysine (55, *via* 54).[291] Alternative syntheses of 1-ACA are routine.[292] Cyclopropanation of aldol condensation products of the chiral glycine equivalent 3,6-dihydro-2H-

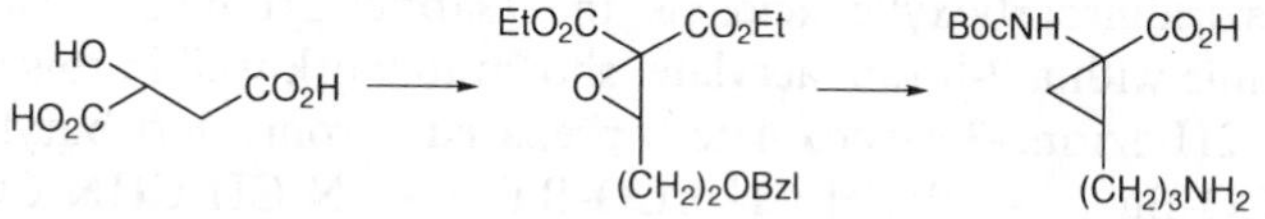

1,4-oxazin-2-one has provided new syntheses of 1-ACA and (−)-allo-norcoronamic and (−)-allo-coronamic acids[293] (see also Refs. 39, 143), and careful assessment of the best catalyst metal complex–ligand cocktail for cyclopropanation leading to 3-phenyl-2,3-methanophenylalanine has been described.[294] Cyclopropanation studies have been reported for 4-methylene-2-phenyloxazolidin-5-one, leading to a mixture of three diastereoisomers when ethyl or t-butyl (dimethylsulfuranylidene)acetate is used as reagent.[295] An erratum referring to an earlier publication on this topic has been published.[296] Chiral imines formed between (−)- or (+)-1-hydroxypinanones and aminoacetonitrile give 1-amino-2-vinylcyclopropanenitrile in no more than 32% e.e. when reacted with (E)- or (Z)-1,4-dichlorobut-2-ene catalysed by a palladium(0)-(R)- or (S)-BINAP complex.[297]

4.4.3 Higher Aminocycloalkanecarboxylic Acids and Saturated Heterocyclic Analogues – 1-Aminocyclopentane-1-carboxylic acids (56) represent a novel glutamate mimic, prepared through a lengthy route including a photoaddition stage.[298] Four racemates of 1-amino-4-hydroxy-2-phenylcyclohexane-1-carboxylic acid emerge from Diels-Alder addition to 4-methylene-2-phenyloxazolidin-5-one and ensuing routine elaboration,[299] and synthesis of (1R,2R)- and (1S,2S)-1-amino-2-phenylcyclohexane-1-carboxylic acids uses an earlier method (Vol. 25, p. 53) and includes a conventional resolution step.[300]

Readily-available azidolactones serve as starting materials for the synthesis of epimers of 2-aminomannofuranose-2-carboxylic esters (57),[301] while conventional general synthetic methods lead to (+)-3-aminotetrahydrofuran-3-carboxylic acid and its thiophen analogue[302] and 2,2,6,6-tetramethylpiperidine-1-oxyl-4-amino-4-carboxylic acid[303] (the latter is used as a nitroxide spin-labelled amino acid in radical reaction studies[304]).

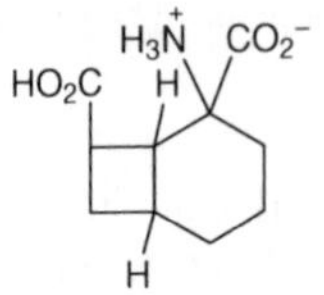

(56) 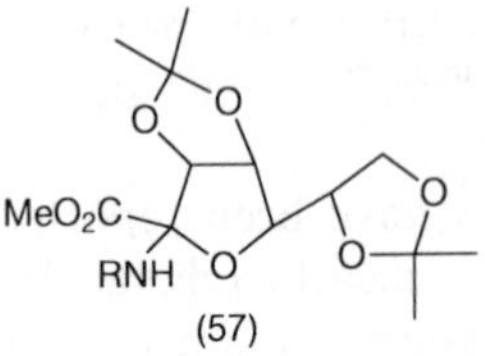   (57)

4.4.4 Aziridinecarboxylic Acids and Azetidinecarboxylic Acids

4.4.4 Aziridinecarboxylic Acids and Azetidinecarboxylic Acids – Conjugate addition of N-Boc-O-benzoylhydroxylamine to an αβ-unsaturated acid that has been amidated with the chiral imidazolidin-2-one (58), followed by NaH-induced cyclization, gives the N-Boc-aziridinecarboxamide.[305] The simplest access to aziridinecarboxylic acids is the Gabriel-Cromwell reaction of a primary amine with a 2-bromoacrylate, shown to work well in this context.[306] tert-Butyl 2H-azirine-3-carboxylate, prepared from tert-butyl acrylate ($CH_2=CHCO_2Bu^t$ → $BrCH_2CHBrCO_2Bu^t$ → $N_3CH_2CHN_3CO_2Bu^t$ → $CH_2=CN_3CO_2Bu^t$ (59), proved to be usable over a period of two hours for Diels-Alder additions leading to other amino acids.[307]

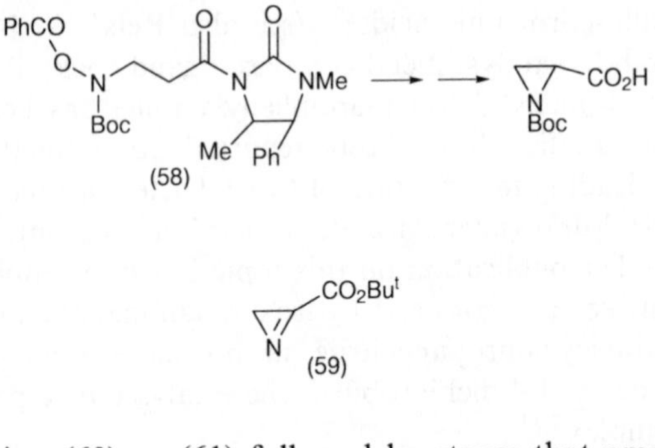

(58)

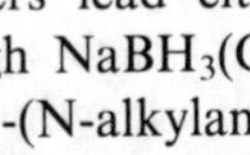

(59)

Photocyclization (60) → (61) followed by stages that amount to standard methodology, leads to (2R)- and (2S)-azetidine-2-carboxylic acids.[308] α-Amino-γ-chloroalkanoate esters lead either to N-alkyl-3,3-dimethylazeti-dine-2-carboxylate esters through $NaBH_3(CN)$ reduction of α-N-alkylimines in refluxing methanol, or to 1-(N-alkylamino)-2,2-dimethylcyclopropane-1-carboxylic acid esters by 1,3-dehydrochlorination.[309]

Diazirinecarboxylic acid esters, prepared from amidines by HOCl oxidation, are useful carbene sources but do not show particular potential for amino acid synthesis.[310]

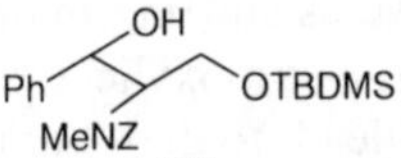

(60) 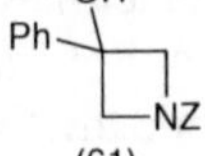   (61)

4.4.5 Pyrrolidine- and Piperidine-carboxylic Acids and Analogues – New syntheses of 4,5-methano-L-proline[311] and its near relative 2,4-methanopyrrolidine-2,4-dicarboxylic acid, prepared through a [2 + 2]photocyclization route and designed as a glutamate analogue,[312] are accompanied in the recent literature by routes to kainoid analogues (62)[313] and (63).[314] The last-mentioned examples involve radical cyclization of an αβ-unsaturated phosphonate, and chain-extension of pinanediol[[(trityloxy)methyl]boronate with LiCHCl$_2$ and ensuing H$_2$O$_2$ oxidation (in very low yield) respectively. (2S,3R,4R)-Diaminoproline is reached from ascorbic acid through conventional amination and

MeO   P(O)(OR)$_2$ CO$_2$Me   N Boc (62)   OH   B(OH)$_2$ N Ac (63)

functional group development,[315] and (2S,3S)-3-hydroxyproline likewise from L-malic acid, cyanide displacement of the methoxy group of an α-methoxypyrrolidine being a key step.[316]

A long route to (2S,4R)-4-carboxyproline starts from the L-cysteine-derived sulfone BocN(Bzl)CH(CH$_2$OBzl)CH$_2$SO$_2$Ph and builds the requisite carbon chain on nitrogen.[317]

(E)-Nitroalkenes participate in [3 + 2]cycloaddition to azomethine ylides derived *in situ* from Schiff bases of methyl glycinate and homologues R^{1}CH=NCHR^{2}CO$_2$Me to give 4-nitroprolines, LiClO$_4$ catalysis giving the endo isomer (64), and AgOAc catalysis favouring exo-addition.[318] Further examples of 1,3-dipolar cycloaddition of these Schiff bases to 4-methylene-2-phenyloxazolidin-5-one demonstrate effective stereocontrol of this route to highly functionalized prolines, now leading to all four stereoisomers of 4-benzamidoproline.[319] This process can be carried out with a tethered dipolarophile and an azomethine ylide.[320] Thermal [3 + 2]cycloaddition of glyoxylate imines BzlON=CHCO$_2$R to protected alkylidenecyclopropanones gives 4,5-substituted prolines, though without offering any obvious generality to the route (Scheme 16).[321] Cycloaddition of [60]fullerene with a glycine Schiff base[322] and resolution through derivatization with a homochiral isocyanate[323] gives enantiopure fullerene-fused prolines; pipecolic acid derivatives (65) are obtained similarly using an azadiene.[324] Fullerene-amino acid reaction products have been reviewed.[325]

R^3 R^2 R^4 CO$_2$Me   R^1 N H R^2 (64)   C$_{60}$ CO$_2$Et   NH (65)

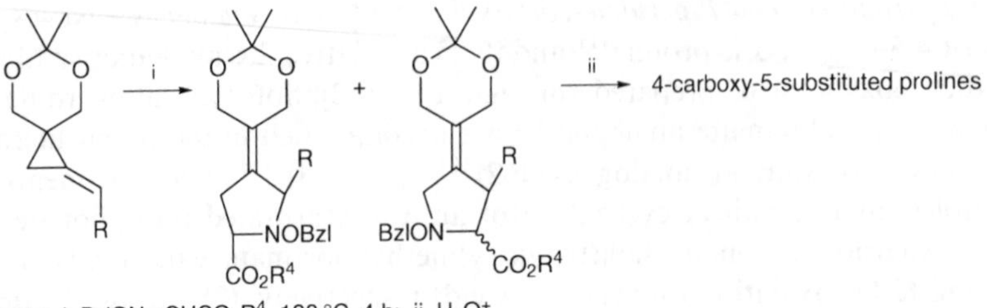

Reagents: i, BzlON=CHCO$_2$R^4, 100 °C, 4 h; ii, H$_3$O$^+$

Scheme 16

Highly-substituted prolines are the inevitable outcome of developing a carboxy substituent on a Boc-protected 2,5-imino-D-glucitol.[326] Polyhydroxylated prolines are accessible from previously-reported azidolactones.[327]

New α-trifluoromethylprolines and pipecolic acids have been prepared in various proportions from $CH_2=CHCH_2NR^1C(CF_3)[(CH_2)_2CH=CH_2]-CO_2Me$ through ruthenium-catalysed ring-closing metathesis.[328] Another cyclization route to prolines, C-H insertion by chiral ester diazoanilides $R^*O_2CC(N_2)$-$CON(Bzl)Ar$ [R* = 2-(1-naphthyl)norbornyl] to give 4-substituted pyrrolidinones, has been illustrated in a synthesis of trans-4-cyclohexyl-L-proline.[329] 4-Alkylpipecolinates can be prepared from 1-acylpyridinium salts through phenyldimethylsilylmethylation, rearrangement, and subsequent oxidative desilylation.[330] Enantiopure 4-piperidinones (66), formed by cycloaddition of chiral 2-aminodienes with imines, have been developed into pipecolic acid derivatives.[331]

Enecarbamates (67) have been developed into β-hydroxy-prolines and -pipecolic acids through epoxidation and cyanohydrin stages.[332]

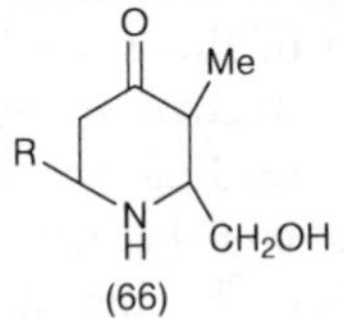

(66) 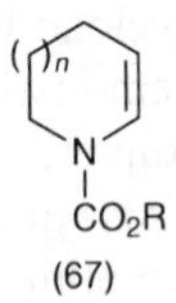 (67)

Imidazolidine-2-carboxylic acid has been prepared in orthogonally protected form from N-Boc-N'-Fmoc-ethylenediamine and glyoxylic acid as a new proline surrogate,[333] and the tetrahydropiperazinecarboxylic acid amide (68) has been prepared by a lengthy route involving asymmetric hydrogenation at the last stage.[334] (3S,5S)-5-Hydroxypiperazic acid, *i.e.* 5-hydroxy-hexahydropyridazine-3-carboxylic acid, has been prepared by the Evans procedure and

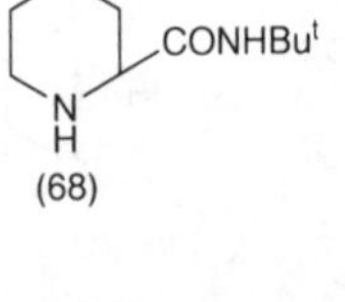

(68)

building on a D-mannitol moiety; this provides material more economically than the current route that calls for a resolution step.[335] Reviews cover piperazic acids[336] and use of the aza-Achmatowicz reaction for the preparation of amides of furan-2-carboxylic acid as starting materials for the synthesis of unusual amino acids.[337]

4.5 Models for Prebiotic Synthesis of Amino Acids – Conventional studies, in which gas mixtures subjected to high energy irradiation generate amino acids, show no signs of losing their fascination. A novel twist to this work is the search for technology for 'fixing' unwanted atmospheric carbon dioxide, and the formation of amino acids, nucleic acid bases, and organic acids through UV irradiation (shorter wavelength than 280 nm) of gaseous moist ammonium carbonate at 80 °C over a magnesium salt is noteworthy.[338] This reaction pathway presumably involves oxalic and oxamic acids. An extraordinary result, which is contradicted by all previous reports of the same experiment, claims that precursors of non-racemic alanine and aspartic acid (15% excess of the L-enantiomer) are formed by synchrotron irradiation of a N_2–CO_2–H_2O mixture.[339] An amorphous film deposited in a reaction vessel containing carbon monoxide and nitrogen held in high energy plasma[340] or irradiated with high energy particles[341] releases amino acids when exposed later to a cold moist atmosphere.

Most of these studies in the past have attempted to model the chemistry of the atmosphere of prebiotic Earth, but events at the surface of the geosphere are also of interest, such as the conversion of aqueous formaldehyde mixed with HCN and NH_3 into α-hydroxyacetonitrile and α-aminoacetonitrile.[342] At first sight the unfavourable equilibrium (99.9:0.1) that is generated is not suited to explain the prebiotic genesis of amino acids, but further processes available only to the latter product (conversion into α-amino acid amides; their reaction with CO_2 to give hydantoins) will shift the equilibrium so that conversion is eventually complete. Prebiotic production of N-carbamoylamino acids is easily envisaged, and so is their generation in space on ice particles under UV irradiation[343] and on carbon powder suspended in a hydrogen–nitrogen mixture (15:85) bombarded with high energy particles.[344] Another study of a role for aqueous formaldehyde that leads to glyoxal sees the generation of alanine and homoserine under mild conditions from ammonia in the presence of a thiol as catalyst.[345] Solid sodium acetate bombarded with high energy nitrogen ions accumulates organic cyanides and releases amino acids through hydrolysis.[346]

Hydrothermal vents on the ocean floor are the currently-favoured environment for prebiotic amino acid synthesis; a persistent objection, that amino acids generated there would be degraded at the high temperatures, has been addressed. The moderately reducing environment that this represents, means that the reaction energetics are as favourable on the sea floor at 100 °C as they are at 18 °C in the oxidizing conditions at the sea surface.[347]

Better focussing of experiments in support of these ideas would be achieved if reliable parameters for young Earth's surface composition could be established. Numerous contributions to a recent Conference have addressed this aspect.[348]

4.6 Synthesis of α-(ω-Halogenoalkyl)-α-Amino Acids – Fluorinated analogues of the common amino acids have been featured regularly in this Section over the years, and also elsewhere in this Chapter. ββ-Difluoro-L-methionine, -L-homocysteine, and -L-homoserine have been prepared starting from 5,6-O-isopropylidene-L-isoascorbic acid,[349] and hexafluoro-L-valine[350] and -leucine[351] have been obtained through a more traditional route (hexafluoroacetone and ethyl bromopyruvate for the latter case in seven steps, including enantioselective reduction of $(CF_3)_2CHCH_2COCO_2Et$).

Novel reagents for the conversion of oxazolines into β-chloro- or iodo-α-amino acids (Scheme 17) are described, giving new life to a well known route from β-hydroxy acids to halogeno-analogues.[352]

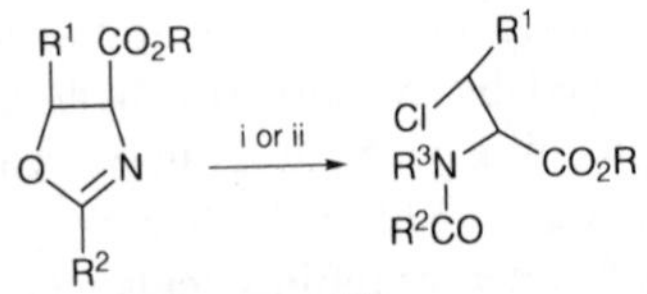

Reagents: i, ROCOCl; ii, NaI, EtCOCl → I in place of Cl

Scheme 17

4.7 Synthesis of α-(ω-Hydroxyalkyl)-α-Amino Acids – The family of β-hydroxy-α-amino acids includes several natural products, not only serine and threonine, and (2R,3R)-p-chloro-3-hydroxytyrosine is also one of the important members, since it is a vancomycin constituent. A synthesis with a conventional basis starts from the α-azido acid derivative $ArCOCH(N_3)CO_2Et$ and exploits the propensity of Baker's yeast to deliver the anti-β-hydroxy-α-azidoester with better than 90%.e.e.[353] Hydrogenation of β-trialkylsilyloxy- and -pivaloyloxy-α-acetamidoacrylates catalysed by a standard rhodium(I)–chiral diphosphine complex offers an alternative enantioselective strategy.[354]

Aldol condensation of an aldose with the anion of a protected glycine or other amino acid is an extension of the standard α-hydroxy-α-amino acid synthesis that facilitates access to polyhydroxylated homologues (Scheme 18).[355] A route from D-mannitol *via* 2-isoxazoline N-oxides also leads to these compounds, including α-alkyl homologues.[356]

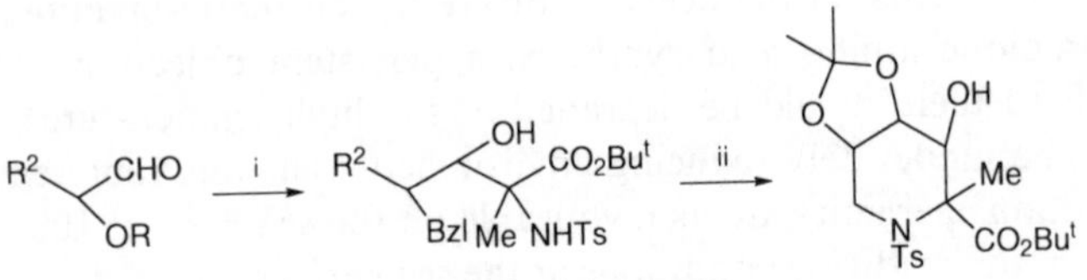

Reagents: i, $TsNHCHMeCO_2Bu^t$, LDA, MX_n; ii, H_2–Pd, then DEAD, PPh_3

Scheme 18

A route from an oxazoline-4-carboxylic ester to a β-hydroxy-α-amino acid (6 M hydrochloric acid at 80–100 °C),[357] reverses a standard heterocyclic synthesis but is a viable general synthesis method since oxazolines can be obtained in other ways, and are open to substitution and manipulation of functional groups.

(4R)-4-Hydroxy-(2S)-amino acids have become more easily available with the finding that kynureninase-catalysis is more generally applicable for trans-aldolization than the single case of kynurenine added to a simple aldehyde.[358]

4.8 Synthesis of N-Substituted α-Amino Acids – Coverage of routine N-substitution reactions of amino acids will be found in Section 6.2.2, and this section is concerned with reaction strategies by which a modified amino group is introduced into a nitrogen-free alkanoic acid synthon.

N-Boc-Hydrazino acids (69), precursors for (3S,4S)-4-hydroxy-2,3,4,5-tetra-hydropyridazine-3-carboxylic acid,[359,360] and 2-(hydroxyimino)propanohy-droxamic acid (HONHCHMeCONHOH), have been prepared by routine methods.[361] The last-mentioned compound is a more effective chelating agent than alanine towards copper(II) and nickel(II) ions.

Further examples (see also Refs. 442, 443) of the use of the N-hydroxy-glycine cation equivalent (70, and its equivalent synthon, 82) in the synthesis of diastereoisomerically-pure N-hydroxyacids have been reported (Scheme 19).[362,363]

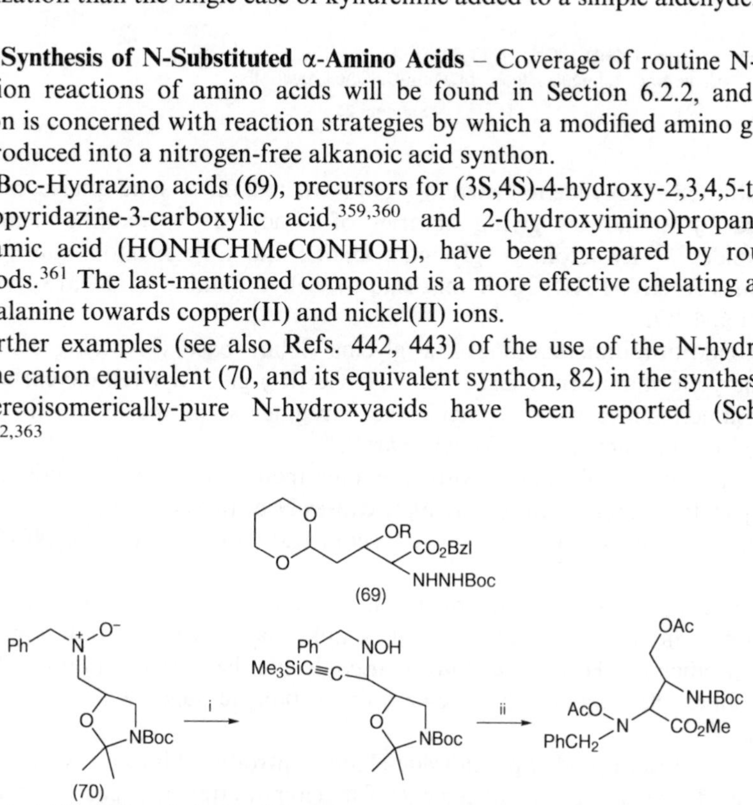

Reagents: i, LiC≡CSiMe$_3$; ii, C≡C → CHCO$_2$Me

Scheme 19

4.9 Synthesis of α-Amino Acids Carrying Unsaturated Aliphatic Side-chains – New synthesis approaches to 'dehydroamino acids', *alias* α-aminoacrylic acid H$_2$NC(= CH$_2$)CO$_2$H and homologues, are reported occasionally, also unusual outcomes of standard syntheses. Dehydration of N-protected β-hydroxy-α-amino acid esters using Boc$_2$O and DMAP gives N-protected N-Boc-dehydroalanines.[364] Oxidative decarboxylation of mono-ethyl acetamidomalo-nates [by lead(IV) acetate with copper(II) acetate; see Vol. 11, p. 25] has been used to prepare dehydroamino acids, with a specifically useful outcome in a preparation of a series of 3-substituted 2,3-dehydroprolines (Scheme 20).[365]

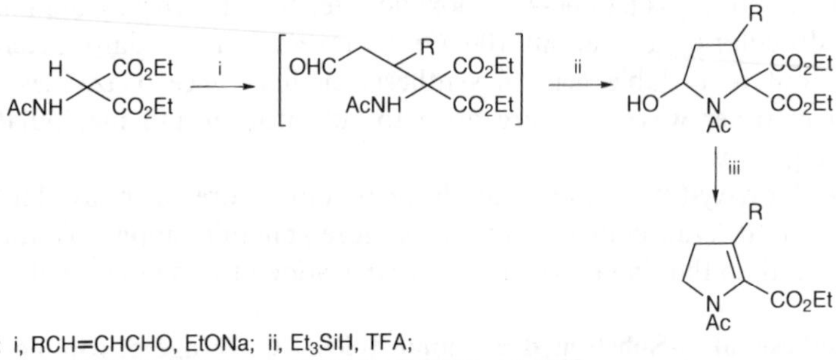

Reagents: i, RCH=CHCHO, EtONa; ii, Et₃SiH, TFA;
iii, aq. NaOH (1 equiv.), then Pb(OAc)₄–Cu(OAc)₂, py, DMF

Scheme 20

2-Substituted 4-alkylidene-oxazolin-5-ones created on a solid phase can be released by aminolysis giving libraries of N-acyl dehydroamino acid N'-alkylamides.[366] Reactions of glycinate anions and α-heteroatom-substituted glycinates $RNHCH(XR)CO_2R$ that lead to 'dehydroamino acids' are covered in Section 4.1.7.

Cycloaddition-dimerization of a glycine Schiff base (involving its azomethine ylide tautomer $R^1CH=NH^+CH^-CO_2R^2$) gives a 1-(alkoxycarbonyl-methyl)imidazolinecarboxylic acid that fragments in basic media into α-dehydroamino acid (R^1 = alkenyl or aryl).[367]

Elimination brought about with threonine methyl ester using acetic anhydride–pyridine at reflux not surprisingly causes N-bis(acetylation), but perhaps unexpectedly leads to a single geometric isomer (Z)-$Ac_2NC(CO_2$-$Me)$=CHMe.[368]

Description of a synthesis of L-vinylglycine using previously reported methods omitted four citations that should have been acknowledged.[369] Stereospecifically ²H-labelled D-propargylglycine has been obtained from nucleophilic cleavage of N-Pmc-aziridinecarboxylic acid with propargyl-lithium.[370]

An α-(γδ-unsaturated aliphatic) side-chain is introduced into an N-protected amino acid by Claisen rearrangement of its enynol ester, induced by forming its lithium enolate.[371] Polyhydroxylated homologues have been prepared by the same strategy.[372]

4.10 Synthesis of α-Amino Acids with Aromatic or Heteroaromatic Groupings in Side-chains – The literature on this topic is split between synthesis starting from the familiar coded amino acids under this heading (this is dealt with in Section 6.3), and applications of the standard general amino acid synthesis strategies.

Some notable achievements do justice to the importance of synthesis targets under this heading, none outshining the total synthesis of vancomycin and eromomycin aglycones.[373] This has required the synthesis of some unusual α-amino acids as described in earlier Volumes of this Specialist Periodical Report, often by standard methods but applied within demanding structural constraints (for example, the enantioselective synthesis of the three atropiso-meric amino acid fragments).[374]

Typical syntheses are illustrated with phenyltris(alanine)s [71; Horner-Wads-worth-Emmons olefination or Heck coupling to connect glycine units, followed by H_2–rhodium(I)/DUPHOS reduction of the resulting tris(dehydroamino acid)],[375] 4-tert-butylphenylalanine (Ref. 154), (R)- and (S)-3,5-dinitro-o-tyro-sine by routine synthesis and enzymic resolution,[376] 3'-amino-L-tyrosine,[377] 4-(4',4'-diphenylbutyl)glutamic acid,[378] naphthyl-substituted N-toluene-p-sulfonylphenylglycines,[379] amidrazonophenylalanine (N-substituted p-amino-amidinophenylalanine N'-methylcyclohexylamide)[380] and other o-, m-, and p-amidinophenylalanine amides showing potential as novel thrombin inhibi-tors,[381] α-methylphenylalanine 4'-phosphonic acid and the cyclic relative (72) as analogues of α-methyl-3-carboxyphenylalanine,[382] and p-boronophenyl-serine and o-carboranylmethyl-3-hydroxytyrosine, prepared from an isocyano-glycine synthon.[383] The binaphthyl (73) and its epimer[384] add new examples of axially chiral bis(α-amino acid)s (Vol. 30, p. 13). 4'-(Diphenylphosphono-methyl)- and 4'-diethylphosphonodifluoromethyl-phenylalanines and β-amino acid homologues have been prepared from 2-benzyl-1,3-propanediols carrying the phosphorus groupings, by lipase-catalysed hydrolysis of the di-acetate, and routine amination steps (DPPA or HN_3 and reduction).[385]

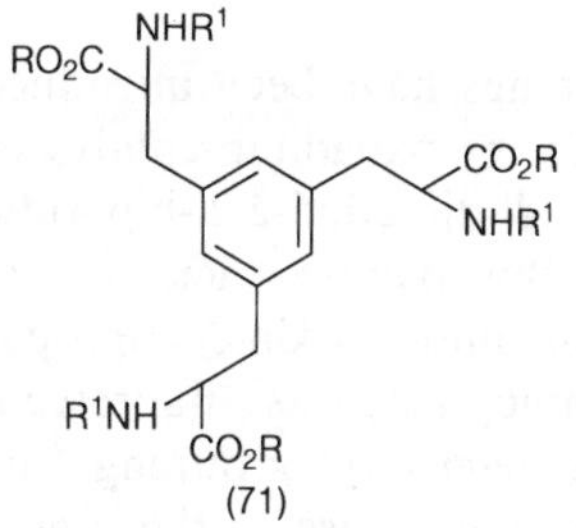

(71)

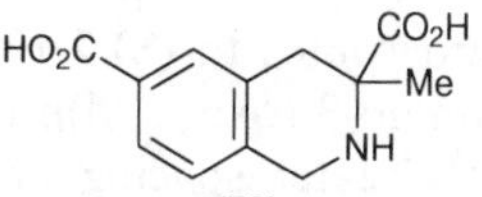

(72)

The interest in α-amino acids carrying heteroaromatic groupings as attach-ments to aliphatic side-chains lies either in their pharmacological potential, or in their use as monomers for the synthesis of peptide nucleic acids (PNAs). α-Amino acids carrying isoxazolyl side chains (74) have been prepared as analogues of AMPA, extending a series of publications that has appeared over several years.[386] α-Substituted-α-(β-triazolylmethyl)glycines have been pre-pared through cycloaddition of alkynes to 2-phenyl-4-azidomethyloxazoline,

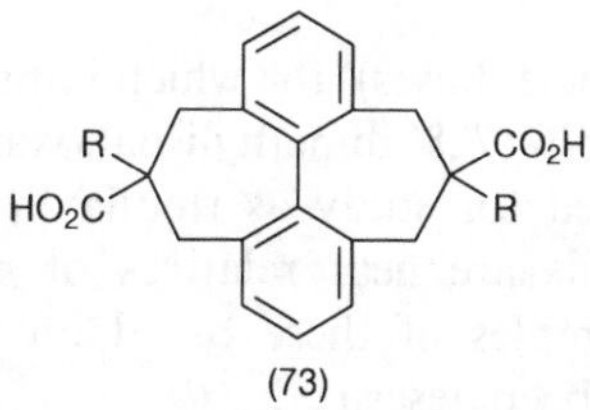

(73)

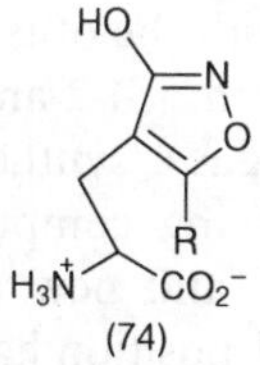

(74)

followed by ring opening and oxidation ($CH_2OH \rightarrow CO_2H$).[387] β-Tetrazolyl analogues have been prepared through addition of 5-aryltetrazoles to N-benzyloxycarbonyl dehydroalanine methyl ester.[388] Other analogues include ring-halogenated histidines starting from imidazoles,[389] β-hydroxy-β-(N'-toluene-p-sulfonylpyrrol-3-yl)alanine, *alias* 3-deaza-β-hydroxyhistidine, using a glycine anion synthon approach, or its Wittig equivalent.[390] Pictet-Spengler cyclization of these products with aldehydes gives pyrrolo[2,3-*c*]tetrahydropyridine-5-carboxylic acid esters which on aromatization with Pd-C give 6-azaindoles. β-Substituted tryptophans have been prepared through ring-opening of 2,3-aziridinotetrahydrofurans with indoles, and elaboration of the reaction products (Scheme 21).[391]

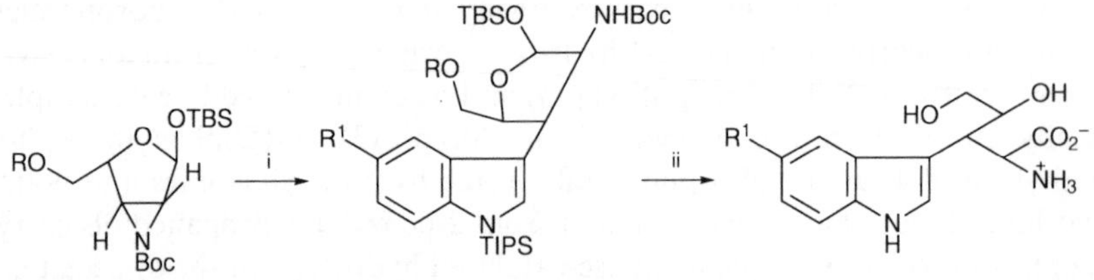

Reagents: i, see text; H_3O^+, then oxidation

Scheme 21

Six-membered heteroaromatic groupings have been introduced into amino acid side-chains using β-iodoalanine (75; by palladium-catalysed condensation with a pyridone triflate),[392] with 4,4'-dimethyl-2,2'-bipyridyl to prepare 2-amino-3-(4'-methyl-2,2'-bipyridin-4-yl)propanoic acid,[393] and similarly straightforward routes to (S)-2-Fmoc-amino-3-(oxin-2-yl)propanoic acid and (S)-2-Fmoc-amino-3-(oxin-5-yl)propanoic acid as fluorescent 8-hydroxy-quinolines.[394] Greater spacing of the heteroaryl grouping from the glycine moiety is seen in butyrines carrying a nucleobase at the γ-position (76, and

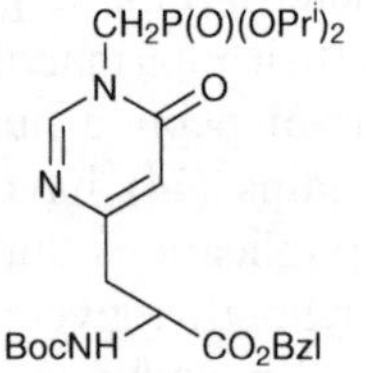

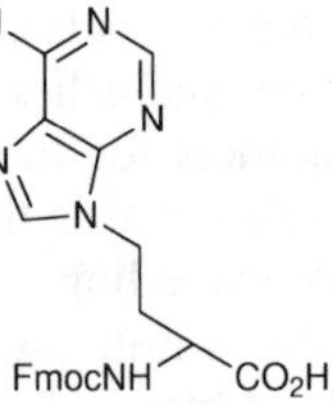

three analogues with the other nucleobases), for which improved syntheses are reported,[395] and in (S)-2-amino-6-(7',8'-dimethylisoalloxazin-10'-yl)hexanoic acid and homologues, synthesized for study as riboflavin analogues.[396] The nucleobase-containing compounds are near relatives of monomers used to prepare PNAs,[397] and new examples of these based on proline carrying a nucleobase at its 4-position have been described.[398]

4.11 Synthesis of α-Amino Acids Carrying Amino Groups, and Related Nitrogen Functional Groups, in Aliphatic Side-chains – β-Aminoalkyl-α-amino acids have become permanent inhabitants of this section, and new syntheses include conversion of tert-butyl crotonates into 2,3-diaminobutanoates through Sharpless asymmetric aminohydroxylation,[399] aminolysis of cyclic sulfamidates and thiohydrolysis of the resulting α-aminosulfamic acids,[400] and Grignard addition to β-aminonitrones derived from L-serine, leading to (2R,3S)- and (2R,3R)-3-substituted αβ-diamino acids.[401] Approaches to novel diamino acid derivatives have led to the squarylium dye (77),[402] and to the L-arginine analogue $H_2NC(=NH)NH(CH_2)_3CH(NHR)B(OH)_2$.[403]

(77)   (78)

4.12 Synthesis of α-Amino Acids Carrying Silicon Functional Groups in Aliphatic Side-chains – A synthesis of γ-silyl-α-amino acids involves a standard development of the aldehyde function of the Garner aldehyde (123) by silylcupration of its ethynyl analogue.[404]

4.13 Synthesis of α-Amino Acids Carrying Phosphorus Functional Groups in Aliphatic Side-chains – Mention has been made, in other Sections, of reports of common amino acids functionalized with phosphorus oxyacid groupings, including analogues in which the carboxy group is replaced [*e.g.* $HO_2C(CH_2)_nCH(NHAc)P(O)(OPh)_2$ as analogues of aspartic and glutamic acids[405]]. The transition state analogue inhibitor (78) of *E. coli* α-glutamylcysteine synthetase is a particularly interesting example of newer thinking in this area.[406]

4.14 Synthesis of α-Amino Acids Carrying Sulfur-, Selenium-, or Tellurium-containing Aliphatic Side-chains – Protected ω-mercapto-α-amino acids have been obtained from ω-bromo-α-azido acids by displacement using a sulfur nucleophile, followed by tin(II) chloride reduction.[407] Derivatives conferring lipophilic properties on S-adenosyl-L-methionine have been prepared.[408]

Sulfinamide and sulfone analogues of (78) were included in the synthesis study referred to in the preceding Section (Ref. 406).

4.15 Synthesis of β-Amino Acids and Higher Homologous Amino Acids – Reviews under this heading include α-hydroxy-β-amino acids.[409]

Asymmetric amination of achiral αβ-unsaturated esters by lithium (S)- or (R)-N-benzyl-N-α-methylbenzylamide continues to serve well for the synthesis of N,N-disubstituted β-amino acid esters (79),[410] including D-β-phenylalanine tert-butyl ester (a constituent of moiramide B and andrimid).[411] Chiral lithium amide bases that are useful in this context have been reviewed,[412] the choice of base for the synthesis of both stereoisomers of β-homothreonine,[413] and other anti-α-alkyl-β-amino acids, by this strategy being (R)-N-(1-phenylethyl)-N-(trimethylsilyl)amine or a homochiral amidocuprate.[414] N-(α-Methylbenzyl)-N-3,4-dimethoxybenzylamine used in this protocol offers the opportunity for selective deprotection.[415] A nine-step route based on this amination approach, to a 2-aminocyclohexanecarboxylic acid derivative, the kelatorphan-like enkephalinase inhibitor (80), starts with tert-butyl cyclohex-1-ene-1-carboxylate,[416] and the same approach provides an essential intermediate in a synthesis of tussilagine and isotussilagine.[417] Another method for asymmetric amination of achiral αβ-unsaturated esters is illustrated with formula 9, Ref. 53.

αβ-Unsaturated pyrazolamides $RCH=CHCONR^1R^2$ (R^1R^2=3,5-dimethyl-pyrazolyl residue) add O-benzylhydroxylamine with high enantioselectivity when a chiral Lewis acid catalyst is used, prepared from $MgBr_2.Et_2O$ and a chiral bis(oxazolyl)cyclopropane.[418]

Conventional amination of lipase-resolved syn-2,3-dihydroxy-3-phenyl-propanoic acid enantiomers by conventional bromination–azidation and epoxyacid pathways gives (2R,3S)-phenylisoserine.[419] 1,2-Dihydroxyindane-2-carboxylic acid esters can be selectively benzylated and substituted by azide, and processed to give the β-aminoindanecarboxylic acid, designed as a conformationally constrained phenylisoserine.[420] Amination *via* azides leading to β-amino acids is also illustrated in Ref. 385.

Lewis acid-promoted addition of chiral imines to enol ethers $CH_2=C(OEt)$-$OSiMe_2Bu^t$ to generate an additional chiral centre leads to (R,R)-β-amino acid esters, although the corresponding process with 1,3-oxazolidines derived from (R)-phenylglycinol yields (R,S)-products.[421] A solid phase version of this process, but with achiral tethered imines, has been explored.[422] Homochiral oxazolidinones used in this way with αβ-unsaturated esters for enantioselective β-amino acid synthesis are, in effect, recyclable chiral reagents.[423] Homochiral N-sulfinylimines (S)-RS(=O)N=CHPh add to the lithium dienolate of methyl propen-3-oate to give (E)-α-ethylidene-α-amino acid esters *via* a cis/trans mixture of the α-vinyl isomer.[424] Catalysis by $ZnCl_2$, $ZnBr_2$, or ScF_3 leads to this outcome, although catalysis is not necessary but absence of catalyst results in the formation of the β-(N-arylsulfinylamino)-α-ethylidenealkanoic acid ester. o-(Nitrophenyl)imines and the lithium enolate of methyl isobutyrate combine to give a β-aryl-αα-dimethyl-β-alanine that is a potentially UV-cleavable resin linker for solid-phase synthesis.[425] Preparations of β-amino-aldehydes *via* β-(toluene-p-sulfinylamino)alkanoate esters prepared in this way, and chain extension to δ-amino acid derivatives, emphasise the usefulness of imines in β-amino acid synthesis.[426] The equivalent Mannich-type process, in which an aldehyde, an amine, and the silyl enol ether are combined under indium trichloride catalysis, proceeds in moderate to good yields and leads to β-aminoketones with simple enol ethers.[427] A mixture of aldehyde, amine and homophthalic anhydride with a Lewis acid catalyst gives a trans-isoquinolonic acid (Scheme 22).[428] C-Acylation of imines with N,N′-carbonyldi-imidazole gives β-enamino acid derivatives [$R^2C(=NR^1)CH_3 \rightarrow R^2C(NHR^1)=CH$-COIm].[429] Imines are reactants for standard β-lactam syntheses; their preparation in immobilized form and conversion into β-lactams is only the start of the problem with solid-phase synthesis but removal of the product from the polymer under acidic and basic conditions has been successful.[430] In what has been described as a new three-component condensation, β-(N-acylamino)-aldehydes $R^1CONHCH(CH_2R^2)CHR^2CHO$ are formed from a primary amide, two equivalents of an aldehyde, and a strong acid catalyst (CF_3SO_3H);[431] this seems to be anything but new, since it is described by the imine alkylation process illustrated in the preceding examples.

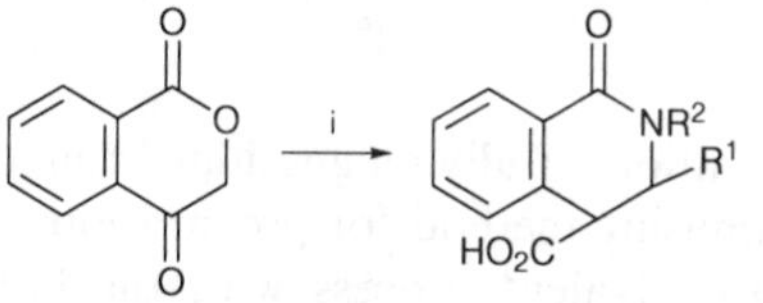

Reagents: i, R^1CHO/R^2NH_2, MeCN, Na_2SO_4, $BF_3 \cdot Et_2O$

Scheme 22

A radical cyclization pathway (Scheme 23) transforms O-benzoylhydroxamic acids into β-lactams,[432] but the other reaction products formed 75% of the total so there is room for improvement. The 'β-lactam synthon method' for β-amino acid synthesis has been illustrated by highly stereo-selective alkylation of 3-trialkylsilyloxy-β-lactams that leads to α-methyliso-serines.[433]

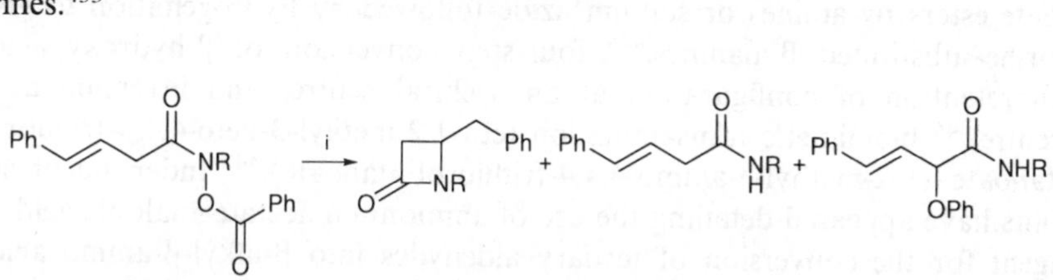

Reagents: i, Bu_3SnH, AIBN

Scheme 23

Attack of N-(benzyloxycarbonylaminomethyl)benzotriazole, an electrophilic aminomethyl synthon, on enolates of chiral 3-acyl-1,3-oxazolidin-2-ones gives 3-aminopropanoic acid derivatives; an alternative approach is to apply the Curtius rearrangement to the same substrate.[434] Dimethylformamide diethyl acetal condenses with α-heteroarylacetic acid esters under microwave irradiation, to give N,N-dimethyl eneamine esters that are readily converted into αβ-unsaturated β-amino acid esters by reaction with a primary amine.[435]

Alternative asymmetric amination approaches to β-amino acids include aminohydroxylation, *e.g.* of (E)-vinylphosphonates $RCH=CHPO_3Et_2$, offering α-hydroxy-β-amino phosphonic acids in moderate e.e.,[436] and a synthesis of (3R)-amino-(2S)-hydroxyheptanoic acid.[437] N-Toluene-p-sulfonyl-aziridine has been identified as an intermediate in a route from N,N-dimethyl-cinnamide to phenylisoserine using this methodology,[438] and aziridinecarboxylic acid derivatives (81) have been used as starting points in routes to 3-amino-2,3-dideoxyhexonic acids.[439] Palladium(0)-catalysed cyclization of (S)-PhCH(NHBz)CH = CHCH$_2$OAc (from L-phenylglycinal) gives (4S,5R)-2,4-diphenyloxazoline-5-carboxylic acid, a precursor for the stereo-isomer of phenylisoserine component of taxol.[440]

(81)   (82)   (83)

Diastereoselective addition of allylmagnesium bromide to an O-(1-phenyl-butyl)aldoxime is a promising method for preparation of a range of protected β-amino acids.[441] The equivalent process with the homochiral nitrone (82) leads to (2S,3R)- and (2S,3S)-3-phenylisoserine after appropriate functional group development, the choice of Lewis acid in the initial addition step determining the stereochemical outcome.[442] Addition of the same nitrone to excess O-methyl-O-tert-butyldimethylsilyl ketene acetal provides entry to the βγ-diamino acid series.[443] Enantioselective and diastereoselective synthesis of syn-β-amino-α-hydroxy acids occurs through addition of toluene-p-sulfonyl isocyanide to chiral α-hydroxy-βγ-unsaturated esters.[444] More conventional amination strategies include: reaction of 3-perfluoroalkyl-3-fluoroprop-2-enoate esters by amines or sodium azide followed by hydrogenation to give fluorine-substituted β-alanines;[445] four-step conversion of β-hydroxy acids with retention of configuration at an α-chiral centre, and inversion at a β-centre;[446] biomimetic transamination (ethyl 2-methyl-3-keto-4,4,4-trifluoro-butanoate → 2-methyl-3-amino-4,4,4-trifluorobutanoate).[447] Independent accounts have appeared detailing the use of ammonium acetate–malonic acid as reagent for the conversion of tertiary aldehydes into β-alkyl-β-amino acids $[R^1R^2R^3CHO \rightarrow R^1C(R^2R^3)CH_2CO_2H]$,[448] and of benzaldehyde into 3-benzoylamino-3-phenylpropanoic acid [used in a preparation of (2R,3S)-

phenylisoserine involving enzymatic resolution and α-iodination (with LiHDMS and I_2) as key steps].[449]

Carboxylation of amines is relatively easy for the preparation of β- and higher homologous amino acids, through cyanohydrin formation of homochiral α-aminoalkanals [(S)-NN-dibenzylphenylalaninal gives (2S,3S)-3-amino-2-hydroxy-4-phenylbutanoic acid;[450] KCN–acyl halide gives α-acyloxy-β-amino acids[451]]. N-Boc-Phenylglycinal adds diethyl phosphite in a moderately diastereoselective fashion to give diethyl 2-N-(Boc-amino)-1-hydroxy-2-phenylethyl phosphonate, a practical benefit being the ease of separation of syn- and anti-isomers by fractional crystallization.[452] α-Trialkylsilyl-β-amino acids are unusually easily desilylated in aqueous media, attributed to anchimeric assistance by amino and carbonyl functions.[453]

Ethyl cyanoacetate is a ready-made source of αα-disubstituted β-amino acids, bis-alkylation by 2,2′-bis(bromomethyl)-1,1′-biphenyl followed by $NaBH_4$–$CoCl_2$ selective reduction of the cyano group leading to αα-disubstituted β-amino acids possessing axial chirality.[454] Allylindium reagents give enamines $RO_2CCR=C(NH_2)CH_2CH=CH_2$.[455] α-Aminonitriles are also suitably- activated in this context, reaction with Reformatzky reagents giving β-amino esters, β-lactams, tertiary amines and cinnamic acid derivatives in an apparently wayward process that can be controlled to give the β-amino esters exclusively.[456] α-Aminoalkylbenzotriazoles, prepared from an aldehyde, an amine and benzotriazole, react with Reformatzky reagents derived from ethyl bromofluoroacetates to give α-fluoro-β-amino acid esters.[457] Acylation of tolyl cyanide $PhCH_2CN$ by the chloroformate of Oppolzer's (S)-sultam and routine ensuing steps leads to (S)-BocNHCH$_2$CHPhCO$_2$H.[458] Another illustration of the use of the tetrahydropyrimidinone chiral synthon (83), prepared from 3-aminobutanoic acid, has been presented; its 5-halogeno-derivative has been used in preparations of enantiomerically pure threonine and allothreonine.[459]

Standard practice is also represented in Arndt-Eistert homologation of N-protected α-amino acids.[460] More than 20 repetitive papers have appeared on this topic in the last four years, including three more. One of the current papers[461] compares the Arndt-Eistert strategy with Evans amination; another[462] deals with the ultrasound-mediated Wolff rearrangement approach. Instead of Wolff rearrangement, an α-amino acid diazoketone can be treated with N,O-dimethylhydroxylamine to give a β-amino acid dimethylhydroxamate.[463] Lossen rearrangement of alkyl succinimidyl carbonates in basic buffers leads to the corresponding β-alanines.[464] Rearrangement of alkoxycarbonylmethyl iminoethers $ArN=CR^FOCH_2CO_2R$ gives β-perfluoroalkyl isoserines, through a promising new general method.[465]

Sequential Birch reduction, 3-alkylation, and catalytic hydrogenation of pyrrole-3-carboxylic acid derivatives give β-substituted β-prolines.[466]

The term 'the γ-amino acids' first brings to mind GABA (γ-aminobutyric acid) and statine as standing at the head of the list, and while the former does not normally stimulate any particular synthesis interest, the latter certainly does. Reductive alkylation of (S)-O-tert-butyldimethylsilyl malimide has been established as the key step in a new synthesis of (3S,4R)-statine,[467] and other non-general routes are available for syntheses of *cis*- and *trans*-(3-substituted vinyl)-γ-aminobutyric acids [by Claisen rearrangement of phthalimides PhtCH$_2$CHCR^{3}CHOHR^2, or by Wittig synthesis starting from 3-formyl-4-phthalimidobutanoic acid],[468] and a synthesis of a GABA analogue [the photogenerated Barton ester radical from N-Fmoc-α-aminoisobutyric acid, trapped by methyl acrylate, gives FmocNHCMe$_2$CH$_2$CH(SPy)CO$_2$Me].[469] Allylic acetates Ph$_2$C=CHCH(OAc)R subjected to palladium-catalysed substitution by methyl cyanoacetate, followed by functional group development, offer an entry to α-substituted GABAs.[470] The useful (−)-isomer of carnitine can be obtained from its (+)-isomer (ca 64% at equilibrium) through the use of d- or l-carnitine dehydrogenases from *Agrobacterium* sp.525a.[471] A short synthesis of another γ-amino acid with physiological activity, R-(−)-baclofen, starts with p-(2-aminoethyl)chlorobenzene, N-malonylation and conversion into the α-diazoester preceding enantioselective ring closure by C-H insertion through the influence of dirhodium(II) tetrakis[(S)-N-phthaloyl tert-leucinate].[472] (S)-Vigabatrin H$_3$N$^+$CH(CH=CH$_2$)(CH$_2$)$_2$CO$_2$$^-$ has been synthesized from butane-1,4-diol, the laborious nature of the route following determination by the authors to use the Sharpless aminohydroxylation.[473] (S,S)-4-Amino-3-hydroxy-5-phenylpentanoic acid has been prepared from 4-benzyl-5-vinyloxazoline, formed from the oxazolidin-2-one derived from L-phenylalanine.[474] γ-Amino α-keto-esters ZNHCHRCH$_2$COCO$_2$R′, prepared from L-α-amino acids *via* β-amino acids, lead to 4-amino-2-hydroxyacids through standard steps including lipase and oxidoreductase resolution to generate the required stereochemistry at C-2.[475]

L-α-Amino acids serve as starting materials for homochiral γ-aminoalkynoic acid esters from aldehydes, through a Corey-Fuchs route -CHO → -CH=CBr$_2$ → -C≡CCO$_2$Me.[476] Urethane-protected N-carboxyanhydrides (UNCAs; see also Ref. 761) give α-alkyl-γ-amino-β-ketoesters through condensation with lithium enolates (84 → 85),[477] and give β-amino α-ketoesters with

(84)   (85)

cyanomethyltriphenylphosphonium chloride.[478] Lactacystin is composed of a cysteine and γ-lactam fragment, and an analogue of the latter has been prepared from a lithium enolate Me$_2$C=C(OMe)OLi and the protected amino aldehyde (86; see Ref. 246),[479] or through a route starting from condensation

(86)

of DL-dimethyl isopropenylmethylmalonate with methyl N-4-methoxybenzyl-glycinate.[480] Other classical organic synthesis procedures are seen in Beckmann ring expansion of dichloroketene – chiral enol ether cycloadducts, to give the lactone of (3S,4S)-4-amino-3-hydroxy-5-phenylpentanoic acid,[481] and ring expansion of β-lactams to γ-lactams with trimethylsilyldiazomethane anion and photolytic Wolff rearrangement.[482]

γ-Amino acids are prototypical dipeptide mimetics, and replacement of the amide bond of a dipeptide by various isosteres is covered elsewhere in this Specialist Periodical Report. αβ-Unsaturated δ-amino acids $PhNHCHArCH_2CR=CHCO_2H$, prepared from an imine and a 1,1-di(tri-methylsilyloxy)-buta-1,3-diene,[483] are one step removed from being described as a dipeptide mimetic, and so are the compounds resulting from a novel 1,6-amino addition reaction to naphthalene followed by electrophilic alkylation.[484] (3R,4S)-3-Tri-isopropylsiloxy-N-Boc-β-lactams start pathways to 4-hydroxy-, 3,4-dihydroxy-, and 3-amino-4-hydroxy-5-amino acids,[485] also obtained from oxazolidine-5-carbaldehydes (Scheme 24).[486] Homochiral vinylogous N-MTS-

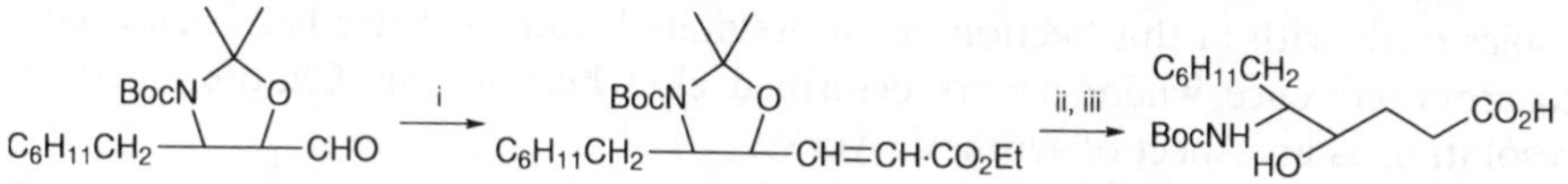

Reagents: i, Ph_3PCHCO_2Et; ii, H_2–Pd/C, then *p*-TsOH, EtOH; iii, *p*-TsOH, thf

Scheme 24

aziridinecarboxylic acids carrying amino acid side-chains undergo ring-opening with TFA to give substituted 5-aminopent-3-enoic acids closely mimicking dipeptides in both structure and stereochemistry.[487] Conformational constraint is introduced by incorporating part of the carbon chain of a δ-amino acid into a cycloalkane grouping, as in piperidine-4-acetic acids[488] and 3-aminomethyl-5-substituted cyclohexanecarboxylic acids.[489] N-Fmoc-2-carboxybenzylamine justifies mention here (since anthranilic acids and related aromatic amino acids are not covered in this Chapter) but only on the basis that it has been proposed as a β-turn-inducing residue for peptide synthesis.[490]

Greater separation of amino and carboxy functions is becoming easier to arrange, with ring-closing metathesis reactions of amides $CH_2=CH(CH_2)_nCONH(CH_2)_mCH=CH_2$ becoming well established (14-membered lactams;[491] lactams of βγ-unsaturated δ-amino acids, from the allylamine prepared from L-phenylalaninal;[492] see also Refs. 907-915, Section 6.3). Other methods are used for 6-aminoalkanoic acids and derivatives, *e.g.* the nega-mycin synthesis intermediate (87) prepared using the L-malic acid-derived synthon (88),[493] the 8-aminononanoic acid, and 9-, and 10-amino homologues that are components of polyazamacrolides in the defence secretion of the ladybird beetle *Epilachna borealis*,[494] the 6-amino-4,6-deoxyheptopyranuronic acid constituent of amipurimycin,[495] 5-aminomethyltetrahydrofuran-1-carboxylic acids (89; given the name C-glycosyl sugar amino acid derivatives) that open up the synthesis of peptide analogues with a pseudopolysaccharide backbone.[496]

(87)   (88)   (89)   (90)   (91)

Compounds with amino and carboxy groups separated over an aromatic ring system, *e.g.* the homochiral [2.2]paracyclophane (90)[497] and the proposed protein β-turn moiety (91),[498] have been prepared.

4.16 Resolution of DL-Amino Acids – The following subsections segregate the topics dealt with in this Section, as in previous Volumes. Care has been taken to cross-reference where papers described elsewhere in this Chapter include resolution as an aspect of synthetic work.

4.16.1 Diastereoisomeric Derivatization and Diastereoisomeric Salt Formation – Classical procedures are represented by DL-leucine–S-(-)-1-phenylethanesulfonic acid mixtures crystallized from MeCN–MeOH (to give the less-soluble L-amino acid salt) or from MeCN–water (to give the less-soluble D-amino acid salt monohydrate);[499] by ethyl DL-methioninate–N-acetyl-D-methionine (less-soluble D,D-salt);[500] by binaphthyl-based 1-aminocycloheptane-1-carboxylic acids showing axial chirality, resolved using L-phenylalanine cyclohexylamide;[501] and by 2-(aminomethyl)cyclopropane-1-carboxylic acids resolved as (R)-pantolactone derivatives.[502] In reverse, the resolution of a DL-alkylamine using N-acetyl-L-leucine may be noted.[503] Kinetic resolution procedures have been employed: methyl DL-4-chlorophenylalaninate–(2S,3S)-tartaric acid–salicylaldehyde from which the derivatized D-amino acid ester crystallizes;[504] the planar-chiral DMAP (92) induces equilibration of 2,4-disubstituted oxazolin-5(4H)-ones with modest e.e.;[505] novel chiral cation exchange resins accomplish the kinetic resolution of cyclic amino acid derivatives.[506] A review covers resolution of amino acids based on asymmetric transformation.[507]

(92)

4.16.2 Enzyme-catalysed Enantioselective Reactions of Amino Acid Derivatives
The propensity of familiar enzymes for handling reactions of non-protein amino acids almost as well as they deal with their normal substrates has enhanced the practical value of this 'resolution' strategy.

Penicillin G acylase-catalysed acylation of the L-enantiomer in DL-phenylglycine is exploited for production of D-phenylglycine,[508] and *Aspergillus* acylase has been applied analogously to the resolution of 7-azatryptophan[509] (see also Refs. 37, 38). An acylase-derivatized medium for use in a centrifugal partition chromatographic mode has been illustrated with N-acetyl-DL-amino acids.[510]

Porcine kidney acylase accepts N-acetyl 4'-substituted (RS)-phenylalanines as substrates.[511] *Candida antarctica* B lipase catalyses the amidation by ammonia of methyl DL-phenylglycinate (47% conversion) giving D-phenylglycinamide in 78% e.e. when the strategy of *in situ* racemization by pyridoxal is employed to maximize the conversion.[512] Substituted DL-hydantoins are accepted as substrates by papain, hydrolysis leading to L-α-amino acids *via* N-carbamoyl derivatives.[513] The more usual protocol with these compounds is demonstrated by conversion by *Agrobacterium tumefaciens* of a DL-p-hydroxyphenylhydantoin into D-p-hydroxyphenylglycine; this involves the cooperation of a racemase, a D-hydantoinase, and an unusual D-selective N-carbamoylamino acid amidohydrolase.[514] In a distantly-related kinetic resolution procedure leading to N-benzyloxycarbonyl-D- and L-serines, acetylation–equilibration of DL-3-(hydroxymethyl)-1,4-benzodiazepin-2-ones has been mediated by immobilized *Mucor miehei* lipase (see Vol. 30, p. 42).[515] A more conventional procedure is represented by the immobilized chymotrypsin–methyl DL-phenylalaninate system.[516]

4.16.3 Chromatographic Resolution, and Applications of Other Physical Methods for Resolution of DL-Amino Acid Derivatives – Newer variations of standard practice are represented by resolution of DL-tryptophan by passage through a porous hollow-fibre membrane in which immobilization of bovine serum albumin has been accomplished using glutaraldehyde,[517] similar results with DL-phenylalanine,[518] and resolution of N-Boc- and -Z-DL-amino acids over a chiral stationary phase (CSP), *viz*. Chiralpak AS.[519] A chiral crown ether stationary phase has been used with modest results for the resolution of cyclic β-amino acids.[520] CSPs have been reviewed.[521]

The binding mechanism for a CSP carrying N-(3,5-dinitrophenyl)-R-phenylglycine[522] and for another standard CSP [(N-3,5-dinitrobenzoyl)-L-leucine as chiral discriminator][523] involves hydrogen bonding and π-stacking interactions; typical applications of CSPs of this type involves resolution of N-(3,5-dimethoxybenzoyl)-DL-amino acid NN-diethylamides over a new CSP derived from (S)-N-(3,5-dinitrobenzoyl)-L-leucine N-phenyl-N-alkylamide [alkyl = $(CH_2)_3Si(OEt)_3$].[524] Reversing the structural features of stationary phase and substrate, resolution of N-(3,5-dinitrobenzoyl-DL-amino acid amides over silica gel derivatized with (R)-4-hydroxyphenylglycine has been demonstrated.[525] Aminopropylsilica derivatized with N-(3,5-dichloro-s-triazinyl)-L-proline tert-butyl ester serves to resolve DL-amino acids as copper(II) complexes on the ligand-exchange chromatography principle.[526] A novel extension of this to ligand-exchange adsorbents imprinted with a polymerizable copper(II) complex of the L-enantiomer of the amino acid to be resolved has been described in an *ACS Symposium Volume* devoted to the topical polymer-imprinting technology.[527] A review of molecule-imprinted polymers for use in analysis has appeared,[528] recent developments including efficient resolution of methyl DL-tryptophanate,[529] and DL-phenylalanine anilides.[530] The principle extends to molecule-imprinted membranes, polymers grafted with the tetrapeptide H-Asp(O-cyclohexyl)-Val-Asn-Glu(O-Bzl)-CH_2-,[531] or with H-Glu(O-Bzl)-Glu-Lys(4-Cl-Z)-Leu-CH_2- (imprinted with N^α-acetyl-L-tryptophan or its N-Boc analogue),[532] which preferentially retain the print isomer and therefore allow the opposite enantiomer to migrate faster. A poly(pyrrole)–poly(vinyl alcohol) colloidal dispersion imprinted with Z-L-aspartic acid preferentially incorporates the D-enantiomer from DL-glutamic acid.[533]

Enantioselective permeation of amino acid perchlorates through a solid membrane formed by grafting $(1 \rightarrow 6)$-2,5-anhydro-3,4-di-O-methyl-D-glucitol on to poly(acrylonitrile) uniformly favours faster permeation by the D-enantiomer, relative rate differences being highest for phenylglycine, lowest for tryptophan. The separation mechanism seems to involve molecular size as a major factor, but best results are not impressive (up to 75% e.e. achieved for phenylglycine).[534]

DL-Kynurenine has been resolved by high speed countercurrent chromatography;[535] other liquid–liquid media capable of resolution of amino acids include chiral liquid membranes.[536]

4.16.4 Host – Guest Resolution – This topic is also covered in Section 7.5, but papers are located here if exploitation of the phenomenon in terms of resolution is involved.

Pronounced enantioselection towards DL-amino acids, in their zwitterionic forms or as K or Na salts, is seen when chiral naphtho-18-crown-6-ethers (93) and aza-analogues derived from α-methyl D-mannopyranoside are contained within a supported liquid membrane separating aqueous media.[537] Protonated heptakis(6-amino-6-deoxy)-β-cyclodextrin binds the anionic form of L-enantiomer of N-acetyl-L-tryptophan, -phenylalanine, -leucine, or -valine more strongly than the D-enantiomer,[538] and permethylated cyclofructans similarly discriminate enantiomers of DL-amino acid ester hydrochlorides.[539]

4.16.5 Prebiotic Resolution of DL-*Amino Acids* – Mechanisms accounting for the presumed predominance of the L-amino acids, destined to be built into proteins at the start of life processes on this planet (and perhaps elsewhere), have largely featured enantioselective destruction, *e.g.* by asymmetric photolysis,[540] or by differential interaction of elementary particles with individual enantiomers.[541] A new idea, that a net natural chiral right-handed helical force field produced by the Earth's orbital chirality could make 'right-handed enantiomers' more stable than their 'left-handed' forms, is asserted to have

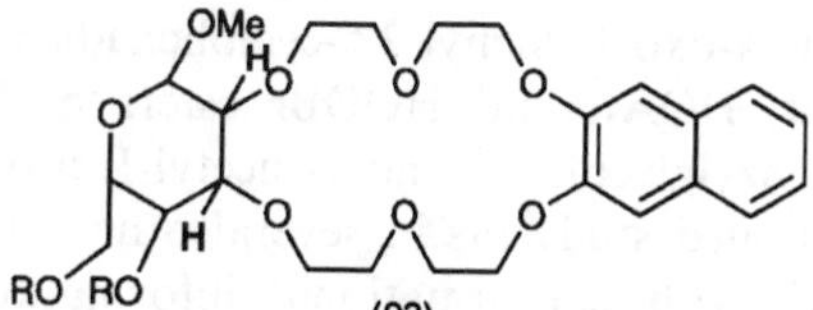

(93)

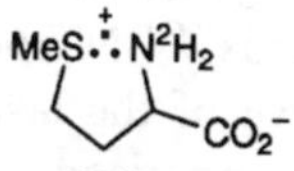

(94)

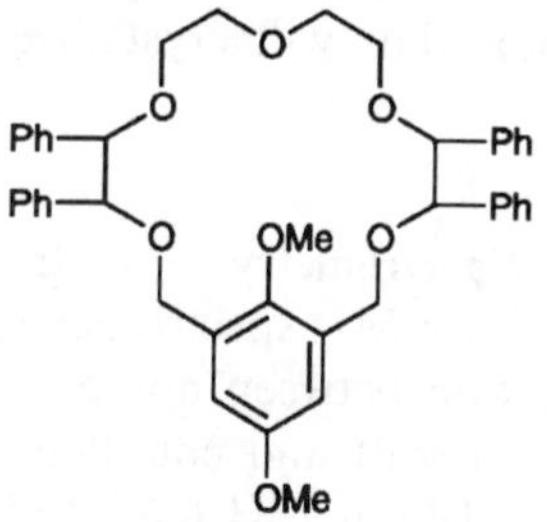

(95)

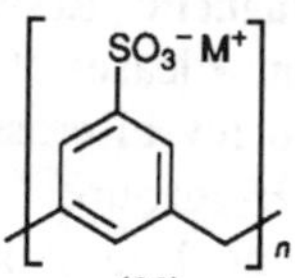

(96)

experimental support (though apparently based on a cyclic argument).[542] Parity-violating energy differences between L- and D-enantiomers of a chiral molecule continue to be claimed to underly the favoured survival of L-amino acids, stabilized by the weak nuclear force.[543]

Selection by RNA of the L-enantiomer from a DL-amino acid, when undergoing aminoacylation *in vitro*, requires the system to be constrained on a surface that mimics a prebiotic monolayer;[544] stereochemical parameters for aminoacylation of RNA have been reviewed.[545]

5 Physico-chemical Studies of Amino Acids

5.1 X-Ray Crystal Analysis of Amino Acids and their Derivatives – Amino acids and their derivatives are covered, in that order, in this Section; a review has appeared of supramolecular networks in amino acid crystals and co-crystals.[546] The continual re-examination of the common amino acids usually accompanies advances in instrumentation or technique, and the 'accelerated' X-ray crystal analysis of DL-proline monohydrate is a significant example,[547] as is DL-aspartic acid at 20 K.[548] More routine work concerns D-phenyl-glycine hydrochloride,[549] oxalic acid salts of DL- and of L-arginine,[550] maleic acid salts of DL- and of L-arginine,[551] the (1:1)-salt of L-histidine with 4,5-imidazoledicarboxylic acid, and the corresponding L-lysine salt,[552] the (1:1)-salt of L-serine with squaric acid[553] and bis(L-asparaginium) hydrogen squa-rate monohydrate,[554] ibotenic acid,[555] and L-carnitine L-tartrate.[556] The electron charge density distribution in X-ray crystal structure of DL-proline hydrate and DL-aspartic acid hydrate has been used to support a discussion of quantum topological theory.[557]

Derivatives assessed by X-ray crystal analysis are mostly chosen so as to be of interest in a biological or a laboratory synthesis context: O-phospho-L-

tyrosine,[558] N-(3,5-di-tert-butyl-4-oxo-1-phenyl-2,5-cyclohexadien-1-yl)-L-iso-
leucine methyl ester,[559] HOBt, HOAt, and HOOBr esters of N-protected
L-amino acids,[560] N-Boc-N-benzylglycine,[561] and N-acetyl-L-prolinamide.[562]
The objective of the last-mentioned study (as in several other cases; Section
5.7) was comparison of data with conformational information obtained
through molecular-orbital calculations. Alkylamines can be chosen to prepare
salts of N,N'-oxalylbis(phenylglycine)s that will crystallize either as monolayers or as bilayers.[563]

5.2 Nuclear Magnetic Resonance Spectrometry – Instrumental advances
continue to give new leases of life to ^{1}H-NMR spectrometry, with a 2D-study
of L-glutamine to reveal cross-relaxation between non-equivalent protons of
the side-chain NH_2 group,[564] chemical shift and coupling constant data for
L-glutamic acid and L-glutamine in ^2H_2O at p^2H 6.6,[565] CRAMPS data for
amino acid crystals to give an accurate measure of the effect of surrounding
magnetic variations on ^{1}H chemical shifts,[566] data for polycrystalline aromatic
amino acids revealing ^{1}H magnetic relaxation and molecular motion at 270–
450 K.[567] Data for tert-butyl N-acetyl-L-prolinate and fulleroproline[568] and
for N-acetyl-L-proline N'-methylamide and the corresponding 4-hydroxy-
proline and 3,3-dimethylproline analogues[569] are aimed at better understanding of the factors that determine the position of the *cis–trans* equilibrium
for the amide bond. The influence of solvent on the position of the tautomeric
equilibrium for N,N-dimethylglycine has been assessed.[570] Detailed NMR
analysis of β-alanine and piperidine-3-carboxylic acid (nipecotic acid) has been
published.[571]

Remarkable instrumental sensitivity has been developed for the estimation
of N-acetylaspartic acid in excised[572] and *in situ* brain tissue.[573] The problem
of peak overlap that undermines *in vivo* analysis of taurine by ^{1}H NMR can be
solved by double quantum filtration.[574]

Commercial samples of glycine can contain both α- and γ-glycine poly-
morphs, which can be detected through their differing ^{13}C NMR shielding
parameters and dynamic properties.[575] Other non-routine NMR studies involving heavier nuclei include ^{31}P NMR estimation of pH-dependent interactions
of O-phosphoserine and -tyrosine with aluminium(III) ions.[576]

A 2-centre 3-electron-bonded radical cation (94) has been identified by
NMR-CIDNP as the intermediate in the photo-oxidation of methionine in
^2H_2O-4-carboxybenzophenone.[577]

^{1}H-NMR assignment of absolute configuration to amines derivatized with
1-fluoro-2,4-dinitrophenyl-5-(R)-phenylethylamine relies on chemical shift
values (see also Ref. 1157).[578]

5.3 Optical Rotatory Dispersion and Circular Dichroism – Spectra of
L-methionine reveal low-lying transitions at ca 262 and 285 nm; this new result
from a CD and MCD study has been explained in terms of singlet–triplet
transitions from the highest filled orbital of a sulfur lone pair.[579]

5.4 Mass Spectrometry – Useful spectra are obtained for underivatized amino acids by CIMS with acetone as reagent, showing $[M+H]^+$, $[M+43]^+$ and $[M+59]^+$ ions,[580] and with dimethyl ether as reagent, showing abundant $[M+13]^+$ and $[M+45]^+$ ions, with $[M+91]^+$ ions being most abundant for serine and threonine.[581] Electrospray mass spectra of threonine or its methyl ester proceed *via* the $[M+H-H_2O]^+$ ion, which is now shown to be a protonated aziridine (protonated hydroxy group, not protonated amino group) rather than dehydrobutyrine as previously supposed.[582] Measurement of ionization energy for 19 L-α-amino acids[583] and collection of data on gas-phase basicities/proton affinities for a broad range of amino acids[584] illustrate routine mass spectrometry studies.

Notable results are being obtained through mass spectrometry of host–guest complexes. β-Cyclodextrin–amino acid complexes give informative spectra,[585] and the fact that the exchange reaction of these complexes that occurs with protonated alkylamines shows different rates for D-amino acid complexes compared with their L-amino acid analogues is a fine demonstration of the information available through modern mass spectrometry.[586] Mass spectra of amino acid complexes formed between the crown ether (95) and L- and D-amino acids[587] and complexes formed with ^{2}H-labelled hosts offer access to useful analytical information, *e.g.* the determination of e.e. data for partly-racemic amino acid samples.[588] The applications of this measurable chiral discrimination towards amino acids and their ester hydrochlorides, arising from the work of Sawada's research group, have been reviewed.[589]

More conventional studies with derivatized amino acids also include access to notable structural information through mass spectrometry: unequivocal distinction between N^τ- and N^π-substituted histidines solves a long-standing problem;[590] distinction between leucine and isoleucine through liquid secondary-ion mass spectrometry of N-acyl derivatives based on the greater relative abundance of $[M+H-(H_2O+CO)]^+$ ions shown by isoleucine;[591] and new information on the mechanism of fragmentation during electrospray mass spectrometry of phenylthiohydantoins derived from α-amino acids.[592] Mass spectra of protonated α-methyl and β-methyl monoesters of aspartic acid have been interpreted in detail.[593]

5.5 Other Spectroscopic Studies of Amino Acids – Increasing numbers of amino acid topics are being identified for infrared spectroscopic study: single crystal L-histidine dihydrogen orthophosphate[594] and corresponding study of glycine, β-alanine, L-histidine, and DL-tryptophan single crystals;[595] non-ionized glycine and its ^{2}H-isotopomer isolated in a low temperature argon matrix[596] and corresponding study of L-alanine;[597] vibrational dynamics study of L-alanine;[598] assignment of vibrational modes for tyrosine through painstaking ^{2}H, ^{13}C, and ^{18}O labelling.[599] The frontier science in this area is infrared cavity ringdown laser absorption spectroscopy, used to show that arginine exists as its neutral tautomer in a supersonic molecular beam.[600] Dissociation kinetics of a dimeric form of protonated methyl N-acetyl-L-alaninate have been determined by black body IR radiative dissociation.[601] N-Boc-L-α-Amino acids (as solids and as CCl_4 solutions) have been studied by near IR spectroscopy.[602] Fourier transform IR (FTIR) spectra of αβ-dehydroamino acids[603] and their N-acetyl N'-methylamides[604] have been interpreted to determine aspects of conformation. FTIR reflection–absorption studies of amphiphilic N-octadecanoyl-L-alanine have revealed the adoption of a regular chiral arrangement for the derivative, when in the form of a surface mono-layer.[605] The temperature dependence of chiral discrimination in an N-hexa-decanoyl-L-alanine monolayer, when it overlays an aqueous zinc(II) salt solution, is particularly clear.[606]

IR-Raman studies are also flourishing, dealing with DL-aspartic acid nitrate monohydrate,[607] acid and alkali salts of glycine,[608] β-alanine,[609] and N-acetyl-L-cysteine and -cystine.[610] Polarized IR-Raman studies are still searching for soluble tasks; gaining the data is routine, *e.g.* for L-threonine crystal samples,[611] but interpretation is problematic.

Microwave spectra for L-prolinamide have been interpreted in terms of conformational equilibria.[612]

Dissociation constants for tyrosine in aqueous media at different ionic strengths have been determined by the standard use of UV spectroscopy.[613] More demanding UV studies deal with assessment of radical formation from short-lived N-chloroamino acids, assisted by EPR data,[614] and UV–resonance Raman spectroscopy of tryptophan and other indoles.[615]

Higher energy radiation produces radicals from glycine (EPR study after 30 keV ion bombardment),[616] and generates photoelectron spectra of representa-tive α-amino acids through X-irradiation.[617]

5.6   Other Physico-chemical Studies of Amino Acids – Sub-sections follow that break up a substantial body of literature into logical topic areas. Selection of papers for citation has not been particularly severe, although much repetitive data-gathering needing only simple equipment can be discerned in the papers cited (especially in the first of the following topic areas). The results may well lead to what can be useful insights.

5.6.1 Measurements for Amino Acid Solutions – Effects of solute on the solubility in aqueous media of DL-α-aminobutyric acid,[618] of glycine,[619] of DL-serine compared with L-serine,[620] L-arginine phosphate,[621] of amino acids in aqueous acetonitrile[622] and in aqueous DMF,[623] and extensions of this field of study[624] can be interpreted in terms of apparent molar volumes, isentropic compressibilities, and activity coefficients. Enthalpies of solution for alanine in aqueous NaBr, KBr, and KI at 298.15 K have been determined.[625] A study of the solubilization of amino acids in cationic reversed micelles has a more practical aim.[626]

Solubility data for argon in aqueous amino acids solutions provide interac-tion coefficients for the solute pairs.[627] Other solute interaction studies invol-ving amino acids include cytosine as a partner,[628] and further development of the extraordinary evidence for chiral recognition by certain N-acetylamino acids (leucine and proline; but not alanine, valine, methionine, or aspartic acid) from calorimetric and nuclear relaxation time mesurements.[629] The results show that interaction energies for DL-pairs differ from those for LL-pairs, thus demonstrating chiral recognition. Calorimetry provides date for enthalpy of transfer of glycine, L-alanine, L-serine and L-proline from water to aqueous glucose at 298.15 K.[630]

Molar volume data have been obtained for aqueous glycine, alanine, and valine[631] and the first two of these amino acids together with β-alanine in aqueous dimethyl sulfoxide,[632] from viscosity coefficients; a related study for aqueous guanidine hydrochloride solutions has been published.[633] Volume properties for glycine have been determined using a new vibrating tub densimeter,[634] and ultrasonic velocity and absorption determinations provide adiabatic compressibilities for glycine, L-alanine, and L-proline.[635]

Dissociation constant data continue to be collected, and redetermined in some cases (first and second dissociation constants for glutamic acid in aqueous media at 298.15 K)[636] for comparison with calculated values.[637] A related study provides the proton binding isotherm for glutamic acid.[638] All the conceivable changes of parameters for such measurements inevitably lead to new data, as in the determination of the variations with ionic strength of acid–base equilibrium constants for glycine in aqueous $NaClO_4$, KCl, and KBr.[639]

5.6.2 Measurements and Studies for Solid Amino Acids – Enthalpies of combustion and formation for enantiomers and racemic forms of aspartic acid[640] and thermal stability data for the same samples and for alanine[641] have been obtained by thermogravimetry and differential thermal analysis. Heat capacities of amino acids, peptides and proteins have been reviewed.[642]

Crystallization studies (L-phenylalanine seeds enhance the growth of an L-glutamic acid polymorph from the racemate;[643] seeding by L-lysine, but not D-lysine, has the same effect[644]) and non-linear optical properties (L-arginine hydrobromide monohydrate[645]) are topics of continuing interest.

5.6.3 Amino Acid Adsorption and Transport Phenomena – Thermodynamics of adsorption of amino acids to silica,[646] and separation and purification of amino acids[647] have been reviewed. Determination of equilibrium isotherms for adsorption of DL-phenylalanine to polymeric adsorbents[648] and adsorption of L-lysine to a strong acid ion-exchanger,[649] and theory aspects for ion exchange and adsorptive parametric pumping for the separation of amino acid mixtures[650] illustrate more fundamental studies, while practical aspects are also covered: use of histidine as a dipolar eluent in comparison with 2,3-diaminopropanoic acid for the chromatography of alkali and alkaline earth cations;[651] isothermal supersaturation of a sulfonic acid cation exchange resin in its hydrogen ion form, by passage of a solution of the sodium salt of an amino acid.[652] The recovery of amino acids from aqueous media by ion exchange has been reviewed.[653]

A flow injection technique has been applied to the determination of diffusion coefficients of glycine, alanine, and proline.[654] Partition of amino acids within a two-phase aqueous poly(ethyleneglycol)–aqueous dipotassium hydrogen phosphate medium,[655] and within a toluene–water–organic acid system,[656] are described in studies that develop a long-running theme. Membrane transport of amino acids is also a well-established topic of study, with nanofiltration (charged amino acids through inorganic membranes[657] and ultrafiltration membranes[658]), permeability of a cation exchanger membrane for leucine,[659] and electrodialysis through bipolar membranes as an amino acid extraction technique.[660] Liquid membranes through which anionic carrier-mediated transport of L-valine can be accomplished feature in the system 1-decanol–bis(2-ethylhexyl)phosphate–water[661] and for reverse micelles of bis(2-ethylhexyl) sulfosuccinate, studied for their potential in selective transport of tryptophan and p-iodophenylalanine.[662] Sulfonate cation exchange in this fashion has been applied to other amino acids.[663]

5.6.4 Host–Guest Studies with Amino Acids – Some of the transport studies of the type described in the preceding section have included crown ethers as complexing agents for amino acids. An example is the role of 18-crown ether-6 in facilitating amino acid transport into chlorinated solvents from water.[664] A crown ether-functionalized nickel salicylaldimine complex [18-crown-6-Ni-(tert-Bu)$_4$-salphen] is effective in transporting amino acids from an aqueous acid phase through a chloroform membrane to pure water,[665] and crown ether complexation assists the partition of amino acids across bi- and triphasic systems.[666] A C$_2$-symmetric bis(18-crown-6) analogue of Troger's base shows moderate enantioselective discrimination towards dimethyl cystinate.[667] Several of these crown ether studies are supported by the collection of fundamental data for binding constants through tritration calorimetry [cryptand(2.2.2) and amino acids in methanol;[668] effect of guanidinium compounds on crown ether–amino acid complexation[669]]. Guanidine-modified cyclodextrins have proved to be effective hosts for tyrosine and O-phosphotyrosine.[670]

Calix[4]arene derivatives continue to prove their potential in this area, and have shown this in a spectacular way; (S)-1,1'-bi-2-naphthyl- and bis(indophenol)-derived calix[4]crown ethers are chromogenic hosts, showing naked eye-detectable chiral recognition of optically-active amines and derivatized amino acids.[671] D-Amino acids (alanine, valine, leucine, tryptophan) induce surface area changes, into monolayers of chiral lipophilic calix[4]resorcinarenes carrying (S)-α-phenylethyl or (1R,2S)-norephedrinyl groupings, that differ from the changes induced by their L-enantiomers.[672] Interaction of a series of calixarenesulfonates (96) with arginine and lysine has been assessed using ^{1}H NMR.[673]

Pyridinium cyclophanes (97) show three times greater binding efficiency towards methyl L-tryptophanate than its D-enantiomer.[674] A zinc complex of a porphyrin carrying an L-threonine grouping exhibits significant chiral recognition towards phenylalanine esters.[675] Novel chiral receptors (98 and its analogue with 1,2-cyclohexyl in place of the 1,2-diphenylethyl moiety) show enantioselective binding of amino acid derivatives.[676] A chiral receptor of a different type, the multichannel taste sensor reported in previous Volumes

(97)   (98)

(Vol. 30, p. 48), has been used to quantify the taste of bitter amino acids[677] and to distinguish D-tryptophan from its L-enantiomer through impedance measurements.[678]

5.6.5 Miscellaneous Physical and Conformational Studies – Methods for determining bond energies for the 20 common amino acids have been reviewed,[679] and existing indices that rank the propensities of these amino acids for conferring β-sheet[680] and α- and β-structures[681] on proteins have been developed further. Conformational characteristics of gem-diamino acids and aminomalonic acids have been reviewed.[682]

Aggregation by the toluene-p-sulfonate of ethyl L-leucinate, leading to gel formation in non-polar solvents, has been visualized at the molecular level as one particular chiral arrangement chosen from a large number of possibilities.[683]

5.7 Molecular Orbital Calculations for Amino Acids – A monograph collects reviews on several topics allied to the applications of molecular orbital (MO) calculations in conformational analysis of amino acids.[684] The primary literature on this topic continues the long series of papers covering gaseous protonated glycine conformers[685] (the general topic of N- versus O-protonation of amino acids in the gas phase,[686,687] as well as N to O proton transfer,[688] has been considered), and an overview of glycine conformations has been published.[689] Intramolecular proton transfer for aqueous glycine[690,691] and an extension of this to include glycine and alanine zwitterions[692] and the stabilization of the alanine zwitterion by four water molecules[693] illustrate MO studies for amino acids in the solution phase. Protonated methionine and its sulfoxide and sulfone have been given similar MO treatment,[694] and the same research group has computed effective charges and proton affinity at carbon atoms in α-amino acids.[695] Interactions of cations with side-chain groupings in the common amino acids have been assessed,[696] and interactions of amino acids with chiral stationary phases (see Sections 4.16.3 and 7.5) have been computed and compared with NMR data.[697]

Considerable effort continues to be expended on calculations of vibrational parameters for the glycine zwitterion,[698] cysteine and serine zwitterions,[699] cysteine and serine hydrochlorides,[700] cysteine and serine zwitterions,[701] and aqueous glutamine.[702] The structure and magnetic properties of glycine radicals in aqueous media at various pH values have been assessed.[703]

Near-edge X-ray absorption features for glycine, phenylalanine, histidine, tyrosine, and tryptophan,[704] and calculated relevant X-ray CD features,[705] have been reported.

Amino acid derivatives have been given MO treatment, continuing the study of N-acylamino acid amides to model the conformational situation of amino acid residues in proteins. Calculations for N-formyl-L-alanine amide[706] and N-formyl-L-serine amide[707] have been presented, similarly for N-acetyl-L-alanine N'-methylamide and its leucine and glutamine analogues.[708] The latter study is unusual since it deals with the conformational behaviour of these derivatives at the water–hexane interface. Calculated conformations of N-acetyl-L-alanine N'-methylamide in homogeneous media concentrate on effects of solvent,[709,710] and on Raman spectroscopic characteristics, vacuum-CD, and Raman optical activity.[711] Calculations for geometrical isomers of N-acetyl 1-aminocyclohexanecarboxylic acid N'-methylamide and its cyclobutane homologue have concentrated on conformational behaviour.[712] N-Thioacetylglycine N'-methylamide conformations revealed through MO calculations continue a study of simple isosteres of the N-acetylamino acid methylamides.[713]

A rare application of MO methods to model a reaction (decomposition of N-chloroamino acids in aqueous media) has been published.[714]

6 Chemical Studies of Amino Acids

6.1 Racemization – Fundamental laboratory assessment of mechanistic details of amino acid racemization is not represented this year, but the dating of fossils continues to be carried out through exploitation of the known kinetics. The ways in which the assumptions involved in applying this method are often overstretched is also becoming better appreciated.

The conclusion that a human tooth dated ca. 3000 BC by ^{14}C methods must have been from a 34 year old individual, based on the D:L-ratio 0.1948 for its aspartic acid content, relies on the assumption that the sample remained at ambient temperature 12.5 °C in the meantime.[715] Fossil bone studies have frequently resorted to amino acid racemization dating, and a useful check on a date obtained in this way is obtained from cystine, methionine and tyrosine levels, which show reduction with time, relative to levels of the other protein amino acids.[716] Dating of younger bone collagen samples has benefited from improved methods of determining enantiomer ratios.[717]

6.2 General Reactions of Amino Acids – The main topics under this heading, which deal with reactions at amino and carboxy groups and at the α-carbon atom of amino acids, fall conveniently into a number of separate sub-sections. The chemistry reviewed here is restricted to coverage of non-routine information, but includes simple new protocols where these offer practical benefits.

6.2.1 Thermal Stability of Amino Acids – A somewhat obscure title, but intended to cover changes, if any, brought about in amino acids through high temperatures. The starting point is the claim that amino acids sublime under partial vacuum at elevated temperatures with racemization or decomposition;[718] this is important in view of the long series of reports stating the opposite that have appeared in the literature over the years. It is further claimed that the trifunctional compounds aspartic acid and serine can be sublimed out from geological and fossil samples, just as well as glycine, alanine, α-aminoisobutyric acid, and valine.

The state of an amino acid sample may determine the effects caused by high temperatures; the formation of short peptides by heating solid amino acid mixtures to 270 °C at 58.6 MPa pressure is assisted by inorganic additives,[719] while glycine in suspension in glycerine at 175–180 °C is almost quantitatively dimerized with loss of water into the cyclic dipeptide, dioxopiperazine.[720] Cyclic dipeptides are formed (ignoring low molecular weight volatile products) by heating alanine and α-aminoisobutyric acid under nitrogen at 500 °C.[721] The isolated yield is 68% for the former but only 1% for the latter because it suffers decarboxylation and loss of isopropylamine to give bicyclic amidines. Valine and leucine held at 500 °C under nitrogen are recovered unchanged to the extent of 10% for the former, and 2% for the latter, releasing alkanoic acids, primary and secondary amides and hydrantoins as well as the other products mentioned above.[722] Amino acids vapourized at 270 °C in the presence of silica gel and pulverized basaltic lava yield bicyclic and tricyclic amidines and cyclic dipeptides, the essential events being dehydrogenation, loss of alkyl groups and dehydration together with significant racemization.[723]

Amino acids chosen for exposure to space conditions on the BIOPAN-1 mission, as samples free from additives, and adsorbed on to clay, were glycine and the mainly racemic amino acids alanine, leucine, valine, aspartic acid, and glutamic acid (as found in the Murchison meteorite); L-tyrosine was also included.[724] There were no detectable D-enantiomers formed, and no other decomposition except partial degradation of aspartic acid and glutamic acid in those samples that were in the free form.

6.2.2 Reactions at the Amino Group – N-Chlorination of amino acids continues to head this subsection to which several research groups have devoted themselves. Decomposition of the products follow unimolecular kinetics[725] and specific attention has been given in this context to N-chlorovaline and N-chloronorvaline.[726] Another oxidative process at nitrogen that has been studied in depth is attack on the glycine anion by methyl and hydroxy radicals, and by the isopropanol carbon-centred radical.[727] Classical deamination protocols are represented for preparation of hydroxy esters from D-leucine[728] and (R)-2-bromo-5-phthalimidopentanoic acid from N^δ-phthalimido-L-ornithine[729] by diazotization, and the conversion of D-phenylalanine into phenylpyruvic acid by a co-immobilized mixture of D-amino acid oxidase and catalase.[730]

The Maillard reaction is another topic that continues to receive in-depth study. It is easy to ascribe this interest to the physiological importance of the reaction, as well as its role in food chemistry, but the fact that the mechanistic challenge that it offers has been only superficially addressed means that there is considerable scope for turning out some worthwhile new knowledge. Reviews from several active research groups published in a Royal Society of Chemistry monograph can be checked for their content through *Chemical Abstracts* (see also Refs. 731, 735, 1020); a typical study is the reaction of glycine with glyoxal that leads to 3-carboxymethyl-1-(2-carboxyethyl)imidazole as well as to Strecker degradation products of the amino acid.[731] Over extended reaction times, a glycine–glucose mixture at physiological temperatures in aqueous media at physiological pH, becomes brown through formation of melanoidins;[732] different amino acids undergo glycation at different rates with galactose under such conditions.[733] Coloured compounds including (99) are

(99)

formed from N-(1-deoxy-D-fructos-1-yl)-L-proline and furan-2-carboxaldehyde (but not from the corresponding L-alanine derivative);[734] these early products in a representative Maillard process arise from acetylformoin, an important intermediate that offers different reaction pathways to other initially-formed compounds. Fluorescent compounds generated from 3-deoxyglucosone, itself derived from initially-formed Amadori compounds, are the result of reactions with arginine and lysine in Maillard processes.[735] Chemiluminescence developed in glycine–glucose solutions (Vol. 30, p. 48) has been studied further.[736]

N-Acylation of amino acids catalysed by 4-dimethylaminopyridine, and reversal of this process, has been reviewed.[737] The use of sodium 2-ethylhexanoate as nonpolar solvent-soluble base in Schotten-Baumann acylation of amino acid amides is beneficial.[738] Enzyme-mediated N-acylation of L-methionine (hog kidney and intestinal aminoacylase) is successful with a range of short-chain alkanoic acids.[739] An unusual approach to N-formylation of an α-amino acid ester [from the Schiff base $Bu^tCH=NCHR^1CO_2R^2$, *via* the oxaziridine (m-chloroperbenzoic acid) and iron(III) sulfate hexahydrate ring-opening] involves non-basic conditions that might be advantageous in certain cases.[740] Intramolecular acyl transfer (100 → 101) occurs slowly (half-life 34

(100)   (101)

days at pH 11) and must be guarded against in the use of these N-acyl-N-(2-aminoethyl)glycines in PNA synthesis.[741]

Resin-tethered N-acylamino acids in which the acyl groups carry an electron-withdrawing group in the α-position have been cleaved from the polymer in the form of a tetramic acid,[742] while 3-acyltetramic acids formed similarly are released from the solid phase under very mild conditions (1 eq KOH in MeOH).[743] N-Alkylamino acids esterified to a solid phase condense with N-α-ketoacylaminoacids to provide a library of (Z)-3-alkylidene-2,5-dioxopiperazines.[744]

3-Chloro-3-(dimethoxyphosphoryl)-1(3H)-isobenzofuranone reacts with amino acids at room temperature to give phthalimido-acids in good yield.[745] Improved selectivity is claimed in optimized procedures for preparations of α-phthalimido and -benzyloxycarbonyl-protected L-arginine.[746]

N-Alkoxycarbonylation using the N-alkoxycarbonylpyrazole (102) tolerates an aqueous medium and does not cause racemization;[747] a preparation of

(102)

Fmoc-amino acids under neutral anhydrous conditions is a little more round-about (first, reflux the amino acid with N-methyl-N-trimethylsilyl-trifluoro-acetamide, add N-Fmoc-succinimide and then desilylate the product with methanol).[748] Selective Boc-protection of piperazinecarboxylic acid can be accomplished with Boc_2O at pH 11.5, the second imino group being arenesul-fonylated in the usual way with di-isopropylethylamine as base.[749] N-Alkoxy-carbonylation[750] and N-acylation[751] of amino acids with a chloroformate or acyl chloride, respectively, is accelerated by the presence of activated zinc. A novel chiral tetra-O-benzylglucosyloxycarbonyl group can be introduced *via* the isocyanato acid ester, and removed under mild conditions using α- and β-glucosidase, after hydrogenolysis of the benzyl groups;[752] another unusual N-protecting group, o-nitromandelyloxycarbonyl, has been shown to be removed from L-glutamic acid by UV irradiation.[753] Acetyl chloride in methanol generates hydrogen chloride that cleaves the N-protecting group from Boc-amino acids; esterification of the carboxy group occurs at the same time.[754] Other new Boc-deprotection procedures include sodium iodide in neutral aqueous media,[755] microwave irradiation of the derivatives adsorbed on silica gel alone[756] and with one equivalent of aluminium chloride,[757] and reaction with $Yb(OTf)_3$ on silica gel.[758] Resin-tethered Boc-peptides that are sensitive to TFA can be deprotected by trimethylsilyl triflate–2,6-lutidine.[759] Fmoc-amino acids tethered to peptide synthesis resin have been converted into N-Boc analogues by either KF with tert-Boc-S-2-mercapto-4,6-dimethylpyri-midine or $KF–Boc_2O$.[760]

UNCAs (see also Ref. 477) react with Grignard reagents to give N-benzyloxycarbonyl-N-acylamino acids, but several side-products are also formed.[761]

N-[2-(N-Protected amino)thioalkanoyl]phthalimides are efficient selective N-thioacylating agents towards β-aminoalkanols, though they are ineffective with hindered amines.[762] Ureas, both symmetrical and unsymmetrical, have been prepared from amino acids using 1,1'-carbonylbis(benzotriazole), in solution and solid phase modes.[763]

N-Arylsulfonyl derivatives are easily prepared represented in the combinatorial synthesis mode for libraries of derivatized L-phenylalanine amides.[764] Toluene-p-sulfonyl group removal from such derivatives has presented a greater problem until recently, and another procedure, but limited to β-amino-α-hydroxyalkanoic esters since it depends on neighbouring group participation by the ester group, involves isopropylidenation with acetone dimethyl acetal–toluene-p-sulfonic acid in toluene at 70 °C (4–6 h).[765] N-Phosphorylation through phosphoramidite–amine exchange in the presence of 1-tetrazole is rapid and efficient,[766] and N-(N-methylaminothiomethanephosphonyl) derivatives show promise, after cyclization, for use in thiopeptide synthesis.[767] A practical N-(di-isopropylphosphoryl)ation procedure involves NaOCl, di-isopropyl phosphite, and NaOH.[768]

N-Arylation occurs without racemization, using aryl halides with copper(I) iodide as catalyst.[769] Mono- and dimethoxytritylation,[770] and protection of an amino acid with hexafluoroacetone prior to N-monomethylation (MeCl with Et$_3$SiH and TFA)[771] or analogous N-phosphinomethylation,[772] have been described. A two-step strategy from an N-Boc- or -Z-amino acid *via* reduction of the oxazolidinone formed with paraformaldehyde gives optically-pure N-methylamino acids (see Ref. 857 for a standard preparation of an oxazolidinone that is the intermediate in this process).[773] N-Benzyl-N-Boc-amino acid esters have been converted into homochiral 2,6-*cis*-disubstituted piperidines, *via* 1,5-aminoalkenols BzlNHCHR^{1}CH=CHCH$_2$CH(OH)R^2.[774] Diastereoselective N-alkylation of α-amino acid esters in which a new chiral centre is created in the resulting alkyl group has been worked out (Scheme 25).[775,776]

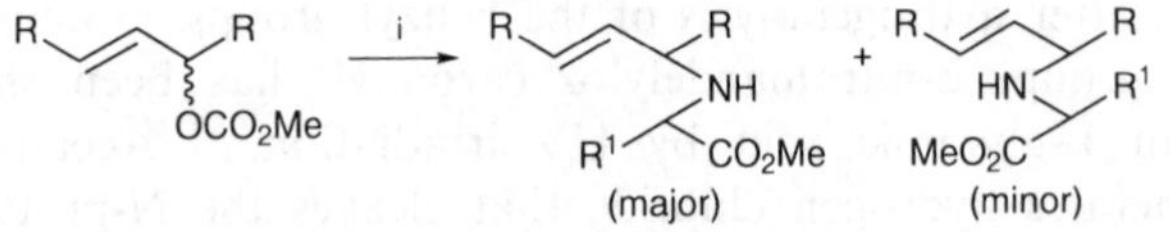

Reagents: i, L-amino acid, [η^3-C$_3$H$_5$PdCl]$_2$, Et$_3$N, CH$_2$Cl$_2$, room temp.

Scheme 25

N-(β-Boc-Aminoethyl)-N-acyl-α-amino acids in which the acyl group carries a nucleobase have been prepared by successive reductive alkylation [BocNHCH$_2$CHO, NaBH$_3$(CN)] and acylation, to give monomers for the preparation of functionalized PNAs.[777]

2,5-Dimethoxyfuran reacts with (S)-phenylglycinol and benzotriazole to give (4S,5R)-5-(benzotriazol-1-yl)-4-phenyl[1,2-a]oxazolopyrrolidine (103) which is easily substituted by Grignard reagents and cleaved to give homochiral 2,5-

(103)

disubstituted pyrrolidines.[778] An oxazolidinone (104) from o-bromobenzalde-hyde, S-lactic acid, and an L-α-amino acid is amenable to radical substitution to give methyl 2-[tributylstannyl)methyl]propenoate after hydrolysis, with modest stereocontrol.[779]

(104)   (105)   (106)

Esters of N-toluene-p-sulfonyl- and -trifluoroacetylamino acids are readily N-allylated using allyl carbonate and a palladium(0) catalyst.[780] N-(3-Pyridazi-namin-3-yl)amino acids are formed with 4-cyano-3-chloropyridazines and removal of the cyano group by catalytic hydrogenolysis (the results are only given for two amino acids, and the hydrogenolysis was not demonstrated).[781] Reductive amination of tert-butyl glycinate with glyoxylic acid, and functional group development to the aldehyde $HCOCH_2N(Boc)CH_2CO_2Bu^t$, provides the starting material for a synthesis of a C-glycosylated peptoid building block through condensation with the aldehyde grouping.[782] The equivalent Michael addition process also leads to the most stable diastereoisomer, illustrated for the addition of isopropyl L-alaninate to ω-nitrostyrene.[783] An unusual version of this category of N-alkylation reaction (amino acids add to 1,2-naphthoquinone-4-sulfonic acid to give coloured derivatives, λ_{max} 480 nm) employs the reagent immobilized on an ion-exchange resin.[784] Bearing in mind the current interest in photosensitized reactions, including amino acids, the observation that their photoalkylation can be brought about by benzophenone is important.[785]

N-Amination is a viable option now that N-Boc-3-trichloromethyloxaziri-dine has been shown to react much faster than aryl analogues to give N^β-Boc-hydrazino-acids, including N^β-Boc-N^β-benzyl-L-glutamic acid, suitable for the preparation of S-piperazic acid and its dihydro-analogue.[786] Bis-Boc-[$^{15}N_2$]Hy-drazine converted into di-tert-butyl azodicarboxylate and thence into [$^{15}N_2$]-S-piperazic acid illustrates standard methodology.[787]

6.2.3 Reactions at the Carboxy Group – Esterification of amino acids can be accomplished in high yield through stirring a suspension in an alkanol over

Amberlyst-15 cation exchange resin in the hydrogen form,[788] or using 2,2-dimethoxypropane, methanol, and a catalytic amount of HCl, the advantage of the latter method being that it is selective for alkanoic acids in the presence of aromatic acid functions.[789] For N-protected amino acids, use of an 2-phenylisopropyl or tert-butyl trichloroacetimidate is advocated, especially useful with β-hydroxy-α-amino acids.[790] Esterification of Boc-glycine to poly-(ethylene glycol), Boc removal, and Schiff base formation have been monitored by ESI-mass spectrometry.[791] Carbodi-imide-mediated esterification of Boc-L-tryptophan to 4-hydroxythiophenol-linked resin and Pictet-Spengler construction of a library of tetrahydro-β-carbolines is followed by release from the resin by a primary amine (see also Ref. 1060).[792] Polymer-bound HOBt mediates the preparation of N-hydroxysuccinimide esters of N-Boc-L-proline.[793]

α-Chymotrypsin-catalysed transesterification of methyl L-tyrosinate gives the fructosyl ester in 63% yield.[794] Fullereneproline esters in toluene can be transesterified by 2,2,2-trifluoroethyl esters of palmitic and butyric acids, using lipases.[795] Porcine liver esterase catalyses the hydrolysis of the pro-S ester group of diethyl α-alkyl-α-(benzyloxycarbonylamino)malonates and enantiodivergent reduction gives α-substituted D-serines.[796] The other major enantioselective reaction topic, ester hydrolysis (but without enzyme catalysis), is represented by lipophilic macrocyclic ligands (105) as catalysts, mediating the hydrolysis of long-chain alkyl esters,[797] and copper(II) complexes of homochiral ligands[798] and sugar-derived surfactants such as N-dodecylmaltobionamide[799] mediating the hydrolysis of p-nitrophenyl esters. Mechanistic aspects are at the heart of these studies (aspects of the field have been reviewed[800]); the general complexity of the topic is illustrated in the last-mentioned research, where it was shown that enhanced enantioselectivity is observed when the pH is changed rapidly at the start of a reaction. Ester hydrolysis protocols ignoring this stereochemical aspect are: lanthanide-ion catalysis of the hydrolysis of simple esters of amino acid derivatives,[801] polymers bearing pyridine side-chains complexed to copper(II) ions as catalysts for the hydrolysis of p-nitrophenyl esters of N-protected α, β, and γ-amino acids,[802] arylpalladium complexes $[Pd(C_6H_4CHMeNMe_2)Cl(py)]$ as catalyst for the hydrolysis of Z-leucine p-nitrophenyl esters,[803]and enhancement of the hydrolysis of ethyl glycinate by CTAB micelles.[804] Transesterification of simple esters leading to hindered esters is catalysed by titanium(IV) ethoxide.[805] Aminolysis of hydroquinone esters of N-protected amino acids is assisted by a bivalent metal oxidant.[806]

Acyl fluorides formed by the reaction of N-tritylamino acids with cyanuric fluoride are already established as powerful acylating agents, and in a broad study their reduction (-COF → -CH$_2$OH), chain extension *via* a phosphorane (→ TrNHCHRCF=CMeCO$_2$Me), and conversion into the trifluoromethyl ketone using Ruppert's reagent, has been demonstrated.[807] Comparison of reactivity of these fluorides with corresponding chlorides has revealed the effectiveness of the chlorides of N-arenesulfonylamino acids, but not the fluorides, for acylation of hindered amines.[808] The use of Fmoc-amino acid chlorides for amination is better assisted by zinc dust rather than an organic base,[809] and the use of Fmoc-α-aminoisobutyroyl chloride for the synthesis of peptides of this highly hindered amino acid uses the potassium salt of hydroxybenzotriazole as co-reactant.[810] Aminolysis of N-protected amino acid esters by amino acids can be usefully mediated by $AlMe_3$.[811] β-Ketoester formation involves reaction of the activated carboxy group with the allyl acetate anion.[812]

Reductive processes applied to the carboxy group and its derivatives typically involve $NaBH_4$–LiCl, applied to protected serine[813] and to amino acids adsorbed on Amberlyst -15 cation exchange resin suspended in THF,[814] to lead to β-amino-alcohols. Fmoc-Amino acids treated to $NaBH_4$ reduction and then Swern oxidation after conversion into the mixed anhydride, or $LiAlH_4$ reduction of Weinreb amides Fmoc-aa-CONMeOMe (-aa- = amino acid residue), give β-amino-aldehydes.[815] Simple N-protected amino acid amides are reduced to amines by BH_3 in a route to spermines starting from β-alanine,[816] but hindered N-phenylethylhomoproline esters could not be reduced by DIBAL-H.[817] The mixed anhydride grouping introduced by isobutyl chloroformate into β-tert-butyl N-Boc-L-aspartate is selectively reduced with $NaBH_4$ to hydroxymethyl.[818]

Reduction of L-tyrosine by BH_3–Me_2S starts a preparation of an oxazolidinone destined to be N-acylated and anchored to a solid phase for use in Evans-type synthesis.[819] N-Protected L-phenylalaninal, and other amino acid aldehydes, can be condensed with a resin-bound Wittig reagent, built upon by normal peptide synthesis and then released as a peptide aldehyde by ozonolysis;[820,821] milder conditions will be needed to make this into a general method. Sodium borohydride reduction of the α-HOBt ester of β-benzyl aspartate gives a quantitative yield of the primary alcohol.[822] Different hydride reagents have been compared in a search for a clean procedure for the reduction of aspartic and glutamic Weinreb amides to aldehydes, leading to the conclusion that $LiAl(O^tBu)_3H$ or the tris(3-ethyl-3-pentyloxy) analogue are best (see also Ref. 957).[823] Condensation of the Weinreb amide of NN-dibenzyl-L-alanine with iodocyclopentene and elaboration leading to the vinylogous amino acid (106) has been reported.[824] N-Benzyloxycarbonyl-L-proline gives the prolinal derivative by the equivalent BH_3–Me_2S reduction followed by either Swern or Dess-Martin periodinane oxidation.[825] The $NaBH_4$–LiCl reagent can reduce an ester grouping in the presence of an α-azido group, thus opening up some novel possibilities for the synthesis of modified amino acids.[826] N-Acyl- or N-thioacyl-N-methyl-α-amino acid esters are fully reduced by $NaBH_4$ in MeOH but the simpler N-H compounds are not.[827] An N-hydroxymethyl group, introduced by treating a Boc-amino acid with formaldehyde, stabilizes the N-Boc-amino acid aldehyde formed by reduction by DIBAL-H, especially against racemization.[828] [See also Refs. 774, 958 for related uses of DIBAL-H.]

The prolinal above has been used to prepare S-(pyrrolidin-2-yl)alkenes, and N-Boc-N-benzyl-L-phenylalaninal has been converted into a 95:5 mixture of BocN(Bzl)CHBzlCH_2CH=CH_2 and BocN(Bzl)CHBzlCH = CHCH_3 *via* the

carbonate BocN(Bzl)CHBzlCH(OCH$_2$CO$_2$Et)CH=CH$_2$.[829] N-Boc-L-leucinal
provides the means of chain construction for a homochiral dihydroisocou-
marin through cycloaddition of the acetylenic ester (107) to a substituted
cyclohexadiene, and elaboration of the adduct.[830] [See also Ref. 476 for
conversion of an amino aldehyde into a α-amino-alkynoic acid.]

Mitsunobu reactions with N-protected β-aminoalkanols, giving O-acyl
derivatives, are best conducted by replacing triphenylphosphine with 1,2-
bis(diphenylphosphino)ethane.[831]

Carbonyl attack by nucleophiles of other types is represented in several
studies. trans-Alkenation is effected with a pathway starting with lithiated
benzylic or allylic benzotriazoles (BtCLiR → BtCRCOR′ → RCH = CHR′).[832]
Reactions of N-protected amino acids esterified to a solid-phase with aqueous
hydroxylamine give corresponding hydroxaminic acids[833] (these have been
used as the source of homochiral acylnitroso compounds for use in Diels-Alder
addition reactions leading to isoxazolidines[834]), and solution reactions of esters
with α-lithiated isocyanides give N-protected 5-(β-aminoethyl)oxazoles.[835] The
equivalent process, reaction of Fmoc-amino acid amides with diazo-esters
RCOC(=N$_2$)CO$_2$Me, gives 2-(β-aminoalkyl)-4-methoxycarbonyl-5-substituted
oxazoles.[836] Pyrazoles and isoxazoles have been prepared from α-aminoalkyl
trimethylsilylethynyl via β-diethylaminoethenyl ketones.[837] Further examples
of solid phase Ugi reactions that convert Boc-amino acids into libraries of
dioxopiperazines, with the beneficial use of cyclohexenyl isonitrile, have been
reported;[838] the same chemistry has been explored in the solution mode,[839]
and L-homoserine gives N-carbamoylmethyl-α-aminobutyrolactones.[840]
α-Hydroxy-β-Boc-amino acid N′-(2-hydroxyethyl)amides have been converted
into oxazolines and thence into α-Boc-amino keto-oxazolines (108) by standard methods.[841] Polymeric aryl hydrazines acylated by an N-protected amino
acid, are oxidized to nitrogen and nitrobenzene by mild oxidants, in the
process of releasing the attached moiety after combinatorial or peptide
synthesis operations.[842] Diazoketones derived from N-toluene-p-sulfonyl-α-
and β-amino acids can react through several different pathways, but the
β-amino acid derivatives give 5-substituted pyrrolidinones in excellent yields
under the conditions of the Wolff rearrangement.[843]

Decarboxylation and electron transfer cyclization have all been established
as outcomes for photolysis of N-phthaloylamino acids, and elimination occurs
with cysteine derivatives.[844] Formation of phthalimido-substituted radicals
from N-phthaloyl-O-tri-isopropylsilyl-DL-threonine by UV-irradiation after
conversion into the Barton ester.[845] Decarboxylation of leucine by hydroxy
radicals generated from the iron(II)–porphyrin–H$_2$O$_2$ reagent system,[846] and

formation of CO_2, NH_3, and alkanoic acids through electrochemical oxida-tion,[847] are further examples of the results of more drastic treatment of amino acids and their derivatives.

α-Aminoketones are acquiring new uses in synthesis, and enantiopure samples have been prepared through Al-Hg reduction of γ-amino-β-ketosul-fones generated through condensation of an N-protected α-amino acid ester with Li_2CHSO_2Tol, followed by alkylation.[848] Treatment of an imidazolide derived from an α-Boc-amino acid, with excess Grignard reagent in the presence of a copper(I) salt, provides a general route to α-aminoketones,[849] as does palladium-catalysed reaction of N-Z-amino acid thiolesters with organo-zinc reagents.[850] Diazoketones (S)-PhtNCHBu^{i}COCH $= N_2$ have provided the first example for this class of compound, of rhodium-catalysed intramolecular C-H insertion.[851] The *threo*-N-methyl-N-benzyloxycarbonyl-L-phenylalanine-derived oxirane (109) has been prepared through established routes; from the bromoketone, or through epoxidation.[852]

6.2.4 Reactions at both Amino and Carboxy Groups – This section mostly covers routes from amino acids to heterocyclic compounds, concentrating on less routine work and leaving β-lactams for coverage elsewhere in this Specialist Periodical Report [except for mention of new examples of the route from β-amino acids to α-amino acids: oxidative ring-expansion of 3-hydroxy-β-lactams to give N-carboxyanhydrides (NCAs) of polyhydroxylated α-amino acids[853]]. Boc-NCAs are also easily prepared from α-amino acids and are effective acylating agents towards alkanols as well as their well-known use in peptide synthesis; sugars in various degrees of protection are aminoacylated by them.[854] Metallacyclic oxidative addition products from NCAs with zerovalent nickel complexes have been described.[855] Other non-routine lactamization studies are published for macrocyclic lactams, which can be formed from 6-amino-7-hydroxyoct-5-enoic acid by way of a ring-contraction rearrange-ment of the lactone.[856]

Oxazolidin-2-one formation from an α-N-Fmoc-amino acid with formalde-hyde is straightforward, and modest improvements are represented in the mild conditions associated with catalysis by $BF_3.Et_2O$ or toluene-p-sulfonic acid on silica gel (see also Ref. 773).[857] An unusual version of this is seen in the sequence (110) to (111), occurring with retention of configuration.[858] More results have been reported for the conversion of 3-(acylamino)benzo[b]furan-2(3H)-ones into 3-arylthio-substituted versions that is accompanied by visible chemiluminescence (see Vol. 30, p. 70).[859]

As usual, this section is dominated by work on oxazoles and oxazolones, the

(110) → (111)

former class being accessible as their 5-trifluoromethyl derivatives, from N-acylamino acids in trifluoroacetic anhydride–pyridine,[860] a reaction medium that promotes an anomalous Dakin-West reaction with an N-(alkoxycarbonyl)proline leading *via* the mesoionic oxazol-5-one to 4-trifluoroacetyl-2,3-dihydropyrroles.[861] 4-Trifluoroacetyl-1,3-oxazolium-5-olates formed from N-alkyl-N-acylglycines with trifluoroacetic anhydride are opened by O-nucleophiles at C-5, and by N-nucleophiles at C-2;[862] phenylhydrazine gives trifluoromethyl-substituted pyrazoles and 1,2,4-triazines.[863] Novel 2-phenyloxazol-5(4H)-one complexes (112) have been prepared.[864] Mesoionic oxazol-4-ones

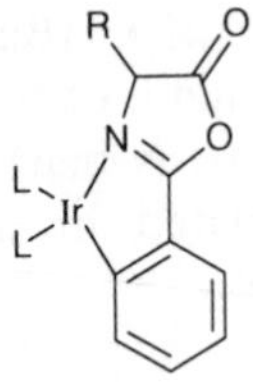

(112)

('isomunchnones') are formed by rhodium(II) catalysed cyclization of α-diazoimides AcN(CHRMe)COC(=N₂)CO₂Me, trapped through additions to dipolarophiles.[865] Oxazoles are formed from the related rhodium(II)-mediated condensation of diazomalonates $(RO_2C)_2C=N_2$ with amino acids; amides R^1CONH_2 give N-acylaminomalonates $(RO_2C)_2CH(CH_2)_2CONHCH(CO_2R)_2$.[866] Further results for coloured 2-arylaminothiazol-5(4H)-ones,[867] and the formation of thiohydantoins *via* oxazol-5(4H)-ones by successive treatment of an α-Boc-amino acid with isobutyl chloroformate and trimethylsilyl isothiocyanate,[868] have been published.

More uses as a chiral reduction catalyst, for the oxazaborolidine (113) prepared from L-threonine,[869] and further examples of boroxazolones (114)[870] and related species, *e.g.* the oxazaborolidinone (115),[871] have been reported.

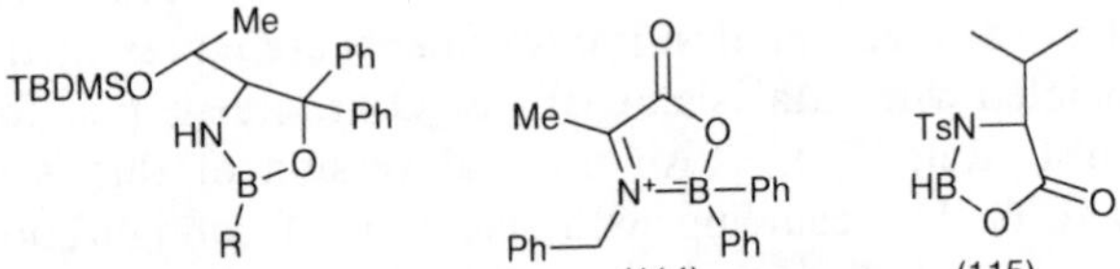

(113)　　　　(114)　　　　(115)

The unexpected displacement of both functional groups from N-toluene-p-sulfonylamino acids by benzene, toluene or p-xylene to give 1,1-diarylalkanes requires long heating at 60–70 °C in the presence of 3 equivalents H_2SO_4,[872] phenylalanine giving 2-arylnaphthalenes.[873] N-Arylsulfonylamino L-acid amides condense with 1,2-dibromoethane to give substituted homochiral oxopiperazines,[874] and larger ring systems can be prepared from β-amino acids as illustrated in a solid-phase synthesis of 1,5-benzodiazepin-2-ones.[875] Hydantoins can be formed by solid-phase reaction of amino acid amides with triphosgene or carbonyldi-imidazole.[876]

The self-condensation of amino acids into dioxopiperazines has been covered in other Sections, and the alternative oligomerization route to generate short peptides under conditions that could have prevailed in the prebiotic environment provides an area of growing interest. Low yields of oligomers are formed from glycine and from glycylglycine adsorbed on silica and alumina through wetting–drying cycles at 80 °C,[877] or on copper(II)-exchanged hectorite.[878] Co-precipitated nickel(II) and iron(II) sulfides suspended in de-oxygenated water at 100 °C under a CO–H_2S or MeSH atmosphere provides a more effective system.[879]

The role of other species capable of acting as reagents rather than catalysts features in several related studies. Oligomerization of aspartic acid as its anion in aqueous solution can be brought about by both carbonyldi-imidazole and 1-ethyl-3-(3-dimethylaminopropyl)carbodi-imide; for simple α-amino acid anions, the former reagent (but not the latter reagent) is effective, the reverse applying for β-amino acids.[880] The impetus given to this topic was the finding that solution self-condensation of amino acids was induced by high salt concentrations, and development of this in terms of finer details: a glycine–alanine mixture induced to oligomerize by copper(II) salts exhibits 'mutual catalysis'; that is, glycine enhances the formation of alanylalanine through a role for the glycylalanylalanine–copper(II) complex.[881]

The benefits of heterogeneous systems in this context are shown by the finding that β-amino acids yield oligomers on mineral surfaces [*e.g.* formation of poly(β-glutamic acid) on hydroxyapatite] but not in free solutions, under the action of a water-soluble carbodi-imide];[882] similar results were found with negatively-charged α-amino acids (glutamic and aspartic acids and O-phosphotyrosine).[883] Polymerization of amino acids on mineral surfaces has been reviewed.[884]

6.2.5 Reactions at the α-Carbon Atom of α-Amino Acids and Higher Homologues – Alkylation of glycine derivatives is standard synthesis methodology (Section 4.1.7), and routes to α-alkyl-α-amino acids are covered in Section 4.3. Therefore this is a small subsection, dealing with α-alkylation of higher homologues; *e.g.* the γ-amino acid (116) with LiHMDS and allyl or benzyl iodide,[885] and γ-lactams and higher homologues using a chiral tetradentate lithium amide in the presence of LiBr.[886]

6.3 Specific Reactions of Amino Acids – Reactions involving primarily the side-chains of common amino acids are covered in this section, which is structured as usual – saturated aliphatic groupings, functionalized aliphatic groupings, aromatic and heteroaromatic side-chains.

Few reagents affect the saturated side-chains of alanine, valine and the leucine isomers in controllable ways, but free-radical bromination, when amino and carboxy groups are suitably protected, yields useful synthons whose preparation and uses have been reviewed.[887] α-(β-Iodoalkyl)glycines and γ-iodo-homologues are best prepared from serine, aspartic acid and glutamic acid, respectively; their conversion into zinc and zinc–copper species, and uses of these in the synthesis of other amino acids, has been reviewed.[888] Examples of such uses include routes to α-(ω-arylalkyl)glycines by palladium-mediated coupling to iodoarenes,[889] to halogenoalkenes and acyl halides,[890] and [2,6-bis(ethoxycarbonyl)pyridin-4-yl]-L-alanines.[891] Organocopper synthons formed with Rieke copper are included in these new studies (Ref. 890), and extension of the organozinc species to include modified aspartic and glutamic acids, *viz.* $BocNHCH(CH_2ZnI)(CH_2)_nCO_2Me$ has opened up routes to substituted β- and γ-amino acids.[892] 2-Amino-4-bromobutanoic acid has been subjected to standard nucleophilic substitution protocols to give expected products, including (S)-azetidine-2-carboxylic acid, hitherto difficult to prepare.[893]

Electrochemical 5-methoxylation of methyl N-Boc-L-prolinate followed by allylation (using allylsilane–BF_3) is the starting point for further development leading to eight-membered lactams.[894] This intermediate has also been used for one of the many epibatidine syntheses that have been reported this year.[895]

N-Benzhydryl α-fulleren[60]ylglycines are now accessible through a novel cyclopropane ring-opening strategy [DBU and CBr_4 in PhCl, followed by $Na(BH_3)CN$] that is unique to a 'methano[60]fullerene amino acid', the cycloadduct (117).[896]

Tautomerization of 'dehydroamino acids', *alias* $\alpha\beta$-unsaturated α-amino acids, accounts for their rapid hydrolysis in neutral aqueous media.[897] Nevertheless, they can be handled without difficulties, *e.g.* in a synthesis of methyl 4-phenylpyroglutamate from N-benzyl-N-α-phenylchloroacetyldehydroalanine through radical intermediates generated by Bu_3SnH-AIBN,[898] in routes to 4-cyanoglutamic acid,[899] to p- and m-phenylenebis(alanine)s (118),[900] and to γ-nitro-α-amino acids by base-catalysed conjugate addition of nitroalkanes,[901] then on to reductive denitration (products of ozonolysis of derived nitronates).[902] N-Acetyl dehydroalanine bound to Wang resin has been shown to undergo Michael addition of 1,2,4-triazole and pyrazole, offering a rapid synthesis of β-heteroaryl-α-amino acids (see also Ref. 998).[903] N-Acyl-N-alk-3-enyl dehydroamino acids participate in tandem radical cyclization to indolizidinones and pyrrolizidinones (Scheme 26).[904]

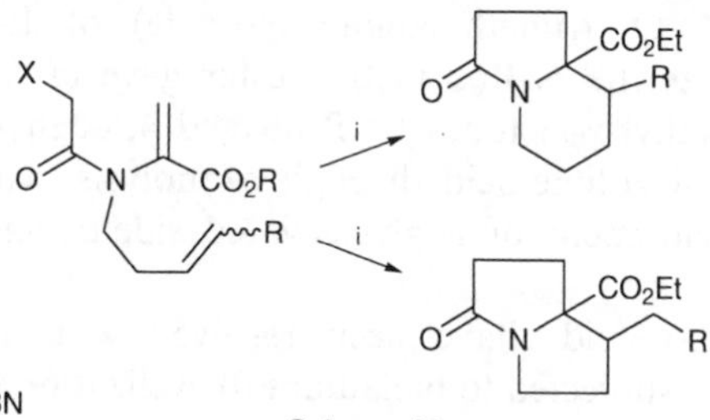

Reagents: i, Bu₃SnH, AIBN

Scheme 26

N-Benzyloxycarbonyl-L-vinylglycine continues to serve in syntheses of other amino acids through standard routes but using currently-favoured methodologies – preparation of [(E)-(2-arylvinyl)]glycines in high enantiomeric purity, by reaction with an iodoarene catalysed by palladium(II) acetate in aqueous Bu_4N–$NaHCO_3$ at 45 °C.[905] Neither dehydroamino acids nor vinylglycines undergo the Grubbs cross-metathesis reactions shown by allyl- and homoallyl-glycines with aryl- and alkyl-substituted alkenes mediated by $(Cy_3P)_2$-$Cl_2Ru=CHPh$,[906] or $RuCl_2$-mediated ring-closing metathesis of αα-bis(ω-alkenyl)glycine derivatives $AcNHC[(CH_2)_nCH=CH_2]CO_2Et$ that leads to 1-aminocycloalken-1-carboxylic acids (119 ; n = 1 or 2),[907,908] proposed for a role as protease inhibitors. The inertness of vinylglycines in this process has been rationalized as a conformation-dependent electronic effect, since N-allyl amides of N-protected vinylglycines were cyclized efficiently.[909] A spectacular Grubbs alkene metathesis route, used for a synthesis of meso-2,6-diaminopimelic acid from the bis(allylglycinate) ester of ethyleneglycol,[910] and for a synthesis of (S,S)-2,7-diaminosuberic acid from the related allylglycine derivative (120),[911,912] also for eneyne metathesis leading to 1-amino-3-vinyl-cyclohex-3-ene-1-carboxylic acid and related constrained amino acids.[913]

(119)

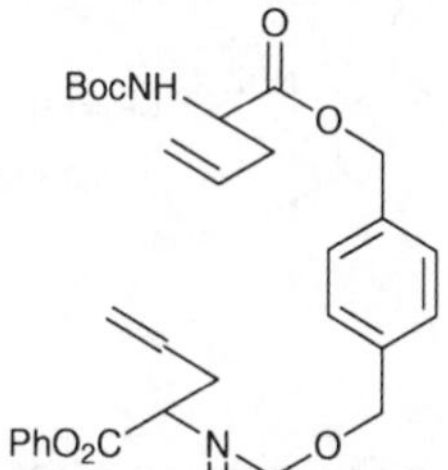

(120)

7-Membered ring lactams have been prepared through a high-speed solid phase route in which N-(N-Boc-allylglycinyl)allylamines are subjected to ring-closing metathesis-promoted cleavage as a key step.[914] Iodocyclization of α-allylglycines TsNH(CO$_2$Me)CH$_2$CH=CHR lacks stereocontrol, leading to *cis*- and *trans*-2,5-disubstituted prolines that undergo aromatization (DBU in DMF) to give corresponding pyrrole 2-carboxylic acids.[915] The equivalent process with alkynylglycines gives 2-substituted-3-iodo-4,5-dehydroprolines.[916] Dihydroxylation (OsO$_4$–trimethylamine N-oxide) of L-α-allylglycine after routine protection as the N-Boc methyl ester gave *cis*- and *trans*-N-Boc-3-amino-5-hydroxymethyl-γ-lactones.[917] Protected 4,5-dehydro-L-leucine is the source of S-2-aminolevulinic acid through ozonolysis, whose keto-group has been used for development of a glycosylated side-chain through an oxime linkage.[918]

ω-Alkynyl-α-amino acids have been resolved with *Pseudomonas putida* aminopeptidase, and subjected to palladium(0) cyclization to give 5-substituted prolines.[919]

α-(ω-Hydroxyalkyl)glycines, particularly L-serine and L-threonine, provide valuable starting points for routes to a wide variety of synthesis targets. A review has appeared covering uses in synthesis of 4,5-disubstituted oxazolin-2-ones formed from D- and L-serine,[920] an illustration being provided for a synthesis of 1,2-aminoalkanols involving organometallic reagents (cerium, magnesium or lithium).[921] Another route to these compounds from the protected D-serine Ph$_2$C=NCH(CH$_2$OTBS)CO$_2$Me also employs organometallic reagents.[922] N-Benzyloxycarbonyl- or -Fmoc-aziridinecarboxylic acid derivatives formed from L-serine undergo Sc(O$_3$SCF$_3$)$_3$-mediated opening by arylindoles to give substituted L-tryptophans.[923] An alternative approach to dehydrotryptophan[924] through the reaction of indole in AcOH with (E)-β-(N-methylamino)dehydroalanine, formed from serine, and similar use of the N-dimethylamino-homologue in preparations of β-aryl and -heteroaryl-dehydroalanines,[925] has been developed further. O-Glycosylation at the end of a route to Flα-antigen is accomplished by TMSOTf- or Cp$_2$ZrCl$_2$-AgClO$_4$-catalysed condensation with the glycosyl trichloracetimidate or fluoride.[926] Preparation of α-acetylglycine derivatives ZNHCH(COMe)COY from L-threonine opens up a synthesis of δ-lactams by aza-annulation with acrylates (121).[927]

(121)

The unique role for D-serine, to offer its side-chain for easy conversion into a carboxy group of a target L-amino acid after functional group changes elsewhere in the molecule, is illustrated in a synthesis of (2S,3S)-β-hydroxy-

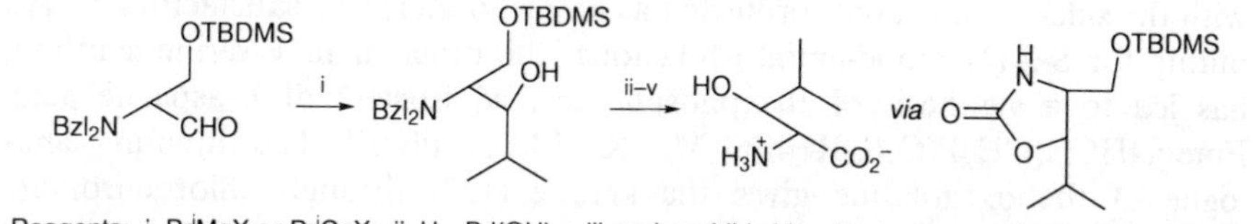

Reagents: i, Pr^{i}MgX or Pr^{i}CeX; ii, H$_2$–Pd(OH)$_2$; iii, carbonyldi-imidazole; iv, Jones oxidation; v, conc. HCl

Scheme 27

leucine (Scheme 27).[928] Cyclic ortho-ester protection of the carboxy group of L-serine, giving an homochiral N-protected serinal equivalent, is part of a route to β-disubstituted serines.[929] D-Serine, converted into the aminodiol derivative TBDMSOCH$_2$CH(NHBoc)CH(OH)Ph (Vol. 30, p. 64), gave chloramphenicol after several routine steps, including p-nitration and dichloroacetylation.[930] The aziridine (122) was prepared by cyclization of protected L-serinol using Ph$_3$P and di-isopropyl azodicarboxylate, and used for the synthesis of N-Boc-D-amino acids through CuBr$_2$.SMe$_2$-catalysed opening with Grignard reagents.[931]

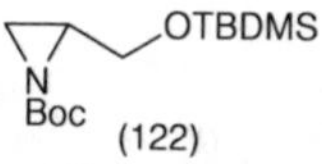

(122)

N,O-Protected L-serinals are useful synthons, and TiCl$_4$- or SnCl$_4$-mediated allyltrimethylsilane addition illustrates a typical chain extension operation;[932] the nature of the protecting groups and the Lewis acid used influences the diastereoselectivity of the process. Most of the serine-based syntheses that build structures on to the carboxy-group continue to employ the double N,O-protection strategy represented in the Garner aldehyde (123): demonstrated in routes to (2R)-F$_3$CCH$_2$CH(NHBoc)CO$_2$H (six steps including use of CF$_3$–TMS for trifluoromethylation),[933] chiral vinyl halides,[934] (2R,3R)-β-hydroxy-aspartic acid *via* (124),[935] D-(+)-erythrosphingosine and its (−)-enantiomer *via* (125),[936] alternatively *via* 3-ketosphinganine (126) from the same starting material,[937] β-diphenylphosphinothio-L-alanine,[938] the aminoalkanol THPOCH$_2$CH(NH$_2$)CH(OH)^{i}Pr in frangulanine synthesis,[939] and novel isoxazol-3- and -5-ylglycines.[940] This Garner aldehyde was not successfully used in a synthesis of S-2-amino-(Z)-3,5-hexadienoic acid due to the need for oxidative treatment in the final stage, but Lajoie's alternative (N-Boc-L-serinal

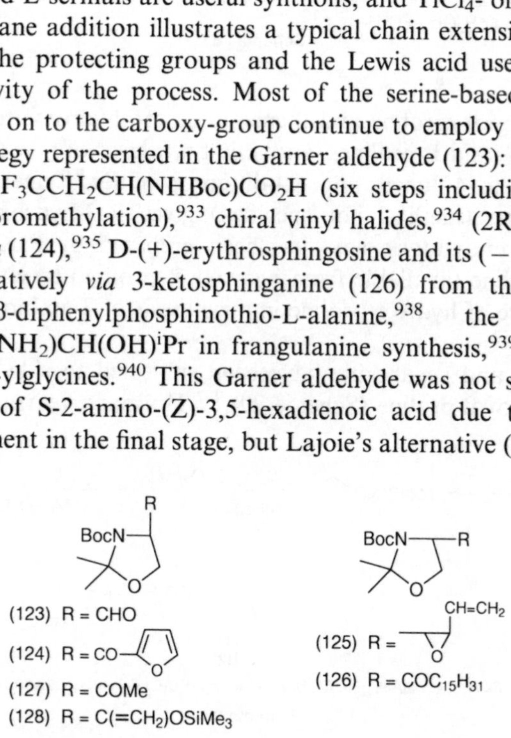

(123) R = CHO
(124) R = CO—
(127) R = COMe
(128) R = C(=CH$_2$)OSiMe$_3$

(125) R =
(126) R = COC$_{15}$H$_{31}$

with the side-chain alkanol protected as an orthoester) was satisfactory.[941] An outing for S-2-(N-Fmoc-amino)-β-lactone, the other main L-serine synthon, has led to a synthesis of the phosphonic acid isostere of L-aspartic acid, $FmocNHCH[CH_2P(O)(OR)_2]CO_2Me$, R = Me or allyl.[942] The threonine analogue of the oxazolidine gives the ketone (127) through chlorochromate oxidation, and this has been used in a multigram synthesis of the silyl enol ether (128), which on condensation with a glycopyranosyl aldehyde gives a 2-amino-5-(β-D-glycopyranosyl)pentanoic acid,[943] and in a route to α- and β-C-galactosylserines.[944]

Hydroxy-L-proline is also a valuable starting material, demonstrated for the synthesis, *via* 3,4-dehydro-L-proline, of L-proline with every methylene group stereospecifically labelled with ^2H,[945] and a synthesis of (2S,3S,4R)-[3,4,5,5,5-$^2H_5$]leucine (Scheme 28),[946] similarly for the synthesis of (2S,3R,4S)-

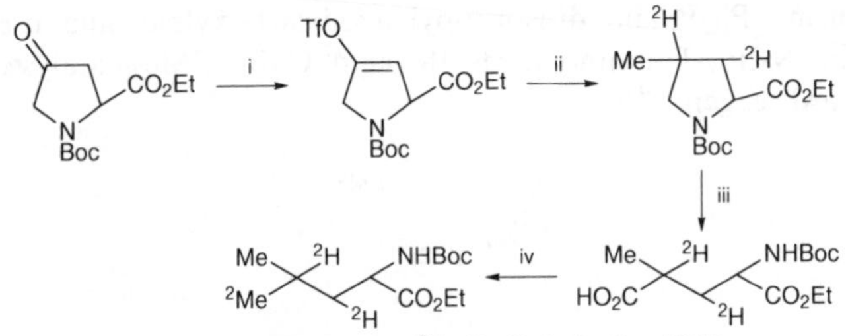

Reagents: i, NaHDMS, Tf₂NPh; ii, Me₂CuMgI, then $^2H_2$–Pt; iii, RuO₄, then LiOH; iv, $CO_2H \rightarrow C^2H_2OH \rightarrow C^2H_2$

Scheme 28

and (2S,3S,4R)-epoxyproline and thence to amino-hydroxyprolines,[947] to *cis*-3,4-dihydroxy-D- and L-prolines, and a first synthesis of (−)-(3S,4R)-dichloroproline,[948] and to 4-fluoro- and 4,4-difluoroprolines.[949] Decarboxylation of hydroxy-L-proline (method not specified) gives R-3-hydroxypyrrolidine.[950] (2S,4S)-2,4-Diaminoglutaric acid has been prepared by RuO₄ oxidation of 4-Boc-aminoproline (available from hydroxy-L-proline) to give the lactam of the target.[951] Use of hydroxyproline in a synthesis of 3-aminopyrrolidines has been described.[952]

Aspartic acid and asparagine have been subjected to side-chain modifications: synthesis of β-proline (Scheme 29);[953] Hofmann rearrangement of $N^α$-

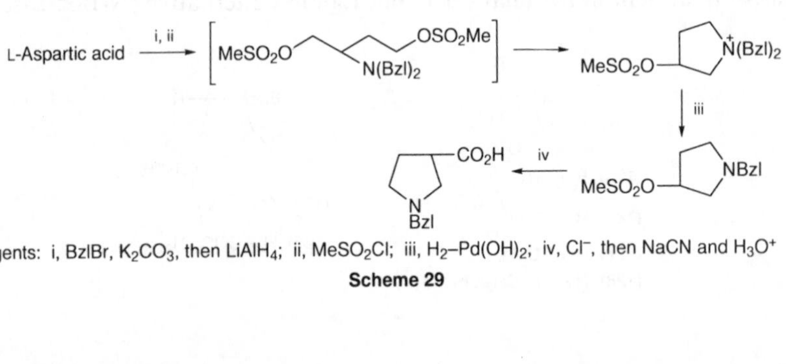

Reagents: i, BzlBr, K₂CO₃, then LiAlH₄; ii, MeSO₂Cl; iii, H₂–Pd(OH)₂; iv, Cl⁻, then NaCN and H₃O⁺

Scheme 29

toluene-p-sulfonyl-L-asparagine giving N^{α}-toluene-p-sulfonyl-L-β-amino-alanine;[954] anti-methallylation;[955] and preparation of chiral oxazin-3-ones.[956] Reactions applied to the general class of α-(ω-carboxyalkyl)glycines include selective reduction of ω-esters by DIBAL-H and chain extension through Wittig reactions on the resulting aldehydes (see also Ref. 823),[957] and corresponding reactions of methyl (S)-2-(di-N-Bocamino)-5-oxopentanoate, leading to α-aminoarachidonic acid;[958] development of the β-aldehyde into indolizidine-9-ones;[959] C-glycosylation of a protected L-aspartic β-aldehyde catalysed by SmI_2,[960] selective microbial protease-catalysed hydrolysis of diesters to give ω-methyl, -ethyl, and isopropyl esters;[961] synthesis of 2-amino-4-aryl-4-oxobutanoic acids and higher homologues through condensation with aryl methyl ethers in liquid HF;[962] development of 2-amino-4-oxo-5-phospho-nopentanoic acid into the carbon isostere of a N^{β}-glucosyl-L-asparagine;[963] cyclization to homochiral 3-aminopyrrolidines and 3-aminopiperidines.[964] The homochiral azetidinone (129) from L-aspartic acid has been chosen to start a synthesis of xemilofiban through extension of its ester grouping to give ethyl β-ethynyl-β-alaninate,[965] and the N-TBS-relative with an aldehyde function in place of the ester grouping has been used in a synthesis of ADDA deriva-tives.[966] The β-lactone (130) from L-aspartic acid has been used to alkylate a triazacyclononane to give an amino acid derivative that offers novel opportu-nities for the synthesis of zinc(II)- and copper(II)-binding peptides.[967]

(129)   (130)

Glutamic acids and their derivatives have been used in Kolbe synthesis of L,L-2,7-diaminosuberic acid,[968] in pyridine synthesis by condensation with alkadienals at 180 °C in aqueous media or in food oils;[969] in Barton radical brominative decarboxylation of α-tert-butyl N-Boc-L-glutamate in syntheses of (2S)- and (2R)-2-amino-4-bromobutanoic acid;[970] in an improved synthesis of (S)-2-amino-5-(aminoxy)pentanoic acid (L-canaline);[971] and in a route to (2R,3R)-3-hydroxyproline.[972]

N-Alkyl-L-pyroglutamic acids have been prepared starting from L-glutamic acid,[973] and amidation of ethyl L- or D-pyroglutamates has been effected with the help of *Candida antarctica* lipase in aqueous media at 60 °C (less than 5 minutes' reaction is needed; at 45 °C the rate for the D-enantiomer is almost unmeasureably slow).[974] Ring cleavage of an N-alkoxycarbonyl pyroglutamate with lithium trimethylsilyldiazomethane provides 6-diazo-5-oxonorleucine.[975]

Uses of L-pyroglutamic acid have been reviewed in the context of the synthesis of amino acids, alkaloids, and antibiotics.[976] New uses for L-pyroglu-tamic acid include routes to L-α-(3-phenylpropyl)glycine, to L-α-(2-benzyl-3-phenylpropyl)glycine, and *trans*-4-benzyl- and *cis*-5-phenyl-L-prolines.[977] L-Pyroglutaminol continues to develop an independent range of applications, including uses in amino acid synthesis: of the echinocandin constituent

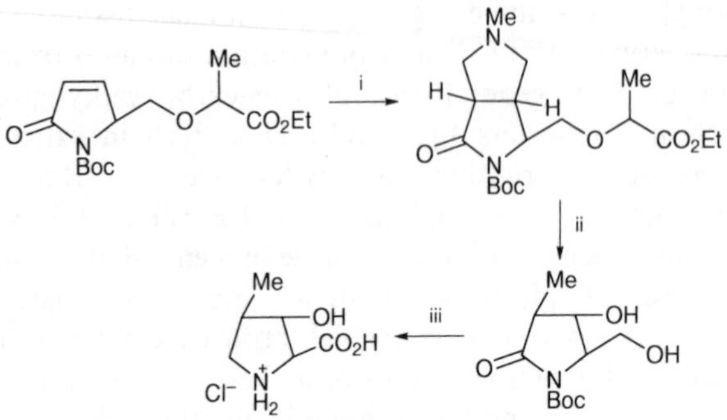

Reagents: i, MeN⁺(O⁻)=CH₂; ii, H₂–Pd(OH)₂; iii, functional group development

Scheme 30

(2S,3S,4S)-3-hydroxy-4-methylproline (Scheme 30);[978] of 3-methoxycarbonyl-methyl-4-substituted L-pyroglutamates as conformationally constrained glutamic acid analogues,[979] and (2S,3S)-3-N-(benzylamino)prolinol and (7S,8R)-7-N-(benzylamino)pyrrolizidin-3-one.[980] The corresponding ketone has been rearranged to 2,3-disubstituted piperidines.[981] The lithium enolate of ethyl N-tert-butoxycarbonyl-L-pyroglutamate starts a route to manzamine alkaloids through quenching with ethyl chloroformate.[982]

Alkylation of N-alkoxycarbonyl-L-pyroglutamic acid at C-4 does not invariably lead to the *trans*-stereochemical outcome, but can be encouraged to give pure *cis*-products by judicious choice of reactants and protecting groups;[983] alkylation at C-2 or C-4 is determined by the nature of the N-protecting group, since this determines the site of enolate formation by LiHMDS; the preparation of 2- and 4-indolylmethylglutamic acids provided the results on which this conclusion was based.[984] An alternative approach to *trans*-4-alkenyl glutamic acids and discussion of methods for preparation of stereoisomers has been published.[985] 4-Cyanomethylation of protected L-pyroglutamic acid, ring-opening and recyclization gives epimeric lactams, reduction giving pyrrolidine lysine mimics.[986] 3-Substitution can be accomplished starting with 3,4-dehydropyroglutamates (Ref.169).

(2S)-4-Methyleneglutamic acid is ideally suited to Diels-Alder and 1,4-ionic addition reactions and radical substitution processes, to give side chain-extended homologues (*e.g.* 131).[987]

Higher amino dicarboxylic acids are represented specifically in a review of the cyclization of α-aminoadipic acid to the β-lactam mediated by *Penicillium chrysogenum*,[988] and in the conversion of the hydroxymethyl analogue of this lactam (prepared from L-lysine) into 5-alkyl derivatives.[989]

Reactions at the sulfur atom of the common sulfur-containing amino acids include disulfide formation of protected cysteines, with dimethyl meso-2,3-dimercaptosuccinate,[990] with S-nitrosocysteine,[991] and through photolysis in the presence of a peroxydiphosphate salt (involving a phosphate radical anion).[992] Cystines protected differently at the four functional groups have been prepared using 3-nitropyridinesulfenyl S-activation.[993] Common oxi-

Me
Me
CO₂H
H₃N⁺ CO₂⁻
(131)

S   S
R²NH R
R²NH R
HO R
OH
R
(132)

R
Ar N S
H
MeO O
(133)

dizing agents have been assessed for their reactions with DL-cysteine; all modify the thiol grouping but some oxidants cause other changes.[994] S-Alkylation of D- or L-cysteine using bromosuccinic acid enantiomers leading to all four stereoisomers of 2-amino-3-[(1,2-dicarboxyethyl)sulfanyl]propanoic acid has been compared with the alternative route employing mercaptosuccinic acid and 2-amino-3-chloropropanoic acid.[995] S-2,2,2-Trifluoroethylation with the iodonium salt $(CF_3SO_2)_2NI(Ph)CH_2CF_3$[996] can be accomplished in aqueous media. Conventional S-alkylation of N-protected L-cysteine with 1,2-dibromoethane, and Grignard addition leads to C_2-symmetric ligands (132) destined for use in conjunction with metal ions to create novel enantioselective catalysts.[997] S-Alkylation of L-cysteine by Merrifield peptide synthesis resin is the prelude to S-oxidation followed by β-elimination leading to a clean synthesis of dehydroalanine derivatives (see also Ref. 903).[998] The reverse process, nucleophilic addition through sulfur to quinone methides in aqueous media[999] and Michael additions to alkynones,[1000] is also applicable, though more reluctantly, to lysine and serine derivatives.

The thiazolidine equivalent of the Garner aldehyde (123; S in place of ring O) has been used in a synthesis of sulfobacins A and B (see also Ref. 179) through aldolization and routine steps.[1001] Ethyl N-formyl-L-cysteinate cyclization by toluene-p-sulfonic acid in boiling toluene gives the (R)-thiazoline-4-carboxylate ester in only 70% e.e.[1002] Cysteine bound through its carboxy group to a resin has been modified at N and S to give a library of thiomorpholin-3-ones after HF cleavage.[1003]

S-Aminoethyl-L-cysteine has been converted through performic acid oxidation into taurine, whereas this reagent gives cysteic acid when applied to cysteine or cystine.[1004] Milder asymmetric S-oxidation of N-Fmoc-S-allyl-L-cysteine [using $Ti(OPr^i)_4/(-)$-diethyl tartrate] gives L-(+)-alliin.[1005] A proposed metabolite of L-histidine catabolism, S-[2-carboxy-1-(1H-imidazol-4-yl)ethyl]-L-cysteine, is a substrate for an enzyme in rat liver homogenate and may have a role in the accumulation of urocanic acid in skin during UV irradiation.[1006] Intramolecular 1,3-cycloaddition of a Schiff base of methyl S-allylcysteinate leads to the bicyclic compound (133).[1007]

L-Methionine shows the standard bivalent sulfur reactions [cyanogen bromide cleavage when bound to peptide synthesis resin, releasing homoserine lactone and adaptable to the generation of a library of analogues;[1008] *Beauveria bassiana* ATCC7159-catalysed oxidation, or H_2O_2 oxidation, of N-phthaloyl-D- or L-methionine to give all four diastereoisomers of the sulfoxide assisted by the ease of separation of the epimers,[1009] and application of H_2O_2 or ozone oxidation for selenoxide formation with DL-selenomethionine;[1010] and conversion of methyl N-Boc-L-methioninate into the [^{18}F]fluoromethyl analogue using XeF_2 with $Bu_4N^{18}F^{1011}$] but the *crème brûlée* approach (amino acids suffering oxidation by a hydrogen–oxygen flame applied to the surface of an aqueous solution) generates other amino acids in an indiscriminate way (2-aminobutanoic acid, homoserine, and glutamic acid are formed from methionine *via* a side-chain carbon radical intermediate).[1012]

Selective Schiff base formation has been investigated for salicylaldehyde with the diamino acids L-2-amino-3-methylaminopropanoic acid, DL-2,3-diaminopropanoic acid and DL-2,4-diaminobutanoic acid.[1013] Ornithine offers easy access to N^ω-alkyl-arginines through reactions with S-alkylisothioureas and iodomethane.[1014] Side-chain tri-alkylsilyl derivatives of N^α-Z-arginine react to give the N^δ,N^ω-bis(ethoxycarbonyl) derivative with ethyl chloroformate.[1015]

Lysine studies have mostly concentrated on models for reactions suffered by its side-chain under physiological conditions, but standard nucleophilic reactions of the ε-amino group have also been reported: hydroxy-deamination (Scheme 31),[1016] and side-chain protection (the N-Ddiv group offers improved

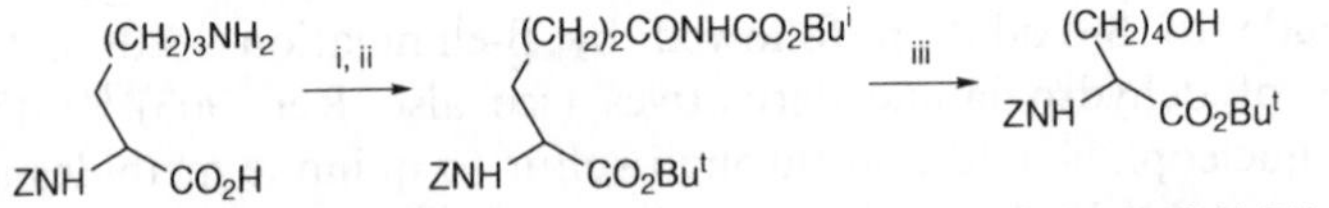

Reagents: i, Bu^tO_2CCl, NaOH, then isobutene, H_2SO_4 (cat.); ii, $NaBrO_3$, RuO_2 (cat.); iii, $NaBH_4$, aq. Pr^iOH

Scheme 31

dimedonyl-type protection[1017]) that is compatible with peptide synthesis protocols. The blue fluorophore (λ_{em} 470 nm) generated between N^α-acetyl-L-lysine and 3-deoxyglucosone by condensation of two molecules of each reactant is a furopyrrolopyridinium salt.[1018] A related study of the fluorophores lipofuscin and ceroid formed from N^α-acetyl-L-lysine and 4-hydroxy-2-nonenal has identified a sequence of reactions starting from the non-fluorescent 2:1-Michael adduct formed between these reactants, cyclization giving 2-hydroxy-3-imino-1,2-dihydropyrrole.[1019] L-Lysine undergoes N^ε-carboxymethylation *via* the Amadori condensation pathway by contact with glyoxal or a reducing sugar (see also Refs. 731, 735);[1020] release of this compound from glycated albumin is greatly accelerated in alkaline media.[1021] Incubation of collagen with glucose under oxidative conditions causes glycation, which eventually leads to the N^ε-carboxymethyl-lysine derivative as well as causing oxidative changes to tyrosine (generation of m-tyrosine, dityrosine, DOPA); the hydroperoxides of valine and leucine are also formed.[1022] Protein crosslinking through lysine residues is well-established, a new crosslink involving condensation of N^ε-groupings with 4-hydroxy-2-alkenal that is produced by lipid peroxidation.[1023] Malondialdehyde reacts with lysine to give a formyl-substituted dihydropyridine rather than the unstable imidopropene Schiff base previously reported.[1024]

The phenylglycine family is unique among aromatic α-amino acids in offering an aryl group for transformation into a carboxy group for the synthesis of other α-amino acids, as in a synthesis of (2S,3R)-N-Boc-3-hydroxyglutamic acid[1025] (see also Scheme 13, Ref. 248).

Side-chain substitution of protected phenylalanine leading to fluorinated analogues[1026] and to 4'-borono-L-phenylalanine by palladium-catalysed cross-coupling with the diboronate 134; R-R = Me_2C-CMe_2[1027] has been reported; the N-Boc-protected 4'-borono-L-phenylalanine can be converted into 4-aryl-L-phenylalanines.[1028] The latter example illustrates a relatively rare approach to 4'-substituted L-phenylalanines for which L-tyrosine is the more usual starting point. Examples are: 4'-borono-L-phenylalanine prepared by palladium-catalysed cross-coupling with the diboronate 134; R-R = Ph_2C-CPh_2;[1029]

$$\begin{array}{c} R\!-\!O \qquad\quad O\!-\!R \\ | \qquad \backslash \quad / \qquad | \\ \;\;\;\; B\!-\!B \\ | \qquad / \quad \backslash \qquad | \\ R\!-\!O \qquad\quad O\!-\!R \end{array}$$

(134)

4'-hydroxymethylphenylalanine prepared by palladium-catalysed hydroformylation of methyl O-trifluoromethanesulfonyl-N-Boc-L-tyrosinate followed by $NaBH_4$ reduction;[1030] and a uridinylated analogue FmocPhe[O-P(O)-(OCH_2CH_2CN)(O-uridinyl)OR[1031]]. Photocyclization of dehydronaphthylalanines gives 1,2-dihydrobenzo[f]quinolinones and benzo[f]isoquinoline.[1032]

(R)-4-Hydroxyphenylglycine has been converted through palladium-catalysed substitutions into 4'-phosphonophenyl-, 4'-(tetrazol-5-yl)- and 4'-carboxy-analogues.[1033] Transformations in the aryl moiety of L-tyrosine derivatives, but not at the OH group, include nitration of the [$^2H_4$]-isotopomer to give 3'-nitro-L-[$^2H_3$]tyrosine,[1034] radioiodination (Chloramine-T and $^{128}I_2$),[1035] routes to 4'-(tert-butoxycarbonylmethoxy)-3'-trimethylsilyloxy-methoxy-L-phenylalanines and 3'-hydroxy- and -tert-butoxycarbonylmethoxy-analogues,[1036] and more extensive changes to give 3'-aryl-5'-phosphonomethyl-L-phenylalanines.[1037] 3'-Hydroxy- and 3'-nitrotyrosines are employed as starting materials in some of these routes, the former (*i.e.* DOPA) being used also for preparation of 6-[^{18}F]fluoroDOPA (by radiofluorodestannylation[1038] and radiofluorodemercuration in a solid phase protocol[1039]). 3'-Nitration and transformation into the 3'-N-acetylamino analogue followed by 6'-nitration and ensuing development leads to 9-substituted 8-membered benzolactams that are important protein kinase C activators.[1040] Intramolecular oxidative coupling of DOPA and 4-fluoro-3-nitrophenylalanine side-chains (Scheme 32)[1041] and mono- and dihalogenated tyrosine using peroxidase (Scheme 33)[1042] has allowed more optimism that current efforts aimed at provision of suitable dityrosines and isodityrosines for syntheses of bastadins 2, 3, and 6 bouvardin will succeed. 4-Hydroxyphenylglycine has been converted into 3,3'-dimers and trimers through treatment with H_2O_2 and horseradish peroxidase.[1043] Simpler oxidative processes with tyrosine itself include the well-known intramolecular attack of the carboxy group on the p-phenolic position,[1044] exploited in synthesis of more complex spirolactams.[1045] DOPA

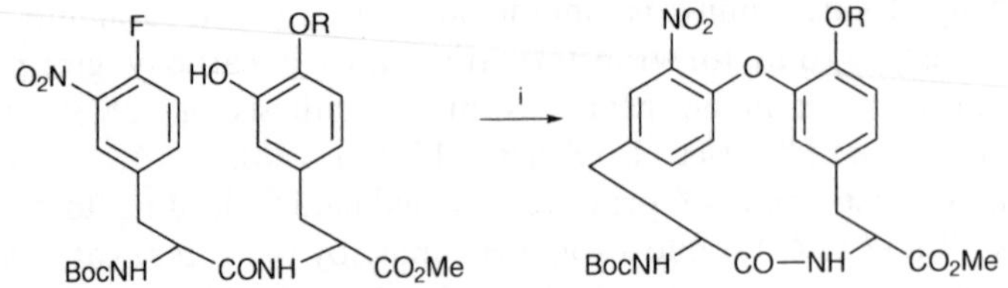

Reagents: i, NaH, thf

Scheme 32

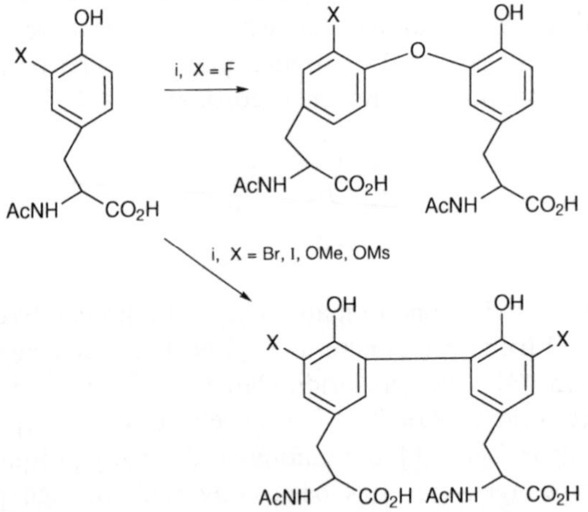

Reagents: i, horseradish peroxidase, H_2O_2, pH 6.0, then $NaHSO_3$

Scheme 33

synthesis is illustrated [hydroxy radicals generated in the iron(II)-EDTA–ascorbic acid cocktail;[1046] horseradish peroxidase-catalysed oxidation enhanced by hydroxycinnamic acid, thus suggesting that tyrosine is substituted *via* its phenoxyl radical[1047]]. Tyrosine partially traps radical intermediates generated in the reaction of peroxynitrite with CO_2.[1048]

Derivatives of tyrosine formed through modification of the phenol group of tyrosine include N-Boc-O-thiophosphono-L-tyrosine[1049] and N-Fmoc-O-(monobenzylphosphono)-L-tyrosine, prepared from its dibenzyl ester.[1050] An arylboronic acid is a suitable reagent for O-arylation of tyrosine and 4-hydroxyphenylglycine;[1051] in the latter case, without the usual problem of racemization. O-Methylation under non-aqueous conditions has been accomplished with dimethyl sulfate and LiOH.[1052] Selectivity in hydrogenolysis of N-benzyloxycarbonyl and nitro-groups on the tyrosine aryl group, leaving O-benzyl in place, can be arranged by having a nitrogen base present.[1053] Reactions of O-(3-pentyl)-tyrosine show that the group offers a new base-resistant side-chain strategy that can be cleaved by HF.[1054]

The mixture of 1- and 3-phosphonylated and 1,3-diphosphonylated histidine derivatives formed by direct reaction was separated by preparative scale chromatography over silica gel.[1055] N^τ-Galactosyl-histidine can be reached with the protected glycos-1-yl bromide as reactant.[1056]

Tryptophan reactions are generally well understood, such as Pictet-Spengler condensation with formaldehyde; but two side-products are also formed in this particular process, the N^{in}-hydroxymethylated product and the dimer (two products linked by a CH_2 group through the indole nitrogen atom).[1057] The tetracyclic ketone (135) is easily prepared from methyl D-tryptophanate and starts a synthesis of norsuveoline,[1058] and the hydroxyimidazoindoline ring system of (−)-asperlicin and (−)-asperlicin C has been created from N^{in}-(N-Z-L-leucinyl)-L-tryptophan.[1059] Tryptophan tethered to a solid phase provides the starting point in a synthesis of indolyl diketopiperazine alkaloids,[1060] and conventional solution reactions for a first synthesis of roquefortine D (136)

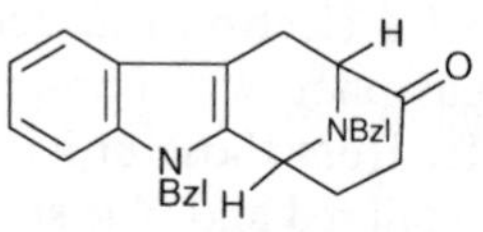

(135) 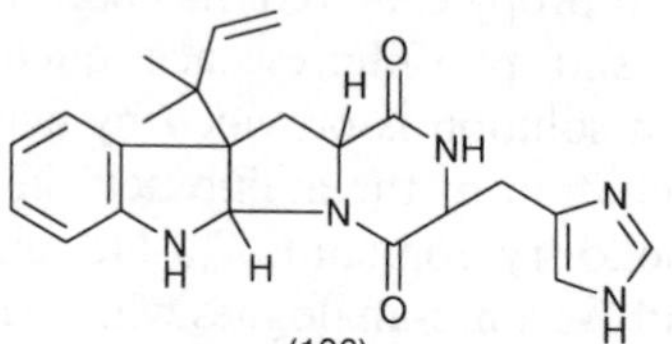   (136)

from L-tryptophan and histidine.[1061] δ,δ'-Dimerization through tryptophan side-chains, the result of an acid-catalysed Mannich-type reaction with an aldehyde, can be reversed in ethanolic HCl at 150 °C in the presence of ethanedithiol.[1062] A continuing series of papers (Vol. 30, p. 71) reveals new reaction pathways for N-acylated or N-alkoxycarbonylated tryptophan derivatives; cyclization or dimerization pathways are favoured (Scheme 34), and bi-indoles are formed, depending on the nature of the side-chain nucleophile.[1063] The related process in which a C-terminal tryptophan-containing dipeptide

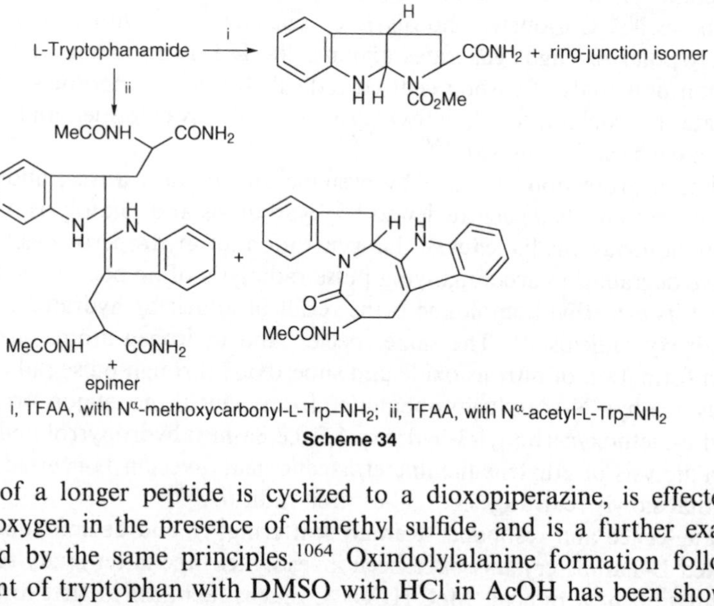

Reagents: i, TFAA, with N^{α}-methoxycarbonyl-L-Trp–NH$_2$; ii, TFAA, with N^{α}-acetyl-L-Trp–NH$_2$

Scheme 34

moiety of a longer peptide is cyclized to a dioxopiperazine, is effected by singlet oxygen in the presence of dimethyl sulfide, and is a further example governed by the same principles.[1064] Oxindolylalanine formation following treatment of tryptophan with DMSO with HCl in AcOH has been shown to involve a chlorotryptophan intermediate.[1065]

6.4 Effects of Electromagnetic Radiation on Amino Acids – Fluorescence studies centred on tyrosine[1066] and dityrosine[1067] relate the data to molecular rotational dynamics. O-Methyl-β-tyrosine shows substantially different fluorescence features compared with those of tyrosine.[1068] A new approach to interpretation of fluorescence spectra of tyrosine and tryptophan and of other compounds carrying the same chromophores has been reported.[1069] Variations in the steady-state fluorescence of N-acetyl-L-tyrosinamide in the presence of N,N-dimethyl- and N-methyl-acetamide and urea, in four different solvents, imply that the amides are effective in dynamic quenching.[1070] A similar study of the effects of acrylamide and iodides on N-acetyl-L-tryptophanamide fluorescence in propyleneglycol has been carried out,[1071] and a new insight on fluorescence and phosphorescence quenching by oxygen of photoexcited tryptophan in solution is provided by evidence for electron transfer gathered for photoexcitation of the amino acid in the gas phase.[1072] Phosphorescence features of solid tryptophan (λ_{max} 442 nm) differ from those of its 4-, 5-, and 6-fluoro-, and 5-bromo-analogues which show small red and blue shifts.[1073]

5-Hydroxytryptophan fluorescence λ_{max} is insensitive to change of solvent, unlike that of tryptophan or 7-azatryptophan.[1074] Phosphorescence features of 5-hydroxytryptophan have been published.[1075]

Photolysis of proline sensitised by Rose Bengal results in quantitative decarboxylation to give 1-pyrroline.[1076] Similar treatment of N^α-acetyltryptophan N′-ethylamide and 5-fluorouracil in water generates the 5-uracilyl cation which gives N^α-acetyl-2-(uracil-5-yl)-tryptophan N′-ethylamide and 2-[1-(2-deoxy-β-D-erythro-pentofuranosyl)-uracil-5-yl]-tryptophan N′-ethylamide.[1077] Alanine or glycine photolysed with ascorbic acid and sodium nitrite yields mono-imino- or monohydrasinyl-ascorbyl radicals formed from intermediate Schiff bases.[1078] Curiously, photolysis of phenylalanine with 205 nm right-circularly-polarized light generates alanine; the D:L-ratio is not changed.[1079] Oxidation potentials of carbon-centred radicals formed by photolysis of ethyl cysteinate hydrochloride, dimethyl cystinate dihydrochloride, and N-Boc-proline have been determined.[1080]

Radiation protection afforded by cysteine and its derivatives, and related sulfur compounds, has been reviewed.[1081] Radiolysis and photolysis of amino acids in aqueous media causes the generation of hydroperoxides,[1082] and oxidative degradation accompanying pulse radiolysis of homocysteine thiolactone and its α-methyl homologue is the result of attack by hydrated electrons and hydroxy radicals.[1083] The same species, and hydrogen atoms, are implicated in formation of nitrous oxide and superoxide through pulse radiolysis of aqueous methyl N-acetyltryptophanate, from which the major product is 1-acetyl-2-methoxycarbonyl-3-hydroxy-1,2,3,8,8a-hexahydropyrroloindole.[1084] Pulse radiolysis of ethylenediaminetetraacetic acid gives an N-centred radical cation that slowly rearranges into C-centred radicals.[1085]

Luminescence and resonance Raman scattering[1086] and deamination[1087] of irradiated L-alanine crystals have been investigated, the latter study revealing the formation of a radical, $MeC \cdot HCO_2{}^-$, that is identical to that formed by pulse radiolysis in aqueous solution. Irradiation of glycine[1088] and threonine[1089] with electrons in the high keV energy region leads to decomposition products that are similar to those seen in radiolysis studies and hot-atom bombardment.

7 Analytical Methods

7.1 Introduction – Several chapters of a recent addition to a standard monograph series are relevant reviews of amino acid analysis methods,[1090] and a review[1091] covers the literature on HPLC and CZE and related methods from 1992. Greater sensitivity that accompanies improved analytical methods requires utmost reliability in sample preparation procedures, and standardization sought for acid hydrolysis of proteins is still not being achieved; for example, microwave heating induces enhanced racemization.[1092] Glycerol added to a protein hydrolysis cocktail leads to under-estimation of aspartic and glutamic acids due to esterification.[1093]

7.2 Gas–Liquid Chromatography – Trimethylsilylation using bis(trimethyl-silyl)trifluoroacetamide (Refs. 1092, 1093) is becoming the more widely used derivatization protocol for preparation of samples for GC-MS, including hydrolysates of samples from oil painting binders[1094] and hydroxypipecolic acids from plants (*Inga* species contain eight compounds of this class).[1095] tert-Butyldimethylsilylation offers identification of amino acids in plant leaves at 1–20 ng levels,[1096] and has been used for estimation of ^{2}H-labelled serine and cysteine.[1097]

N-Methoxycarbonyl amino acid esters are also favoured derivatives in recent studies, of [1-^{13}C]valine,[1098] and of pyroglutamic acid.[1099] This last study incorporates a two-column GLC protocol, the second column being a CSP for enantiomer separation prior to MS analysis.

Attomole levels can be assessed for 3-nitrotyrosine after reduction and conversion into the N-heptafluorobutyroyl derivative of its propyl ester,[1100] and N-pentafluoropropionyl analogues of representative amino acids have been employed for enantiomer analysis over L-Chirasil-Val.[1101]

Standard GLC methods for the estimation of homocysteine compare favourably with results of a novel commercial ELISA-based method.[1102]

7.3 Ion-exchange Chromatography – Routine studies using standard amino acid analysis methodology, illustrated for the special case of homocysteine and cysteine in clinical samples,[1103] are not covered. Some standard HPLC analytical protocols employ ion-exchangers as stationary phases, as in the quantitation of glycosylated hydroxylysines as their N-Fmoc derivatives.[1104]

7.4 Thin-layer Chromatography – Advantages of TLC methods still remain clear, for estimation of the enantiomeric purity of ^{14}C and ^{3}H-labelled L-amino acids (as dansyl derivatives with a β-cyclodextrin-containing mobile phase) using reverse isotope dilution analysis,[1105] and for PTH analysis.[1106]

TLC methods for the estimation of lysine, homoserine, and threonine have been reviewed.[1107]

7.5 High-performance Liquid Chromatography – Estimation of components in mixtures of underivatized amino acids separated by HPLC calls for specific structural characteristics in the amino acids to which detectors will respond. These usually rely on UV absorption for aromatic and heteroaromatic compounds (tryptophan and kynurenine,[1108] detection of direct interaction between L-phenylalanine and DNA[1109]) and 3-nitrotyrosine. For the last-mentioned amino acid, of interest for its possible presence in human brain tissue, errors in identification can arise due to other constituents with very similar retention times,[1110] and a new post-column process that is specific for this amino acid, UV photolysis conversion into DOPA, followed by electro-chemical detection has been advocated.[1111] Electrochemical detection is already used in HPLC analysis of DOPA-containing samples,[1112] and is also standard practice for homocysteine[1113] and cysteinesulfinic acid and hypo-taurine,[1114] and for amino acids in nanolitre samples from brain, as their tert-butyl thiol derivatives.[1115] A special situation allows scintillation counting detection for radiolabelled leucine and phenylalanine.[1116]

The protein crosslinking amino acids that turnover causes to be present in blood and excreted in urine are useful markers of cirrhosis and liver fibrosis (desmosine and hydroxylysylpyridinoline)[1117] and markers of bone formation and resorption[1118] (these include pyridinoline and deoxypyridinoline;[1119] see also Ref. 1191), as well as collagen crosslinks pentosidine and other fluorescent pyridinium-containing amino acids,[1120] have continued to receive special attention.[1121] A notable chemiluminescence detection method for deoxypyridi-noline is emerging, and compares well with ELISA.[1122] Carnitine and acylcar-nitines are diagnostic markers for inborn errors of fatty acid oxidation, and an HPLC-MS protocol has been developed for their quantitation in physiological samples,[1123,1124] a technique also used to quantitate and identify two side-products in a synthesis of 4'-boronophenylalanine using two different synthesis pathways.[1125] ICP-MS detection for selenium analogues of common amino acids has been demonstrated.[1126]

Amino acids are derivatized prior to HPLC analysis in the majority of cases, to achieve the required sensitivity. The familiar protocols continue to be used; pre-column derivatization using phthaldialdehyde (OPA) with 2-mercapto-ethanol (D-phenylalanine in urine,[1127] dimethylated arginines,[1128] and sixteen common amino acids[1129]) or with 3-mercaptopropionic acid (glutamine in rat inestinal mucosa,[1130] representative amino acids,[1131] and histidine using ion-pair HPLC[1132]) remains the most favoured. Results with the OPA–3-mercap-topropionic acid reagent have been compared with those for OPA–N-acetyl-L-cysteine as far as stability of fluorescence of the derivatives are concerned, and both offer reliable quantitation;[1133] the OPA variants work equally well for analysis of aspartic and glutamic acids.[1134] The particular situation for OPA derivatization of N-acetyl-S-ethyl-L-cysteine in urine (after post-column de-acetylation using acylase),[1135] and selenocysteine[1136] and selenomethionine[1137] has been addressed, including, in the last-mentioned study, comparison with the corresponding use of naphthalene-1,2-dialdehyde as reagent (giving more intensely fluorescent derivatives) and β-cyclodextrin as mobile phase additive leading to enantiomer separation.

Fluorescent derivatives are formed using alternative protocols: homocysteine analysis using 7-fluorobenzofurazone-4-sulfonic acid[1138] and an equivalent commercial kit,[1139] 4-fluoro-7-nitro-2,1,3-benzoxadiazole (for analysis of opines),[1140] and new L-proline-substituted members of this class [4-(2-carboxypyrrolidin-1-yl)-7-nitro-2,1,3-benzoxadiazole and its 7-(N,N-dimethylaminosulfonyl) analogue] and the N-methyl-L-alanine relative [4-(N-1-carbethoxethyl-N-methyl)amino-7-nitro-2,1,3-benzoxadiazole and its 7-(N,N-dimethylaminosulfonyl) analogue].[1141] These form diastereoisomeric derivatives with partly racemic samples, to permit the determination of isomer ratios for amino acids with high sensitivity associated with the easily measured fluorescence (λ_{ex} 469 nm; λ_{em} 535 and 569 nm).

N-(2,4-Dinitrophenyl)amino acids (DNP-amino acids) have been used to explore a novel instrumental HPLC technique, toroidal-coil countercurrent flow.[1142] N-Dansyl-DL-amino acids have been resolved using human serum albumin as stationary phase; some benefit is seen in incorporating perchlorate salts in the buffer.[1143] Amino acids carrying standard N-protecting groups (Fmoc, Boc, TFA, Pmc) have been used to test protein-based HPLC stationary phases, including macrocyclic antibiotics.[1144] Synthetic CSPs, *e.g.* silica carrying aromatic substituents as part of a chiral grouping (137),[1145] are also

O_2N—⟨⟩—CONH ~(CH$_2$)$_3$—Si—O—

NO$_2$ (137)

usually tested in the same way, while macrocyclic hosts, a new example being (+)-(18-crown-6)-2,3,11,12-tetracarboxylic acid,[1146] have been tried with underivatized amino acids (with poor resolution; though rather better results are achieved with amino acid monoalkylamides). Benzophenone Schiff bases of DL-amino acids can be resolved over [(R,R-Whelk-OI], the L-enantiomer travelling fastest.[1147] Dansylation has been chosen for quantitation of S-methyl-, S-prop-2-enyl- and S-prop-1-enyl-L-cysteine sulfoxides from eight different members of the *Allium* species.[1148] Coloured derivatives formed by dabsylation can be detected at 780 femtomole levels using non-linear laser absorbance detection;[1149] phosphorescence developed by dabsylation of threonine and tyrosine has been assessed for its exploitation in amino acid analysis.[1150] Other conventional fluorophores can be introduced into amino acids using N-(O-succinimidyl)-α-(9-acridine)acetate[1151] or 2-(9-anthryl)ethyl chloroformate,[1152] which give tagging groups with fluoresecence characteristics comparable with those of the Fmoc group.

Derivatization through N-phenylthiocarbamoylation by phenyl isothiocyanate seems to require fewer papers to be published on its applications, now that the resulting PTAs have entered routine use for amino acid analysis.[1153] The separation of enantiomers of PTHs formed from PTA-peptides in the Edman sequencing protocol has been accomplished on CSPs to which β-cyclodextrin is bonded;[1154] this requires the use of non-racemizing conditions (BF₃ – HCl/MeOH) for the protocol leading to PTHs. A fluorescence-generating chiral relative, R-(−)-4-(3-isothiocyanatopyrrolidin-1-yl)-7-(N,N-dimethylaminosulfonyl)-2,1,3-benzoxadiazole successfully, amalgamates several of the established fluorophores into one reagent, giving derivatives showing λ_{ex} 460 nm, λ_{em} 550 nm;[1155] the results that are achievable through HPLC of the derivatives were demonstrated for DL-amino acids in yoghourt (11.9% D-glutamic acid, 27.6% D-aspartic acid, and 56.7% D-alanine). The fluorescence-generating reagents (1R,2R)- and (1S,2S)-N-[(2-isothiocyanato)cyclohexyl]-6-methoxy-4-quinolinylamide (λ_{ex} 333 nm, λ_{em} 430 nm) have been introduced for the same purpose.[1156] The classical diastereoisomer-generating Marfey's reagent (1-fluoro-2,4-dinitrophenyl-5-D-leucinamide or the L-enantiomer) has been used in the same way ('advanced Marfey's method') for absolute configurational assignments to amino acids based on elution sequence data[1157] (see also Ref. 578).

Methods for HPLC analysis that are applicable specifically to sulfur and selenium-containing amino acids include bromobimane derivatization of cysteine and homocysteine[1158] and similar fluorophore tagging for a homocysteine assay,[1159] S-pyridinylation of homocysteine,[1160] and S-nitrosation of cysteine and N-acetylcysteine by treatment with excess sodium nitrite.[1161] The nitrosothiols are easily detected, since the chromophore absorbs at 333 nm; or it can be cleaved by a mercury(II) salt, the nitric oxide being trapped with sulfanilamide to give a stable azo-dye.[1162]

7.6 Capillary Zone Electrophoresis (CZE), and Related Analytical Methods –

Reviews of capillary electrochromatography [a hybrid of capillary electrophoresis (CE) and HPLC][1163] and chiral CE with respect to its control through inclusion-complexation by chiral selectors[1164] have appeared.

CZE analysis studies of underivatized amino acids include S-[2-carboxy-1-(1-imidazol-4-yl)ethyl]cysteine, a putative histidine metabolite (see Ref. 1006),[1165] cysteine and homocysteine,[1166] and o-, m-, and p-fluoro- and -hydroxy-DL-phenylalanines (by LE-MEKC with copper(II)-4-hydroxy-L-proline as chiral selector).[1167]

The range of derivatives that are familiar from HPLC studies are used also in CE analysis. DNP-DL-Amino acids can be analysed rapidly by CZE,[1168] and have been resolved by CZE using cyclic hexapeptides as chiral selectors.[1169] Resolution of dansyl-DL-amino acids, with β-cyclodextrin as chiral selector[1170] and over molecule-imprinted copolymers of vinylpyridine and methacrylic acid,[1171] and dabsyl-amino acids,[1172] are also represented. The last-mentioned study was a test of fibre-optic thermal lens detected that allows working with microlitre volumes, also a feature of the estimation of L-arginine in marine snail neurones at 50 attomole levels after fluoresceamine derivatization.[1173] A mixture of common PTHs can be analysed efficiently through CMEK separation.[1174]

Fluorescent tagging in conjunction with CZE analysis leads to high sensitivity, fluorescamine giving better results than OPA in this respect.[1175] Fluorescein isothiocyanate derivatization of homocysteine generates only modest fluorescence yield in laser-induced fluorescence detection,[1176] a technique used also for quantitation of derivatized glutamic acid and glutamine.[1177] Derivatization with R-(−)-4-(3-isothiocyanatopyrrolidin-1-yl)-7-nitro-2,1,3-benzoxadiazole or its S-(+)-enantiomer permits CZE determination of enantiomer ratios.[1178]

7.7 Other Analytical Methods – Total D-amino acids in plasma can be estimated through treatment of samples with immobilized D-amino acid oxidase followed by measurement of chemiluminescence generated with peroxyoxalate.[1179]

7.8 Assays for Specific Amino Acids – A monograph[1180] and a review of biosensors[1181] cover topics that have provided most of the literature for this Section over the years. Immobilized L-glutamate oxidase and L-glutamate dehydrogenase are at the heart of a novel instrumental configuration for online monitoring of L-glutamic acid production,[1182] a need that has also been met in an independent study leading to a similar biosensor.[1183] Biosensor principles applying to L-glutamic acid and L-glutamine have been outlined,[1184] and two amperometric biosensors responding to L-alanine and pyruvic acid have been developed.[1185] An amperometric biosensor set up for assays of specific L-amino acid can be tuned to discriminate for another L-amino acid through electronic circuit modifications.[1186]

A laboratory version of the chemistry underlying these instruments is seen in an L-glutamic acid assay that employs L-glutamate oxidase and L-glutamic acid–pyruvic acid transaminase leading to α-ketoglutaric acid and H_2O_2, the latter being converted into a chromophoric derivative that is assayed spectrophotometrically.[1187]

An ELISA using monoclonal anti-nitrotyrosine antibodies permitting 3-nitrotyrosine quantitation illustrates another approach to specific amino acid assay.[1188]

Routine colorimetry for cysteine has been used in a novel context, applied after acylase treatment of a cysteine–N-acetylcysteine mixture whose cysteine content was already determined.[1189] The Udenfriend tyrosine fluorimetric assay has been updated,[1190] and chemiluminescence measurements (immunoassay for deoxypyridinoline[1191] and for the DOPA–$KMnO_4$ system[1192]) continue to show valuable characteristics that may lead to robust biosensors and chemical sensors. Fluorescence generated by addition of thiols to maleimides has been related to cysteine concentration.[1193]

Electrochemical detection of nitric oxide released by copper(I)-catalysed homolysis of S-nitrosocysteine employs conventional apparatus that permits quantitative assay of this derivative;[1194] a warning has appeared that nitric oxide microsensors (porphyrinic membranes exploiting oxidative electrochemistry[1195]) respond equally to tyrosine since both species have the same oxidation potential. A graphite paste membrane electrode, used for potentiometric assay of L-proline, has been rendered enantioselective by impregnation with 2-hydroxy-3-trimethylammoniopropyl-β-cyclodextrin.[1196]

References

1.   M.J. Larkin, *Chem. and Ind.*, 1999, 488 (correspondence, *ibid.*, p. 490).
2.   *Practical Approach in Chemistry: Amino Acid Derivatives*, ed. G.C. Barrett, Oxford University Press, 1999.
3.   A.S. Bommarius, M. Schwann, and K. Drauz, *J. Mol. Catal. B: Enzymol.*, 1998, **5**, 1.
4.   Z. Feng and C. Yin, *Huaxue Tongbao*, 1998, 16 (*Chem. Abs.*, 1999, **130**, 1417).
5.   W.M. Bandaranayake, *Nat. Prod. Rep.*, 1998, **15**, 159.
6.   M.G. Moloney, *Nat. Prod. Rep.*, 1998, **15**, 205; see also *Idem, ibid.*, 1999, **16**, 485.
7.   H. Homma and K. Imai, *Bunseki*, 1998, 266 (*Chem. Abs.*, 1998, **129**, 64984).
8.   L.P. Osman, S.C. Mitchell, and R.H. Waring, *Sulfur Rep.*, 1997, **20**, 155.
9.   M. Saffran, *Biochem. Educ.*, 1998, **26**, 216.
10. B.R.T. Simoneit, R.E. Summons, and L.L. Jahnke, *Origins Life Evol. Biosphere*, 1998, **28**, 475.
11. R.R. Burky, D.L. Kirner, R.E. Taylor, P.E. Hare, and J.R. Southon, *Radiocarbon*, 1998, **40**, 11.
12. K. Ujikawa, *Rev. Port. Farm.*, 1998, 48, 23 (*Chem. Abs.*, 1998, **129**, 200062).
13. R.J. Capon, S.P.B. Ovenden, and T. Dargaville, *Aust. J. Chem.*, 1998, **51**, 169.
14. S.D. Bull, S.G. Davies, R.M. Parkin, and F. Sanchez-Sancho, *J. Chem. Soc., Perkin Trans. I*, 1998, 2313.
15. N.A. Kocharova, E.V. Vinogradov, S.A. Borisova, A.S. Shashkov, and Y.A. Knivel, *Carbohydr. Res.*, 1998, **309**, 131.
16. D. Enneking, I.M. Delaere, and M.E. Tate, *Phytochemistry*, 1998, **48**, 643.
17. H.-P. Zhang, H. Kakeya, and H. Osada, *Tetrahedron Lett.*, 1998, **39**, 6947.
18. R. Jacobsen, E.C. Jiminez, M. Grilley, M. Watkins, D. Hillyard, L.J. Cruz, and M.B. Olivera, *J. Pept. Res.*, 1998, **51**, 173.
19. H. Uemoto, Y. Yahiro, H. Shigemori, M. Tsuda, T. Takao, Y. Shimonishi, and J. Kobayashi, *Tetrahedron*, 1998, **54**, 6719.
20. S. Takahashi, T. Matsunaga, C. Hasegawa, H. Saito, D. Fujita, F. Kiuchi, and Y. Tsuda, *Chem. Pharm. Bull.*, 1998, **46**, 1527.
21. S. Hatanaka, Y. Niimura, K. Takishima, and J. Sugiyama, *Phytochemistry*, 1998, **49**, 573.
22. C.-J. Li, F.J. Schmitz, and M. Kelly-Borges, *J. Nat. Prod.*, 1998, **61**, 387.
23. A. Evidente, R. Capasso, A. Cutignano, O. Taglialatela-Seafati, M. Vurro, M.C. Zonno, and A. Motta, *Phytochemistry*, 1998, **48**, 1131.
24. H. Wang, J.B. Gloer, D.T. Wicklow, and P.F. Dowd, *J. Nat. Prod.*, 1998, **61**, 804.
25. S. Kodani, K. Ishida, and M. Murakami, *J. Nat. Prod.*, 1998, **61**, 854.
26. M.V.R. Reddy, M.K. Harper, and D.J. Faulkner, *Tetrahedron*, 1998, **54**, 10649.
27. B. Picur, M. Lisowski, and I.Z. Siemion, *Lett. Pept. Sci.*, 1998, **5**, 183.

28. B.-S. Yun, I.-J. Ryoo, I.-K .Lee, and I.-D Yoo, *Tetrahedron Lett.*, 1998, **39**, 993.
29. S. Kajiyama, H. Kanzaki, H. Kawazu, and A. Kobayashi, *Tetrahedron Lett.*, 1998, **39**, 3737.
30. K. Umezawa, K. Nakazawa, T. Uemura, Y. Ikeda, S. Kondo, H. Naganawa, N. Kinoshita, H. Hashizume, M. Hamada, T. Takeuchi, and S. Ohba, *Tetrahedron Lett.*, 1998, **39**, 1389.
31. G.A. Kleter, J.J.M. Damen, J.J. Kettenes van den Bosch, R.A. Bank, J.M. te Koppele, J.R. Veraart, and J.M. ten Cate, *Biochim.Biophys.Acta*, 1998, **1381**, 179.
32. A. Dabrowska, W. Wiczk, and L. Lankiewicz, *Wiad.Chem.*, 1998, **52**, 1.
33. M. Ayoub, A. Brunissen, H. Josien, A. Loffet, G. Chassaing, and S. Lavielle, *Actual.Chim.Ther.*, 1996, **22**, 83.
34. F.A. Davis, P. Zhou, and B.-C. Chen, *Chem.Soc.Rev.*, 1998, **27**, 13.
35. A.Dondini, *Synthesis*, 1998, 1681.
36. K. Wisniewski, A.S. Koldziejczyk, and B. Falkiewicz, *J.Pept.Sci.*, 1998, **4**, 1; *Wiad.Chem.*, 1998, **52**, 243.
37. A.V. Osorio Lozada and C.M. Fuertes Ruiton, *Bol.Soc.Quim.Peru*, 1998, **64**, 24.
38. M. Bois-Choussy and J. Zhu, *J.Org.Chem.*, 1998, **63**, 5662.
39. V. Atlan, S. Racouchot, M. Rubin, C. Bremer, J. Ollivier, A. de Meijere, and J. Salaun, *Tetrahedron: Asymmetry*, 1998, **9**, 1131.
40. M. Lakhrissi and Y. Chapleur, *Tetrahedron Lett.*, 1998, **39**, 4659.
41. D.S. Brown, J.M. Revill, and R.E. Shute, *Tetrahedron Lett.*, 1998, **39**, 8533.
42. C. Anne, M.-C. Fournie-Zaluski, B.P. Roques, and F. Cornille, *Tetrahedron Lett.*, 1998, **39**, 8973.
43. T. Bach and C. Korber, *Tetrahedron Lett.*, 1998, **39**, 5015.
44. B. Merla, H.-J. Grumbach, and N. Risch, *Synthesis*, 1998, 1609.
45. Y. Wang, G. Ruan, and J. Fu, *Huaxue Tongbao*, 1998, 29 (*Chem.Abs.*, 1999, **130**, 81819).
46. V.K. Aggarwal, J.K. Barrell, J.M. Worrall, and R. Alexander, *J.Org.Chem.*, 1998, **63**, 7128.
47. R.V. Hoffman, M.C. Johnson, and J.F. Okonya, *Tetrahedron Lett.*, 1998, **39**, 1283.
48. R.D. Crouch, M.S. Holden, and T.M. Weaven, *Chem.Educ.*, 1998, **3** (online computer file; *Chem.Abs.*, 1998, **129**, 95022).
49. P. Coutrot, C. Grison, and F. Coutrot, *Synlett*, 1998, 393.
50. C. Yuan, X. Zhang, W. Luo, and Z. Yao, *Heteroat.Chem.*, 1998, **9**, 139.
51. M. Yamada, N. Nagashima, J. Hasegawa, and S. Takahashi, *Tetrahedron Lett.*, 1998, **39**, 9019.
52. E. Coulon, M.C. de Andrade, V. Ratovelomanana-Vidal, and J.-P. Genet, *Tetrahedron Lett.*, 1998, **39**, 6467.
53. S.-K. Chung, T.-H. Jeong, and D.-H. Kang, *J.Chem.Soc., Perkin Trans. I*, 1998, 969.
54. P. Bravo, G. Cavicchio, M. Crucianelli, A. Poggiali, A. Volonterio, and M. Zanda, *J.Chem.Res., Synop.*, 1998, 666.
55. U. Tedebark, M. Meldal, L. Panza, and K. Bock, *Tetrahedron Lett.*, 1998, **39**, 1815.
56. J.T. Lundquist and T.A. Dix, *Tetrahedron Lett.*, 1998, **39**, 775.
57. J. Lin, S. Liao, and V.J. Hruby, *Tetrahedron Lett.*, 1998, **39**, 3117.
58. R.N. Ben, A. Orellana, and P. Arya, *J.Org.Chem.*, 1998, **63**, 4817; P. Arya, R.N. Ben, and H. Qin, *Tetrahedron Lett.*, 1998, **39**, 6131.

59. C.A.M. Afonso, *Synth. Commun.*, 1998, **28**, 261.
60. C.A. Celatka and J.S. Panek, *Chemtracts*, 1998, **11**, 836.
61. S.G. Nelson, *Tetrahedron: Asymmetry*, 1998, **9**, 357; see also K. Fuji and T. Kawabata, *Chem-Eur.J.*, 1998, **4**, 373.
62. S. Fukuzawa and Y. Hongo, *Tetrahedron Lett.*, 1998, **39**, 3521.
63. S. Caddick, K. Jenkins, N. Treweeke, S.X. Candeias, and C.A.M. Afonso, *Tetrahedron Lett.*, 1998, **39**, 2203.
64. C.L. Gibson, K. Gillon, and S. Cook, *Tetrahedron Lett.*, 1998, **39**, 6733.
65. K.H. Ahn, S.-K. Kim, and C. Ham, *Tetrahedron Lett.*, 1998, **39**, 6323.
66. A.B. Charette, A. Gagnon, M. Janes, and C. Mellon, *Tetrahedron Lett.*, 1998, **39**, 5147.
67. K.L. Reddy and K.B. Sharpless, *J. Am. Chem. Soc.*, 1998, **120**, 1207.
68. P. O'Brien, S.A. Osborne, and D.D. Parker, *Tetrahedron Lett.*, 1998, **39**, 4099.
69. B. Tao, G. Schlingloff, and K.B. Sharpless, *Tetrahedron Lett.*, 1998, **39**, 2507.
70. K.R. Dress, L.J. Goossen, H. Liu, D.M. Jerina, and K.B. Sharpless, *Tetrahedron Lett.*, 1998, **39**, 7669.
71. F. Benedetti, F. Berti, and S. Norbedo, *Tetrahedron Lett.*, 1998, **39**, 7971 .
72. M. Zhao, J. Li, Z. Song, R. Desmond, D.M. Tschaen, E.J.J. Grabowski, and P.J. Reider, *Tetrahedron Lett.*, 1998, **39**, 5323.
73. Y. Langlois, *Curr. Org. Chem.*, 1998, **2**, 1.
74. J. Armbruster, S. Grabowski, T. Ruch, and H. Prinzbach, *Angew. Chem. Int. Ed.*, 1998, **37**, 2242.
75. A.G.M. Barrett, D.C. Braddock, P.W.N. Christian, D. Pilipauskas, A.J.P. White, and D.J. Williams, *J. Org. Chem.*, 1998, **63**, 5818.
76. S. Sabelle, D. Lucet, T. Le Gall, and C. Mioskowski, *Tetrahedron Lett.*, 1998, **39**, 2111.
77. P. Dauban and R.H. Dodd, *Tetrahedron Lett.*, 1998, **39**, 5739.
78. K. Jahnisch, *GIT Labor.-Fachz.*, 1998, **42**, 232.
79. K. Jahnisch, *Schriftenr. Nachwachsende Rohst.*, 1998, **10**, 324 (*Chem. Abs.*, 1999, **130**, 110577).
80. S.-K. Choi and W.-K. Lee, *Heterocycles*, 1998, **48**, 1917.
81. J.-W. Chang, J.H. Bae, S.-H. Shin, C.S. Park, D. Choi, and W.-K. Lee, *Tetrahedron Lett.*, 1998, **39**, 9193.
82. K. Uneyama and T. Kato, *Tetrahedron Lett.*, 1998, **39**, 587.
83. S. Kacker, J.S. Kim, and A. Sen, *Angew. Chem. Int. Ed.*, 1998, **37**, 1251.
84. T. Iwama, T. Kataoka, O. Muraoka, and G. Tanabe, *J. Org. Chem.*, 1998, **63**, 8355.
85. T. Satoh, M. Ozawa, K. Takano, and M. Kudo, *Tetrahedron Lett.*, 1998, **39**, 2345.
86. M. Lange and K. Undheim, *Tetrahedron*, 1998, **54**, 5337; K. Undheim, *Lett.-Pept. Sci.*, 1998, **5**, 227.
87. S.D. Bull, A.N. Cherulga, S.G. Davies, W.O. Moss, and R.M. Parkin, *Tetrahedron*, 1998, **54**, 10379.
88. M. Lange and P.M. Fischer, *Helv. Chim. Acta*, 1998, **81**, 2053.
89. P. Kuemminger and K. Undheim, *Tetrahedron: Asymmetry*, 1998, **9**, 1183.
90. A.H. Fauq, C. Cherif-Ziani, and E. Richelson, *Tetrahedron: Asymmetry*, 1998, **9**, 2333.
91. M. Ohba, Y. Nishimura, M. Imasho, T. Fujii, J. Kubanek, and R.J. Andersen, *Tetrahedron Lett.*, 1998, **39**, 5999.
92. H. Hasegawa and Y. Shinohara, *J. Chem. Soc., Perkin Trans. I*, 1998, 243.

93. S. Zhao, T. Gan, P. Yu, and J.M. Cook, *Tetrahedron Lett.*, 1998, **39**, 7009.
94. K. Hammer and K. Undheim, *Tetrahedron: Asymmetry*, 1998, **9**, 2359.
95. S. Krikstolaityte, K. Hammer, and K. Undheim, *Tetrahedron Lett.*, 1998, **39**, 7595.
96. K. Hammer, C. Romming, and K. Undheim, *Tetrahedron*, 1998, **54**, 10837.
97. B. Moller and K. Undheim, *Tetrahedron*, 1998, **54**, 5789.
98. M. Oba, T. Terauchi, Y. Owazi, Y. mai, I. Motoyama, and K. Nishiyama, *J.Chem.Soc., Perkin Trans. I*, 1998, 1275.
99. T. Abellan, C. Najera, and J.M. Sansano, *Tetrahedron: Asymmetry*, 1998, **9**, 2211.
100. S. Sano, M. Takebayashi, T. Miwa, T. Ishii, and Y. Nagao, *Tetrahedron: Asymmetry*, 1998, **9**, 3611.
101. S. Sano, T. Miwa, X.-K. Liu, T. Ishii, T. Takehisa, M. Shiro, and Y. Nagao, *Tetrahedron: Asymmetry*, 1998, **9**, 3615.
102. S.D. Bull, S.G. Davies, and W.O. Moss, *Tetrahedron: Asymmetry*, 1998, **9**, 321.
103. G. Porzi, S. Sandri, and P. Verrocchio, *Tetrahedron: Asymmetry*, 1998, **9**, 119; G. Porzi, and S. Sandri, *ibid.*, p. 3411.
104. S.D. Bull, S.G. Davies, S.W. Epstein, and J.V.A. Ouzman, *Chem.Commun.*, 1998, 659.
105. S.D. Bull, S.G. Davies, S.W. Epstein, and J.V.A. Ouzman, *Tetrahedron: Asymmetry*, 1998, **9**, 2795.
106. S.D. Bull, S.G. Davies, S.W. Epstein, M.A. Leech, and J.V.A. Ouzman, *J.Chem.Soc., Perkin Trans. I*, 1998, 2321.
107. G. Kardassis, P. Brungs, and E. Steckhan, *Tetrahedron*, 1998, **54**, 3471.
108. L.M. Harwood, S.A. Anslow, I.D. MacGilp, and M.G.B. Drew, *Tetrahedron: Asymmetry*, 1998, **9**, 4007.
109. R. Chinchilla, N. Galinda, and C. Najera, *Tetrahedron: Asymmetry*, 1998, **9**, 2769.
110. D. Alker, G. Hamblett, L.M. Harwood, S.M. Robertson, D.J. Watkin, and C.E. Williams, *Tetrahedron*, 1998, **54**, 6089.
111. L.M. Harwood and S.M. Robertson, *Chem.Commun.*, 1998, 2641.
112. D. Alker, L.M. Harwood, and C.E. Williams, *Tetrahedron Lett.*, 1998, **39**, 475.
113. Y. Aoyagi and R.M. Williams, *Synlett*, 1998, 1099.
114. J.D. Scott, T.N. Tippie, and R.M. Williams, *Tetrahedron Lett.*, 1998, **39**, 3659.
115. H. Fretz, *Tetrahedron*, 1998, **54**, 4849.
116. M.P. Bertrand, L. Feray, R. Nouguier, and L. Stella, *Synlett*, 1998, 780.
117. G. Kardassis, P. Brungs, C. Nothhelfer, and E. Steckhan, *Tetrahedron*, 1998, **54**, 3479.
118. V. Joerres, H. Keul, and H. Hoecker, *Macromol.Chem.Phys.*, 1998, **199**, 825.
119. S.W. Baldwin, B.G. Young, and A.T. McPhail, *Tetrahedron Lett.*, 1998, **39**, 6819.
120. P. Karoyan, S. Sagan, G. Clodic, S. Lavielle, and G. Chassaing, *Bioorg.Med.-Chem.Lett.*, 1998, **8**, 1369.
121. C.G. Knight, *Lett.Pept.Sci.*, 1998, **5**, 1.
122. A. Martin, G. Chassaing, and A.Vanhove, *Isot.Environ.Health Stud.*, 1996, **32**, 15.
123. E. Koltai, A. Alexin, G. Rutkai, and E. Toth-Sarudy, *J.Labelled Compd.Radiopharm.*, 1998, **41**, 977.
124. S.N. Lodwig and C.J. Unkefer, *J.Labelled Compd.Radiopharm.*, 1998, **41**, 983.
125. A. Lopez and R. Pleixats, *Tetrahedron: Asymmetry*, 1998, **9**, 1967.

126. M. Nishihara, K. Takatori, N. Wada, S. Toyama, and M. Kajiwara, *Igaku*, 1998, **8**, 6 (*Chem.Abs.*, 1998, **129**, 136461).
127. M. Haddad, H. Imogaie, and M. Larcheveque, *J.Org.Chem.*, 1998, **63**, 5680.
128. R.P. Frutos and D.M. Spero, *Tetrahedron Lett.*, 1998, **39**, 2475.
129. D.M. Hodgson, A.J. Thompson, and S. Wadman, *Tetrahedron Lett.*, 1998, **39**, 3357; erratum, *ibid.*, p. 8003.
130. M. Matsushita, H. Maeda, and M. Kodama, *Tetrahedron Lett.*, 1998, **39**, 3749.
131. M. Foldfield and N.P. Botting, *J.Labelled Compd.Radiopharm.*, 1998, **41**, 29.
132. M. Beller, M. Eckert, and E.W. Holla, *J.Org.Chem.*, 1998, **63**, 5658; M. Beller, M. Eckert, H. Geissler, B. Napierski, H.-P. Rebenstock, and E.W. Holla, *Chem.- Eur.J.*, 1998, **4**, 935.
133. M. Beller, M. Eckert, and F. Vollmuller, *J.Mol.Catal.A: Chem.*, 1998, **135**, 23.
134. J. Li, L. Li, N. Zhang, H. Li, W. Han, T. Jin, and Y. Lu, *Hecheng Huaxue*, 1997, **5**, 106 (*Chem.Abs.*, 1998, **128**, 295009).
135. D.C. Horwell, M.J. McKiernan, and S.D. Osborne, *Tetrahedron Lett.*, 1998, **39**, 8729.
136. A. Boeijen, A.W. Kruijtzer, and R.M.J. Liskamp, *Bioorg.Med.Chem.Lett.*, 1998, **8**, 2375.
137. S.-H. Lee, S.-H. Chung, and Y.-S. Lee, *Tetrahedron Lett.*, 1998, **39**, 9469.
138. S. Kobayashi and T. Busujima, *Chem.Commun.*, 1998, 981.
139. A. Heydari, P. Fatemi, and A.A. Alizadeh, *Tetrahedron Lett.*, 1998, **39**, 3049.
140. E.F. Kogut, J.C. Thoen, and M.A. Lipton, *J.Org.Chem.*, 1998, **63**, 4684.
141. F.A. Davis and D.L. Fanelli, *J.Org.Chem.*, 1998, **63**, 1981.
142. C. Vergne, J.-P. Bouillon, J. Chastanet, M. Bois-Choussy, and J.Zhu, *Tetrahedron: Asymmetry*, 1998, **9**, 3095.
143. A. Fadel and A. Khesrani, *Tetrahedron: Asymmetry*, 1998, **9**, 305.
144. M.S. Sigman and E.N. Jacobsen, *J.Am.Chem.Soc.*, 1998, **120**, 4901.
145. I.K. Ugi, *Proc.Est.Acad.Sci., Chem.*, 1998, **47**, 107 (*Chem.Abs.*, 1999, **130**, 109755).
146. I. Ugi, W. Horl, C. Hanusch-Kompa, T. Schmid, and E. Herdtweck, *Heterocycles*, 1998, **47**, 965.
147. C. Hulme, J. Peng, B. Louridas, P. Menard, P. Krolikowski, and N.V. Kumar, *Tetrahedron Lett.*, 1998, **39**, 8047.
148. S.W. Kim, Y.S. Shin, and S. Ro, *Bioorg.Med.Chem.Lett.*, 1998, **8**, 1665.
149. S.W. Kim, S.M. Bauer, and R.W. Armstrong, *Tetrahedron Lett.*, 1998, **39**, 6993.
150. C.D. Floyd, L.A. Harnett, A. Miller, S. Patel, L. Saroglou, and M. Whittaker, *Synlett*, 1998, 637.
151. S. Patel, L. Saroglou, C.D. Floyd, A. Miller, and M. Whittaker, *Tetrahedron Lett.*, 1998, **39**, 8333.
152. T. Ziegler, R. Schlomer, and C. Koch, *Tetrahedron Lett.*, 1998, **39**, 5957.
153. N.A. Petasis and I.A. Zavialov, *J.Am.Chem.Soc.*, 1998, **120**, 11798.
154. J. Jiang, R.B. Miller and J.C. Tolle, *Synth.Commun.*, 1998, **28**, 3015.
155. R.X. Yuan and Y.-W. Zhang, *Hecheng Huaxue*, 1998, **6**, 106 (*Chem.Abs.*, 1998, **129**, 28189).
156. R. Braslau and M.O. Anderson, *Tetrahedron Lett.*, 1998, **39**, 4227.
157. N. Abe, F. Fujisaki, and K. Sumoto, *Chem.Pharm.Bull.*, 1998, **46**, 142.
158. J.E. Semple, *Tetrahedron Lett.*, 1998, **39**, 6645.
159. V.N. Kobzareva, L.I. Deiko, O.S. Vasileva, V.M. Beristovitskaya, and G.A. Berkova, *Russ.J. Org.Chem.*, 1997, **33**, 1180.

160. G. Tang, S. Wang, A. Wu, and G. Jiang, *Hejishu*, 1998, **21**, 304 (*Chem. Abs.*, 1998, **129**, 203222).

161. I.K. Mosevich, O.F. Kuznetsova, O.S. Fedorova, and M.V. Korsakov, *Radiokhimiya*, 1997, **39**, 552.

162. F. Tanaka, M. Oda, and I. Fujii, *Tetrahedron Lett.*, 1998, **39**, 5057.

163. T. Miura, M. Fujii, K. Shingu, I. Koshimizu, J. Naganoma, T. Kajimoto, and Y. Ida, *Tetrahedron Lett.*, 1998, **39**, 7313.

164. V.P. Vassilev, E.E. Simanek, M.R. Wood, and C.-H. Wong, *Chem. Commun.*, 1998, 1865.

165. G.K. Hsiao and D.G. Hangauer, *Synthesis*, 1998, 1043.

166. B. Sauvagnat, F. Lamaty, R. Lazaro, and J. Martinez, *Tetrahedron Lett.*, 1998, **39**, 821.

167. K.W. Laue and G. Haufe, *Synthesis*, 1998, 1453.

168. G. Haufe, K.W. Laue, M.U. Triller,Y. Takeuchi, and N. Shibata, *Tetrahedron*, 1998, **54**, 5929.

169. G. Guillena, B. Mancheno, C. Najera, J.Ezquerra, and C. Pedregal, *Tetrahedron*, 1998, **54**, 9447.

170. E. Dominguez, M.J. O'Donnell, and W.L. Scott, *Tetrahedron Lett.*, 1998, **39**, 2167.

171. M.J. O'Donnell, F. Delgado, C. Hostettler, and R. Schwesinger, *Tetrahedron Lett.*, 1998, **39**, 2167.

172. S.P. Chavan and M.S. Venkatraman, *Tetrahedron Lett.*, 1998, **39**, 6745.

173. E.J. Corey, M.C. Noe, and F. Xu, *Tetrahedron Lett.*, 1998, **39**, 5347.

174. A. Lopez, M. Moreno-Manas, R. Pleixats, and A. Roglans, *An. Quim. Int. Ed.*, 1997, **93**, 355.

175. R. Fornicola, E. Oblinger, and J. Montgomery, *J. Org. Chem.*, 1998, **63**, 3528.

176. L. Meyer, J.-M. Poirier, P. Duhamel, and L. Duhamel, *J. Org. Chem.*, 1998, **63**, 8094.

177. A. Solladie-Cavallo and T. Nsenda, *Tetrahedron Lett.*, 1998, **39**, 2191.

178. A. Solladie-Cavallo, J. Schwarz, and C. Mouza, *Tetrahedron Lett.*, 1998, **39**, 3861.

179. N. Irako and T. Shioiri, *Tetrahedron Lett.*, 1998, **39**, 5793, 5797; erratum, *ibid.*, p.7623.

180. A.S. Sagiyan, S.Z. Sagyan, G.L. Grigoryan, T.F. Saveleva, Y.N. Belokon, and S.K. Grigoryan, *Khim. Zh. Arm.*, 1997, **50**, 149.

181. A.S. Sagiyan, S.M. Dzhamgaryan, G.L. Grigoryan, S.R. Kagramanyan, G.T. Ovsepyan, S.K. Grigoryan, and Y.N. Belokon, *Khim. Zh. Arm.*, 1997, **50**, 75.

182. Y.N. Belokon, K.A. Kochetkov, T.D. Churkina, N.S. Ikonnikov, A.A. Chesnokov, O.V. Larionov, V.S. Parmar, R. Kumar, and H.B. Kagan, *Tetrahedron: Asymmetry*, 1998, **9**, 851.

183. A.S. Sagiyan, A.E. Avfetisyan, S.M. Dzhamgaryan, L.R. Dzhilavyan, E.L. Gyulumyan, L.B. Danielyan, S.K. Grigoryan, G.L. Grigoryan, and Y.N. Belokon, *Khim. Zh. Arm.*, 1996, **49**, 146 106 (*Chem. Abs.*, 1998, **128**, 308718); A.S. Sagiyan, A.E. Avfetisyan, S.M. Dzhamgaryan, L.R. Dzhilavyan, V.I. Tadarov, S.K. Grigoryan, and Y.N. Belokon, *Khim. Zh. Arm.*, 1996, **49**, 153; A.S. Sagiyan, S.K. Grigoryan, S.M. Dzhamgaryan, G.L. Grigoryan, and Y.N. Belokon, *Khim. Zh. Arm.*, 1997, **50**, 142.

184. Y.N. Belokon, K.A. Kochetkov, T.D. Churkina, A.A. Chesnokov, V.V. Smirnov, N.S. Ikonnikov, and S.A. Orlova, *Russ. Chem. Bull.*, 1998, **47**, 74.

185. S. Collet, P. Bauchat, R. Danion-Bougot, and D. Danion, *Tetrahedron: Asymmetry*, 1998, **9**, 2121.

186. S. Kotha, E. Brahmachary, and N. Sreenivasachary, *Tetrahedron Lett.*, 1998, **39**, 4095.

187. W.J. Drury, D. Ferraris, C. Cox, B. Young, and T. Lectka, *J.Am.Chem.Soc.*, 1998, **120**, 11006.

188. K. Mikami, T. Yajima, and M. Kaneko, *Amino Acids*, 1998, **14**, 311.

189. D. Ferraris, B. Young, C. Cox, W.J. Drury, T. Dudding, and T. Lectka, *J.Org.Chem.*, 1998, **63**, 6090.

190. S. Yao, X. Fang, and K.A. Jorgensen, *Chem.Commun.*, 1998, 2547.

191. C. Agami, D. Bihan, L. Hamon, and C. Puchot-Kadouri, *Tetrahedron*, 1998, **54**, 10309; C. Agami, D. Bihan, L. Hamon, C. Puchot-Kadouri, and M. Lusinchi, *Eur.J. Org.Chem.*, 1998, 2461; for a review, see C. Agami, F. Couty, and C. Puchot-Kadouri, *Synlett*, 1998, 449.

192. E. Hagiwara, A. Fujii, and M .Sodeoka, *J.Am.Chem.Soc.*, 1998, **120**, 2474.

193. S. Bromidge, P.C. Wilson, and A. Whiting, *Tetrahedron Lett.*, 1998, **39**, 8905.

194. M. Barreau, A. Commercon, S. Mignani, D. Mouysset, P. Perfetti, and L. Stella, *Tetrahedron*, 1998, **54**, 11501.

195. N. Gautier and R.H. Dodd, *Synth.Commun.*, 1998, **28**, 3769.

196. W. Wu and Y. Zhang, *Tetrahedron: Asymmetry*, 1998, **9**, 1441.

197. J.M. McIntosh, E.J. Kiser, and Z. Tian, *Can.J. Chem.*, 1998, **76**, 147.

198. J. Rancourt, V. Gorys, and E. Jolicoeur, *Tetrahedron Lett.*, 1998, **39**, 5339.

199. J. Lee, W.-B. Choi, J.E. Lynch, R.P. Volante, and P.J. Reider, *Tetrahedron Lett.*, 1998, **39**, 3679.

200. D. Seebach, M. Hoffmann, A.R. Sting, J.N. Kinkel, M. Schulte, and E. Kusters, *J.Chromatogr. A*, 1998, **796**, 299.

201. M. Hoffmann, S. Blank, D. Seebach, E. Kusters, and E. Schmid, *Chirality*, 1998, **10**, 217.

202. D. Seebach and M. Hoffmann, *Eur.J. Org.Chem.*, 1998, 1337.

203. E.J. Iwanowicz, M.F. Malley, and K. Smith, *Synth.Commun.*, 1998, **28**, 3711.

204. E.J. Iwanowicz, P. Blomgren, P.T.W. Cheng, K. Smith, W.F. Lau, Y.Y. Pan, H.H. Gu, M.F. Malley, and J.Z. Gougoutas, *Synlett*, 1998, 664.

205. S.G. Pyne and K. Schafer, *Tetrahedron*, 1998, **54**, 5709.

206. Y. Yamamoto, S. Onuki, M. Yumoto, and N. Asao, *Heterocycles*, 1998, **47**, 765.

207. G. Guillena and C. Najera, *Tetrahedron: Asymmetry*, 1998, **9**, 1125.

208. G.H.P. Roos, *S.Afr.J. Chem.*, 1998, **51**, 7.

209. A.G. Moglioni, E. Garcia-Exposito, G.Y. Moltrasio, and R.M. Ortuno, *Tetrahedron Lett.*, 1998, **39**, 3593.

210. R.S. Hoerrner, D. Askin, R.P. Volante, and P.J. Reider, *Tetrahedron Lett.*, 1998, **39**, 3455.

211. J.E. Beecher and D.A. Tirrell, *Tetrahedron Lett.*, 1998, **39**, 3927.

212. H. Imogai, G. Bernadinelli, C. Graenicher, M. Moran, J.-C. Rossier, and P. Mueller, *Helv.Chim.Acta*, 1998, **81**. 1754.

213. T. Fuchss and R.R. Schmidt, *Synthesis*, 1998, 753.

214. M. Brandl, S.I. Kozhushkov, S. Braese, and A. de Meijere, *Eur.J. Org.Chem.*, 1998, 453.

215. I. Coldham, M.L. Middleton, and P.L. Taylor, *J.Chem.Soc., Perkin Trans. I*, 1998, 2817.

216. U. Kazmaier and C. Schneider, *Tetrahedron Lett.*, 1998, **39**, 817.

217. J.M. Percy, M.E. Prime, and M.J. Broadhurst, *J.Org.Chem.*, 1998, **63**, 8049.

218. R.A. Tommasi, W.M. Macchia, and D.T. Parker, *Tetrahedron Lett.*, 1998, **39**, 5947.
219. H.-J. Kreuzfeld and C. Dobler, *J. Mol. Katal. A: Chem.*, 1998, **136**, 105.
220. J. Zhao, W. Tan, and S. Yang, *Fenzi Cuihua*, 1997, **11**, 421 (*Chem. Abs.*, 1998, **128**, 230651).
221. S.A. Laneman, D.E. Froen, and D.J. Ager, *Chem. Ind. (New York)*, 1998, **75**(Catalysis of Organic Reactions), 525.
222. U. Schmidt, H.W. Krause, G. Oehme, M. Michalik, and C. Fischer, *Chirality*, 1998, **10**, 564.
223. T. Imamato, J. Watanabe, Y. Wada, H. Masuda, H. Yamada, H. Tsuruta, S. Matsukawa, and K. Yamaguchi, *J. Am. Chem. Soc.*, 1998, **120**, 1635.
224. K.C. Nicolaou and K. Namoto, *Chem. Commun.*, 1998, 1757.
225. R. Selke, J. Holz, A. Riepe, and A. Borner, *Chem.-Eur. J.*, 1998, **4**, 769.
226. S. Qiao and G.C. Fu, *J. Org. Chem.*, 1998, **63**, 4168.
227. J.J.A. Perea, A.Borner, and P.Knochel, *Tetrahedron Lett.*, 1998, **39**, 8073.
228. A.Q. Mi, R.L. Lou, Y.Z. Jiang, J.G. Deng, Y. Qih, F.M. Fu, W.H. Hu, and A.S.C. Chan, *Synlett*, 1998, 847.
229. G. Zhu and X. Zhang, *J. Org. Chem.*, 1998, **63**, 3183.
230. G. Zhu, A.L. Casalnuovo, and X. Zhang, *J. Org. Chem.*, 1998, **63**, 8100.
231. R. Kuwano, S. Okuda, and Y.Ito, *Tetrahedron: Asymmetry*, 1998, **9**, 2773.
232. M.J. Burk and F. Bienewald, *Transition Met. Org. Synth.*, 1998, **2**, 13; U. Nagel and J. Albrecht, *Top. Catal.*, 1998, **5**, 3; V.A. Slovinskaya, G. Cipens, A. Strautina, D. Sile, E.K. Korchagova, I. Liepina, and E. Lukevics, *Latv. Kim. Z.*, 1998, 3.
233. S.-S. Lin, O. Miyawaki, and K. Nakamura, *Biosci., Biotechnol., Biochem.*, 1997, **61**, 2029.
234. P. Pialis and B.A. Saville, *Enzyme Microb. Technol.*, 1998, **22**, 261.
235. B. Sookkheo, S. Phutakul, S.-T. Chen, and K.-T. Wang, *J. Chin. Chem. Soc. (Taipei)*, 1998, **45**, 525 (*Chem. Abs.*, 1998, **129**, 276257).
236. V.A. Abelyan and E.G. Afrikyan, *Prikl. Biokhim. Mikrobiol.*, 1997, **33**, 620.
237. S.-Y. Huang and S.-Y. Chen, *J. Biotechnol.*, 1998, **62**, 95.
238. A.H. Sassi, L. Fauvart, A.M. Deschamps, and J.M. Lebeault, *Biochem. Eng. J.*, 1998, **1**, 85.
239. P.P. Taylor, D.P. Pantaleone, R.F. Senkpeil, and I.G. Fotheringham, *Trends Biotechnol.*, 1998, **16**, 412.
240. N.J. Grinter, *Chem. Tech.*, 1998, **28**, 33.
241. Y. Ikenaka and S. Takahashi, *Barosaiensu to Indasutori*, 1998, **56**, 759 (*Chem. Abs.*, 1999, **130**, 65277).
242. M. Nampoothiri and A. Pandey, *Process Biochem.*, 1998, **33**, 147.
243. J.F. Morot-Gaudry, in *Assimilation Azote Plants*, ed. J.F. Morot-Gaudry, Institut National de la Recherche Agronomique, Paris, 1997, p.199.
244. S. Kobayashi, T. Furuta, T. Hayashi, M. Nishijima, and K. Hamada, *J. Am. Chem. Soc.*, 1998, **120**, 908 (for an erratum, see *ibid.*, p.4256).
245. B.M. Trost and C.B. Lee, *J. Am. Chem. Soc.*, 1998, **120**, 6818.
246. E.J. Corey, W. Li, and G.A. Reichard, *J. Am. Chem. Soc.*, 1998, **120**, 2330.
247. S.H. Kang and H.-S. Jun, *Chem. Commun.*, 1998, 1929; S.H. Kang, H.-S. Jun, and J.-H. Youn, *Synlett*, 1998, 1045.
248. G. Veeresa and A. Datta, *Tetrahedron Lett.*, 1998, **39**, 119.
249. K. Kato, C.Y. Chen, and H. Akita, *Synthesis*, 1998, 1527.
250. D. Zhang and M.J. Miller, *J. Org. Chem.*, 1998, **63**, 755.

251. S.S. Klair, H.R. Mohan, and T. Kitahara, *Tetrahedron Lett.*, 1998, **39**, 89.
252. M. Sendzik, W. Guarnieri, and D. Hoppe, *Synthesis*, 1998, 1287.
253. K. Hashimoto, S. Higashibayashi, and H. Shirahama, *Heterocycles*, 1997, **46**, 581.
254. K. Tanaka and H. Sawanishi, *Tetrahedron*, 1998, **54**, 10029.
255. H. Akita, C.Y. Chen, and K. Kato, *Tetrahedron*, 1998, **54**, 11011.
256. J. Cossy, M. Cases, and D. Gomez Pardo, *Synlett*, 1998, 507.
257. I. Collado, J. Ezquerra, A.I. Mateo, and A. Rubio, *J. Org. Chem.*, 1998, **63**, 1995.
258. A. Rubio, J. Ezquerra, A. Escribano, M.J. Remuinan, and J.J. Vaquero, *Tetrahedron Lett.*, 1998, **39**, 2171.
259. O. Miyata, Y. Ozawa, I. Ninomiya, K. Aoe, H. Hiramatsu, and T. Naito, *Heterocycles*, 1997, **46**, 321.
260. M.D. Bachi and A. Melman, *Pure Appl. Chem.*, 1998, **70**, 259.
261. J.E. Baldwin, A.M. Fryer, G.J. Pritchard, M.R. Spyvee, R.C. Whitehead, and M.E. Wood, *Tetrahedron Lett.*, 1998, **39**, 707; *Tetrahedron*, 1998, **54**, 7465.
262. Review: K. Rueck-Braun, *Nachr. Chem., Tech. Lab.*, 1998, **46**, 1182.
263. C. Cativiela and M.D. Diaz-De-Villegas, *Tetrahedron: Asymmetry*, 1998, **9**, 3517.
264. J.C. Ruble and G.C. Fu, *J. Am. Chem. Soc.*, 1998, **120**, 11532.
265. B.M. Trost and X. Ariza, *Angew. Chem. Int. Ed.*, 1997, **36**, 2635.
266. M.J. O'Donnell, Z. Fang, X. Ma, and J.C. Huffmann, *Heterocycles*, 1997, **46**, 617.
267. F. Alonso, S.G. Davies, A.S. Elend, and J.L. Haggitt, *J. Chem. Soc., Perkin Trans. I*, 1998, 257.
268. M. Alias, C. Cativiela, M.D. Diaz-De-Villegas, J.A. Galvez, and Y. Lapena, *Tetrahedron*, 1998, **54**, 14963.
269. S. Wenglowsky and L.S. Hegedus, *J. Am. Chem. Soc.*, 1998, **120**, 12468.
270. A. Carloni, G. Porzi, and S. Sandri, *Tetrahedron: Asymmetry*, 1998, **9**, 2987.
271. E. Juaristi, H. Lopez-Ruiz, D. Madrigal, Y. Ramirez-Quiros, and J. Escalante, *J. Org. Chem.*, 1998, **63**, 4706.
272. H. Shao, J.K. Rueter, and M. Goodman, *J. Org. Chem.*, 1998, **63**, 5240.
273. R. Grandel and U. Kazmaier, *Eur. J. Org. Chem.*, 1998, 409.
274. P. Pravo, M. Crucianelli, B. Vergani, and M. Zanda, *Tetrahedron Lett.*, 1998, **39**, 7771.
275. M. Goodman, J. Zhang, P. Gantzel, and E. Benedetti, *Tetrahedron Lett.*, 1998, **39**, 9589.
276. A.P. Kozikowski, D. Steensma, M. Varasi, S. Pshenichkin, E. Surina, and J.T. Wroblewski, *Bioorg. Med. Chem. Lett.*, 1998, **8**, 447.
277. M. Carda, J. Murga, S. Rodriguez, F. Gonzalez, E. Castillo, and J.A. Marco, *Tetrahedron: Asymmetry*, 1998, **9**, 1703; see also J.A. Marco, M. Carda, J. Murga, R. Portoles, E. Falomir, and J. Lex, *Tetrahedron Lett.*, 1998, **39**, 3237.
278. A.B. Charette and C. Mellon, *Tetrahedron*, 1998, **54**, 10525.
279. N. Evrard-Todeschi, J. Gharbi-Benarous, A. Cosse-Barbi, G. Thirot, and J.-P. Girault, *J. Chem. Soc., Perkin Trans. II*, 1997, 2677.
280. A.S. Demir, C. Tanyli, A. Cagir, M.N. Tahir, and D. Ulku, *Tetrahedron: Asymmetry*, 1998, **9**, 1035.
281. S. Hannour, M.-L. Roumestant, P. Viallefont, C. Riche, J. Martinez, A. El Hallaoui, and F. Ouazzani, *Tetrahedron: Asymmetry*, 1998, **9**, 2329.
282. A. Shibuya, A. Sato, and T. Taguchi, *Bioorg. Med. Chem. Lett.*, 1998, **8**, 1979.
283. I. Collado, J. Ezquerra, A. Mazon, C. Pedregal, B. Yruretagoyena, A.E. Kingston, R. Tomlinson, R.A. Wright, B.G. Johnson, and D.D. Schoepp, *Bioorg.-*

Med.Chem.Lett., 1998, **8**, 2849; P.L. Ornstein, T.J. Bleisch, M.B. Arnold, R.A. Wright, B.G. Johnson, and D.D. Schoepp, *Bioorg.Med.Chem.Lett.*, 1998, **8**, 346; P.L. Ornstein, T.J. Bleisch, M.B. Arnold, R.A. Wright, B.G. Johnson, J.P. Tizzano, D.R. Helton, M.J. Kallmann, D.D. Schoepp, and M. Herin, *Bioorg.-Med.Chem.Lett.*, 1998, **8**, 358.

284. P.L. Ornstein, M.B. Arnold, T.J. Bleisch, R.A. Wright, W.J. Wheeler, and D.D. Schoepp, *Bioorg.Med.Chem.Lett.*, 1998, **8**, 1919.

285. R. Pellicciari, G. Costantino, E. Giovagnoni, L. Mattoli, I. Brabet, and J.-P. Pin, *Bioorg.Med.Chem.Lett.*, 1998, **8**, 1569.

286. H. Tsujishima, K. Nakatani, K. Shimamoto, Y. Shigeri, N. Yumoto, and Y. Ohfune, *Tetrahedron Lett.*, 1998, **39**, 1193.

287. Y. Ohfune, T. Kan, and T. Nakajima, *Tetrahedron*, 1998, **54**, 5207.

288. J. Eustache, A. Grob, C. Lam, O. Sellier, and G. Schulz, *Bioorg.Med.Chem.Lett.*, 1998, **8**, 2961.

289. X. Zhu and P. Gan, *Synth.Commun.*, 1998, **28**, 3159.

290. A. Hercouet, N. Godbert, and M. Le Corre, *Tetrahedron: Asymmetry*, 1998, **9**, 2233.

291. D. Lim and K. Burgess, *J.Org.Chem.*, 1997, **62**, 9382.

292. C. Balsamini, A. Bedini, G. Spadoni, G. Tarzia, A. Tontini, W. Balduini, and M. Cimini, *Farmaco*, 1998, **53**, 181.

293. R. Chinchilla, L.R. Falvello, N. Galindo, and C. Najera, *Tetrahedron: Asymmetry*, 1998, **9**, 2223.

294. D. Moye-Sherman, M.B. Welch, J. Reibenspies, and K. Burgess, *Chem.Commun.*, 1998, 2377.

295. S.G. Pyne, K.Schafer, B.W. Skelton, and A.H. White, *Aust.J. Chem.*, 1998, **51**, 127.

296. S.G. Pyne, K. Schafer, B.W. Skelton, and A.H. White, *Chem.Commun.*, 1998, 1607 (refers to *idem.*, *ibid.*, 1997, 2667).

297. P. Dorizon, G. Su, G. Ludvig, L. Nikitina, J. Ollivier, and J. Salauen, *Synlett*, 1998, 483.

298. A.P. Kozikowski, G.L. Araldi, J. Flippen-Anderson, C. George, S. Pshenichkin, E. Surina, and J.T. Wroblewski, *Bioorg.Med.Chem.Lett.*, 1998, **8**, 925.

299. A. Avenoza, J.H. Busto, J.M. Peregrina, and C. Cativiela, *An.Quim.Int.Ed.*, 1998, **94**, 50.

300. A. Avenoza, J.H. Busto, C. Cativiela, J.M. Peregrina, and F. Rodriguez, *Tetrahedron*, 1998, **54**, 11659.

301. J.C. Esterez, J.W. Burton, R.J. Esterez, H. Ardron, M.R. Wormald, R.A. Dwek, D. Brown, and G.W.J. Fleet, *Tetrahedron: Asymmetry*, 1998, **9**, 2137.

302. K. Lavrador, D. Guillerm, and G. Guillerm, *Bioorg.Med.Chem.Lett.*, 1998, **8**, 1629.

303. C. Toniolo, M. Crisma, and F. Formaggio, *Biopolymers*, 1998, **47**, 153.

304. B. Pispisa, A. Palleschi, L. Stella, M. Venanzi, and C. Toniolo, *J.Phys.Chem.B*, 1998, **102**, 7890.

305. G. Cardillo, L. Gentilucci, I.R. Bastardas, and A. Tolomelli, *Tetrahedron*, 1998, **54**, 8217.

306. S.N. Filigheddu and M. Taddei, *Tetrahedron Lett.*, 1998, **39**, 3857.

307. M.J. Alves and T.L. Gilchrist, *Tetrahedron Lett.*, 1998, **39**, 7579.

308. P. Wessig and J. Schwartz, *Helv.Chim.Acta*, 1998, **81**, 1803.

309. N. De Kimpe, M. Boeykens, and D. Tourwe, *Tetrahedron*, 1998, **54**, 2619.

310. R.A. Moss and D.C. Merrer, *Tetrahedron Lett.*, 1998, **39**, 8067.

311. S. Hanessian, U. Reinhold, M. Saulnier, and S. Claridge, *Bioorg. Med. Chem. Lett.*, 1998, **8**, 2123.
312. C.S. Esslinger, H.P. Koch, M.O. Kavanaugh, D.P. Philips, A.R. Chamberlin, C.M. Thompson, and R.J. Bridges, *Bioorg. Med. Chem. Lett.*, 1998, **8**, 3101.
313. Y. Yuasa, N. Fujimaki, T. Yokomatsu, J. Ando, and S. Shibuya, *J. Chem. Soc., Perkin Trans. I*, 1998, 3577.
314. D. Matteson and J. Lu, *Tetrahedron: Asymmetry*, 1998, **9**, 2423.
315. M.E. Pfeifer and J.A. Robinson, *Chem. Commun.*, 1998, 1977.
316. J.-O. Durand, M. Larcheveque, and Y. Petit, *Tetrahedron Lett.*, 1998, **39**, 5743.
317. Q. Wang, A. Sasaki, and P. Potier, *Tetrahedron Lett.*, 1998, **39**, 5755.
318. M. Ayerbe, A. Anieta, F.P. Cossio, and A. Linden, *J. Org. Chem.*, 1998, **63**, 1795.
319. S.G. Pyne, J. Safaei, A.K. Schafer, A. Javidan, B.W. Skelton, and A.H. White, *Aust. J. Chem.*, 1998, **51**, 137.
320. S. Najdi, K.-H. Park, M.M. Olmstead, and M.J. Kurth, *Tetrahedron Lett.*, 1998, **39**, 1685; Y.-D. Gong and M.J. Kurth, *ibid.*, p. 3379.
321. S. Yamago, M. Nakamura, X.Q. Wang, M. Yanagawa, S. Tokumitzu, and E. Nakamura, *J. Org. Chem.*, 1998, **63**, 1694.
322. S.-H. Wu, W.-Q. Sun, D.-W. Zhang, L.-H. Shu, H.-M. Wu, J.-F. Xu, and X.-F. Lao, *J. Chem. Soc., Perkin Trans I*, 1998, 1733.
323. X. Tan, D.I. Schuster, and S.R. Wilson, *Tetrahedron Lett.*, 1998, **39**, 4187.
324. M. Ohno, S. Kojima, Y. Shirakawa, and S. Eguchi, *Heterocycles*, 1997, **46**, 49.
325. M.E. Volpin, Z.N. Parnes, and V.S. Romanova, *Russ. Chem. Bull.*, 1998, **47**, 1021.
326. I. McCort, A. Dureault, and J.-C. Depezay, *Tetrahedron Lett.*, 1998, **39**, 4463.
327. D.D. Long, S.M. Frederiksen, D.G. Marquess, A.L. Lane, D.J. Watkin, D.A. Winkler, and G.W.J. Fleet, *Tetrahedron Lett.*, 1998, **39**, 6091.
328. S.N. Osipov, M. Picquet, A.F. Kolomiets, C. Bruneau, and P.H. Dixneuf, *Chem. Commun.*, 1998, 2053.
329. A.G.H. Wee, B. Liu, and D.D. McLeod, *J. Org. Chem.*, 1998, **63**, 4218.
330. A. Nazih, M.-R. Schneider, and A. Mann, *Synlett*, 1998, 1337.
331. J. Barlenga, F. Aznar, C. Valdes, and C. Ribas, *J. Org. Chem.*, 1998, **63**, 3918.
332. C.H. Sugisaki, P.J. Carroll, and C.R.D. Correia, *Tetrahedron Lett.*, 1998, **39**, 3413.
333. L. Rene, L. Yaouancq, and B. Badet, *Tetrahedron Lett.*, 1998, **39**, 2569.
334. K. Rossen, P.J. Pye, L.M. DiMichele, R.P. Volante, and P.J. Reider, *Tetrahedron Lett.*, 1998, **39**, 6823.
335. K.J. Hale, N. Jogiya, and S. Manaviazar, *Tetrahedron Lett.*, 1998, **39**, 7163.
336. M.A. Ciufolini and N. Xi, *Chem. Soc. Rev.*, 1998, **27**, 437.
337. M.A. Ciufolini, C.Y.W. Hermann, Q. Dong, T. Shimizu, S. Swaminathan, and N. Xi, *Synlett*, 1998, 105.
338. S. Kihara, K. Maeda, T. Hori, and T. Fujinaga, *Stud. Surf. Sci. Catal.*, 1998, **114**, 189.
339. Y. Utsumi and J.-I. Takahashi, *Jpn. J. Appl. Phys., Part 2*, 1998, **37**, L1268 (*Chem. Abs.*, 1999, **130**, 625721).
340. S. Miyakawa, H. Tamura, A.B. Sawaoka, and K. Kobayashi, *Appl. Phys. Lett.*, 1998, **72**, 990.
341. K. Kobayashi, T. Kaneko, T. Saito, and T. Oshima, *Origins Life Evol. Biosphere*, 1998, **28**, 155.
342. J. Taillades, I. Benzelin, L. Garrel, V. Tabacik, C. Bied, and A. Commeyras, *Origins Life Evol. Biosphere*, 1998, **28**, 61.
343. W.H. Sorrell, *Astrophys. Space Sci.*, 1997, **253**, 27.

344. F.M. Devienne, C. Barnabe, M. Couderc, and G. Ourisson, *C.R. Acad.Sci., Ser. IIc: Chim.*, 1998, **1**, 435.

345. A.L. Weber, *Origins Life Evol.Biosphere*, 1998, **28**, 259.

346. X. Wang, J. Han, C. Shao, and Z. Yu, *Viva Origino*, 1998, **26**, 109.

347. J.P. Amend and E.L. Shock, *Science*, 1998, **281**, 1659.

348. D.J. Des Marais, *Astron.Soc.Pac.Conf.Ser.*, 1998, **148**(Origins), 415 (*Chem.Abs.* 1999, **130**, 48736; see also adjacent abstracts).

349. K. Li, C. Leriche, and H.-W. Liu, *Bioorg.Med.Chem.Lett.*, 1998, **8**, 1097.

350. M.K. Eberle, R. Keese, and U. Stoeckli-Evans, *Helv.Chim.Acta*, 1998, **81**, 182.

351. C. Zhang, C. Ludin, M.K. Eberle, U. Stoeckli-Evans, and R. Keese, *Helv.Chim. Acta*, 1998, **81**, 174.

352. A. Laaziri, J. Uziel, and S. Juge, *Tetrahedron: Asymmetry*, 1998, **9**, 437.

353. N.W. Fadnavis, S.K. Vadivel, M. Sharfuddin, and U.T. Bhalereo, *Tetrahedron: Asymmetry*, 1998, **9**, 4003.

354. R. Kuwano, S. Okuda, and Y. Ito, *J.Org.Chem.*, 1998, **63**, 3499.

355. R. Grandel, U. Kazmaier, and F. Rominger, *J.Org.Chem.*, 1998, **63**, 4524.

356. E. Marotta, M. Baravelli, L. Maini, P. Righi, and G. Rosini, *J.Org.Chem.*, 1998, **63**, 8235.

357. H. Suga, K. Ikai, and T. Ibata, *Tetrahedron Lett.*, 1998, **39**, 869.

358. T. Miura, N. Masuo, Y. Fusamae, T. Kajimoto, and Y. Ida, *Synlett*, 1998, 631.

359. D.L. Boger and G. Schuele, *J.Org.Chem.*, 1998, **63**, 6421.

360. L.J. Wilson, M. Li, and D.E. Portlock, *Tetrahedron Lett.*, 1998, **39**, 5135.

361. A. Dobosz, I.O. Fritsky, A. Karaczyn, H. Kozlowski, T.Y. Sliva, and J. Swiatek-Kozlowska, *J.Chem.Soc., Dalton Trans.*, 1998, 1089.

362. P. Merino, E. Castillo, S. Franco, F.L. Merchan, and T. Tejero, *J.Org.Chem.*, 1998, **63**, 2371.

363. P. Merino, S. Franco, F.L. Merchan, and T. Tejero, *J.Org.Chem.*, 1998, **63**, 5627.

364. P.M.T. Ferreira, H.L.S. Maia, and L.S. Monteiro, *Tetrahedron Lett.*, 1998, **39**, 9575.

365. S. Osada, T. Fumoto, H. Kodama, and M. Kondo, *Chem.Lett.*, 1998, 675.

366. J.N. Kyranos and J.C. Hogan, *Anal.Chem.*, 1998, **70**, 389A.

367. P.W. Groundwater, T. Sharif, A. Arany, D.E. Hibbs, M.B. Hursthouse, I. Garnett, and M. Nyerges, *J.Chem.Soc., Perkin Trans. I*, 1998, 2837; P.W. Groundwater, T. Sharif, A. Arany, D.E. Hibbs, M.B. Hursthouse, and M. Nyerges, *Tetrahedron Lett.*, 1998, **39**, 1433.

368. W.A. Nugent and J.E. Feaster, *Synth.Commun.*, 1998, **28**, 1617.

369. R. Baddorey, C. Cativiela, M.D. Diaz-De-Villegas, and J.A. Galvez, *Synthesis*, 1998, 454.

370. N.J. Church and D.W. Young, *J.Chem.Soc., Perkin Trans. I*, 1998, 1475.

371. F.L. Zumpe and U. Kazmaier, *Synlett*, 1998, 434.

372. U. Kazmaier and C. Schneider, *Synthesis*, 1998, 1321.

373. D.A. Evans, M.R. Wood, B.W. Trotter, T.I. Richardson, J.C. Barrow, and J.L. Katz, *Angew.Chem.Int.Ed.*, 1998, **37**, 2700; K.C. Nicolau, S. Natarajan, H. Li, N.F. Jain, R. Hughes, M.E. Solomon, J.M. Ramanjulu, C.N.C. Boddy, and M. Takayanaji, *ibid.*, p.2708; K.C. Nicolau, N.F. Jain, S. Natarajan, R. Hughes, M.E. Solomon, H. Li, J.M. Ramanjulu, M. Takayanaji, A.E. Koumbis, and T. Bando, *ibid.*, p.2714; K.C. Nicolau, M. Takayanaji, N.F. Jain, S. Natarajan, A.E. Koumbis, T. Bando, and J.M. Ramanjulu, p.2717.

374. D.A. Evans, C.J. Dinsmore, P.S. Watson, M.R. Wood, T.I. Richardson, B.W. Trotter, and J.L. Katz, *Angew.Chem.Int.Ed.*, 1998, **37**, 2704.

375.　A. Ritzen, B. Basu, A. Wallberg, and T. Frejd, *Tetrahedron: Asymmetry*, 1998, **9**, 3491.

376.　G. Sun, M. Slavica, N.J. Uretsky, L.J. Wallace, G. Shams, D.M. Weinstein, J.C. Miller, and D.D. Miller, *J.Med.Chem.*, 1998, **41**, 1034.

377.　T. Boxus, R. Touillaux, G. Dive, and J. Marchand-Brynaert, *Bioorg.Med.Chem. Lett.*, 1998, **6**, 1577.

378.　A. Escribano, J. Ezquerra, C. Pedregal, A. Rubio, B. Yruretagoyena, S.R. Baker, R.A. Wright, B.G. Johnson, and D.D. Schoepp, *Bioorg.Med.Chem.Lett.*, 1998, **8**, 765.

379.　A.B. Charette and P. Chua, *Mol. Online*, 1998, **2**, 63.

380.　K. Lee, S.Y. Huang, S. Hong, C.Y. Hong, C.-S. Lee, Y. Shin, S. Saagsoo, M. Yun, Y.J. Yoo, M. Kang, and Y.S. Oh, *Bioorg.Med.Chem.Lett.*, 1998, **6**, 869.

381.　D.R. Kent, W.L. Cody, and A.M. Doherty, *J.Pept.Res.*, 1998, **52**, 201.

382.　D. Ma, Z. Ma, A.P. Kozikowski, S. Pshenichkin, and J.T. Wroblewski, *Bioorg. Med.Chem.Lett.*, 1998, **6**, 2447.

383.　T. Nakano, M. Nishiaki, M. Kirihata, M. Takagaki, and K. Ono, *Kyoto Daigaku Genshiro Jikkensho Gakujutsu Koenkai Hobunshu*, 1998, **32**, 73 (*Chem.Abs.*, 1998, **128**, 267767).

384.　M. Tichy, L. Ridvan, M. Budesinsky, J. Zavada, J. Podlaha, and I. Cisarova, *Collect.Czech.Chem.Commun.*, 1998, **63**, 211.

385.　T. Yokomatsu, T. Minowa, T. Murano, and S. Shibuya, *Tetrahedron*, 1998, **54**, 9341.

386.　N. Skjaerback, L. Brehm, T.N. Johansen, L.M. Hansen, B. Nielsen, B. Ebert, K.K. Soby, T.B. Stensbol, E. Falch, and P. Krogsgaard-Larsen, *Bioorg.Med. Chem.*, 1998, **6**, 119.

387.　F. Zaid, S. El Haiji, A. El Hallaoui, A. Elachqar, A. Kerbal, M.L. Roumestant, and P. Viallefont, *Prep.Biochem.Biotechnol.*, 1998, **28**, 137; F. Zaid, S. El Haiji, A. El Hallaoui, A. Elachqar, A. Kerbal, M.L. Roumestant, and P. Viallefont, *ibid.*, p.155.

388.　A. Alami, A. El Hallaoui, A. Elachqar, S. El Haiji, M.L. Roumestant, and P. Viallefont, *Prep.Biochem.Biotechnol.*, 1998, **28**, 167.

389.　R. Jain, B. Avramovitch, and L.A. Cohen, *Tetrahedron*, 1998, **54**, 3235.

390.　J.-F. Rousseau and R.H. Dodd, *J.Org.Chem.*, 1998, **63**, 2731.

391.　B. Hofmann, P. Dauban, J.-P. Biron, P. Potier, and R.H. Dodd, *Heterocycles*, 1998, **46**, 473.

392.　J.-M. Fu and A.L. Castelhano, *Bioorg.Med.Chem.Lett.*, 1998, **8**, 2813.

393.　K.J. Kise and B.E. Bowler, *Tetrahedron: Asymmetry*, 1998, **9**, 3319.

394.　G.K. Walkup and B. Imperiali, *J.Org.Chem.*, 1998, **63**, 6727.

395.　P. Ciapetti, A. Mann, A. Schoenfelder, M. Taddei, E. Trifilieff, I. Canet, and J.L. Canet, *Lett.Pept.Sci.*, 1997, **4**, 341.

396.　T. Carell, H. Schmid, and M. Reinhard, *J.Org.Chem.*, 1998, **63**, 8741.

397.　Review: P.E. Nielsen, *Pure Appl.Chem.*, 1998, **70**, 105.

398.　G. Lowe, T. Vilaivan, and M.S. Westwell, *Bioorg.Chem.*, 1997, **25**, 321.

399.　H. Han, J. Yoon, and K.D. Janda, *J.Org.Chem.*, 1998, **63**, 2045.

400.　B.M. Kim and S.M. So, *Tetrahedron Lett.*, 1998, **39**, 5381.

401.　P. Merino, A. Lanaspa, F.L. Merchan, and T. Tejero, *Tetrahedron: Asymmetry*, 1998, **9**, 629.

402.　T. Mikayama, H. Matsuoka, K. Uehara, I. Shimizu, and S. Yoshikawa, *Biopolymers*, 1998, **47**, 179.

403. C. Lebarbier, F. Carreaux, B. Carboni, and J.L. Boucher, *Bioorg.Med.Chem. Lett.*, 1998, **8**, 2573.
404. G. Reginato, A. Mordini, and M. Valacchi, *Tetrahedron Lett.*, 1998, **39**, 9545.
405. R. Hamilton, B. Walker, and B.J. Walker, *Bioorg.Med.Chem.Lett.*, 1998, **8**, 1655.
406. N. Tokutake, J. Hiratake, M. Katoh, T. Irie, H. Kato, and J.Oda, *Bioorg.Med. Chem.*, 1998, **6**, 1935.
407. J.T. Lundquist, E.E. Bullesbach, and T.A. Dix, *Tetrahedron: Asymmetry*, 1998, **9**, 2739.
408. O. Juanes, P. Goya, and A. Martinez, *J.Heterocycl.Chem.*, 1998, **35**, 727.
409. R. Andruszkiewicz, *Pol.J. Chem.*, 1998, **72**, 1.
410. D. Ma and J. Jiang, *Tetrahedron: Asymmetry*, 1998, **9**, 575.
411. S.G. Davies and D.J. Dixon, *J.Chem.Soc., Perkin Trans. I*, 1998, 2635.
412. P. O'Brien, *J.Chem.Soc., Perkin Trans. I*, 1998, 1439.
413. M. Koerner, M. Findeisen, and N. Sewald, *Tetrahedron Lett.*, 1998, **39**, 3463.
414. N. Sewald, K.D. Hiller, M. Koerner, and M. Findeisen, *J.Org.Chem.*, 1998, **63**, 7263.
415. S.G. Davies and O. Ichihara, *Tetrahedron Lett.*, 1998, **39**, 6045.
416. S.G. Davies and D.J. Dixon, *J.Chem.Soc., Perkin Trans. I*, 1998, 2629.
417. D. Ma and J. Zhang, *Tetrahedron Lett.*, 1998, **39**, 9067.
418. M.P. Sibi, J.J. Shay, M. Liu, and C.P. Jasperse, *J.Am.Chem.Soc.*, 1998, **120**, 6615.
419. D. Lee and M.-J. Kim, *Tetrahedron Lett.*, 1998, **39**, 2163.
420. L. Barboni, C. Lambertucci, R. Ballini, G. Appendino, and E. Bombardelli, *Tetrahedron Lett.*, 1998, **39**, 7177.
421. K. Higashiyama, H. Kyo, and H. Takahashi, *Synlett*, 1998, 489.
422. S. Kobayashi and Y. Aoki, *Tetrahedron Lett.*, 1998, **39**, 7345.
423. T. Ishikawa, K. Nagai, T. Kudoh, and S. Saito, *Synlett*, 1998, 1291.
424. J.L. Garcia Ruano, I. Fernandez, M. del P. Catalina, J.A. Hermoso, J. Sanz-Aparicio, and M. Martinez-Ripoll, *J.Org.Chem.*, 1998, **63**, 7157.
425. S.M. Sternson and S.L. Schreiber, *Tetrahedron Lett.*, 1998, **39**, 7451.
426. F.A. Davis and J.M. Szewczyk, *Tetrahedron Lett.*, 1998, **39**, 5951.
427. T.-P. Loh and L.-L. Wei, *Tetrahedron Lett.*, 1998, **39**, 323.
428. N. Yu, L. Bourel, B. Deprez, and J.-C. Gesquiere, *Tetrahedron Lett.*, 1998, **39**, 829.
429. S. Fustero, M.G. de la Torre, V. Jofre, R.P. Carlon, A. Navarro, A.S. Fuentes, and J.S. Carrio, *J.Org.Chem.*, 1998, **63**, 8825.
430. V. Molteni, R. Annunziata, M. Cinquini, F. Cozzi, and M. Benaglia, *Tetrahedron Lett.*, 1998, **39**, 1257.
431. C.M. Marson and A. Fallah, *Chem.Commun.*, 1998, 83.
432. A.J. Clark and J.L. Peacock, *Tetrahedron Lett.*, 1998, **39**, 1265.
433. I. Ojima, T. Wang, and F. Delaloge, *Tetrahedron Lett.*, 1998, **39**, 3663.
434. E. Arvanitis, H. Ernst, A.A. Ludwige D'Souza, A.J. Robinson, and P.B. Wyatt, *J.Chem.Soc., Perkin Trans. I*, 1998, 521; erratum, *ibid.*, p.1459.
435. Z. Dahmani, M. Rahmouni, R. Brugifou, J.P. Bazureau, and J. Hamelin, *Tetrahedron Lett.*, 1998, **39**, 8453.
436. G. Cravotto, G.B. Giovenzana, R. Pagliarin, G. Palmisano, and M. Sisti, *Tetrahedron: Asymmetry*, 1998, **9**, 745.
437. S.J. Keding, N.A. Dales, S. Lim, D. Beaulieu, and D.H. Rich, *Synth.Commun.*, 1998, **28**, 4463.

438. A.E. Rubin and K.B. Sharpless, *Angew. Chem. Int. Ed.*, 1997, **36**, 2637.
439. C. Joergensen, C. Pedersen, and I. Soetofte, *Synthesis*, 1998, 325.
440. K.Y. Lee, Y.-H. Kim, M.-S. Park, and W.-H. Ham, *Tetrahedron Lett.*, 1998, **39**, 8129.
441. C.J. Moody and J.C.A. Hunt, *Synlett*, 1998, 733.
442. P. Merino, E. Castillo, S. Franco, F.L. Merchan, and T. Tejero, *Tetrahedron*, 1998, **54**, 12301.
443. P. Merino, S. Franco, F.L. Merchan, and T. Tejero, *Tetrahedron Lett.*, 1998, **39**, 6411.
444. H. Sugimura, M. Miura, and N. Yamada, *Tetrahedron: Asymmetry*, 1998, **9**, 4089.
445. M.S. Ozer, S. Thiebaut, C. Gerardin-Charbonnier, and C. Selve, *Synth. Commun.*, 1998, **28**, 2429.
446. Y. Jin and D.H. Kim, *Synlett*, 1998, 1189.
447. V.A. Soloshonok, I.V. Soloshonok, V.P. Kukhar, and V.K. Svedas, *J. Org. Chem.*, 1998, **63**, 1878.
448. L. Lazar, T. Martinek, G. Bernath, and F. Fulop, *Synth. Commun.*, 1998, **28**, 219.
449. G. Cardillo, L. Gentilucci, A. Tolomelli, and C. Tomasini, *J. Org. Chem.*, 1998, **63**, 2351.
450. N. Shibita, E. Itoh, and S. Terashima, *Chem. Pharm. Bull.*, 1998, **46**, 733.
451. A. Fassler, G. Bold, and H. Steiner, *Tetrahedron Lett.*, 1998, **39**, 4925.
452. A.E. Wroblewski and D.G. Piotrowska, *Tetrahedron*, 1998, **54**, 8123.
453. M.N. Greco, H.M. Zhong, and B.E. Maryanoff, *Tetrahedron Lett.*, 1998, **39**, 4959.
454. A. Gaucher, F. Bintein, M. Wakselman, and J.-P. Mazaleyrat, *Tetrahedron Lett.*, 1998, **39**, 575.
455. N. Fujiwara and Y. Yamamoto, *Tetrahedron Lett.*, 1998, **39**, 4729.
456. C. Dartiguelongue, S. Payan, O. Duval, L.M. Gomes, and R.D. Waigh, *Bull. Soc. Chim. Fr.*, 1997, **134**, 769.
457. A.R. Katritzky, D.A. Nichols, and M. Qi, *Tetrahedron Lett.*, 1998, **39**, 7063.
458. R. Ponsinet, G. Chassaing, and S. Lavielle, *Tetrahedron: Asymmetry*, 1998, **9**, 865.
459. G. Cardillo, L. Gentilucci, A. Tolomelli, and C. Tomasini, *J. Org. Chem.*, 1998, **63**, 3458.
460. E.P. Ellmerer-Mueller, D. Broessner, N. Maslouh, and A. Tako, *Helv. Chim. Acta*, 1998, **81**, 59.
461. C. Guichard, S. Abele, and D. Seebach, *Helv. Chim. Acta*, 1998, **81**, 187.
462. A. Mueller, C. Vogt, and N. Sewald, *Synthesis*, 1998, 837.
463. D. Limal, A. Quesnel, and J.-P. Briand, *Tetrahedron Lett.*, 1998, **39**, 4239.
464. S. Zalipsky, *Chem. Commun.*, 1998, 69.
465. K. Uneyama, J. Hao, and H. Amii, *Tetrahedron Lett.*, 1998, **39**, 4079.
466. T.J. Donohoe, P.M. Guyo, R.R. Harji, M. Helliwell, and R.P.C. Cousins, *Tetrahedron Lett.*, 1998, **39**, 3075.
467. P.Q. Huang, J.L. Ye, Z. Chen, Y.P. Ruan, and J.X. Gao, *Synth. Commun.*, 1998, **28**, 417.
468. L. Serfass and P.J. Casara, *Bioorg. Med. Chem. Lett.*, 1998, **8**, 2599.
469. P. Garner, J.T. Anderson, S. Dey, W.J. Youngs, and K. Galat, *J. Org. Chem.*, 1998, **63**, 5732.
470. C.J. Martin, D.J. Rawson, and J.M.J. Williams, *Tetrahedron: Asymmetry*, 1998, **9**, 3723.

471. S. Setyahadi, E. Harada, N. Mori, and Y. Kitamoto, *J. Mol. Catal. B: Enzymol.*, 1998, **4**, 205.
472. M. Anada and S. Hashimoto, *Tetrahedron Lett.*, 1998, **39**, 79.
473. S. Chandrasekhar and S. Mohapatra, *Tetrahedron Lett.*, 1998, **39**, 6415.
474. G.R. Cook and P.S. Shanker, *Tetrahedron Lett.*, 1998, **39**, 3405.
475. A. Sutherland and C.L. Willis, *J. Org. Chem.*, 1998, **63**, 7764.
476. G. Reginato, A. Mordini, A. Capperucci, A. Degl'Innocenti, and S. Manganiello, *Tetrahedron*, 1998, **54**, 10217.
477. M. Paris, J.-A. Fehrentz, A. Heitz, and J. Martinez, *Tetrahedron Lett.*, 1998, **39**, 1569.
478. M. Paris, C. Pothion, C. Michalak, J. Martinez, and J.-A. Fehrentz, *Tetrahedron Lett.*, 1998, **39**, 6889.
479. E.J. Corey and W.-D. Z. Li, *Tetrahedron Lett.*, 1998, **39**, 7475; E.J. Corey, W. Li, and T. Nagamitsu, *Angew. Chem., Int. Ed.*, 1998, **37**, 1676.
480. E.J. Corey and W.-D. Z. Li, *Tetrahedron Lett.*, 1998, **39**, 8043.
481. A. Kanazawa, S. Gillet, P. Delair, and A.E. Greene, *J. Org. Chem.*, 1998, **63**, 4660.
482. D.-C. Ha, S. Kang, C.-M. Chung, and H.-K. Lim, *Tetrahedron Lett.*, 1998, **39**, 7541.
483. M. Bellassoued, R. Ennigrou, R. Gil, and N. Leasen, *Synth. Commun.*, 1998, **28**, 3955.
484. M. Shimano and A. Matsuo, *Tetrahedron*, 1998, **54**, 4787.
485. I. Ojima, H. Wang, T. Wang, and E.W. Ng, *Tetrahedron Lett.*, 1998, **39**, 923.
486. M. Pasto, A. Moyeno, M.A. Pericas, and A. Riera, *Tetrahedron Lett.*, 1998, **39**, 1233.
487. H. Tamamura, *Chem. Commun.*, 1997, 2327.
488. A.K. Ghosh and K. Krishnan, *Tetrahedron Lett.*, 1998, **39**, 947.
489. B. Schmidt and C. Kuehn, *Synlett*, 1998, 1240.
490. J.-H. Sun and W.F. Daneker, *Synth. Commun.*, 1998, **28**, 4525.
491. W.P.D. Golding, A.S. Hodder, and L. Weiler, *Tetrahedron Lett.*, 1998, **39**, 4955.
492. H. Sauriat-Dorizon and F. Guibe, *Tetrahedron Lett.*, 1998, **39**, 6711.
493. M. Shimizu, A. Morita, and T. Fujisawa, *Chem. Lett.*, 1998, 467.
494. F.C. Schroder, J.J. Farmer, S.R. Smedley, T. Eisner, and J. Meinwald, *Tetrahedron Lett.*, 1998, **39**, 6625.
495. P. Garner, J.H. Yoo, R. Sarabu, V.O. Kennedy, and W.J. Youngs, *Tetrahedron*, 1998, **54**, 9303.
496. M.D. Smith, D.D. Long, T.D.W. Claridge, G.W.J. Fleet, and D.G. Marquess, *Chem. Commun.*, 1998, 2039; D.D. Long, M.D. Smith, D.G. Marquess, T.D.W. Claridge, and G.W.J. Fleet, *Tetrahedron Lett.*, 1997, **39**, 9293.
497. A. Pelter, R.A.N.C. Crump, and H. Kidwell, *Tetrahedron: Asymmetry*, 1997, **8**, 3873.
498. F. Jean, E. Buisine, O. Melnyk, H. Drobecq, B. Odaert, M. Hugues, G. Lippens, and A. Tartar, *J. Am. Chem. Soc.*, 1998, **120**, 6076.
499. R. Yoshioka, K. Okamura, S. Yamada, K. Aoe, and T. Da-te, *Bull. Chem. Soc. Jpn.*, 1998, **71**, 1109.
500. M. Yokota, Y. Takahashi, A. Sato, N. Kubota, F. Masumi, and H. Takeuchi, *Chem. Eng. Sci.*, 1998, **53**, 1473.
501. J.-P. Mazaleyrat, A. Boutboul, Y. Lebars, A. Gaucher, and M. Wakselman, *Tetrahedron: Asymmetry*, 1998, **9**, 2701.
502. R.K. Duke, R.D. Allan, M. Chebib, J.R. Greenwood, and G.A.R. Johnston, *Tetrahedron: Asymmetry*, 1998, **9**, 2533.

503. S. Brown, A.M. Jordan, N.J. Lawrence, R.G. Pritchard, and A.T. McGown, *Tetrahedron Lett.*, 1998, **39**, 3559.

504. C.A. Maryanoff, L. Scott, R.D. Shah, and F.J. Villani, *Tetrahedron: Asymmetry*, 1998, **9**, 3247.

505. J. Liang, J.C. Ruble, and G.C. Fu, *J.Org.Chem.*, 1998, **63**, 3154.

506. M.D. Weingarten, K. Sekanina, and W.C. Still, *J.Am.Chem.Soc.*, 1998, **120**, 9112.

507. X. Guo, Q. Liu, H. Bo, and F. Zhang, *Huaxue Yanjiu Yu Yingyong*, 1998, **10**, 164 (*Chem.Abs.*, 1999, **130**, 95784).

508. A. Bossi, M. Cretich, and P.G. Righetti, *Biotechnol.Bioeng.*, 1998, **60**, 454.

509. L. Lecointe, V. Rolland-Fulcrand, M.L. Roumestant, P. Viallefont, and J. Martinez, *Tetrahedron: Asymmetry*, 1998, **9**, 1753.

510. J.L. den Hollander, B.I. Stribos, M.J. van Buel, K.C.A.M. Luyben, and L.A.M. van der Wielen, *J.Chromatogr.B: Biomed.Appl.*, 1998, **711**, 223.

511. C.J. Easton and J.B. Harper, *Tetrahedron Lett.*, 1998, **39**, 5269.

512. M.A.P.J. Hacking, M.A. Wegman, J. Rops, F. van Rantwijk, and R.A. Sheldon, *J.Mol.Catal.B: Enzymol.*, 1998, **5**, 155.

513. R. Rai and V. Taneja, *Biochem.Biophys.Res.Commun.*, 1998, **244**, 889.

514. C.J. Hartley, S. Kirchmann, S.G. Burton, and R.A. Dorrington, *Biotechnol.Lett.*, 1998, **20**, 707.

515. A. Avdagic and V. Sunjic, *Helv.Chim.Acta*, 1998, **81**, 85.

516. L. Liu, P. Yang, and R. Zhuo, *Zhongguo Yiyao Gongye Zazhi*, 1997, **28**, 342 (*Chem.Abs.*, 1998, **128**, 270851).

517. M. Nakamura, S. Kiyohara, K. Saito, K. Sugita, and T. Sugo, *J.Chromatogr.A*, 1998, **822**, 53; I. Koguma, M. Nakamura, K. Saito, K. Sugita, S. Kiyohara, and T. Sugo, *Kagaku Kogaku Ronbunshu*, 1998, **24**, 458 (*Chem.Abs.*, 1998, **129**, 161824).

518. H. Escalante, A.I. Alonso, I. Ortiz, and A. Irabien, *Sep.Sci.Technol.*, 1998, **33**, 119.

519. B.-H. Kim and W. Lee, *Bull.Korean Chem.Soc.*, 1998, **19**, 289.

520. A. Peter, G. Torok, and F. Fulop, *J.Chromatogr.Sci.*, 1998, **36**, 311.

521. D.W. Armstrong, *J.Chin.Chem.Soc.*, 1998, **45**, 581.

522. W. Golkiewicz and B. Polak, *Chem.Anal.*, 1998, **43**, 591.

523. E. Horvatti, L. Kocsis, R.L. Frost, B. Hren, and L.P. Szabo, *Anal.Chem.*, 1998, **70**, 2766.

524. M.H. Hyun, S.J. Lee, and J.-J. Ryoo, *Bull.Korean Chem.Soc.*, 1998, **19**, 1105.

525. M.H. Hyun and C.-S. Min, *Chirality*, 1998, **10**, 592.

526. M. Wachsmann and H. Bruckner, *Chromatographia*, 1998, **47**, 637.

527. F.H. Arnold, S. Striegler, and V. Sundaresan, *ACS Symp.Ser.*, 1998, **703**(Molecular and Ionic Recognition with Imprinted Polymers), 109.

528. P.A.G. Cormack and K. Mosbach, *Am.Biotechnol.Lab.*, 1998, **16**, 47.

529. M. Yoshida, K. Uezu, M. Goto, S. Furusaki, and M. Takagi, *Chem.Lett.*, 1998, 925.

530. P. Sajouz, M. Kele, G. Zhong, B. Sellergren, and G. Guichon, *J.Chromatogr.A*, 1998, **810**, 1.

531. M. Yoshikawa, T. Fujisawa, J. Izumi, T. Kitao, and S. Sakamoto, *Sen'i Gakkaishi*, 1998, **54**, 77 (*Chem.Abs.*, 1998, **128**, 230682).

532. M. Yoshikawa, T. Fujisawa, J. Izumi, T. Kitao, and S. Sakamoto, *Anal.Chim. Acta*, 1998, **365**, 59.

533. Z. Chen, T. Kiyonaga, and T. Nagaoka, *Bunseki Kagaku*, 1998, **47**, 519.

534. T. Satoh, Y. Tanaka, K. Yokota, and T. Kakuchi, *React.Funct.Poly.*, 1998, **37**, 293.

535. H. Shinomiya, Y. Kabasawa, and Y. Ito, *J.Liq.Chromatogr.Relat.Technol.*, 1998, **21**, 135.

536. J.K. Kim and W.S. Kim, *J.Ind.Eng.Chem.(Seoul)*, 1998, **4**, 105; for a review, see S.Tone, *Kagaku Kogaku*, 1998, **62**, 93 (*Chem.Abs.*, 1998, **128**, 115205).

537. M. Pietraszkiewicz, M. Kozbial, and O. Pietraszkiewicz, *Enantiomer*, 1997, **2**, 319; *J.Membr. Sci.*, 1998, **138**, 109.

538. T. Kitae, T. Nakayama, and K. Kano, *J.Chem.Soc., Perkin Trans.II*, 1998, 207.

539. M. Sawada, Y. Takai, M. Shizuma, Y. Takai, T. Takeda, H. Adachi, and T. Uchiyama, *Chem.Commun.*, 1998, 1453.

540. W.A. Bonner, *EXS*, 1998, **85**(D-Amino Acids in Sequences of Secreted Peptides), 159 (*Chem.Abs.*, 1998, **129**, 340877).

541. W. Wang, *Theor.Biophys.Biomath., Proc.Int.Symp.*, Eds. L. Luo, Q. Li, and W.Lee, Inner Mongolia University Press, Hohhor, People's Republic of China, 1997, 141 (*Chem.Abs.*, 1998, **129**, 327319).

542. Y.J. He, F. Qi, and S.C. Qi, *Med.Hypotheses*, 1998, **51**, 125 (*Chem.Abs.*, 1998, **129**, 327407).

543. R. Zanasi and P. Lazzeretti, *Chem.Phys.Lett.*, 1998, **286**, 240.

544. J.M. Bailey, *FASEB J.*, 1998, **12**, 503; cf. J.M. Bailey, *Biochem.Soc.Trans.*, 1998, **26**, 5262.

545. M.Yarus, *J.Mol.Evol.*, 1998, **47**, 109.

546. B.Schade and J.-H. Fuhrhop, *New J.Chem.*, 1998, **22**, 97.

547. T.Koritsanszky, R.Flaig, D.Zobel, H.-G. Krane, W.Morgenroth, and P.Luger, *Science*, 1998, **279**, 356.

548. R.Flaig, T.Koritsanszky, D.Zobel, and P.Luger, *J.Am.Chem.Soc.*, 1998, **120**, 2227.

549. S.Ravichandran, J.K. Dattagupta, and C.Chakrabarti, *Acta Crystallogr., Sect.C: Cryst.Struct.Commun.*, 1998, **C54**, 499.

550. N.R. Chandra, M.M. Prabu, J.Venkatraman, S.Suresh, and M.Vijayan, *Acta Crystallogr., Sect.B: Struct.Sci.*, 1998, **B54**, 257

551. R.Ravishankar, R.Nagasuma, and M.Vijayan, *J.Biomol.Struct.Dyn.*, 1998, **15**, 1093.

552. C.H. Gorbitz and J.Husdal, *Acta Chem.Scand.*, 1998, **52**, 218.

553. T.Koler, R.Stahl, H.Preut, P.Bleckmann, and V.Radomirska, *Z.Kristallogr. – New Cryst.Struct.*, 1998, **213**, 169.

554. T.Koler, R.Stahl, H.Preut, L.Koniczek, P.Bleckmann, and V.Radomirska, *Z.Kristallogr. – New Cryst.Struct.*, 1998, **213**, 167.

555. L.Brehm, K.Frydenvang, P.Krogsgaard-Larsen, and T.Liljefors, *Struct.Chem.*, 1998, **9**, 149.

556. H.Schmidbaur, A.Schier, and A.Bayler, *Z.Naturforsch., B: Chem.Sci.*, 1998, **53**, 788.

557. R.Latz and R.Boese, *Chem.Unserer Zeit*, 1998, **32**, 84.

558. T.Suga, C.Inubashi, and N.Okabe, *Acta Crystallogr., Sect.C: Cryst.Struct. Commun.*, 1998, **C54**, 83.

559. W.Hiller, M.Neumayer, A.Rieker, M.H. Khalifa, and G.Jung, *Z.Kristallogr. – New Cryst.Struct.*, 1998, **213**, 188.

560. M.Crisma, G.Valle, V.Moretto, F.Formaggio, C.Toniolo, and F.Albericio, *Lett. Pept.Sci.*, 1998, **5**, 247.

561. M.Doi, K.Kinoshita, A.Asano, R.Yoneda, T.Kurihara, and T.Ishida, *Acta Crystallogr., Sect.C: Cryst.Struct.Commun.*, 1998, **C54**, 1164.
562. G.Raabe, A.Sudeikat, and R.W. Woody, *Z.Naturforsch., A: Phys.Sci.*, 1998, **53**, 61.
563. M.Akazome, T.Takahashi, and K.Ogura, *Tetrahedron Lett.*, 1998, **39**, 4839.
564. N.Jurinac, Z.Zolnai, and S.Macura, *Mol.Phys.*, 1998, **95**, 833.
565. V.Govindaraju, V.J. Basus, G.B. Matson, and A.A. Maudsley, *Magn.Reson. Med.*, 1998, **39**, 1011.
566. H.Kimura, K.Nakamura, A.Eguchi, H.Sugisawa, K.Deguchi, K.Ebisawa, E.Suzuki, and A.Shoji, *J.Mol.Struct.*, 1998, **447**, 247.
567. V.R. Reddy, Y.M.K. Reddy, P.N. Reddy, and B.P.N. Reddy, *Indian J.Pure Appl.Phys.*, 1997, **35**, 653.
568. A.Bianco, V.Lucchini, M.Maggini, M.Prato, G.Scorrano, and C.Toniolo, *J.Pept.Sci.*, 1998, **4**, 364.
569. E.Beausoleil, R.Sharma, S.W. Michnick, and W.D. Lubell, *J.Org.Chem.*, 1998, **63**, 6572.
570. A.D. Headley, R.E. Corona, and E.T. Cheung, *J.Phys.Org.Chem.*, 1997, **10**, 898.
571. F.Gregoire, S.H. Wei, E.W. Streed, K.A. Brameld, D.Fort, L.J. Hanely, J.D. Walls, W.A. Goddard, and J.D. Roberts, *J.Am.Chem.Soc* , 1998, **120**, 7537.
572. Y.Assif and Y.Cohen, *J.Magn.Reson.*, 1998, **131**, 69.
573. O.Gonen, A.K. Viswanathan, I.Catalaa, J.Babb, J.Udupa, and R.I. Grossman, *Magn.Reson.Med.*, 1998, **40**, 684.
574. D.L. Hardy and T.J. Norwood, *J.Magn.Reson.*, 1998, **133**, 70.
575. M.J. Potrzebowski, P. Tekely, and Y. Dusausay, *Solid State Nucl.Magn.Reson.*, 1998, **11**, 253.
576. E. Kiss, A. Lakatos, I. Banyai, and T. Kiss, *J.Inorg.Biochem.*, 1998, **69**, 145.
577. M. Goez, J. Rozwadowski, and B. Marciniak, *Angew.Chem.Int.Ed.*, 1998, **37**, 628; M. Goez and J. Rozwadowski, *J.Phys.Chem.A*, 1998, **102**, 7945.
578. K. Harada, Y. Shimizu, and K. Fujii, *Tetrahedron Lett.*, 1998, **39**, 6245.
579. P. Faller and M. Vasak, *Inorg.Chim.Acta*, 1998, **272**, 150.
580. S. Prabhakar, P. Krishna, and M. Vairamani, *Rapid Commun.Mass Spectrom.*, 1997, **11**, 1945.
581. E.P. Burrows, *J.Mass Spectrom.*, 1998, **33**, 221.
582. R.A.J. O'Hair and G.E. Reid, *Rapid Commun.Mass Spectrom.*, 1998, **12**, 999.
583. Z. Guo and Q. An, *Hebei Daxue Xuebao Ziran Kexueban*, 1997, **17**, 73 (*Chem.Abs.*, 1998, **128**, 192904).
584. A.G. Harrison, *Mass Spectrom.Rev.*, 1997, **16**, 201.
585. W.-X. Jun, Z.-Q. Liu, F.-R. Song, and S.-Y. Liu, *Gaodeng Xuexiao Huaxue Xubao*, 1998, **19**, 708 (*Chem.Abs.*, 1999, **129**, 161823).
586. J. Ramirez, F. He, and C.B. Lebrilla, *J.Am.Chem.Soc.*, 1998, **120**, 7387.
587. M. Sawada, Y. Takai, H. Yamada, J. Nishida, T. Kaneda, R. Arakawa, M. Okamoto, K. Hirose, T. Tanaka, and K. Naemura, *J.Chem.Soc., Perkin Trans. II*, 1998, 701.
588. M. Sawada, Y. Takai, H. Yamada, M. Sawada, H. Yamaoka, T. Azuma, T. Fujioka, Y. Kawai, and T. Tanaka, *Chem.Commun.*, 1998, 1569.
589. M. Shizuma, *J.Mass Spectrom.Soc.Jpn.*, 1998, **46**, 211 (*Chem.Abs.*, 1998, **129**, 272420).
590. F. Turecek, J.L. Kerwin, R. Xu, and K.J. Kramer, *J.Mass Spectrom.*, 1998, **33**, 392.

591. P. Krishna, S. Prabhahkar, and M. Vairamani, *Rapid Commun. Mass Spectrom.*, 1998, **12**, 1429.

592. T. Yalcin, W. Gabryelski, and L. Li, *J. Mass Spectrom.*, 1998, **33**, 543; W.Gabryelski, R.W. Purves, and L.Li, *Int.J. Mass Spectrom.*, 1998, **176**, 213.

593. A.G. Harrison and Y.-P. Tu, *J. Mass Spectrom.*, 1998, **33**, 532.

594. E. Espinoza, B. Wyncke, F. Brehat, X. Gerbaux, S. Vientemillas, and E. Molins, *Infrared Phys. Technol.*, 1997, **38**, 449.

595. G. Dovbeshko and L. Berezhinsky, *J. Mol. Struct.*, 1998, **450**, 121.

596. S.G. Stepanian, I.D. Reva, E.D. Radchenko, M.T.S. Rosado, M.L.T.S. Duarte, R. Fausto, and L. Adamowicz, *J. Phys. Chem. A*, 1998, **102**, 1041.

597. S.G. Stepanian, I.D. Reva, E.D. Radchenko, and L. Adamowicz, *J. Phys. Chem. A*, 1998, **102**, 4623.

598. H.N. Bordallo, M. Barthes, and J. Eckert, *Physica B*, 1997, **241**, 1138.

599. C. Berthomieu, C. Boullais, J.-M. Neumann, and A. Boussac, *Biochim. Biophys. Acta*, 1998, **1365**, 112.

600. C.J. Chapo, J.B. Paul, R.A. Provencal, K. Roth, and R.J. Saykally, *J. Am. Chem. Soc.*, 1998, **120**, 12956.

601. R.A. Jockusch and E.R. Williams, *J. Phys. Chem. A*, 1998, **102**, 4543.

602. K. de Wael, C. Bruyneel, and T. Zeegers-Huyskens, *Spectrosc. Lett.*, 1998, **31**, 283.

603. M.A. Broda and B. Rzeszotarska, *Lett. Pept. Sci.*, 1998, **5**, 441.

604. M.A. Broda, B. Rzeszotarska, L. Smelka, and G. Pietrzynski, *J. Pept. Res.*, 1998, **52**, 72.

605. X. Du, B. Shi, and Y. Liang, *Langmuir*, 1998, **14**, 3631.

606. F. Hoffmann, H. Huehnerfuss, and K.J. Stine, *Langmuir*, 1998, **14**, 4525.

607. B.J.M. Rajkumar, V. Ramakrishnan, and R.K. Rajaram, *Spectrochim. Acta, Part A*, 1998, **54A**, 1527.

608. M.T.S. Rosado, M.L.T.S. Duarte, and R. Fausto, *Vib. Spectrosc.*, 1998, **16**, 35.

609. L.I. Berezhinsky, G.I. Dovbeshko, M.P. Lisitsa, and G.S. Litvinov, *Spectrochim. Acta, Part A*, 1998, **54A**, 349.

610. M. Picquart, Z. Abedinzadeh, L. Grajcar, and M.H. Baron, *Chem. Phys.*, 1998, **228**, 279.

611. B.L. Silva, P.T.C. Friere, F.E.A. Melo, I. Guedes, M.A.A. Silva, J.M. Filo, and A.J.D. Moreno, *Braz. J. Phys.*, 1998, **28**, 19.

612. K.A. Kuhls, C.A. Centrone, and M.J. Tubergen, *J. Am. Chem. Soc.*, 1998, **120**, 10194.

613. J.-Z. Yang and Z.-F. Zhang, *Talanta*, 1998, **45**, 947.

614. C.L. Hawkins and M.J. Davies, *J. Chem. Soc., Perkin Trans II*, 1998, 1937.

615. M. Matsuno and H. Takeuchi, *Bull. Chem. Soc. Jpn.*, 1998, **71**, 851.

616. W. Huang, J. Han, X. Wang, Z. Yu, and Y. Zhang, *Nucl. Instrum. Methods Phys. Res., Sect. B*, 1998, **140**, 137.

617. X. Li and X. Su, *Fenxi Huaxue*, 1999, **27**, 47 (*Chem. Abs.*, 1999, **130**, 110597).

618. A. Soto, A. Arce, M.K. Khoshkbarchi, and J.H. Vera, *Biophys. Chem.*, 1998, **73**, 77.

619. A. Soto, A. Arce, and M.K. Khoshkbarchi, *Biophys. Chem.*, 1998, **74**, 165.

620. Y.K. Lim and R.B. Lal, *J. Mater. Sci. Lett.*, 1998, **17**, 1363.

621. M.K. Khoshkbarchi, A.M. Soto-Campos, and J.H. Vera, *J. Solution Chem.*, 1997, **26**, 941.

622. K. Gekko, E. Ohmae, K. Kameyama, and T. Takagi, *Biochim. Biophys. Acta*, 1998, **1387**, 195.

623. X. Ren, X. Hu, R. Liu, and H. Zong, *J.Chem.Eng.Data*, 1998, **43**, 700.
624. A. A. Pradhan and J. H. Vera, *Fluid Phase Equilib.*, 1998, **152**, 121.
625. Y. Lu, T. Bai, W. Xie, and J. Lu, *Thermochim.Acta*, 1998, **319**, 11.
626. M.M. Cardoso, M.J. Barradas, M.T. Carrondo, K.H. Kroner, and J.G. Crespo, *Bioseparation*, 1998, **7**, 65.
627. V.K. Abrasimov and R.V. Chumakova, *Zh.Fiz.Khim.*, 1998, **72**, 994.
628. P.V. Lapshev and O.V. Kulikov, *Russ.J. Coord.Chem.*, 1998, **24**, 449.
629. S. Andini, G. Castronuovo, V. Elia, and F. Velleca, *J.Chem.Soc., Faraday Trans.*, 1998, **94**, 1271.
630. Y. Lou and R. Lin, *Thermochim.Acta*, 1998, **316**, 145.
631. M.N. Islam and A.A. Khan, *J.Bangladesh Chem.Soc.*, 1996, **9**, 49.
632. U.N. Dash and N.N. Pasupalak, *Indian J.Chem., Sect.A: Inorg., Bio-inorg., Phys., Theor., Anal.Chem.*, 1997, **36A**, 834.
633. Z. Yan, J. Wang, H. Zheng, and D. Liu, *J.Solution Chem.*, 1998, **27**, 473.
634. A.W. Hakin, D.C. Daisley, L. Delgado, J.L. Liu, R.A. Marriott, J.L. Marty, and G. Tompkins, *J.Chem.Thermodyn.*, 1998, **30**, 583.
635. D. Ragouramane and A.S. Rao, *Indian J.Chem., Sect.A: Inorg., Bio-inorg., Phys., Theor., Anal.Chem.*, 1998, **37A**, 659.
636. J.I. Partanen, P.M. Juusola, and P.O. Minkkinen, *Acta Chem.Scand.*, 1998, **52**, 198.
637. J.I. Partanen, *Ber.Bunsen-Ges.*, 1998, **102**, 855.
638. H.A. Saroff, *Anal.Biochem.*, 1998, **257**, 71.
639. P. Alonso, J.L. Barriada, P. Rodriguez, I. Brandariz, and M.E. Sastre de Vicente, *J.Chem.Eng.Data*, 1998, **43**, 876.
640. I. Contineanu and D.I. Marchidan, *Rev.Roum.Chem.*, 1997, **42**, 605.
641. S. Calin, G. Chilom, and C. Telea, *Rev.Roum.Chem.*, 1998, **43**, 101.
642. G.I. Makhatadze, *Biophys.Chem.*, 1998, **71**, 133.
643. M. Kitamura and T. Ishizu, *J.Cryst.Growth*, 1998, **192**, 225.
644. D.K. Kondepudi and M. Culha, *Chirality*, 1998, **10**, 238.
645. S. Mukerji and T. Kar, *Mater.Res.Bull.*, 1998, **33**, 619.
646. V.A. Basiuk, *Surfactant Sci.Ser.*, 1998, **75**(Biopolymers at Interfaces), 55.
647. T. Kawakita, *Kagaku Kogaku*, 1998, **62**, 391.
648. S. Diez, A. Leitao, L. Ferreira, and A. Rodrigues, *Sep.Purif.Technol.*, 1998, **13**, 25.
649. X. Yan, L. Zhu, R. Zhang, and Y. Yang, *Huadong Ligong Daxue Xuebao*, 1998, **24**, 312 (*Chem.Abs.*, 1998, **129**, 216872).
650. G. Simon, L. Hanak, G. Grevillot, T. Szanya, and G. Marton, *Magy.Kem.Lapja*, 1998, **53**, 309; *idem., ibid.*, p.379.
651. P. Hajos, *J.Chromatogr.A*, 1997, **789**, 141.
652. D. Muraviev, *Langmuir*, 1998, **14**, 4169.
653. Z.I. Kuvaeva, *Vestsi Akad.Navuk Belarusi, Ser.Khim.Navuk*, 1997, 6 (*Chem.Abs.*, 1998, **128**, 164467).
654. I. Mohammed, G. Zou, and J. Zhu, *Nanjing Daxue Xuebao*, 1997, **33**, 369 (*Chem.Abs.*, 1998, **128**, 292423).
655. C. Grossmann, R. Tintinger, J. Zhu, and G. Maurer, *Fluid Phase Equilib.*, 1997, **137**, 20; see also K. Mishima, K. Matsuyama, S. Oka, Y. Taruta, S. Takarabe, and M. Nagatani, *Fukuoka Daigaku Kogaku Shuho*, 1998, **60**, 207 (*Chem.Abs.*, 1998, **129**, 149236).
656. N.A. Kelly, M. Lukhezo, B.G. Reuben, L.J. Dunne, and M.S. Verrall, *J.Chem.-Technol.Biotechnol.*, 1998, **72**, 347.

657. C. Martin-Orue, S. Bonhallab, and A. Garem, *J. Membr. Sci.*, 1998, **142**, 225.
658. J.M.K. Timmer, M.P.J. Speelmans, and H.C. Van der Horst, *Sep. Purif. Technol.*, 1998, **14**, 133.
659. M. Minagawa and A. Tanioka, *J. Colloid Interface Sci.*, 1998, **202**, 149.
660. H. Grib, L. Bonnal, J. Sandeaux, R. Sandeaux, C. Gavach, and N. Mameri, *J. Chem. Technol. Biotechnol.*, 1998, **73**, 64.
661. M.H. Yi, S.J. Nam, and S.T. Chung, *Korean J. Chem. Eng.*, 1997, **14**, 263.
662. M. Hebrant and C. Tondre, *Anal. Sci.*, 1998, **14**, 109; erratum, *idem.*, *ibid.*, p.874.
663. V.S. Soldatov, Z.I. Kuraeva, V.A. Bychkova, and L.A. Bodopyanova, *Zh. Fiz. Khim.*, 1998, **72**, 136.
664. D.O. Popescu, L. Mutihac, and T. Constantinescu, *Rev. Roum. Chim.*, 1997, **42**, 907.
665. D.T. Rosa and D. Coucouvanis, *Inorg. Chem.*, 1998, **37**, 2328.
666. L. Mutihac, H.-J. Buschmann, C. Bala, and R. Mutihac, *An. Quim. Int. Ed.*, 1997, **93**, 332.
667. A.P. Hansson, P.-O. Norrby, and K. Warnmark, *Tetrahedron Lett.*, 1998, **39**, 4565.
668. H.J. Buschmann, E. Schollmeyer, and L. Mutihac, *J. Inclusion Phenom. Mol. Recognit. Chem.*, 1998, **30**, 21; *Thermochim. Acta*, 1998, **313**, 189.
669. M. Czekalla, A. Stephan, B. Habermann, J. Trepte, K. Gloe, and F.P. Schmidtchen, *Thermochim. Acta*, 1998, **313**, 137.
670. E.S. Cotner and P.J. Smith, *J. Org. Chem.*, 1998, **63**, 1737.
671. Y. Kubo, N. Hirota, S. Maeda, and S. Tokita, *Anal. Sci.*, 1998, **14**, 183.
672. M. Pietraszkiewicz, P. Prus, and W. Fabianowski, *Pol. J. Chim.*, 1998, **72**, 1068.
673. N. Douteau-Guevel, A.W. Coleman, J.-P. Morel, and N. Morel-Desrosiers, *J. Phys. Org. Chem.*, 1998, **11**, 693.
674. J.A. Gavin, N. Deng, M. Alcala, and T.E. Mallouk, *Chem. Mater.*, 1998, **10**, 1937.
675. G. Luo, H. Liu, J. Huang, X. Peng, and J. Li, *Zhongshan Daxue Xuebao, Ziran Kexueban*, 1997, **36**, 125 (*Chem. Abs.*, 1998, **128**, 308723).
676. H. Tye, C. Eldred, and M. Wills, *J. Chem. Soc., Perkin Trans. II*, 1998, 457.
677. T. Nagamori and K. Toko, *Res. Rep. Inf. Sci. Electr. Eng. Kyushi Univ.*, 1998, **3**, 101 (*Chem. Abs.*, 1998, **129**, 106095).
678. H. Chibvongodze, H. Akiyama, T. Matsuno, and K. Toko, *Res. Rep. Inf. Sci. Electr. Eng. Kyushi Univ.*, 1998, **3**, 45 (*Chem. Abs.*, 1998, **129**, 106183).
679. M. Spiliopoulous, *Chem. Chron., Genike Ekdose*, 1998, **60**, 149.
680. T.S. Niwa and A. Ogino, *THEOCHEM*, 1997, **419**, 155.
681. B. Jiang, T. Guo, L.-W. Peng, and Z.-R. Sun, *Biopolymers*, 1998, **45**, 35.
682. J. Piuggali and J.A. Subirana, *Biopolymers*, 1998, **45**, 149.
683. K. Hanabusa, H. Kobayashi, M. Suzuki, M. Kimura, and H. Shirai, *Colloid Polym. Sci.*, 1998, **276**, 252.
684. L. Schafer, S.Q. Newton, and X. Jiang, in *Molecular Orbital Calculations for Biological Systems*, ed. A.-M. Sapse, Oxford University Press, New York, 1998, p.181.
685. K. Zhang and A. Chung-Phillips, *J. Comput. Chem.*, 1998, **19**, 1862; *J. Phys. Chem. A*, 1998, **102**, 3625.
686. I.A. Topol, S.K. Burt, M. Toscano, and N. Russo, *THEOCHEM*, 1998, **430**, 41.
687. P. Perez and R. Contreras, *Chem. Phys. Lett.*, 1998, **293**, 239.
688. J.R. Sambrano, A.R. de Sousa, J.J. Queralt, J. Andres, and E. Longo, *Chem. Phys. Lett.*, 1998, **294**, 1.

689. M. Kieninger, S. Suhai, and O.N. Ventura, *THEOCHEM*, 1998, **433**, 193.
690. M. Nagaoka, N. Okuyama-Yoshida, and T. Yamabe, *J.Phys.Chem.A*, 1998, **102**, 8202.
691. I. Tunon, E. Silla, C. Millot, M.T.C. Martins-Costa, and M.F. Ruiz-Lopez, *J.Phys.Chem.A*, 1998, **102**, 8673.
692. F.R. Tortonda, J.-L. Pascual-Ahuir, E. Silla, I. Tunon, and F.J. Ramirez, *J.Chem.Phys.*, 1998, **109**, 592.
693. E. Tajkhorshid, K.J. Jalkanen, and S. Suhai, *J.Phys.Chem.B*, 1998, **102**, 5899.
694. Y.A. Borisov, Y.A. Zolotarev, E.V. Laskatelev, and N.F. Myasoedov, *Russ.Chem.Bull.*, 1998, **47**, 1442.
695. Y.A. Zolotarev, E.V. Laskatelev, S.G. Rosenberg, Y.A. Borisov, and N.F. Myasoedov, *Russ.Chem.Bull.*, 1997, **46**, 1536.
696. O. Donini and D.F. Weaver, *J.Comput.Chem.*, 1998, **19**, 1515.
697. F. Betschinger, J. Libman, C.E. Felder, and A. Shanzer, *Chirality*, 1998, **10**, 396.
698. D. Chakraborty and S. Manogaran, *Chem.Phys.Lett.*, 1998, **294**, 56.
699. D. Chakraborty and S. Manogaran, *THEOCHEM*, 1998, **429**, 31.
700. D. Chakraborty and S. Manogaran, *THEOCHEM*, 1998, **429**, 13.
701. P. Tarakeshwar and S. Manogaran, *THEOCHEM*, 1997, **417**, 255.
702. F.J. Ramirez, I. Tunon, and E. Silla, *J.Phys.Chem.B*, 1998, **102**, 6290.
703. N. Rega, M. Cossi, and V. Barone, *J.Am.Chem.Soc.*, 1998, **120**, 5273.
704. V. Carravetta, O. Plashkevych, and H. Agren, *J.Chem.Phys.*, 1998, **109**, 1456.
705. O. Plashkevych, V. Carravetta, O. Vantras, and H. Agren, *Chem.Phys.*, 1998, **232**, 49.
706. A.M. Rodriguez, H.A. Baldoni, F. Surire, R.N. Vazquez, G. Zamarbide, R.D. Enriz, O. Farkas, A. Perczel, M.A. McAllister, L.L. Torday, J.G. Papp, and I.G. Csizmadia, *THEOCHEM*, 1998, **455**, 275.
707. A. Perczel, O. Farkas, I. Jakli, and I.G. Csizmadia, *THEOCHEM*, 1998, **455**, 315.
708. G. Chipot and A. Pohorille, *J.Phys.Chem.B*, 1998, **102**, 281.
709. M. Scarsi, J. Apostolakis, and A. Caflisch, *J.Phys.Chem.B*, 1998, **102**, 3637.
710. N. Gresh, G. Tiraboschi, and D.R. Salahub, *Biopolymers*, 1998, **45**, 405.
711. W.-G. Han, K.J. Jalkanen, M. Elstner, and S. Suhai, *J.Phys.Chem.B*, 1998, **102**, 2587.
712. S.N. Rao, S. Profeta, and V.N. Balaji, *Protein Pept.Lett.*, 1998, **5**, 109.
713. D.R. Artis and M.A. Lipton, *J.Am.Chem.Soc.*, 1998, **120**, 12200.
714. J.J. Queralt, V.S. Safont, V. Moliner, and J. Andres, *Chem.Phys.*, 1998, **229**, 125.
715. S. Ohtani and T. Yamamoto, *Bull.Kanagawa Dent.Coll.*, 1998, **26**, 22 (*Chem.Abs.*, 1998, **129**, 216158).
716. J. Csapo, Z. Csapo-Kiss, and J. Csapo, *Trends Anal.Chem.*, 1998, **17**, 140.
717. T.C. O'Connell, R.E.M. Hedges, and G.J. van Klinken, *Ancient Biomol.*, 1997, **1**, 215.
718. D.P. Glavin and J.L. Bada, *Anal.Chem.*, 1998, **70**, 3119.
719. K. Gamoh and N. Yamasaki, *Bunseki Kagaku*, 1998, **47**, 303 (*Chem.Abs.*, 1998, **129**, 1930).
720. T. Wei and S. Gu, *Xibei Shifan Daxue Xuebao, Ziran Kexueban*, 1998, **34**, 93 (*Chem.Abs.*, 1998, **129**, 216909).
721. V.A. Basiuk, R. Navarro-Gonzalez, and E.V. Basiuk, *J.Anal.Appl.Pyrolysis*, 1998, **45**, 89.
722. V.A. Basiuk, *J.Anal.Appl.Pyrolysis*, 1998, **47**, 127.

723. V.A. Basiuk, R. Navarro-Gonzalez, and E.V. Basiuk, *Origins Life Evol. Biosphere*, 1998, **28**, 167.
724. B. Barbier, A. Chabin, D. Chaput, and A. Brack, *Planet. Space Sci.*, 1998, **46**, 391.
725. J.J. Queralt, V.S. Safont, V. Moliner, and J. Andres, *Chem. Phys. Lett.*, 1998, **283**, 213.
726. J. Franco, X. Rodriguez, J.J. Lamelas, and J.M. Antelo, *Afinidad*, 1998, **55**, 213.
727. M. Bonifacic, I. Stefanic, G.L. Hug, D.A. Armstrong, and K.-D. Asmus, *J. Am. Chem. Soc.*, 1998, **120**, 9930.
728. M.N. Qabar, J.P. Meara, M.D. Ferguson, C. Lum, H.-O. Kim, and M. Kahn, *Tetrahedron Lett.*, 1998, **39**, 5895.
729. M.J. Bunegar, V.C. Dyer, A.P. Green, G.G. Gott, C.M. Jaggs, C.J. Lock, B.J.V. Mead, W.R. Spearing, P.D. Tiffin, N. Tremayne, and M. Woods, *Org. Process Res. Dev.*, 1998, **2**, 334.
730. R. Fernandez-Lafuente, V. Rodriguez, and J.M. Guisan, *Enzyme Microb. Technol.*, 1998, **23**, 28.
731. J. Velisek, T. Davidek, and J. Davidek, *Spec. Publ. Roy. Soc. Chem.*, 1998, **223**(Maillard Reactions in Food and Medicine), 204 (see also, *e.g.*, *Chem. Abs.*, 1998, **129**, 290382-290387).
732. R. Vasiliauskaite, *Chimija*, 1998, 75.
733. S. Ramakrishnan, K.N. Sulochana, and R. Punitham, *Indian J. Biochem. Biophys.*, 1997, **34**, 518.
734. T. Hofmann, *J. Agric. Food Chem.*, 1998, **46**, 3918.
735. F. Hayase, N. Nagashima, T. Koyama, S. Sagara, and Y. Takahashi, in Ref. 731, p. 204 (see also, *e.g.*, *Chem. Abs.*, 1998, **129**, 312213 and 315199).
736. V.L. Voeikov and V.I. Naletov, *Proc. SPIE-Int. Soc. Opt. Eng.*, 1998, **3252**, 140.
737. U. Ragnarsson and L. Grehn, *Acc. Chem. Res.*, 1998, **31**, 494.
738. J. Fitt, K. Prasad, O. Repic, and T.J. Blacklock, *Tetrahedron Lett.*, 1998, **39**, 6991.
739. A. Ferjancic-Biagini, T. Giardina, M. Reynier, and A. Puigserver, *Biocatal. Biotransform.*, 1997, **15**, 313.
740. T. Giard, D. Benard, and J.C. Plaquevent, *Synthesis*, 1998, 297.
741. M. Eriksson, L. Christensen, J. Schmidt, G. Haaima, L. Orgel, and P.E. Nielsen, *New J. Chem.*, 1998, **22**, 1055.
742. B.A. Kulkarni and A. Ganesan, *Tetrahedron Lett.*, 1998, **39**, 4369.
743. T.T. Romoff, L. Ma, Y. Wang, and D.A. Campbell, *Synlett*, 1998. 1341.
744. W.-R. Li and S.-Z. Peng, *Tetrahedron Lett.*, 1998, **39**, 7373.
745. J. Kehler and E. Breuer, *Synthesis*, 1998, 1419.
746. H.A. Moynihan and W. Yu, *Synth. Commun.*, 1998, **28**, 17.
747. C. Kashima, S. Tsuruoka, and S. Mizuhara, *Tetrahedron*, 1998, **54**, 14679.
748. S.P. Raillard, A.D. Mann, and T.A. Baer, *Org. Prep. Proced. Int.*, 1998, **30**, 183.
749. J.G. Breitenbucher, C.R. Johnson, M. Haight, and J.C. Phelan, *Tetrahedron Lett.*, 1998, **39**, 1295.
750. J.S. Yadav, G.S. Reddy, M.M. Reddy, and H.M. Meshram, *Tetrahedron Lett.*, 1998, **39**, 3259.
751. H.M. Meshram, G.S. Reddy, M.M. Reddy, and J.S. Yadav, *Tetrahedron Lett.*, 1998, **39**, 4103.
752. T. Kappes and H. Waldmann, *Carbohydr. Res.*, 1998, **305**, 341.
753. F.M. Rossi, M. Margulis, C.-M. Tang, and J.P.Y. Kao, *J. Biol. Chem.*, 1997, **272**, 32933.

754. A. Nudelman, Y. Bechno, E. Falb, B. Fischer, B.A. Wexler, and A. Nudelman, *Synth.Commun.*, 1998, **28**, 471.

755. J. Ham, K. Choi, J. Ko, H. Lee, and M. Jung, *Protein Pept.Lett.*, 1998, **5**, 257.

756. J.G. Siro, J. Martin, J.L. Garcia-Navio, M.J. Remuinan, and J.J. Vaquero, *Synlett*, 1998, 147.

757. D.S. Bose and V. Lakshminarayana, *Tetrahedron Lett.*, 1998, **39**, 5631.

758. H. Kotsuki, T. Ohishi, T. Araki, and K. Arimura, *Tetrahedron Lett.*, 1998, **39**, 4869.

759. A.J. Zhang, D.H. Russell, J. Zhu, and K. Burgess, *Tetrahedron Lett.*, 1998, **39**, 7439.

760. R.L.E. Furlan and E.G. Mata, *Tetrahedron Lett.*, 1998, **39**, 6421.

761. C. Pothion, J.A. Fehrentz, P. Chevallet, A. Loffet, and J. Martinez, *Lett.-Pept.Sci.*, 1997, **4**, 241.

762. C.T. Brain, A. Hallett, and S.Y. Koo, *Tetrahedron Lett.*, 1998, **39**, 127.

763. J.W. Nieuwenhuijzen, P.G.M. Conti, H.C.J. Ottenheijm, and J.T.M. Linders, *Tetrahedron Lett.*, 1998, **39**, 7811.

764. S.W. Kim, C.Y. Hong, K. Lee, E.J. Lee, and J.S. Koh, *Bioorg.Med.Chem.Lett.*, 1998, **8**, 735.

765. S. Chandrasekhar and S. Mohopatra, *Tetrahedron Lett.*, 1998, **39**, 695.

766. C.P. Chow and C.E. Berkman, *Tetrahedron Lett.*, 1998, **39**, 7471.

767. S. Alibert, D. Crestia, F. Dujols, and M. Mulliez, *Tetrahedron Lett.*, 1998, **39**, 8841.

768. K.M.J. Brands, K. Wiedbrauk, J.M. Williams, U.-H. Dolling, and P.J. Reider, *Tetrahedron Lett.*, 1998, **39**, 9583.

769. D. Ma, Y. Zhang, J. Yao, S. Wu, and F. Tao, *J.Am.Chem.Soc.*, 1998, **120**, 12459.

770. S. Matysiak, T. Boldicke, W. Tegge, and R. Frank, *Tetrahedron Lett.*, 1998, **39**, 1733.

771. J. Spengler and K. Burger, *Synthesis*, 1998, 67.

772. J. Spengler and K. Burger, *J.Chem.Soc., Perkin Trans. I*, 1998, 2091.

773. G.V. Reddy, G.V. Rao, and D.S. Iyengar, *Tetrahedron Lett.*, 1998, **39**, 1985.

774. T. Xin, S. Okamoto, and F. Sato, *Tetrahedron Lett.*, 1998, **39**, 6927.

775. B.M. Trost, T.L. Calkins, C.Oertelt, and J.Zambrano, *Tetrahedron Lett.*, 1998, **39**, 1713.

776. M.E. Humphries, B.P. Clark, and J.M.J. Williams, *Tetrahedron: Asymmetry*, 1998, **9**, 749.

777. A. Puschl, S. Sforza, G. Haaima, O. Dahl, and P.E. Nielsen, *Tetrahedron Lett.*, 1998, **39**, 4707.

778. A.R. Katritzky, X.-L. Cui, B. Yang, and P.J. Steel, *Tetrahedron Lett.*, 1998, **39**, 1697.

779. L. Giraud and P. Renaud, *J.Org.Chem.*, 1998, **63**, 9162.

780. F.L. Zumpe and U. Kazmaier, *Synlett*, 1998, 1199.

781. S. Chayer, E.M. Essassi, and J.-J. Bourguignon, *Tetrahedron Lett.*, 1998, **39**, 841.

782. M.A. Dechantsreiter, F. Burkhart, and H. Kessler, *Tetrahedron Lett.*, 1998, **39**, 253.

783. M. Knollmueller, L. Gaischin, M. Ferencic, M. Noe-Letschnig, U. Girreser, P. Gaertner, K. Mereiter, and C.R. Noe, *Monatsh.Chem.*, 1998, **129**, 1025.

784. M. Vela, J. Saurina, and S. Hernandez-Casson, *Anal.Lett.*, 1998, **31**, 313.

785. E. Deseke, Y. Nakatani, and G. Ourisson, *Eur.J. Org.Chem.*, 1998, 243.

786. J. Vidal, J.-C. Hannachi, G. Hourdin, J.-C. Mulatier, and A. Collet, *Tetrahedron Lett.*, 1998, **39**, 8845.

787. J.M. Herbert, *J.Labelled Compd.Radiopharm.*, 1998, **41**, 859.
788. R.C. Anand and Vimal, *Synth.Commun.*, 1998, **28**, 1963.
789. A. Rodriguez, M. Nomen, B.W. Spur, and J.J. Godfroid, *Tetrahedron Lett.*, 1998, **39**, 8563.
790. J. Thierry, C. Yue, and P. Potier, *Tetrahedron Lett.*, 1998, **39**, 1557.
791. B. Sauvignat, C. Enjalbal, F. Lamaty, R. Lazaro, J. Martinez, and J.-L. Aubagnac, *Rapid Commun.Mass Spectrom.*, 1998, **12**, 1024.
792. P.P. Fantauzzi and K.M. Yager, *Tetrahedron Lett.*, 1998, **39**, 1291.
793. K.G. Dendrinos and A.G. Kalivretenos, *Tetrahedron Lett.*, 1998, **39**, 1321; *J.Chem.Soc., Perkin Trans. I*, 1998, 1463.
794. X. Cameleyre, J.M. Taconnat, A. Guibert, and D. Combes, *Bioprocess Eng.*, 1998, **18**, 361.
795. S. Schergna, T. Da Ros, P. Linda, C. Ebert, L. Gardossi, and M. Prato, *Tetrahedron Lett.*, 1998, **39**, 7791.
796. S. Sano, K. Hayashi, T. Miwa, T. Ishii, M. Fujii, H. Mima, and Y. Nagao, *Tetrahedron Lett.*, 1998, **39**, 5571.
797. J. You, X. Yu, X. Li, Q. Yan, and R. Xie, *Tetrahedron: Asymmetry*, 1998, **9**, 1197.
798. F. Bertoncin, F. Mancin, P. Scrimin, P. Tecilla, and U. Tonellato, *Langmuir*, 1998, **14**, 975.
799. T. Kida, K. Isogawa, W. Zhang, Y. Nakatsuji, and I. Ikeda, *Tetrahedron Lett.*, 1998, **39**, 4339.
800. F. Tanaka, K. Kinoshita, R. Tanimura, and I. Fujii, *Chemtracts*, 1997, **10**, 1039.
801. T. Takarada, R. Takahashi, M. Yashiro, and M. Komiyama, *J.Phys.Org.Chem.*, 1998, **11**, 41.
802. P. Scrimin, P. Tecilla, and U. Tonellato, *Eur.J. Org.Chem.*, 1998, 1143.
803. A.D. Ryabov, G.M. Kazankov, S.A. Kurzeev, P.V. Samuleev, and V.A. Polyakov, *Inorg.Chim.Acta*, 1998, **280**, 57.
804. H.S. Rama and D.P. Kumar, *Colloids Surf., A*, 1998, **142**, 41.
805. P. Krasik, *Tetrahedron Lett.*, 1998, **39**, 4223.
806. G. Reischl, M. El-Mobayed, R. Beisswenger, K. Regier, C. Maichle-Moessmer, and A. Rieker, *Z.Naturforsch., B: Chem.Sci.*, 1998, **53**, 765.
807. G. Karigiannis, C. Athanassopoulos, P. Mamos, N. Karamanos, D. Papaioannou, and G.W. Francis, *Acta Chem.Scand.*, 1998, **52**, 1144.
808. L.A. Carpino, D. Ionescu, A. El-Faham, P. Henklein, H. Wenschuh, M. Bienert, and M. Beyermann, *Tetrahedron Lett.*, 1998, **39**, 241.
809. H.N. Gopi and V.V.S. Babu, *Tetrahedron Lett.*, 1998, **39**, 9769.
810. V.V.S. Babu and H.N. Gopi, *Tetrahedron Lett.*, 1998, **39**, 1049.
811. S.F. Martin, M.P. Dwyer, and C.L. Lynch, *Tetrahedron Lett.*, 1998, **39**, 1517.
812. R.V. Hoffman and J. Tao, *Tetrahedron Lett.*, 1998, **39**, 4195.
813. R. Benhida, M. Devys, J.-L. Fourrey, F. Lecubin, and J.-S. Sun, *Tetrahedron Lett.*, 1998, **39**, 6167.
814. R.C. Anand and Vimal, *Tetrahedron Lett.*, 1998, **39**, 917.
815. J.J. Wen and C.M. Crews, *Tetrahedron: Asymmetry*, 1998, **9**, 1855.
816. G. Karigiannis, P. Mamos, G. Balayiannis, I. Katsoulis, and D. Papaioannou, *Tetrahedron Lett.*, 1998, **39**, 5117.
817. A. Bardou, J.-P. Celerier, and G. Lhommet, *Tetrahedron Lett.*, 1998, **39**, 5189.
818. C. Dallaire and P. Arya, *Tetrahedron Lett.*, 1998, **39**, 5129.
819. C.W. Phoon and C. Abell, *Tetrahedron Lett.*, 1998, **39**, 2655.
820. B.J. Hall and J.D. Sutherland, *Tetrahedron Lett.*, 1998, **39**, 6593.

821. M. Paris, A. Heitz, V. Guerlavais, M. Cristau, J.-A. Fehrentz, and J. Martinez, *Tetrahedron Lett.*, 1998, **39**, 7287.
822. R.P. McGeary, *Tetrahedron Lett.*, 1998, **39**, 3319.
823. M. Paris, C. Pothion, A. Heitz, J. Martinez, and J.-A. Fehrentz, *Tetrahedron Lett.*, 1998, **39**, 1341.
824. S.A. Hart, M. Sabat, and F.A. Etzkorn, *J. Org. Chem.*, 1998, **63**, 7580.
825. J.M. Gardiner and S.E. Bruce, *Tetrahedron Lett.*, 1998, **39**, 1029.
826. H.-O. Kim, *Synth. Commun.*, 1998, **28**, 1713.
827. A. Roy, N.C. Bar, B. Achari, and S.B. Mandal, *Indian J. Chem., Sect. B: Org. Chem., Incl. Med. Chem.*, 1998, **37B**, 644.
828. S.I. Hyun and Y.G. Kim, *Tetrahedron Lett.*, 1998, **39**, 4299.
829. S. Matsuda, D.K. An, S. Okamoto, and F. Sato, *Tetrahedron Lett.*, 1998, **39**, 7513.
830. A.K. Ghosh and J. Cappiello, *Tetrahedron Lett.*, 1998, **39**, 8803.
831. I.A. O'Neil, S. Thompson, C.L. Murray, and S.B. Kalindjian, *Tetrahedron Lett.*, 1998, **39**, 7787.
832. A.R. Katritzky, D. Cheng, and J. li, *J. Org. Chem.*, 1998, **63**, 3438.
833. S. Dankwardt, *Synlett.*, 1998, 761.
834. P.F. Vogt, M.J. Miller, M.J. Mulvihill, S. Ramurthy, G.C. Savela, and A.R. Ritter, *Enantiomer*, 1997, **2**, 367.
835. M. Ohba, H. Kubo, S. Seto, T. Fujii, and H. Ishibashi, *Chem. Pharm. Bull.*, 1998, **46**, 860.
836. M. Falorni, G. Dettori, and G. Giacomelli, *Tetrahedron: Asymmetry*, 1998, **9**, 1419.
837. M. Falorni, G. Giacomelli, and A.M. Spanedda, *Tetrahedron: Asymmetry*, 1998, **9**, 3039.
838. C. Hulme, M.M. Morrissette, F.A. Volz, and C.J. Burns, *Tetrahedron Lett.*, 1998, **39**, 1113.
839. C. Hanusch-Komppa and I. Ugi, *Tetrahedron Lett.*, 1998, **39**, 2725.
840. S.J. Park, G. Keum, S.B. Kang, H.Y. Koh, Y. Kim, and D.H. Lee, *Tetrahedron Lett.*, 1998, **39**, 7109.
841. D. Dunn and S. Chatterjee, *Bioorg. Med. Chem. Lett.*, 1998, **8**, 1273.
842. C.R. Millington, R. Quarrell, and G. Lowe, *Tetrahedron Lett.*, 1998, **39**, 7201.
843. J. Wang and Y. Hou, *J. Chem. Soc., Perkin Trans. I*, 1998, 1919.
844. A.G. Griesbeck, J. Hirt, W. Kramer, and P. Dallakian, *Tetrahedron*, 1998, **54**, 3169; A.G. Griesbeck, A. Henz, W. Kramer, P. Wamser, K. Peters, and E.-M. Peters, *Tetrahedron Lett.*, 1998, **39**, 1549; for reviews see A.G. Griesbeck, *EPA Newsletter*, 1998, **62**, 3; and A.G. Griesbeck, *Chimia*, 1998, **52**, 272.
845. A. Stojanovic, P. Renaud, and K. Schenk, *Helv. Chim. Acta*, 1998, **81**, 268.
846. J. Guitton, F. Tinardon, R. Lamrini, P. Lacan, M. Desage, and A. Francina, *Free Radical Biol. Med.*, 1998, **25**, 340.
847. K. Ogura, M. Kobayashi, M. Nakayama, and Y. Miho, *J. Electroanal. Chem.*, 1998, **449**, 101.
848. S. Sengupta, D.S. Sarma, and S. Mondal, *Tetrahedron*, 1998, **54**, 9791.
849. B.F. Bonini, M. Comes-Franchini, M. Fochi, G. Mazzanti, A. Ricci, and G. Varchi, *Synlett.*, 1998, 1013.
850. H. Tokuyama, S. Yokoshima, T. Yamashita, and T. Fukuyama, *Tetrahedron Lett.*, 1998, **39**, 3189.
851. S. Sengupta and D. Das, *Synth. Commun.*, 1998, **28**, 403.
852. C. Beier, E. Schaumann, and G. Adiwidjaja, *Synlett.*, 1998, 41.

853. C. Palomo, M. Oiarbide, A. Esnal, A. Landa, J.I. Miranda, and A. Linden, *J.Org.Chem.*, 1998, **63**, 5838.
854. D. Cabaret, M. Wakselman, A. Kaba, L. Ilungu, and C. Chany, *Synth.Commun.*, 1998, **28**, 2713.
855. T.J. Deming, *J.Am.Chem.Soc.*, 1998, **120**, 4240.
856. S. Derrer, N. Feeder, S.J. Teat, J.E. Davies, and A.B. Holmes, *Tetrahedron Lett.*, 1998, **39**, 9309.
857. M.A. Blaskovich and M. Kahn, *Synthesis*, 1998, 379.
858. A.G. Brewster, C.S. Frampton, J. Jayatissa, M.B. Mitchell, R.J. Stoodley, and S.Vohra, *Chem.Commun.*, 1998, 299.
859. B. Matuszczak, *J.Prakt.Chem./Chem.-Ztg.*, 1998, **340**, 20.
860. M. Kawase, H. Miyamae, and T. Kurihara, *Chem.Pharm.Bull.*, 1998, **46**, 749.
861. M. Kawase, M. Hirabayashi, H. Koiwai, K. Yamamoto, and H. Miyamae, *Chem.Commun.*, 1998, 641.
862. M. Kawase, H. Koiwai, S. Saito, and T. Kurihara, *Tetrahedron Lett.*, 1998, **39**, 6189.
863. M. Kawase, H. Koiwai, A. Yamano, and H. Miyamae, *Tetrahedron Lett.*, 1998, **39**, 663.
864. W. Bauer, M. Prem, K. Polborn, K. Suenkel, W. Steglich, and W. Beck, *Eur.J. Inorg.Chem.*, 1998, 485.
865. A. Padwa and M. Prein, *Tetrahedron*, 1998, **54**, 6957.
866. M.C. Bagley, R.T. Buck, S.L. Hind, and C.J. Moody, *J.Chem.Soc., Perkin Trans. I*, 1998, 591.
867. H. Matsunaga, T. Santa, T. Iida, T. Fukushima, H. Homma, and K. Imai, *Analyst*, 1997, **122**, 931.
868. P.D. Ribiero, R.S.M. Borges, P.R.R. Costa, E.W. Alves, and O.L.T. Machado, *Protein Pept.Lett.*, 1998, **5**, 251.
869. T. Fujisawa, Y Onogawa, and M. Shimizu, *Tetrahedron Lett.*, 1998, **39**, 6019.
870. A. Gonzalez, J. Granell, J.F. Piniella, and A. Alvarez-Larena, *Tetrahedron*, 1998, **54**, 13313.
871. S. Kiyooka, H. Maeda, M.A. Hena, M. Uchida, C.-S. Kim, and M. Horiike, *Tetrahedron Lett.*, 1998, **39**, 8287.
872. M.R. Seong, H.J. Lee, and J.N. Kim, *Tetrahedron Lett.*, 1998, **39**, 6219.
873. M.R. Seong, H.N. Song, and J.N. Kim, *Tetrahedron Lett.*, 1998, **39**, 7101.
874. N. Mohamed, U. Bhatt, and G. Just, *Tetrahedron Lett.*, 1998, **39**, 8213.
875. M.K. Schwarz, D. Tumelty, and M.A. Gallop, *Tetrahedron Lett.*, 1998, **39**, 8397.
876. A. Nefzi, J.M. Ostresh, M. Giulianotti, and R.A. Houghten, *Tetrahedron Lett.*, 1998, **39**, 8199.
877. J. Bujdak and B.M. Rode, *React.Kinet.Catal.Lett.*, 1997, **62**, 281; H.L. Son, Y. Suwannachot, J. Bujdak and B.M. Rode, *Inorg.Chim.Acta*, 1998, **272**, 89.
878. T.L. Porter, M.P. Eastman, M.E. Hagerman, L.B. Price, and R.F. Shand, *J.Mol.Evol.*, 1998, **47**, 373
879. C. Huber and G. Wachtershauser, *Science*, 1998, **281**, 670.
880. R. Liu and L.E. Orgel, *Origins Life Evol.Biosphere*, 1998, **28**, 47.
881. Y. Suwannachot and B.M. Rode, *Origins Life Evol.Biosphere*, 1998, **28**, 79.
882. R. Liu and L.E. Orgel, *Origins Life Evol.Biosphere*, 1998, **28**, 245.
883. A.R. Hill, C. Bohler, and L.E. Orgel, *Origins Life Evol.Biosphere*, 1998, **28**, 235.
884. L.E. Orgel, *Origins Life Evol.Biosphere*, 1998, **28**, 227.
885. H.A. Kelly, R. Bolton, S.A. Brown, S.J. Coote, M. Dowle, U. Dyer, H. Finch,

D. Golding, A. Lowdon, J. McLaren, J.G. Montana, M.R. Owen, N.A. Pegg, B.C. Ross, R. Thomas, and D.A. Walker, *Tetrahedron Lett.*, 1998, **39**, 6979.

886. J. Matsuo, S. Kobayashi, and K. Koga, *Tetrahedron Lett.*, 1998, **39**, 9723.

887. C.J. Easton and C.A. Hutton, *Synlett.*, 1998, 457.

888. S. Gair and R.F.W. Jackson, *Curr. Org. Chem.*, 1998, **2**, 527.

889. R.F.W. Jackson, R.J. Moore, C.S. Dexter, J. Elliott, and C.E. Mowbray, *J. Org. Chem.*, 1998, **63**, 7875.

890. R.F.W. Jackson, J.L. Fraser, N. Wishart, B. Porter, and M.J. Wythes, *J. Chem. Soc., Perkin Trans. I*, 1998, 1903.

891. B. Schmidt and D.K. Ehlert, *Tetrahedron Lett.*, 1998, **39**, 3999.

892. C.S. Dexter and R.F.W. Jackson, *Chem. Commun.*, 1998, 75.

893. P. Ciapetti, A. Mann, A. Shoenfelder, and M. Taddei, *Tetrahedron Lett.*, 1998, **39**, 3843.

894. L.M. Beal and K.D. Moeller, *Tetrahedron Lett.*, 1998, **39**, 4639.

895. D.L.J. Clive and V.S.C. Yeh, *Tetrahedron Lett.*, 1998, **39**, 4789.

896. G.A. Burley, P.A. Keller, S.G. Pyne, and G.E. Ball, *Chem. Commun.*, 1998, 2539.

897. S.-P. Lu and A.H. Lewin, *Tetrahedron*, 1998, **54**, 15097.

898. S.R. Baker, A.F. Parsons, and M. Wilson, *Tetrahedron Lett.*, 1998, **39**, 2815.

899. C. Dugave, J. Cluzeau, A. Menez, M. Gaudry, and A. Marquet, *Tetrahedron Lett.*, 1998, **39**, 5775.

900. A. Ritzen, B. Basu, S.K. Chattopadhyay, F. Dossa, and T. Frejd, *Tetrahedron: Asymmetry*, 1998, **9**, 503.

901. M.J. Crossley, Y.M. Fung, J.J. Potter, and A.W. Stamford, *J. Chem. Soc., Perkin Trans. I*, 1998, 1113.

902. M.J. Crossley, Y.M. Fung, E. Kyriakopoulos, and J.J. Potter, *J. Chem. Soc., Perkin Trans. I*, 1998, 1123.

903. M. Barbaste, V. Rolland-Fulcrand, M.-L. Roumestant, P. Viallefont, and J. Martinez, *Tetrahedron Lett.*, 1998, **39**, 6287.

904. S.R. Baker, A.F. Parsons, J.-F. Pons, and M. Wilson, *Tetrahedron Lett.*, 1998, **39**, 7197.

905. T. Itaya and Y. Hozumi, *Chem. Pharm. Bull.*, 1998, **46**, 1094.

906. S.C.G. Biagini, S.E. Gibson, and S.P. Keen, *J. Chem. Soc., Perkin Trans. I*, 1998, 2485.

907. S. Kotha and N. Sreenivasachary, *Bioorg. Med. Chem. Lett.*, 1998, **8**, 257.

908. A.S. Ripka, R.S. Bohacek, and D.H. Rich, *Bioorg. Med. Chem. Lett.*, 1998, **8**, 357.

909. J.-M. Campagne and L. Ghosez, *Tetrahedron Lett.*, 1998, **39**, 6175; A.D. Abell, J. Gardiner, A.J. Phillips, and W.T. Robinson, *ibid.*, p.9563.

910. Y. Gao, P. Lane-Bell, and J.C. Vederas, *J. Org. Chem.*, 1998, **63**, 2133.

911. R.M. Williams and J. Liu, *J. Org. Chem.*, 1998, **63**, 2130.

912. D.J. O'Leary, S.J. Miller, and R.H. Grubbs, *Tetrahedron Lett.*, 1998, **39**, 1689.

913. S. Kotha, N. Sreenivasachary, and E. Brahmachary, *Tetrahedron Lett.*, 1998, **39**, 2805.

914. A.D. Piscopio, J.F. Miller, and K. Koch, *Tetrahedron Lett.*, 1998, **39**, 2667.

915. D.W. Knight, A.L. Redfern, and J. Gilmore, *Synlett.*, 1998, 731.

916. D.W. Knight, A.L. Redfern, and J. Gilmore, *Chem. Commun.*, 1998, 2207.

917. A. Girard, C. Greck, and J.P. Genet, *Tetrahedron Lett.*, 1998, **39**, 4259.

918. L.A. Marcaurelle and C.R. Bertozzi, *Tetrahedron Lett.*, 1998, **39**, 7279; L.A. Marcaurelle, E.C. Rodriguez, and C.R. Bertozzi, *ibid.*, 8417.

919. L.B. Wolf, K.C.M.F. Tjen, F.P.J.T. Rutjes, H. Hiemstra, and H.E. Schoemaker, *Tetrahedron Lett.*, 1998, **39**, 5081.

920. M.C. Di Giovanni, D. Misiti, G. Zappia, and G. Delle Monache, *Gazz.Chim. Ital.*, 1997, **127**, 475.
921. A.G.H. Wee and F. Tang, *Can.J. Chem.*, 1998, **76**, 1070.
922. S.A. Mitchell, B.D. Oates, H. Ravazi, and R. Polt, *J.Org.Chem.*, 1998, **63**, 8837; H. Ravazi and R. Polt, *Tetrahedron Lett.*, 1998, **39**, 3371.
923. Y.L. Bennani, G.-D. Zhu, and J.C. Freeman, *Synlett.*, 1998, 754.
924. T. Nakazawa, M. Ishii, H.-J. Musiol, and L. Moroder, *Tetrahedron Lett.*, 1998, **39**, 1381.
925. R. Toplak, J.Svete, and B. Stanovnik, in *Zb.Ref.Posvetovanja Slov.Kem.Dnavi*, Eds. P. Glavic and D. Brodnjak-Voncina, Maribor, Slovenia, 1997, p.19 (*Chem.Abs.*, 1999, **130**, 110169).
926. X.-T. Chen, D. Sames, and S.J. Danishefsky, *J.Am.Chem.Soc.*, 1998, **120**, 7760.
927. P. Benovsky, G.A. Stephenson, and J.R. Stille, *J.Am.Chem.Soc.*, 1998, **120**, 2493.
928. T. Laieb, J. Chastanet, and J. Zhu, *J.Org.Chem.*, 1998, **63**, 1709.
929. M.A. Blaskovich, G. Evindar, N.G.W. Rose, S. Wilkinson, Y. Luo, and G.A. Lajoie, *J.Org.Chem.*, 1998, **63**, 3631; erratum, *ibid.*, p.4560.
930. G. Veeresa and A. Datta, *Tetrahedron Lett.*, 1998, **39**, 8503.
931. J.M. Travins and F.A. Etzkorn, *Tetrahedron Lett.*, 1998, **39**, 9389.
932. J. Jurczak and P. Prokopowicz, *Tetrahedron Lett.*, 1998, **39**, 9835.
933. F.L. Qing, S. Peng, and C.-M. Hu, *J.Fluorine Chem.*, 1998, **88**, 79.
934. E. Branquet, P. Meffre, P. Durand, and F.Le Goffic, *Synth.Commun.*, 1998, **28**, 613.
935. P. Merino, S. Franco, F.L. Merchan, and T. Tejero, *Molecules*, 1998, **3**, 26.
936. C. Hertweck, W. Boland, and H. Goerls, *Chem.Commun.*, 1998, 1955.
937. R.V. Hoffman and J. Tao, *Tetrahedron Lett.*, 1998, **39**, 3953.
938. A.M. Porte, W.A.. van der Donk, and K. Burgess, *J.Org.Chem.*, 1998, **63**, 5262.
939. S.P. East and M.M. Joullie, *Tetrahedron Lett.*, 1998, **39**, 7211.
940. M. Falorni, G. Giacomelli, and E. Spanu, *Tetrahedron Lett.*, 1998, **39**, 9241.
941. S. Cameron and B.P.S. Khambay, *Tetrahedron Lett.*, 1998, **39**, 1987.
942. F.A. Lohse and R. Felber, *Tetrahedron Lett.*, 1998, **39**, 2067.
943. A. Dondoni, A. Massi, and A. Marra, *Tetrahedron Lett.*, 1998, **39**, 6601.
944. A. Dondini, A. Marra, and A. Massi, *Chem.Commun.*, 1998, 1741.
945. M. Oba, T. Terauchi, and K. Nishiyama, *Jpn.J. Deuterium Sci.*, 1998, **7**, 31.
946. M. Oba, T. Terauchi, A. Miyakawa, H. Kamo, and K. Nishiyama, *Tetrahedron Lett.*, 1998, **39**, 1595.
947. J.K. Robinson, V. Lee, T.D.W. Claridge, J.E. Baldwin, and C.J. Schofield, *Tetrahedron*, 1998, **54**, 981.
948. K.K. Schumacher, J. Jiang, and M.M. Joullie, *Tetrahedron: Asymmetry*, 1998, **9**, 47.
949. L. Demange, A. Manez, and C. Dugave, *Tetrahedron Lett.*, 1998, **39**, 1169.
950. S. Murahashi, H. Ohtake, and Y. Imada, *Tetrahedron Lett.*, 1998, **39**, 2765.
951. K.-I. Tanaka and H. Sawanishi, *Tetrahedron: Asymmetry*, 1998, **9**, 71.
952. A. Corruble, J.-Y. Valnot, J. Maddaluno, and P. Duhamel, *J.Org.Chem.*, 1998, **63**, 8266.
953. C. Thomas, U. Ohnmacht, M. Niger, and P. Gmeiner, *Bioorg.Med.Chem.Lett.*, 1998, **8**, 2885; C. Thomas, F. Orecher, and P. Gmeiner, *ibid.*, 1491.
954. J.S. Amato, C. Bagner, R.J. Cvetovich, S. Gomolka, F.W. Hartner, and R. Reamer, *J.Org.Chem.*, 1998, **63**, 9533.
955. S. Hanessian, R. Margarita, A. Hall, and X. Luo, *Tetrahedron Lett.*, 1998, **39**, 5883.

956. B.B. Lohray, S. Baskaran, B.Y. Reddy, and K.S. Rao, *Tetrahedron Lett.*, 1998, **39**, 6555.
957. J.M. Padron, G. Kokotos, T. Martin, T. Markidis, W.A. Gibbons, and V.S. Martin, *Tetrahedron: Asymmetry*, 1998, **9**, 3381.
958. G. Kokotos, J.M. Padron, T. Martin, W.A. Gibbons, and V.S. Martin, *J.Org.Chem.*, 1998, **63**, 3741.
959. F. Gosselin and W.D. Lubell, *J.Org.Chem.*, 1998, **63**, 7463.
960. D. Urban, T. Skrydstrup, and J.-M. Beau, *Chem.Commun.*, 1998, 955.
961. T. Miyazawa, M. Ogura, S. Nakajo, and T. Yamada, *Biotechnol.Tech.*, 1998, **12**, 431.
962. M.A. Bednarek, *J.Pept.Res.*, 1998, **52**, 195.
963. R.M. Werner, L.M. Williams, and J.T. Davis, *Tetrahedron Lett.*, 1998, **39**, 9135.
964. S.-H. Moon and S. Lee, *Synth.Commun.*, 1998, **28**, 3919.
965. M.L. Boys, *Tetrahedron Lett.*, 1998, **39**, 3449.
966. D.J. Cundy, A.C. Donohue, and T.D. McCarthy, *Tetrahedron Lett.*, 1998, **39**, 5125.
967. P. Rossi, F. Felluga, and P. Scrimin, *Tetrahedron Lett.*, 1998, **39**, 7159.
968. J. Hiebl, M. Blanka, A. Guttman, H. Kollmann, K. Leitner, G. Mayrhofer, F. Rovenszky, and K. Winkler, *Tetrahedron*, **54**, 2059.
969. Y.-S. Kim and C.-T. Ho, *J.Food Lipids*, 1998, **5**, 173.
970. P. Ciapetti, M. Faloni, A. Mann, and M. Taddei, *Mol. Online*, 1998, **2**, 86.
971. I. Lang, N. Donze, P. Garrouste, P. Dumy, and M. Mutter, *J.Pept.Sci.*, 1998, **4**, 72.
972. O. Poupardin, C. Greck, and J.-P. Genet, *Synlett.*, 1998, 1279.
973. S. Marchalin, K. Kadlecikova, N. Bar, and B. Decroix, *Synth.Commun.*, 1998, **28**, 3619.
974. S. Conde, P. Lopez-Sarrano, and A. Martinez, *Biotechnol.Lett.*, 1998, **20**, 261.
975. I.G.C. Coutts and R.E. Saint, *Tetrahedron Lett.*, 1998, **39**, 3243.
976. M.B. Smith, *Alkaloids: Chem.Biol.Perspect.*, 1998, **12**, 229.
977. J. van Betsbrugge, W.Van Den Nest, P. Verheyden, and D. Tourwe, *Tetrahedron*, 1998, **54**, 1753.
978. N. Langlois, *Tetrahedron: Asymmetry*, 1998, **9**, 1333.
979. J. Dyer, S. Keeling, and M.G. Moloney, *Chem.Commun.*, 1998, 461.
980. N. Langlois and M.-O. Radom, *Tetrahedron Lett.*, 1998, **39**, 857.
981. O. Calvez, A. Chiaroni, and N. Langlois, *Tetrahedron Lett.*, 1998, **39**, 9447.
982. K.M.J. Brands and L.M. DiMichele, *Tetrahedron Lett.*, 1998, **39**, 1677.
983. J.-D. Charrier, J.E.S. Duffy, P.B. Hitchcock, and D.W. Young, *Tetrahedron Lett.*, 1998, **39**, 2199.
984. M.F. Brana, M. Garranzo, and J. Perez-Castells, *Tetrahedron Lett.*, 1998, **39**, 6569.
985. S. Hanessian and R. Margarita, *Tetrahedron Lett.*, 1998, **39**, 5887.
986. P.J. Murray, I.D. Starkey, and J.E. Davies, *Tetrahedron Lett.*, 1998, **39**, 6721.
987. J.M. Receveus, J. Guiramand, M. Rescaseus, M.-L. Roumestant, P. Viallefont, and J. Martinez, *Bioorg.Med.Chem.Lett.*, 1998, **8**, 127.
988. C.M. Henricksen, J. Nielsen, and J. Villadsen, *J.Antibiot.*, 1998, **51**, 99.
989. C.E. Mills, T.D. Heightman, S.A. Hermitage, M.G. Moloney, and G.A. Woods, *Tetrahedron Lett.*, 1998, **39**, 1025.
990. Y. Li, D.E. Carter, and E.A. Mash, *Synth.Commun.*, 1998, **28**, 2057.
991. A.P. Dicks, E. Li, A.P. Munro, H.R. Swift, and D.L.H. Williams, *Can.J. Chem.*, 1998, **76**, 789.

992. M.R. Kumar and M. Adinarayana, *Indian J.Chem., Sect.A: Inorg., Bioinorg., Phys., Theor., Anal.Chem.*, 1998, **37A**, 346.
993. S. Zuberi, A. Glen, R.C. Hider, and S.S. Bansal, *Tetrahedron Lett.*, 1998, **39**, 7567.
994. J. Darkwa, C.M. Simoyi, and H. Reuben, *J.Chem.Soc., Faraday Trans.*, 1998, **94**, 1971.
995. T. Shiraiwa, M. Ohkubo, M. Kubo, H. Miyazaki, M. Takehata, H. Izawa, K. Nakagawa, and H. Kurokawa, *Chem.Pharm.Bull.*, 1998, **46**, 1364.
996. D. DesMarteau and V. Montanari, *Chem.Commun.*, 1998, 2241.
997. M. Kossenjans and J. Martens, *Tetrahedron: Asymmetry*, 1998, **9**, 1409.
998. M. Yamada, T. Miyajima, and H. Horikawa, *Tetrahedron Lett.*, 1998, **39**, 289.
999. J.L. Bolton, S.B. Turnipseed, and J.A. Thompson, *Chem.-Biol.Interact.*, 1997, **107**, 185.
1000. G.T. Crisp and M.J. Millan, *Tetrahedron*, 1998, **54**, 637, 649.
1001. H. Takikawa, S. Muto, D. Nozawa, A. Kayo, and K. Mori, *Tetrahedron Lett.*, 1998, **39**, 6931.
1002. F. Almqvist, D. Guillaume, S.J. Huttgren, and G.R. Marshall, *Tetrahedron Lett.*, 1998, **39**, 2295.
1003. A. Nefzi, M. Guilianotti, and R.A. Houghten, *Tetrahedron Lett.*, 1998, **39**, 3671.
1004. T.W. Thannhauser, R.W. Sherwood, and H.A. Scheraga, *J.Protein Chem.*, 1998, **17**, 37.
1005. I. Koch and M. Keusgen, *Pharmazie*, 1998, **53**, 668.
1006. M. Kinuta, H. Shimizu, N. Masuoka, J. Ohta, W.-B. Yao, and T. Ubuka, *Amino Acids*, 1997, **13**, 163.
1007. A.M.T.D.P.V. Cabral, A.M.D'A. Rocha Gansalves, and T.M.V.D. Pinho e Melo, *Molecules*, 1998, **3**, 60.
1008. D.-H. Ko, D.J. Kim, C.S. Lyu, I.K. Min, and H.-S. Moon, *Tetrahedron Lett.*, 1998, **39**, 297.
1009. H.L. Holland and F.M. Brown, *Tetrahedron: Asymmetry*, 1998, **9**, 535.
1010. H.A. Zainal, W.R. Wolf, and R.M. Waters, *J.Chem.Technol.Biotechnol.*, 1998, **72**, 38.
1011. K. Hatano, *Chem.Pharm.Bull.*, 1998, **46**, 1337.
1012. S. Nomoto, A. Shimoyama, and S. Shiraishi, *Tetrahedron Lett.*, 1998, **39**, 1009; S. Nomoto, A. Shimoyama, S. Shiraishi, T. Seno, and D. Sahara, *Biosci., Biotechnol., Biochem.*, 1998, **62**, 643.
1013. S. Mahmood, M.A. Malik, M. Motevalli, P.B. Nunn, and P. O'Brien, *Tetrahedron*, 1998, **54**, 5721.
1014. K.J. Kennedy, T.L. Simandan, and T.A. Dix, *Synth.Commun.*, 1998, **28**, 741; B.-C. Chen, S. Shiu, and D.-Y. Yang, *J.Chin.Chem.Soc.*, 1998, **45**, 549 (*Chem.Abs.*, 1998, **129**, 276258).
1015. H.A. Moynihan and W.Yu, *Tetrahedron Lett.*, 1998, **39**, 3349; *Synth.Commun.*, 1998, **28**, 17.
1016. C.R. Nevill and P.T. Angell, *Tetrahedron Lett.*, 1998, **39**, 5671.
1017. S.R. Chhabra, B. Hothi, D.J. Evans, P.D. White, B.W. Bycroft, and W.C. Chan, *Tetrahedron Lett.*, 1998, **39**, 1603.
1018. M. Prabhakaram and V.V. Mossine, *Prep.Biochem.Biotechnol.*, 1998, **28**, 319.
1019. L. Tsai, P.A. Szweda, O. Vinogradova, and L.I. Szweda, *Proc.Natl.Acad.Sci. U.S.A.*, 1998, **95**, 7975.
1020. A. Ruttkat and H.F. Erbersdobler, in Ref.731, p. 94.

1021. R. Nagai, K. Ikeda, Y. Kawasaki, H. Sano, M. Yoshida, T. Araki, S. Ueda, and S. Horiuchi, *FEBS Lett.*, 1998, **425**, 355.

1022. S. Fu, M.-X. Fu, J.W. Baynes, S.R. Thorpe, and R.T. Dean, *Biochem.J.*, 1998, **330**, 233.

1023. G. Xu and L.M. Sayre, *Chem.Res.Toxicol.*, 1998, **11**, 247.

1024. D.A. Slatter, M. Murray, and A.J. Bailey, *FEBS Lett.*, 1998, **421**, 180.

1025. G. Veeresa and A. Datta, *Tetrahedron Lett.*, 1998, **39**, 3069.

1026. J.P. Duneau, N. Garnier, G. Cremel, G. Nuhlans, P. Hubert, D. Genest, M. Vincent, J. Gallay, and M. Genest, *Biophys.Chem.*, 1998, **73**, 109.

1027. C. Malan and C. Morin, *J.Org.Chem.*, 1998, **63**, 8019.

1028. F. Firooznia, C. Gude, K. Chan, and Y. Satoh, *Tetrahedron Lett.*, 1998, **39**, 3985.

1029. H. Nakamura, M. Fujiwara, Y. Yamamoto, *J.Org.Chem.*, 1998, **63**, 7529.

1030. E. Morera, G. Ortar, and A. Varani, *Synth.Commun.*, 1998, **28**, 4279.

1031. D. Filippov, E. Kuyl-Yeheskiely, G.A. van der Marel, G.I. Tesser, and J.H. van Boom, *Tetrahedron Lett.*, 1998, **39**, 3597.

1032. K. Kubo, Y. Ishii, T. Sakurai, and M. Makino, *Tetrahedron Lett.*, 1998, **39**, 4083.

1033. D. Ma and H. Tian, *J.Chem.Soc., Perkin Trans. I*, 1998, 3493.

1034. E. Schwedhelm, J. Sandmann, and D. Tsikas, *J.Labelled Compd.Radiopharm.*, 1998, **41**, 773.

1035. K. Farah and N. Farouk, *J.Labelled Compd.Radiopharm.*, 1998, **41**, 255.

1036. T.R. Burke, Z.-J. Yao, H. Zhao, G.W.A. Milne, L. Wu, Z.-Y. Zhang, and J.H. Voigt, *Tetrahedron*, 1998, **54**, 9981.

1037. W. Mueller, M. Baenziger, and P. Kipfer, *Helv.Chim.Acta*, 1998, **81**, 729.

1038. F. Dolle, S. Demphel, F. Hinnen, D. Fournier, F. Vaufrey, and C. Crouzel, *J.Labelled Compd.Radiopharm.*, 1998, **41**, 105.

1039. C.P. Szajek, M.A. Channing, and W.C. Eckelman, *Appl.Radiat.Isot.*, 1998, **49**, 795.

1040. D. Ma and W. Tang, *Tetrahedron Lett.*, 1998, **39**, 7369.

1041. A. Bigot and J. Zhu, *Tetrahedron Lett.*, 1998, **39**, 551.

1042. Z.-W. Guo, K. Machiya, G.M. Salamonczyk, and C.J. Sih, *J.Org.Chem.*, 1998, **63**, 4269; Y.-A. Ma, Z.-W. Guo, and C.J. Sih, *Tetrahedron Lett.*, 1998, **39**, 9357.

1043. Z.-W. Guo, K.Machiya, Y.-A. Ma, and C.J. Sih, *Tetrahedron Lett.*, 1998, **39**, 5679.

1044. A. Hutinec, A. Ziogas, M. El-Mobayed, and A. Rieker, *J.Chem.Soc., Perkin Trans. I*, 1998, 2201.

1045. N.A. Braun, M.A. Ciufolini, K.Peters, and E.-M. Peters, *Tetrahedron Lett.*, 1998, **39**, 4667.

1046. G. Cohen, S. Yakushin, and D. Dembiec-Cohen, *Anal.Biochem.*, 1998, **263**, 232.

1047. U. Takahama and K. Yoshitama, *J.Plant.Res.*, 1998, **111**, 97.

1048. H. Zhang, G.L. Squadrito, and W.A. Pryor, *Nitric Oxide*, 1997, **1**, 301.

1049. E.-K. Kim, H. Choi, and E.-S. Lee, *Arch.Pharmacol.Res.*, 1998, **21**, 330.

1050. B.K. Handa and C.J. Hobbs, *J.Pept.Sci.*, 1998, **4**, 138.

1051. D.A. Evans, J.L. Katz, and T.R. West, *Tetrahedron Lett.*, 1998, **39**, 2937.

1052. A. Basak, M.K. Nayak, and A.K. Chakraborti, *Tetrahedron Lett.*, 1998, **39**, 4883.

1053. H. Sajiki, H. Kuno, and K. Hirota, *Tetrahedron Lett.*, 1998, **39**, 7127.

1054. J. Bodi, Y. Nishiuchi, H. Nishio, T. Inui, and T. Kimura, *Tetrahedron Lett.*, 1998, **39**, 7117.

1055. P.G. Besant, L. Byrne, G. Thomas, and P.V. Attwood, *Anal.Biochem.*, 1998, **258**, 372.

1056. M.M.W. McLachlan and C.M. Taylor, *Tetrahedron Lett.*, 1998, **39**, 3055.
1057. K. Iterbeke, G. Laus, P. Verheyden, and D. Tourwe, *Lett.Pept.Sci.*, 1998, **5**, 121.
1058. T. Wang, P. Lu, J. Li, and J.M. Cook, *Tetrahedron Lett.*, 1998, **39**, 8009.
1059. F. He, B.M. Foxman, and B.B. Snider, *J.Am.Chem.Soc.*, 1998, **120**, 6417.
1060. A. van Loevezin, J.H. van Maarseveen, K. Stegman, G.M. Visser, and G.-J. Koomen, *Tetrahedron Lett.*, 1998, **39**, 4737.
1061. W.-C. Chen and M.M. Joullie, *Tetrahedron Lett.*, 1998, **39**, 8401.
1062. B. Biggs, A.L. Presley, and D.L. van Vranken, *Bioorg.Med.Chem.*, 1998, **6**, 975.
1063. P.B. Holst, U. Anthoni, C. Christophersen, S. Larsen, P.H. Nielsen, and A. Puschl, *Acta Chem.Scand.*, 1998, **52**, 683.
1064. U. Anthoni, C. Christophersen, P.H. Nielsen, M.W. Christoffersen, and D. Sorensen, *Acta Chem.Scand.*, 1998, **52**, 958.
1065. M. Van de Weert, F. M. Lagerwerf, J. Haverkamp, and W. Heerma, *J.Mass Spectrom.*, 1998, **33**, 884.
1066. G.S. Harms, S.W. Pauls, J.F. Hedstrom, and C.K. Johnson, *J.Fluoresc.*, 1997, **7**, 273.
1067. G.S. Harms, S.W. Pauls, J.F. Hedstrom, and C.K. Johnson, *J.Fluoresc.*, 1997, **7**, 283.
1068. W. Wiczk, K. Lankiewicz, C. Czaplewski, S. Oldziej, K. Stachowiak, A. Michniewicz, B. Micewicz, and A. Liwo, *J.Fluoresc.*, 1997, **7**, 257.
1069. V.I. Emelyanenko and E.A. Burshtein, *J.Appl.Spectrosc.*, 1998, **65**, 372.
1070. J.K. Lee and R.T. Ross, *J.Phys.Chem.B*, 1998, **102**, 4612.
1071. B. Zelent, J. Kusba, I. Gryczynski, M.L. Johnson, and J.R. Lakowicz, *Biophys.Chem.*, 1998, **73**, 53.
1072. R. Weinkauf and J. Schiedt, *Photochem.Photobiol.*, 1997, **66**, 569.
1073. C. McCaul and R.D. Ludescher, *Proc.SPIE-Int.Soc.Opt.Eng.*, 1998, **3256**(Advances in Optical Biophysics), 263.
1074. J. Guharay, B. Sengupta, and P.K. Sengupta, *Spectrochim.Acta, Part A*, 1998, **54A**, 185.
1075. P. Cioni, L. Erijman, and G.B. Strambini, *Biochem.Biophys.Res.Commun.*, 1998, **248**, 347.
1076. K. Endo, K. Hirayama, Y. Aota, K. Seya, H. Asakura, and K. Hisamichi, *Heterocycles*, 1998, **47**, 865.
1077. L. Celewicz, *J.Photochem.Photobiol.*, 1998, **43**, 66.
1078. C. Lagercrantz, *Acta Chem.Scand.*, 1998, **52**, 37.
1079. K. Nakagawa, T. Mochida, T. Okamoto, S. Saijoh, S. Ueji, T. Amakawa, T. Yamada, and H. Onuki, *Kobe Daigaku Hattatsu Kagakubu Kenkyu Kiyo*, 1998, **5**, 709 (*Chem.Abs.*, 1998, **129**, 105693).
1080. M. Jonsson and H.-B. Kraatz, *J.Chem.Soc.,Perkin Trans. II*, 1997, 2673.
1081. C.J. Koch, in *Radioprotectors*, Eds. E.A. Bump and K. Malaker, CRC Publishing Co., Boca Raton, Florida, 1998, p.25.
1082. S. Robinson, R. Bevan, J. Lunec, and H. Griffiths, *FEBS Lett.*, 1998, **430**, 297.
1083. N. Getoff, S. Solar, and G. Lubec, *Life Sci.*, 1998, **63**, 1469.
1084. X. Fang, F. Jin, H. Jin, and C. von Sonntag, *J.Chem.Soc., Perkin Trans. II*, 1998, 259.
1085. B. Hobel and C. von Sonntag, *J.Chem.Soc., Perkin Trans. II*, 1998, 509.
1086. E. Winkler, P. Etchegoin, A. Fainstein, and C. Fainstein, *Phys.Rev.B: Condens.-Matter Mater.Phys.*, 1998, **57**, 13477.
1087. Z.P. Zagorski and K. Sehested, *J.Radioanal.Nucl.Chem.*, 1998, **232**, 139.
1088. W. Huang, Z. Yu, and Y. Zhang, *Chem.Phys.*, 1998, **237**, 223.

1089. W. Huang, Z. Yu, and Y. Zhang, *Nucl. Instrum. Methods, Phys. Res., Sect. B*, 1998, **134**, 202.

1090. A.J. Smith, *Methods Enzymol.*, 1998, **289**(Solid Phase Peptide Synthesis), 419.

1091. I. Molnar-Perl, *J. Chromatogr. Libr.*, 1998, **60**(Advanced Chromatographic and Electromigration Methods in Biosciences), 415.

1092. M. Weiss, M. Manneberg, J.-F. Juranville, H.-W. Lahm, and M. Fountoulakis, *J. Chromatogr., A*, 1998, **795**, 263.

1093. J.-F. Juranville, B. Poeschl, G. Oesterhelt, H.-J. Schoenfeld, and M. Fountoulakis, *Amino Acids*, 1998, **15**, 253.

1094. M.P. Colombini, R. Fuoco, A. Giacomelli, and B. Muscatello, *Stud. Conserv.*, 1998, **43**, 33.

1095. G.C. Kite and M.J. Hughes, *Phytochem. Anal.*, 1997, **8**, 294.

1096. Z. Chen, P. Landman, T.D. Colmer, and M.A. Adams, *Anal. Biochem.*, 1998, **259**, 203.

1097. S.M. Liu and S. Figliomeni, *Rapid Commun. Mass Spectrom.*, 1998, **12**, 1199.

1098. W. Kulik, J.A.N. Meesterburrie, C. Jakobs, and K. de Meer, *J. Chromatogr., B: Biomed. Appl.*, 1998, **710**, 37.

1099. M. Heil, F. Podebrad, T. Beck, A. Mosandl, A.C. Sewell, and H. Bohles, *J. Chromatogr., B: Biomed. Appl.*, 1998, **714**, 119.

1100. J.R. Crowley, K. Yarasheski, C. Leeuwenburgh, J. Turk, and J.W. Heinecke, *Anal. Biochem.*, 1998, **259**, 127.

1101. J.C. Laubscher, *Schweiz. Lab.-Z.*, 1998, **55**, 144.

1102. J. Pietzsch and A. Pixa, *Clin. Chem.*, 1998, **44**, 1781.

1103. A. Briddon, *Amino Acids*, 1998, **15**, 235.

1104. R.A. Bank, B. Beckman, R. Tenni, and J.M. Te Koppele, *J. Chromatogr., B: Biomed. Appl.*, 1997, **703**, 267.

1105. J.W. Lefevre, N.J. Bonzagni, L.L. Chappell, D.J. Clement, J.R. Albro, and D.M. Lezynski, *J. Labelled Compd. Radiopharm.*, 1998, **41**, 477.

1106. R. Bhushan and R. Agarwal, *Biomed. Chromatogr.*, 1998, **12**, 322.

1107. I.I. Malakhova, V.D. Krasikov, E.V. Degterev, E.V. Kuznetsov, and B.V. Tyaglov, *Biotekhnologiya*, 1996, 27 (*Chem. Abs.*, 1998, **128**, 179374).

1108. B. Widner, W.R. Werner, H. Schennach, H. Wachter, and D. Fuchs, *Clin. Chem.*, 1997, **43**, 2424.

1109. K.H. Schulpis, G.A. Karikas, and G. Kokotos, *Clin. Chem.*, 1998, **44**, 178.

1110. H. Kaur, L. Lyras, P. Jenner, and B. Halliwell, *J. Neurochem.*, 1998, **70**, 2220.

1111. H. Liu, C.T. Duda, T. Huang, W.O. Aruda, and P.T. Kissinger, *J. Chromatogr., A*, 1998, **818**, 69; H. Liu, T. Huang, C.B. Kissinger, and P.T. Kissinger, *J. Chromatogr., B: Biomed. Appl.*, 1998, **713**, 289.

1112. Y. Qu, L. Moons, and F. Vandesande, *J. Chromatogr., B: Biomed. Appl.*, 1997, **704**, 351.

1113. B.P. Solomon and C.T. Duda, *Curr. Sep.*, 1998, **17**, 3.

1114. T. Togawa, A. Ohsawa, K. Kawanabe, and S. Tanabe, *J. Chromatogr., B: Biomed. Appl.*, 1997, **704**, 83.

1115. J.S. Soblosky, L.L. Colgin, C.M. Parrish, J.F. Davidson, and M.E. Carey, *J. Chromatogr., B: Biomed. Appl.*, 1998, **712**, 31.

1116. L.S. Ahmed, H. Moorehead, C.A. Leitch, and E.A. Liechty, *J. Chromatogr., B: Biomed. Appl.*, 1998, **710**, 27.

1117. N.H. Afdhal, A.P. Keaveny, T.B. Cohen, D.P. Nunes, N. Maldonado, M. O'Brien, and P.J. Stone, *J. Hepatol.*, 1997, **27**, 993 (*Chem. Abs.*, 1998, **128**, 138283).

1118. M. Power, *Biochem. Soc. Trans.*, 1998, **26**, 45.

1119. J. Weng, Y. Liao, and B. Yu, *Sepu*, 1997, **15**, 521 (*Chem. Abs.*, 1998, **128**, 138187).

1120. R.A. Bank, B. Beekman, N. Verzijl, J.A.D.M. de Roos, A.N. Sakkee, and J.M. Te Koppele, *J. Chromatogr., B: Biomed. Appl.*, 1997, **703**, 37.

1121. I.T. James and D. Perrett, *J. Chromatogr., A*, 1998, **798**, 159.

1122. M.J. Seibel, H.W. Woitge, B. Auler, C.Kissling, and R.Ziegler, *Clin. Lab.*, 1998, **44**, 129; *idem., ibid.*, p. 269.

1123. C. Tallarico, S. Pace, and A. Longo, *Rapid Commun. Mass Spectrom.*, 1998, **12**, 403.

1124. E. Schmidt-Sommerfeld, P.J. Bobrowski, D. Penn, W.J. Rhead, R.J.A. Wanders, and M.J. Bennett, *Pediatr. Res.*, 1998, **44**, 210.

1125. J. Vahatalo, J. Tuominen, J. Kokkonen, O. Kriz, S.-L. Karonen, and M. Kallio, *Rapid Commun. Mass Spectrom.*, 1998, **12**, 1118.

1126. S.M. Bird, H. Ge, P.C. Uden, J.F. Tyson, E. Block, and E. Denoyer, *J. Chromatogr., A*, 1997, **789**, 349.

1127. P. Khan, *Anal. Commun.*, 1998, **35**, 37.

1128. D. Tsikas, W. Junker, and J.C. Frolich, *J. Chromatogr., B: Biomed. Appl.*, 1998, **705**, 174.

1129. G. Noctor and C.H. Foyer, *Anal. Biochem.*, 1998, **264**, 98.

1130. G. Shang, Y. Jiang, H. Tang, X. Fan, S. Wang, G. Dong, and J. Wang, *Sepu*, 1997, **15**, 474 (*Chem. Abs.*, 1998, **128**, 125453).

1131. D. Mu, J. Li, Z. Liu, G. Li, J. Liu, and Z. Zhou, *Fenxi Ceshi Xuebao*, 1998, **17**, 18 (*Chem. Abs.*, 1999, **130**, 49325).

1132. N. Takeda, K. Matsuhisa, K. Hasebe, N. Kitajima, and T. Miura, *J. Chromatogr., B: Biomed. Appl.*, 1998, **718**, 235.

1133. I. Molnar-Perl and I. Bozar, *J. Chromatogr., A*, 1998, **798**, 37.

1134. J. Kehr, *J. Chromatogr., B: Biomed. Appl.*, 1997, **708**, 27.

1135. M. Eskinja, G. Lamprecht, G. Scherer, and E.R. Schmidt, *J. Chromatogr., B: Biomed. Appl.*, 1997, **704**, 159.

1136. C. Hammel, A. Kyriakopoulos, U. Rosick, and D. Behne, *Analyst*, 1997, **122**, 1359.

1137. S.P. Mendez, E.B. Gonzalez, H.L. Fernandez Sanchez, and A.S. Medel, *J. Anal. At. Spectrom.*, 1998, **13**, 893.

1138. M. Troebs, T. Renner, and G. Scherer, *Chromatographia*, 1998, **48**, 506.

1139. V.C. Dias, F.J. Bamforth, M. Tesonovic, M.E. Hyndman, H.G. Parsons, and G.S. Cembrowski, *Clin. Chem.*, 1998, **44**, 2199.

1140. D. Zhang, R. Shelby, M.A. Savka, Y. Dessaux, and M. Wilson, *J. Chromatogr., A*, 1998, **813**, 247.

1141. S. Al-Kindy, T. Santa, T. Fukushima, H. Homma, and K. Imai, *Biomed. Chromatogr.*, 1998, **12**, 276.

1142. K. Matsuda, S. Matsuda, and Y. Ito, *J. Chromatogr., A*, 1998, **808**, 95.

1143. E. Peyrin and Y.C. Guillaume, *Chromatographia*, 1998, **48**, 431.

1144. J.B. Esquivel, C. Sanchez, and M.J. Fazio, *J. Liq. Chromatogr. Relat. Technol.*, 1998, **21**, 777.

1145. W.H. Pirkle and W. Lee, *Bull. Korean Chem. Soc.*, 1998, **19**, 1277 (*Chem. Abs.*, 1999, **130**, 95796).

1146. M.H. Hyun, J.-S. Jin, and W. Lee, *J. Chromatogr., A*, 1998, **822**, 155.

1147. W. Lee, *Bull. Korean Chem. Soc.*, 1998, **19**, 715; *Anal. Lett.*, 1997, **30**, 2791.

1148. K.S. Yoo and L.M. Pike, *Sci. Hortic.*, 1998, 75, 1.

1149. Z. Wu and W.G. Tong, *J. Chromatogr., A*, 1998, **805**, 63.

1150. L. Vallorani, F. Palma, R. De Bellis, G. Piccoli, C. Sacconi, L. Cucchiarini, and V. Stocchi, *Anal. Biochem.*, 1998, **258**, 376.
1151. J.-M. You, X.-J. Fan, Q. Zhu, and Y. Su, *Anal. Chim. Acta*, 1998, **367**, 69.
1152. J. Bjorklund, S. Einarsson, A. Engstrom, A. Grzegorczyk, H.-D. Becker, and B. Josefsson, *J. Chromatogr., A*, 1998, **798**, 1.
1153. X. Kuang and J. Wu, *Sichuan Daxue Xuebao, Ziran Kexueban*, 1998, **35**, 491 (*Chem. Abs.*, 1999, **130**, 107052).
1154. T. Iida, H. Matsunaga, T. Santa, T. Fukushima, H. Homma, and K. Imai, *J. Chromatogr., A*, 1998, **813**, 267.
1155. D. Jin, K. Nagakura, S. Murofushi, T. Miyahara, and T. Toyo'oka, *J. Chromatogr., A*, 1998, **822**, 215.
1156. O.P. Kleidernigg and W. Lindour, *J. Chromatogr., A*, 1998, **795**, 251.
1157. K. Fujii, *LC/MS No Jissai*, 1996, 154 (*Chem. Abs.*, 1999, **130**, 106999); K. Fujii, T. Shimoya, Y. Ikai, H. Oka, and K. Harada, *Tetrahedron Lett.*, 1998, **39**, 2579.
1158. A. Pastore, R. Massoud, C. Motti, A. Lo Russo, G. Fucci, C. Cortese, and G. Federici, *Clin. Chem.*, 1998, **44**, 825.
1159. Z.K. Shihabi, M.E. Hinsdale, and P.E. Lantz, *J. Liq. Chromatogr., Relat. Technol.*, 1998, **21**, 2715.
1160. E. Kaniowska, G. Chwatko, R. Glowacki, P. Kubalczyk, and E. Bald, *J. Chromatogr., A*, 1998, **798**, 27.
1161. D. Tsikas, J. Sandmann, M. Ikic, J. Fauler, D.O. Stichtenoth, and J.C. Frolich, *J. Chromatogr., B: Biomed. Appl.*, 1998, **708**, 55.
1162. R.K. Goldman, A.A. Vlessis, and D.D. Trunkey, *Anal. Biochem.*, 1998, **259**, 98.
1163. K.D. Attria, N.W. Smith, and C.H. Turnbull, *J. Chromatogr., B: Biomed. Appl.*, 1998, **717**, 341.
1164. S. Fanali, *J. Chromatogr., A*, 1997, **792**, 227; see also H. Nishi, *ibid.*, p.327.
1165. M. Kinuta, H. Shimiz, and T. Ubuka, *J. Chromatogr., A*, 1998, **802**, 73.
1166. S. Giorgieri, A. D'Antuono, K. Panak, O. Ruiz, and L. Diaz, *J. Capillary Electrophor.*, 1997, **4**, 257.
1167. Z. Chen, J.-M. Liu, K. Uchiyama, and T. Hobo, *J. Chromatogr., A*, 1998, **813**, 369.
1168. Z. Shen and L. Wu, *Zhonghua Yikue Jianyan Zazhi*, 1998, **21**, 151 (*Chem. Abs.*, 1999, **130**, 22384).
1169. M. Chiari, V. Desperati, E. Manera, and R. Longhi, *Anal. Chem.*, 1998, **70**, 4967.
1170. C.-F. Tsai and H.-M. Chang, *J. Agric. Food Chem.*, 1998, **46**, 979; H.-M. Chang, C.-F. Tsai, and C.-F. Li, *ibid.*, p.4598.
1171. J.-M. Lin, K. Uchiyama, and T. Hobo, *Chromatographia*, 1998, **47**, 625.
1172. B.S. Seidel and W. Faubel, *J. Chromatogr., A*, 1998, **817**, 223.
1173. P.D. Floyd, L.L. Moroz, R. Gillette, and J.V. Sweedler, *Anal. Chem.*, 1998, **70**, 2243.
1174. H. Wang, H. Hu, S. Shang, T. Ding, J. Gu, and R. Fu, *Fenxi Huaxue*, 1998, **26**, 1161 (*Chem. Abs.*, 1999, **130**, 1933).
1175. R. Zhu and W.T. Kok, *J. Chromatogr., A*, 1998, **814**, 213.
1176. E. Causse, R. Terrier, S. Champagne, M. Nertz, P. Valdiguie, R. Salvayre, and F. Couderc, *J. Chromatogr., A*, 1998, **817**, 181.
1177. S. Tucci, C. Pinto, J. Gogo, P. Rada, and L. Hernandez, *Clin. Biochem.*, 1998, **37**, 143.
1178. Y.-M. Liu, M. Schneider, C.M. Sticha, T. Toyooka, and J.V. Sweedler, *J. Chromatogr., A*, 1998, **800**, 345.
1179. M. Wado, N. Kuroda, S. Akiyama, and K. Nakashima, *Anal. Sci.*, 1997, **13**, 945.

1180. U.E. Spichiger-Keller, *Chemical Sensors and Biosensors for Medical and Biological Applications*, VCH, Weinheim, 1997.
1181. V.V. Sorochinskii and B.I. Kurganov, *Prikl. Biokhim. Microbiol.*, 1997, **33**, 579.
1182. T. Yao, S. Suzuki, T. Nakahara, and H. Nishino, *Talanta*, 1998, **45**, 917.
1183. O. Niwa, T. Horiuchi, R. Kurita, H. Tabei, and K. Torimitsu, *Anal. Sci.*, 1998, **14**, 947.
1184. L. Bucsis, *Labor-Praxis*, 1998, **22**, 84, 86.
1185. M.Gilis, H. Durliat, and M. Comtat, *Anal. Chim. Acta*, 1997, **355**, 235.
1186. S. Nakata, Y. Hirata, R. Takitani, and K. Yoshikawa, *Chem. Lett.*, 1998, 401.
1187. E. Valero and F. Garcia-Carmona, *Anal. Biochem.*, 1998, **259**, 265.
1188. J.C.A. ter Steege, L. Koster-Kamphuis, E.A. van Straaten, P.P. Forget, and W.A. Buurman, *Free Radical Biol. Med.*, 1998, **25**, 953.
1189. V. Ogwu and G. Cohen, *Free Radical Biol. Med.*, 1998, **25**, 362.
1190. V.B. Gavrilov, A.V. Lychkovskii, E.P. Shostack, and S.V. Konev, *J. Appl. Spectrosc.*, 1998, **65**, 379.
1191. T.G. Rosano, R.T. Peaston, H.G. Bone, H.W. Woitge, R.M. Francis, and M.J. Seibel, *Clin. Chem.*, 1998, **44**, 2126.
1192. M. Yang, L. Li, M. Feng, and J. Lu, *Yaowu Fenxi Zazhi*, 1998, **18**, 41 (*Chem. Abs.*, 1998, **129**, 13275).
1193. A.P. De Silva, H.Q.N. Gunaratne, and T. Gunnlaugsson, *Tetrahedron Lett.*, 1998, **39**, 5077.
1194. S. Pfeiffer, A. Schrammel, K. Schmidt, and B. Mayer, *Anal. Biochem.*, 1998, **258**, 68.
1195. R. Stingele, D.A. Wilson, R.J. Traystman, and D.F. Hanley, *Am. J. Physiol.*, 1998, **274**, H1698.
1196. R.-I. Stefan, J.F. Van Staden, and H.Y. Aboul-Enein, *Anal. Lett.*, 1998, **31**, 1787.

2
Peptide Synthesis

BY DONALD T. ELMORE

1 Introduction

Rather fewer reviews have been located this year compared with the previous Report.[1] Some reviews[2][6] relate to several sections of this Report whereas others are cognate to particular sections as follows: Section 2.1,[7] Section 2.4,[8,9] Section 2.5,[10] Section 2.6,[11][26] Section 2.7,[27,28] Section 3.3,[29,30] Section 3.5,[31][34] Section 3.7,[35] Section 3.9,[36][39] and Section 3.10.[40][43]

2 Methods

2.1 Amino-group Protection – The Boc derivatives of the stereoisomeric 4-fluoro- and the 4-difluoro-prolines have been synthesized and converted into the Fmoc derivatives, but no peptides were described.[44] Fmoc derivatives of amino acids and dipeptides linked to Wang resin are easily converted into the corresponding Boc analogues using either KF/Boc-S-2-mercapto-4,6-dimethylpyrimidine or KF/(Bu^{t}OCO)$_2$O.[45] Although the need for this transformation may not be instantly apparent, the authors suggest that the procedure is valuable since Boc peptides can be detached from resins used for SPPS under mild conditions using Me$_3$SnOH. Boc groups can be removed from substrates that are attached to Rink's amide resin by treatment with CF$_3$SO$_3$SiMe$_3$/2,6-lutidine.[46] Boc groups can also be cleanly removed by reaction with NaI under neutral conditions.[47] The cyclohexyloxycarbonyl (Choc) group is suitable for N-protection since yields are high and stereochemical purity is retained in peptide synthesis.[48] The group is resistant to treatment with CF$_3$SO$_3$SiMe$_3$/CF$_3$CO$_2$H/PhSMe at 0 °C for 1 hr but is cleaved by HF. It is a useful orthogonal group in conjunction with Bzl- or ClZ-functions.

An efficient method for the preparation of Fmoc amino acids has been described.[49] Amino acids are treated with CF$_3$CONMeSiMe$_3$ in refluxing CH$_2$Cl$_2$ until a clear solution is obtained. Fmoc-Suc is added and the solution is stirred at room temperature to give the silylated Fmoc acid which is not isolated but treated with MeOH to give the Fmoc amino acid. Fmoc and 2-(4-nitrophenyl-sulfenyl)ethoxycarbonyl (Nsc) groups have been compared in the SPPS of some difficult sequences.[50] Little difference was found by examina-

Amino Acids, Peptides and Proteins, Volume 31

tion of the HPLC profiles of the products. $CaCl_2$ dramatically increases the lifetime of the Fmoc group in alkaline solutions in aqueous Pr^iOH, but it has little effect on hydrolytic reactions.[51] Consequently, it is possible to saponify Me or Bzl esters of Fmoc peptides without the risk of removing the Fmoc group. Perhaps more importantly, Fmoc peptides can be released after SPPS using an *N*-acylurea-based linker and $CaCl_2$-catalysed hydrolysis. The converse objective of removing an Fmoc group without affecting a thioester group in a peptide can be achieved using a mixture of 1-methylpyrrolidine (25% v/v), hexamethyleneimine (2% v/v) and HOBt (2% v/v) in a mixture of 1-methyl-pyrrolid-5-one and Me_2SO (1:1 v/v).[52] The 1-methylpyrrolidine is a sufficiently strong base to remove the Fmoc group, hexamethyleneimine is an efficient nucleophile for scavenging dibenzofulvene and HOBt efficiently suppresses aminolysis by the last reagent. Using this technique, the authors synthesized a partial sequence of verotoxin by SPPS starting from Fmoc-Gly-SCMe$_2$CH$_2$-CONH-resin. The synthesis of peptides containing β-amino acids is currently of interest so the synthesis of Fmoc derivatives of β-amino acids from the corresponding α-analogues by the classical Arndt-Eistert method of homologation is a useful technique.[53]

Perhaps interest in the use of phthaloyl protection of amino groups will be revived by the introduction of a new reagent, 3-chloro-3-(dimethoxyphosphoryl)-isobenzofuran-1(3H)-one for introducing the group[54] (Scheme 1). The potentiality of the Dde group for *N*-protection has been studied. An unprotected

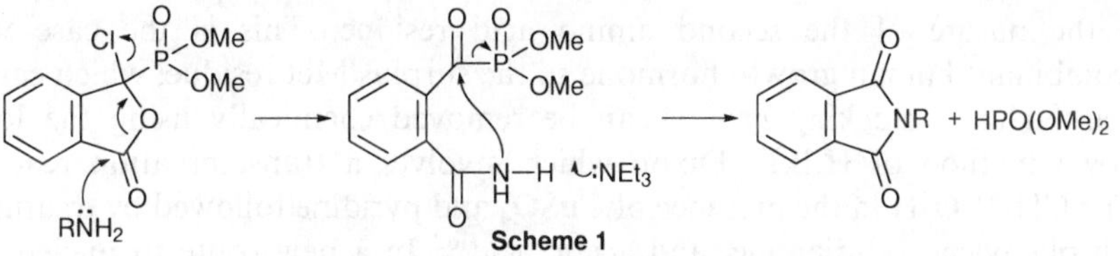

Scheme 1

ε-NH$_2$ group can acquire a Dde group from another Lys residue or from a α-Dde substituent by a rearrangement on a resin during the removal of an ε-Fmoc protecting group.[55] It has not been determined whether this process is intra- or inter-molecular. Several new variants of the Dde group with alkyl substituents at the exocyclic alkene position have been synthesized and shown to be fairly resistant to this rearrangement.[56] Moreover the new derivatives are completely resistant to 20% piperidine. Hydrazinolysis of Dde groups is troublesome in presence of Alloc groups. The latter are concurrently reduced both in solution and on a solid support thus preventing subsequent deprotection of side chains protected by Alloc groups. Fortunately, addition of allyl alcohol circumvents this side reaction so that the Dde and Alloc groups are completely orthogonal both in solution synthesis and SPPS.[57] As a further extension of the Dde method of protection, 2-acetyl-4-nitro-indane-1,3-dione has been examined as a reagent.[58] The resultant *N*-1-(4-nitro-1,3-dioxoindan-2-ylidene)ethyl (Nde) derivatives are stable to acids and secondary and tertiary bases. Deprotection is effected by 2% N_2H_4 at room temperature. Importantly, amino-acid derivatives do not cyclize to oxazolones and therefore do not racemize by this route. Some examples of urethane groups that are removed enzymically have been described. The 4-acetoxybenzyloxycarbonyl (AcOZ) group is removed by either acetylesterase or lipase and has been used for the synthesis of lipopeptides.[59] N-Protection of amino acids with the tetrabenzylglucosyloxycarbonyl (BGloc) group can be effected by reaction of α-isocyanato acid allyl esters with 2,3,4,6-tetrabenzylglucose followed by removal of the allyl group.[60] A mixture of α- and β-anomers is produced. The BGloc-amino acids were coupled with amino acid esters affording BGloc-dipeptide esters. The benzyl ether groups were removed by hydrogenolysis and the glycoside links were cleaved using a mixture of β-glucosidase from almonds and α-glucosidase from baker's yeast at pH 5.5 and 37 °C. Although this might appear to be more tedious than the use of Boc or Fmoc protection, there might be a useful application since the N-glycosyl peptide product might be easily purified by affinity chromatography on a support containing immobilized 3-aminophenylboronic acid. The 2-(phenylsulfonyl)ethoxycarbonyl (Psc) group has been proposed for the protection of the ε-amino group of Lys during solution-phase synthesis.[61] It is stable during coupling and during removal of Boc and 2,2,2-trichloroethyl groups. The Psc group is finally removed with piperidine. Salmon calcitonin (1–32) was synthesized using this procedure.

Recombinant proteins contain N-terminal Met due to the ATG initiation codon. This residue may be difficult to remove with aminopeptidase depending on the nature of the second amino acid residue. This is the case with recombinant human growth hormone so the surplus Met residue, which can be regarded as a blocking group, can be removed chemically using the long-known method of H.B.F. Dixon which involves a transamination reaction using $CHOCO_2H$ in the presence of $CuSO_4$ and pyridine followed by treatment with phenylene-1,2-diamine, and acetic acid.[62] In a new route to nucleotide-peptide hybrids, amino groups were blocked by Alloc groups.[63]

2.2 Carboxyl-group Protection – There is very little new work to report in this area. Amino acids are esterified in high yield and purity by stirring with Amberlyst-15 in appropriate alcohols.[64] Tertiary alkyl esters can be prepared from amino acids and the required alcohol using 4-dimethylaminopyridine (DMAP) and scandium triflate and $Pr^iN=C=NPr^i$ in CH_2Cl_2 at -8 °C.[65] This could be a useful method for attaching the C-terminal residue of a forthcoming peptide to a trityl resin ready for SPPS. The 2-(phenylsulfonyl)ethyl (Pse) group has been proposed to protect the side chains of Asp and Glu.[66] Some calcitonin fragments were synthesized by this method.

Proteinases from several bacteria catalyse highly regioselective hydrolysis of diesters of N-acylated derivatives of Glu, Asp, α-aminoadipic and α-suberic acids to produce ω-monoesters.[67] Tetrabutylammonium hydroxide is effective in hydrolysing polypeptide esters with minimal enantiomerization at the C-terminus.[68] This method is especially suitable for nonpolar or long peptide esters that are insoluble in most solvents. In the synthesis of lipopeptides, the carboxyl group was protected with the allyl group and deprotection was achieved using Pd(0) catalysis in the presence of Pd[PPh$_3$]$_4$ (0.01 equivalents) and morpholine (1.2 equivalents) at room temperature for 30–60 min.[69]

2.3 Side-chain Protection – The 3-pentyl (Pen) group has been proposed as a new protecting group for the phenolic function of the Tyr side chain.[70] It is resistant to bases such as 20% (v/v) piperidine/CHONMe$_2$ and to 50% CF$_3$CO$_2$H/CH$_2$Cl$_2$, but is readily cleaved by HF without the formation of significant amounts of alkyltyrosines. In the SPPS of pseudopeptides containing -CH$_2$NH- groups, removal of Trt groups from the Asn side chain is incomplete if the latter is close to the ψ-peptide bond.[71] Either a longer period of deprotection or the replacement of Trt by Me-Trt is recommended to overcome this problem. The basicity of the ψ-peptide bond is blamed for this behaviour. 1-Aminotetralins (1) have been examined as reagents for the

(1)

generation of protected carboxamide groups of Asn and Gln in peptide synthesis.[72] All the Gln derivatives were deprotected in 24 h using CF$_3$CO$_2$H/CH$_2$Cl$_2$/PhOMe at 25 °C but only the derivatives with X = OMe, Y = Z = H and X = Z = H, Y = OMe were deprotected by this treatment. In contrast, B(CF$_3$CO$_2$) deprotected all derivatives. Treatment of α-N-Z-Arg-OH with Me$_3$SiCl and tertiary base afforded δ-N,ω-N,O-tris(trialkylsilyl)-α-N-Z-Arg-OH. Reaction of this with RCOOCl gave δ-N,ω-N-bis(alkyloxycarbonyl)-α-N-Z-arginines from which the Z group can be removed by hydrogenation.[73] These potentially useful derivatives have not yet been reported as intermediates in peptide synthesis. In a study of the synthesis of peptide-oligonucleotide conjugates in which a Hse residue is the site of attachment of a nucleotide chain and in which a His residue follows the substituted Hse residue, protection of the imidazole ring is not required during elongation of the nucleotide chain using the phosphite triester approach.[74] The reduction of Met(O) residues with NH$_4$I in CF$_3$CO$_2$H in peptides containing His, Tyr, Trp or cysteine residues has been studied.[75] His and Tyr residues are stable in this process but cysteine residues are oxidized to cystine and Trp residues undergo dimerization to form 2-indolyl-2'-indolenine derivatives. Addition of Me$_2$S increases the rate of reduction of Met(O) residues and minimizes the extent of modification of Trp residues. The use of N^{in}-For-Trp avoids this side reaction completely. In a further study of the synthesis of peptides containing Trp and Met,[76] it was shown that the use of Fmoc/Trt chemistry gave purer products than when Fmoc/Bu^t chemistry was used.

2.4 Disulfide Bond Formation – The search for specific methods for forming disulfides from cysteinyl peptides continues to occupy a number of research teams. For example, of four methods used to form the disulfide bonds in calcitonin, namely aerial oxidation, Me_2SO/CF_3CO_2H, Me_2SO/H_2O and $K_3Fe(CN)_6$, the last was fastest and afforded the best yield.[77] In contrast, in the synthesis of human and murine stromal cell-derived factor-1s, which contains two disulfide bonds and a Trp residue that is sensitive to oxidation, a stepwise procedure involving $CF_3SO_3Ag/Me_2SO/HCl$ aq. and aerial oxidation was successfully used.[78] Again, spontaneous oxidation of a palindromic peptide containing a cysteine residue at the *N*-terminus and another just before the *C*-terminus gave up to three products, namely a cyclic monomer, an antiparallel dimer and a trimer with two parallel and one antiparallel chains.[79] The relative amounts of each product depended on the helical content of the peptide which was determined by the solvent. The parallel dimer, which was not formed by this method, had to be made by spontaneous oxidation of a mono-Acm derivative. A new reagent (2) for the synthesis of cyclic disulfides has been designed.[80] It consists of Ellman's reagent attached through two sites on a suitable insoluble support such as either Sephadex™ modified with PEG/polystyrene or controlled-pore glass (Scheme 2). A considerable excess of reagent is used in dilute acid; pH 2.7 gave the fastest reaction. The method was successfully used to make several disulfides. It was found that when the desired peptide contains two disulfide bonds, the major product contains correctly paired components of the disulfide bonds. No doubt this methodology will be subjected to a searching examination in view of its obvious potential. Solid-phase methodology has also been used to effect the oxidation of cysteinyl peptides on the resin.[81,82] This has the advantage that intermolecular coupling of two immobilized peptide chains is almost impossible. There is no obvious reason, however, why it should not be possible to couple one immobilized peptide chain to another peptide that is in the surrounding solution phase. The synthesis of circulin B and cyclopsychotride, which both contain 31 amino-acid residues and three disulfide bonds, is perhaps the most ambitious project undertaken in this area of peptide synthesis.[83]

A method has been described for the attachment of a carbohydrate moiety to a protein by forming a disulfide bond from the thiol group of a cysteine residue in the protein and a thiol group in the carbohydrate.[84] Thus (3) reacts with RSH to give (4) and if RSH represents a protein such as bovine serum albumin (chosen because it has only one thiol group), the product resembles an *N*-glycoprotein with the carbohydrate moiety attached to Asn. Fluorescence-quenched libraries of peptides containing interchain disulfide bonds have been assembled for studying the specificity and mechanism of action of protein disulfide isomerases.[85]

2.5 Peptide Bond Formation – There has been further research to determine the limitations of and to optimize the use of *N*-protected amino acid halides for forming peptide bonds. For example, although Fmoc amino acid fluorides are excellent for coupling moderately hindered amino acids (*e.g.* Aib to Aib),

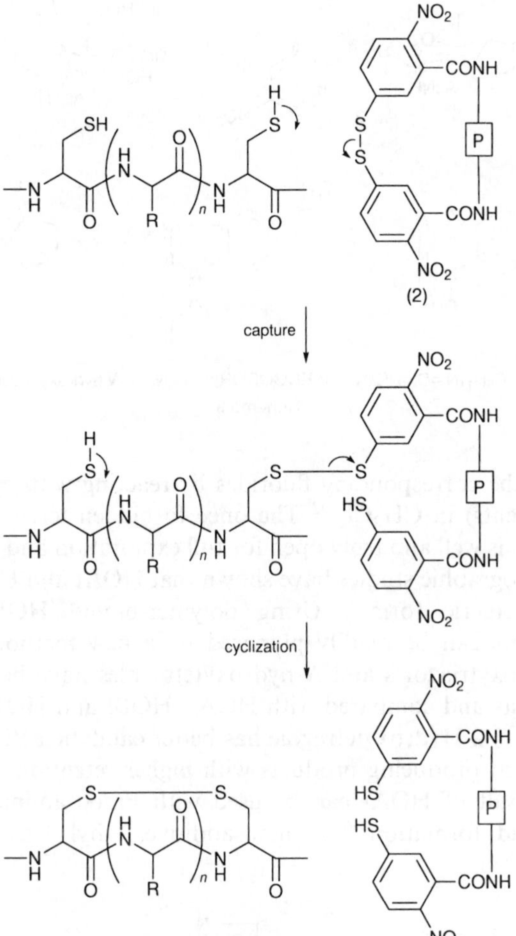

Scheme 2

they are not suitable for more hindered sytems (*e.g.* Aib to MeAib).[86] Again, although urethane-protected amino acid chlorides are more reactive than fluorides, they are also unsuitable for hindered systems because of competing formation of oxazolones.[86] The authors recommend using arenesulfonyl protection and Fmoc amino acid chlorides. Several peptides containing Aib have been synthesized using Fmoc amino acid chlorides in the presence of KOBt.[87] Since no additional base was required, it was found that the duration of coupling could be safely extended with improvement of yields. This procedure was extended to the synthesis of peptides of α,α-dialkylamino acids.[88] The use of Fmoc peptidyl chlorides has even been tried.[89] The coupling of Fmoc amino acid chlorides has been shown to be catalysed by the addition of Zn dust without the addition of base.[90] Coupling of Fmoc amino acid chlorides has also been effected in the presence of AgCN without added base.[91] Reaction is fast and free from enantiomerization. *N*-Trityl amino acids can be

(3)   (4)

Reagents: i, RCO$_2$H, C$_6$H$_{11}$N=C=NC$_6$H$_{11}$ or RCOOCOR, C$_6$H$_5$N; ii, *N*-hydroxysuccinimide
Scheme 3

converted into the corresponding fluorides by reacting with cyanuric fluoride (2 mole equivalents) in CH$_2$Cl$_2$.[92] The once forbidden territory of couplings with acyl halides is well and truly open for full exploration and exploitation.

X-ray crystallographic studies have shown that HOAt and HOBt exist in the 1-hydroxy tautomeric form.[93] Using polymer-bound HOBt, *N*-hydroxysuccinimide esters can be readily prepared by a new method (Scheme 3).[94] Various *N*-hydroxytriazoles and *N*-hydroxytetrazoles have been examined as potential catalysts and compared with HOAt, HOBt and HODhbt in peptide bond formation.[95] 2-Hydroxytetrazole has better catalytic activity than HOAt but is no better at producing products with higher retention of chiral purity. The potassium salt of HOAt can be used with Fmoc amino acid chlorides for peptide bond formation.[96] A new additive, ethyl 1-hydroxy-1H-1,2,3-

(5)

triazole-4-carboxylate (HOCt)(5), has been synthesized and assessed as a potential reagent in peptide coupling.[97] Apart from some difficulties in synthesizing the reagent, Fmoc amino acid esters of HOCt were very hygroscopic. Synthesis of the Ala-His system using τ-Trt protection on the imidazole ring caused negligible loss of chiral purity if coupling was carried out at 0 °C in a sonic bath. A more extensive comparison with HOAt and HOBt is required before a definitive assessment of the value of HOCt can be given. 1-β-Naphthalenesulfonyloxybenzotriazole (NSBt) has been used as a coupling agent for peptide synthesis in the presence of fluorinated alcohols as cosolvents.[98] *N*-Protected peptide azides have enjoyed a special place in the history of peptide synthesis because chiral purity is well preserved during coupling, but the instability of acyl azides requires the use of a low temperature and that means that coupling is necessarily slow. Generation of the acyl azide in the presence of HOAt or HOCt leads to the formation of the relevant reactive ester. The latter can react with a variety of nucleophiles such as peptides, alcohols, thiols and hydroxylamine to give higher yields than those obtained by the classical azide procedure.[99,100] This procedure, which has been given the rather forbidding name of transfer active ester condensation (TAEC), promises to be particularly useful for coupling sizeable segments of proteins as well as providing a convenient route to a range of *C*-terminal derivatives of peptides.

The use of onium reagents for peptide synthesis has been further studied. An improved synthesis of bis(tetramethylene)fluoroformamidinium hexafluoro-phosphate has been reported.[101] A very detailed study[102] of the use of onium coupling reagents revealed that (a) these usually give better results than carbodiimides, (b) pyrrolidine derivatives are generally preferred to piperidine derivatives, (c) although phosphonium salts are slightly less reactive than aminium or uronium salts, they are preferred for the coupling of sterically hindered *N*-terminal amino acids in order to avoid the formation of guanidine derivatives derived from the *C*-terminal moiety. *O*-Pentafluorophenyluronium salts have been used[103] for the synthesis of a glycopeptide sequence from epithelial cadherin 1.

Functionally active cell permeable peptides can be produced by the ligation of two peptide modules. The *N*-terminal portion of the final peptide comprises a cell permeable peptide (CPP) and this is coupled to a functional domain

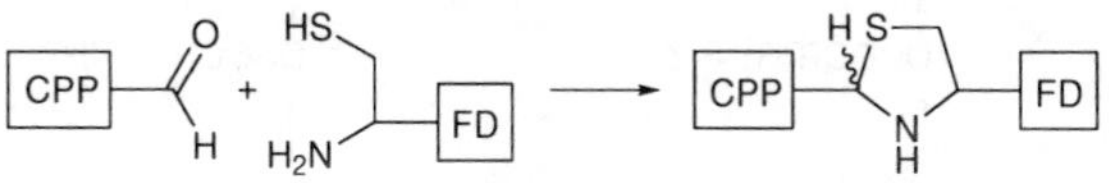

Scheme 4

(FD) that possesses an *N*-terminal Cys residue (Scheme 4).[104] In a similar vein, parathyroid hormones have been assembled from two peptides (Scheme 5).[105] The *N*-terminal residue of the *C*-terminal moiety is homocysteine and this is converted into methionine by *S*-methylation after coupling of the fragments. Boc-Cys(Npys)-OH and Z-Cys(Npys)-OH have been prepared and placed at the *N*-terminus of an otherwise unprotected peptide which was then coupled to another peptide destined to be the *N*-terminal moiety of the complete peptide.[106]

An interesting new method for peptide coupling uses an oxidative method for activating a phenyl ester.[107] 2,6-Di-t-butyl hydroquinone is acylated with an *N*-protected amino acid to give (6) and this is oxidized, preferably by indirect electrolysis, to give (7) which is sufficiently reactive to couple with an amino ester to give a peptide with no detectable enantiomerization. An intermediate (8) was isolated and the structure of the alanine derivative was proved by X-ray crystallography. Spirolactones as typified by (8) behave as reactive esters and afford dipeptide derivatives when treated with an amino acid ester.[108] The tyrosine moiety of the product contains a 4-hydroxy-

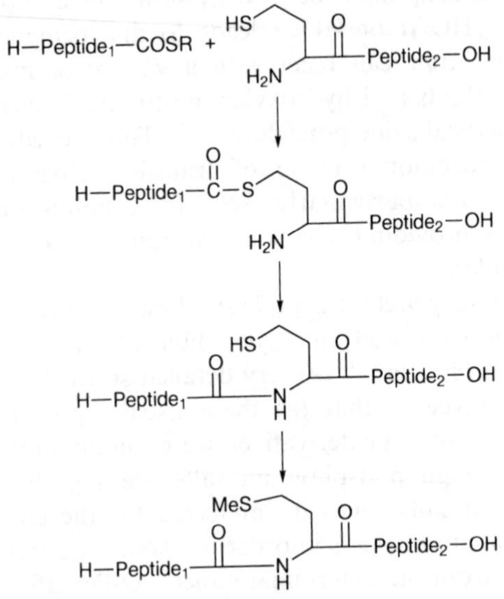

Scheme 5

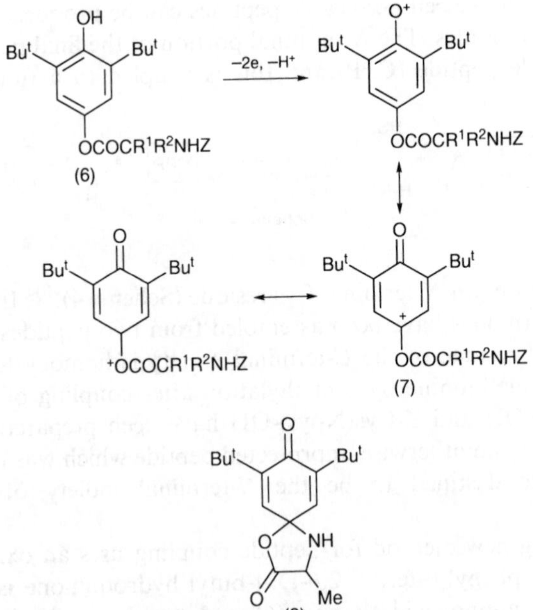

cyclohexa-2,5-diene-1-one system and this can be aromatized by reduction with 5,6-O-isopropylidene-L-ascorbic acid.

The precursor molecule of the *Aequorea* green fluorescent protein, a molecule containing 238 amino acid residues, has been constructed by the ligation of 26 segments.[109] This demonstrates that chemical synthesis has caught up with genetic engineering in the ability to synthesize large proteins. Reaper, a peptide of 65 residues that contains cysteine has been assembled from two segments using a peptide thioester in the presence of AgCl.[110] A transmembrane protein has also been synthesized by the ligation of two solubilized segments, a useful ploy to overcome the problems associated with the coupling of rather insoluble peptides containing a high proportion of hydrophobic amino acids.[111]

This section concludes with a miscellany of unusual or new methods of peptide bond synthesis. An unprotected amino acid or peptide destined to form the *C*-terminal moiety of the completed peptide is preactivated with AlMe$_3$ and then treated with the methyl ester of the *N*-terminal unit (Scheme 6).[112] The preferred solvent is ClCH$_2$CH$_2$Cl. Reaction is sluggish at room

$$\text{H—Xaa—OH} \xrightarrow{\text{i, ii}} \text{RCO—Xaa—OH}$$

Reagents: i, AlMe$_3$; ii, RCO$_2$Me

Scheme 6

temperature but proceeds readily at 80°C affording moderate yields. For example, coupling of Boc-Phe-OMe and H-Val-OH gave only 50% of protected dipeptide. The modified Ugi reaction using Schiff bases from amino and ketonic components can be used to synthesize tripeptides of α,α-diphenylglycine (Dph).[113] Very crowded tripeptides such as Z-Dph-Dph-Dph-OMe were made. *N*-Carboxy anhydrides of polyhydroxylated α-amino acids have been synthesized and are intended as intermediates for making directly linked peptidyl nucleoside antibiotics.[114] With eyes focused outside the laboratory and on the distant past, it has been shown that montmorillonite clay enhances the salt-induced formation of tripeptides from glycine and/or alanine.[115] The clay appears to stabilize peptides against their destruction by hydrolysis. It was postulated that this kind of process could have occurred on primitive Earth. Amino acids were also shown to form low yields of dipeptides when heated under pressure without water.[116]

2.6 Peptide Synthesis on Macromolecular Supports and Methods of Combinatorial Synthesis – A new procedure for the synthesis of aminomethylpolystyrene and 4-methylbenzhydrylamine polystyrene resins starting from the synthons (9) has been described.[117] A Merrifield resin has been generated on SynPhase^{TM} crowns and used for SPPS.[118] A new resin based on methylbenzhydrylamine has been reported[119] from which a peptide product containing a *C*-terminal secondary amide group can be detached by mild acidolysis. The resin (10) is reacted with a primary amine before construction of the peptide is commenced. This has the advantage that a wide variety of secondary amides is accessible. A convenient method for preparing 2-chlorotrityl chloride resin has also been reported.[120] A new resin (11a) that is an analogue of trityl chloride was designed to allow attachment of an amino acid allyl ester at the amino group.[121] The allyl group was removed and the free carboxy group was coupled to H-Phe-OMe. The dipeptide ester was

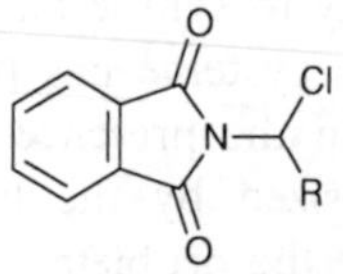

(9) R = H, 4-Me₆H₄

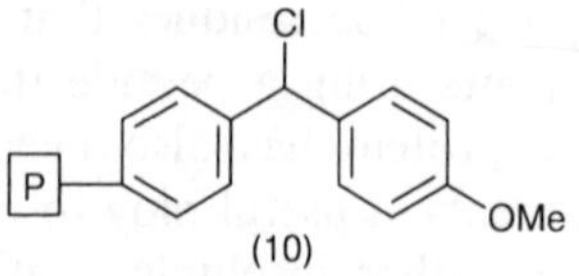

(10)

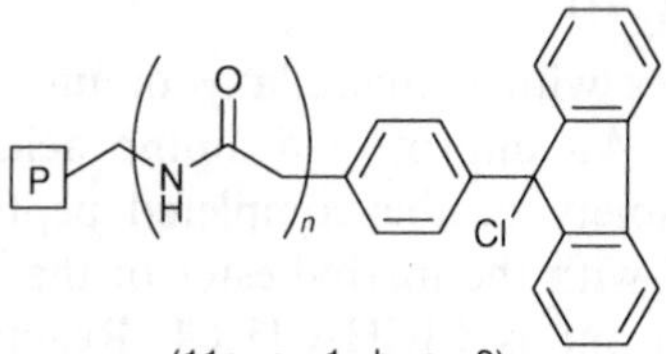

(11a, *n* = 1; b, *n* = 0)

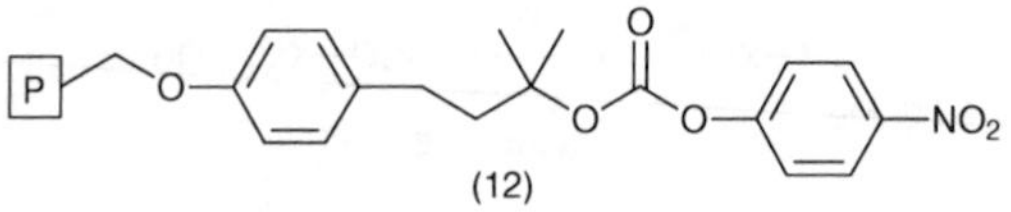

(12)

detached with 95% CF_3CO_2H. A very similar support (11b) has been produced.[122] Another resin (12) is a 4-nitrophenyl carbonate ester which reacts with the Li salt of an amino acid.[123] This was converted into a series of dipeptide amides. Linear poly-4- and poly-2-vinylpyridines and copolymers of these with divinylbenzene have been used as an efficient solid phase for the synthesis of acid anhydrides and amides by pyridine-catalysed acyl transfer to carboxylic acids.[124] A chemically inert hydrophilic resin has been prepared by radical polymerization of PEG that is substituted by a suitable styrene derivative (*e.g.* 13).[125] The resin is chemically inert but swells in water and so combinatorial libraries of products can be exposed to enzymes in aqueous solution while still attached to the resin. This could be a new way of effecting enzyme-catalysed fragment coupling of peptides. Moreover these resins are stable to Lewis acids because they do not contain benzylic ether groups. The search for better linkers has proceeded vigorously. An improved synthesis of 2-(N^{im}-Boc-5-methylimidazol-4-yl)-2-hydroxyacetic acid, a safety-catch which allows direct release of peptide acids into aqueous buffers, has been designed.[126] Linkers based on 10,11-dihydro-5H-dibenzo[*a,d*]cyclohepten-5-one (14) have been designed for the synthesis of peptide amides.[127] A novel and versatile type of linker involves an aryl hydrazine moiety; cleavage requires a Cu(II) catalyst, a base and a nucleophile.[128] The base serves a dual function, proton abstraction and complexation of copper. Coupling of a 4-hydrazinobenzoic acid to an amino-substituted resin was slow so the former was first linked to an amino acid which is more reactive in conjugating to an insoluble amine and also acts as an internal reference amino acid. A typical linker (15) attached to a support is prepared for peptide synthesis by removing the Fmoc group. After peptide construction is complete, detachment is brought about by oxidation with Fehling's solution in the presence of a nucleophile (water, alcohol or amine) and a tertiary base. A thiol obviously cannot be the

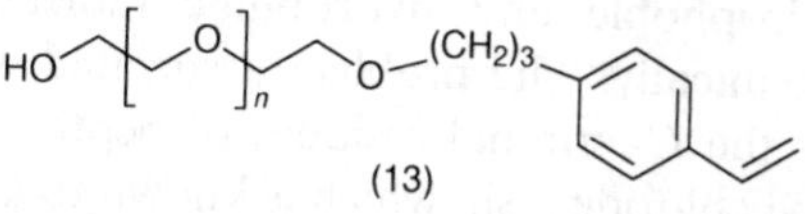

(13)

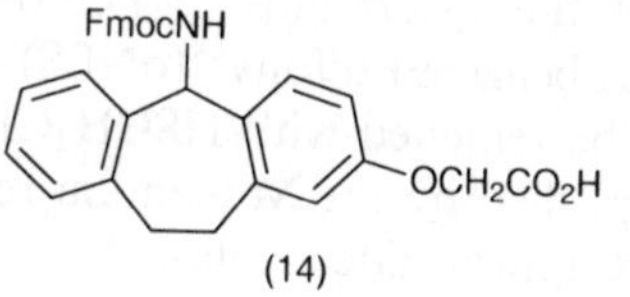

(14)

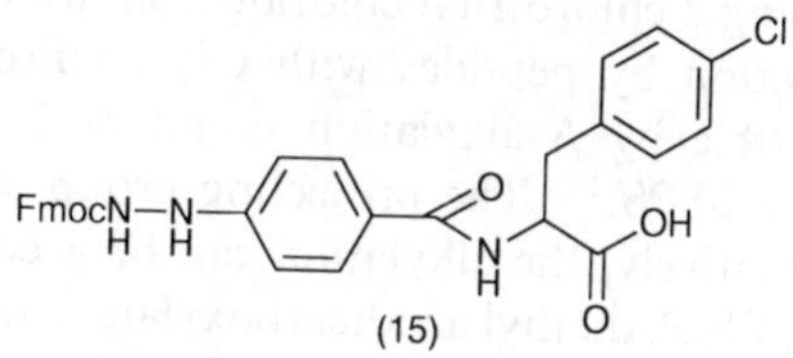

(15)

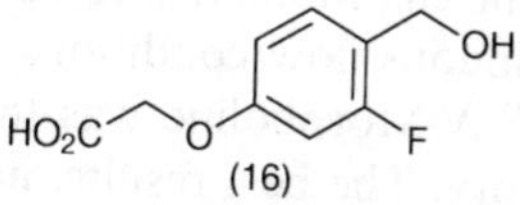

(16)

nucleophile which is unfortunate since thiol esters are invaluable in fragment couplings. Development of the PAL type of linker involves reductive amination of 5-(4-formyl-3,5-dimethoxy-phenoxy)valeric acid with the α-amino group of the intended *C*-terminal acid which has its carboxy group esterified. The other carboxy group can then be used to couple to the insoluble support. Before that, the secondary amine group is protected with a Fmoc group. SPPS is then conventional with the exception that the *C*-terminal amino acid can be coupled to a new nucleophile before detachment from the resin. Alternatively, the peptide can be cyclized while it is still attached to the resin.[129] A simple linker was used to synthesize octreotide.[130] 4-Carboxybenzaldehyde reacted with Fmoc threoninol to form the 4-carboxybenzacetal and this was coupled to an amino resin and the peptide assembled. An alternative method for attaching a *C*-terminal alcohol involves the use of 3,4-dihydro-2H-pyran-2-carboxylate as a bifunctional linker.[131] This was also used to synthesize octreotide. The Dde protecting group has been used in the SPPS of the spider toxin pseudoargiopinine III.[132] A allyl protected Trp linker analogous to a known Lys side chain linker can be used to effect cyclization of a peptide before detachment from the support.[133] By using fluorinated analogues of three known linkers (*e.g.* 16), NMR spectroscopy can allow monitoring of several stages of SPPS,[134] such as attachment of the linker to the support, coupling of the first residue to the support and ultimate detachment of the product.

Incorporation of either a strongly basic or acidic tail during SPPS of a hydrophobic peptide is reported to facilitate synthesis without resorting to protection of backbone -NH- groups.[135,136] One group[135] recommends inserting a 4-hydroxymethylbenzoic acid between the solubilizing and the hydrophobic group to be constructed:

$$H_2N\text{-hydrophobic peptide-}COOCH_2C_6H_4CO\text{-solubilising peptide}$$

At the end of the synthesis, the required peptide is detached by mild aqueous basic hydrolysis of the ester link. The other group[136] does not use a spacer to separate the hydrophobic and hydrophilic moieties and detaches the required peptide enzymically. The problem associated with the formation of diketopiperazines by the *C*-terminal residues of peptoids has been addressed by using 2-chlorotrityl chloride resin which is known to avoid diketopiperazine formation by peptides with Gly or Pro at the *C*-terminus.[137] Peptoids are accessible by *N*-alkylation of an *N*-(2-nitrobenzene)sulfonyl (oNBS) group during SPPS.[138] The protecting group can be removed with $HSCH_2CH_2OH$. Alternatively, the alkylation can be accomplished by the Mitsunobu reaction using Ph_3P, diethyl azodicarboxylate and the appropriate alcohol.[139]

There have been several studies of technical difficulties and tactical improvements in SPPS. A study of the enantiomerization of Fmoc-Ser(Bu^t)-OH during SPPS under standard continuous-flow conditions revealed that the choice of tertiary base was crucial.[140] *N*-Morpholine was totally unacceptable whereas 2,4,6-collidine was satisfactory. The best results, however, were obtained with HATU as coupling agent and with either HOAt or HOBt as additive. These results agree with those of several other workers but particularly of Carpino's group. In an evaluation of potential orthogonal protecting groups, it was found that monomethyltrityl groups were not as easily cleaved on hydrophilic supports as had been expected.[141] The more labile monomethoxytrityl group was preferred. A nonacidolytic method for detaching Boc-peptides that are linked to polystyrene resin by a phenacyl ester moiety uses Me_3SnOH.[142,143] The method is compatible with the use of *N*-Boc and *O*-Bzl groups. A study has been made of 4-cresol as a reversible acylium ion scavenger during HF cleavage after SPPS.[144] Contrary to previous reports, it is claimed that 4-cresol and 4-thiocresol predominantly form aryl esters. Whereas cresyl esters are stable at $0\,°C$, they undergo the Fries rearrangement to give aryl ketones at $5-20\,°C$. 4-Thiocresyl esters behave differently and undergo further addition of 4-thiocresol to form ketene-bisthioacetals and trithioorthoesters. Further 4-cresyl esters at Glu side chains are susceptible to amidation and fragmentation in the presence of mild bases. This side reaction can be avoided by using H_2O_2-catalysed hydrolysis which converts cresyl adducts into free carboxylic acids in near quantitative yield. In the synthesis of peptides of α-hydroxymethylserine (Hms), the two hydroxy groups can be protected by an *O,O*-isopropylidene group using Fmoc chemistry.[145] The 2-chlorotrityl polystyrene resin is recommended in order to avoid diketopiperazine formation if Hms is *C*-terminal and to avoid $N{\rightarrow}O$ acyl shift if prolonged acid exposure is required. Peptide aldehydes can be made on a PAM resin because of its stability to acids and bases. The majority of the peptide is assembled conventionally and then an amino acid dimethylacetal is added and this effects aminolysis and peptide detachment from the support.[146] An alternative approach to the synthesis of peptide aldehydes involves attachment of a Wittig or Wittig-Horner reagent on an MBHA resin and reaction with an *N*-protected aminoaldehyde. This process is followed by assembly of the rest of the peptide chain and ozonolysis. An aldehyde group is generated on the same residue that originally possessed such a group before the Wittig synthesis.[147] By using 3-mercaptopropionamide-PEG-(poly-*N,N*-dimethylacrylamide) copolymer, a peptide thioester can be assembled on the resin and deprotected *in situ* ready for fragment condensation.[148] Attempted SPPS from δ-amino acids has been plagued by the tendency for cyclization to lactam to occur. This was the exclusive route when trying to synthesize peptides of 2-(aminomethyl)phenyl acetic acid.[149] This problem was neatly solved by converting the δ-amino group into an azido group and attaching the azido acid to a Wang chloride resin. The azido group was activated by treatment with tributylphosphine forming an iminophosphorane and reaction was then allowed to proceed with an unsymmetrical acid anhydride derived from the protected amino acid intended to form the other moiety of the required peptide. Reaction at 0–20 °C afforded a very good yield of product. The construction of peptide thioesters on a resin enables the coupling of peptide fragments and the synthesis of *N*-alkylated peptide amides although the latter reaction is rather sluggish. *N*-Aryl amides are not conveniently accessible by this method. The discovery that Ag$^+$-assisted aminolysis proceeds rapidly has changed this situation dramatically.[150] Careful adjustment of the ratio of amine to silver salt is required in order to minimize the extent of enantiomerization.

Several new or improved analytical techniques have been reported. A record of possible changes in peptide conformation especially aggregation processes has been obtained by determining the Fourier transform IR (FTIR) spectrum on the resin after each coupling step.[151] A fragment of HIV proteinase (80–99) was synthesized by this method. FTIR and Raman spectroscopy have also been used to identify suitably derivatized peptides.[152] Incorporation of an amino acid that contains a >C=O group as well as an amino acid that contains a thiol group provides sites for the installation of a donor and acceptor pair in one step.[153] This modificaton of a peptide permits structural studies using fluorescence resonance energy transfer (FRET). Determination of the loading of resin during SPPS with Fmoc amino acids has usually been achieved by deprotection with piperidine. It is now claimed that 1,8-diazabicyclo [5.4.0]undec-7-ene (DBU) is the preferred reagent since piperidine yields a dibenzofulvene-piperidine adduct whereas DBU gives rise to unassociated dibenzofulvene.[154] Moreover sample preparation has been automated with this latter method. An alternative method involves determination of the basic groups on the resin using an ion-selective electrode.[155] Free basic groups are protonated with $HClO_4$, then ClO_4^- ions are quantitatively released by base for potentiometric determination. The progress of liquid-phase synthesis on soluble polymer supports can be followed by MALDI-TOF spectroscopy.[156] An apparatus ('domino blocks') has been designed for manual and semi-automatic parallel SPPS.[157] Resin-bound peptoids can be sequenced by Edman degradation followed by identification of the 3-phenyl-2-thiohydantoin by chromatography.[158]

Combinatorial synthesis of peptides continues to attract much attention. A new hybrid resin allows products to be detached in a stepwise manner for screening.[159] This technique is combined with Edman sequencing on single resin beads. Some new cyclic scaffolds have been designed.[160-162] The last two of these permit construction of different molecules. A novel concept in combinatorial synthesis involves the preparation of omission libraries in which one amino acid is omitted in all positions of a peptide.[163] This approach is claimed to lead rapidly to the identification of those amino acids that make an important contribution to the manifestation of biological activity. The use of membrane supports in SPPS permits other conveniences.[164] The membrane can be cut into pieces and combined into groups and index patterns can be printed on the membrane allowing direct identification of compounds.

Some examples of the chemical aspects of combinatorial synthesis of peptide libraries conclude this section. A library of 27 tripeptide aldehydes has been prepared.[165] (*cf.* ref. 147). A library of peptides containing biphenyl amino acids has been produced and tested as antagonists against the vitronectin receptor.[166] A library of β-peptoids composed of N-substituted β-amino-propionic acid oligomers has been assembled.[167] Libraries of small peptides were generated by the Ugi 4-component route.[168,169] Both solution and solid-phase methodologies were used.[168] The specificity pockets of two homologous SH3 domains were explored with a library of peptides produced by split-pool synthesis.[170] A library of peptides related to Leu-enkephalin was assembled on a support prepared by coupling aminomethyl(polystyrene) and PEG.[171] The peptides could be detached under very mild conditions with 1% CF_3CO_2H. A new method of detaching members of a peptide library involved the use of Fmoc-Met-OH as the C-terminal residue for all peptides and, when the library was complete, detachment was effected with CNBr when all members had Hse lactone at the *C*-terminus.[172] Ramage's group have installed the tetrabenzo [*a,c,g,i*]fluorenyl-17-methoxycarbonyl (Tbfmoc) group as the common substituent at the N-terminus of a completed peptide library.[173] This permitted purification of individual members by chromatography on graphitized carbon. A library of 1000 tripeptides has been synthesized and screened with a molecular tweezer receptor (17).[174] The tweezer showed 95% selectivity for Val at the *C*-terminus and 40% selectivity for Glu(OBu^t) at the *N*-terminus. Another method for screening a peptide library involves radiolabelling of a suitable protein probe.[175] [^{14}C]-HCHO was used to label the probe. This technique could be made more sensitive by radioiodination as used in radioimmunoassays.

2.7 Enzyme-mediated Synthesis and Semisynthesis – There is still interest in using immobilized or chemically modified enzymes in order to diminish the chance of enzyme denaturation by organic solvent and/or autolysis of the enzyme during peptide coupling. Papain deposited on celite has been used for coupling N-phenylacetyl- or N-mandelyl-glycine esters to nucleophiles.[176] Trypsin in which statistically 8 of the 14 lysine ε-amino groups had been blocked was found to be more stable than the native enzyme in the range 30–70 °C.[177] The modified enzyme had a decreased rate of autolysis and displayed higher activity in some solvents. For example, it gave higher yields of Bz-Arg-Leu-NH$_2$ in 95% MeCN, but zero yield was obtained in 95% CHONMe$_2$. Immobilized trypsin gave Z-Lys-Gly-OMe (80%) while immobilized thermolysin gave a quantitative yield of Z-Ser-Leu-OMe in

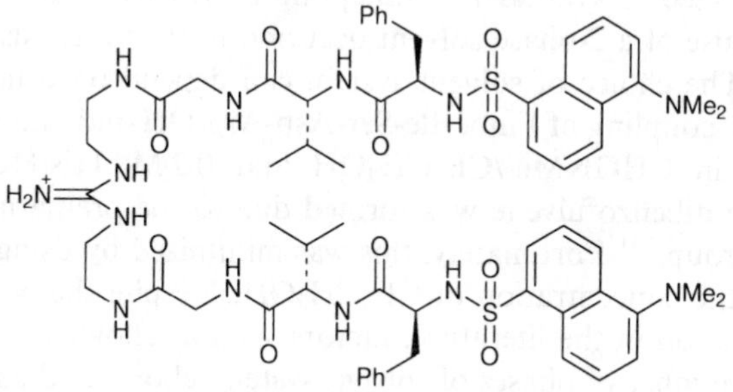

(17)

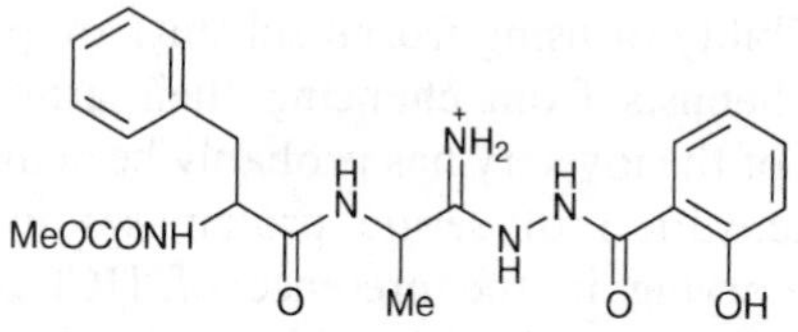

(18)

aqueous organic solvent.[178] Commercial crosslinked thermolysin has been used to synthesize small peptide derivatives.[179] Semisynthesis of human insulin from porcine insulin lacking the *C*-terminal residue in the B chain was effected by incubation of the substrate with H-Thr(Bu^t)-OBu^t in the presence of immobilized trypsin.[180]

Investigation of reaction conditions for enzyme-catalysed peptide synthesis continues to focus on the choice of solvent. In recent years, systems containing little or no water or frozen aqueous solutions appear to be favoured, but there is still considerable effort being applied to the study of all kinds of systems. Thermolysin gives satisfactory yields of protected small peptides in t-amyl alcohol containing 6%(v/v) water.[181,182] Leu-enkephalin was also synthesized using α-chymotrypsin in CH$_2$Cl$_2$. Thirteen homogeneous aqueous solvents were used in the synthesis of Z-Phe-Phe-OMe in the presence of thermolysin.[183] The enzyme was strongly activated by MeCN, Me$_2$CO, CHONMe$_2$, tetrahydrofuran and pyridine, but the degree of activation does not appear to be related to the physical parameters of solvents. A range of systems including frozen solutions was used for the α-chymotrypsin-catalysed synthesis of eledoisin and LH-RH.[184] Best results were obtained with aqueous buffers, especially boric acid-borate or ammonium acetate buffers. The products precipitated from solution thus preventing hydrolysis. Thermolysin displayed spectral changes in alcoholic solvents.[185] Mixtures of good hydrogen bond donors such as CF$_3$CH$_2$OH or (CF$_3$)$_2$CHOH and an acceptor (*e.g.* CHONMe$_2$) are not as suitable for peptide synthesis using trypsin as had previously been claimed and fluorescence measurements on the enzyme in such solvents indicated that Trp residues had become more exposed.[186] For the esterification of *N*-acetyltyrosine using α-chymotrypsin, the best solvent is claimed to be acetone containing 10% v/v of the alcohol and 3% v/v water.[187] Similar conditions were used for the pronase-catalysed[188] and the carboxypeptidase-catalysed[189] synthesis of dipeptide derivatives. An example of the advantageous use of a 2-phase solvent occurred in the synthesis of CHO-Asp-Phe-OMe.[190] The choice of solvent system can depend on other factors. For example, in the coupling of Fmoc-Ile-Ser-Asp-Arg-OH and amino components using trypsin in $CHONMe_2/CF_3CH_2OH$ and 0.2 M Tris/HCl buffer (pH 8.0–8.9), some dibenzofulvene was formed due to the premature detachment of the Fmoc group.[191] Fortunately, this was minimized by using 1 M buffer or by increasing the concentration of CF_3CH_2OH. Despite the wealth of experimental information in the literature, factors such as choice of enzyme, choice of substrates, number of phases of solvent system, choice and concentration of organic solvent, chemical nature, molarity and pH of buffer, temperature (particularly the possibility of using frozen solvent) are probably disincentives influencing organic chemists from chancing their arm in enzyme-catalysed methodology. A little of the mystery has probably been dispersed by the report that the dramatic activation of serine proteinases in nonaqueous media resulting from freeze-drying in the presence of HCl is not related to the relaxation of potential substrate diffusional limitation but is due to an intrinsic enzyme activation.[192] Pretreatment of enzyme with crown ethers can effect a large increase in catalytic activity. A typical result is reported in the synthesis of Ac-Phe-Phe-NH_2 using subtilisin.[193]

The use of surface-active additives such as bis(2-ethylhexyl) sodium sulfosuccinate to form reverse micelles with *e.g.* isooctane seems to have fallen out of favour but some small peptide derivatives required for the synthesis of RGD peptides have been obtained by this method.[194]

Surprisingly, the use of frozen aqueous solutions seems to be mainly confined to Jakubke and co-workers. Granted that the yield enhancement due to using frozen solutions can be very variable,[195] but the yield does not appear to depend on whether native or some chemically modified form of enzyme is used.[196]

Noncoded amino acids can be incorporated as P_1 components employing an enzyme such as papain of rather low specificity. The kinetically controlled approach with a methyl ester as acyl donor is used.[197] A related problem occurs with α-chymotrypsin operating on esters of an amino acid for which the enzyme has a low specificity. The solution is to use an ester group that confers higher activity as an acylating agent. Thus, α-chymotrypsin can couple Ala in the P_1 position using CF_3CH_2 or NH_2COCH_2 esters.[198] The latter ester reacts 133 times faster than the methyl ester. This approach presumably is only successful with enzymes that catalyse reactions by the ping pong mechanism. Peptides with His at P_1 and Lys at $P_{1\alpha}$ can be synthesized using trypsin[199] but Lys derivatives with no protecting group on the ε-NH_2 group could be hydrolysed albeit slowly by trypsin. Improved yields were obtained when the Lys side chain was protected. A better solution is to use an unprotected Lys derivative and α-chymotrypsin as catalyst. Presumably, the absence of a protecting group facilitates binding to the enzyme. A further improvement was obtained when the synthesis was carried out in frozen solution. Peptide synthesis can be achieved using inverse substrates. Trypsin normally requires a protonated strong base in the side chain of the residue destined to be at the P_1 site. An amino acid with an uncharged side chain can occupy this position, however, if the acylating substrate is either a 4-guanidinophenyl or a 4-(guanidinomethyl)phenyl ester.[200] This type of synthesis was mentioned in the previous Report. The importance of substrate structure is well illustrated by a paper describing the splicing of two haemoglobin fragments by V8 proteinase in the presence of n-PrOH.[201] Hbα (24–30) and Hbα (31–40) are the shortest segments that can be spliced in this way. By replacing Glu^{27} and Arg^{31} in the two substrates it was shown that these were essential and it was concluded that (i, i+4) carboxylate-guanidine interaction drives the formation of an α-helix and this is further assisted by the organic solvent. The authors describe the phenomemon as a molecular trap for splicing. Once formed, the α-helix imparts some resistance to hydrolysis of the Glu^{30}-Arg^{31} bond. No doubt, further examples of molecular trapping will come to light as enzyme-catalysed fragment coupling is attempted at sites where secondary structure is generated with the formation of a new peptide bond.

This section ends with a small collection of experiments with less common enzymes including mutant form of proteinases and nonproteolytic enzymes. Carboxypeptidase Y has been used to effect the semisynthesis of peptide amides with either Asp or Glu amide derivatives at the C-terminus.[202] Enantioselective synthesis of amino acid amides is possible using *Candida antarctica* B lipase (Novozyme 435).[203] Di- and tri-peptides have been synthesized using the aspartic proteinases cardosin A and B from *Cynara cardunculus L.*[204] An RGD tripeptide has been synthesized using chymopapain as catalyst.[205] Using the Gln19Glu mutant of papain, incubation of MeOCO-Phe-NHCHMeC≡N and salicylhydrazide gives the amidrazone (18).[206] The nitrile reacts with the thiol group of papain to form the thioimidate adduct which is then the target for nucleophilic attack by salicyl hydrazide. Genetically engineered mutants of the neutral proteinase from *B. stearothermophilus* have been examined as potential catalysts for the synthesis of Z-Asp-Met-OMe.[207] Ile140Pro and Asp141Pro greatly stabilized the enzyme in $CHOMe_2$ and these substantially enhanced the yield of peptide. Orange flavedo peptide amidase converts several peptides with a free carboxy group into the corresponding amide in NH_4CO_3.[208] Porcine pancreatic lipase is capable of using D-amino acid esters or amides as occupants of the $P_{1\alpha}$ position in the synthesis of dipeptide derivatives.[209] In a new synthesis of nucleopeptides, the amino group of adenine or cytosine was protected with a phenylacetyl group.[210] This group was removed by penicillin acylase. In the same synthesis, the carboxy group at the C-terminus of the peptide moiety was protected as a choline ester. This was removed by butyrylcholine esterase.

2.8 Miscellaneous Reactions Related to Peptide Synthesis – Trifluoromethyl trimethylsilane, the Ruppert reagent, reacts additively with N-substituted oxazolidin-5-ones.[211] Mild acidic hydrolysis of adducts that have an electron-releasing substituent at C_2 affords N-substituted α-aminotrifluoromethyl ketones and these can be coupled with amino acid fluorides giving access to peptidic trifluoromethyl ketones, which are well known proteinase inhibitors (see Section 3.3). During the SPPS of a peptide containing a Lys residue with an unprotected ε-amino group close to an Arg residue with the guanidino group bearing a tosyl group, treatment with isonicotinoyl 4-nitrophenyl carbonate resulted in substitution of an iNoc group on the guanidino group.[212] A potentially important and simple route to peptides containing an ω-guanidino group involves guanidination of an ω-amino group with a new reagent, N,N'-Boc$_2$-N''-triflylguanidine (Scheme 7).[213,214] This technique could well

Reagent: i, RNH$_2$

Scheme 7

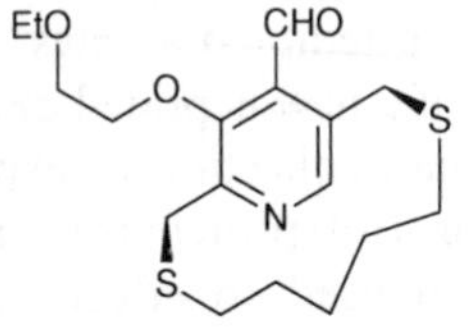

(19)

prove to be preferable to the use of protected guanidino acids especially since peptides of homoarginine and norarginine should be readily accessible and also because hydroxy groups can be replaced by guanidino groups.

Electrophilic iodination of aromatic groups can be effected on solid phases using IPy$_2$BF$_4$ as reagent.[215] It will be interesting to see if this method can be used for labelling peptides and proteins for use in radioimmunoassay. Peptides containing Tyr(SO$_3$H) residues can be obtained by either direct reaction with CHONMe$_2$-SO$_3$ complex or by using Fmoc-Tyr(SO$_3$Na) in the synthesis.[216] Peptides related to CCK were synthesized to test the method. It is possible to trifluoroethylate thiol and ε-amino groups in amino acids and peptides in aqueous solution using the novel iodonium salt (CF$_3$SO$_2$)$_2$NI(Ph)CH$_2$CF$_3$.[217] For example, glutathione was converted into S-trifluoroethyl glutathione. Peptides containing a β-hydroxyamino acid can be cyclized under Mitsunobu conditions leading to a constrained structure that resembles a reverse turn.[218] Using Bu^{t}Li, N-Boc-β-tripeptide esters can be converted into the Li enolate of the C-terminal residue which can then be selectively alkylated with a variety of alkyl halides.[219] Peptides containing a central residue of N-acylated aminomalonic acid can be C-alkylated under mildly basic conditions.[220] Similar reactions occur with peptides containing a residue of α-cyanoglycine. These peptides with an acidic CH-group can also add to Michael acceptors in the presence of catalytic amounts of alkoxide bases. Stereoselective and N-terminal selective α-alkylation of peptides can be achieved using an N-terminal activator (19) which also functions as a chiral auxiliary.[221] Interestingly, the stereoselectivity in the presence of Li$^+$ ions is the opposite to that when either another alkali cation or no ionic additive is present. Finally, in an exciting series of experiments,[222] a replicating peptide system involving auto- and cross-catalysis has been described. The peptides K1 and K2 couple to give K1K2 autocatalytically at pH 7.5 in 2M $NaClO_4$. E1 and E2 similarly afford E1E2 at pH 4.0.

K1 Ac-KLYALKEKLGALKEKL-COSR
K2 H-CLKEKLGALKEKLYALKE-$CONH_2$
E1 Ac-ELYALEKELGALEKELA-COSR
E2 H-CLEKELGALEKELYALEK-$CONH_2$

Cross catalysis is illustrated by the coupling of E1 and E2 in the presence of K1K2. Further activity in this area is to be expected from those interested in the synthesis of naturally occurring peptides and proteins and from those who seek plausible answers to questions about the origin of life. In addition, to those interested in biochemical kinetics, there is the prospect of an infinite number of substrates to study. Oh, to be young again!

3 Appendix: A list of Syntheses Reported mainly in 1998

Peptide/protein	*Ref.*
3.1 Natural Peptides, Proteins and Partial Sequences	
ACTH	
Protected fragments	223
Urocortin	224
Amyloid protein	
Fragments labelled with tritium	225
Antamanide (ex *Amanita phalloides*)	
Linear and cyclic analogues	226
Antibacterial peptides	
Mycobactin analogues	227
Moiramide B and andrimid, active against MRSA	228
A biotinylated gramicidin	229
Structure–function studies on antibacterial peptides	230
Analogues of trichogin GA IV	231
Zervamicin-IIB	232
Head-to-tail dimers of vancomycin	233
Blood-clotting factors	
Glycopeptide fragment (447–460) of protein S	234
Thrombin-sensitive module and EGF module of protein S	235
Peptide that inhibits platelet aggregation *in vitro*	236
Orally active fibrinogen receptor antagonists	237
Bradykinin	
Antagonists	238,239

3.2 Sequential Oligo- and Poly-peptides

3.3 Enzyme Substrates and Inhibitors

3.4 Conformations of Synthetic Peptides

3.5   Glycopeptides

3.9   Miscellaneous Peptides

3.10 Purification Methods

References

1. D.T. Elmore, *Specialist Periodical Report: Amino Acids, Peptides and Proteins*, 1998, **30**, 111.
2. G.C. Barrett and D.T. Elmore, *Amino Acids and Peptides*, Cambridge University Press, Cambridge, 1998.
3. M.C. Fitzgerald and S.B.H. Kent in *Bioorg. Chem.: Pept. Proteins*, ed. S.M. Hecht, Oxford Univ. Press: New York, 1998, p. 65. and p. 480.
4. J. Wilken and S.B.H. Kent, *Curr. Opin. Biotechnol.*, 1998, **9**, 412.
5. M. Sandberg, L. Eriksson, J. Jonsson, M. Sjöstroem and S. Wold, *J. Med. Chem.*, 1998, **41**, 2481.
6. V.J. Hruby and J.-P. Meyer in *Bioorg. Chem.: Pept. Proteins*, ed. S.M. Hecht, Oxford Univ. Press: New York, 1998, p. 27 and p. 473.
7. R.S. Givens, J.F.W. Weber, A.H. Jung and C.-H. Park, *Methods Enzymol.*, 1998, **291**, 1.
8. E.V. Kudryavtseva, M.V. Sidorova and R.P. Evstigneeva, *Usp. Khim.*, 1998, **67**, 611.
9. L.-Y. Wang, H.-P. Pan and Z.-Y. Chen, *Youji Huaxue*, 1998, **18**, 576.
10. S. Aimoto and T. Kawakami, *Baiosaiensu to Indasutori*, 1998, **56**, 445.
11. G.B. Fields, in *Molecular Biomethods Handbook*, eds. R. Rapley and J.M. Walker, Humana, Totowa, N.Y., 1998, p. 527.
12. M. Meisenbach, H. Echner and W. Voelter, *Chim. Oggi*, 1998, **16**, 67.
13. B. Seligman, M. Lebl and K.S. Lam, in *Comb. Chem. Mol. Diversity Drug Discovery*, eds. E.M. Gordon and J.F. Kerwin, Wiley-Liss, New York, N.Y., 1998, p. 39.
14. K.M. Youssef, *Saudi Pharm. J.*, 1998, **6**, 1.
15. R.M.J. Liskamp, *Pharmacochem. Libr.*, 1997, **28**, 291.
16. D.J. Hammond, *Chromatographia*, 1998, **47**, 475.
17. F. Al-Obeidi, V.J. Hruby and T.K. Sawyer, *Mol. Biotechnol.*, 1998, **9**, 205.
18. J.M. Ostresh, B. Doerner and R.A. Houghten, *Methods Mol. Biol.*, 1998, **87**, 41.
19. K.S. Lam and M. Lebl, *Methods Mol. Biol.*, 1998, **87**, 1. K.S. Lam, *Methods Mol. Biol.*, 1998, **87**, 7.
20. C.T. Dooley and R.A. Houghten, *Methods Mol. Biol.*, 1998, **87**, 13.
21. A.A. Vergilio and J.A. Ellman, in *Comb. Chem. Mol. Diversity Drug Discovery*, eds. E.M. Gordon and J.F. Kerwin, Wiley-Liss, New York, N.Y. 1998, p. 133.
22. O. Seitz, *Angew. Chem., Int. Ed.*, 1998, **37**, 3109.
23. C.K. Jayawickreme, S.P. Jayawickreme and M.R. Lerner, *Methods Mol. Biol.*, 1998, **87**, 107.
24. A. Kramer and J. Schneider-Mergener, *Methods Mol. Biol.*, 1998, **87**, 25.
25. R.H. Hoess, in *Comb. Chem. Mol. Diversity Drug Discovery*, eds. E.M. Gordon and J.F. Kerwin, Wiley-Liss, New York, N.Y., 1998, p. 389.
26. J.W. Jacobs, D.V. Patel, Z. Yuan, C.P. Holmes, J. Schullek, V.V. Antonenko, J.R. Grove, N. Kulikov, D. Maclean, M. Navre, C. Nguyen, L. Shi, A. Sundaram and S.A. Sundberg, in *Comb. Chem. Mol. Diversity Drug Discovery*, eds. E.M. Gordon and J.F. Kerwin, Wiley-Liss, New York, N.Y. 1998, p. 111.

27. M. Erbeldinger, X. Ni and P.J. Halling, *Enzyme Microb.Technol.*, 1998, **23**, 141.
28. V.K. Svedas and A.I. Beltser, *Ann.N.Y.Acad.Sci.*, 1998, **864**, 524.
29. E.A. Lunney and C. Humblet, *Methods Princ.Med.Chem.*, 1998, **6**, 37.
30. Y. Kiso, *Yuki Gosei Kagaku Kyokaishi*, 1998, **56**, 896.
31. G.-J. Boons and R.L. Polt, *Carbohydr.Chem.*, 1998, p. 223.
32. J.-X. Xu and J.-H. Yang, *Hecheng Huaxue*, ed. G.-J. Boons, Blackie, London, 1998, **6**, 248.
33. N.V. Bovin, *Glycoconjugate J.*, 1998, **15**, 431.
34. H. Kunz and M. Schultz, *Chim.Oggi*, 1998, **16**, 45.
35. D.C. Hancock, N.J. O'Reilly and G.I. Evan, *Methods Mol.Biol.*, 1998, **80**, 69.
36. M. Sisido, T. Matsubara and H. Shinohara in *Novel Trends in Electroorganic Synthesis*, ed. S. Torii, Springer: Tokyo, 1998, p.413.
37. R. Sreekumar, M. Ikebe and F.S. Fay, *Methods Enzymol.*, 1998, **291**, 78.
38. H. Bayley, C.-Y. Chang, W.T. Miller, B. Niblack and P. Pan, *Methods Enzymol.*, 1998, **291**, 117.
39. M.D. Fletcher and M.M. Campbell, *Chem.Rev.*, 1998, **98**, 763.
40. G.M. McLaughlin, K.W. Anderson and D.K. Hauffe, *Chem.Anal.*, 1998, **146**, 637.
41. H.E. Schwartz, A. Guttman and A. Vinther, in *Capillary Electrophoresis* (2nd edn.), ed. P. Camilleri, C.R.C.: Boca Raton, Fl., 1998, p. 363.
42. J. Carlsson, F. Batista-Viera and L. Ryden, in *Protein Purification* (2nd edn.), eds. J.C. Janson and L. Ryden, Wiley-Liss, N.Y., 1998, p. 343.
43. G.N. Okafo, in *Capillary Electrophoresis* (2nd edn.), ed. P. Camilleri, C.R.C.: Boca Raton, Fla., 1998, p. 183.
44. L. Demange, A. Ménez and C. Dugave, *Tetrahedron Lett.*, 1998, **39**, 1169.
45. R.L.E. Furlán and E.G. Mata, *Tetrahedron Lett.*, 1998, **39**, 6421.
46. A.J. Zhang, D.H. Russell, J. Zhu and K. Burgess, *Tetrahedron Lett.*, 1998, **39**, 7439.
47. J. Ham, K. Choi, J. Ko, H. Lee and M. Jung, *Protein Pept.Lett.*, 1998, **5**, 257.
48. G. Mezó, N. Mihala, G. Kóczán and F. Hudecz, *Tetrahedron*, 1998, **54**, 6757.
49. S.P. Raillard, A.D. Mann and T.A. Baer, *Org.Prep.Proced.Int.*, 1998, **30**, 183.
50. A.N. Sabirov, Y.-D. Kim, H.-J. Kim and V.V. Samukov, *Protein Pept.Lett.*, 1998, **5**, 57.
51. R. Pascal and R. Sola, *Tetrahedron Lett.*, 1998, **39**, 5031.
52. X. Li, T. Kawakami and S. Aimoto, *Tetrahedron Lett.* 1998, **39**, 8669.
53. E.P. Ellmerer-Müller, D. Brössner, N. Maslouh and A. Takó, *Helv.Chim. Acta*, 1998, **81**, 59.
54. J. Kehler and E Breuer, *Synthesis*, 1998, 1419.
55. K. Augustyns, W. Kraas and G. Jung, *J.Pept.Res.*, 1998, **51**, 127.
56. S.R. Chhabra, B. Hothi, D.J. Evans, P.D. White, B.W. Bycroft and W.-C. Chan, *Tetrahedron Lett.*, 1998, **39**, 1603.
57. B. Rohwedder, Y. Mutti, P. Dumy and M. Mutter, *Tetrahedron Lett.*, 1998, **39**, 1175.
58. B. Kellam, B.W. Bycroft, W.C. Chan and S.R. Chhabra, *Tetrahedron*, 1998, **54**, 6817.
59. E. Naegele, M. Schelhaas, N. Kuder and H. Waldmann, *J.Am.Chem.Soc.*, 1998, **120**, 6889.
60. T. Kappes and H. Waldmann, *Carbohydr.Res.*, 1998, **305**, 341.
61. Y.S. Lee, H.J. Lee, P.I. Pozdnyakov, V.V. Samukov and H.J. Kim, *Bull. Korean Chem.Soc.*, 1998, **19**, 696.

62. O. Nishimura, M. Suenaga, H. Ohmae, S. Tsuji and M. Fujino, *Chem. Commun.*, 1998, 1135.
63. A. Sakakura and Y. Hayakawa, *Nucleic Acids Symp. Ser.*, 1998 **39**, 25.
64. R.C. Anand and Vimal, *Synth. Commun.*, 1998, **28**, 1963.
65. H. Zhao, A. Pendri and R.B. Greenwald, *J. Org. Chem.*, 1998, **63**, 7559.
66. Y.S. Lee, H.J. Lee, P.I. Pozdnyakov, V.V. Samukov and H.J. Kim, *Bull. Korean Chem. Soc.*, 1998, **19**, 717.
67. T. Miyazawa, M. Ogura, S. Nakajo and T. Yamada, *Biotechnol. Tech.*, 1998, **12**, 431.
68. A.F. Abdel-Magid, J.H. Cohen, C.A. Maryanoff, R.D. Shah, F.J. Villani and F. Zhang, *Tetrahedron Lett.*, 1998, **39**, 3391.
69. T. Schmittberger, A Cotté and H. Waldmann, *Chem. Commun.*, 1998, 937.
70. J. Bódi, Y. Nishiuchi, H. Nishio, T. Inui and T. Kimura, *Tetrahedron Lett.*, 1998, **39**, 7117.
71. A. Quesnel and J-P. Briand, *J. Pept. Res.*, 1998, **52**, 107.
72. P.M. Gitu, A.O. Yusuf and B.M. Bhatt, *Bull. Chem. Soc. Ethiop.*, 1998, **12**, 35; P.M. Gitu, A.O. Yusuf, V.O. Ogutu and B.M. Bhatt, *Int. J. BioChemiPhysics*, 1998, **6 & 7**, 7.
73. H.A. Moynihan and W. Yu. *Tetrahedron Lett.*, 1998, **39**, 3349.
74. M. Beltrán, E. Pedroso and A. Grandas, *Tetrahedron Lett.*, 1998, **39**, 4115.
75. M. Vilaseca, E. Nicolás, F. Capdevila and E. Giralt, *Tetrahedron*, 1998, **54**, 15273.
76. K. Barlos, D. Gatos and S. Koutsogianni, *J. Pept. Res.*, 1998, **51**, 194.
77. H. Pan, L. Wang and Z. Chen, *Zhongguo Yaowu Huaxue Zazhi*, 1997, **7**, 294.
78. H. Tamamura, F. Matsumoto, K. Sakano, A. Otaka, T. Ibuka and N. Fujii, *Chem. Commun.*, 1998, 151.
79. M. Royo, M.A. Contreras, E. Giralt, F. Albericio and M. Pons, *J. Am. Chem. Soc.*, 1998, **120**, 6639.
80. I. Annis, L. Chen and G. Barany, *J. Am. Chem. Soc.*, 1998, **120**, 7226.
81. H. Tamamura, T. Ishihara, H. Oyake, M. Imai, A. Otaka, T. Ibuka, R. Arakaki, H. Nakashima, T. Murakami, M. Waki, A. Matsumoto, N. Yamamoto and N. Fujii, *J. Chem. Soc., Perkin Trans. 1*, 1998, 495.
82. D. Limal, J.-P. Briand, P. Dalbon and M. Jolivet, *J. Pept. Res.*, 1998, **52**, 121.
83. J.P. Tam and Y.-A. Lu, *Protein Sci.*, 1998, **7**, 1583.
84. W.M. Macindoe, A.H. van Oijen and G.-J. Boons, *Chem. Commun.*, 1998, 847.
85. J.C. Spetzler, V. Westphal, J.R. Winther and M. Meldal, *J. Pept. Sci.*, 1998, **4**, 128.
86. L.A. Carpino, D. Ionescu, A. El-Faham, P. Henklein, H. Wenschuh, M. Bienert and M. Beyerman, *Tetrahedron Lett.*, 1998, **39**, 241.
87. V.V.S. Babu and H.N. Gopi, *Tetrahedron Lett.*, 1998, **39**, 1049.
88. K. Ananda, H.N. Gopi and V.V.S. Babu, *Lett. Pept. Sci.*, 1998, **5**, 277.
89. K. Gayathri, H.N. Gopi and V.V.S. Babu, *Indian J. Chem.*, 1998, **37B**, 151.
90. H.N. Gopi and V.V.S. Babu, *Tetrahedron Lett.*, 1998, **39**, 9769.
91. V.V.S. Babu and K. Gayathri, *Indian J. Chem.*, 1998, **37B**, 1109.
92. G. Karygiannis, C. Athanassopoulos, P. Mamos, N. Karamanos, D. Papaioannou and G.W. Francis, *Acta Chem. Scand.*, 1998, **52**, 1144.
93. M. Crisma, G. Valle, V. Moretto, F. Formaggio, C. Toniolo and F. Albericio, *Lett. Pept. Sci.*, 1998, **5**, 247.
94. K.G. Dendrinos and A.G. Kalivretenos, *Tetrahedron Lett.*, 1998, **39**, 1321.

95. J.C. Spetzler, M. Meldal, J. Felding, P. Vedsø and M. Bergtrup, *J.Chem. Soc., Perkin Trans. 1*, 1998, 1727.
96. H.N. Gopi and V.V.S. Babu, *Indian J.Chem.*, 1998, **37B**, 394.
97. L. Jiang, A. Davison, G. Tennant and R. Ramage, *Tetrahedron*, 1998, **54**, 14233.
98. S.K. Khare, G. Singh, K.C. Agarwal and B. Kundu, *Protein Pept.Lett.*, 1998, **5**, 171.
99. P. Wang, R. Layfield, M. Landon, R.J. Mayer and R. Ramage, *Tetrahedron Lett.*, 1998, **39**, 8711.
100. P. Wang, K.T. Shaw, B. Whigham and R. Ramage, *Tetrahedron Lett.*, 1998, **39**, 8719.
101. A. El-Faham, *Chem.Lett.*, 1998, 671.
102. F. Albericio, J.M. Bofill, A. El-Faham and S.A. Kates, *J.Org.Chem.*, 1998, **63**, 9678.
103. J. Habermann and H. Kunz, *J.Prakt.Chem./Chem.-Ztg.*, 1998, **340**, 233.
104. L. Zhang, T.R. Torgerson, X.-Y. Liu, S. Timmons, A.D. Colosia, J. Hawiger and J.P. Tam, *Proc.Natl.Acad.Sci., U.S.A.*, 1998, **95**, 9184.
105. J.P. Tam and Q. Yu, *Biopolymers*, 1998, **46**, 319.
106. H. Huang and R.I. Carey, *J.Pept.Res.*, 1998, **51**, 290.
107. G. Reischl, M. El-Mobayed, R. Beisswenger, K. Regier, C. Maichle-Mössmer and A. Rieker, *Z.Naturforsch., B: Chem.Sci.*, 1998, **53**, 765.
108. A. Hutinec, A. Ziogas, M. El-Mobayed and A. Rieker, *J.Chem.Soc., Perkin Trans.1*, 1998, 2201.
109. Y. Nisiuchi, T. Inui, H. Nishio, J. Bódi, T. Kimura, F.T. Tsuji and S. Sakakibara. *Proc.Natl.Acad.Sci., U.S.A.*, 1998, **95**, 13549.
110. T. Kawakami, S. Yoshimura and S. Aimoto, *Tetrahedron Lett.*, 1998, **39**, 7901.
111. D.R. Englebretsen, C.T. Choma and G.T. Robillard, *Tetrahedron Lett.*, 1998, **39**, 4929.
112. S.F. Martin, M.P. Dwyer and C.L. Lynch, *Tetrahedron Lett.*, 1998, **39**, 1517.
113. T. Yamada, Y. Omote, Y. Yamanaka, T. Miyazawa and S. Kuwata, *Synthesis*, 1998, 991.
114. C. Palomo, M. Oiarbide, A. Esnal, A. Landa, J.I. Miranda and A. Linden, *J.Org.Chem.*, 1998, **63**, 5838.
115. H.L. Son, Y. Suwannachot, J. Bujdak and B.M. Rode, *Inorg.Chim.Acta*, 1998, **272**, 89.
116. K. Gamoh and N. Yamasaki, *Bunseki Kagaku*, 1998, **47**, 303.
117. J.H. Adams, R.M. Cook, D. Hudson, V. Jammalamadaka, M.H. Lyttle and M.F. Songster, *J.Org.Chem.*, 1998, **63**, 3706.
118. S. Wendeborn, R. Beaudegnies, K.H. Ang and J.N. Maeji, *Biotechnol. Bioeng.*, 1998, **61**, 89.
119. D.S. Brown, J.M. Revill and R.E. Shute, *Tetrahedron Lett.*, 1998, **39**, 8533.
120. G. Orosz and L.P. Kiss, *Tetrahedron Lett.*, 1998, **39**, 3241.
121. K.H. Bleicher and J.R. Wareing, *Tetrahedron Lett.*, 1998, **39**, 4587, 4591.
122. B. Henkel and E. Bayer, *Tetrahedron Lett.*, 1998, **39**, 9401.
123. R. Léger, R. Yen, M.W. She, V.J. Lee and S.J. Hecker, *Tetrahedron Lett.*, 1998, **39**, 4171.
124. J.-G. Rodriguez, R. Martin-Villamil and S. Ramos, *New J.Chem.*, 1998, **22**, 865.
125. J. Buchardt and M. Meldal, *Tetrahedron Lett.*, 1998, **39**, 8695.
126. G. Panke and R. Frank, *Tetrahedron Lett.*, 1998, **39**, 17.
127. M. Noda, *Chem.Pharm.Bull.*, 1998, **46**, 1157.

128. C.R. Millington, R. Quarrell and G. Lowe, *Tetrahedron Lett.*, 1998, **39**, 7201.

129. K.J. Jensen, J. Alsina, M.F. Songster, J. Vágner, F. Albericio and G. Barany, *J.Am.Chem.Soc.*, 1998, **120**, 5441.

130. Y.-T. Wu, H.-P. Hsieh, C.-Y. Wu, H.-M. Yu, S.-T. Chen and K.-T. Wang, *Tetrahedron Lett.*, 1998, **39**, 1783.

131. H.-P. Hsieh, Y.-T. Wu and S.-T. Chen, *Chem.Commun.*, 1998, 649.

132. S.R. Chhabra, A.N. Khan and B.W. Bycroft, *Tetrahedron* Lett., 1998, **39**, 3585.

133. A.T. Dabre, T.S. Yokum and M.L. McLaughlin, *Proc. - NOBCChE*, 1997, **24**, 177.

134. A. Svensson, K.-E. Bergquist, T. Fex and J. Kihlberg, *Tetrahedron Lett.*, 1998, **39**, 7193.

135. C.T. Choma, G.T. Robillard and D.R. Englebretsen, *Tetrahedron Lett.*, 1998, **39**, 2417.

136. B.D. Larsen and A. Holm, *J.Pept.Res.*, 1998, **52**, 470.

137. C. Anne, M.-C. Fournié-Zaluski, B.P. Roques and F. Cornille, *Tetrahedron Lett.*, 1998, **39**, 8973.

138. S.C. Miller and T.S. Scanlan, *J.Am.Chem.Soc.*, 1998, **120**, 2690.

139. J.F. Reichwein and R.M.J. Liskamp, *Tetrahedron Lett.*, 1998, **39**, 1243.

140. A. Di Fenza, M. Tancredi, C. Galoppini and P. Rovero, *Tetrahedron Lett.*, 1998, **39**, 8529.

141. S. Matysiak, T. Böldicke, W. Tegge and R. Frank, *Tetrahedron Lett.*, 1998, **39**, 1733.

142. R.L.E. Furlán, E.G. Mata and O.A. Mascaretti, *J.Chem.Soc., Perkin Trans.1*, 1998, 355.

143. R.L.E. Furlán, E.G. Mata, O.A. Mascaretti, C. Peña and M.P. Coba, *Tetrahedron*, 1998, **54**, 13023.

144. L.P. Miranda, A. Jones, W.D.F. Meutermans and P.F. Alewood, *J.Am. Chem.Soc.*, 1998, **120**, 1410.

145. M. Stasiak and M.T. Leplawy, *Lett.Pept.Sci.*, 1998, **5**, 449.

146. D. Lelièvre, H. Chabane and A. Delmas, *Tetrahedron Lett.*, 1998, **39**, 9675.

147. M. Paris, A. Heitz, V. Guerlavais, M. Cristau, J.-A. Fehrentz and J. Martinez, *Tetrahedron Lett.*, 1998, **39**, 7287.

148. J.A. Camerero, G.J. Cotton, A. Adeva and T.W. Muir, *J.Pept.Res.*, 1998, **51**, 303.

149. Z. Tang and J.C. Pelletier, *Tetrahedron Lett.*, 1998, **39**, 4773.

150. K. Kaljuste and J.P. Tam, *Tetrahedron Lett.*, 1998, **39**, 9327.

151. B. Henkel and E. Bayer, *J.Pept.Sci.*, 1998, **4**, 461.

152. S.S. Rahman, D.J. Busby and D.C. Lee, *J.Org.Chem.*, 1998, **63**, 6196.

153. L.A. Marcaurelle and C.R. Bertozzi, *Tetrahedron Lett.*, 1998, **39**, 7279.

154. W.S. Newcomb, T.L. Deegan, W. Miller and J.A. Porco, *Biotechnol. Bioeng.*, 1998, **61**, 55.

155. M. Pátek, S. Bildstien and Z. Flegelová, *Tetrahedron Lett.*, 1998, **39**, 753.

156. R. Thirlner, M. Meisenbach, H. Echner, A. Weiler, A. Al-Qawasmeh, W. Voelter, U. Korff and W. Schmitt-Sody, *Rapid Commun.Mass Spectrom.*, 1998, **12**, 398.

157. V. Krchňák and V. Padĕra, *Bioorg.Med.Chem.Lett.*, 1998, **8**, 3261.

158. A. Boeijen and R.M.J. Liskamp, *Tetrahedron Lett.*, 1998, **39**, 3589.

159. H.S. Hiemstra, W.E. Benckhuijsen, R. Amons, W. Rapp and J.W. Drijfhout, *J.Pept.Sci.*, 1998, **4**, 282.

160. D. Mink, S. Mecozzi and J. Rebek, *Tetrahedron Lett.*, 1998, **39**, 5709.
161. B.R. Neustadt, E.M. Smith, T. Nechuta and Y. Zhang, *Tetrahedron Lett.*, 1998, **39**, 5317.
162. M. Falorni, G. Giacomelli, L. Mameli and A. Porcheddu, *Tetrahedron Lett.*, 1998, **39**, 7607.
163. E. Câmpian, M.L. Peterson, H.H. Saneii and A. Furka, *Bioorg. Med. Chem. Lett.*, 1998, **8**, 2357.
164. F. Dittrich, W. Tegge and R. Frank, *Bioorg. Med. Chem. Lett.*, 1998, **8**, 2351.
165. B.J. Hall and J.D. Sutherland, *Tetrahedron Lett.*, 1998, **39**, 6593.
166. B.R. Neustadt, E.M. Smith, N. Lindo, T. Nechuta, A. Bronnenkant, A. Wu, L. Armstrong and C. Kumar, *Bioorg. Med. Chem. Lett.*, 1998, **8**, 2395.
167. B.C. Hamper, S.A. Kolodziej, A.M. Scates, R.G. Smith and E. Cortez, *J. Org. Chem.*, 1998, **63**, 708.
168. S.W. Kim, Y.S. Shin and S. Ro, *Bioorg. Med. Chem. Lett.*, 1998, **8**, 1665.
169. S.W. Kim, S.M. Bauer and R.W. Armstrong, *Tetrahedron Lett.*, 1998, **39**, 6993.
170. T.M. Kapoor, A.H. Andreotti and S.L. Schreiber, *J. Am. Chem. Soc.*, 1998, **120**, 23.
171. N.J. Wells, M. Davies and M. Bradley, *J. Org. Chem.*, 1998, **63**, 6430.
172. D.-H. Ko, D.J. Kim, C.S. Lyu, I.K. Min and H.-s. Moon, *Tetrahedron Lett.*, 1998, **39**, 297.
173. R. Ramage, H.R. Swenson and K.T. Shaw, *Tetrahedron Lett.*, 1998, **39**, 8715.
174. M. Davies, M. Bonnat, F. Guillier, J.D. Kilburn and M. Bradley, *J. Org. Chem.*, 1998, **63**, 8696.
175. K. Mondorf, D.B. Kaufman and R.G. Carbonell, *J. Pept. Res.*, 1998, **52**, 526.
176. M. Fité, G. Alvaro, P. Clapés, J. López-Santin, M.D. Benaiges and G. Caminal, *Enzyme Microb. Technol.*, 1998, **23**, 199.
177. A. Murphy and C. O'Fagain, *Biotechnol. Bioeng.*, 1998, **58**, 364.
178. M.D. Romero, J. Aguado, M.J. Guerra, G. Alvaro, R. Navarro and E. Rubio, *Stud. Surf. Sci. Catal.*, 1997, **108**, 657.
179. M. Baust-Timpson and P. Seufer-Wasserthal, *Spec. Chem.*, 1998, **18**, 300.
180. Y. Zhao and J. Yang, *Huaxi Yike Daxue Xuebao*, 1998, **29**, 275.
181. G.-W. Xing, G.-L. Tian and Y.-H. Ye, *J. Pept. Res.*, 1998, **52**, 300.
182. Y.-H. Ye, G.-L. Tian, G.-W. Xing, D.-C. Dai, G. Chen and C.-X. Li, *Tetrahedron*, 1998, **54**, 12585.
183. M.N. Alam, K. Tadasa, T. Maeda, H. Kawase and H. Kayahara, *Shinshu Daigaku Nogakubu Kiyo*, 1998, **35**, 51.
184. P. Bjorup, J.L. Torres, P. Adlercreutz and P. Clapés, *Bioorg. Med. Chem.*, 1998, **6**, 891.
185. M.N. Alam, K. Tadasa and H. Kayahara, *Biotech. Tech.*, 1998, **12**, 115.
186. M.P. Bemquerer, C.W. Liria, K. Kitagawa, M.T.M. Miranda and M. Taminaga, *J. Pept. Res.*, 1998, **51**, 29.
187. K. Laszlo and L.M. Simon, *Prog. Biotechnol.*, 1998, **15**, 713.
188. M. Lobell and M.P. Schneider, *J. Chem. Soc., Perkin Trans. 1*, 1998, 319.
189. A. Vertesi and L.M. Simon, *J. Biotechnol.*, 1998, **66**, 75.
190. Y. Murakami, T. Yoshida and A. Hirata, *Biotechnol. Lett.*, 1998, **20**, 767.
191. C.W. Liria, M.P. Bemquerer and M.T.M. Miranda, *Tetrahedron Lett.*, 1998, **39**, 4207.
192. B.A. Bedell, V.V. Mozhaev, D.S. Clark and J.S. Dordick, *Biotechnol. Bioeng.*, 1998, **58**, 654.

193. D.J. van Unen, I.K. Sakodinskaya, J.F.J. Engbersen and D.N. Reinhoudt, *J.Chem.Soc., Perkin Trans.1*, 1998, 3341.
194. Y.-X. Chen, X.-Z. Zhang, K. Zheng, S.-M. Chen, Q.-C. Wang and X.-X. Wu, *Enzyme Microb.Technol.*, 1998, **23**, 243.
195. M. Hänsler and H.-D. Jakubke, *Enzyme Microb.Technol.*, 1998, **22**, 617.
196. M. Hänsler, S. Gerisch and H.-D. Jakubke, *Pharmazie*, 1998, **53**, 135.
197. T. Miyazawa, K. Hamahara, M. Matsuoka, Y. Shindo and T. Yamada, *Biotechnol.Lett.*, 1998, **20**, 389.
198. T. Miyazawa, K. Tanaka, E. Ensatsu, R. Yanagihara and T. Yamada, *Tetrahedron Lett.*, 1998, **39**, 997.
199. K. Beck-Piotraschke and H.-D. Jakubke, *Tetrahedron: Assymetry*, 1998, **9**, 1505.
200. H. Sekizaki, K. Itoh, E. Toyota and K. Tanizawa, *Chem.Pharm.Bull.*, 1998, **46**, 846.
201. G. Sahni, S.A. Khan and A.S. Acharya, *J.Protein Chem.*, 1998, **17**, 669.
202. D.B. Henriksen, M. Rolland, K. Breddam and O. Buchardt, *Protein Pept. Lett.*, 1998, **5**, 141.
203. M.A.P.J. Hacking, M.A. Wegman, J. Rops, F. van Rantwijk and R.A. Sheldon, *J.Mol.Catal.B:Enzym.*, 1998, **5**, 155.
204. A.C. Sarmento, L. Silvestre, M. Barros and E. Pires, *J.Mol.Catal.B: Enzym.*, 1998, **5**, 327.
205. J.-E. So, B.-K. Bae and B.-G. Kim, *Ann.N.Y.Acad.Sci.*, 1998, **864**, 327.
206. E. Dufour, W. Tam, D.K. Nagler, A.C. Storer and R. Menard, *FEBS Lett.*, 1998, **433**, 78.
207. S. Nakamura and S. Nakai, *Nahrung*, 1998, **42**, 131.
208. V. Čeřovský and M.-R. Kula, *Angew.Chem., Int.Ed.*, 1998, **37**, 1885.
209. J.-E. So, S.-H. Kang and B.-G. Kim, *Enzyme Microb.Technol.*, 1998, **23**, 211.
210. V. Jungmann and H. Waldmann, *Tetrahedron Lett.*, 1998, **39**, 1139.
211. M.W. Walter, R.M. Adlington, J.E. Baldwin and C.J. Schofield, *J.Org. Chem.*, 1998, **63**, 5179.
212. O. Melnyk, P. Chaurand, C. Rommens, H. Drobecq, J.M. Wieruszeski, B. Spengler and H. Gras-Masse, *J.Pept.Res.*, 1998, **51**, 188.
213. K. Feichtinger, C. Zapf, H.L. Sings and M. Goodman, *J.Org.Chem.*, 1998, **63**, 3804.
214. K. Feichtinger, H.L. Sings, T.J. Baker, K. Matthews and M. Goodman, *J.Org.Chem.*, 1998, **63**, 8432.
215. G. Arsequell, G. Espuña, G. Valencia, J. Barluenga, R.P. Carlón and J.M. González, *Tetrahedron Lett.*, 1998, **39**, 7393.
216. S. Futaki, *Yakugaku Zasshi*, 1998, **118**, 493.
217. D.D. DesMarteau and V. Montanari, *Chem.Commun.*, 1998, 2241.
218. K. Wisniewski, A.S. Koldziejezyk and B. Falkiewicz, *J.Pept.Sci.*, 1998, **4**, 1.
219. T. Hintermann, C. Mathes and D. Seebach, *Eur.J.Org.Chem.*, 1998, 2379.
220. T. Matt and D. Seebach, *Helv.Chim.Acta*, 1998, **81**, 1845.
221. K. Miyashita, H. Iwaki, K. Tai, H. Murafuji and T.Imanishi, *Chem.Commun.*, 1998, 1987.
222. S. Yao, I. Ghosh, R. Zutshi and J. Chmielewski, *Nature*, 1998, **396**, 447; *Angew.Chem., Int.Ed.*, 1998, **37**, 478.
223. H. Wang, M. Zhao and S. Peng, *Zhongguo Yaowu Huaxue Zazhi*, 1998, **8**, 54.
224. K. Iguchi, T. Yamaki, H. Kadoyama, T. Mochizuki, M. Hoshino, Y. Futai and G. Lee, *ACTH Relat.Pept.*, 1997, **8**, 53.

225. E.C. Gulyas, K. Soos, J. Varga, G. Toth and B. Penke, *J. Labelled Compd. Radiopharm.*, 1998, **41**, 763.
226. P. Ruzza, A. Calderan, M. Carrara, T. Tancredi and G. Borin, *Curr. Top. Pept. Protein Res.*, 1997, **2**, 21.
227. Y. Xu and M.J. Miller, *J. Org. Chem.*, 1998, **63**, 4314.
228. S.G. Davies and D.J. Dixon, *J. Chem. Soc., Perkin Trans. 1*, 1998, 2635.
229. E. Suarez, E. De, G. Molle, R. Lazaro and P. Viallefont, *J. Pept. Sci.*, 1998, **4**, 371.
230. Y. Lu, H. Wang and X. Qu, *Shengwu Gongcheng Xuebao*, 1997, **13**, 35.
231. V. Monaco, E. Locardi, F. Formaggio, M. Crisma, S. Mammi, E. Peggion, C. Toniolo, S. Rebuffat and B. Bodo, *J. Pept. Res.*, 1998, **52**, 261.
232. A. Ogrel, A. Ogrel, S. Ogrel, V. Shvets and J. Raap, *Lett. Pept. Sci.*, 1998, **5**, 175.
233. T. Staroske and D.H. Williams, *Tetrahedron Lett.*, 1998, **39**, 4917.
234. B. Holm, S. Linse and J. Kihlberg, *Tetrahedron*, 1998, **54**, 11995.
235. T.M. Hackeng, P.E. Dawson, S.B.H. Kent and J.H. Griffin, *Biopolymers*, 1998, **46**, 53.
236. C.V. Ramesh, R. Jayakumar and R. Puvanakrishnan, *Peptides*, 1998, **19**, 1695.
237. S.I. Klein, B.F. Molino, M. Czekaj, C.J. Gardner, V. Chu, K. Brown, R.D. Sabatino, J.S. Bostwick, C. Kasiewski, R. Bentley, V. Windisch, M. Perrone, C.T. Dunwiddie and R.J. Leadley, *J. Med. Chem.*, 1998, **41**, 2492.
238. I. Daffix, M. Amblard, G. Berge, P. Dodey, D. Pruneau, J.-L. Paquet, C. Fouchet, R.-M. Franck, E. Defrene, J.-M. Luccarini, P. Belichard and J. Martinez, *J. Pept. Res.*, 1998, **52**, 1.
239. M. Lange, A.S. Cuthbertson, R. Towart and P.M. Fischer, *J. Pept. Sci.*, 1998, **4**, 289.
240. M. Mizuno, I. Muramoto, T. Kawakami, M. Seike, S. Aimoto, K. Haneda and T. Inazu, *Tetrahedron Lett.*, 1998, **39**, 55.
241. H.-P. Pan, L.-Y. Wang, Z.-Y. Chen, F. Wang and H.-X. Wang, *Zhongguo Shengwu Huaxue Yu Fenzi Shengwu Xuebao*, 1998, **14**, 463.
242. B. Yang, H. Dong, Z. Han, Z. Ding, X. Ma and Q. Zhang, *Junshi Yixue Kexueyuan Yuankan*, 1997, **21**, 247.
243. L. Wang, Z. Chen and H. Pan, *Zhongguo Yaowu Huaxue Zazhi*, 1997, **7**, 287.
244. S. Saha, D.J.J. Waugh, P. Zhao, P.W. Abel and D.D. Smith, *J. Pept. Res.*, 1998, **52**, 112.
245. B.E. Kaplan, L.J. Hefta, R.C. Blake, K.M. Swiderek and J.E. Shively, *J. Pept. Res.*, 1998, **52**, 249.
246. S. Vunnam, P. Juvvadi, K.S. Rotondi and R.B. Merrifield, *J. Pept. Res.*, 1998, **51**, 38.
247. G. Pagani Zecchini, I. Torrini, M. Paglialunga Paradisi, G. Lucente, G. Mastropietro and S. Spisani, *Amino Acids*, 1998, **14**, 301.
248. S. Spisani, M.E. Ferretti, E. Fabbri, S. Alvoni and G. Cavicchioni, *Protein Pept. Lett.*, 1998, **5**, 155.
249. I. Torrini, G. Mastropietro, G. Pagani Zecchini, M. Paglialunga Paradisi, G. Lucente and S. Spisani, *Arch. Pharm.*, 1998, **331**, 170.
250. G. Cavicchioni, L.G. Monesi, M.E. Ferretti, E. Fabbri, O. Rizzuti and S. Spisani, *Arch. Pharm.*, 1998, **331**, 368.
251. M. Amblard, M. Rodriguez, M.-F. Lignon, M.-C. Galas, N. Bernad, A. Aumelas and J. Martinez, *Eur. J. Med. Chem.*, 1998, **33**, 171.
252. B. Bellier, S. Da Nascimento, H. Meudal, E. Gincel, B.P. Roques and C. Garbay, *Bioorg. Med. Chem. Lett.*, 1998, **8**, 1419.
253. E.A. Jefferson, E. Locardi and M. Goodman, *J. Am. Chem. Soc.*, 1998, **120**, 7420.

254. M. Goodman, M. Bhumralkar, E.A. Jefferson, J. Kwak and E. Locardi, *Biopolymers*, 1998, **47**, 127.

255. J. Rivier, J. Gulyas, A. Corrigan, V. Martinez, A.G. Craig, Y. Taché, W. Vale and C. Rivier, *J. Med. Chem.*, 1998, **41**, 5012.

256. S.C. Koerber, J. Gulyas, S.L. Lahrichi, A. Corrigan, A.G. Craig, C. Rivier, W. Vale and J. Rivier, *J. Med. Chem.*, 1998, **41**, 5002.

257. X. Lauth, A. Nesin, J.-P. Briand, J.-P. Roussel and C. Hetru, *Insect Biochem.-Mol. Biol.*, 1998, **28**, 1059.

258. A.J. Pfizenmayer, M.D. Vera, X. Ding, D. Xiao, W.-C. Chen and M.M. Joullié, *Bioorg. Med. Chem. Lett.*, 1998, **8**, 3653.

259. J.J. Wen and C.M. Crews, *Tetrahedron Lett.*, 1998, **39**, 779.

260. G.R. Pettit, J.K. Srirangam, J. Barkoczy, M.D. Williams, M.R. Boyd, E. Hamel, R.K. Pettit, F. Hogan, R. Bai, J.-C. Chapuis, S.C. McAllister and J.M. Schmidt, *Anti-Cancer Drug Des.*, 1998, **13**, 243.

261. J. Schmitt, M. Bernd, B. Kutscher and H. Kessler, *Bioorg. Med. Chem. Lett.*, 1998, **8**, 385.

262. C. Thomas, U. Ohnmacht, M. Niger and P. Gmeiner, *Bioorg. Med. Chem. Lett.*, 1998, **8**, 2885.

263. E. Cassano, C. Galoppini, L. Giusti, M. Hamdan, M. Macchia, M.R. Mazzoni, E. Menchini, S. Pegoraro and P. Rovero, *Folia Biol.*, 1998, **44**, 11.

264. S. Pegoraro, S. Fiori, S. Rudolph-Bohner, T.X. Watanabe and L. Moroder, *J. Mol. Biol.*, 1998, **284**, 779.

265. J. Sakaki, T. Murata, Y. Yuumoto, I. Nakamura and K. Hayakawa, *Bioorg. Med. Chem. Lett.*, 1998, **8**, 2247.

266. J. Sakaki, T. Murata, Y. Yuumoto, I. Nakamura, T. Frueh, T. Pitterna, G. Iwasaki, K. Oda, T. Yamamura and K. Hayakawa, *Bioorg. Med. Chem. Lett.*, 1998, **8**, 2241.

267. H. Chen, A.L. Pyluck, M. Janik and N.S. Sampson, *Biopolymers*, 1998, **47**, 299.

268. M. Shiozaki, N. Deguchi, T. Ishikawa, H. Haruyama, Y. Kawai and M. Nishijima, *Tetrahedron Lett.*, 1998, **39**, 4497.

269. N.S. Sturm, Y. Lin, S.K. Burley, J.L. Krstenansky, J.-M. Ahn, B.Y. Azizeh, D. Trivedi and V.J. Hruby, *J. Med. Chem.*, 1998, **41**, 2693.

270. A. Graul, X. Rabasseda and J. Castaner, *Drugs Future*, 1998, **23**, 1057.

271. S. Rahimipour, L. Weiner, P.B. Shrestha-Dawadi, S. Bittner, Y. Koch and M. Fridkin, *Lett. Pept. Sci.*, 1998, **5**, 421.

272. M. Keramida, J.M. Matsoukas, G. Agelis, D. Panagiotopoulos, J. Cladas, D. Pati, G.J. Moore and H.R. Habibi, *Lett. Pept. Sci.*, 1998, **5**, 305.

273. V. Cavallaro, P. Thompson and M. Hearn, *J. Pept. Sci.*, 1998, **4**, 335.

274. L. Yang, G. Morriello, Y. Pan, R.P. Nargund, K. Barakat, K. Prendergast, K. Cheng, W.W.-S. Chan, R.G. Smith and A.A. Patchett, *Bioorg. Med. Chem. Lett.*, 1998, **8**, 759.

275. M. Ankersen, N.L. Johansen, K. Madsen, B.S. Hansen, K. Raun, K.K. Nielsen, H. Thøgersen, T.K. Hansen, B. Peschke, J. Lau, B.F. Lundt and P.H. Andersen, *J. Med. Chem.*, 1998, **41**, 3699.

276. M. Ota, Y. Shimizu, K. Tonosaki and Y. Ariyoshi, *Biopolymers*, 1998, **46**, 65.

277. A. Lombardi, F. Nastri, M. Sanseverino, O. Maglio, C. Pedone and V. Pavone, *Inorg. Chim. Acta*, 1998, 275.

278. B. Karawajczyk, J. Wysocki, D. Kunikowska, Z. Makiewicz, R. Glosnicka, A. Korzeniowski, J. Gorski and G. Kupryszewski, *Pol. J. Chem.*, 1998, **72**, 1017.

279. J. Hlaváček, R. Tykva, B. Bennettová and T. Barth, *Bioorg. Chem.*, 1998, **26**, 131.

280. D. Konopinska, H. Bartosz-Bechowski, M. Kuczer, G. Rosinski, I. Janssen and A. De Loof, *Lett.Pept.Sci.*, 1998, **5**, 391.

281. Y. Ye, M.-H. Xu, X.-T. Zhang, S.-Y. Yao and S.-Q. Zhu, *Shengwu Huaxue Yu Shengwu Wuli Xuebao*, 1998, **30**, 114.

282. J. Shi, J. Fang, Z. Cheng, C. Cheng and B. Sun, *Disi Junyi Daxue Xuebao*, 1998, **19**, 520.

283. F. Azam and C.M. Bladon, *Tetrahedron Lett.*, 1998, **39**, 6377.

284. S.Y. Kassim, L.M. Restrepo and A.G. Kalivretenos, *J.Chromatogr.*, 1998, **816**, 11.

285. M. Maeda, Y. Izuno, K. Kawasaki, Y. Kaneda, Y. Mu, Y. Tsutsumi, S. Nakagawa and T. Mayumi, *Chem.Pharm.Bull.*, 1998, **46**, 347.

286. Y. Nishiyama, T. Yoshikawa, T. Mori, K. Kurita, K. Hojo, K. Kawasaki, Y. Tsutsumi and T. Mayumi, *Kichin.Kitosan Kenkyu*, 1998, **4**, 218.

287. M. Maeda, K. Kawasaki, Y. Mu, H. Kamada, Y. Tsutsumi, T.J. Smith and T. Mayumi, *Biochem.Biophys.Res.Commun.*, 1998, **248**, 485.

288. P. Mayer-Fligge, J. Volz, U. Kruger, E. Sturm, W. Gernandt, K.P. Schafer and M. Przybylski, *J.Pept.Sci.*, 1998, **4**, 355.

289. H.N. Shroff, C.F. Schwender, A.D. Baxter, F. Brookfield, L.J. Payne, N.A. Cochran, D.L. Gallant and M.J. Briskin, *Bioorg.Med.Chem.Lett.*, 1998, **8**, 1601.

290. M. Kohmura and Y. Ariyoshi, *Biopolymers*, 1998, **46**, 215.

291. W.M.M. Schaaper, R.A.H. Adan, T.A. Posthuma, J. Oosterom, W.-H. Gispen and R.H. Meloen, *Lett.Pept.Sci.*, 1998, **5**, 205.

292. V.J. Hruby, G. Han and M.E. Hadley, *Lett.Pept.Sci.*, 1998, **5**, 117.

293. M. Ota and Y. Ariyoshi, *Biosci. Biotechnol. Biochem.*, 1998, **62**, 2043.

294. A. Lombardi, B. D'Agostino, A. Filippelli, C. Pedone, M.G. Matera, M. Falciani, M. De Rosa, F. Rossi and V. Pavone, *Bioorg.Med.Chem.Lett.*, 1998, **8**, 1735.

295. A. Lombardi, B. D'Agostino, F. Nastri, L.D. D'Andrea, A. Filippelli, M. Falciani, F. Rossi and V. Pavone, *Bioorg.Med.Chem.Lett.*, 1998, **8**, 1153.

296. F. Sebestyen, G. Szendrei, M. Mak, M. Doda, E. Illyes, G. Szokan, K. Kindla, W. Rapp, P. Szego, E. Campian and A. Furka, *J.Pept.Sci.*, 1998, **4**, 294.

297. D. Terriere, K. Chavatte, M. Ceusters, D. Tourwe and J. Mertens, *J.Labelled Compd.Radiopharm.*, 1998, **41**, 19.

298. D. Tourwe, J. Mertens, M. Ceusters, L. Jeannin, K. Iterbeke, D. Terriere, C. Chavatte and R. Boumon, *Tumour Targeting*, 1998, **3**, 41.

299. T.A. Gudasheva, T.A. Voronina, R.U. Ostovskaya, N.I. Zaitsheva, N.A. Bondarenko, V.K. Briling, L.S. Asmakova, G.G. Rozantsev and S.B. Seredenin, *J.Med.Chem.*, 1998, **41**, 284.

300. M. Kuczer, G. Rosinski, J. Issberner, R. Osborne and D. Konopinska, *Lett.-Pept.Sci.*, 1998, **5**, 387.

301. J. Leprince, P. Gandolfo, J.-L. Thoumas, C. Patte, J.-L. Fauchère, H. Vaudry and M.-C. Tonon, *J.Med.Chem.*, 1998, **41**, 4433.

302. R.J. Nachman, G. Moyna, H.J. Williams, S.S. Tobe and A.I. Scott, *Bioorg.Med.-Chem.*, 1998, **6**, 1379.

303. G. Calo, R. Guerrini, R. Bigoni, A. Rizzi, C. Bianchi, D. Regoli and S. Salvadori, *J.Med.Chem.*, 1998, **41**, 3360.

304. T.-W. Lee, S.-R. Chang, S.-T. Chen and Z.-T. Tsai, *Appl.Radiat.Isot.*, 1998, **49**, 1581.

305. T.W. Spradau, W.B. Edwards, C.J. Anderson, M.W. Welch and J.A. Katzenellenbogen, *Nucl.Med.Biol.*, 1998, **26**, 1.

306. P.M. Smith-Jones, B. Stolz, R. Albert, G. Ruser, U. Briner, H.R. Macke and C. Bruns, *Nucl. Med. Biol.*, 1998, **25**, 181.

307. E. Masiukiewicz and B. Rzeszotarska, *Chem. Pharm. Bull.*, 1998, **46**, 1672.

308. H.N. Gopi and V.V.S. Babu, *J. Indian Chem. Soc.*, 1998, **75**, 511.

309. K. Shreder, L. Zhang, T. Dang, T.L. Yaksh, H. Umeno, R. DeHaven, J. Daubert and M. Goodman, *J. Med. Chem.*, 1998, **41**, 2631.

310. K. Shreder, L. Zhang and M. Goodman, *Tetrahedron Lett*, 1998, **39**, 221.

311. H.I. Mosberg, J.C. Ho and K. Sobczyk-Kojiro, *Bioorg. Med. Chem. Lett.*, 1998, **8**, 2681.

312. M. Čudić, J. Horvat, M. Elofsson, K.-E. Bergquist, J. Kihlberg and S. Horvat, *J. Chem. Soc., Perkin Trans. 1*, 1998, 1789.

313. I. Bobrova, N. Abissova, N. Mishlakova, G. Rozentals and G. Chipens, *Eur. J. Med. Chem.*, 1998, **33**, 255.

314. J. Olczak, K. Kaczmarek, I. Maszczynska, M. Lisowski, D. Stropova, V.J. Hruby, H.I. Yamamura, A.W. Lipkowski and J. Zabrocki, *Lett. Pept. Sci.*, 1998, **5**, 437.

315. D. Winkler, N. Sewald, K. Burger, N.N. Chung and P.W. Schiller, *J. Pept. Sci.*, 1998, **4**, 496.

316. H. Han, J. Yoon and K.D. Janda, *Bioorg. Med. Chem. Lett.*, 1998, **8**, 117.

317. M. Horikawa, Y. Shigeri, N. Yumoto, S. Yoshikawa, T. Nakajima and Y. Ohfune, *Bioorg. Med. Chem. Lett.*, 1998, **8**, 2027.

318. J.A.W. Kruijtzer, L.J.F. Hofmeyer, W. Heerma, C. Versluis and R.M.J. Liskamp, *Chem.–Eur. J.*, 1998, **4**, 1570.

319. B. Kellam, B. Drouillat, G. Dekany, M.S. Starr and I. Toth, *Int. J. Pharm.*, 1998, **161**, 55.

320. G. Li, W. Haq, L. Xiang, B.-S. Lou, R. Hughes, I.A. De Leon, P. Davis, T.J. Gillespie, M. Romanowski, X. Zhu, A. Misicka, A.W. Lipkowski, F. Porreca, T.P Davis, H.I. Yamamura, D.F. O'Brien and V.J. Hruby, *Bioorg. Med. Chem.-Lett.*, 1998, **8**, 555.

321. I. Kertesz, G. Balboni, S. Salvadori, L.H. Lazarus and G. Toth, *J. Labelled Compd. Radiopharm.*, 1998, **41**, 1083.

322. P.W. Schiller, R. Schmidt, G. Weltrowska, I. Berezowska, T.M.-D. Nguyen, S. Dupuis, N.N. Chung, C. Lemieux, B.C. Wilkes and K.A. Carpenter, *Lett. Pept. Sci.*, 1998, **5**, 209.

323. Y. Okada, M. Tsukatani, H. Taguchi, T. Yokoi, S.D. Bryant and L.H. Lazarus, *Chem. Pharm. Bull.*, 1998, **46**, 1374.

324. K.M. Sivanandaiah, V.V.S. Babu and S.C. Shankaramma, *Indian J. Chem.*, 1998, **37B**, 760.

325. T. Naqvi, R. Raghubir, W. Haq, A. Tripathi, G.K. Patnaik and K.B. Mathur, *Neuropeptides*, 1998, **32**, 333.

326. A. Calderan, P. Ruzza, B. Ancona, L. Cima, P. Giusti and G. Borin, *Lett. Pept. Sci.*, 1998, **5**, 71.

327. L. Leelasvatanakij and J.V. Aldrich, *Warasan Phesatchasat*, 1996, **23**, 22.

328. R. Guerrini, A. Capasso, M. Marastoni, S.D. Bryant, P.S. Cooper, L.H. Lazarus, P.A. Temussi and S. Salvadori, *Bioorg. Med. Chem.*, 1998, **6**, 57.

329. S.U. Kook and K.N. Son, *J. Korean Chem. Soc.*, 1998, **42**, 214.

330. S. Liao, M.D. Shenderovich, Z. Zhang, L. Maletinska, J. Slaninova and V.J. Hruby, *J. Am. Chem. Soc.*, 1998, **120**, 7393.

331. H. Susaki, K. Suzuki, M. Ikeda, H. Yamada and H.K. Watanabe, *Chem. Pharm. Bull.*, 1998, **46**, 1530.

332. O. Nishimura, T. Moriya, M. Suenaga, Y. Tanaka, T. Itoh, N. Koyama, R. Fujii, S. Hinuma, C. Kitada and M. Fujino, *Chem.Pharm.Bull.*, 1998, **46**, 1490.

333. H. Fukuda, K. Irie, A. Nakahara, K. Oie, H. Ohigashi and P.A. Wender, *Tetrahedron Lett.*, 1998, **39**, 7943.

334. A. Dal Pozzo, L. Muzi, M. Moroni, R. Rondanin, R. De Castiglione, P. Bravo and M. Zanda, *Tetrahedron*, 1998, **54**, 6019.

335. T. Boxus, R. Touillaux, G. Dive and J. Marchand-Brynaert, *Bioorg.Med. Chem.*, 1998, **6**, 1577.

336. P.P. De Laureto, E. Scaramella, V. De Filippis, O. Marin, M.G. Doni and A. Fontana, *Protein Sci.*, 1998, **7**, 433.

337. H. Fujii, N. Nishikawa, H. Komazawa, M. Suzuki, M. Kojima, I. Itoh, A. Obata, K. Ayukawa, I. Azuma and I. Saiki, *Clin.Exp.Metastasis*, 1998, **16**, 94.

338. D.A. Annis, O. Helluin and E.N. Jacobsen, *Angew.Chem., Int.Ed.*, 1998, **37**, 1907.

339. B.T. Houseman and M. Mrksich, *J.Org.Chem.*, 1998, **63**, 7552.

340. D. Lim, D. Moye-Sherman, I. Ham, S. Jin, K. Burgess and J.M. Scholtz *Chem.Commun.*, 1998, 2375.

341. B. Arshava, S.-F. Liu, H. Jiang, M. Breslav, J.M. Becker and F. Naider, *Biopolymers*, 1998, **46**, 343.

342. C. Gilon, M. Huenges, B. Mathä, G. Gellerman, V. Hornik, M. Afargan, O. Amitay, O. Ziv, E. Feller, A. Gamliel, D. Shohat, M. Wanger, O. Arad and H. Kessler, *J.Med.Chem.*, 1998, **41**, 919.

343. T.-A. Tran, R.-H. Mattern, M. Afargan, O. Amitay, O. Ziv, B.A. Morgan, J.E. Taylor, D. Hoyer and M. Goodman, *J.Med.Chem.*, 1998, **41**, 2679.

344. S.J. Hocart, R. Jain, W.A. Murphy, J.E. Taylor, B. Morgan and D.H. Coy, *J.Med.Chem.*, 1998, **41**, 1146.

345. A. Nagy, A.V. Schally, G. Halmos, P. Armatis, R.-Z. Cai, V. Csernus, M. Kovacs, M. Koppan, K. Szepeshazi and Z. Kahan, *Proc.Natl.Acad.Sci.*, *U.S.A.* 1998, **95**, 1794.

346. I. Wojciechowska, B. Kochanska, E. Stelmanska, N. Knap, Z. Mackiewicz and G. Kupryszewski, *Pol.J.Chem.*, 1998, **72**, 2098.

347. Y. Tong, Y.M. Fobian, M. Wu, N.D. Boyd and K.D. Moeller, *Bioorg.Med. Chem.Lett.*, 1998, **8**, 1679.

348. P. Karoyan, S. Sagan, G. Clodic, S. Lavielle and G. Chassaing, *Bioorg. Med.Chem.Lett.*, 1998, **8**, 1369.

349. D.A. Kirby, W. Wang, M.C. Gershengorm and J.E. Rivier, *Peptides*, 1998, **19**, 1679.

350. E. Mahé, P. Vossen, H.W. Van den Hooven, D. Le-Nguyen, J. Vervoort and P.J.G.M. De Wit, *J.Pept.Res.*, 1998, **52**, 482.

351. B.Z. Dolimbek and M.Z. Atassi, *Chem.Nat.Compd.*, 1998, **33**, 485.

352. S.T. Moe, D.L. Smith, Y. Chien, J.L. Raszkiewicz, L.D. Artman and A.L. Mueller, *Pharm.Res.*, 1998, **15**, 31.

353. H. Saito, E. Yuri, M. Miyazawa, Y. Itagaki, T. Nakajima and M. Miyashita, *Tetrahedron Lett.*, 1998, **39**, 6479.

354. H. Nishio, T. Inui, Y. Nishiuchi, C.L.C. De Medeiros, E.G. Rowan, A.L. Harvey, E. Katoh, T. Yamazaki, T. Kimura and S. Sakakibara, *J.Pept.Res.*, 1998, **51**, 355.

355. X. Wang, S. Liang and Z. Luo, *Shengming Kexue Yanjiu*, 1998, **2**, 87.

356. W. Lee, N.R. Krishnat, G.M. Anatharamaiah and H.C. Cheung, *Bull.Korean Chem.Soc.*, 1998, **19**, 57.

357. H. Mostafavi, K. Adermann, S. Austermann, M. Raida, M. Meyer and W.G. Forssmann, *Biomed.Pept., Proteins Nucleic Acids*, 1995, **1**, 255.

358. A.M. Felix, Z. Zhao, T. Lambros, M. Ahmad, W. Liu, A. Daniewski, J. Michalewsky and E.P. Heimer, *J.Pept.Res.*, 1998, **52**, 155.

359. D.L. Rabenstein, J. Russell and J. Gu, *J.Pept.Res.*, 1998, **51**, 437.

360. M. Dettin, C. Scarinci, C. Zanotto, A. Cabrelle, A. De Rossi and C. Di Bello, *J.Pept.Res.*, 1998, **51**, 110.

361. J.H. Viles, S.U. Patel, J.B.O. Mitchell, C.M. Moody, D.E. Justice, J. Uppenbrink, P.M. Doyle, J. Harris, P.J. Sadler and J.M. Thornton, *J.Mol. Biol.*, 1998, **279**, 973.

362. B. Pengo, F. Formaggio, M. Crisma, C. Toniolo, G.M. Bonora, Q.B. Broxterman, J. Kamphuis, M. Saviano, R. Iacovino, F. Rossi and E. Benedetti, *J.Chem.Soc., Perkin Trans.2*, 1998, 1651.

363. T. Nakato, M. Yoshitake, K. Matsubara, M. Tomida and T. Kakuchi, *Macromolecules*, 1998, **31**, 2107.

364. H. Kanazawa, *Mol.Cryst.Liq.Cryst.Sci.Technol.*, Sect A, 1998, **313**, 205.

365. T.M. Cooper, L.V. Natarajan and C.G. Miller, *Polym.Prepr.*, 1998, **39**, 760.

366. M.D. Smith, T.D.W. Claridge, G.E. Tranter, M.S.P. Sansom and G.W.J. Fleet, *Chem.Commun.*, 1998, 2041.

367. M.D. Smith, D.D. Long, D.G. Marquess, T.D.W. Claridge and G.W.J. Fleet, *Chem.Commun.*, 1998, 2039.

368. W. Hoffmueller, M. Maurus, K. Severin and W. Beck, *Eur.J.Inorg.Chem.*, 1998, 729.

369. M. Yu and T.J. Deming, *Macromolecules*, 1998, **31**, 4739.

370. A.C. Birchall, *Chem.Commun.*, 1998, 1335.

371. S. Hashimoto, K. Yamamoto, T. Yamada and Y. Nakamura, *Heterocycles*, 1998, **48**, 939.

372. L.J. Twyman, A.E. Beezer, R. Esfand, B.T. Mathews and J.C. Mitchell, *J.Chem.Res., Synop.*, 1998, 758.

373. D.B. O'Sullivan, E. Murphy, T.P. O'Connell, E.A. Murphy and J.P.G. Malthouse, *Biochem.Soc.Trans.*, 1998, **26**, S67.

374. M. Renatus, W. Bode, R. Huber, J. Stürzebecher and M.T. Stubbs, *J.Med. Chem.*, 1998, **41**, 5445.

375. A. Jaskiewicz, A. Lesner, J. Rozycki, G. Kupryszewski and K. Rolka, *Pol.J. Chem.*, 1998, **72**, 2537.

376. W. Lu, M.A. Starovasnik, and S.B.H. Kent, *FEBS Lett.*, 1998, **429**, 31.

377. R. Kasher, G. Bitan, C. Halloun and C. Gilon, *Lett.Pept.Sci.*, 1998, **5**, 101.

378. D.S. Jackson, S.A. Fraser, L.-M. Ni, C.-M. Kam, U. Winkler, D.A. Johnson, C.J. Froelich, D. Hudig and J.C. Powers, *J.Med.Chem.*, 1998, **41**, 2289.

379. J.F. Lynas, P. Harriott, A. Healy, M.A. McKervey and B. Walker, *Bioorg.Med.-Chem.Lett.*, 1998, **8**, 373.

380. R.T. Lum, M.G. Nelson, A. Joly, A.G. Horsma, G. Lee, S.M. Meyer, M.M. Wick and S.R. Schow, *Bioorg.Med.Chem.Lett.*, 1998, **8**, 209.

381. L. Guy, J. Vidal, A. Collet, A. Amour and M. Reboud-Ravaux, *J.Med.Chem.*, 1998, **41**, 4833.

382. L. Wang and F. Xu, *Zhongguo Yaoke Daxue Xuebao*, 1998, **29**, 153.

383. R.J. Cregge, S.L. Durham, R.A. Farr, S.L. Gallion, C.M. Hare, R.V. Hoffman, M.J. Janusz, H.-O. Kim, J.R. Koehl, S. Mehdi, W.A. Metz, N.P.

Peet, J.T. Pelton, H.A. Schreuder, S. Sunder and C. Tardif, *J.Med.Chem.*, 1998, **41**, 2461.

384. A. Amour, M. Reboud-Ravaux, E. De Rosny, A. Abouabdellah, J.-P. Begue, D. Bonnet-Delpon and M. Le Gall, *J.Pharm.Pharmacol.*, 1998, **50**, 593.

385. J.P. Burkhart, S. Mehdi, J.R. Koehl, M.R. Angelastro, P. Bey and N.P. Peet, *Bioorg.Med.Chem.Lett.*, 1998, **8**, 63.

386. J.M. Fevig, J. Buriak, J. Cacciola, R.S. Alexander, C.A. Kettner, R.M. Knabb, J.R. Pruitt, P.C. Weber and R.R. Wexler, *Bioorg.Med.Chem.Lett.*, 1998, **8**, 301.

387. J.V. Duncia, J.B. Santella, C.A. Higley, M.K. Vanatten, P.C. Weber, R.S. Alexander, C.A. Kettner, J.R. Pruitt, A.Y. Liauw, M.L. Quan, R.M. Knabb and R.R. Wexler, *Bioorg.Med.Chem.Lett.*, 1998, **8**, 775.

388. G. Liu, S. Mu, L. Yun, Z. Ding, Y. Cong and Z. Yin, *Zhongguo Yaowu Huaxue Zazhi*, 1998, **8**, 14.

389. V. De Filippis, D. Quarzago, A. Vindigni, E. di Cera and A. Fontana, *Biochemistry*, 1998, **37**, 13507.

390. K. Lee, S.Y. Hwang, M. Yun and D.S. Kim, *Bioorg.Med.Chem.Lett.*, 1998, **8**, 1683.

391. V. Caciagli, F. Cardinali, F. Bonelli and P. Lombardi, *J.Pept.Sci.*, 1998, **4**, 327.

392. W.J. Hoekstra, B.L. Hulshizer, D.F. McComsey, P. Andrade-Gordon, J.A. Kauffman, M.F. Addo, D. Oksenberg, R.M. Scarborough and B.E. Maryanoff, *Bioorg.Med.Chem.Lett.*, 1998, **8**, 1649.

393. T. Takeuchi, A. Böttcher, C.M. Quezada, M.I. Simon, T.J. Meade and H.B. Gray, *J.Am.Chem.Soc*, 1998, **120**, 8555.

394. W.C. Lumma, K.M. Witherup, T.J. Tucker, S.F. Brady, J.T. Sisko, A.M. Naylor-Olsen, S.D. Lewis, B.J. Lucas and J.P. Vacca, *J.Med.Chem.*, 1998, **41**, 1011.

395. B. Kundu, M. Bauser, J. Betschinger, W. Kraas and G. Jung, *Bioorg.Med. Chem.Lett.*, 1998, **8**, 1669.

396. S.W. Kim, C.Y. Hong, K. Lee, E.J. Lee and J.S. Koh, *Bioorg.Med.Chem. Lett.*, 1998, **8**, 735.

397. C.V. Ramesh, R. Jayakumar and R. Puvanakrishnan, *Protein Pept.Lett.*, 1998, **5**, 147.

398. R. Hamilton, B. Walker and B.J. Walker, *Bioorg.Med.Chem.Lett.*, 1998, **8**, 1655.

399. Y. Tsuda, K. Wanaka, M. Tada, S. Okamoto, A. Hijikata-Okunomiya and Y. Okada, *Chem.Pharm.Bull.*, 1998, **46**, 452.

400. M. Eda, A. Ashimori, F. Akahoshi, T. Yoshimura, Y. Inoue, C. Fukaya, M. Nakajima, H. Fukuyama, T. Imada and N. Nakamura, *Bioorg.Med.Chem.Lett.*, 1998, **8**, 913.

401. M. Eda, A. Ashimori, F. Akahoshi, T. Yoshimura, Y. Inoue, C. Fukaya, M. Nakajima, H. Fukuyama, T. Imada and N. Nakamura, *Bioorg.Med.Chem. Lett.*, 1998, **8**, 919.

402. S.E. Webber, K. Okano, T.L. Little, S.H. Reich, Y. Xin, S.A. Fuhrman, D.A. Matthews, R.A. Love, T.F. Hendrickson, A.K. Patick, J.W. Meador, R.A. Ferre, E.L. Brown, C.E. Ford, S.L. Binford and S.T. Worland, *J.Med.Chem.*, 1998, **41**, 2786.

403. P.S. Dragovich, S.E. Webber, R.E. Babine, S.A. Fuhrman, A.K. Patick, D.A. Matthews, C.A. Lee, S.H. Reich, T.J. Prins, J.T. Marakovits, E.S. Littlefield, R. Zhou, J. Tikhe, C.E. Ford, M.B. Wallace, J.W. Meador, R.A. Ferre, E.L. Brown, S.L. Binford, J.E.V. Harr, D.M. DeLisle and S.T. Worland, *J.Med.Chem.*, 1998, **41**, 2806.

404. P.S. Dragovich, S.E. Webber, R.E. Babine, S.A. Fuhrman, A.K. Patick, D.A. Matthews, S.H. Reich, J.T. Marakovits, T.J. Prins, R. Zhou, J. Tikhe, E.S. Littlefield, T.M. Bleckman, M.B. Wallace, T.L. Little, C.E. Ford, J.W. Meador, R.A. Ferre, E.L. Brown, S.L. Binford, D.M. DeLisle and S.T. Worland, *J. Med. Chem.*, 1998, **41**, 2819.

405. J.-s. Kong, S. Venkatraman, K. Furness, S. Nimkar, T.A. Shepherd, Q.M. Wang, J. Aubé and R.P. Hanzlik, *J. Med. Chem.*, 1998, **41**, 2579.

406. M. Llinàs-Brunet, M. Bailey, G. Fazal, S. Goulet, T. Halmos, S. Laplante, R. Maurice, M. Poirier, M.-A. Poupart, D. Thibeault, D. Wernic and D. Lamarre, *Bioorg. Med. Chem. Lett.*, 1998, **8**, 1713.

407. M. Llinàs-Brunet, M. Bailey, R. Déziel, G. Fazal, V. Gorys, S. Goulet, T. Halmos, R. Maurice, M. Poirier, M.-A. Poupart, J. Rancourt, D. Thibeault, D. Wernic and D. Lamarre, *Bioorg. Med. Chem. Lett.*, 1998, **8**, 2719.

408. P.R. Bonneau, C. Plouffe, A. Pelletier, D. Wernic and M.-A. Poupart, *Anal. Biochem.*, 1998, **255**, 59.

409. M. Bouygues, M. Medou, G. Quéléver, J.C. Chermann, M. Camplo and J.L. Kraus, *Bioorg. Med. Chem. Lett.*, 1998, **8**, 277.

410. S. Yao, R. Zutshi and J. Chmielewski, *Bioorg. Med. Chem. Lett.*, 1998, **8**, 699.

411. A. Asagarasu, N. Takayanagi and K. Achiwa, *Chem. Pharm. Bull.*, 1998, **46**, 867.

412. N.M. Patel, F. Bennett, V.M. Girijavallabhan, B. Dasmahapatra, N. Butkiewicz and A. Hart, *Bioorg. Med. Chem. Lett.*, 1998, **8**, 931.

413. A. Asagarasu, T. Uchiyama and K. Achiwa, *Chem. Pharm. Bull.*, 1998, **46**, 697.

414. S.F. Martin, G.O. Dorsey, T. Gane, M.C. Hillier, H. Kessler, M. Baur, B. Mathä, J.W. Erickson, T.N. Bhat, S. Munshi, S.V. Gulnik and I.A. Topol, *J. Med. Chem.*, 1998, **41**, 1581.

415. A. Fässler, G. Bold and H. Steiner, *Tetrahedron Lett.*, 1998, **39**, 4925.

416. G. Bold, A. Fässler, H.-G. Capraro, R. Cozens, T. Klimkait, J. Lazdins, J. Mestan, B. Poncioni, J. Roesel, D. Stover, M. Tintelnot-Blomley, F. Acemoglu, W. Beck, E. Boss, M. Eschbach, T. Huerlimann, E. Masso, S. Roussel, K. Ucci-Stoll, D. Wyss and M. Lang, *J. Med. Chem.*, 1998, **41**, 3887.

417. M. Alterman, M. Bjoersne, A. Muehlman, B. Classon, I. Kvarnstroem, H. Danielson, P.-O. Markgren, U. Nillroth, T. Unge, A. Hallberg and B. Samuelsson, *J. Med. Chem.*, 1998, **41**, 3782.

418. M.T. Reetz, C. Merk and G. Mehler, *Chem. Commun.*, 1998, 2075.

419. M. Marastoni, F. Bortolotti, S. Salvadori and R. Tomatis, *Arzneim.-Forsch.*, 1998, **48**, 709.

420. M. Bouygues, M. Medou, J.-C. Chermann, M. Camplo and J.-L. Kraus, *Eur. J. Med. Chem.*, 1998, **33**, 445.

421. P. Garrouste, M. Pawlowski, T. Tonnaire, S. Sicsic, P. Dumy, E. De Rosny, M. Reboud-Ravaux, P. Fulcrand and J. Martinez, *Eur. J. Med. Chem.*, 1998, **33**, 423.

422. R.J. Broadbridge and M. Akhtar, *Chem. Commun.*, 1998, 1449.

423. H.R. Wiltshire, K.J. Prior, J. Dhesi, F. Trach, M. Schlageter and H. Schonenberger, *J. Labelled Compd. Radiopharm.*, 1998, **41**, 1103.

424. T. Lee, G.S. Laco, B.E. Torbett, H.S. Fox, D.L. Lerner, J.H. Elder and C.-H. Wong, *Proc. Natl. Acad. Sci., U.S.A.*, 1998, **95**, 939.

425. R. Xing and R.P. Hanzlik, *J. Med. Chem.*, 1998, **41**, 1344.

426. R. Tripathy, Z.-Q. Gu, D. Dunn, S.E. Senadhi, M.A. Ator and S. Chatterjee, *Bioorg. Med. Chem. Lett.*, 1998, **8**, 2647.

427. M. Tao, R. Bihovsky, G.J. Wells and J.P. Mallamo, *J.Med.Chem.*, 1998, **41**, 3912.

428. T. Yasuma, S. Oi, N. Choh, T. Nomura, N. Furuyama, A. Nishimura, Y. Fujisawa and T. Sohda, *J.Med.Chem.*, 1998, **41**, 4301.

429. G.M. Dubowchik and R.A. Firestone, *Bioorg.Med.Chem.Lett.*, 1998, **8**, 3341.

430. G.M. Dubowchik, K. Mosure, J.O. Knipe and R.A. Firestone, *Bioorg.Med. Chem.Lett.*, 1998, **8**, 3347.

431. J.J. Peterson and C.F. Meares, *Bioconj. Chem.*, 1998, **9**, 618.

432. J. Billson, J. Clark, S.P. Conway, T. Hart, T. Johnson, S.P. Langston, M. Ramjee, M. Quibell and R.K. Scott, *Bioorg.Med.Chem.Lett.*, 1998, **8**, 993.

433. K.A. Scheidt, W.R. Roush, J.H. McKerrow, P.M. Selzer, E. Hansell and P.J. Rosenthal, *Bioorg.Med.Chem.*, 1998, **6**, 2477.

434. W.R. Roush, F.V. González, J.H. McKerrow and E. Hansell, *Bioorg.Med. Chem.Lett.*, 1998, **8**, 2809.

435. S. Bajusz, I. Fauszt, K. Németh, E. Barbarás, A. Juhász and M. Patthy, *Bioorg.Med.Chem.Lett*, 1998, **8**, 1477.

436. D.S. Karanewsky, X. Bai, S.D. Linton, J.F. Krebs, J. Wu, B. Pham and K.J. Tomaselli, *Bioorg.Med.Chem.Lett.*, 1998, **8**, 2757.

437. J. Litera, J. Weber, I. Krizova, I. Pichova, J. Konvalinka, M. Fusek and M. Soucek, *Coll.Czech.Chem.Commun.*, 1998, **63**, 541.

438. M. Kratzel, R. Hiessböck and A. Bernkop-Schnürch, *J.Med.Chem.*, 1998, **41**, 2339.

439. C.D. Carroll, H. Patel, T.O. Johnson, T. Guo, M. Orlowski, Z.-M. He, C.L. Cavallaro, J. Guo, A. Oksman, I.Y. Gluzman, J. Connelly, D. Chelsky, D.E. Goldberg and R.E. Dolle, *Bioorg.Med.Chem.Lett.*, 1998, **8**, 2315.

440. C.D. Carroll, T.O. Johnson, S. Tao, G. Lauri, M. Orlowski, I.Y. Gluzman, D.E. Goldberg and R.E. Dolle, *Bioorg.Med.Chem.Lett*, 1998, **8**, 3203.

441. A.S. Ripka, R.S. Bohacek and D.H. Rich, *Bioorg.Med.Chem.Lett.*, 1998, **8**, 357.

442. R.M. McConnell, C. Patterson-Goss, W. Godwin and B. Stanley, *J.Org. Chem.*, 1998, **63**, 5648.

443. G.L. Jung, P.C. Anderson, M. Bailey, M. Baillet, G.W. Bantle, S. Berthiaume, P. Lavallee, M. Llinàs-Brunet, B. Thavonekham, D. Thibeault and B. Simoneau, *Bioorg.Med.Chem.*, 1998, **6**, 2317.

444. N.-J. Hong, Y.-A. Park and K.-N. Son, *Bull.Korean Chem.Soc.*, 1998, **19**, 189.

445. D.E. Levy, F. Lapierre, W. Liang, W. Ye, C.W. Lange, X. Li, D. Grobelny, M. Casabonne, D. Tyrrell, K. Holme, A. Nadzan and R.E. Gelardy, *J.Med. Chem.*, 1998, **41**, 199.

446. S. Patel, L. Saroglou, C.D. Floyd, A. Miller and M. Whittaker, *Tetrahedron Lett.*, 1998, **39**, 8333.

447. D. Krumme, H. Wenzel and H. Tschesche, *FEBS Lett.*, 1998, **436**, 209.

448. I.C. Jacobson, P.G. Reddy, Z.R. Wasserman, K.D. Hardman, M.B. Covington, E.C. Arner, R.A. Copeland, C.P. Decicco and R.L. Magolda, *Bioorg.Med. Chem.Lett.*, 1998, **8**, 837.

449. C.D. Floyd, L.A. Harnett, A. Miller, S. Patel, L. Saroglou and M. Whittaker, *Synlett*, 1998, 637.

450. M. Renil, M. Ferreras, J.M. Delaisse, N.T. Foged and M. Meldal, *J.Pept.Sci.*, 1998, **4**, 195.

451. E.P. Johnson, W.R. Cantrell, T.M. Jenson, S.A. Miller, D.J. Parker, N.M. Reel, L.G. Sylvester, R.J. Szendroi, K.J. Vargas, J. Xu and J.A. Carlson, *Org.Process Res.Dev.*, 1998, **2**, 238.

452. H. Chen, *Zhongguo Yaowu Huaxue Zazhi*, 1997, **7**, 175.
453. J.S. Warmus, T.R. Ryder, J.C. Hodges, R.M. Kennedyand K.D. Brady, *Bioorg.-Med.Chem.Lett.*, 1998, **8**, 2309.
454. C. Nguyen, J, Blanco, J.-P. Mazaleyrat, B. Krust, C. Callebaut, E. Jacotot, A.G. Hovanessian and M. Wakselman, *J.Med.Chem.*, 1998, **41**, 2100.
455. M. Yamada, C. Okagaki, T. Higashijima, S. Tanaka, T. Ohnuki and T. Sugita, *Bioorg.Med.Chem.Lett.*, 1998, **8**, 1537.
456. J. Lin, P.J. Toscano and J.T. Welch, *Proc.Natl.Acad.Sci., U.S.A.*, 1998, **95**, 14020.
457. M. Tandon, M. Wu, T.P. Begley, J. Myllyharju, A. Pirskanen and K. Kivirikko, *Bioorg.Med.Chem.Lett.*, 1998, **8**, 1139.
458. E.J. Corey and W.-D.Z. Li, *Tetrahedron Lett.*, 1998, **39**, 7475.
459. Y.-J. Hu, P.T.R. Rajagopalan and D. Pei, *Bioorg.Med.Chem.Lett.*, 1998, **8**, 2479.
460. S. Petursson and J.E. Baldwin, *Tetrahedron*, 1998, **54**, 6001.
461. J.E. Baldwin, R.M. Adlington, N.P. Crouch and I.A. Pereira, *J.Labelled Compd.Radiopharm.*, 1998, **41**, 1145.
462. T. Xu, H. Khanna and J.K. Coward, *Bioorg.Med.Chem.*, 1998, **6**, 1821.
463. V. Ferro, L. Weiler and S.G. Withers, *Carbohydr.Res.*, 1998, **306**, 531.
464. Z.-B. Zheng, S. Nagai, N. Iwanami, A. Kobayashi, M. Hijikata, S. Natori and U. Sankawa, *Chem.Pharm.Bull.*, 1998, **46**, 1950.
465. Z. Li, S.L. Yeo, C.J. Pallen and A. Ganesan, *Bioorg.Med.Chem.Lett.*, 1998, **8**, 2443.
466. H.D. Ly, S.L. Clugston, P.B. Sampson and J.F. Honek, *Bioorg.Med.Chem. Lett.*, 1998, **8**, 705.
467. S. Sasaki, T. Hashimoto, N. Obana, H. Yasuda, Y. Uehara and M. Maeda, *Bioorg.Med.Chem.Lett.*, 1998, **8**, 1019.
468. R.J. Cox, J.A. Schouten, R.A. Stentiford and K.J. Wareing, *Bioorg.Med. Chem.Lett.*, 1998, **8**, 945.
469. S.J. deSolms, E.A. Guiliani, S.L. Graham, K.S. Koblan, N.E. Kohl, S.D. Mosser, A.I. Oliff, D.L. Pompliano, E. Rands, T.H. Scholz, C.M. Wiscount, J.B. Gibbs and R.L. Smith, *J.Med.Chem.*, 1998, **41**, 2651.
470. R.K. Jain, D.A. Sarracino and C. Richert, *Chem.Commun.*, 1998, 423.
471. M. Gobbo, L. Biondi, F. Filira, F. Formaggio, M. Crisma, R. Rocchi, C. Toniolo, Q.B. Broxterman and J. Kamphuis, *Lett.Pept.Sci.*, 1998, **5**, 105.
472. D. Dieudonné, A. Gericke, C.R. Flach, X. Jiang, R.S. Farid and R. Mendelsohn, *J.Am.Chem.Soc.*, 1998, **120**, 792.
473. K. Kirshenbaum, A.E. Barron, R.A. Goldsmith, P. Armand, E.K. Bradley, K.T.V. Truong, K.A. Dill, F.E. Cohen and R.N. Zuckermann, *Proc.Natl. Acad.Sci., U.S.A.*, 1998, **95**, 4303.
474. L. Williams, K. Kather and D.S. Kemp, *J.Am.Chem.Soc.*, 1998, **120**, 11033.
475. D.C. Tahmassebi and T. Sasaki, *J.Org.Chem.*, 1998, **63**, 728.
476. J.S. Johansson, B.R. Gibney, J.J. Skalicky, A.J. Wand and P.L. Dutton, *J.Am.Chem.Soc.*, 1998, **120**, 3881.
477. T.Z. Rizvi, M. Petukhov, Y. Tatsu and S. Yoshikawa, *J.Chem.Soc.Pak.*, 1998, **20**, 137.
478. C.-F. Chang and M.H. Zehfus, *Biopolymers*, 1998, **46**, 181.
479. D. Gani, A. Lewis, T. Rutherford, J. Wilkie, I. Stirling, T. Jenn and M.D. Ryan, *Tetrahedron*, 1998, **54**, 15793.

480. M. Allert, M. Kjellstrand, K. Broo, A. Nilsson and L. Baltzer, *J.Chem. Soc., Perkin Trans.2*, 1998, 2271.

481. G. Impellizzeri, G. Pappalardo, R. Purrello, E. Rizzarelli and A.M. Santoro, *Chem.–Eur.J.*, 1998, **4**, 1791.

482. R. Gratias, R. Konat, H. Kessler, M. Crisma, G. Valle, A. Polese, F. Formaggio, C. Toniolo, Q.B. Broxterman and J. Kamphuis, *J.Am.Chem. Soc.*, 1998, **120**, 4763.

483. Y.T. Scott, M.G. Bursavich, T. Gauthier, R.P. Hammer and M.L. McLauglin, *Chem.Commun.*, 1998, 1801.

484. W.M. Wolf, M. Stasiak, M.T. Leplawy, A. Bianco, F. Formaggio, M. Crisma and C. Toniolo, *J.Am.Chem.Soc.*, 1998, **120**, 11558.

485. A. Aubry, D. Bayeul, H. Bruckner, N. Schiemann and E. Benedetti, *J.Pept. Sci.*, 1998, **4**, 502.

486. S.R. Raghothama, S.K. Awasthi and P. Balaram, *J.Chem.Soc., Perkin Trans. 2*, 1998, 137.

487. A.J. Maynard, G.J. Sharman and M.S. Searle, *J.Am.Chem.Soc.*, 1998, **120**, 1996.

488. C.R. Das, S. Raghothama and P. Balaram, *J.Am.Chem.Soc.*, 1998, **120**, 5812.

489. F. Jean, E. Buisine, O. Melnyk, H. Drobecq, B. Odaert, M. Hugues, G. Lippens and A. Tartar, *J.Am.Chem.Soc.*, 1998, **120**, 6076.

490. D. Ranganathan, V. Haridas, S. Kurur, A. Thomas, K.P. Madhusudanan, R. Nagaraj, A.C. Kunwar, A.V.S. Sarma and I.L. Karle, *J.Am.Chem.Soc.*, 1998, **120**, 8448.

491. M. Zouikri, A. Vicherat, A. Aubry, M. Marraud and G. Boussard, *J.Pept. Res.*, 1998, **52**, 19.

492. R. Vanderesse, V. Grand, D. Limal, A. Vicherat, M. Marraud, C. Didierjean and A. Aubry, *J.Am.Chem.Soc.*, 1998, **120**, 9444.

493. I.G. Jones and M. North, *Lett.Pept.Sci.*, 1998, **5**, 171.

494. D.E. Hibbs, M.B. Hursthouse, I.G. Jones, W. Jones, K.M.A. Malik and M. North, *J.Org.Chem.*, 1998, **63**, 1496.

495. J. Klose, A. Ehrlich and M. Bienert, *Lett.Pept.Sci.*, 1998, **5**, 129.

496. M.E. Pfeifer and J.A. Robinson, *Chem.Commun.*, 1998, 1977.

497. F. Eblinger and H.-J. Schneider, *Chem.Commun.*, 1998, 2297.

498. S. Yinlin, H. Yongliang, W. Qi and L. Luhua, *Protein Pept.Lett.*, 1998, **5**, 259.

499. Y. Feng, Z. Wang, S. Jin and K. Burgess, *J.Am.Chem.Soc.*, 1998, **120**, 10768.

500. T. Kortemme, M. Ramirez-Alvarado and L. Serrano, *Science*, 1998, **281**, 253.

501. D. Seebach, S. Abele, T. Sifferlen, M. Hänggi, S. Gruner and P. Seiler, *Helv.Chim.Acta*, 1998, **81**, 2218.

502. C.O. Ogbu, M.N. Qabar, P.D. Boatman, J. Urban, J.P. Meara, M.D. Ferguson, J. Tulinsky, C. Lum, S. Babu, M.A. Blaskovich, H. Nakanishi, F. Ruan, B. Cao, R. Minarik, T. Little, S. Nelson, M. Nguyen, A. Gall and M. Kahn, *Bioorg.Med.-Chem.Lett.*, 1998, **8**, 2321.

503. L. Mayne, S.W. Englander, R. Qiu, J. Yang, Y. Gong, E.J. Spek and N.R. Kallenbach, *J.Am.Chem.Soc.*, 1998, **120**, 10643.

504. J.D. Hartgerink, T.D. Clark and M.R. Ghadiri, *Chem.–Eur.J.*, 1998, **4**, 1367.

505. M.E. Polaskova, N.J. Ede and J.N. Lambert, *Austr.J.Chem.*, 1998, **51**, 535.

506. W.-J. Wu and D.P. Raleigh, *J.Org.Chem.*, 1998, **63**, 6689.

507. S. Kimura, K. Fujita, Y. Miura, G. Xu, Y. Muraji, T. Kidchob and Y. Imanishi, *Curr.Top.Pept.Protein Res.*, 1997, **2**, 137.

508. Y. Takahashi, A. Ueno and H. Mihara, *Chem.–Eur.J.*, 1998, **4**, 2475.

509. G. Tuchscherer, C. Lehmann and M. Mathieu, *Angew.Chem., Int.Ed.*, 1998, **37**, 2990.

510. S. Sakamoto, A. Ueno and H. Mihara, *J.Chem.Soc., Perkin Trans.2*, 1998, 2395.

511. T. Ishizu, M. Fujiwara, A. Yagi and S. Noguchi, *Chem.Pharm.Bull.*, 1998, **46**, 690.

512. M. Keller, C. Sager, P. Dumy, M. Schutkowski, G.S. Fischer and M. Mutter, *J.Am.Chem.Soc.*, 1998, **120**, 2714.

513. R. Zhang, F. Brownewell and J.S. Madalengoitia, *J.Am.Chem.Soc.*, 1998, **120**, 3894.

514. D. Ranganathan, S. Kurur, K.P. Madhusudanan, R. Roy and I.L. Karle, *J.Pept.Res.*, 1998, **51**, 297.

515. D. Seebach, S. Abele, K. Gademann, G. Guichard, T. Hintermann, B. Jaun, J.L. Matthews, J.V. Schreiber, L. Oberer, U. Hommel and H. Widmer, *Helv.Chim. Acta*, 1998, **81**, 932.

516. T. Hintermann, K. Gademann, B. Jaun and D. Seebach, *Helv.Chim.Acta*, 1998, **81**, 983.

517. S. Hanessian, X. Luo, R. Schaum and S. Michnick, *J.Am.Chem.Soc.*, 1998, **120**, 8569.

518. V. Brecx, P. Verheyden and D. Tourwe, *Lett.Pept.Sci.*, 1998, **5**, 67.

519. J. Broddefalk, K.-E. Bergquist and J. Kihlberg, *Tetrahedron*, 1998, **54**, 12047.

520. E.C. Rodriguez, L.A. Marcaurelle and C.R. Bertozzi, *J.Org.Chem.*, 1998, **63**, 7134.

521. P. Laurent, L. Hennig, K. Burger, W. Hiller and M. Neumayer, *Synthesis*, 1998, 905.

522. D. Cabaret, M. Wakselman, A. Kaba, L. Ilunga and C. Chany, *Synth. Commun.*, 1998, **28**, 2713.

523. P.R. Hansen, C.E. Olsen and A. Holm, *Bioconjugate Chem.*, 1998, **9**, 126.

524. M. Bengtsson, J. Broddefalk, J. Dahmen, K. Henriksson, J. Kihlberg, H. Lonn, B.R. Srinivasa and K. Stenvall, *Glycoconjugate J.*, 1998, **15**, 223.

525. L.A. Marcaurelle, E.C. Rodriguez and C.R. Bertozzi, *Tetrahedron Lett.*, 1998, **39**, 8417.

526. M.A. Dechantsreiter, F. Burkhart and H. Kessler, *Tetrahedron Lett.*, 1998, **39**, 253.

527. U. Tedebark, M. Meldal, L. Panza and K. Bock, *Tetrahedron Lett.*, 1998, **39**, 1815.

528. P. Arya, K.M.K. Kutterer, H. Qin, J. Roby, M.L. Barnes, J.M. Kim and R. Roy, *Bioorg.Med.Chem.Lett.*, 1998, **8**, 1127.

529. M. Ledvina, J. Jezek, D. Saman and V. Hribolova, *Collect.Czech.Chem. Commun.*, 1998, **63**, 590.

530. I.P. Chikailo, V.Ya. Gorishnii and M.N. Bidyuk, *Farm.Zh.*, 1996, 112.

531. Z.-F. Wang and J.-C. Xu, *Tetrahedron*, 1998, **54**, 12597.

532. H.-J. Kohlbau, J. Tschakert, R.A. Al-Qawasmeh, T.A. Nizami, A. Malik and W. Voelter, *Z. Naturforsch.*, 1998, **53B**, 753.

533. S.A. Hitchcock, C.N. Eid, J.A. Aikins, M. Zia-Ebrahimi and L.C. Blaszczak, *J.Am.Chem.Soc.*, 1998, **120**, 1916.

534. C.N. Eid, M.J. Nesler, M. Zia-Ebrahimi, C.-Y.E. Wu, R. Yao, K. Cox and J. Richardson, *J.Labelled Compd.Radiopharm.*, 1998, **41**, 705.

535. C.-Y. Tsai, W.K.C. Park, G. Weitz-Schmidt, B. Ernst and C.-H. Wong, *Bioorg.Med.Chem.Lett.*, 1998, **8**, 2333.

536. T.F.J. Lampe, G. Weitz-Schmidt and C.-H. Wong, *Angew. Chem., Int. Ed.*, 1998, **37**, 1707.

537. G. Baisch and R. Ohrlein, *Carbohydr. Res.*, 1998, **312**, 61.

538. M.M. Palcic, H. Li, D. Zanini, R.S. Bhella and R. Roy, *Carbohydr. Res.*, 1998, **305**, 433.

539. X. Zeng, T. Murata, H. Kawagishi, T. Usui and K. Kobayashi, *Biosci. Biotechnol. Biochem.*, 1998, **62**, 1171.

540. S. Kamiya and K. Kobayashi, *Macromol. Chem. Phys.*, 1998, **199**, 1589.

541. A.E. Zemlyakov, V.O. Kur'yanov, V.N. Tsikalova and V.Ya. Chirva, *Chem. Nat. Compd.*, 1998, **33**, 568.

542. T. Inazu, K. Haneda and M. Mizuno, *Yuki Gosei Kagaku Kyokaishi*, 1998, **56**, 210.

543. L. Biondi, F. Filira, M. Gobbo, E. Pavin and R. Rocchi, *J. Pept. Sci.*, 1998, **4**, 58.

544. I. Zigrovic, J. Kidric and S. Horvat, *Glycoconjugate J.*, 1998, **15**, 563.

545. L. Andersson, G. Stenhagen and L. Baltzer, *J. Org. Chem.*, 1998, **63**, 1366.

546. P. Braun, G.M. Davies, M.R. Price, P.M. Williams, S.J.B. Tendler and H. Kunz, *Bioorg. Med. Chem.*, 1998, **6**, 1531.

547. M. Ledvina, D. Zyka, J. Jezek, T. Trnka and D. Saman, *Collect. Czech. Chem. Commun.*, 1998, **63**, 577.

548. S. Ando, J.-I. Aikawa, Y. Nakahara and T. Ogawa, *J. Carbohydr. Chem.*, 1998, **17**, 633.

549. N.J. Snyder, R.D.G. Cooper, B.S. Briggs, M. Zmijewski, D.L. Mullen, R.E. Kaiser and T.I. Nicas, *J. Antibiot.*, 1998, **51**, 945.

550. C. Mouton, F. Tillequin, E. Seguin and C. Monneret, *J. Chem. Soc., Perkin Trans. 1*, 1998, 2055.

551. J. Habermann and H. Kunz, *Tetrahedron Lett.*, 1998, **39**, 265.

552. J. Habermann and H. Kunz, *Tetrahedron Lett.*, 1998, **39**, 4797.

553. J.Y. Roberge, X. Beebe and S.J. Danishefsky, *J. Am. Chem. Soc.*, 1998, **120**, 3915.

554. E. Meinjohanns, M. Meldal, H. Paulsen, R.A. Dwek and K. Bock, *J. Chem. Soc., Perkin Trans. 1*, 1998, 549.

555. Y. Nakahara, Y. Nakahara, Y. Ito and T. Ogawa, *Carbohydr. Res.*, 1998, **309**, 287.

556. H. Yuasa, Y. Kamata, S. Kurono and H. Hashimoto, *Bioorg. Med. Chem. Lett.*, 1998, **8**, 2139.

557. P.M. St.Hilaire, T.L. Lowary, M. Meldal and K. Bock, *J. Am. Chem. Soc.*, 1998, **120**, 13312.

558. T. Xu, R.M. Werner, K.-C. Lee, J.C. Fettinger, J.T. Davis and J.K. Coward, *J. Org. Chem.*, 1998, **63**, 4767.

559. K. Witte, O. Seitz and C.-H. Wong, *J. Am. Chem. Soc.*, 1998, **120**, 1979.

560. K. Haneda, T. Inazu, M. Mizuno, R. Iguchi, K. Yamamoto, H. Kumagai, S. Aimoto, H. Suzuki and T. Noda, *Bioorg. Med. Chem. Lett.*, 1998, **8**, 1303.

561. C. Unverzagt, *Carbohydr. Res.*, 1998, **305**, 423.

562. K. Yamamoto, K. Fujimori, K. Haneda, M. Mizuno, T. Inazu and H. Kumagai, *Carbohydr. Res.*, 1998, **305**, 415.

563. I.L. Deras, K. Takegawa, A. Kondo, I. Kato and Y.C. Lee, *Bioorg. Med. Chem. Lett.*, 1998, **8**, 1763.

564. A. Tholey, R. Pipkorn, M. Zeppezauer and J. Reed, *Lett. Pept. Sci.*, 1998, **5**, 263.

565. L. Galzigna, N. Domergue and A. Previero, *J. Pept. Res.*, 1998, **52**, 15.

566. N. Mora and J.-M. Lacombe, *Curr. Top. Pept. Protein Res.*, 1997, **2**, 169.

567. M. Ueki, M. Goto, J. Okumura and Y. Ishii, *Bull. Chem. Soc. Jpn.*, 1998, **71**, 1887.

568. C. Mathé, C. Périgaud, G. Gosselin and J.-L. Imbach, *J.Org.Chem.*, 1998, **63**, 8547.

569. B.K. Handa and C.J. Hobbs, *J.Pept.Sci.*, 1998, **4**, 138.

570. J.H. Lai, T.H. Marsilje, S. Choi, S.A. Nair and D.G. Hangauer, *J.Pept. Res.*, 1998, **51**, 271.

571. P. Furet, B. Gay, G. Caravatti, C. Garcia-Echeverria, J. Rahuel, J. Schoepfer and H. Fretz, *J.Med.Chem.*, 1998, **41**, 3442.

572. H.-Y. Lu, N.-J. Zhang, X. Chen, H. Fu and Y.-F. Zhao, *Synth.Commun.*, 1998, **28**, 1727.

573. P.A. Lohse and R. Felber, *Tetrahedron Lett.*, 1998, **39**, 2067.

574. S. Chen and J.K. Coward, *J.Org.Chem.*, 1998, **63**, 502.

575. P.H. Dorff, G. Chiu, S.W. Goldstein and B.P. Morgan, *Tetrahedron Lett.*, 1998, **39**, 3375.

576. K. Ikeda, K. Miyajima and K. Achiwa, *Yuki Gosei Kagaku Kyokaishi*, 1998, **56**, 567.

577. B. Hemmer, C. Pinilla, J. Appel, J. Pascal, R. Houghten and R. Martin, *J.Pept.Res.*, 1998, **52**, 338.

578. J.R. Appel, G.D. Campbell, J. Buencamino, R.A. Houghten and C. Pinilla, *J.Pept.Res.*, 1998, **52**, 346.

579. E. Krambovitis, G. Hatzidakis and K. Barlos, *J.Biol.Chem.*, 1998, **273**, 10874.

580. A. Murray, D.I.R. Spencer, S. Missailidis, G. Denton and M.R. Price, *J.Pept.Res.*, 1998, **52**, 375.

581. V. Krsmanovic, J.M. Biquard, M. Sikorska-Walker, J.F. Whitfield, J.P. Durkin, I. Cosic, C. Desgranges, M.-A. Trabaud, A. Achour and M.T.W. Hearn, *J.Pept.Res.*, 1998, **52**, 410.

582. J.M. Lozano, F. Espejo, D. Diaz, L.M. Salazer, J. Rodriguez, C. Pinzón, J.C. Caloo, F. Guzmán and M.E. Patarroyo, *J.Pept.Res.*, 1998, **52**, 457.

583. P.J. Cachia, L.M.G. Glasier, R.R.W. Hodgins, W.Y. Wong, R.T. Irvin and R.S. Hodges, *J.Pept.Res.*, 1998, **52**, 289.

584. A. Bianco, C. Zabel, P. Walden and G. Jung, *J.Pept.Sci.*, 1998, **4**, 471.

585. J.Y. Lee, Y.J. Chung, Y.-S. Bae, S.H. Ryu and B.H. Kim, *J.Chem.Soc., Perkin Trans.1*, 1998, 359.

586. G. Aldrian-Herrada, A. Rabie, R. Wintersteiger and J. Brugidou, *J.Pept. Sci.*, 1998, **4**, 266.

587. A. Kreimeyer and P. Marliere, *Nucleosides Nucleotides*, 1998, **17**, 2339.

588. U. Diederichsen, *Angew.Chem., Int.Ed.*, 1998, **37**, 2273.

589. E. Uhlmann, A. Peyman, G. Breipohl and D.W. Will, *Angew.Chem., Int.Ed.*, 1998, **37**, 2796.

590. V.A. Efimov, M.V. Choob, A.A. Buryakova and O.G. Chakhmakhcheva, *Nucleosides Nucleotides*, 1998, **17**, 1671.

591. T.E. Goodwin, R.D. Holland, J.O. Lay, and K.D. Raney, *Bioorg.Med.Chem. Lett.*, 1998, **8**, 2231.

592. S. Peyrottes, B. Mestre, F. Burlina and M.J. Gait, *Tetrahedron*, 1998, **54**, 12513.

593. C. Dalliare and P. Arya, *Tetrahedron Lett.*, 1998, **39**, 5129.

594. A.C. Van der Laan, I. Van Amsterdam, G.I. Tesser, J.H. Van Boom and E. Kuyl-Yeheskiely, *Nucleosides Nucleotides*, 1998, **17**, 219.

595. C.N. Tetzlaff, I. Schwope, C.F. Bleczinski, J.A. Steinberg and C. Richert, *Tetrahedron Lett.*, 1998, **39**, 4215.

596. A. Püschl, S. Sforza, G. Haaima, O. Dahl and P.E. Nielsen, *Tetrahedron Lett.*, 1998, **39**, 4707.

597. A.H. Krotz, S. Larsen, O. Buchardt, M. Eriksson and P.E. Nielsen, *Bioorg.Med. Chem.*, 1998, **6**, 1983.

598. P. Clivio, D. Guillaume, M.-T. Adeline, J. Hamon, C. Riche and J.-L. Fourrey, *J.Am.Chem.Soc.*, 1998, **120**, 1157.

599. X. Li, C. Huang, Y. Wang, Y. Chen, L. Zhang, J. Lu and L. Zhang, *Curr.Sci.*, 1998, 74, 624.

600. U. Diederichsen and H.W. Schmitt, *Angew.Chem., Int.Ed.*, 1998, **37**, 302.

601. A.M. Shalaby, W.M. Basyouni and K.A.M. El-Bayouki, *J.Chem.Res., Synop.*, 1998, 134.

602. D.K. Alargov, Z. Naydenova, K. Grancharov, P. Denkova and E. Golovinsky, *Monatsh.Chem.*, 1998, **129**, 756.

603. B.D. Gildea, S. Casey, J. MacNeill, H. Perry-O'Keefe, D. Sørensen and J.M. Coull, *Tetrahedron Lett.*, 1998, **39**, 7255.

604. T. Miura, M. Fujii, K. Shingu, I. Koshimizu, J. Naganoma, T. Kajimoto and Y. Ida, *Tetrahedron Lett.*, 1998, **39**, 7313.

605. S. Soukchareum, J. Haralambidis and G. Tregear, *Bioconjugate Chem.*, 1998, **9**, 466.

606. A. Maruyama, H. Watanabe, A. Ferdous and T. Akaike, *Nucleic Acids Symp. Ser.*, 1997, **37**, 225.

607. B. Armitage, T. Koch, H. Frydenlund, H. Orum and G.B. Schuster, *Nucleic Acids Res.*, 1998, **26**, 715.

608. A.C. van der Laan, P. Havenaar, R.S. Oosting, E. Kuyl-Yeheskiely, E. Uhlmann and J.H. van Boom, *Bioorg.Med.Chem.Lett.*, 1998, **8**, 663.

609. J.G. Harrison and S. Balasubramanian, *Nucleic Acids Res.*, 1998, **26**, 3136.

610. A. Domling, *Nucleosides Nucleotides*, 1998, **17**, 1667.

611. M. Yamada, T. Miyajima and H. Horikawa, *Tetrahedron Lett.*, 1998, **39**, 289.

612. S.A. Burrage, T. Raynham and M. Bradley, *Tetrahedron Lett.*, 1998, **39**, 2831.

613. R. Vijayaraghavan, P. Kumar, S. Dey and T.P. Singh, *J.Pept.Res.*, 1998, **52**, 89.

614. U.A. Ramagopal, S. Ramakumar, R.M. Joshi and V.S. Chauhan, *J.Pept.Res.*, 1998, **52**, 208.

615. H. Feng, J. Zhong, G. Liu and W. Li, *Zhengzhou Liangshi Xueyuan Xuebao.*, 1998, **19**, 15.

616. A.E. Zemlyakov, *Chem.Nat.Compd.*, 1998, **34**, 78.

617. T. Mikayama, H. Matsuoka, K. Uehara, I. Shimizu and S. Yoshikawa, *Biopolymers*, 1998, **47**, 179.

618. J.E. Semple, *Tetrahedron Lett.*, 1998, **39**, 6645.

619. M.L. Falck-Pedersen and K. Undheim, *Acta Chem.Scand.*, 1998, **52**, 1327.

620. A.R. Butler, H.H. Al-Sa'doni, I.L. Megson and F.W. Flitney, *Nitric Oxide*, 1998, **2**, 193.

621. J. Vinsova, K. Waisser, Z. Odlerova, *Folia Pharm.Univ.Carol.*, 1998, **21/22**, 83.

622. M.E. Vol'pin, Z.N. Parnes and V.S. Romanova, *Russ.Chem.Bull.*, 1998, **47**, 1021.

623. Y. Takeuchi, M. Kamezaki, K. Kirihara, G. Haufe, K.W. Laue and N. Shibata, *Chem.Pharm.Bull.*, 1998, **46**, 1062.

624. G. Porzi, S. Sandri and P. Verrocchio, *Tetrahedron: Asymmetry*, 1998, **9**, 119.

625. S. Booth, E.N.K. Wallace, K. Singhal, P.N. Bartlett and J.D. Kilburn, *J.Chem.Soc., Perkin Trans.1*, 1998, 1467.

626. C. Gennari, C. Longari, S. Ressel, B. Salom and A. Mielgo, *Eur.J.Org. Chem.*, 1998, 945.

627. C. Gennari, C. Longari, S. Ressel, B. Salom, U. Piarulli, S. Ceccarelli and A. Mielgo, *Eur.J.Org.Chem.*, 1998, 2437.

628. G. Guichard, S. Abele and D. Seebach, *Helv.Chim.Acta*, 1998, **81**, 187.
629. J.L. Matthews, K. Gademann, B. Jaun and D. Seebach, *J.Chem.Soc., Perkin Trans.1*, 1998, 3331.
630. S. Abele, G. Guichard and D. Seebach, *Helv.Chim.Acta*, 1998, **81**, 2141.
631. D.J. Cundy, A.C. Donohue and T.D. McCarthy, *Tetrahedron Lett.*, 1998, **39**, 5125.
632. C. Fernández-García, K. Prager, M.A. McKervey, B. Walker and C.H. Williams, *Bioorg.Med.Chem.Lett.*, 1998, **8**, 433.
633. M. Stasiak, W.M. Wolf and M.T. Leplawy, *J.Pept.Sci.*, 1998, **4**, 46.
634. K.J. Kise and B.E. Bowler, *Tetrahedron:Asymmetry*, 1998, **9**, 3319.
635. T. Chiba, N. Takahasi and Y. Takasugi, *Akita Kogyo Koto Senmon Gakko Kenkyu Kiyo*, 1998, **33**, 55.
636. L. Révész, F. Bonne, U. Manning and J.-F. Zuber, *Bioorg.Med. Chem.Lett.*, 1998, **8**, 405.
637. K. Burger, K. Mutze, W. Hollweck and B. Koksch, *Tetrahedron*, 1998, **54**, 5915.
638. S. Feng, X. Xu and H. Zhang, *Yaoxue Xuebao*, 1998, **33**, 429.
639. S.L. Belagali and M. Himaja, *Indian J.Heterocycl.Chem.*, 1998, **8**, 11.
640. S.L. Belagali, K.H. Kumar, P. Boja and M. Himaja, *Indian J.Chem.*, 1998, **37B**, 370.
641. H. Tamamura, M. Waki, M. Imai, A. Otaka, T. Ibuka, K. Waki, K. Miyamoto, A. Matsumoto, T. Murakami, H. Nakashima, N. Yamamoto and N. Fujii, *Bioorg.Med.Chem.*, 1998, **6**, 473.
642. H. Tamamura, R. Arakaki, H. Funakoshi, M. Imai, A. Otaka, T. Ibuka, H. Nakashima, T. Murakami, M. Waki, A. Matsumoto, N. Yamamoto and N. Fujii, *Bioorg.Med.Chem.*, 1998, **6**, 231.
643. O. Nyeki, A. Rill, I. Schon, A. Orosz, J. Schrett, L. Bartha and J. Nagy, *J.Pept.Sci.*, 1998, **4**, 486.
644. G. Czerwinski, N.I. Tarasova and C.J. Michejda, *Proc.Natl.Acad.Sci., U.S.A.*, 1998, **95**, 11520.
645. F. Itoh, Y. Nishikimi, A. Hasuoka, Y. Yoshioka, K. Yukishige, S. Tanida and T. Aono, *Chem.Pharm.Bull.*, 1998, **46**, 255.
646. S.D. Abbott, L. Gagnon, M. Lagraoui, S. Kadhim, G. Attardo, B. Zacharie and C.L. Penney, *J.Med.Chem.*, 1998, **41**, 1909.
647. G.J. Pacofsky, K. Lackey, K.J. Alligood, J. Berman, P.S. Charifson, R.M. Crosby, G.F. Dorsey, P.L. Feldman, T.M. Gilmer, C.W. Hummel, S.R. Jordan, C. Mohr, M. Rodriguez, L.M. Shewchuk and D.D. Sternbach, *J.Med. Chem.*, 1998, **41**, 1894.
648. K.-I. Nunami, M. Yamada and R. Shimizu, *Bioorg.Med.Chem.Lett.*, 1998, **8**, 2517.
649. M. Kogiso, S. Ohnishi, K. Yase, M. Masuda and T. Shimizu, *Langmuir*, 1998, **14**, 4978.
650. K.D. Roberts, J.N. Lambert, N.J. Ede and A.M. Bray, *Tetrahedron Lett.*, 1998, **39**, 8357.
651. D.S. Jones, C.A. Gamino, M.E. Randow, E.J. Victoria, L. Yu and S.M. Coutts, *Tetrahedron Lett.*, 1998, **39**, 6107.
652. L. Yu, Y. Lai, J.V. Wade and S.M. Coutts, *Tetrahedron Lett.*, 1998, **39**, 6633.
653. T. Wakamiya, M. Kamata, S. Kusumoto, H. Kobayashi, Y. Sai, I. Tamai and A. Tsuji, *Bull.Chem.Soc.Jpn.*, 1998, **71**, 699.
654. T. Lescrinier, C. Hendrix, L. Kerremans, J. Rozenski, A. Link, B. Samyn, A. Van

Aerschot, E. Lescrinier, R. Eritja, J. Van Beeumen and P. Herdewijn, *Chem.-Eur.J.*, 1998, **4**, 425.

655. T. Sakthivel, I. Toth and A.T. Florence, *Pharm.Res.*, 1998, **15**, 776.

656. C. Boeckler, B. Frisch and F. Schuber, *Bioorg.Med.Chem.Lett.*, 1998, **8**, 2055.

657. O. Melnyk, M. Bossus, D. David, C. Rommens and H. Gras-Masse, *J.Pept. Res.*, 1998, **52**, 180.

658. Y. Crozet, J.J. Wen, R.O. Loo, P.C. Andrews and A.F. Spatola, *Mol. Diversity*, 1998, **3**, 261.

659. D. Ranganathan, V. Haridas and I.L. Karle, *J.Am.Chem.Soc.*, 1998, **120**, 2695.

660. P. Stefanowicz, R. Zbozine and I.Z. Siemon, *Lett.Pept.Sci.*, 1998, **5**, 329.

661. B. Chen, G. Bestetti, R.M. Day and A.P.F. Turner, *Biosens.Bioelectron.*, 1998, **13**, 779.

662. M. Takahashi, A. Ueno, T. Uda and H. Mihara, *Bioorg.Med.Chem.Lett.*, 1998, **8**, 2023.

663. S. De Luca, G. Bruno, R. Fattorusso, C. Isernia, C. Pedone and G. Morelli, *Lett.Pept.Sci.*, 1998, **5**, 269.

664. F. Chillemi, P. Francescato, A. Fraccari and I. Galatulas, *Anticancer Res.*, 1998, **18**, 757.

665. M. Horikawa, T. Nakajima and Y. Ohfune, *Synlett*, 1998, 609.

666. R.J. Bergeron, W.R. Weimar, R. Mueller, C.O. Zimmerman, B.H. McCosar, H. Yao and R.E. Smith, *J.Med.Chem.*, 1998, **41**, 3888.

667. T. Carell, H. Schmid and M. Reinhard, *J.Org.Chem.*, 1998, **63**, 8741.

668. J. Lehmann, A. Linden and H. Heimgartner, *Tetrahedron*, 1998, **54**, 8721.

669. A. Bianco, M. Maggini, S. Mondini, A. Polese, G. Scorrano, C. Toniolo and D.M. Guldi, *Proc.-Electrochem.Soc.*, 1998, **98**, 1145.

670. R.J. Cregge, T.T. Curran and W.A. Metz, *J.Fluorine Chem.*, 1998, **88**, 71.

671. D.A. Pearce, G.K. Walkup and B. Imperiali, *Bioorg.Med.Chem.Lett.*, 1998, **8**, 1963.

672. D.D. Long, S.M. Frederiksen, D.G. Marquess, A.L. Lane, D.J. Watkin, D.A. Winkler and G.W.J. Fleet, *Tetrahedron Lett.*, 1998, **39**, 6091.

673. I. Lang, N. Donze, P. Garrouste, P. Dumy and M. Mutter, *J. Pept.Sci.*, 1998, **4**, 72.

674. A. Pohlmann, D. Guillaume, J.-C. Quirion and H.-P. Husson, *J.Pept.Res.*, 1998, **51**, 116.

675. C.J. Stankovic, M.S. Plummer and T.K. Sawyer, *Adv.Amino Acid Mimetics Peptidomimetics*, 1997, **1**, 127.

676. J.W. Guiles, C.L. Lanter and R.A. Rivero, *Angew.Chem., Int.Ed.*, 1998, **37**, 926.

677. T. Yoshida and T. Okada, *Chromatography*, 1998, **19**, 376.

678. J. Barbosa, I. Toro and V. Sanz-Nebot, *J.Chromatogr., A*, 1998, **823**, 497.

679. T.R. Londo, T. Kehoe, S.A. Kates, S. Hantman and N.F. Gordon, *Lett. Pept.Sci.*, 1998, **5**, 285.

680. S. Gebauer, S. Friebe, G. Scherer, G. Gubitz and G.-J. Krauss, *J.Chromatogr.Sci.*, 1998, **36**, 388.

681. P.A. D'Agostino, J.R. Hancock, L.R. Provost, P.D. Semchuk and R.S. Hodges, *J.Chromatogr.*, 1998, **800**, 89.

682. C. Miller and J. Rivier, *J.Pept.Res.*, 1998, **51**, 444.

683. S. Sabah and G.K.E. Gerhard, *J.Microcolumn Sep.*, 1998, **10**, 255.

684. M.A. Shogren-Knaak and B. Imperiali, *Tetrahedron Lett.*, 1998, **39**, 8241.

3
Analogue and Conformational Studies on Peptides, Hormones and Other Biologically Active Peptides

BY ANAND S. DUTTA

1 Introduction

The subject matter included this year is broadly similar to that included last year.[1] Most of the publications covered in this chapter were published in 1998. However, some of the 1997 publications (not covered last year) have been included. This is especially so in the case of peptides not discussed last year due to space restrictions. No work published in patents or in unrefereed form (such as conference proceedings) has been included. Non-peptide ligands acting on the peptide receptors have again been included in sections on biologically active peptides. As in the 1997 chapter, due to space limitations, the structure-activity studies on non-peptide series of compounds are not described in detail. Only the more potent compounds from each series are highlighted to give an idea about the structural types displaying the desired activity and pharmaco-kinetic profile. As last year, a small section dealing with the advances in formulation and delivery technology has been included. In addition, work on some of the peptides not discussed last year has been included in the miscellaneous peptides section. In the first two sections below (peptide backbone modifications and di-, tri-peptide mimetics and cyclic peptides), only examples from those peptides are included which are not covered in the sections on the biologically active peptides. Throughout this chapter, amino acids are referred to by their three letter codes following standard nomenclature. For the naturally occurring L-amino acids, no stereochemistry is specified in the text.

2 Peptide Backbone Modifications and Di-, Tri-peptide Mimetics

2.1 Aza, Iminoaza and Reduced Aza Peptides – Solid phase and solution phase methods for the synthesis of azapeptides have been reported. Peptides containing azaglycine located terminally or within the backbone [*e.g.* Ac-Phe-Ala-AzGly-Leu-NH$_2$, Phe-Gly-AzGly-Ala and Ac-Aib-AzGly-NH$_2$] were prepared using MBHA, PAM and Rink amide resins.[2] PAM resin has also been

Amino Acids, Peptides and Proteins, Volume 31
© The Royal Society of Chemistry, 2000

used for the synthesis of aza-, iminoaza- and reduced aza-peptide homologues using the Boc-strategy.[3] Synthesis of the hydrazinocarbonyl peptide-PAM precursor was carried out by coupling the N-Boc-aza-amino acid chloride (*e.g.* Boc-NH-NMe-COCl), obtained by the action of triphosgene on the corresponding N-Boc-hydrazine, to the growing peptide chain. From the same hydrazinocarbonyl peptide-PAM precursor, the coupling of either a Boc-amino acid or a Boc-amino aldehyde gives rise to an aza-peptide or an iminoaza-peptide containing the C^α-CO-NH-N^α-CO-NH-C^α or C^α-CH=N-N^α-CO-NH-C^α surrogate of the peptide motif, respectively. Reduction of the latter by $NaBH_3CN$ leads to a reduced aza-peptide containing the C^α-CH_2-NH-N^α-CO-NH-C^α moiety.[3] These modifications have been introduced in position 1–2 of the benzodiazepine-like decapeptide [Tyr-Leu-Gly-Tyr-Leu-Glu-Gln-Leu-Leu-Arg]. Using the thiophosgene method, larger peptides based on the main immunogenic region (Trp-Asn-Pro-Ala-Asp-Tyr-Gly-Gly-Ile-Lys) of the *Torpedo californica* electric organ acetylcholine receptor were synthesised using combinations of solution phase and solid phase methods. Two of the analogues, Trp-AzAsn-Pro-Ala-Asp-Tyr-Gly-Gly-Ile-Lys and Trp-Asn-Pro-AzAla-Asp-Tyr-Gly-Gly-Ile-Lys, were tested against mAb6 monoclonal antibody raised against the receptor. The AzAsn analogue did not bind to the antibody whereas the AzAla analogue retained about 30% of the antibody binding affinity.[4]

Structural analysis has been carried out in solution by H^1 NMR and IR spectroscopy and in the solid state by X-ray diffraction using model pseudopeptides like Piv-Proψ[CH_2NH]Ala-NH^{i}Pr, Piv-Proψ[CH_2NZ]Ala-NH^{i}Pr, Piv-Proψ[CH_2NH]NH^{i}Pr, Piv-Proψ[$CH_2N^+H_2$]Ala-NH^{i}Pr, Piv-Proψ[$CH_2N^+H_2$] NH^{i}Pr, Piv-Proψ[CH_2O]Gly-NH^{i}Pr, Piv-Proψ[CH=N]Az-Ala-NH^{i}Pr, Boc-Proψ[CH=N]AzaGly-NH$_2$ and Piv-Proψ[CH_2NH]AzAla-NH^{i}Pr.[5] Substitution of AzPro for proline in the octapeptide Thr-Thr-Ser-Ala-Pro-Thr-Thr-Ser (motif present in the peptide backbone of MUC5AC mucin) was analysed in terms of conformational changes and O-glycosylation. In DMSO solution, AzPro prevents β-turn formation in which it would occupy the i+1 position, and therefore behaves quite opposite to Pro, whereas both AzPro and Pro can support a β-turn in the i+2 position with a *cis* disposition of the preceding tertiary amide function. The structural modifications do not prevent O-glycosylation taking place at the same specific site, but it occurs at a reduced rate.[6]

2.2 ψ[CH_2NH]-Aminomethylene, Hydroxyethylene, Dihydroxyethylene and Hydroxyethylamine Analogues

– Reductive amination on resins derivatised with linkers **1** or **2** on the aminomethylated or chloromethylated supports

(1) R = OMe or H   (2) R = OMe or H

indicated that, depending on the linker, spacer and the solid support, yields from 10% to 75% were obtained under exactly the same reaction conditions. Within one type of solid support and linker, those with the valeric acid spacer reacted faster and gave higher yields.[7] Synthetic routes to hydroxyethylene, dihydroxyethylene and hydroxyethylamine have been reported.[8,9]

2.3 ψ[CO-N(alkyl)] Analogues – Solid phase synthesis of N-alkylglycine oligomers on 2-chlorotritylchloride resin was carried out by using alternative reactions of bromoacetic acid and different amines.[10] This strategy avoids diketopiperazine formation in peptides bearing a glycine or a proline residue at the C-terminus. Similar procedures using Wang resin were not successful. A series of thiol containing peptides [*e.g.* HS-CH$_2$-CO-N(CH$_2$Ph)-CH$_2$CO-Gly (K_i 15 μM against thermolysin)] was synthesised as inhibitor of zinc metallopeptidases. For this purpose, an additional coupling step of bromoacetic acid and a subsequent substitution of the bromine atom by potassium thioacetate was achieved on the peptoid N(CH$_2$Ph)-CH$_2$CO-Gly.[10]

N-Alkyl substituted amino acid methyl esters using N-*p*-nitro-benzenesulfonyl protected amino acid methyl esters were prepared. The increased acidity of the sulfonamide NH allowed alkylation using alkylbromide and potassium carbonate.[11] Deprotection of the N-alkyl-sulfonamides (thiophenol and potassium carbonate) led to N-alkyl amino acids. Studies using several coupling agents indicated that HOAt based couplings gave the best results for the synthesis of dipeptides containing N-alkyl amino acids (*e.g.* Boc-Phe-N-alkylPhe-OMe). Several N-alkyl Leu-enkephalin analogues [Boc-Tyr(*t*-butyl)-Gly-Gly-Phe-N-EtLeu-OMe, Boc-Tyr(*t*-butyl)-Gly-Gly-N-EtPhe-Leu-OMe, Boc-Tyr(*t*-butyl)-Gly-N-EtGly-Phe-Leu-OMe, Boc-Tyr(*t*-butyl)-N-EtGly-Gly-Phe-Leu-OMe] were synthesised on Tentagel resin. *o*-Nitrobenzenesulfonylamide gave better results than *p*-Nitrobenzenesulfonylamide group. No biological activity data are given.

2.4 ψ[CH$_3$C=CH], ψ[CF$_3$C=CH], ψ[CON=S(Ph)=O], ψ[CH=CH-SO$_2$NH] and ψ[CSNH] Analogues – Syntheses of methyl- and (trifluoromethyl)alkene peptide isosteres (like **3** and **4**) and sulfoximine-containing pseudopeptides of type **5** are reported.[12,13] Conformational properties of the compounds were investigated by X-ray or NMR analysis. Chiral vinylogous sulfonamidopeptides (**6**) were synthesised on Tentagel resin employing (S)- and (R)-N-Boc-vinylogous sulfonyl chlorides as building blocks.[14] Vinylogous sulfonamidopeptide libraries were synthesised by the 'split-mix synthesis' method. Taking advantage of the acidic character of the sulfonamides, conditions were developed to alkylate the sulfonamide nitrogen atom so as to reduce the acidity of the monomers and of the oligomers and increase the diversity and *in vivo* bioavailability. The conformations of four model pseudodipeptides, Ac-ψ[CSNH]-Gly-NHMe, Ac-Gly-ψ[CSNH]-NHMe, Ac-ψ[CSNH]-Ala-NHMe, and Ac-Ala-ψ[CSNH]-NHMe, were studied by high-level *ab initio* calculations. Whereas the conformations of the C-terminal thioamides were close to those of the corresponding peptides, the N-terminal thioamides displayed markedly different conformational behaviour. The changes in the conformational profile of thioamide-containing peptides appear to result from a combination of the decreased hydrogen bonding-accepting ability and increased size of sulfur versus oxygen and lengthening of the C=S bond in the thioamide as compared to the C=O bond in an amide.[15]

2.5 Rigid Di-, Tri-peptide and Turn Mimetics – Syntheses of several conformationally constrained amino acids (**7-12**) have been reported.[16-18] Some of the proline-derivatives may be considered as Pro-Leu (**8**), Pro-Lys (**9**), Pro-Arg (**10**) and Pro-Glu (**11**) based dipeptide mimetics. A number of these including **12** (R = Boc-NH-CH$_2$- or COOH) can be used in the design of peptidomimetics with defined and predictable conformations. In addition to these amino acid based structures several other non-peptidic templates (**13–18**) have also been designed for incorporation in peptides.[19-23] The synthesis of model tri- and tetrapeptide analogues based on the constrained 10-membered lactam **16** (R = Boc-Phe or Boc-Ala-Phe; R′ = Phe-OMe or Met-NH$_2$) have been reported. Conformational preferences of some of these peptide derivatives have been investigated by spectroscopic techniques. Other peptides containing turn-mimetic structures include compounds **19, 20, 21** (X = NH, S, O, n = 0–3, R^1 and R^2 = side chains of Ile, Ser, Thr, Val, Leu, Asn, Glu, Lys), **22** and **23**.[24-26] Compounds **22** and **23** are both based on the tetrapeptide motif, Asn-Pro-Asn-Ala, present in the immunodominant central portion of the circumsporozoite surface protein of the malaria parasite *Plasmodium falciparum*. The template in these peptides was designed to stabilise β-turns in the peptide loop and to allow its conjugation to T-cell epitopes in a multiple antigen-peptide.[26]

Synthesis and structural analysis of a series of cyclic helical peptides wherein ring-closing metathesis is used to incorporate a carbon-carbon tether between amino acid side chains is reported.[27] In peptides like Boc-Val-Ala-Leu-Aib-Val-Ala-Leu-OMe, the Aib residue has been shown to stabilise a helical structure. Analogues of this peptide [Boc-Val-Ser(CH$_2$-CH = CH$_2$)-Leu-Aib-Val-Ser(CH$_2$-CH = CH$_2$)-Leu-OMe and Boc-Val-hSer(CH$_2$-CH = CH$_2$)-Leu-Aib-Val-hSer(CH$_2$-CH = CH$_2$)-Leu-OMe] were reacted with the olefin meta-

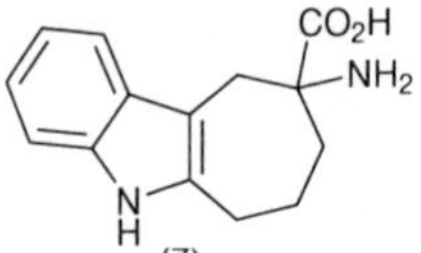

(7) 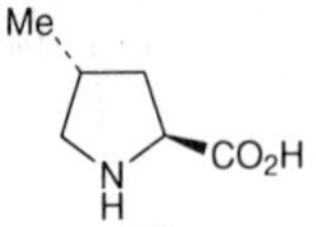 (8) 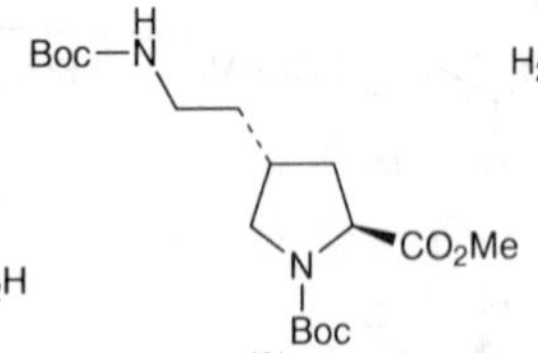 (9) 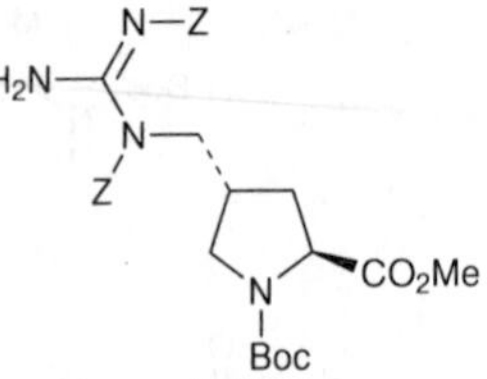 (10)

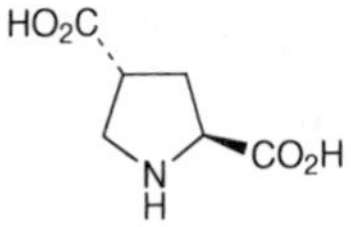

(11) 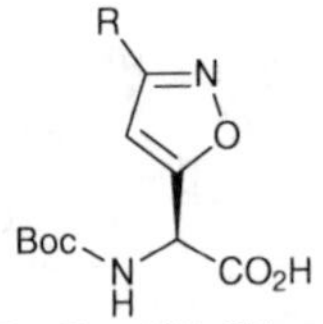 (12) R = Boc–NH–CH$_2$ or CO$_2$H (13) 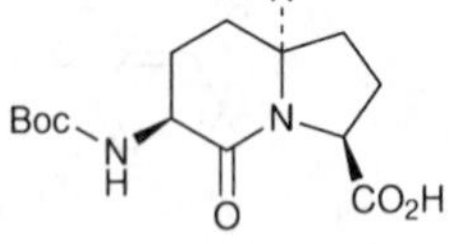 (14)

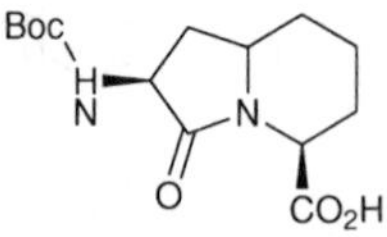

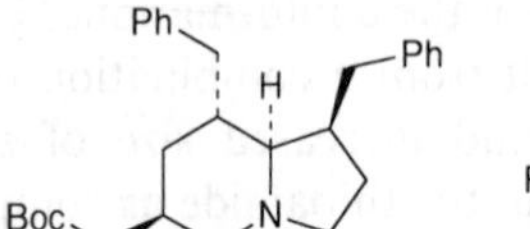

(15) 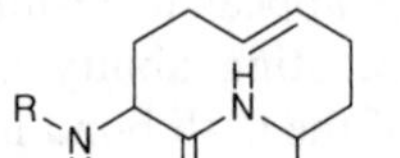 (16) 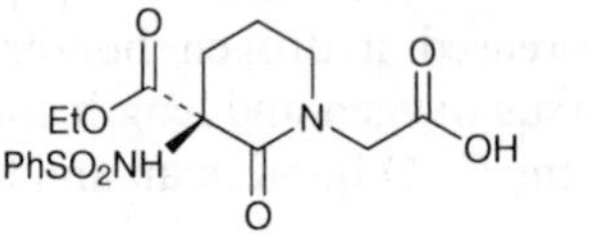 (17)

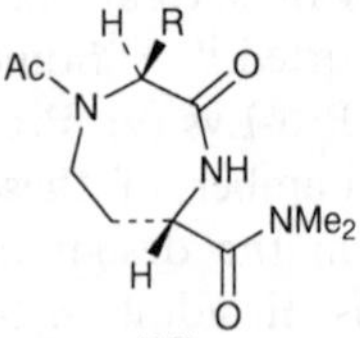

(18) 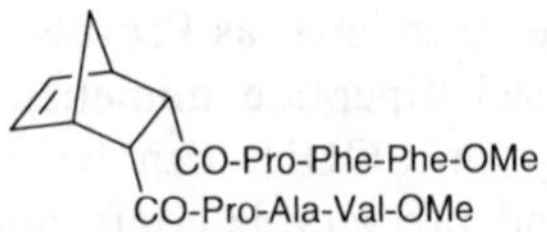 (19) 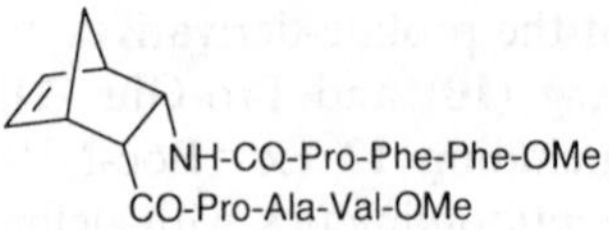 (20)

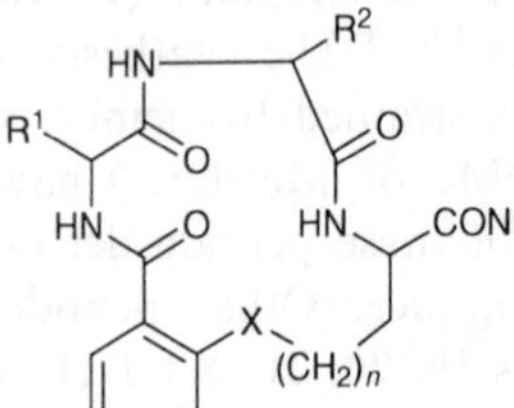

(21) 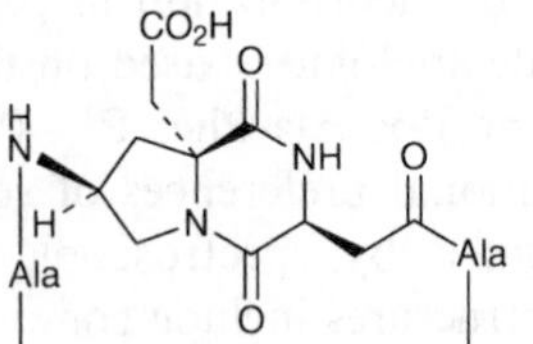 (22) 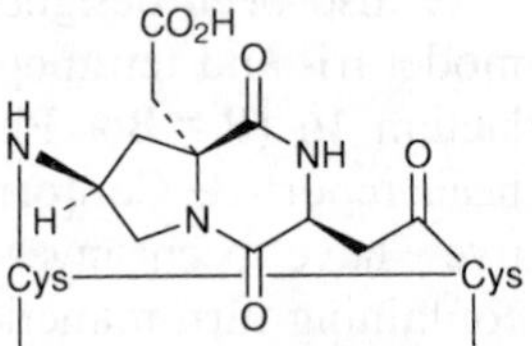 (23)

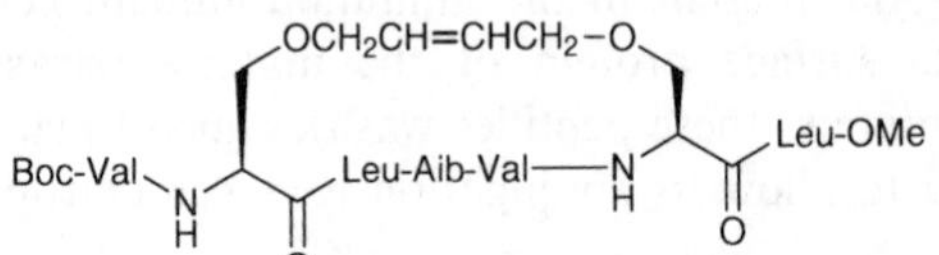

(24)

thesis catalyst [(PCy$_3$)$_2$Cl$_2$Ru=CHPh] to give the cyclic peptides (24) and the corresponding homoserine analogue. Catalytic hydrogenation gave the reduced versions of the serine and homoserine derivatives.

3   Cyclic Peptides

Cyclic peptide analogues of biologically active peptides are included in the sections dealing with individual peptides (Section 4). Sequences of other cyclic peptides isolated from natural sources and their biological activities are presented here. Side reactions in the synthesis of backbone to backbone cyclic peptides requiring N^α-substituted amino acid derivatives have been reported. The use of N-carboxymethyl amino acids can lead to diketopiperazine formation by intramolecular aminolysis which occurs despite the tert-butyl protection of the carboxy group.[28] This side reaction can be prevented by a very short deprotection time for the Fmoc group, by elongation of the N-carboxyalkyl chain or by forming the backbone (lactam) bridge before Fmoc removal, but not by the use of DBU or additives.

A general method was described for the synthesis of thioether cyclic peptides (e.g. 25).[29] The thioether linkage was formed through an intramolecular substitution of the chloro group of β-chloroalanine with the thiol group of cysteine. Using a similar procedure, thioether-linked cyclic peptide libraries involving a cyclisation step between a C-terminal cysteine side chain and an N-terminal bromoacetyl group were prepared.[30] Cleavage of peptides like BrCH$_2$CO-Asp(OBu^t)-Asp(OBu^t)-Ile-Lys(Boc)-Cys(methoxytrityl or trityl)-Lys(X)-NH-Rink resin from the solid support under acid conditions leads to cysteine deprotection and peptide cyclisation occurs in the same process. Conditions for the synthesis of i to i+4 side chain-to-side chain (Lys → Glu and Glu → Lys) linked cyclic peptides related to hypoglycaemic analogues of human growth hormone hGH(6-13) [Fmoc-Lys-Phe-D-Ala-Pro-Glu-Gly, Fmoc-Glu-Phe-D-Ala-Pro-Lys-Gly, Fmoc-Lys-Phe-D-Ala-Pro-Glu-Leu, Fmoc-Gly-Lys-Phe-D-Ala-Pro-Glu-Gly and Fmoc-Gly-Glu-Phe-D-Ala-Pro-Lys-Gly attached to the MBHA-resin] have been examined.[31] Cyclisation reactions of the partially protected linear peptides were both coupling reagent as well as sequence dependent, with competing inter-chain oligomerisation predominating in some cases. Protection with the bulky Fmoc group of the amino acid residues immediately adjacent to the side chain-deprotected Lys and Glu residues, which participate in the cyclisation reaction, enhanced the rate of lactam formation. Solid-phase and solution phase syntheses of cyclic depsipeptides like AM-toxins [c(DehydroAla-Ala-O-CH(CHMe$_2$)-CO-NHCH(CH$_2$)$_3$-Ph(p-X)] (X = OMe, H or OH)] and analogues, fragments of pseudotetrapeptide didemnin M [Pyr-Glnψ[COO]Ala-Pro], arenastatin A (26, a cytotoxic peptide isolated from the Okinawan sponge *Dysidea arenaria*) and a cyclic peptide analogue of dolastatin 10 (27) are reported.[32-35] Dolastatin 10 analogue 27 (less potent than dolastatin) inhibited growth of HT-29 and L1210 cells (IC$_{50}$ values 60–110 μM) and tubulin polymerisation (IC$_{50}$ 39 μM). The linear peptide dolastatin 10 and four analogues were fungicidal for American Type Culture Collection strains and clinical isolates (including fluconazole-resistant strains) of *Cryptococcus neoformans*.[36]

Chemical structures of several biologically active cyclic peptides [comoramides A (28) and B (29), mayotamides A and B (30, R = -CHMe-CH$_2$Me or

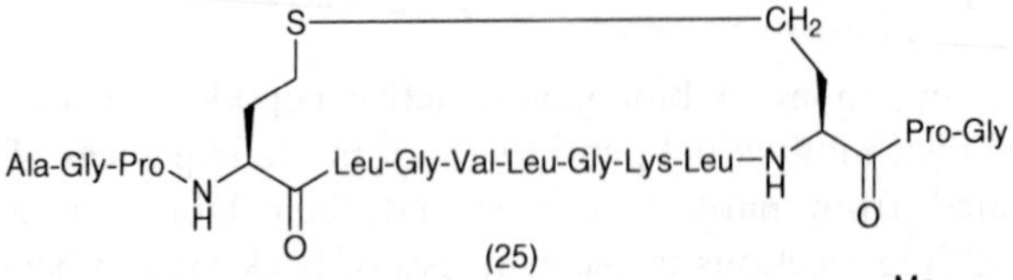

(25)

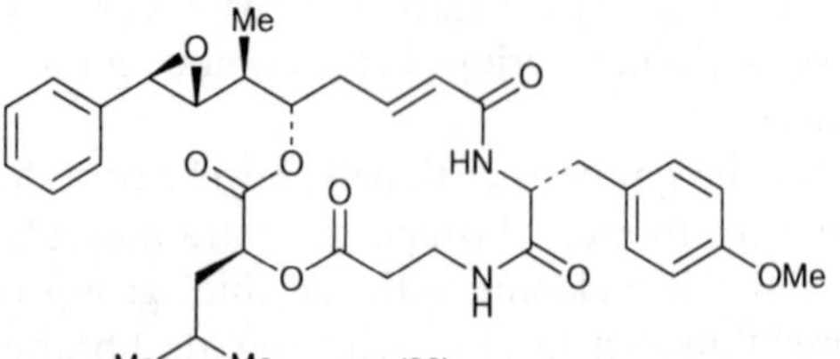

Me   Me   (26)

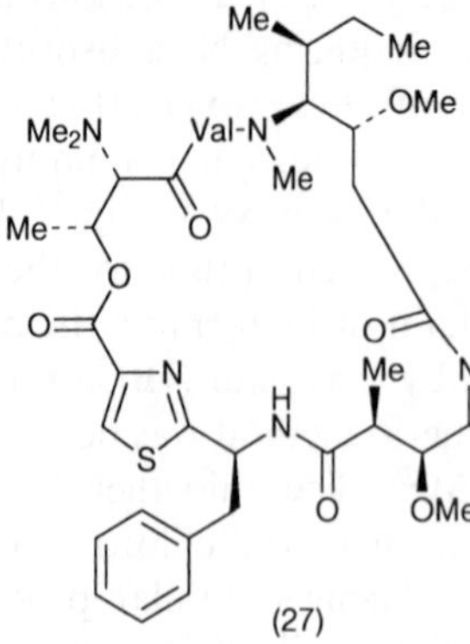

(27)

-CHMe$_2$), polyoxypeptin, ustiloxin D and F (**31**, R = -CHMe$_2$ or -Me), hibispeptin A (**32**), axinellins A and B and cyclolinopeptide A] isolated from various sources have been reported.[37-43] Polyoxypeptin (inducer of apoptosis in human pancreatic carcinoma cells), a hexadepsipeptide isolated from *Streptomyces* culture broth, contains all unnatural amino acids (N-hydroxyvaline, 3-hydroxy-3-methylproline, 5-hydroxyhexahydropyridazine-3-carboxylic acid, N-hydroxyalanine, piperazic acid and N-hydroxyleucine).[38] Hibispeptin A was isolated from the root bark of *Hibiscus syriacus* Linne (Malcaceae), which has been used as antipyretic, anthelmintic and antifungal agent in the Orient.[41] Comoramides A and B and mayotamides A and B exhibited cytotoxicity (IC$_{50}$ 5-10 µg ml^{-1}) against several tumour cells (A549, HT29 and MEL-28). Usliloxin F inhibited microtubule assembly with an IC$_{50}$ value of 10.3 µM.[39] Axinellins A [c(Phe-Pro-Asn-Pro-Phe-Thr-Ile)] and B [c(Phe-Pro-Thr-Leu-Val-Pro-Trp-Pro)], isolated from the marine sponge *Axinella carteri*, exhibited *in vitro* antitumour activity against human non-small-cell-lung-carcinoma lines (NSCLC-NG) (IC$_{50}$ values of 3.0 to 7.3 µg ml^{-1}).[42] The immunosuppressive cyclolinopeptide A [c(Val-Pro-Pro-Phe-Phe-Leu-Ile-Ile-Leu)] inhibited calcium-dependent activation of T lymphocytes comparably to the actions of cyclosporin A and FK506. The concentration required for complete inhibition, however, is 10-times higher than that of cyclosporin A.[43] Biological activities of four cyclic peptides, c(Phe-Pro), c(Tyr-Pro), c(Trp-Pro) and c(Trp-Trp) were reported. Cyclic peptides c(Trp-Pro) and c(Trp-Trp) blocked calcium channels and c(Tyr-Pro) and c(Trp-Pro) blocked delayed-rectifier potassium channels. All the peptides increased the expression of alkaline phosphatase and showed concentration-dependent antibacterial properties (MIC values 6-130 nM against Gram-positive and Gram-negative bacteria).[44]

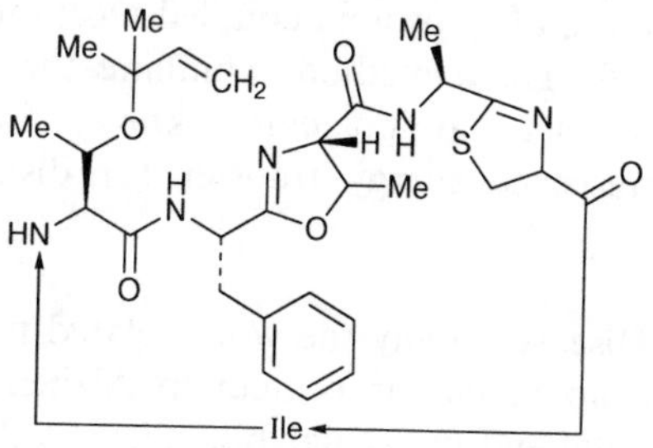

(28)

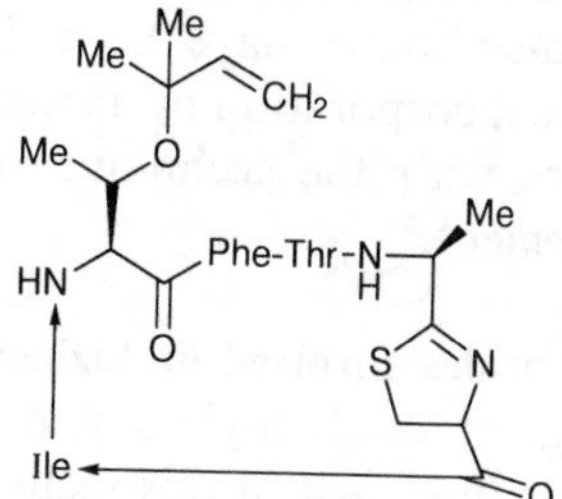

(29)

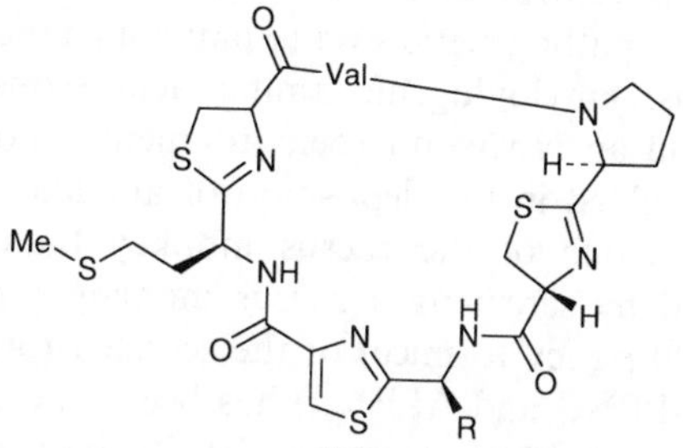

(30)

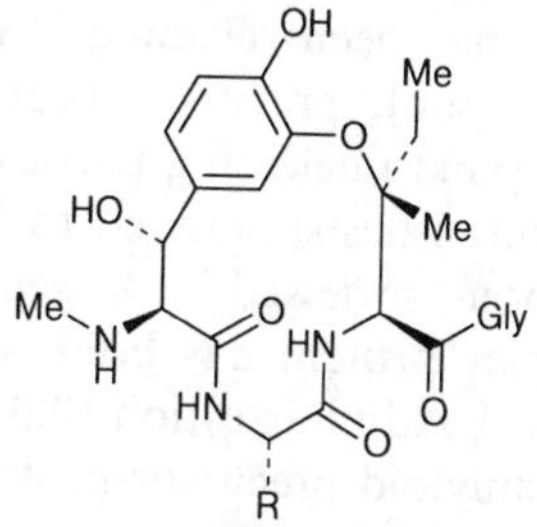

(31)

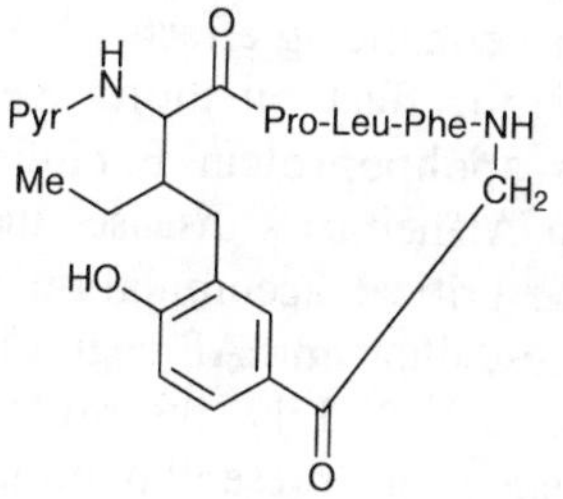

(32)

4 Biologically Active Peptides

General reviews on peptides and their non-peptidyl mimetics in endocrinology, role of angiogenesis in hypertension, antiangiogenic tumour therapy and pathological significance of chemokine receptors have appeared.[45-49] Various aspects, including diversity of receptor-ligand interactions and mechanisms of agonist activation, of G-protein-coupled receptors have been reviewed.[50,51] The role of orphan G-protein-coupled receptors as drug targets has also been reviewed.[52] The use of approaches like reverse molecular pharmacology which involves cloning and expression of orphan G-protein-coupled receptors in mammalian cells and screening these cells for a functional response to cognate or surrogate agonists present in biological extract preparations, peptide libraries, and compound collections and functional genomics involving the use of 'humanised' yeast cells may lead to novel receptor ligands. Once activating ligands are identified they can be used as pharmacological tools to explore receptor function and relationship to disease.[52] Fluorescence approaches to investigate ligand recognition and structure of G protein-coupled receptors in native membranes have been developed. These methods combine the biosynthetic incorporation of unnatural fluorescent amino acids at known sites in receptors with the technique of fluorescence energy transfer for distance measurement.[53]

4.1   Peptides Involved in Alzheimer's Disease – Only the work related to the involvement of β-amyloid and β-amyloid precursor protein in Alzheimer's disease is discussed in this section. Genetic and molecular biological evidence suggesting that the peptide Aβ42 is central to the etiology of Alzheimer's disease has been reviewed.[54] *In vivo* pathogenetic events [amyloid precursor protein pool, precursor proteolysis, amyloidogenic amino acid sequences, fibrillogenic nucleating particles, and an *in vivo* microenvironment conducive to fibrillogenesis] believed to be involved in the deposition of amyloids have also been reviewed.[55] A genomic clone of the rhesus monkey β-amyloid precursor protein has been isolated by screening a rhesus monkey genomic library.[56] Administration (0.05–5000 pg icv to mice) of the secreted forms of the β-amyloid precursor protein (APP^S_{751} and APP^S_{695}) has been reported to display memory-enhancing effects and block learning deficits induced by scopolamine.[57] The memory enhancing effects of APP^S were blocked by anti-APP^S antisera. APP^S had no effect on motor performance or exploratory activity. In normal brains apolipoprotein E binds and sequesters Aβ, preventing its aggregation. In Alzheimer's disease, the impaired apolipoprotein E-Aβ binding leads to the critical accumulation of Aβ, facilitating plaque formation.[58] The role of presenilins and Aβ degrading enzymes in Alzheimer's disease has been investigated.[59-63] Like the mutations identified within the β-amyloid precursor protein gene, presenilin mutations cause the increased generation of a highly neurotoxic variant of amyloid β-peptide. Mutations in two familial AD-linked genes, presenilins 1 and 2, selectively increase the production of Aβ42 in cultured cells and the brains of transgenic mice, and gene deletion of presenilin 1 shows that it is required for normal γ-secretase cleavage of the β-amyloid precursor protein to generate Aβ. Presenilin proteins are proteolytically processed to an N-terminal similar to 30-kDa and a C-terminal similar to 20-kDa fragment that form a heterodimeric complex which is resistant to proteolytic degradation, whereas the full-length precursor is rapidly degraded. The presenilin 1 heterodimeric complex can be attacked by proteinases of the caspase superfamily. Changes in fragment generation/degradation might be important for Alzheimer's disease-associated pathology.[60,61] Other enzymes like cathepsins D and B are also thought to be involved in the amyloidogenic processing of amyloid precursor protein.[62] Insulin-degrading enzyme (a thiol metalloendopeptidase) that degrades peptides such as insulin, glucagon, and atrial natriuretic peptide also degrades Aβ. Degradation of both endogenous and synthetic Aβ was completely inhibited by the competitive insulin-degrading enzyme substrates like insulin. In addition to its ability to degrade Aβ, insulin-degrading enzyme activity was found to be associated with oligomerisation of synthetic Aβ at physiological levels in the conditioned media of cultured cells; this process, which may be initiated by insulin-degrading enzyme-generated proteolytic fragments of Aβ, was prevented by the insulin-degrading enzyme inhibitors.[63] Thus insulin-degrading enzyme may be capable of down-regulating the levels of secreted Aβ extracellularly.

Structural, biological and neurotoxicity studies on amyloid β peptide and fragments have been reported. NMR data, used to resolve the fibril structure in aggregated β-amyloid peptide, indicated that the Aβ peptide stacks as a directly aligned parallel β-sheet.[64] Using CD and NMR spectroscopic techniques, the solution structure of residues 1-28 of the amyloid β peptide when bound to micelles was investigated. The results showed that micelles prevent formation of the toxic β-sheet structure for the 1-28 region.[65] Synthetic amyloid like structures have been developed using tripeptide derivatives. Several of the derivatives like **33** formed an aggregate not only in water but

(33)

also in some nonpolar organic solvents.[66] Small diffusible Aβ oligomers derived from Aβ(1-42) have been suggested to be responsible for neurological dysfunction well in advance of cellular degeneration and, therefore, impaired synaptic plasticity and associated memory dysfunction during early stage Alzheimer's disease may be due to these oligomers acting upon particular neural signal transduction pathways before the amyloid fibril formation.[67] The neurotoxicity of Aβ(25-35) has been associated with the haemolytic activity of the peptide when it forms β-sheet structure and aggregates. In its monomeric random coil structure form, Aβ(25-35) does not perturb lipid membranes significantly, and exhibits no haemolytic activity.[68] Because of its sequence similarity to tachykinins, the Aβ(25-35) [Gly-Ser-Asn-Lys-Gly-Ala-Ile-Ile-Gly-Leu-Met], Aβ(25-35)-NH$_2$ and their Nle[35] and Phe[31] analogues were investigated for the tachykinin-like activities. In ligand displacement studies on tachykinin NK$_1$ receptors, only the Phe[31] analogue [Gly-Ser-Asn-Lys-Gly-Ala-Phe-Ile-Gly-Leu-Met] showed activity comparable to that of genuine tachykinins.[69]

Incubation of Aβ1-40 for 7 days in the presence of a β-sheet breaker pentapeptide Leu-Pro-Phe-Phe-Asp (equimolar or 20-fold molar excess) produced 53.5 and 84.1% inhibition of fibrillogenesis, respectively. Under the same conditions Aβ1-42 fibrillogenesis was inhibited 34.2 and 71.9%. The pentapeptide also dissolved pre-formed fibrils. Co-injection with Aβ1-42 with the pentapeptide reduced fibril formation in a rat brain model of amyloidosis.[70] Small peptides containing His-His-Gln-Lys [Aβ(13-16)] inhibited Aβ(1-42) cell binding as well as plaque induction of neurotoxicity in human microglia. *In vivo* experiments confirmed that the His-His-Gln-Lys peptide reduces rat brain inflammation elicited after infusion of Aβ peptides or implantation of native plaque fragments.[71]

4.2 Antimicrobial Peptides – Reviews on cationic antibiotic peptides, two-component signal transduction as a target for microbial anti-infective therapy and structure-function relationships of antimicrobial peptides have been published.[72-74] A number of antimicrobial peptides of plants and animals are cationic (containing Lys and Arg residues) amphipathic molecules and, because of their antibacterial, anti-endotoxic, antibiotic-potentiating or anti-fungal properties, are being developed for use as a novel class of antimicrobial agents and as the basis for making transgenic disease-resistant plants and animals. The two-component signal transduction system combines signal recognition, signal transduction, and gene activation in a two-protein system and consists of a sensor histidine kinase and a response regulator. The sensor kinase is the primary signal transduction protein that interacts directly with a signal ligand or with a receptor that binds to the signal ligand.[73] The bactericidal activities of four peptides, cecropin P1, magainin II, indolicidin, and ranalexin (all reported previously), were compared against >200 clinical isolates of Gram-positive and Gram-negative aerobic bacteria.[75] The peptides exhibited different *in vitro* activities and rapid time-dependent killing. Some of the antimicrobial peptides have been reported to display other biological properties. For example, tumourigenicity of cecropin expressing bladder carcinoma derived cell clones in nude mice was greatly reduced.[76] Similarly, a proline-arginine-rich antimicrobial peptide PR-39 and a biologically active fragment were found to bind to a number of cytoplasmic proteins and alter mammalian cell gene expression and behaviour.[77]

4.2.1 Antibacterial Peptides – As in previous years, new antibacterial peptides from natural sources have been isolated and SAR and mechanism of action studies on several other antibacterial peptides have been reported. Six peptides (including the known hypotensive peptide caerulin) were isolated from the dorsal glands of the tree frog *Litoria genimaculata* and named maculatins [Gly-Leu-Phe-Gly-Val-Leu-Ala-Lys-Val-Ala-Ala-His-Val-Val-Pro-Ala-Ile-Ala-Glu-His-Phe-NH₂ (maculatin 1.1), Phe-Gly-Val-Leu-Ala-Lys-Val-Ala-Ala-His-Val-Val-Pro-Ala-Ile-Ala-Glu-His-Phe-NH₂, Gly-Leu-Phe-Gly-Val-Leu-Ala-Lys-Val-Ala-Ser-His-Val-Val-Pro-Ala-Ile-Ala-Glu-His-Phe-Gln-Ala-NH₂, Gly-Phe-Val-Asp-Phe-Leu-Lys-Lys-Val-Ala-Gly-Thr-Ile-Ala-Asn-Val-Val-Thr-NH₂ and Gly-Leu-Leu-Gln-Thr-Ile-Lys-Glu-Lys-Leu-Glu-Ser-Leu-Ala-Lys-Gly-Ile-Val-Ser-Gly-Ile-Gln-Ala-NH₂. Maculatin 1.1 showed the most pronounced activity, particularly against Gram-positive organisms (MIC values 3–100 μg ml⁻¹).[78] Two peptide toxins with antimicrobial activity, lycotoxins I (Ile-Trp-Leu-Thr-Ala-Leu-Lys-Phe-Leu-Gly-Lys-His-Ala-Ala-Lys-His-Leu-Ala-Lys-Gln-Gln-Leu-Ser-Lys-Leu-NH₂) and II (Lys-Ile-Lys-Trp-Phe-Lys-Thr-Met-Lys-Ser-Ile-Ala-Lys-Phe-Ile-Ala-Lys-Glu-Gln-Met-Lys-Lys-His-Leu-Gly-Gly-Glu-NH₂), were identified from venom of the wolf spider *Lycosa carolinensis*. Both peptides showed antibacterial and antifungal activities at micromolar concentrations.[79] In response to epidermal injury, *Pavasilurus asotus* (a catfish) secreted a strong antimicrobial peptide into the epithelial mucosal layer. Eighteen of the 19 residues of the peptide, parasin I (Lys-Gly-Arg-Gly-Lys-Gln-Gly-Gly-Lys-Val-Arg-Ala-Lys-Ala-Lys-Thr-Arg-Ser-Ser) were identical to the N-terminal of buforin I, a 39-residue antimicrobial peptide derived from the N-terminal of toad histone H2A. Parasin I showed antimicrobial activity against several Gram-negative and Gram-positive bacteria and fungi (MIC values 1-4 μg ml^{-1}) (12–100 times more potent than magainin 2), against a wide spectrum of micro-organisms, without any haemolytic activity.[80] Proenkephalin-A derived peptides (PEAP), natural enkelytin [Ser^{221,223} bisphosphorylated PEAP$_{209-237}$, Phe^{209}-Ala-Glu-Pro-Leu-Pro-Ser^{215}-Glu-Glu-Glu-Gly-Glu-Ser^{221}-Tyr-Ser^{223}-Lys-Glu-Val-Pro-Glu-Met-Glu^{230}-Lys-Arg-Tyr-Gly-Gly-Phe-Met^{237}] and natural bisphosphorylated (Ser^{221,223}) PEAP$_{209-239}$, known as peptide B [-Arg^{238}-Phe^{239}], inhibited the growth of *M. luteus* at a concentration of 0.2 μM, but were unable to inhibit that of *E. coli* in the concentration range 0.2–3 μM. Synthetic unphosphorylated PEAP$_{209-237}$ and some of its fragments, 209-220, 224-237, and 233-237 did not show significant antibacterial activity.[81] Various other antibacterial peptides obtained from the South American frog *Phylomedusa bicolor*, seeds of pokeweed (*Phytolacca americana*) and human lung epithelium have been reported.[82-84]

Antibacterial and haemolytic activities of crabrolin (Phe-Leu-Pro-Leu-Ile-Leu-Arg-Lys-Ile-Val-Thr-Ala-Leu-NH$_2$), present in the venom of the hornet *Vespa crabro*, and analogues are reported.[85] The [Leu^3, Pro^6] analogue was similar in potency to the parent peptide but the [Lys^3, Pro^6] analogue was about 2-fold less potent. [Ala^3]-crabrolin, Fmoc-crabrolin and Fmoc-[Lys^3, Pro^6]-crabrolin showed improved antibacterial activity compared to the parent peptides. Fmoc-[Lys^3, Pro^6] analogue was the most potent analogue (MICs 11, 15, 7.5 and 90 μg ml^{-1} against *E. coli*, *S. aureus*, *B. subtilis* and *P. putida*, respectively). Although Fmoc-crabrolin showed significant haemolytic activity, the other analogues were somewhat less potent as haemolytic agents. Antibacterial, haemolytic and antitumour activities of cecropin A-magainin 2 and cecropin A-melittin hybrid peptides have been reported.[86,87] The cecropin A(1-8) and magainin 2(1-12) hybrid peptide [Lys-Trp-Lys-Leu-Phe-Lys-Lys-Ile-Gly-Ile-Gly-Lys-Phe-Leu-His-Ser-Ala-Lys-Lys-Phe-NH$_2$] and a few analogues, [Thr^{18,19}]-, [Ala^{12}]- and [Leu^{16}]-hybrids, retained significant antimicrobial activity and were much less haemolytic (0–11% haemolysis at 100 μg ml^{-1}). In comparison, a hybrid peptide of cecropin(1-8) and melittin [Lys-Trp-Lys-Leu-Phe-Lys-Lys-Ile-Gly-Ile-Ala-Val-Leu-Lys-Val-Thr-Thr-Gly-NH$_2$] was somewhat less potent as an antibacterial agent and displayed significant haemolytic activity (51% at 100 μg ml^{-1}).[86] Cecropin A(1-8) and magainin(1-12) hybrid analogues [Ala^{12}, Val^{13}, Lys^{15}, Val^{16}, Leu^{17}, Gly^{20}]-, [Thr^{18,19}]-, [Ala^{12}]-, [Leu^{16}]-, [Ala^{12}, Leu^{16}]-, [Lys^{15}, Leu^{16}]-, [Ala^6, Leu^{16}]- and [Lys^4, Leu^{6,16}]- were investigated for antitumour activities using small cell lung cancer cell lines (NCI-H69, NCI-H128 and NCI-H146). Greater antitumour activity was observed when the residues 16, 18 and 19 of the peptide were hydrophobic (Leu or Val), basic (Lys) and basic (Lys), respectively.[87] Hybrids

[Ala^{12}, Val^{13}, Lys^{15}, Val^{16}, Leu^{17}, Gly^{20}]-, [Leu^{16}]- and [Lys^{15}, Leu^{16}]- with these properties had higher antitumour activity (IC$_{50}$ 2-4 µM), but were relatively less haemolytic. Residue 12 was related to haemolytic activity rather than antitumour activity. Increase in amphipathicity of the cecropin A(1-8) and magainin(1-12) hybrid (by changes at positions 4 and 5 or 6) enhanced haemolytic activity without significant change in antitumour activity. *In vitro* antimicrobial activity of cationic peptide MSI-78 (a magainin analogue) against various clinical isolates has been reported.[88]

To study the function of the N-terminal loop and disulfide bridges, N-terminal loop deleted tenecin 1, reduced tenecin 1 and tenecin 1 (an inducible antibacterial protein secreted in the larvae of *Tenebrio molitor*) were synthesised. The loop deleted tenecin and reduced tenecin 1 did not show antibacterial activity.[89] An active fragment of tenecin 1 has been identified.[90] The C-terminal β-sheet domain [tenecin(29-43), Arg-Ser-Gly-Gly-Tyr-Cys-Asn-Gly-Lys-Arg-Val-Cys-Val-Cys-Arg-NH$_2$] showed activity against fungi as well as Gram-positive and Gram-negative bacteria, whereas tenecin 1, the native protein, showed activity only against Gram-positive bacteria. Truncated fragments, Tyr-Cys-Asn-Gly-Lys-Arg-Val-Cys-Val-Cys-Arg-NH$_2$ and Cys-Asn-Gly-Lys-Arg-Val-Cys-Val-Cys-Arg-NH$_2$, were similar in potency to tenecin (29-43). Further deletions from the N- or C-terminal ends of tenecin(34-43) or protection of the cysteine residues by acetamidomethyl groups resulted in loss of antimicrobial activity. The mechanism of action of buforin II (21-amino acid antimicrobial peptide) was compared with magainin 2, which has a pore-forming activity on the cell membrane. Buforin II killed *E. coli* without lysing the cell membrane but magainin 2 lysed the cell to death under the same conditions.[91] Labelled buforin II was found to penetrate the cell membrane and accumulate inside *E. coli*. It bound to DNA and RNA of the cells over 20 times more strongly than magainin 2. Thus buforin II inhibits the cellular functions by binding to DNA and RNA of cells after penetrating the cell membranes, resulting in the rapid cell death, which is quite different from that of magainin 2 even though they are structurally similar α-helical peptides.

4.2.2 Antifungal Peptides – Chemical structures of several antifungal peptides like **34–36** isolated from natural sources have been reported.[92-94] The cyclo-depsipeptides W493 B (**34**) and an analogue containing a Val in place of Ile, isolated from a culture broth of *Fusarium* sp., showed antifungal activity (MIC values 1–10 µg ml^{-1}) against *Venturia inaequalis*, *Monilinia mali* and *Cochliobolus miyabeanus*. Isomeric linear peptides, aspergillamides A and B possessing dehydrotryptamine functionalities, were isolated from the mycelium of a cultured marine fungus of the genus *Aspergillus*. Aspergillamide A (**35**) showed modest *in vitro* cytotoxicity (IC$_{50}$ 16 µg ml^{-1}) toward the human colon carcinoma cell line HCT-116. Antifungal cyclic depsipeptide, cyclolithistide A isolated from a marine sponge, *Theonella swinhoei*, contains unnatural amino acids 4-amino-3,5-dihydroxyhexanoic acid, formyl-leucine, and chloroisoleucine.[94]

Using combinatorial libraries consisting of various amino acid sequences, a

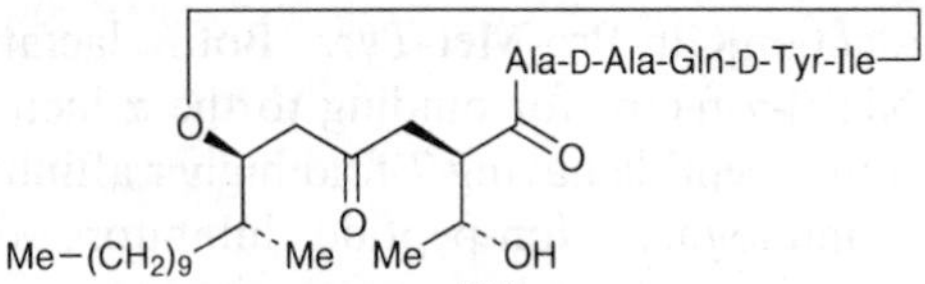

Ala-D-Ala-Gln-D-Tyr-Ile

Me−(CH₂)₉ Me Me OH

(34)

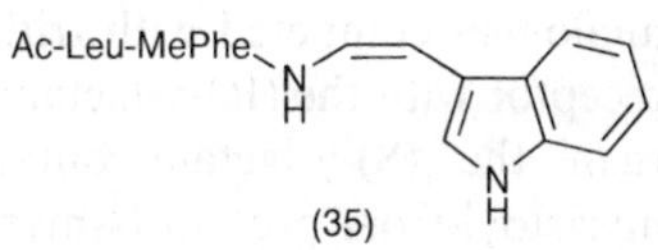

Ac-Leu-MePhe N
H

(35)

Cl

Me Me

N—MeLeu-Ala
H

HO N
H

Me Me

NH

O Me H O

Gln-alloThr-Phe-Gly-Nva

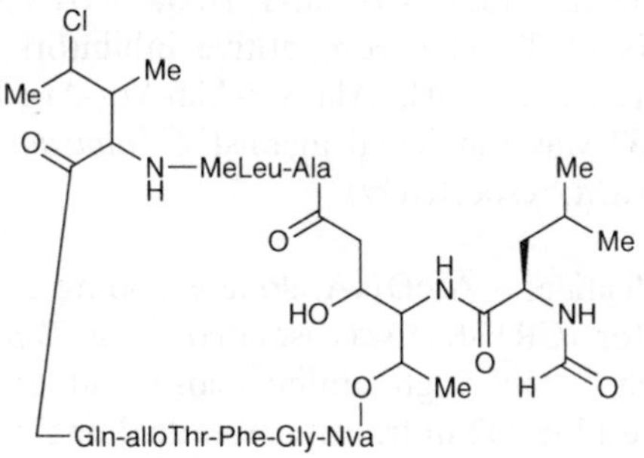

(36)

decapeptide (Lys-Lys-Val-Val-Phe-Lys-Val-Lys-Phe-Lys) (KSL) was found to be active against the *Candida albicans* membrane and showed no significant haemolytic activity up to a concentration of 25 µg ml⁻¹.[95] This peptide irreversibly inhibited the growth of *C. albicans* and showed a broad range of antibacterial activity but no haemolytic activity. Synthetic analogues [Val⁵]-, [Phe⁷]- and [Leu³,⁴,⁷]-KSL showed similar potency against various Gram-positive and Gram-negative bacteria and fungi. [Pro⁵]-KSL which contained a Pro (an α-helix breaker) did not show activity at concentrations up to 100 µg ml⁻¹. A truncated derivative of the natural amphibian skin peptide dermaseptin S3-(1-16)-NH₂ [Ala-Leu-Trp-Lys-Asn-Met-Leu-Lys-Gly-Leu-Gly-Lys-Leu-Ala-Gly-Lys-NH₂] inhibited completely the growth of *Saccharomyces cerevisiae* at a concentration of 8.63 µg ml⁻¹.[96] Neuropeptide Y and N-terminally truncated analogues were reported to be fungicidal against *Candida albicans* (MICs, approximately 1 µM).[97] In comparison to NPY (MIC, approximately 10–12 µM), NPY(13-36), [Lys¹³]NPY(13-36), [Thr¹³]NPY(13-36), [Asn¹²]NPY(12-36), [Lys¹²]NPY(12-36), [Thr¹²]NPY(12-36), [Arg¹¹]NPY (11-36), [Asn¹¹]NPY(11-36), and [Lys¹¹]NPY(11-36) (MICs 1.0–2.0 µM), and [Asp¹³]NPY(13-36), [Asp¹²]NPY(12-36), [Glu¹²]NPY(12-36) and [Glu¹¹]NPY (11-36) were more potent (3–7 µM). N-Terminally acetylated or succinylated analogues of [Asn¹²]NPY(12-36), [Thr¹²]NPY(12-36) and [Asp¹²]NPY(12-36) were 3–10-fold less potent as antifungal peptides than the parent peptides containing a free N-terminal amino group. However, [Lys¹²]NPY(12-36) and its N-acetyl and N-succinyl derivatives were nearly equipotent.

Analogues of the *Saccharomyces cerevisiae* α-factor (tridecapeptide mating pheromone, Trp-His-Trp-Leu-Gln-Leu-Lys-Pro-Gly-Gln-Pro-Met-Tyr) incorporating a (R or S)-γ-lactam conformational constraint in place of the Pro-Gly at residues 8 and 9 of the peptide have been reported.[98] Analogue Trp-His-Trp-Leu-Gln-Leu-Lys[(R)-γ-lactam]Gln-Pro-Nle-Tyr was more potent than

Trp-His-Trp-Leu-Gln-Leu-Lys[(S)-α-lactam]Gln-Pro-Met-Tyr. Both lactam analogues competed with tritiated [Nle^{12}]-α-factor for binding to the α-factor receptor with the (R)-γ-lactam-containing peptide having 7-fold higher affinity than the (S)-γ-lactam-containing homologue. Non-peptide inhibitors of myristoyl-CoA-protein N-myristoyl transferase have been synthesised starting from the octapeptide substrate Gly-Leu-Tyr-Ala-Ser-Lys-Leu-Ser-NH$_2$. Both (R) and (S) isomers of **37** were competitive inhibitors with respect to the octapeptide substrate Gly-Asn-Ala-Ala-Ser-Ala-Arg-Arg-NH$_2$ (K_i values 9–20 μM). Compounds **37** was fungicidal against *C. albicans* and *C. neoformans* (MFCs 250 and 200 μM, respectively).

4.3 ACTH/CRF Peptides – A cDNA clone encoding corticotropin-releasing factor type 1 receptor (CRF-R$_1$) was isolated from *Tupaia belangeri* with a PCR-based approach.[100] Only eight amino acids (residues 3, 4, 6, 35, 36 and 39 in the N-terminus, residue 232 in transmembrane domain 4 and residue 410 in the C-terminus) differed between this receptor and hCRF-R$_1$. Binding studies using HEK293 cells stably transfected with tCRF-R$_1$ showed that the CRF agonists ovine CRF, human/rat CRF, urocortin and sauvagine (K_d values 0.37–1.28 nM) were bound with significantly higher affinities than the CRF antagonist astressin (K_d = 12.4 nM). The CRFR$_1$ and CRFR$_2$ regions (present in the second and third extracellular domains) important for the binding of rat/human CRF also effect the binding of urocortin and sauvagine, two other members of the CRF peptide family.[101] A fourth region in the third extra-cellular domain, Asp^{254}, was important for sauvagine but not CRF or urocortin binding. In addition, His^{199} and Met^{276} (3rd and 5th transmembrane domain, respectively) were important for binding the non-peptide high-affinity CRFR$_1$ antagonist NBI 27914.

(37)   (38)

SAR studies on urotensin I fragments using cultured rat pituitary cells are reported.[102] The potency of urotensin I(1-41) was about one seventh that of rat CRF on a molar basis. Urotensin I fragments, 1-36, 4-36, 6-36 and 1-19 were 70–700-fold less potent than urotensin(1-41) and urotensin 9-36 and 17-36 were even weaker agonists. The activity of urotensin(1-36) was weaker than urotensin(1-41), suggesting that the C-terminal 37-41 sequence was required to express the full ACTH-release activity, although each of four C-terminal fragments, 24-36, 24-41, 29-36 and 29-41, exhibited no activity.

Conformationally constrained agonist and antagonist analogues of hCRF have been reported.[103-105] Successive c(i to i+3)[Lys^i-Glu^{i+3} and a Glu^i-Lys^{i+3}] bridges were introduced from residues 4 to 14 of the agonist [Ac-Pro^4, D-Phe^{12}, Nle^{21,38}]hCRF(4-41) and related compounds.[103] While c(30-33) [Ac-Leu^8, D-Phe^{12}, Nle^{21}, Glu^{30}, Lys^{33}, Nle^{38}]hCRF(8-41) was the shortest analogue of CRF to be equipotent to CRF (70% intrinsic activity), the corresponding linear analogue was 120 times less potent. Addition of one amino acid at the N-terminus gave a 5-fold more potent analogue c(30-33) [Ac-Ser^7, D-Phe^{12}, Nle^{21}, Glu^{30}, Lys^{33}, Nle^{38}]hCRF(7-41). Other more potent c(30-33) analogues included [D-Phe^{12}, Nle^{21}, Glu^{30}, Lys^{33}, Nle^{38}]hCRF(5-41), [Ac-Pro^5, D-Phe^{12}, Nle^{21}, Glu^{30}, Lys^{33}, Nle^{38}]hCRF(5-41), [D-Phe^{12}, Nle^{21}, Glu^{30}, Lys^{33}, Nle^{38}]hCRF(6-41), [Ac-Ile^6, D-Phe^{12}, Nle^{21}, Glu^{30}, Lys^{33}, Nle^{38}]hCRF(6-41) and [D-Phe^{12}, Nle^{21}, Glu^{30}, Lys^{33}, Nle^{38}]hCRF(7-41).

D-Amino acids were incorporated within the cycles formed between residues 20 and 23 and residues 30 and 33 in CRF agonists and antagonists.[104] In comparison to the agonist analogue c(20-23)[Ac-D-Pro^4, D-Phe^{12}, Glu^{20}, Nle^{21,38}, Lys^{33}]hCRF$_{(4-41)}$, several monocyclic and bicyclic analogues like c(20-23)[Ac-Pro^4, D-Phe^{12}, Glu^{20}, Nle^{21,38}, D-Ala^{22}, Lys^{33}]-, c(30-33)[Ac-Pro^4, D-Phe^{12}, Nle^{21,38}, Glu^{30}, Lys^{33}]-, c(30-33)[Ac-Pro^4, D-Phe^{12}, Nle^{21,38}, Glu^{30}, D-His^{32}, Lys^{33}]-, c(30-33)[Ac-Pro^4, D-Phe^{12}, Nle^{21,38}, Glu^{30}, D-Ala^{31}, Lys^{33}]- and bicyclic(20-23, 30-33)[Ac-Pro^4, D-Phe^{12}, Glu^{20}, Nle^{21,38}, D-Ala^{22}, Lys^{23}, Glu^{30}, D-His^{32}, Lys^{33}]-hCRF$_{(4-41)}$ were less potent. Several other analogues of carp urotensin and sauvagine were also less potent. In c(30-33)[D-Phe^{12}, Nle^{21,38}, Glu^{30}, Lys^{33}]hCRF$_{(12-41)}$ (astressin) series of antagonists, only the D-His^{32} analogue was somewhat more potent than the parent peptide. All the other analogues, [D-Ala^{31}], [D-Ala^{31}, D-His^{32}], [D-Nal^{32}], [D-Glu^{32}] and [D-Arg^{32}], were less potent than astressin. Two of the bicyclic antagonists, c(20-23, 30-33)[D-Phe^{12}, Glu^{20}, Nle^{21,38}, Lys^{23}, Glu^{30}, Lys^{33}]- and c(20-23, 30-33)[D-Phe^{12}, Glu^{20}, Nle^{21,38}, D-Ala^{22}, Lys^{23}, Glu^{30}, Lys^{33}]hCRF$_{(12-41)}$, also showed reduced antagonist potency.[104] Extending the N-terminus by additional amino acids gave N-acetyl-c(30-33)[D-Phe^{12}, Nle^{21,38}, Glu^{30}, Lys^{33}]hCRF$_{(9-41)}$ which was more potent than astressin.[105] Other analogues like c(30-33)[D-Phe^{12}, Nle^{21,38}, Glu^{30}, Lys^{33}]hCRF$_{(11-41)}$, c(30-33)[D-Phe^{12}, Nle^{21,38}, Glu^{30}, Lys^{33}]hCRF$_{(10-41)}$, c(30-33)[D-Phe^{12}, Nle^{21,38}, Glu^{30}, Lys^{33}]hCRF$_{(9-41)}$, their D-His^{32} and N-acetyl derivatives were less potent. Some of the analogues containing an α-methyl Leu residue in position 27 {c(30-33)[D-Phe^{12}, Nle^{21,38}, α-MeLeu^{27}, Glu^{30}, D-His^{32}, Lys^{33}]hCRF$_{(11-41)}$, c(30-33)[D-Phe^{12}, Nle^{21,38}, α-MeLeu^{27}, Glu^{30}, Lys^{33}]hCRF$_{(9-41)}$ and N-acetyl-c(30-33)[D-Phe^{12}, Nle^{21,38}, α-MeLeu^{27}, Glu^{30}, D-His^{32}, Lys^{33}]hCRF$_{(9-41)}$}, though less potent in the *in vitro* assays, showed increased duration of action in an *in vivo* assay when administered intravenously at a dose of 100 μg kg^{-1}.

Receptor selective analogues of CRF were obtained based on the sauvagine and urocortin sequence.[106] In comparison with astressin, which exhibited a similar affinity to rCRFR$_1$ and mCRFR$_{2(}$ (K_d 4.0–5.7 nM), [D-Phe^{11}, His^{12}] sauvagine$_{11-40}$, [D-Leu^{11}]sauvagine$_{11-40}$, [D-Phe^{11}]sauvagine$_{11-40}$ and sauvagine$_{11-40}$ bound, respectively, with a 110-, 80-, 68- and 54-fold higher affinity to $mCRFR_{2\beta}$ than to $rCRFR_1$. The truncated analogues of rat urocortin, $[D\text{-}Phe^{11}]rUcn_{11\text{-}40}$, $[D\text{-}Phe^{11}, Glu^{12}]rUcn_{11\text{-}40}$, $c(29\text{-}32)[D\text{-}Phe^{11}, Glu^{29}, Lys^{32}]rUcn_{11\text{-}40}$ and $[D\text{-}Leu^{11}, Glu^{12}]rUcn_{11\text{-}40}$, displayed modest preference (2-7-fold) for binding to $mCRFR_{2\beta}$. $[D\text{-}Phe^{11}, His^{12}]$sauvagine$_{11\ 40}$ (anti-sauvagine-30) was the most potent and selective ligand to suppress agonist-induced adenylate cyclase activity in human embryonic kidney cells expressing $mCRFR2\beta$. Corticotropin-inhibiting peptide [a 32 amino acid peptide identical to the fragment 7-38 of human ACTH(1-39)], isolated from human pituitary extracts, was an antagonist at ACTH receptors.[107] It inhibited ACTH-stimulated glucocorticoid secretion of dispersed rat adrenocortical cells and raised basal aldosterone secretion from fresh suspensions of rat zona glomerulosa cells.

Non-peptide antagonists acting at the CRF_1 receptor (38) have been reported.[108] Compound 38 showed a K_i value of 5 nM in a binding assay. The corresponding 2-chlorophenyl, 4-chlorophenyl and 2,4,6-trimethylphenyl analogues were less potent (K_i values 15-93 nM).

4.4 Angiotensin II Analogues and Non-peptide Angiotensin II Receptor Ligands

– Various aspects of angiotensin II type 2 receptor, including the regulation of AT_2-receptor expression, its cellular localisation, its pathological role in cardiovascular and kidney diseases, have been reviewed.[109] SAR studies on angiotensin II analogues modified in positions 1 and 8 were studied using human myometrium (AT_2 receptor).[110] The pharmacological profile of the receptors was typical of an angiotensin AT_2 receptor with the following order of affinities: [angiotensin III $\geqslant$ angiotensin II > angiotensin I > PD123319 > angiotensin(1-7) > angiotensin(1-6) $\approx$ angiotensin IV >> losartan]. In comparison to angiotensin II, the position 1 modified analogues [Asn^1]-, [Glu^1]-, [Gly^1]-, [Sar^1]-, [Me$_2$Gly^1]-, [Me$_3$Gly^1]-, [Ac^1]-, [Ac-Gly^1]-, [Me$_3$Ser^1]-, [Ac-Ser^1]-, [Lac^1]-, [Ser^1]-, [Cys(Acm)^1]-, [Cysteic acid^1]-, [Aib^1]-, [Lys^1]-, [p-benzoylPhe^1]- and [N$_3$-benzoyl^1]-angiotensin II analogues were more potent (1.5–15-fold) at the AT_2 receptor preparation, [Me$_2$Gly^1]-, [Me$_3$Gly^1]- and [Me$_3$Ser^1]- being the most potent (>1000-fold) of the series. The position 8 modified analogues [Phe^8]- and [D-Phe^8]- were 28–44-fold more potent than angiotensin II. The positions 1 and 8 modified analogues [Sar^1, Ala^8]-, [Sar^1, Ile^8]-, [Sar^1, Leu^8]-, [Sar^1, Met^8]- and [Sar^1, Phe^8]-angiotensin II were 28–345-fold more potent than angiotensin II. The most potent of the analogues, [Sar^1, Phe^8]-angiotensin II, was 345-fold more potent than angiotensin II at the AT_2 receptor preparation. When the aromatic ring of the phenylalanine residue in position 8 was replaced or substituted many of the analogues, [Sar^1, D-Phe^8]-, [Sar^1, Trp^8]-, [Sar^1, Tyr^8]-, [Sar^1, Phe(S-Acm)^8]-, [Sar^1, N$_3$-D-Phe^8]- and [Sar^1, Br$_5$Phe^8]-angiotensin II, were less potent than [Sar^1]-angiotensin II. Only [Sar^1, Phe(p-Cl)^8]-, [Sar^1, Phe(Br)^8]-, [Sar^1, Phe(N$_3$)^8]-, [Sar^1, Me$_5$Phe^8]- and [Sar^1, Cl$_5$Phe^8]-angiotensin II were 2–5-fold more potent at the AT_2 receptors.

Publications on the chemical and biological profiles of a number of non-peptide angiotensin receptor antagonists have appeared.[111-114] Metabolites of the angiotensin II antagonist tasosartan (39) were identified. One of these (40)

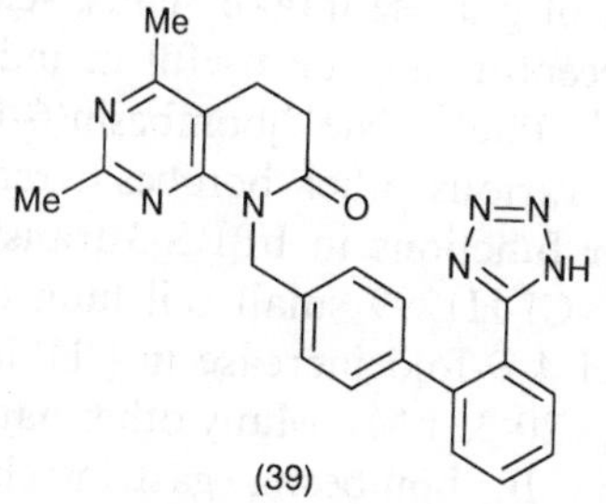

(39)

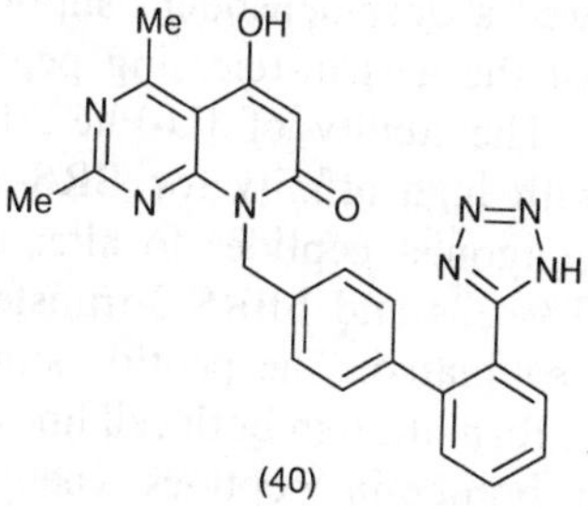

(40)

which contains a relatively acidic hydroxyl group was not absorbed as well as **39** after oral administration.

4.5 Bombesin/Neuromedin Analogues – Work on bombesin receptor antagonists has been reviewed.[115] Four subtypes of bombesin receptor have been identified (GRP receptor, neuromedin B receptor, the orphan receptor bombesin receptor subtype 3 and bombesin receptor subtype 4). The role of individual bombesin receptors is under investigation and selective ligands for these receptor subtypes are being synthesised. A series of 'balanced' neuromedin-B preferring (BB_1)/gastrin-releasing peptide preferring (BB_2) receptor ligands as exemplified by PD 176252 (**41**) were synthesised starting from an α-methyl-tryptophan derivative (PD165929) reported earlier to be a BB_1 selective (BB_1 K_i 6.3 nM and BB_2 K_i >10.000 nM) compound.[116] PD176252 (**41**) displaying a BB_2 receptor affinity of 1 nM whilst retaining subnanomolar (0.17 nM) BB_1 receptor affinity was a competitive antagonist at both the receptor subtypes.

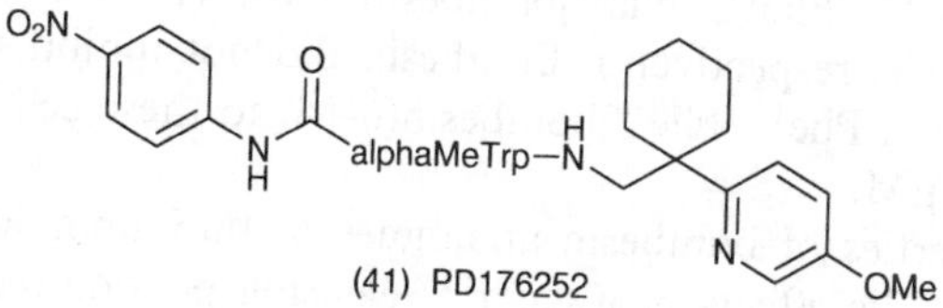

(41) PD176252

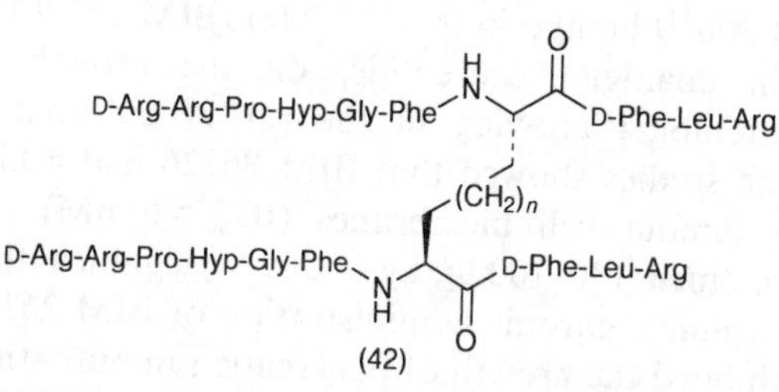

(42)

The role of bombesin receptor subtypes in food intake has been explored.[117,118] Mice lacking a functional gastrin-releasing peptide receptor gene were shown to develop and reproduce normally and showed no gross phenotypic abnormalities. However, peripheral administration of bombesin to such mice had no effect on the suppression of glucose intake, whereas normal mice showed a dose-dependent suppression of glucose intake. Thus, selective agonists of the gastrin-releasing peptide receptor may be useful in inducing satiety.[117] The ability of [D-Phe^6, β-Ala^{11}, Phe^{13}, Nle^{14}]bombesin(6-14), a peptide with high affinity for BRS-3, and various other bombesin receptor agonist/antagonist peptides to alter cellular functions in hBRS-3-transfected BALB 3T3 cells and hBRS-3-transfected NCI-H1299 small cell lung cancer cells was evaluated. The peptide stimulated 4–9-fold increase in [^{3}H]inositol phosphate formation in both cell lines (EC$_{50}$s 20–35 nM). Many other naturally occurring bombesin peptides {neuromedin B, bombesin, gastrin-releasing peptide, phyllolitorin, litorin, [Phe^{13}]-bombesin, [D-Phe^6]-bombesin(6-13) propylamide, [D-Phe^6, Phe^{13}]-bombesin(6-13)propylamide and Ac-neuromedin B(3-10)} were not active in this assay.[118] Neuromedin B receptor-specific antagonist [D-Nal-Cys-Tyr-D-Trp-Lys-Val-Cys-Nal-NH$_2$, cyclic peptide] inhibited hBRS-3 receptor activation in a competitive fashion (K_i 0.5 μM).

Nonapeptide ^{125}I-[D-Tyr^6, β-Ala^{11}, Phe^{13}, Nle^{14}]bombesin(6-14) has been shown to be a high affinity ligand for all the four receptor subtypes.[119] In rat pancreatic acini containing only GRP receptor and BB$_4$ transfected BALB cells, both [D-Phe^6, β-Ala^{11}, Phe^{13}, Nle^{14}]bombesin(6-14) and bombesin inhibited the binding of ^{125}I-[D-Tyr^6, β-Ala^{11}, Phe^{13}, Nle^{14}]bombesin(6-14) and ^{125}I-[Tyr^4]bombesin (a GRP receptor ligand), respectively, in a dose-dependent manner. In neuromedin B receptor transfected BALB cells, both neuromedin B and [D-Phe^6, β-Ala^{11}, Phe^{13}, Nle^{14}]bombesin(6-14) inhibited the binding of ^{125}I-[Tyr^0]neuromedin B (a neuromedin B receptor ligand) and ^{125}I-[D-Tyr^6, β-Ala^{11}, Phe^{13}, Nle^{14}]bombesin(6-14), respectively, in a dose-dependent manner. In BRS-3 transfected BALB 3T3 cells, or hBRS-3-transfected H1299 cells, [D-Phe^6, β-Ala^{11}, Phe^{13}, Nle^{14}]bombesin(6-14) inhibited the binding of ^{125}I-[D-Tyr^6, β-Ala^{11}, Phe^{13}, Nle^{14}]bombesin(6-14) (IC$_{50}$ 5 and 9 nM at H1299 and 3T3 cells, respectively). Bombesin did not inhibit the binding of ^{125}I-[D-Tyr^6, β-Ala^{11}, Phe^{13}, Nle^{14}]bombesin(6-14) to these cells, even up to a concentration of 1 μM.

Biological properties of bombesin analogues in tumour models have been investigated.[120,121] The effects of a gastrin-releasing peptide receptor antagonist, [D-F$_5$ Phe^6, D-Ala^{11}] bombesin (6-13)-OMe) (BIM 26226), and the long-acting somatostatin analogue, lanreotide, on the growth of an acinar pancreatic adenocarcinoma growing in the rat or cultured *in vitro* were investigated. Binding studies showed that BIM 26226 had a high affinity for GRP receptors in tumour cell membranes (IC$_{50}$ = 6 nM). While gastrin-releasing peptide treatment (30–100 μg kg^{-1} day^{-1}) significantly increased the pancreatic tumour volume, chronic administration of BIM 26226 and lanreotide significantly inhibited the growth of pancreatic tumours stimulated or not by gastrin-releasing peptide (GRP), as shown by a reduction in tumour volume. In cell cultures, both BIM 26226 and lanreotide (1 μM) inhibited [H^3]thymidine incorporation in tumour cells induced or not by GRP.[120] Incorporation of ^{3}H-thymidine in NCI-H69 human small cell lung cancer cells was also inhibited by a number of other bombesin [Pyr-Trp-Ala-Val-Gly-D-Phe-Leu-Leu-NH$_2$, Gln-Trp-Ala-Val-Gly-D-Ser-Leu-Leu-NH$_2$, and Gln-Trp-

Ala-Val-Gly-D-Ser-cycloLeu-Leu-NHCH$_3$] (IC$_{50}$ values 84–124 μM) and substance P analogues [D-MePhe-D-Trp-Phe-D-Trp-Leuψ(CH$_2$NH)Leu-NH$_2$, D-Tyr-D-Trp-Phe-D-Trp-Leuψ(CH$_2$NH)Leu-NH$_2$ and D-Tyr(Et)-D-Trp-Phe-D-Trp-Leu((CH$_2$NH)Leu-NH$_2$] (IC$_{50}$ values 2-9 μM).[121] Some smaller fragments like Leuψ(CH$_2$NH)Leu-NH$_2$, D-Trp-Leuψ(CH$_2$NH)Leu-NH$_2$, Phe-D-Trp-Leuψ(CH$_2$NH)Leu-NH$_2$ and D-Trp-Phe-D-Trp-Leuψ(CH$_2$NH)Leu-NH$_2$ were also effective in inhibiting ^{3}H-thymidine incorporation in NCI-H69 human SCLC cells (IC$_{50}$ values 12–42 μM).

4.6 Bradykinin Analogues – Bradykinin-related peptides generated in the plasma of six sarcopterygian species after a treatment with trypsin were characterised as [Tyr^1, Gly^2, Ala^7, Pro^8]- (lungfish), [Phe^1, Ile^2, Leu^5]- (amphiuma), [Val^1, Thr^6]- (bullsnake and coachwhip), [Leu^2, Thr^6]- (Gila monster) and [Thr^6]-bradykinin (Gray's monitor).[122] Bolus intra-arterial injections of synthetic lungfish bradykinin (100 nmol) into an anaesthetised dog produced no change in the arterial pressure or blood flow in the bronchial artery. A 14 kDa bradykinin binding protein was purified from inflammatory cells like human blood neutrophils and peripheral blood mononuclear cells. The protein was also present on a variety of bradykinin responsive human cell types, *i.e.* CCD-16Lu lung fibroblasts, HL60 promyelocytes, U937 myelomonocytes and Jurkat T lymphocytes.[123] The conformational profile of five bradykinin analogues, c(Gly-Thi-D-Tic-Oic-Arg), c(Gly-Ala-D-Tic-Oic-Arg) and c(Abu-Ala-Ser-D-Tic-Oic-Arg) (antagonists of kinin-induced rabbit jugular vein and rabbit aorta smooth muscle contraction) and c(Abu-D-Phe-Ala-D-Tic-Oic-Arg) and Thi-Ser-D-Tic-Oic-Arg (non-binders) was assessed by computational methods. Comparison of the common conformations among antagonists and at the same time not found in the non-binders provided a putative bioactive conformation.[124]

Total retro-inverso analogues of bradykinin [H$_2$N-D-Arg<D-Pro<D-Pro<Gly<D-Phe<D-Ser<D-Pro<D-Phe<D-Arg-H], the B$_{2a}$-selective kinin antagonist D-Arg^0[Hyp^3, D-Phe^7, Leu^8]-bradykinin [H$_2$N-Arg<D-Arg<D-Pro<D-Pro<Gly<D-Phe<D-Ser<Phe<D-Leu<D-Arg-H, HO-Arg<D-Arg <D-Pro<D-Pro<Gly<D-Phe<D-Ser<Phe<D-Leu<D-Arg-H and C$_6$H$_5$CH$_2$-HN-Arg<D-Arg<D-Pro<D-Pro<Gly<D-Phe<D-Ser<Phe<D-Leu<D-Arg-H], angiotensin II [H$_2$N-D-Asp<D-Arg<D-Val<D-Tyr<D-Ile<D-His<D-Pro<D-Phe-H] and the AT II antagonist Saralasin [H$_2$N-Gly<D-Arg<D-Val<D-Tyr<D-Val<D-His<D-Pro<D-Ala-H] were prepared by solid-phase synthesis.[125] Total retro-inverso analogues of D-Arg^0[Hyp^3, D-Phe^7, Leu^8]-bradykinin bound to the kidney medulla B$_{2a}$ bradykinin receptor (K_d values 4–64 μM). Total retro-inverso analogues of bradykinin, AT II and Saralasin, did not bind to either the B$_{2a}$ bradykinin receptor or the rat AT$_{1a}$ angiotensin receptor, respectively. Three peptidomimetics of D-Arg^0[Hyp^3, D-Phe^7, Leu^8]-bradykinin were weak inhibitors of ACE. Bradykinin analogues with dehydrophenylalanine (ΔPhe) or its ring substituted analogues at position 5, {[ΔPhe^5]-, [ΔPhe(2-Me)^5]-, [ΔPhe(2,5-Me)^5]-, [ΔPhe(4-I)^5]- and [ΔPhe(4-F)^5]-bradykinin} were partial antagonists.[126] Ring substitutions by methyl groups

or iodine reduce both the agonist and antagonist activities. Only substitution by fluorine gives a high potency. Incorporation of ΔPhe into different representative antagonists with key modifications at position 7 {[ΔPhe^5, D-Phe^7]-, [ΔPhe^5, D-Tic^7, Oic^8]-, [Hyp^3, ΔPhe^5, D-Tic^7, Oic^8]- and [D-Arg^0, Hyp^3, ΔPhe^5, D-Tic^7, Oic^8]-bradykinin} also gave less potent antagonists. Only the combination of ΔPhe at position 5 with D-Phe at position 7 increases the antagonist potency on guinea pig ileum. Some other analogues like [ΔPhe^5, D-MePhe^7]-, [MePhe^2, ΔPhe^5]- and [MePhe^2, ΔPhe^5, MePhe^7]-bradykinin were agonists. However, except for [MePhe^2, (Phe^5)-bradykinin, the agonist activity compared to [D-MePhe^7]-bradykinin was much reduced in the rat uterus preparation. Based on the cross-linked dimeric peptides reported earlier, some new dimeric bradykinin antagonist peptides [H-D-Arg-Arg-Pro-Hyp-Gly-Phe]$_2$-X-[D-Phe-Leu-Arg-OH]$_2$ were synthesised (X corresponds to a L,L-2,7-diaminosuberic or L,L-2,9-diaminosebacic acid residue, respectively).[127] The pA$_2$ values for compounds **42a** (n = 2) and **42b** (n = 4) in a guinea pig ileum contraction assay were 5.9 and 6.8, respectively. Both the compounds were less potent antagonists than Hoe-140 (pA$_2$ 8.4).

A number of non-peptide antagonists of bradykinin have been reported.[128-133] One of the antagonists (NPC18884) (**43**) containing three arginine residues given intraperitoneally or orally inhibited bradykinin-induced leukocytes influx (ID$_{50}$ value of 63 nmol kg^{-1} and 141 nmol kg^{-1}, respectively) and the exudation induced by bradykinin.[128] The effects of **43** lasted for up to 4 h and were selective for the bradykinin B$_2$ receptors; at similar doses it had no significant effect against the inflammatory responses induced by des-Arg(9)-bradykinin, histamine or substance P. In another series of non-peptide antagonists of bradykinin, potent antagonists were obtained by chemical modifications starting from the random screening lead **44**. Several of the compounds (**45, 46**) were active in animal models when given by the oral route.[129-132] FR173657 inhibited bradykinin-induced contractions of the rabbit isolated iris sphincter muscle mediated by tachykinin release from trigeminal afferent neurones (pK$_B$ 7.9). Bradykinin-induced nociceptive behavioural responses and blood pressure reflexes in anaesthetised rats were also inhibited by the non-peptide derivative **45**. In the blood pressure assay, the non-peptide was much less potent (ID$_{50}$ 1.1 μmol kg^{-1}) than the peptidic B$_2$ antagonist Hoe-140 {D-Arg^0-[Hpro^3, Thi^5, D-Tic^7, Oic^8]-bradykinin} (ID$_{50}$ 8.5 nmol kg^{-1}).[133] Another nonpeptide antagonist LF160335 (**47**) displaced [^{3}H]-bradykinin binding to membrane preparations from CHO cells expressing the cloned human B$_2$ receptor, INT 407 cells and human umbilical vein with K values of 0.84, 1.26 and 2.34 nM, respectively.[134] LF160335 had no affinity for the cloned human kinin B$_1$ receptor stably expressed in 293 cells. LF160335 inhibited the concentration-contraction curve to bradykinin in the human umbilical vein giving a pA$_2$ value of 8.30.

4.7 Cholecystokinin Analogues – The role of various mediators involved in acid secretion (gastrin, CCK, histamine, somatostatin, enterogastrones and CGRP) is discussed.[135] Mutations within the cholecystokinin-B/gastrin re-

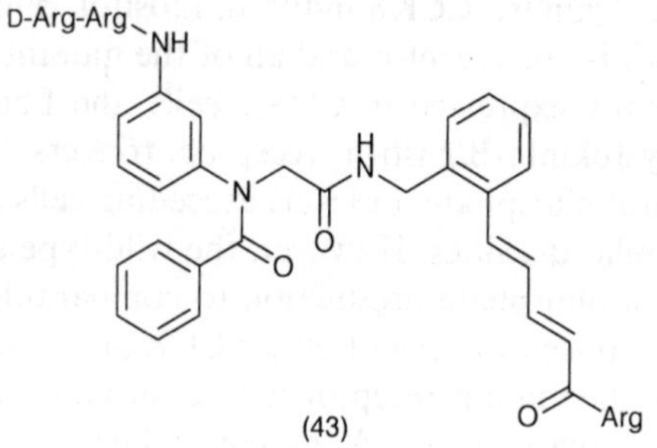

(43)

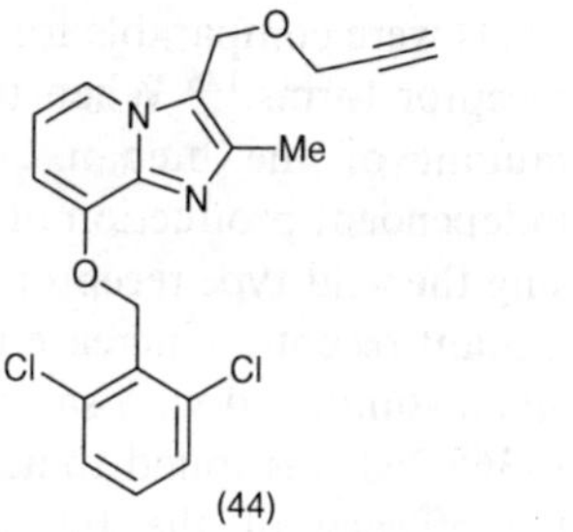

(44)

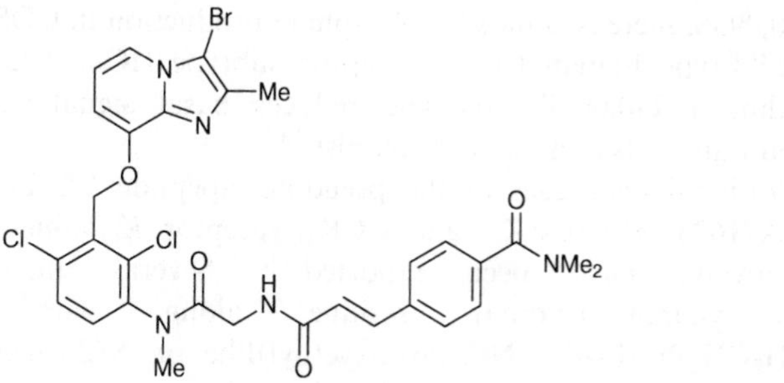

(45) FR173657

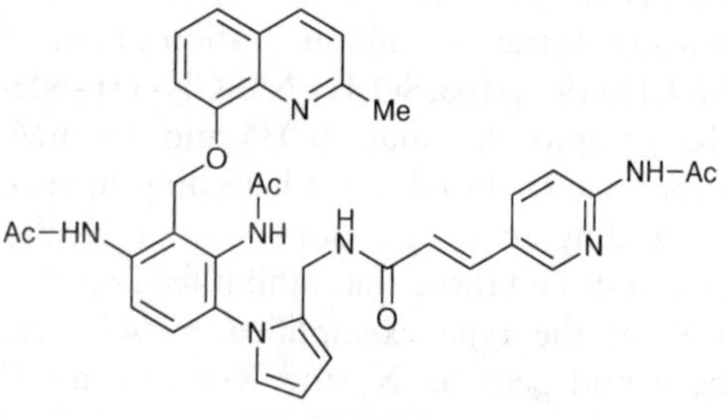

(46) FR193517

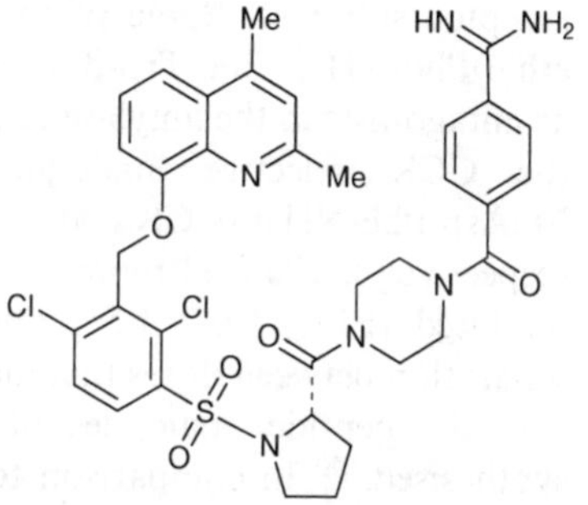

(47)

ceptor ligand pocket were shown to modify the functions of nonpeptide agonists and antagonists.[136,137] Alanine substitutions within the CCK_B receptor binding pocket altered the affinities and/or functional activities of L-365,260 (CCK_B receptor antagonist), YM022 (antagonist) and L-740,093S (partial agonist). For each of the nonpeptide ligands examined, a distinct series of mutations altered the affinity, suggesting that each ligand possessed a characteristic pattern of interactions within the CCK_B receptor pocket. For example, $Trp^{346}Ala$ mutation converted YM022 to a partial agonist from an antagonist on the wild-type receptor. In the case of L-740093S, $Ser^{219}Ala$, $Val^{349}Ala$ mutations increased the efficacy and $Thr^{111}Ala$, $Trp^{346}AlaAsn^{353}Leu$ mutations decreased the efficacy of the compound when compared with the efficacy at the wild-type receptor. $Ser^{219}Ala$, $Val^{349}Ala$, $Asn^{353}Leu$ and $Ser^{379}Ala$ mutations increased the efficacy of L-365,260. $Asn^{353}Leu$ and $Ser^{379}Ala$ mutations converted this antagonist into a partial or a full agonist.

In contrast to the non-peptide ligands, CCK8-induced inositol phosphate levels were comparable for the wild-type receptor and all of the mutant CCK_B receptor forms.[136] When transiently expressed in COS-7 cells, the $Leu^{325}Glu$ mutant of the human cholecystokinin-B/gastrin receptor triggers agonist independent production of inositol phosphate, to levels exceeding cells expressing the wild-type receptor at similar densities. However, the wild-type and the mutant receptors increase inositol phosphate production to comparable levels when stimulated with saturating concentrations of either CCK_8 or gastrin-17. L-365,260 was found to activate the mutant receptor with approximately half the efficacy of the full agonist (gastrin-17). A closely related analogue, L-740,093S, increased inositol phosphate production in COS-7 cells expressing the wild-type human CCK_B receptor subtype (IC_{50} 2.45 nM). The corresponding L-740,093R analogue reduces basal signalling of the mutated receptor and acts as an inverse agonist.[137]

Additional analogues of the pseudotetrapeptide CCK_A selective agonists like A71623 (**48**) (CCK_A and CCK_B receptor K_i values 18 and 700 nM, respectively) have been reported.[138] Several analogues containing Lys(o-tolylaminocarbonyl) residue along with N-benzyl-β-Ala, $N(CH_2CH_2Ph)$-β-Ala, N(2-carboxyethyl)Phe or N(2-carboxyethyl)Phe residues in place of the MePhe residue showed much weaker affinity at both the receptor subtypes. Some of the analogues [Boc-Trp-Lys(Tac)-N-(2-carboxyethyl)Phe-NH_2, Boc-Trp-β-hLys(Tac)-N-(2-carboxyethyl)Phe-NH_2] behaved as antagonists in the amylase release assay. Intranasal administration of one of the CCK_A-selective analogue ARL15849 [Hpa(SO_3H)-Nle-Gly-Trp-Nle-MeAsp-Phe-NH_2] (CCK_A and CCK_B receptor K_i values 0.034 and 224 nM, respectively), 100-fold more potent than CCK_8 in inhibit 3-h feeding in rats, inhibited (ED_{50} 5.0 ± 1.5 μg kg^{-1}) feeding in beagle dogs with a greater separation between doses that induce emesis and those that inhibit feeding.[139]

Cyclic peptide analogues of CCK of the type exemplified by **49** were synthesised.[140] In comparison to the parent peptide (X_1 and X_2 = Lys, n = 1) (IC_{50} values 300 and 1.4 nM, respectively, at the rat pancreatic acini and guinea pig brain membranes; amylase secretion EC_{50} 3 nM), all the other analogues (X_1 and X_2 = Lys, Orn or Dab or X_1 = Orn and X_2 = Lys, n = 2–4) (IC_{50} values >800 and 180 nM, respectively, at the rat pancreatic acini and guinea pig brain membranes, amylase secretion EC_{50} >50 nM) were much less potent in the receptor binding and amylase release assays. One of the ornithine containing analogue, (X_1 and X_2 = Orn and n = 4), showed no agonist activity at a very high dose (10 μM) and another similar analogue, (X_1 and X_2 = Orn and n = 3), showed partial agonist activity (40% maximum release at 10 μM).

Based on the biochemical and pharmacological profiles of Boc-c(D-Asp-Tyr(SO_3Na)-Nle-D-Lys)-Trp-Nle-Asp-Phe-NH_2 and Boc-Tyr(SO_3Na)-gNle-mGly-Trp-MeNle-Asp-Phe-NH_2 CCK_B receptor agonists (K_i values 1.01 and 0.31 nM, respectively) indicating the existence of two CCK_B receptor subsites, analogues were synthesised which may be more selective.[141] While RB360 (**50**) (CCK_B K_i 18 nM) has a CCK_{B1} profile with anxiogenic-like effects in the elevated plus-maze test at a dose of 30 μg kg^{-1} and is inactive in the Y-maze

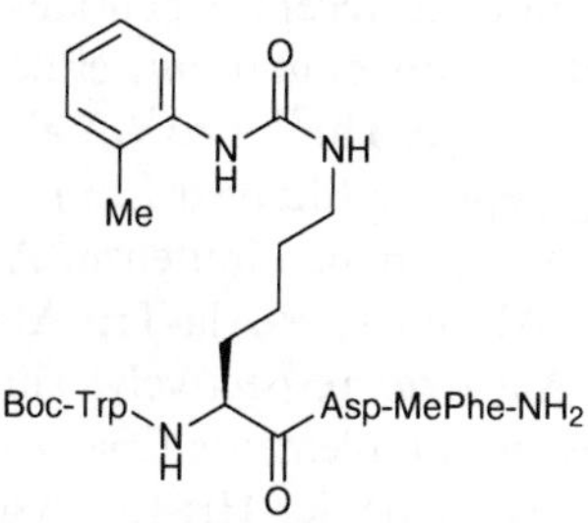

(48) A71623

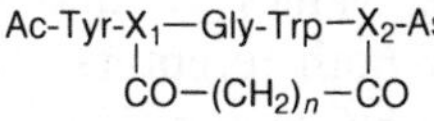

Ac-Tyr-X$_1$—Gly-Trp—X$_2$-Asp-Phe-NH$_2$
$\overset{|}{\underset{}{CO}}$—(CH$_2$)$_n$—$\overset{|}{CO}$
(49)

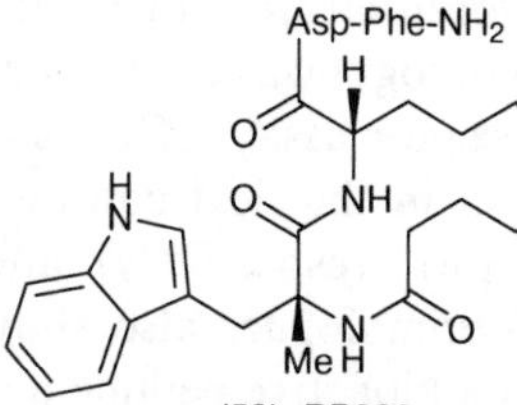

(50) RB360

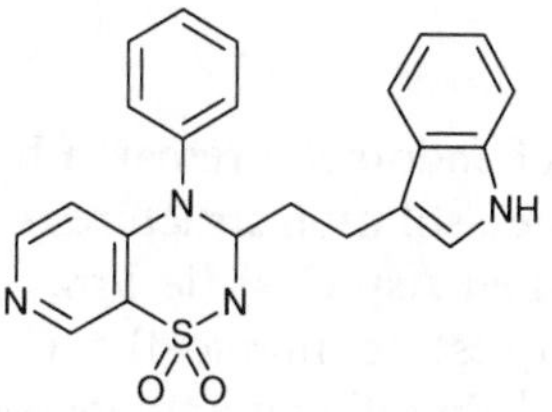

(51)

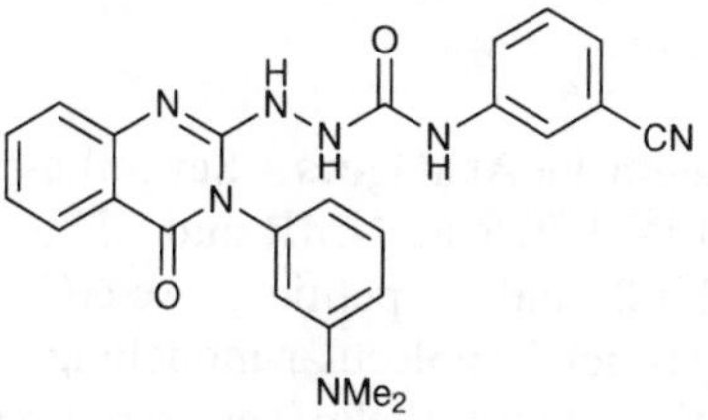

(52)

test, RB400 (HOOC-CH$_2$-CO-Trp-MeNle-Asp-Phe-NH$_2$) (CCK$_B$ K_i 0.42 nM), like BC264, seems to be a specific CCK$_{B2}$ agonist, able to increase attention and/or memory processes in the Y-maze test at 0.3 and 1 mg kg^{-1} dose levels (no anxiogenic-like effect at 10 mg kg^{-1}). Non-peptide CCK antagonists like 51 and 52 have been reported.[142-143] Compound 51 exhibited a binding affinity for CCK$_A$ and CCK$_B$ receptors in the µM range (45 and 4.8 µM, respectively). The CCK$_B$ receptor antagonist 52 was orally active (22% bioavailability) in the elevated rat X-maze test.

4.8 Complement-related Peptide/Non-peptide Analogues – The role of complement and complement inhibitors in autoimmune and inflammatory diseases has been reviewed.[144] The interactions of C5a [74 residue glycoprotein derived from the fifth component (C5) of the complement system upon proteolytic activation] with the N-terminal domain of the C5a receptor were examined by use of recombinant human C5a molecules and peptide fragments Met[1]-Asn-Ser-Phe-Asn-Tyr-Thr-Thr-Pro-Asp[10]-Tyr-Gly-His-Tyr-Asp-Asp-Lys-Asp-Thr-Leu[20]-Asp-Leu-Asn-Thr-Pro[25]-Val-Asp-Lys-Thr-Ser[30]-Asn-Thr-Leu-Arg, acetyl-His-Tyr-Asp[15]-Asp-Lys-Asp-Thr-Leu[20]-Asp-Leu-Asn-Thr-Pro[25]-Val-Asp-Lys-Thr-Ser[30]-Asn-Thr-Leu-Arg, and acetyl-Thr-Leu[20]-Asp-Leu-Asn-Thr-Pro[25]-Val-Asp-Lys-Thr-Ser[30]-Asn-amide derived from human C5a receptor.[145]

The results indicated that all three receptor peptides responded to the binding of C5a through the 21-30 region containing either hydrophobic, polar, or positively charged residues such as Thr^{24}, Pro^{25}, Val^{26}, Lys^{28}, Thr^{29}, and Ser^{30}.

Three ileum-contracting peptides (derived from the tryptic digest of serum albumin) [bovine, human and porcine albutensin A sequences: Ala-Leu-Lys-Ala-Trp-Ser-Val-Ala-Arg, Ala-Phe-Lys-Ala-Trp-Ala-Val-Ala-Arg, and Ala-Phe-Lys-Ala-Trp-Ser-Leu-Ala-Arg, respectively] showed homology with the COOH-terminal sequences of complements C3a α-Ala-Arg-Ala-Ser-His-Leu-Gly-Leu-Ala-Arg) and C5a α-Ile-Ser-His-Lys-Asp-Met-Gln-Leu-Gly-Arg), which are essential for their activities.[146] Porcine albutensin A (highest sequence homology) was confirmed to act through both C3a and C5a receptors by a radioreceptor assay (IC_{50} values 2.3 and 10 μM, respectively) and cross-desensitisation in the ileal contraction. In addition, bovine (IC_{50} values 110 and >2000 μM, respectively) and human (IC_{50} values 75 and 670 μM, respectively) homologues also showed affinity for both receptors. This study suggests that a bioactive peptide acting through both C3a and C5a receptors is released by the proteolytic cleavage of serum proteins other than complement components.

4.9 Endothelin Analogues – Several aspects of endothelin research have been reviewed.[147-149] Using NMR and CD techniques, solution structures of human endothelin-2 and a peptide, Ac-c(Cys-His-Leu-Asp-Cys)-Ile-Trp, designed (computer-aided molecular-modelling techniques) to mimic the C-terminal hexapeptide of endothelin were reported.[150,151] Endothelin antagonists, based on the D-Phe-Val dipeptide structure, acting at both ET_A and ET_B receptor subtypes have been reported.[152] N-Methyl-[4-isoxazolyl]-D-Phe-Val derivative IRL3461 (**53**) (obtained by additional SAR studies on IRL2500 reported earlier) displayed similar potency at both ET_A and ET_B receptor subtypes (K_i 1.8 and 1.2 nM, respectively). Many other analogues containing a 2-naphthyl-alanine, ethylglycine, Met, Leu, Ile, Cha, or Thr in place of Val were 2-4-fold more potent at the ET_B receptor. Compounds in the tryptophan series (**54**)

(53) IRL3461

(54)

including IRL2500 (R = Ph) containing 2-pyridyl, 2-furyl, 2-thienyl, 3-thienyl, 3-isoxazolyl or 5-isoxazolyl substituents were all >100-fold more potent at the ET_B receptors (ET_B K_i values 0.21-4.4 nM, ET_A K_i values 45-2300 nM).

A number of publications on the non-peptide antagonists of endothelin acting at ET_A, ET_B or both the receptor subtypes have been reported.[153-160] Examples of some of the structurally different antagonists (**55-60**) are shown below. The selective ET_A receptor antagonist **55** (A-216546; K_i 0.46 nM for ET_A and 13000 nM for ET_B) showed 48% oral bioavailability in rats. T-0201 (**60**) (K_i 0.015 and 41 nM at ET_A and ET_B receptors, respectively) (pA_2 values 9.0 in isolated rat aorta, 6.8 in the isolated rat trachea and 5.7 in the isolated rabbit pulmonary artery) inhibited the pressor response to exogenous big ET-1 after both i.v. and p.o. administration in a dose dependent manner (0.01–1.0 mg kg^{-1}). The inhibitory effect of orally administered **60** on big ET-1-induced pressor response lasted for 4 h at 0.1 mg kg^{-1} and for 8 h at 1 mg kg^{-1}.

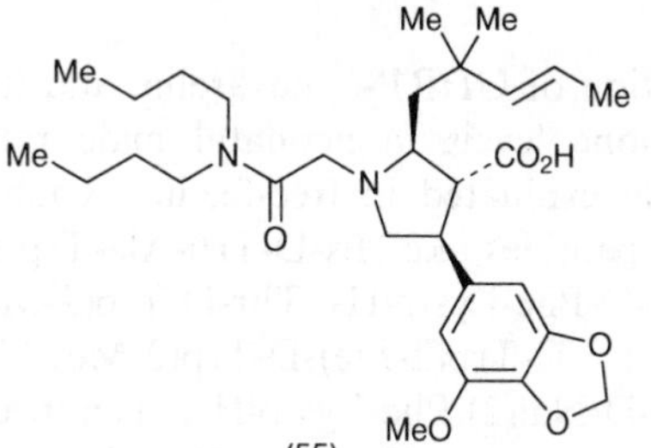

(55) 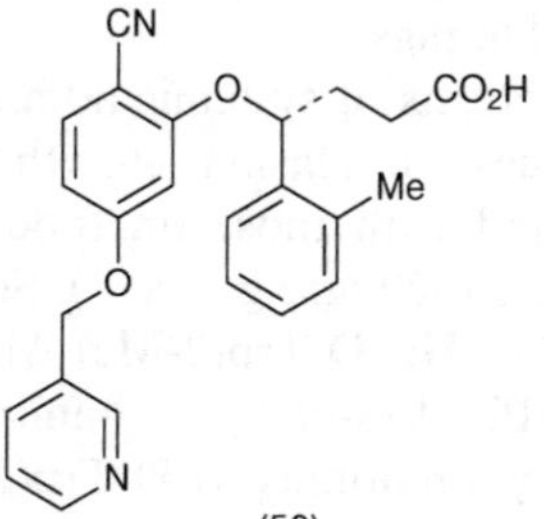 (56) 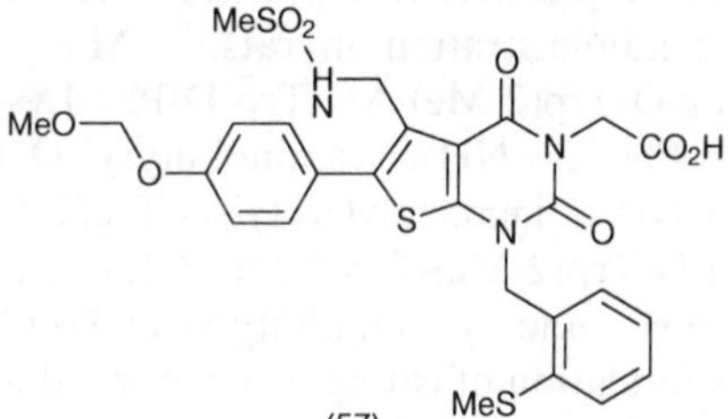 (57) 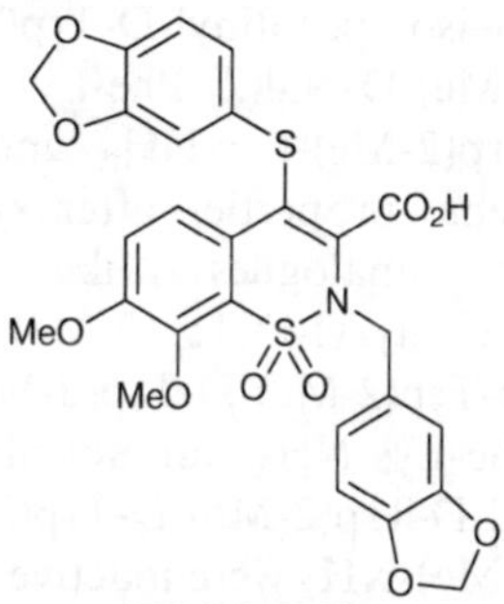 (58) PD164800 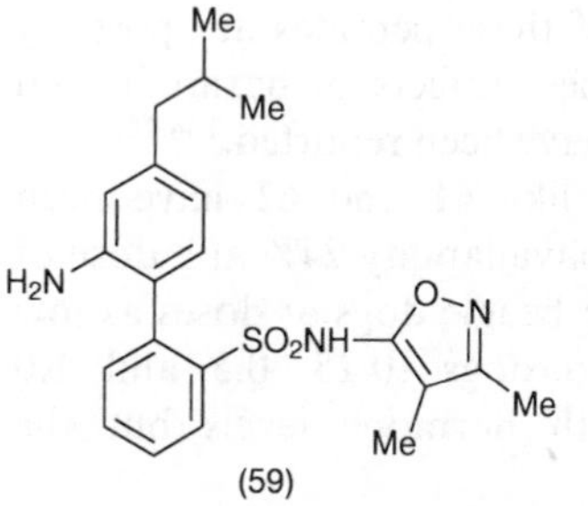 (59) 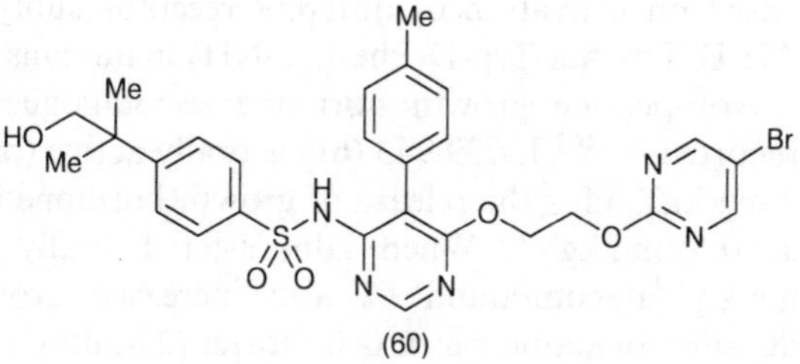 (60)

4.10 Growth Hormone-releasing Peptide and Non-peptide Analogues – Peptidomimetic growth hormone secretagogues have been reviewed.[161] New analogues of a growth hormone releasing peptide hexarelin have been reported.[162,163] One of these peptides, ipamorelin (Aib-His-D-Nal(2)-D-Phe-Lys-NH₂), obtained by deleting the middle -Ala-Trp- dipeptide from a similar heptapeptide Ala-His-D-Nal(2)-Ala-Trp-D-Phe-Lys-NH$_2$, was comparable to the hexapeptide His-D-Trp-Ala-Trp-D-Phe-Lys-NH$_2$ in several *in vitro* and *in vivo* growth hormone releasing tests. Two of the other analogues Ala-His-D-Nal(2)-D-Phe-Lys-NH$_2$ and D-Ala-His-D-Nal(2)-D-Phe-Lys-NH$_2$ were >5-fold less potent than ipamorelin in releasing growth hormone from rat pituitary cells. A pharmacological profiling using growth hormone-releasing peptide and growth hormone-releasing hormone antagonists demonstrated that ipamorelin as GHRP-6 stimulates growth hormone release *via* a growth hormone-releasing peptide-like receptor.[162] In contrast to both GHRP-6 and GHRP-2, administration of ipamorelin did not cause an increase in plasma level of ACTH and cortisol in pigs.

The effects of systemic administration of GHRP-6, hexarelin, and its novel analogues on plasma growth hormone levels in neonatal male rats were compared with those on food intake evaluated in free-feeding young-adult male rats (320 µg kg^{-1} s^{-1}). Several peptides like His-D-Trp-Ala-Trp-D-Phe-Lys-NH$_2$, His-D-Trp(2-Me)-Ala-Trp-D-Phe-Lys-NH$_2$, Thr-D-Trp(2-Me)-Ala-Trp-D-Phe-Lys-NH$_2$, γ-aminobutyryl-D-Trp(2-Me)-D-Trp(2-Me)-Phe-Lys-NH$_2$, γ-aminobutyryl-D-Trp(2-Me)-D-Nal(2)-Phe-Lys-NH$_2$, isonipecotinyl-D-Nal(2)-D-Trp-Phe-Lys-NH$_2$, isonipecotinyl-D-Nal(2)-D-Nal(2)-Phe-Lys-NH$_2$, isonipecotinyl-D-Trp(2-Me)-D-Trp-Phe-Lys-NH$_2$, isonipecotinyl-D-Trp(2-Me)-D-Nal(2)-Phe-Lys-NH$_2$, γ-aminobutyryl-D-Trp(2-Me)-D-Trp(2-Me)-Trp(2-Me)-Lys-NH$_2$ and Aib-D-Trp(2-Me)-D-Trp(2-Me)-NH$_2$ display orexigenic properties after systemic administration in rats.[163] Many other similar analogues like D-Thr-D-Trp(2-Me)-Ala-Trp-D-Phe-Lys-NH$_2$, γ-aminobutyryl-D-Trp(2-Me)-D-Trp-Phe-Lys-NH$_2$, γ-aminobutyryl-D-Trp(2-Me)-D-Trp(2-Me)-D-Trp(2-Me)-Lys-NH$_2$, imidazolylacetyl-D-Trp(2-Me)-D-Trp-Phe-Lys-NH$_2$, aminoisobutyryl-D-Trp(2-Me)-Trp(2-Me)-NH$_2$, γ-aminobutyryl-D-Trp(2-Me)-D-Trp(2-Me)-NH$_2$ and γ-aminobutyryl-D-Trp(2-Me)-Trp(2-Me)-NH$_2$ were inactive. The stimulation of eating behaviour induced by the peptides was independent of the effects that these compounds exert on growth hormone release. Thus the two effects of these peptides are possibly based on activation of different receptor subtypes. Effects of hexarelin and Ala-D-Trp-Ala-Trp-D-Phe-Lys-NH$_2$ in humans have been reported.[164,165]

Non-peptide growth hormone secretagogues like **61** and **62** have been reported.[166,167] L-739,943 (**61**) is orally active (bioavailability 24% at a dose of 2 mg kg^{-1}) for the release of growth hormone in beagle dogs at doses as low as 0.5 mg kg^{-1}. When administered orally to dogs (0.25, 0.5 and 1.0 mg kg^{-1}), compound **62** also increased growth hormone levels but the duration of action was much shorter (2 hours).

SAR studies on [Nle^{27}]human growth hormone-releasing hormone(1-29)-NH$_2$ have been reported.[168] Single point alanine replacements of each amino acid by alanine led to the identification of [Ala^8]-, [Ala^9]-, [Ala^{15}]-, [Ala^{22}]- and [Ala^{28}, Nle^{27}]-hGH-RH(1-29)-NH$_2$ as being 2-6 times more potent than hGH-RH(1-40)-OH (standard) *in vitro*. Nearly complete loss in potency was seen for [Ala^1]-, [Ala^3]-, [Ala^5]-, [Ala^6]- [Ala^{10}]-, [Ala^{11}]-, [Ala^{13}]-, [Ala^{14}]- and [Ala^{23}]-,

(61) L-739,943   (62)

whereas [Ala^{16}]-, [Ala^{18}]-, [Ala^{24}]-, [Ala^{25}]-, [Ala^{26}]- and [Ala^{29}]- yielded equipotent analogues and [Ala^{7}]-, [Ala^{12}]-, [Ala^{17}]-, [Ala^{20}]-, [Ala^{21}]- and [Ala^{27}]- gave weak agonists with potencies 15-40% that of the standard. The multiple-alanine substituted peptides [MeTyr^{1}, Ala^{15,22}, Nle^{27}]- and [MeTyr^{1}, Ala^{8,9,15,22,28}, Nle^{27}]-hGH-RH(1-29)-NH$_2$ released growth hormone 26 and 11 times, respectively, more effectively than the standard *in vitro*. Application of lactam constraints in the C-terminus of GHRH(1-29)-NH$_2$ identified c(25-29)[MeTyr^{1}, Ala^{15}, D-Asp^{25}, Nle^{27}, Orn^{29}]-hGH-RH(1-29)-NH$_2$, containing an optimum 19-membered ring in this region of the molecule, as one of the more potent (17-times more potent than the standard) peptides. Equally effective was an [i-(i+3)] constraint yielding the 18-membered ring c(25-28)[MeTyr^{1}, Ala^{15}, Glu^{25}, Nle^{27}, Lys^{28}]-hGH-RH(1-29)-NH$_2$ which was 14-times more potent than the standard. Nude mice bearing xenografts of DU-145 human androgen-independent prostate carcinoma were treated with a growth hormone-releasing hormone antagonist MZ-5-156 (20 μg per animal, s.c. twice a day) for eight weeks.[169] After 8 weeks of therapy, final volume and weight of tumours in mice treated with MZ-5-156 {[Ph-CH$_2$CO-Tyr^{1}, D-Arg^{2}, Phe(p-Cl)^{6}, Abu^{15}, Nle^{27}, Agm^{29}]hGH-RH(1-29)} were significantly decreased, and serum IGF-1 showed a significant reduction. The levels of IGF-II in tumour tissue (treated animals) were also reduced by 77%.

4.11 Integrin-related Peptide and Non-peptide Analogues – Various aspects of cell adhesion molecules are reviewed.[170-172] Iminodiacetic acid diamide dimer, trimer and tetramer libraries have been synthesised for probing protein-protein interactions.[173] Chemistry, biology, pharmacokinetic and clinical details of some of the IIb/IIIa antagonists in development [Eptifibatide, α-S-CH$_2$-CH$_2$-CO-Arg-Gly-Asp-Trp-Pro-Cys-NH$_2$) (disulfide bridge containing cyclic peptide), Sibrafiban (**63**, Ro-48-3657) and Orbofiban (**64**, SC-511)] have been published.[174-176]

4.11.1 IIb/IIIa Antagonists – As in previous years, most of the work in the integrin area has been concentrated in the field of peptide/non-peptide IIb/IIIa antagonists as platelet aggregation inhibitors. New potent inhibitors of platelet aggregation, acanthin I and II, were purified from the venom of a common death adder (*Acanthophis antarcticus*). The inhibitors (IC$_{50}$ 4-12 nM against

(63) Sibrafiban, Ro-48-3657   (64) Orbofiban, SC-511

collagen- and ADP-induced platelet aggregation) are basic proteins and exhibit phospholipase enzyme activity.[177] An atrolysin E disintegrin domain has been isolated from crotalid snake venom and shown to contain a Met-Val-Asp sequence instead of an Arg-Gly-Asp sequence.[178] Nevertheless, the protein is a potent inhibitor of collagen- and ADP-stimulated platelet aggregation (IC$_{50}$ 4-8 nM, respectively. A cyclised synthetic peptide, Ac-c(Cys-Arg-Val-Ser-Met-Val-Asp-Arg-Asn-Asp-Asp-Thr-Cys)-NH$_2$, which represents the sequence of the atrolysin disintegrin domain non-Arg-Gly-Asp loop, was demonstrated to be a weak inhibitor of platelet aggregation (IC$_{50}$ ~ 2 mM).

Cyclic peptides like c(Arg-Gly-Asp-D-Phe-β-Ala), c(Arg-Gly-Asp-D-Phe-(R)-β-Leu), c(Arg-Gly-Asp-(S)-β-Phe-Val) and c(Arg-Gly-Asp-D-Phe-Val-α-Ala) inhibited ADP- and thrombin-receptor activating peptide-mediated platelet aggregation with IC$_{50}$ values in the region of 9-140 μM. The most potent compound of the series was c(Arg-Gly-Asp-(S)-β-Phe-Val) (IC$_{50}$ 9-10 μM). Cyclic peptide c(Arg-Gly-Asp-(S)-β-Phe-Val-Gly) was much less potent.[179] Cyclic peptide containing a diazaethylene glycol derivative, c(Arg-Gly-Asp-NH-CH$_2$-CH$_2$-O-CH$_2$-CH$_2$-NHCO-CH$_2$-CH$_2$-CO), showed a weak inhibitory activity toward the fibrinogen or fibronectin binding to their respective integrin receptors (IC$_{50}$ values 14-65 μM) and was also weakly active in the dog platelet-rich plasma anti-aggregatory assay (IC$_{50}$ value 73 μM).[180] A heptapeptide Leu-Ser-Ala-Arg-Leu-Ala-Phe (reported earlier), designed to bind to residues 315-321 of α$_{IIb}$ which is adjacent to presumed fibrinogen γ chain binding site on α$_{IIb}$ was shown to activate platelets by an unknown mechanism and causes aggregation.[181] To develop polymer membranes suitable for supporting *in vitro* cultivation of mammalian cells, several Arg-Gly-Asp peptidomimetics were prepared starting from an ortho-amino-tyrosine template and various ω-amino acid derivatives. The most flexible compounds, including **65**, have shown a biological activity similar to that of Arg-Gly-Asp-Ser in the platelet aggregation test.[182] The compound **65** was covalently fixed on the surface of a poly(ethylene terephthalate) membrane.

(65)

A number of non-peptide analogues based on the Arg-Gly-Asp sequence have been reported.[183-200] Selected examples of such compounds, all inhibiting platelet aggregation after oral administration (some as ester prodrugs), include **66-75**. Several of the diaminopropionic acid derivatives like XV459 (**66**) and XV454 (methylphenylsulfonyl group in place of n-butyloxycarbonyl group in **66**) inhibited human platelet aggregation induced by ADP, thrombin receptor agonist peptide or collagen, bound to both activated and unactivated human platelets with high affinity and showed oral activity in animal models.[183-187] The methyl ester prodrug of **66** (XU065) showed antiplatelet effects in dogs after oral administration (100% inhibition at 1.6 mg kg^{-1}, up to 5 h). Some compounds like **71** were active against $\alpha_{IIb}\beta_3$ and $\alpha_v\beta_3$ (inactive against the $\alpha_v\beta_5$ integrin) and inhibited bovine fibroblast growth factor-induced angiogenesis in chick chorioallantoic membrane assay.[194] Several oligocarbamate ligands like **73** for IIb/IIIa were discovered using a combinatorial library approach.[200] Several other compounds unrelated to the Arg-Gly-Asp sequence have been reported as thrombin-induced platelet aggregation inhibitors (**74**) and fibrinogen receptor antagonists (**75**).[201-203]

4.11.2 $\alpha_v\beta_3$ Antagonists – Role of the integrin $\alpha_v\beta_3$ in a tumour growth has been discussed.[204-206] Recent reports suggest that tumour-associated blood

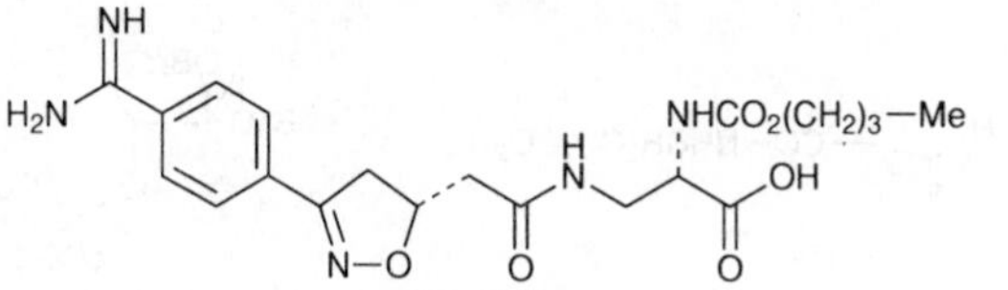

(66) XV459

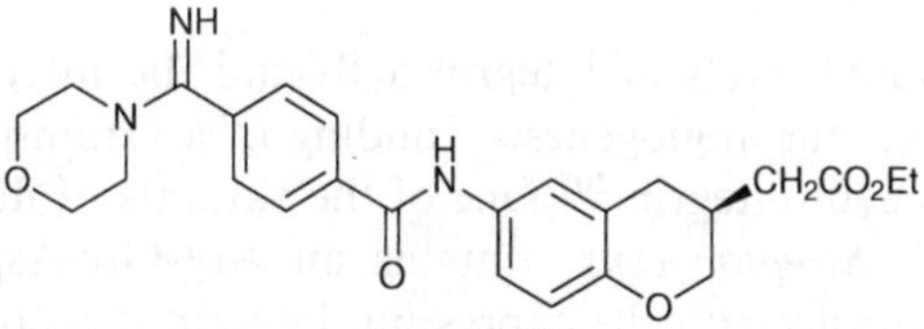

(67) MS-180

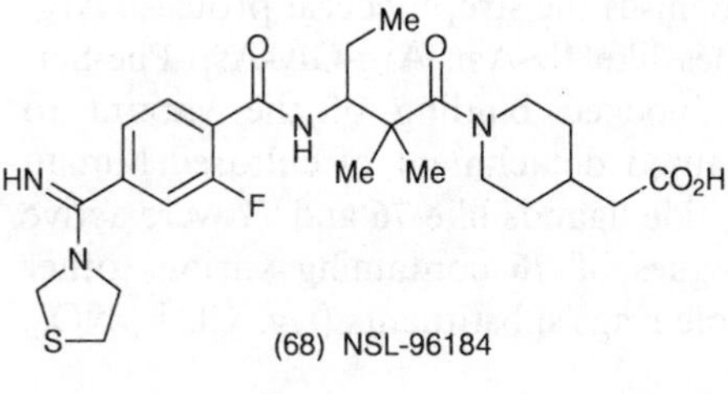

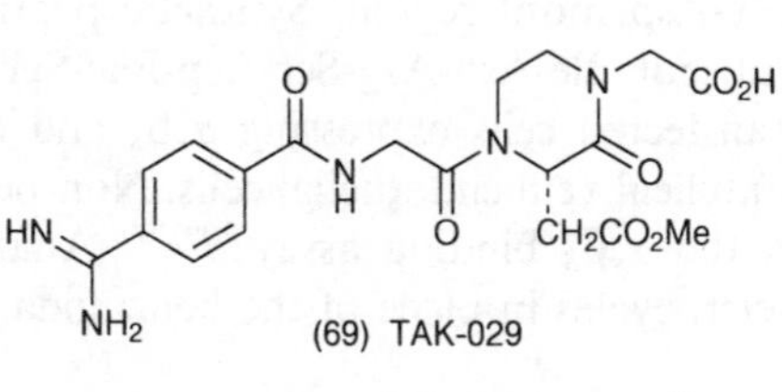

(68) NSL-96184

(69) TAK-029

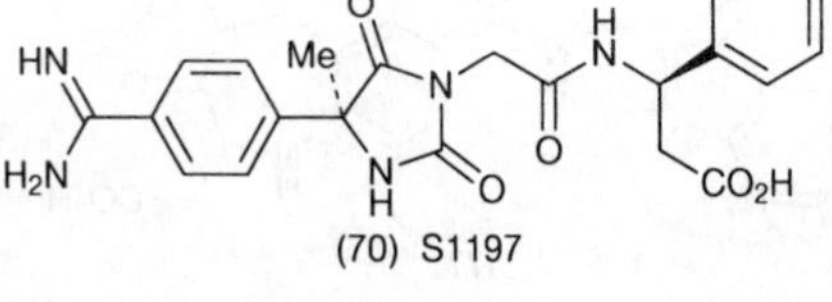

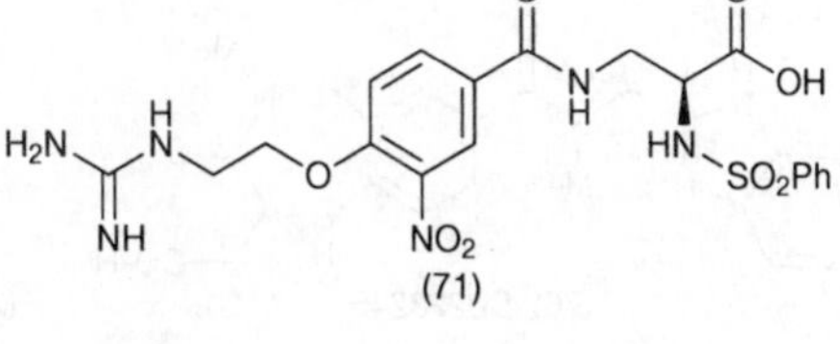

(70) S1197

(71)

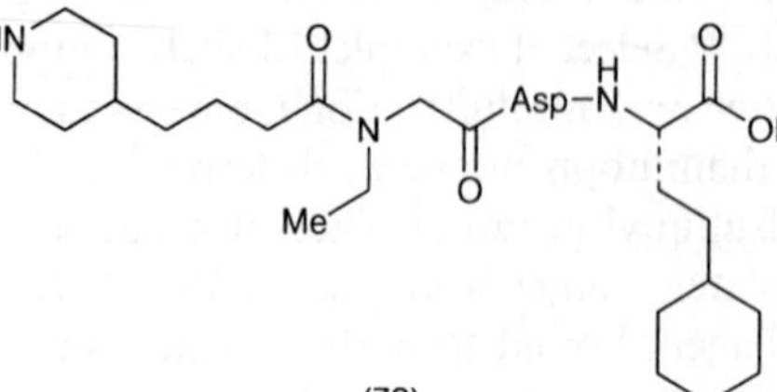

(72)

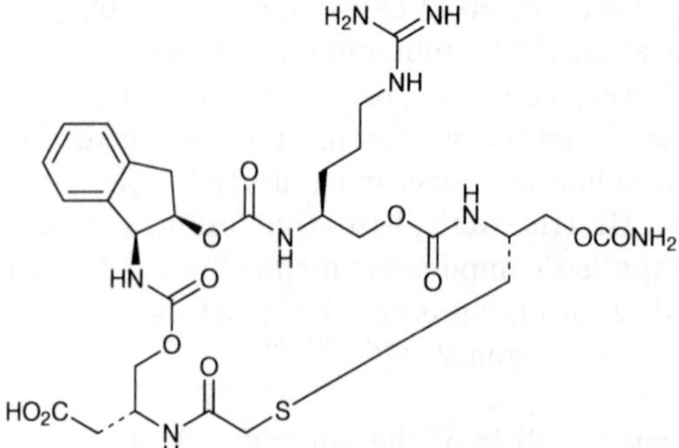

(73)

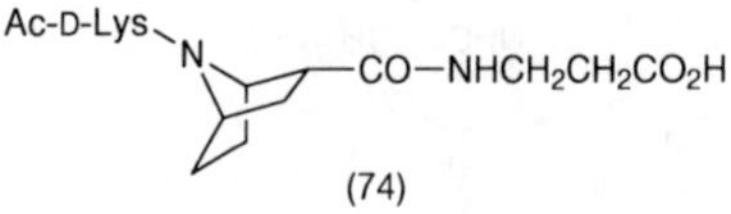

(74)

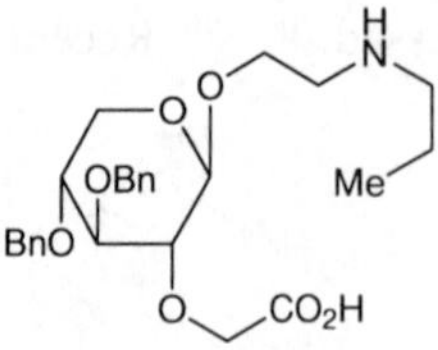

(75)

vessels express elevated levels of integrin $\alpha_v\beta_3$ and the integrin may play an important role in tumour angiogenesis. Binding of a Gram-positive virulence factor also requires $\alpha_v\beta_3$ integrin.[207] One of the variants of human pathogenic bacterium group A *Streptococcus* contains an Arg-Gly-Asp sequence. The variant bound to transfected cells expressing integrin $\alpha_v\beta_3$ or $\alpha_{IIb}\beta_3$, and the binding was blocked by a mAb that recognises the streptococcal protease Arg-Gly-Asp motif region. Synthetic peptides like Ile-Asn-Arg-Gly-Asp-Phe-Ser, but not Ile-Asn-Arg-Ser-Asp-Phe-Ser, blocked binding of the variant to transfected cells expressing $\alpha_v\beta_3$ and caused detachment of cultured human umbilical vein endothelial cells. Non-peptide ligands like **76** and **77** were active in the $\alpha_v\beta_3$ binding assay.[208,209] Analogues of **76** containing various other heterocycles in place of the benzimidazole ring, substituents (*e.g.* Cl, F, NO$_2$,

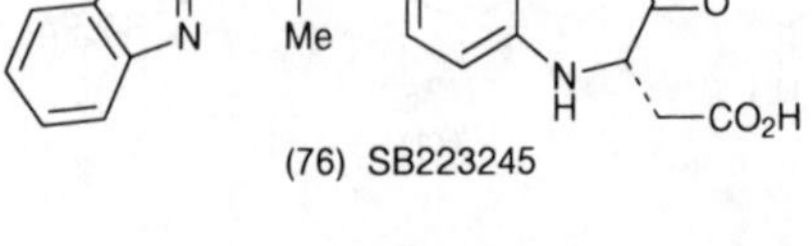

(76) SB223245

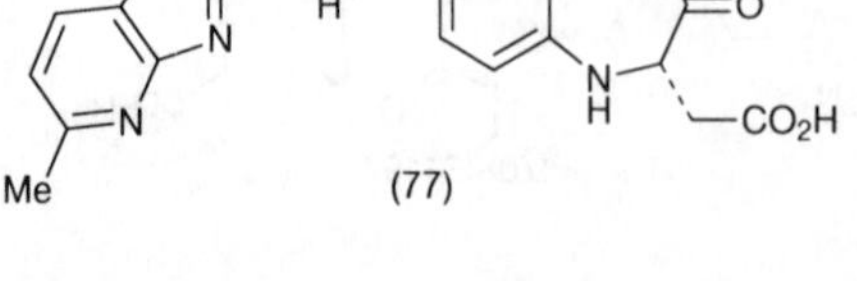

(77)

CN, OCH$_3$) into the benzimidazole ring or other substituents in place of the methyl group on the amide groups resulted in compounds which retained activity in the $\alpha_v\beta_3$ binding assay. The imidazopyridine analogue **77** (K_i 45 nM in the $\alpha_v\beta_3$ binding assay) showed efficacy in an animal model of restenosis. Like the benzimidazole derivative **76**, compound **77** showed poor oral bioavailability (<10%).

4.11.3 $\alpha_4\beta_1$ and $\alpha_5\beta_1$ Antagonists – Role of $\alpha_4\beta_1$ integrin in monocyte recruitment in atherogenesis and in the development of experimental autoimmune encephalomyelitis has been reported.[210,211] A small linear binding epitope in domain 1 of VCAM-1 (contained within the region Arg[36]-Thr-Gln-Ile-Asp-Ser-Pro-Leu-Asn[44]) has been used to design cyclic peptides which could block $\alpha_4\beta_1$ binding to VCAM-1. The cyclic peptides c(Thr-Gln-Ile-Asp-Ser-Pro-Asn-Gly), c(Thr-Gln-Ile-Asp-Ser-Pro-Ala-Gly), c(Thr-Gln-Ile-Asp-Ser-Pro-Asn-Ala), c(Thr-Gln-Ile-Asp-Ser-Pro-Ala-Ala), c(Ala-Gln-Ile-Asp-Ser-Pro-Asn-Ala), c(Thr-Ala-Ile-Asp-Ser-Pro-Asn-Ala), c(Thr-Gln-Ala-Asp-Ser-Pro-Asn-Ala), c(Thr-Gln-Ile-Asp-Ala-Pro-Asn-Ala), c(Thr-Gln-Ile-Asp-Ala-Ala-Asn-Ala), c(Thr-Ala-Ile-Asp-Ala-Pro-Asn-Ala), c(Thr-Ala-Ile-Ala-Ala-Pro-Asn-Ala), c(Thr-Gln-Ile-Ala-Ala-Pro-Asn-Ala), c(Val-Gln-Ile-Asp-Ser-Pro-Asn-Ala), c(Val-Gln-Ile-Asp-Ala-Pro-Asn-Ala), c(Thr-Pro-Gly-Asp-Ser-Pro-Asn-Ala) and c(Thr-Gln-Ile-Asp-Ser-Pro-Asn-Arg) were active in the micromolar range (IC$_{50}$ values 4.3-500 μM).[212] The most potent peptides of the series were c(Thr-Ala-Ile-Asp-Ala-Pro-Asn-Ala), c(Val-Gln-Ile-Asp-Ser-Pro-Asn-Ala) and c(Val-Gln-Ile-Asp-Ala-Pro-Asn-Ala) (IC$_{50}$ values 4-6 μM). Using a library of small molecule β-turn mimetics, tyrosine derivatives like **78** and **79** (IC$_{50}$ values 5-8 μM) were identified as inhibitors of $\alpha_4\beta_1$ integrin binding.[213] Analogues of **78** containing Nle or Leu side chains in place of the Glu side chain were much less potent (IC$_{50}$ values >100 μM).

(78)   (79)

Integrin $\alpha_5\beta_1$ is a widely distributed cell surface receptor for the extracellular matrix glycoprotein fibronectin. It appears to play a significant role during the interaction of human newborn smooth muscle cell lines with fibrin.[214] A heptapeptide Arg-Arg-Glu-Thr-Ala-Trp-Ala is reported as a ligand for integrin $\alpha_5\beta_1$, which blocks $\alpha_5\beta_1$-mediated cell adhesion to fibronectin. The binding site for this peptide on $\alpha_5\beta_1$ has been identified using inhibitory monoclonal antibodies and site-directed mutagenesis.[215] A cyclic peptide containing this sequence (Cys-Arg-Arg-Glu-Thr-Ala-Trp-Ala-Cys) had little effect on the binding of most anti-$\alpha 5$ and anti-$\beta 1$ mAbs to $\alpha_5\beta_1$ but blocked binding of the anti-α_5 mAb 16. The cyclic peptide also acted as a competitive inhibitor of the binding of Arg-Gly-Asp-containing fibronectin fragments to $\alpha_5\beta_1$.

4.11.4 Mac-1/ICAM-1 Antagonists – ICAM-1 [consisting of five Ig-like domains (D1-D5), a short transmembrane region, and a small carboxyl-terminal cytoplasmic domain] is a cell surface, transmembrane molecule that is rapidly up-regulated by cytokine stimulation, enhancing adhesion of leukocytes to endothelial cells at sites of infection or injury. ICAM-1 is also used by various pathogens, such as common cold human rhinoviruses, coxsackievirus A21, and the malarial parasite *Plasmodium falciparum*. Ligands for ICAM-1 include leukocyte function-associated antigen (LFA-1, CD11a/CD18), macrophage-1 antigen (Mac-1, CD11b/CD18, $\alpha_M\beta_2$) and fibrinogen. ICAM-1 does not possess an Arg-Gly-Asp motif, but has a larger, more extended binding surface. Mac-1 (CD11b/CD18, $\alpha_M\beta_2$) has been implicated in intimal thickening after angioplasty or stent implantation. Administration of M1/70, an anti-CD11b blocking mAb, to rabbit immediately before, and every 48 h for 3, 6 and 14 days after, iliac artery balloon denudation or deeper stent-induced injury inhibited Mac-1-mediated fibrinogen binding *in vitro*, reduced leukocyte recruitment more than 2-fold 3, 6 and 14 days after injury.[216] Neointimal growth 14 days after injury was markedly attenuated by treatment with M1/70. Crystal structure of a 190-residue fragment of intracellular adhesion molecule-1 has been reported.[217] Three-dimensional atomic structure of the two amino-terminal domains (D1 and D2) of ICAM-1 has also been determined.[218]

Peptide inhibitors of the integrin Mac-1 with its receptor, intercellular adhesion molecule-1 (ICAM-1), have been identified. Peptides derived from the complementarity determining regions of three antibodies (44aacb, MY904, and 118.1) shown to block Mac-1-mediated cell adherence showed weak blocking activity at 10–100 μM. By using phage display of peptide libraries based on the 118.1 CDR peptide with five residues randomised, weak Mac-1 antagonists were identified.[219] The peptides Tyr-Thr-Phe-Thr-Asn-Tyr-Trp-Ile-Asn-Trp-Val-Lys-Gln,   Tyr-Thr-Phe-Ser-Asn-Tyr-Trp-Ile-Glu-Trp-Val-Lys-Gln, Glu-Trp-Ile-Gly-Tyr-Ile-Asp-Pro-Tyr-Tyr-Gly-Gly-Ile-Thr-Tyr-Asn-Gln-Ile-Phe-Lys-Gly-Lys-Ala and Ala-Val-Tyr-Phe-Cys-Ala-Arg-Gly-Gly-Ile-Ile-Thr-Thr-Ala-His-Tyr-Phe-Asp-Tyr-Trp-Gly-Gln-Gly were weak inhibitors of Mac-1/ICAM-1 interaction (IC$_{50}$s 33–100 μM).

4.12 LHRH Analogues – Chemistry, molecular biology and utility of LHRH agonists and antagonists has been reviewed.[220-222] Additional details on several of the agonist (avorelin, MF6001) and antagonist (abarelix and cetrorelix) analogues in development have been reported.[223-225] Avorelin (Meterelin), [D-2-Me-Trp^6, Pro^9-NHEt]LHRH, was formulated in polylactic glycolic acid to afford protracted and continuous release of the peptide from subcutaneous implants. Two different formulations (10 and 15 mg) were effective in reducing the testosterone levels in both men and dogs to the castrate levels for a period of 6 months.[223] A second isoform of GnRH, [His^5, Trp^7, Tyr^8] GnRH-I (GnRH-II) was shown to be present in the brain of the mouse, rat and human.[226]

Analogues of [Tyr(OMe)^5]-LHRH with modifications at positions 6, 9 and 10 are reported.[227] In comparison to LHRH and [Tyr-(OMe)^5]-LHRH (K_d values 3.44 and 1570 nM, respectively, in the rat pituitary binding assay], two of the analogues, [Tyr-(OMe)^5, D-Arg^6, Aze^9]- and [Tyr-(OMe)^5, D-Ser^6, Aze^9-NHEt]-LHRH, were similar in potency to LHRH. In the goldfish pituitary gonadotropin-release assay, the D-Arg containing analogues [Tyr-(OMe)^5, D-Arg^6, Aze^9]- and [Tyr-(OMe)^5, D-Arg^6, Aze^9-NHEt]-LHRH were about 10-40-fold more potent than LHRH. Cytotoxic analogues of LHRH have been prepared. The analogues contained a 2-β-alanyl-3-chloro-1,4-naphthoquinone or a 2-(5-aminopentanoic acid)-1,4-naphthoquinone moiety attached to the side chain amino of the Lys^6 residue of [D-Lys^6]-LHRH.[228]

Non-peptide antagonists of LHRH have been discovered by using a directed screening approach based on the Tyr^5-Gly-Leu-Arg^8 region of LHRH.[229] Further chemistry on one of the early leads (80) [67% inhibition of [^{125}I]leuprorelin at 20 μM] led to a potent antagonist 81 (T-98475). In the binding

(80)   (81) T-98475

assay using cloned human receptor and membrane fractions of monkey and rat pituitaries (IC$_{50}$s 0.2, 4 and 60 nM against human, monkey and rat receptors, respectively), 81 was as potent as and [D-Leu^6, Pro^9-NHEt]-LHRH. The *in vivo* activity (suppression of plasma LH levels) of 81 was investigated in castrated male cynomolgus monkeys. Oral administration of 81 (60 mg kg^{-1}) exhibited >70% inhibition of plasma LH levels 8 hours after administration of the compound. The inhibitory effect lasted for more than 10 hours at this dose.

4.13 α-MSH Analogues – Work on melanocortin receptors and the role of MSH peptides in pigmentation, obesity and cardiovascular regulation has been reviewed.[230-232] The agouti-related protein gene, involved in the regulation of melanocortin receptors expressed in the central nervous system, is implicated in the control of feeding behaviour. Its overexpression in transgenic animals results in obesity and diabetes. Both agouti-related protein and a truncated form act as competitive antagonists of α-MSH at MC$_3$ and MC$_4$ receptors.[233] Alanine-scanning mutagenesis was performed on the agouti protein carboxyl terminus to locate residues important for melanocortin receptor binding inhibition. Replacement of the agouti residues Asp^{108}, Arg^{116} Phe^{118} and Phe^{117} to alanine has significant effect on the binding to melanocortin receptors.[234] A synthetic C-terminal human agouti-related protein(83-132)-NH$_2$ fragment [Ser-Ser-Arg-Arg-Cys-Val-Arg-Leu-His-Glu-Ser-Cys-Leu-Gly-Gln-Gln-Val-Pro-Cys-Cys-Asp-Pro-Cys-Ala-Thr-Cys-Tyr-Cys-Arg-Phe-Phe-Asn-Ala-Phe-Cys-Tyr-Cys-Arg-Arg-Leu-Gly-Thr-Ala-Met-Asn-Pro-Cys Ser-Arg-Thr-NH$_2$, containing many disulfide bridges] was found to inhibit (K 0.7 nM) α-MSH response in a *Xenopus laevis* dermal melanophore cell preparation. Fragments(25-51)-NH$_2$ and (54-82)-NH$_2$ were inactive.[235] When administered icv into rats, the C-terminal (83-132)-NH$_2$ fragment increased food intake over a 24-h period.[236] The hyperphagia was similar to that seen when synthetic MC$_3$ and MC$_4$ cyclic peptide antagonist SHU9119. Two other disulfide bridge containing cyclic peptides HS014 [Cys-Glu-His-D-Nal-Arg-Trp-Gly-Cys-Pro-Pro-Lys-Asp-NH$_2$] and HS024 (c[Ac-Cys^3, Nle^4, Arg^5 D-Nal^7, Cys-NH$_2^{11}$] α-MSH-(3-11)) (both MC$_4$ receptor antagonists) significantly increased food intake in rats. Peptide HS014 showed a number of behavioural effects but HS024 was inactive in elevated plus-maze and open field experiments on rats.[237-239]

Cyclic α-MSH(1-13) and (4-10) lactam analogues [Ac-Ser-Tyr-Ser-Nle c(Asp-His-D-Phe-Arg-Trp-Lys)-Gly-Pro-Val-NH$_2$, Ac-Nle-c(Asp-His-D-Nal-Arg-Trp-Lys)-NH$_2$ and Ac-Nle-c(Asp-His-D-Tyr-Arg-Trp-Lys)-NH$_2$] and α-MSH(1-13) disulfide bridge containing peptide [Ac-Ser-Tyr-Ser-c(Cys-Gly-His-D-Phe-Arg-Trp-Cys)-Lys-Pro-Val-NH$_2$] were tested for their selectivity for the rat MC$_4$ receptor.[240] Both the larger peptides, Ac-Ser-Tyr-Ser-c(Cys-Gly-His-D-Phe-Arg-Trp-Cys)-Lys-Pro-Val-NH$_2$ and Ac-Ser-Tyr-Ser-Nle c(Asp-His-D-Phe-Arg-Trp-Lys)-Gly-Pro-Val-NH$_2$, were about 20-50-fold more potent at the rMC$_4$ receptors (EC$_{50}$s 0.04-0.05 nM) in comparison to the rMC$_3$ and mMC$_5$ receptors (EC$_{50}$s 1-3 nM). One of the smaller peptides, Ac-Nle-c(Asp-His-D-Tyr-Arg-Trp-Lys)-NH$_2$, was similar in potency to the larger peptides at the rMC$_4$ receptors (EC$_{50}$ 0.07 nM) but was much less potent at the other receptor subtypes (EC$_{50}$s 80 and >300 nM, respectively, at the rMC$_3$ and mMC$_5$ receptors). The other heptapeptide Ac-Nle-c(Asp-His-D-Nal-Arg-Trp-Lys)-NH$_2$ was an agonist at the mMC$_5$ receptor subtype (EC$_{50}$ 2 nM) and an antagonist at the rMC$_3$ and rMC$_4$ receptor subtypes.

In [Cys^4, D-Nal^7, Cys^{11}]MSH(4-11)-NH$_2$ (HS964) (K_i 1460, 281, 23 and 164 nM at the MC$_1$, MC$_3$, MC$_4$ and MC$_5$ receptors, respectively) series of cyclic peptides, replacement of the D-Nal^7 residue by D-Phe, D-Cha and D-Phe(p-benzoyl) residues led to poorly active compounds at all the four receptors (K values 1-10 μM).[241] Similarly, replacement of the Gly^{10} by Asp led to a poorly active compound at all the four receptors (K_i values 5-91 μM). Replacement of Glu^5 in the cyclic peptide by Arg led to [Cys^4, Arg^5, D-Nal^7, Cys^{11}]MSH(4-11)-NH$_2$, which was more potent than the parent peptide in all the four receptor preparations (K_i 200, 8, 10 and 42 nM at the MC$_1$, MC$_3$, MC$_4$ and MC$_5$ receptors, respectively). The SAR results indicated that Glu^5 and D-Nal^7 were important for the potency and MC$_4$ receptor selectivity in this series of compounds. In comparison to one of the more potent MC_4 receptor ligands reported earlier, [Nle^4, Asp^5, D-Nal^7, Lys^{10}]MSH(4-10)-NH_2, (amide bond between the Asp^5 and Lys^{10} side chains) (SHU9119)] (K_i values 0.714, 1.20, 0.36 and 1.12 nM at the MC_1, MC_3, MC_4 and MC_5 receptors, respectively), HS964 was about >50-fold less potent at the MC_4 receptors but the selectivity profile of the two compounds was quite different. The most potent and selective compound at the MC_4 receptor subtypes [Cys^4, D-Nal^7, Cys^{11}]MSH(4-11)-Pro-Pro-Lys-Asp-NH_2 (HS014) antagonised cAMP stimulation induced by α-MSH at the MC_3 and MC_4 receptor transfected cells. However, for the MC_1 and MC_5 receptors, the peptide was shown to increase intracellular levels of cAMP, but without reaching maximum levels. Thus the compound appears to be an antagonist at the MC_3 and MC_4 receptors and partial agonist at the MC_1 and MC_5 receptor subtypes.[241]

Somatostatin-based compounds like D-Phe-c(Cys-Tyr-D-Trp-Orn-Thr-Pen)-Thr-NH_2, D-Phe-c(Cys-His-D-Phe-Arg-Trp-Pen)-Thr-NH_2, D-Phe-c(Cys-His-D-Phe-Arg-Trp-Cys)-Thr-NH_2, D-Phe-c(hCys-His-D-Phe-Arg-Trp-Cys)-Thr-NH_2, D-Phe-c(Asp-His-D-Phe-Arg-Trp-Lys)-Thr-NH_2 and Ac-Nle-c(Asp-His-D-Phe-Arg-Trp-Lys)-NH_2 were α-MSH agonists in a frog skin assay.[242] Two of the more potent compounds, D-Phe-c(hCys-His-D-Phe-Arg-Trp-Cys)-Thr-NH_2 and D-Phe-c(Asp-His-D-Phe-Arg-Trp-Lys)-Thr-NH_2, were nearly equipotent to α-MSH. Several other analogues, D-Phe-c(Cys-Tyr-D-Trp-Arg-Thr-Pen)-Thr-NH_2, D-Phe-c(Cys-Tyr-D-Trp-Arg-Trp-Pen)-Thr-NH_2, D-Phe-c(Cys-Phe-D-Trp-Arg-Trp-Pen)-Thr-NH_2 and D-Phe-c(Cys-His-D-Phe(p-I)-Arg-Trp-Cys)-Thr-NH_2, were moderately potent antagonists of α-MSH in a frog skin assay. The ^{99m}Tc- and ^{188}Re complexes of [$Cys^{4,10}$, D-Phe^7]-α-MSH(4-13) and [$Cys^{3,4,10}$, D-Phe^7]-α-MSH(3-13) were synthesised and shown to be stable. *In vivo*, the ^{99m}Tc-[$Cys^{3,4,10}$, D-Phe^7]-α-MSH(3-13) complex exhibited significant tumour uptake and retention and was effective in imaging melanoma in a murine-tumour model system.[243] Binding affinity on B16 F1 murine melanoma cells was reduced to 1% of its original level after Re incorporation into the cyclic [$Cys^{4,10}$, D-Phe^7]-α-MSH(4-13) analogue. Two-dimensional NMR studies on the complexes are reported.

4.14 MHC Class I and II Analogues – Different aspects of peptide presentation by the major histocompatibility complex class I and class II are reviewed.[244-247] Class I molecules are cell surface proteins that present foreign peptides as targets for cytotoxic T cell killing and elimination. Bound peptides eluted from class I molecules have a size preference for 8, 9, and 10 amino acids. Sequence alignment studies using sequences of 181 binding phage and 129 non-binding phage indicated that approximately 82% of the sequences contained anchor residues, either Tyr or Phe in the expected positions (positions 3 or 5) and approximately 80% contained anchor residues Leu, Val, Ile, Met in the eighth position. About 18% of the sequences contained no anchor residues in positions 3 and 5 and of those 18%, approximately 32% registered as strong binders. Using this data computational analysis technique was shown to identify strong binders.[248] In the case of MHC class II, foreign protein antigens must be broken down within endosomes or lysosomes to generate suitable peptides that will form complexes with class II molecules for presentation to T cells. An asparagine-specific cysteine endopeptidase has been shown to be involved in the processing of the antigen. Asparagine-containing peptides like Fmoc-Ala-Glu-Asn-Lys-NH$_2$ and Fmoc-Lys-Asn-Asn-Glu-NH$_2$ inhibited B-cell asparaginyl endopeptidase processing of microbial tetanus toxin antigen *in vitro*. *In vivo*, these inhibitors slow tetanus toxin antigen presentation to T cells, whereas pre-processing of tetanus toxin antigen with asparaginyl endopeptidase accelerates its presentation, indicating that this enzyme performs a key step in tetanus toxin antigen processing.[249]

MHC class I ligands containing an N-hydroxy amide bond [Ser-Ile-Ile-Asn-Phe-Gluψ[CO(NOH)]Gly-Leu, Ser-Ile-Ile-Asn-Pheψ[CO(NOH)]Gly-Lys-Leu, Ser-Ile-Ile-Asnψ[CO(NOH)]Gly-Glu-Lys-Leu, Ser-Ile-Ileψ[CO(NOH)]Gly-Phe-Glu-Lys-Leu, Ser-Ileψ[CO(NOH)]Gly-Asn-Phe-Glu-Lys-Leu, Serψ[CO-(NOH)]Gly-Ile-Asn-Phe-Glu-Lys-Leu and ψ[CO(NOH)]Gly-Ile-Ile-Asn-Phe-Glu-Lys-Leu], designed from the natural epitope Ser-Ile-Ile-Asn-Phe-Glu-Lys-Leu, were investigated for binding to the class I molecule H-2K^b and to induce T cell responses.[250] Binding to the MHC molecule was diminished by the N-hydroxy group at positions 2 and 3 of the oligomer and improved in the case of positions 4, 5, 6 and 7. No change was seen for position 1. The efficacy of T cell stimulation was strongly reduced by the modification of all positions except for position 1. A complete loss of activity was found for the N-hydroxy variant in positions 4 and 6. N-hydroxyamide containing peptides displayed an enhanced stability to enzymatic degradation. Poly-N-acylated amines, as a new class of synthetic non-peptide ligands for the murine MHC class I molecule H-2K^b, were developed by combinatorial approaches on the basis of the ovalbumin-derived peptide epitope Ser-Ile-Ile-Asn-Phe-Glu-Lys-Leu.[251] More potent compounds identified included **82** and **83**.

Ser-Ile-Ile ... Gly-Leu   Ser-Ile-Ile ... Leu

(82)   (83)

Several MHC class II binding peptides have been reported.[251-255] The interactions of an N-terminal peptide Ac-1-11 [Ac-Ala-Ser-Gln-Lys-Arg-Pro-Ser-Gln-Arg-His-Gly] of myelin basic protein, known to induce autoimmune encephalomyelitis in H-2^u and (H-2^u × H-2^s) mice but not in H-2^s mice, were studied. Two polymorphic residues that differ between I-A^u and I-A^s, Tyr^{28}β and Thr^{28}β, and one conserved residue, Glu^{74}β, confer specific binding of Ac-1-11 to I-A^u. A fourth residue, Arg^{70}β in I-A^u, affects both peptide binding and T cell recognition.[252] Another peptide from human myelin basic protein (residues 85-99; Glu-Asn-Pro-Val-Val-His-Phe-Phe-Lys-Asn-Ile-Val-Thr-Pro-Arg), previously reported to bind to purified HLA-DR2 and recognised by human myelin basic protein-specific T cell clones, was used in crystallisation studies directed towards understanding the T-cell receptor recognition and signalling.[253]

Type II collagen has been implicated in rheumatoid arthritis, an autoimmune disease associated with the HLA-DR4 and DR1 alleles. The DR4- and DR1-restricted immunodominant T cell epitope in this protein corresponds to amino acids 251-273. MHC and T cell contacts in collagen II 261-273 [Ala-Gly-Phe-Lys-Gly-Glu-Gln-Gly-Pro-Lys-Gly-Glu-Pro] were defined.[254] Each of the amino acid residues in the peptide was replaced by alanine residues and relative binding of the analogues to DRB1*0401 was determined. In comparison to the parent peptide (IC$_{50}$ 180 nM), Ala^2, Ala^5, Ala^7, Ala^8, Ala^{10}, Ala^{11} and Ala^{12} were more potent (IC$_{50}$ values 21-115 nM). Replacement of the Phe^3 by Ala, Asn and Ser, and Glu^6 by Lys and Arg led to inactive compounds (IC$_{50}$ values >35,000 nM). A few other analogues (Ala^4, Ala^6, Ala^9, Ala^{13} and Tyr^3) retained some of the binding affinity (IC$_{50}$ values 429-2022 nM).

Incorporation of nonpeptide β-strand peptidomimetic based on the 3,5,5-pyrrolin-4-one scaffold in the influenza virus hemagglutinin peptide fragment Pro-Lys-Tyr-Val-Lys-Gln-Asn-Thr-Leu-Lys-Leu-Ala-Thr in place of the middle tetrapeptide Val-Lys-Gln-Asn led to compound **84**. Affinity binding experiments revealed that pyrrolinone-peptide hybrid **84** was a ligand for HLA-DR1, having an IC$_{50}$ of 137 nM, compared with 89 nM for the HA306-318 peptide and 176 nM for the control peptide (Pro-Lys-Tyr-Gly-Leu-Leu-Leu-Thr-Leu-Lys-Leu-Ala-Thr).[256]

(84)

4.15 Neuropeptide Y (NPY) Analogues – In addition to other activities associated with neuropeptide Y, the peptide has been involved in food intake and it is this aspect which has been investigated in more detail recently using various antagonists of neuropeptide Y. Although the role of individual receptor subtypes in the feeding behaviour has not been fully established, two of the receptor subtypes (Y$_1$ and/or Y$_5$) were claimed to be more closely associated with the food intake.[257-259] Several antagonists of NPY including 1229U91 [(Ile-Glu-Pro-Dpr-Tyr-Arg-Leu-Arg-Tyr-NH$_2$)$_2$], BIBP3226 [(N^2-(diphenylacetyl)-N-[(4-hydroxyphenyl)methyl]-D-arginine amide], BIBO3304 **(85)** (IC$_{50}$ values 0.38 and 0.72 nM, respectively for human and rat Y$_1$ receptors and >1000 nM for the human Y$_2$, human and rat Y$_4$ and Y$_5$

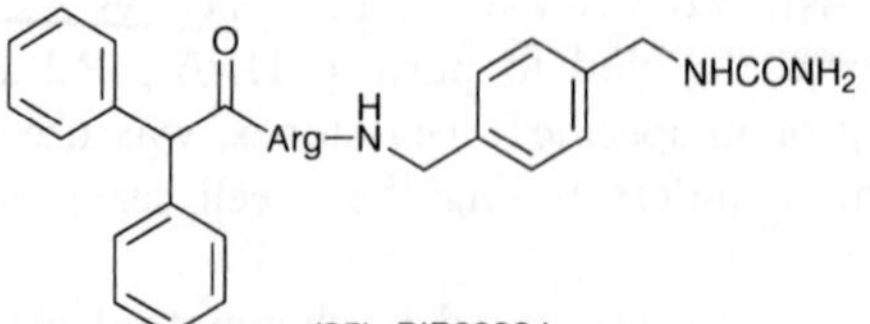

(85) BIB03304

receptors) inhibited food intake in various animal models after icv administration.[260] [262] Repeated central administration of Y_5 antisense oligodeoxynucleotides also decreased spontaneous food intake and subsequently resulted in a significant weight loss.[263]

Based on a series of indole-based NPY antagonists and dual CCK_B and histamine H_2 receptor antagonist reported previously, new NPY antagonists (**86, 87**) were synthesised.[264,265] One of the more potent benzimidazole derivatives **86** and other similar analogues were antagonists at the Y_1 receptor and showed <30% inhibition of the binding of peptide YY at human Y_2, Y_4 and Y_5 receptors at a concentration of 1 µM. Similarly, compound **87** competitively inhibited [I^{125}] peptide YY binding to Y_1 receptors in human neuroblastoma SK-N-MC cells (K_i of 6.4 nM), while it had no effect on [I^{125}]PYY binding to Y_2 or Y_5 receptors even at 1 µM.

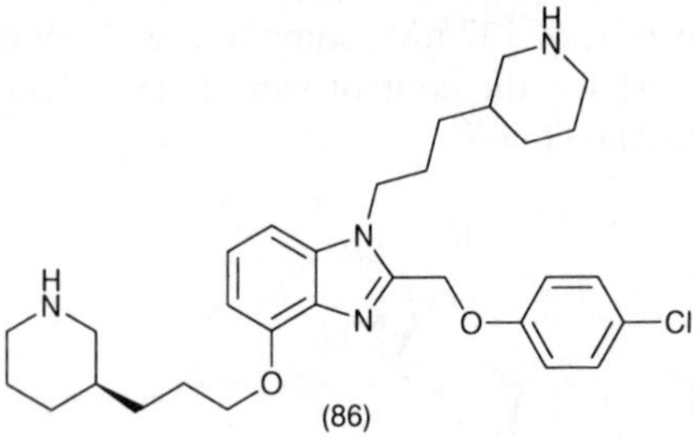

(86)

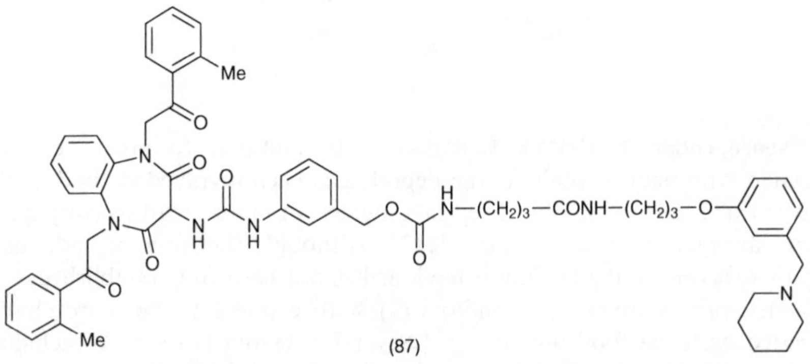

(87)

4.16 Opioid (Enkephalin, β-Casomorphin, Morphiceptin, Deltorphin and Dynorphin) Peptides – Reviews on new δ-opioid antagonists, orphanin FQ/

nociceptin [Phe-Gly-Gly-Phe-Thr-Gly-Ala-Arg-Lys-Ser-Ala-Arg-Lys-Leu-Ala-Asn-Gln] and neuropeptide FF [Phe-Leu-Phe-Gln-Pro-Gln-Arg-Phe-NH$_2$] and their role in pain and analgesia have appeared.[266-268] A new peptide, Tyr-Ile-Phe-His-Leu-Met-Asp-NH$_2$ [Ile^2 analogue of Met-deltorphin], was identified from cDNA encoding a precursor of dermorphin from the skin of *Pachymedusa dacnicolor* and, based on some earlier work, the second residue was predicted to be D-allo-Ile.[269] Analogues of Tyr-D-allo-Ile-Phe-His-Leu-Met-Asp-NH$_2$ containing D-amino acids in position 2 were investigated for δ agonist activity. The K_i values for the D-Met^2, D-aIle^2, D-Ile^2, D-Nle^2 and D-Val^2 analogues at the δ receptor were 9.6, 54, 24. 3.8 and 34 nM, respectively, and at the μ receptor 1630, 452, 1021, 314 and 293 nM, respectively. A large-scale synthesis of Met-enkephalin (150 g) using a solution phase segment condensation [1+(2+2)] method has been reported.[270] Conformations of two cyclic enkephalin analogues, dansyl-c[D-Dab-Gly-Trp-Leu] and dansyl-c[D-Dab-Gly-Trp-D-Leu], and relationship between solid-state X-ray structure and opioid peptide activity of various opiate peptides have been reported.[271,272]

A number of new enkephalin linear and cyclic peptide analogues have been reported.[273-281] Leu-enkephalin analogues containing a 1,5-disubstituted tetrazole ring, Tyr-D-Ala-Gly-Pheψ[CN$_4$]-Leu-NH$_2$, Tyr-D-Ala-Gly-Pheψ[CN$_4$]-Leu, Tyr-D-Ala-Gly-Phe-Leuψ[CN$_4$]-CH$_3$, Tyr-Gly-Glyψ[CN$_4$]-Phe-Leu-NH$_2$ and Tyr-Gly-Glyψ[CN$_4$]-Phe-Leu, were weakly active (IC$_{50}$s 9-1351 nM) at μ and δ receptors. Tyr-Gly([CN$_4$]-Gly-Phe-Leu, was inactive in binding assays.[273] Leu-enkephalin analogue **88** containing substituted serine instead of Gly^2 was a potent agonist at δ-opioid receptors (IC$_{50}$s 0.25, 0.01 and 6310 nM at μ, δ and κ receptors, respectively), and was 10 times more potent than Leu-enkephalin.[274] The corresponding 5-membered aminocyclopentane derivative was less potent at all the receptor subtypes. [Aib^2]-, [(S)-α-MeSer^2]- and [(R)-α-MeSer^2]-enkephalin were also less potent than **88**. Branched analogues of Tyr-D-Orn-Gly-Phe containing Pro, Leu, Asn or Met residues linked to the side-chain amino group of D-Orn were compared to those of the linear peptides containing these amino acids in position 5. Tyr-D-Orn(Pro)-Gly-Phe and Tyr-D-Orn(Pro)-Gly-Phe-NH$_2$ were more potent than Tyr-D-Orn-Gly-Phe-Pro in guinea pig ileum and mouse vas deferens preparations. Similar results were seen in the case of Tyr-D-Orn(Leu)-Gly-Phe, Tyr-D-Orn(Leu)-Gly-Phe-NH$_2$, Tyr-D-Orn(Asn)-Gly-Phe, Tyr-D-Orn(Asn)-Gly-Phe-NH$_2$ and Tyr-D-Orn(Met)-Gly-Phe analogues.[275] Several analogues of Leu-enkephalin (*e.g.* **89**, **90**) N-alkylated at either the Gly^2 or Gly^2, Gly^3 residues with a 6-deoxy-D-galactose moiety were synthesised by employing N-glycated glycine as the building block. The relative populations of the *cis* and *trans* isomers in these compounds were estimated by NMR.[276]

A cyclic peptide enkephalin analogue (**91**) was nearly equipotent at μ and δ receptor subtypes and about 150-fold less potent at the κ receptor subtypes. The peptide showed analgesic activity in an *in vivo* thermal escape assay (ED$_{50}$ 0.027 μg) in the rat when administered intrathecally.[277] Analogues of δ- and μ-selective cyclic peptides, Tyr-c[D-Cys-Phe-D-Pen] (K_i μ 51.5 nM and K_i δ

0.74 nM) and Tyr-c[D-Cysα-CH$_2$-CH$_2$-)Phe-D-Pen]-NH$_2$ (containing an S-CH$_2$-CH$_2$-S ring) (K_i μ 0.29 nM and K_i δ 24.8 nM), containing Phe in place of Tyr[1] were 5-50-fold less potent than the parent peptides. The K_i values for Phe-c[D-Cys-Phe-D-Pen] were 610 and 42.4 nM and for Phe-c[D-Cysα-CH$_2$-CH$_2$-)-Phe-D-Pen]-NH$_2$ 1.36 nM and 1020 nM, respectively, at μ and δ receptor subtypes.[278] Several p-substituted Phe[4] analogues of the δ$_1$-selective antagonist c[D-pen[2], Ala[3], D-Pen[5]]-enkephalin were evaluated for their brain binding and *in vitro* pharmacological effects. [Phe(p-F)[4]]-, [Phe(p-Cl)[4]]-, [Phe(p-Br)[4]]-, [Phe(p-I)[4]]- and [Phe(p-NO$_2$)[4]]- analogues of c[D-pen[2], Ala[3], D-Pen[5]]-enkephalin were all more potent at the δ-receptors (mouse vas deferens IC$_{50}$s 3.0-11 nM).[279] In comparison the IC$_{50}$s in the α-receptor preparation (guinea pig ileum) were 4600-54000 nM. The [Phe(p-Cl)[4]]- and [Phe(p-Br)[4]]-analogues were the most δ-selective (14000-fold). In the rat brain binding assay the [Phe(p-F)[4]]- analogue shows the highest affinity for the brain δ-receptors (IC$_{50}$ 0.55 nM). The other three analogues [Phe(p-Cl)[4]]-, [Phe(p-Br)[4]]- and [Phe(p-NO$_2$)[4]]- of c[D-pen[2], Ala[3], D-Pen[5]]-enkephalin all show IC$_{50}$ values similar to that of c[D-pen[2], D-Pen[5]]-enkephalin (4.5 nM). One of the analogues of [D-Pen[2], D-Pen[5]]enkephalin containing 4,4-difluoro-2-aminobutyric acid (Dfa), [D-Pen[2], D-Dfa[3], D-Pen[5]]enkephalin, was a more potent δ agonist (IC$_{50}$s >10,000 and 17.9 nM, respectively, in the guinea pig ileum and mouse vas deference) than the corresponding L-Dfa analogue (IC$_{50}$s >10,000 and 155 nM, respectively, in the μ and δ receptor preparations).[280]

Piperazine, piperazinone and pyrazinone-based opiate ligands (**92-96**) have been reported.[281-283] One of the compounds (**93**), based on a similar compound **94** (IC$_{50}$ values 9.7 and 0.31 nM at μ and δ receptors, respectively) reported earlier showed considerable selectivity (IC$_{50}$ values 17000 and 8.4 nM at μ and δ receptors, respectively). The corresponding analogue with phenolic OH replaced by OMe was much less potent and selective [IC$_{50}$ values >8000 and

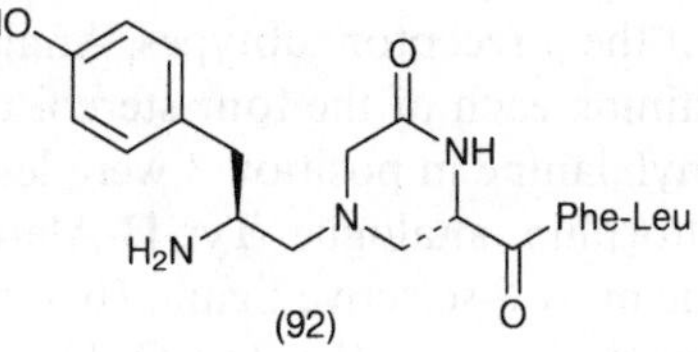

(92)

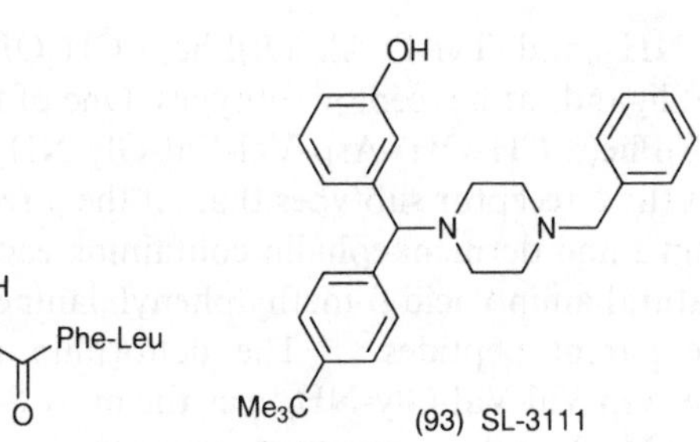

(93) SL-3111

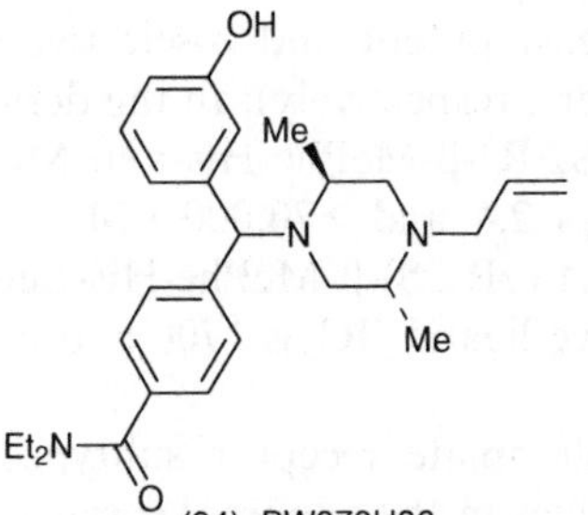

(94) BW373U86

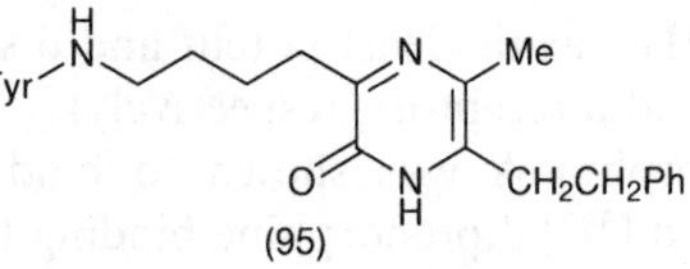

(95)

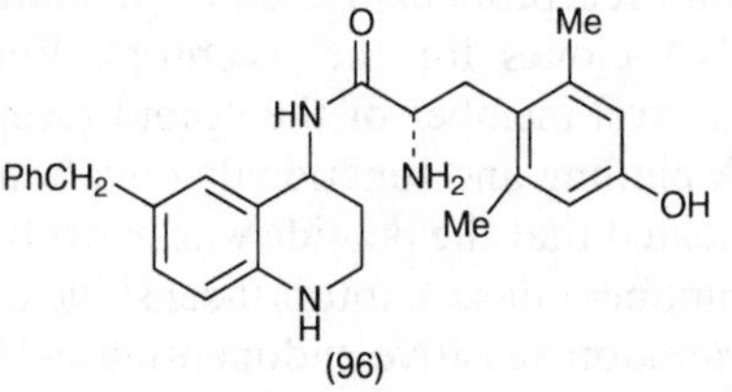

(96)

1800 nM at μ and δ receptors, respectively]. Analogue of **94** (BW373U86) containing an OMe group in place of the phenolic hydroxyl was more potent and selective [IC$_{50}$ values 2500 and 1.06 nM at μ and δ receptors, respectively] than **93**. Pyrazinone derivative **95** exhibited strong binding to the μ-receptor with a K_i value of 55.8 nM and to the δ-opioid receptor with a K_i value of 2165 nM. In comparison, the corresponding compound containing a phenyl group in place of the phenethyl group was about 2-fold more potent at the δ receptors (K_i 1092 nM) and about 5-fold less potent at the μ receptors (K_i 267 nM).[283] A non-peptidic 2′,6′-dimethyl-L-Tyr derivative (**96**) was more potent at the μ receptor subtypes.[284]

Hydrophilic analogues of enkephalin and deltorphin have been synthesised by incorporating a hydroxymethyl group into the α-position of phenylalanine or tyrosine residues of these peptides.[285] Introduction of the (R) or (S) Tyr(α-CH$_2$OH) residues in enkephalin analogues led to much less potent compounds [(S)-Tyr(α-CH$_2$OH)-D-Ala-Gly-Phe-Leu-NH$_2$, (S)-Tyr(α-CH$_2$OH)-D-Ala-Gly-Phe-Leu, (R)-Tyr(α-CH$_2$OH)-D-Ala-Gly-Phe-Leu-NH$_2$ and (R)-Tyr(α-CH$_2$OH)-D-Ala-Gly-Phe-Leu] at both δ and μ receptors. However, incorporation of (R)- or (S)-Phe(α-CH$_2$OH) residues in position 3 of enkephalin [Tyr-D-Ala-Gly-Phe-Leu-(S)(α-CH$_2$OH) and Tyr-D-Ala-Gly-Phe-Leu-(R)(α-CH$_2$OH)] and deltorphin [Tyr-D-Ala-(S)Phe(α-CH$_2$OH)-Asp-Val-

Val-Gly-NH$_2$ and Tyr-D-Ala-(R)Phe(α-CH$_2$OH)-Asp-Val-Val-Gly-NH$_2$] led
to potent ligands at δ receptor subtypes. One of the deltorphin analogues, Tyr-
D-Ala-(S)Phe(α-CH$_2$OH)-Asp-Val-Val-Gly-NH$_2$, was about 10,000-fold more
potent at the δ receptor subtypes than at the μ receptor subtypes. Analogues of
deltorphin I and dermenkephalin containing each of the four stereoisomers of
the unnatural amino acid β-methylphenylalanine in position 3 were less potent
than the parent peptides.[286] The deltorphin analogue Tyr-D-Ala-(2S,3R)-
β-MePhe-Asp-Val-Val-Gly-NH$_2$ was the most δ-selective ligand (IC$_{50}$s 2.5 and
>72,000 nM, δ and μ receptors, respectively) and the Tyr-D-Ala-(2R,3R)-
β-MePhe-Asp-Val-Val-Gly-NH$_2$ was the least potent and δ-selective ligand
(IC$_{50}$s 530 and >20,000 nM, δ and μ receptors, respectively). In the dermenke-
phalin series of compounds, Tyr-D-Met-(2S,3R)-β-MePhe-His-Leu-Met-Asp-
NH$_2$ was the most δ-selective ligand (IC$_{50}$s 2.4 and >70,000 nM, δ and μ
receptors, respectively) and the Tyr-D-Met-(2R,3S)-β-MePhe-His-Leu-Met-
Asp-NH$_2$ was the least potent and δ-selective (IC$_{50}$s 1700 and >50,000
nM, δ and μ receptors, respectively).

Dynorphin A was shown to bind to all opiate receptor subtypes.[287] It
displaced [^{3}H]-diprenorphine binding (K_i values in the nanomolar range) from
μ, δ and κ human opioid receptors expressed on mammalian cells transfected
with each of the cDNA clones for the receptors. Furthermore, an opioid
receptor-like receptor (novel member of the opioid receptor gene family) also
displayed dynorphin A binding and functional activation. Degradation studies
on dynorphin(1-8) indicated that the peptide was hydrolysed by three enzymes,
amastatin-sensitive aminopeptidase, captopril-sensitive dipeptidyl carboxypeptidase I and phosphoramidon-sensitive endopeptidase-24.11, in both ileal and
striatal membranes.[288] Thus, endogenously released dynorphin(1-8) may act
either through dynorphin(1-8) itself on κ receptors or through Tyr-Gly-Gly-
Phe-Leu-Arg on μ or δ receptors depending on both the peptidase activities
and the receptor type densities at the target synaptic membrane. Reports on
the conformational analysis of endomorphin-1 (Tyr-Pro-Trp-Phe-NH$_2$), (μ
receptor agonist), performed using multidimensional NMR and molecular
modelling techniques, and pharmacological characterisation of endomorphin-
1 and -2 in mouse brain have appeared.[289,290] Both endomorphins competed
for μ_1 and μ_2 receptor sites and neither showed affinity for δ or κ_1 receptors.
However, the two endomorphins displayed reasonable affinities for κ_3 binding
sites, with K_i values 20-30 nM. Endomorphin analgesia was blocked by
naloxone, as well as the μ-selective antagonist β-funaltrexamine and naloxona-
zine. Finally, the endomorphins inhibited gastrointestinal transit.

Several analogues of dermorphin (H-Tyr-D-Ala-Phe-Gly-Tyr-Pro-Ser-NH$_2$)
have been reported.[291-294] O-Glycopeptide analogues, [Tyr(β-D-Glc)^5]- and
[Ser(β-D-Glc)^5, Tyr^7)]-dermorphin, exhibited analgesic (80–90% of morphine)
and antidiarrhoeal activity (50% of dermorphin). Analogues containing
3-aminoTyr, 3-nitroTyr, 4-aminoPhe, 3-(uracilyl-1)alanine, 3-(thyminyl-1)
alanine and 3-(6-methyluracilyl-1)alanine residues in place of Tyr^1 showed low
binding affinity with respect to μ- and δ-receptors (IC$_{50}$s >750 nM) and
exhibited low antinociceptive activity *in vivo* in the tail-pinch assay (1% of

dermorphin). Only one of the analogues containing 3-(thyminyl-1)alanine residue in position 1 demonstrated about 10% of the dermorphin activity in different tests when administered intracisternally in mice. Additional analogues of the N-terminal tetrapeptide analogue of dermorphin (Tyr-D-Arg-Phe-βAla-NH$_2$) were prepared by replacing the aromatic residues at positions 1 or/and 3 by unnatural or constrained amino acids.[294] In comparison to the parent peptide [IC$_{50}$s 4.3 and >10,000 nM at μ and δ receptors, respectively], most of the analogues, Tyr-D-Arg-Phe-βAla, Tyr-D-Arg-Cha-βAla-NH$_2$, Htc-D-Arg-Cha-βAla-NH$_2$, MeTyr-D-Arg-Cha-βAla-NH$_2$, Tyr-D-Arg-Aic-βAla-NH$_2$, Tyr-D-Arg-Tic-βAla-NH$_2$ and Tyr-D-Arg-MePhe-βAla-NH$_2$, were either less potent or less selective for the μ receptor subtype. Compounds like Tyr-D-Arg-Cha-βAla-NH$_2$, MeTyr-D-Arg-Cha-βAla-NH$_2$, and Tyr-D-Arg-Aic-βAla-NH$_2$ which retained activity at the μ receptors (IC$_{50}$s 2.5–9.7 nM) also showed enhanced potency at the δ receptors (IC$_{50}$ values 54, 400 and 4100 nM, respectively).

The neuropeptide nociceptin [Phe-Gly-Gly-Phe-Thr-Gly-Ala-Arg-Lys-Ser-Ala-Arg-Lys-Leu-Ala-Asn-Gln] (also known as orphanin FQ) and nocistatin (Thr-Glu-Pro-Gly-Leu-Glu-Glu-Val-Gly-Glu-Ile-Glu-Gln-Lys-Gln-Leu-Gln) are processed from the same 176 amino acids precursor prepronociceptin.[295] Nociceptin, an endogenous ligand for the orphan opioid-like receptor, induces both hyperalgesia and allodynia (pain induced by innocuous tactile stimuli) when administered intrathecally and nocistatin blocks nociceptin induced hyperalgesia and allodynia, and attenuates pain evoked by prostaglandin E$_2$. Nocistatin and the C-terminal(10-17) fragment (Glu-Ile-Glu-Gln-Lys-Gln-Leu-Gln) were the most potent peptides in inhibiting nociceptin-induced allodynia (ED$_{50}$s 0.715 and 0.125 pg per mouse). Further truncation to hexapeptide Glu-Gln-Lys-Gln-Leu-Gln significantly reduced the potency of the peptide (ED$_{50}$ 14.2 pg). The pentapeptides Gln-Lys-Gln-Leu-Gln and Glu-Gln-Lys-Gln-Leu were not active at a dose of 500 pg per mouse. A pseudopeptide analogue of nociceptin, [Phe$^1\psi$(CH$_2$NH)Gly^2]nociceptin-(1-13)-NH$_2$, which acts as an antagonist in peripheral assays showed agonist activity after spinal administration.[296,297] The noxious evoked activity of the neurones was inhibited by [Phe$^1\psi$(CH$_2$NH)Gly^2]nociceptin(1-13)-NH$_2$, which was as potent as nociceptin itself. The pseudopeptide also inhibited morphine analgesia *in vivo*, an effect parallel to that of nociceptin. The anti-opioid actions of nociceptin are not blocked by [Phe$^1\psi$(CH$_2$NH)Gly^2]-nociceptin-(1-13)-NH$_2$.

New analogues of a δ-opioid antagonist H-Tyr-Tic-Phe-Phe were reported.[298,299] Several β-methyl amino acid containing analogues were prepared. [(2S,3R)-β-MePhe^3]-Analogue (K_i values >10,000 and 0.38 nM at μ and δ receptors, respectively; guinea pig ileum IC$_{50}$ >10,000 nM, mouse vas deferens Ke 0.192 nM) was among the most potent δ-antagonists. In the D-β-methyl amino acid series, the [D-β-MeTic^2] analogues were δ-selective antagonists whereas [D-Tic^2]-NH$_2$ analogue was a δ-agonist. Analogues displaying partial agonism included (2S,3R)-β-MeTyr-Tic-Phe-Phe-NH$_2$ and Tyr-Tic-Phe-(2R,3R)-β-MePhe-NH$_2$. Three mixed μ agonist/δ antagonists

[Dmt-Ticψ[CH$_2$NH]Phe-Phe-NH$_2$,   Dmt-c(D-Orn-D-Nal-D-Pro-Gly)   and
Dmt-Tic-NH(CH$_2$)$_3$Ph] were also discovered.

A combinatorial library (6,250,000 tetrapeptides) approach has been used to
identify opiate μ, δ, and κ receptor ligands.[300] Examples of compounds active
at the μ receptor included Tyr-D-Nva-Gly-Nal-NH$_2$, Tyr-D-Nva-Gly-Trp-
NH$_2$, Tyr-D-Tyr-Ala-Nal-NH$_2$, Tyr-D-Nva-Phe-Trp-NH$_2$, Tyr-D-Nva-Ala-
Nal-NH$_2$, Tyr-D-Tyr-Gly-Trp-NH$_2$, Tyr-D-Tyr-Ala-Trp-NH$_2$, Tyr-D-Arg-
Ala-Nal-NH$_2$, Tyr-D-Nle-Gly-Nal-NH$_2$, Tyr-D-Nva-Phe-Nal-NH$_2$, Tyr-D-
Nva-Ala-Trp-NH$_2$, Tyr-D-Nle-Ala-Nal-NH$_2$, Tyr-D-Arg-Ala-Trp-NH$_2$, Tyr-
D-Nva-Trp-Trp-NH$_2$, Tyr-D-Nle-Ala-Trp-NH$_2$ and Tyr-D-Nle-Phe-Trp-NH$_2$
(K_i values 0.4–2.7 nM). Examples of compounds active at the δ receptor
included Tyr-D-Tyr-Gly-Trp-NH$_2$, Trp-D-Tyr-Abu-Arg-NH$_2$, Trp-D-Tyr-
Nva-Arg-NH$_2$, Trp-D-Tyr-Met-Arg-NH$_2$, Tyr-D-Nva-Gly-Trp-NH$_2$, Tyr-D-
Tyr-Nva-Arg-NH$_2$, Tyr-D-Trp-Gly-Trp-NH$_2$, Tyr-D-Nva-Cha-Arg-NH$_2$,
Trp-D-Tyr-Cha-Arg-NH$_2$ and Tyr-D-Tyr-Cha-Trp-NH$_2$, (K_i values 3–120
nM). Some of the δ receptor active compounds were up to 10-fold more potent
at the δ receptor than at the μ receptor. Examples of compounds active at the κ
receptor included D-Phe-D-Phe-D-Nle-D-Arg-NH$_2$, D-Nle-D-Nal-D-Ile-D-
Arg-NH$_2$, D-Nle-D-Nal-D-Nle-D-Arg-NH$_2$, D-Phe-D-Phe-D-Ile-D-Arg-NH$_2$,
D-Phe-D-Nal-D-Ile-D-Arg-NH$_2$, D-Nle-D-Phe-D-Nle-D-Arg-NH$_2$, D-Phe-D-
Nal-D-Nle-D-Arg-NH$_2$, D-Nle-D-Phe-D-Ile-D-Arg-NH$_2$, D-Phe-D-Phe-D-
Nle-D-Cha-NH$_2$, D-Nle-D-Nal-D-Nle-D-Cha-NH$_2$, D-Phe-D-Nal-D-Nle-D-
Cha-NH$_2$, D-Phe-D-Nal-D-Ile-D-Cha-NH$_2$, D-Nle-D-Nal-D-Ile-D-Cha-NH$_2$
and D-Phe-D-Phe-D-Ile-D-Cha-NH$_2$ (K_i values 1.2-64 nM).

4.17 Somatostatin Analogues – The role of somatostatin and its receptors in
various forms of cancer, gastrointestinal tract, restenosis after coronary
interventions and pathophysiology of rheumatoid arthritis has been published.[301-309] Cytotoxic hybrid analogues (*e.g.* **97**) of octapeptides [D-Phe-

CH$_2$OCO(CH$_2$)$_3$CO-D-Phe-c(Cys-Tyr-D-Trp-Lys-Val-Cys)-Thr-NH$_2$

(97)

c(Cys-Tyr-D-Trp-Lys-Val-Cys)-Trp-NH$_2$] and [D-Phe-c(Cys-Tyr-D-Trp-Lys-
Val-Cys)-Thr-NH$_2$] were prepared by linking the peptides to doxorubicin or
pyrrolinodoxorubicin. In the binding assay, the hybrid peptides were >10-fold
less potent than the parent peptide. The hybrid peptides were active in
inhibiting proliferative activity of various tumour cell lines and also demonstrated activity in animal models of breast and prostate cancers.[310,311]

Various cyclic peptide analogues of somatostatin have been reported. A backbone-cyclic somatostatin analogue (**98**) showed high selectivity to the subtype 5 receptor (IC_{50} 67 nM). The IC_{50} values against $sstr_{1\text{-}4}$ were >1000 nM. Peptide **98** was stable against enzymes present in rat renal homogenate and human serum.[312] The cyclic peptide inhibited bombesin-and caerulein-induced amylase and lipase release from the pancreas without inhibiting growth hormone or glucagon release. In comparison, sandostatin ($sstr_2$, -3 and -5 ligand) was a potent inhibitor of glucagon and growth hormone release. Binding studies on a α-benzyl-o-aminomethylphenylacetic acid linked cyclic peptide **99** using SST_2 receptor showed the (*S*) isomer to be six times more

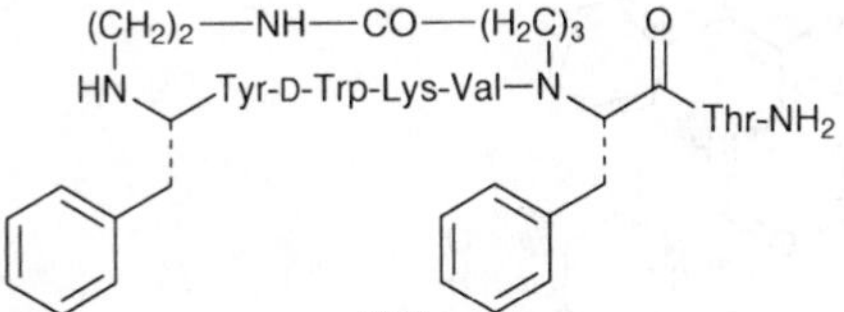

(CH2)2—NH—CO—(H2C)3
HN
Tyr-D-Trp-Lys-Val—N
O
Thr-NH2

(98)

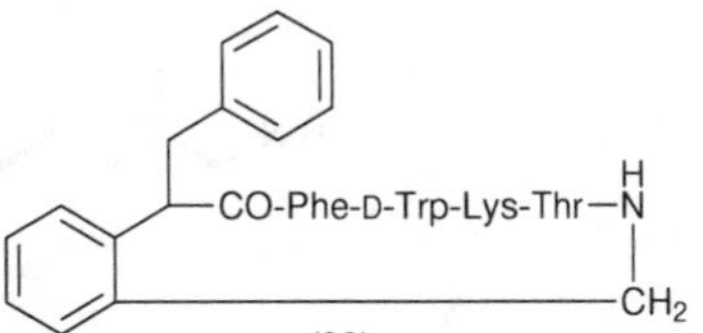

CO-Phe-D-Trp-Lys-Thr—N
H
CH2

(99)

potent (K_i 5.1 nM) than the (*R*) isomer.[313] N- to C-terminal cyclic peptides containing an N-benzylglycine (**100**, R = H) residue are reported.[314,315] The peptide **100** (K_i hsst1 >1000, $hsst_2$ 6.98, $hsst_3$ 253, $hsst_4$ >1000, $hsst_5$ 100 nM) and the βMe-analogues (**100**, R = Me) were more selective agonists at the $hsst_2$ receptor subtype. *In vivo*, the analogues (100 μg kg^{-1}) selectively inhibited growth hormone release in rats but had no effect on the inhibition of insulin at the same dose. N-Terminally modified cyclic peptide analogues like **101** were stable to degradation in rats after subcutaneous injection.[316]

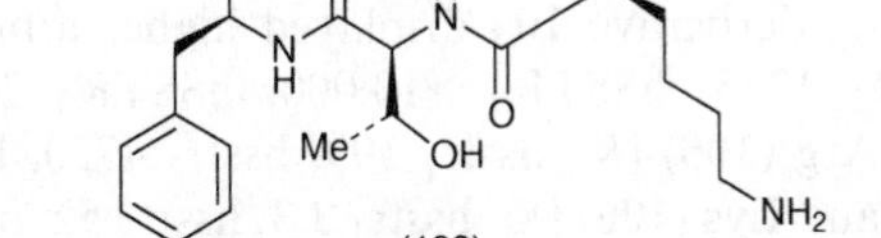

R
N
O
NH
Me
OH
NH2

(100)

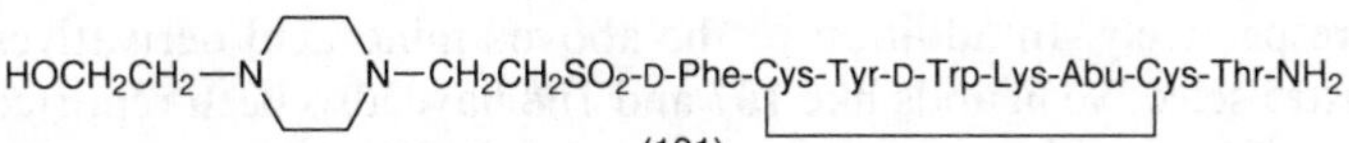

HOCH2CH2—N
N—CH2CH2SO2-D-Phe-Cys-Tyr-D-Trp-Lys-Abu-Cys-Thr-NH2

(101)

Several non-peptide, receptor selective agonist analogues of somatostatin have been reported. Several amino acid or dipeptide derivatives like **102-108** were obtained by combinatorial screening approaches.[317-319] The Nal derivative **102** was selective for the $hsstr_1$ subtype [K_i, $sstr_1$ 1.4, $sstr_2$ 1875, $sstr_3$ 2240,

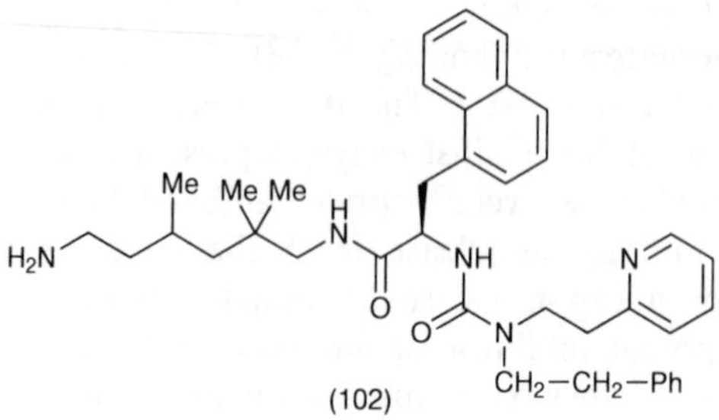

(102)

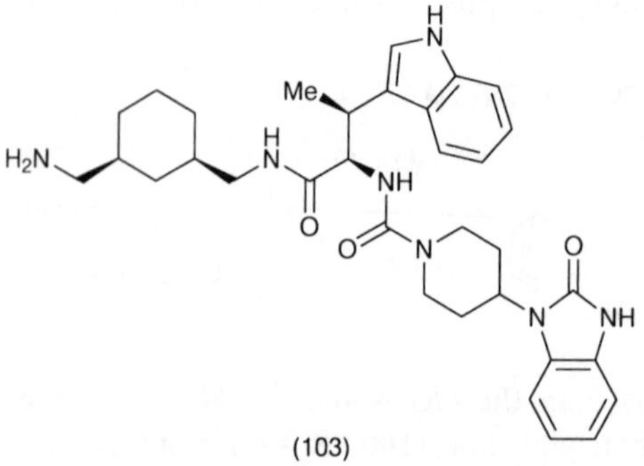

(103)

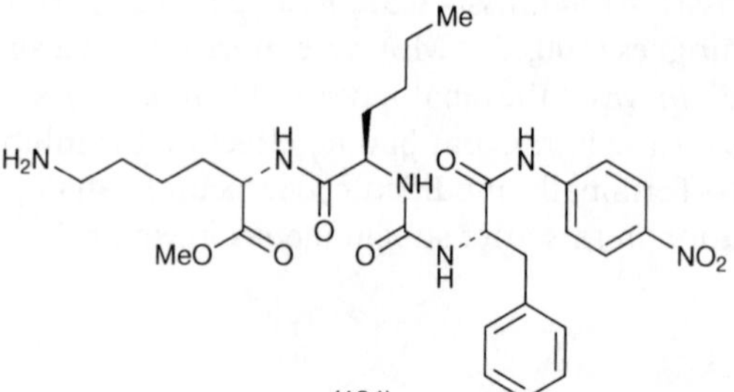

(104)

sstr$_4$ 170, sstr$_5$ 3600 nM] and the Trp derivative **103** was selective for the hsstr$_2$ subtype [K_i, hsstr$_1$ 2760, hsstr$_2$ 0.05, hsstr$_3$ 729, hsstr$_4$ 310, hsstr$_5$ 4260 nM]. The Nle-Lys dipeptide derivative **104** displayed higher affinity for the hsstr$_3$ subtype [K_i, hSSTR$_1$ 1255, hSSTR$_2$ >10,000, hSSTR$_3$ 24, hSSTR$_4$ 8650, hSSTR$_5$ 1200]. The Arg (**105**) [K_i, hsstr$_1$ 199, hsstr$_2$ 4720, hsstr$_3$ 1280, hsstr$_4$ 0.7, hsstr$_5$ 3880 nM] and Lys (**106**) [K_i, hsstr$_1$ 3.3, hsstr$_2$ 52, hsstr$_3$ 64, hsstr$_4$ 82, hsstr$_5$ 0.4 nM] derivatives were more selective for the hsstr$_4$ and hsstr$_5$ receptor subtypes, respectively. In addition to the above amino acid derivatives, other sstr4 and sstr5 selective ligands like **107** and **108** have also been reported.[320-322] The K_i values for the thiourea **107** derivative (sstr$_2$ 621 and sstr$_4$ 6 nM) and the corresponding urea derivative (sstr$_2$ 4200 nM and sstr$_4$ 14 nM) showed at least 100-fold selectivity for the sstr$_4$ subtype against the sstr$_2$ subtype.

Several antagonists of somatostatin have been reported.[323-325] An all D-amino acid hexapeptide antagonist Ac-D-His-D-Phe-D-Ile-D-Arg-D-Trp-D-Phe-NH$_2$ was obtained by screening a synthetic hexapeptide library. This

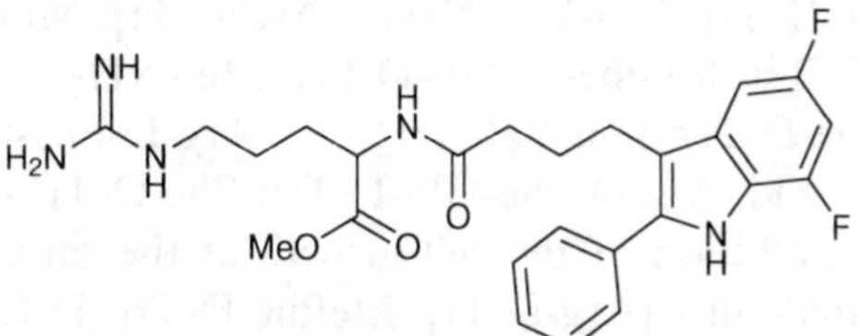

(105)

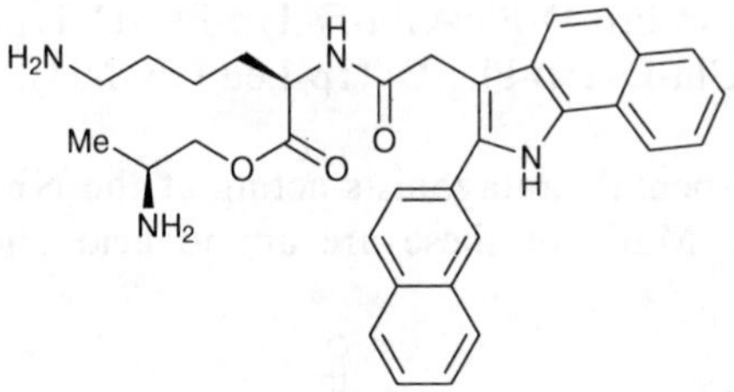

(106)

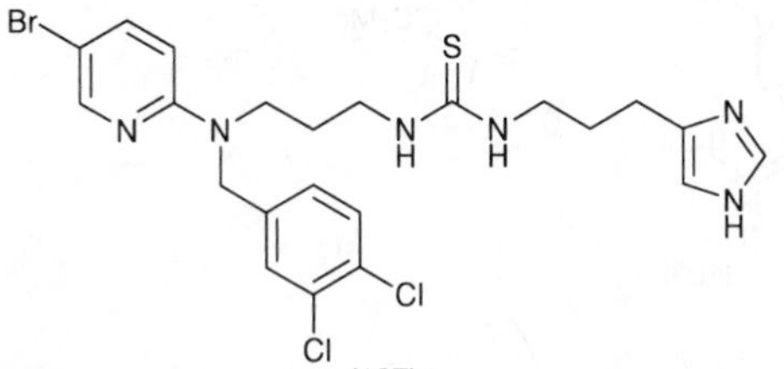

(107)

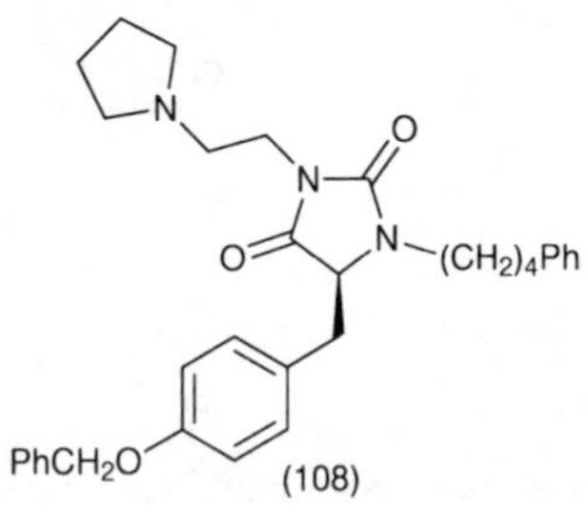

(108)

D-hexapeptide bound somatostatin receptor type 2 (K_i 172 nM), blocked somatostatin inhibition of adenylate cyclase *in vitro* (IC_{50} 5.1 μM), and induced growth hormone release when given alone (50 μg iv) to anaesthetised rats with or without pre-treatment with a long-acting SRIF agonist. A cyclic peptide antagonist Nal-c[D-Cys-Pal-D-Trp-Lys-Val-Cys]-Nal-NH_2 was selective for hsstr$_2$ with an affinity of 75 nM.

4.18 Tachykinin (Substance P and Neurokinins) Analogues – Residues in transmembrane domains believed to be part of the presumed main ligand-binding pocket of the NK$_1$ receptor were probed by alanine substitution and introduction of residues with larger and/or chemically distinct side chains. Binding affinity measurements for peptide agonists (substance P, substance P-OMe, eledoisin and neurokinin A) and several non-peptide antagonists (*e.g.* CP96,345, CP99,994, RP67,580, RPR100,893, and CAM4092, LY303,870, FK888 and SR140,333) indicated that the peptides and certain non-peptides do not use this pocket as part of their binding site in the NK$_1$ receptor.[326]

Analogues of substance P were investigated for their effects on the growth of small cell lung cancer cells.[327,328] The most potent analogues D-Pro-Lys-Pro-D-Trp-Gln-D-Trp-Phe-D-Trp-Leu-Leu-NH_2 and D-Pro-Lys-Pro-D-Phe-Gln-D-Trp-Phe-D-Trp-Leu-Leu-NH_2 inhibited growth of H-69 SCLC cell line (92 and 61% respectively) at a concentration of 25 μM. Many other analogues,

Arg-D-Trp-MePhe-D-Trp-Leu-Met-NH$_2$,   Arg-D-Trp-MePhe-D-Trp-Leu-D-Met-NH$_2$,   Arg-D-Trp-MePhe-D-Trp-D-Leu-Met-NH$_2$,   D-Pro-Lys-Pro-D-Phe-Gln-D-Trp-Phe-D-Trp-Leu-Val-NH$_2$,   Arg-D-Phe-Gln-D-Trp-Phe-D-Trp-Leu-Leu-NH$_2$ and Arg-D-Phe-Gln-D-Trp-Phe-D-Trp-Leu-Gly-NH$_2$, resulted in 30–40% inhibition of the cell growth at the same concentration. A number of other analogues [Arg-D-Trp-MePhe-D-Trp-D-Leu-Gly-NH$_2$, Arg-D-Trp-MePhe-D-Trp-Leu,   D-Pro-Lys-Pro-D-Phe-Gln-D-Trp-Phe-D-Trp-Leu-Gly-NH$_2$, Ac-Lys-Pro-D-Phe-Gln-D-Trp-Phe-D-Trp-Leu-Leu-NH$_2$ and Ac-Lys-Pro-D-Phe-Gln-D-Trp-Phe-D-Trp-Leu-Gly-NH$_2$] gave less than 20% inhibition at 25 µM.

A number of non-peptide antagonists acting at the NK$_1$ or NK$_2$ receptors have been reported. Many of these are amino acid and dipeptide deriva-

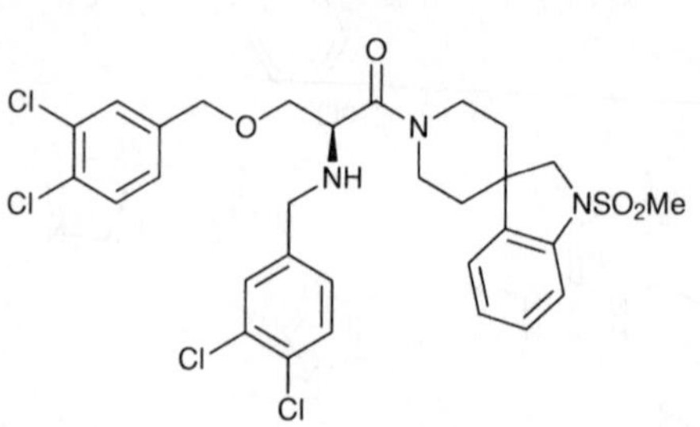

(109)

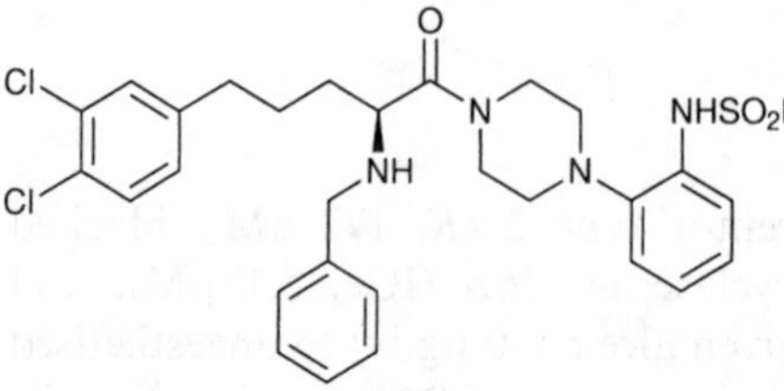

(110)

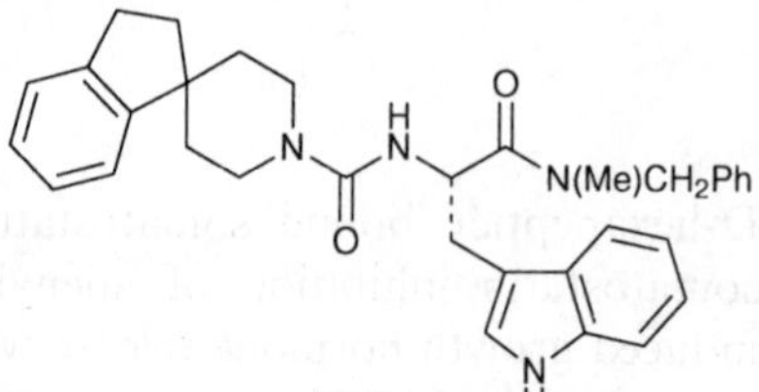

(111)

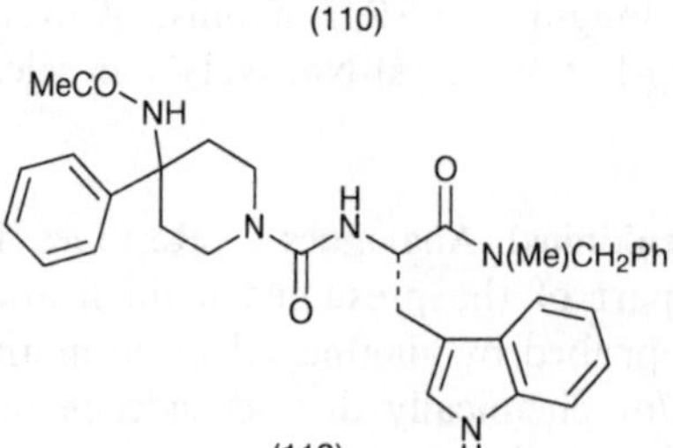

(112)

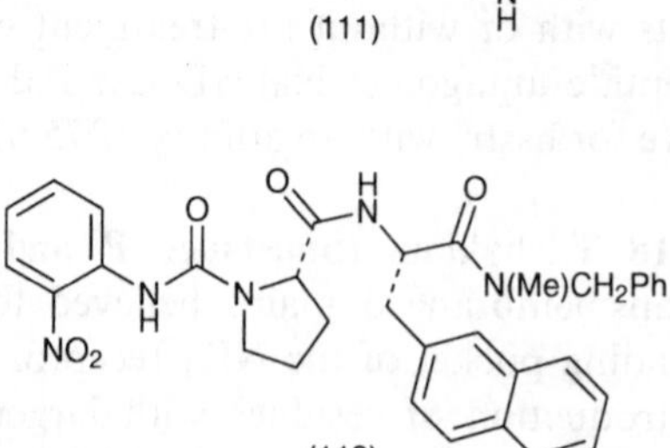

(113)

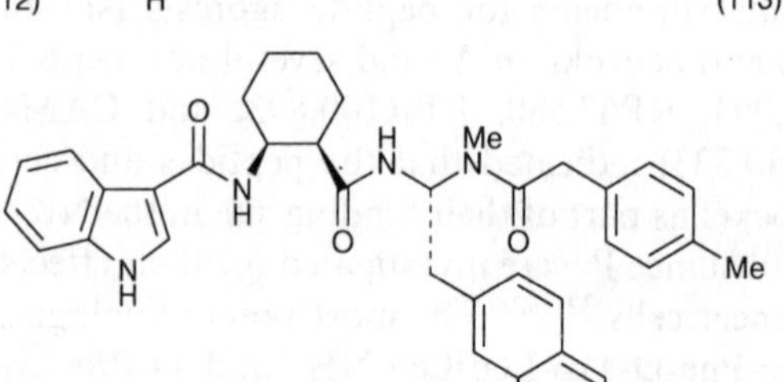

(114)

tives.[329-331] In the serine derivative **109** (hNK$_1$ IC$_{50}$ 1.7 nM), the ether linkage was not essential for activity. Compound **110** (hNK$_1$ IC$_{50}$ 4.8 nM) was similar in binding affinity to **109**. Both NK$_1$ and NK$_2$ tachykinin receptor antagonists were obtained in the case of tryptophan derivatives.[331] For example, compound **111** displayed IC$_{50}$ values of 56 and 27 nM at hNK$_1$ and hNK$_2$ receptors, respectively. A similar compound (**112**) was nearly equipotent to **111** at the NK$_1$ receptors (IC$_{50}$ 77 nM) but much less potent at the hNK$_2$ receptors (IC$_{50}$ 928 nM). The dipeptide based compounds include **113** and **114**. Both of these were NK$_1$ receptor antagonists.[332-334] Compound **113** inhibited [^{3}H]-substance P binding to the human NK$_1$ receptor (IC$_{50}$ 0.62 nM) but showed much lower affinity at rat NK$_1$ receptors (IC$_{50}$ 451 nM) and human NK$_2$ and NK$_3$ receptors (K_i 0.52 and 3.4 μM, respectively). In anaesthetised guinea-pigs i.v. administered **113** antagonised [Sar^9]SP sulfone-evoked bronchoconstriction (70% reduction at 0.4 mg kg^{-1}). In an analgesic model (Carrageenan-induced hyperalgesia in guinea pigs), compound **113** showed 68% inhibition at a dose of 30 mg kg^{-1} when administered orally.

Examples of other non-peptide tachykinin antagonists include compounds **115-119**. The morpholine derivative **115** (hNK$_1$ receptor antagonist, IC$_{50}$ 0.09 nM) was active in an inflammation model after oral administration in the guinea pig (IC$_{50}$ 0.008 and 1.8 mg kg^{-1} after 1 and 24 h, respectively).[335] Like **115**, several other compounds like **116-118** were also NK$_1$ receptor antagonists.[336-338] 4-Alkylpiperidine derivatives like **119** were NK$_2$ receptor antagonists which exhibited oral activity in a model of dyspnea in guinea pigs.[339]

(115)   (116)   (117)

(118)   (119)

4.19 Thyrotropin-releasing Hormone Analogues – A new receptor subtype (50% identical to that of rat TRH receptor gene reported previously) of thyrotropin-releasing hormone was isolated from rat brain cDNAs.[340,341] Although the binding of TRH was similar at both the receptor subtypes, the active metabolite of TRH, c(His-Pro) showed no specific binding to the new receptor. Three of the recently identified naturally occurring peptides, pGlu-Glu-Pro-NH$_2$, pGlu-Phe-Pro-NH$_2$ and pGlu-Gln-Pro-NH$_2$, (similar structures to TRH) were shown to increase levels of triiodothyronine (T3) and to a lesser extent tetraiodothyronine (T4) in the circulation of male and female mice.[342] Tetracyclic peptidomimetics of TRH like **120** were found to be partial agonists. Even at maximally effective concentrations the compounds produced 28–47% response.[343] A chemical brain-targeting system, in which a redox 1,4-dihydropyridine-pyridinium function serves as a targeting moiety and a cholesteryl ester contributes to the improved penetration across the blood-brain barrier, was synthesised for a centrally active TRH analogue, pGlu-Leu-Pro-NH$_2$. Treatment with the chemical targeting system (**121**) significantly improved memory-related behaviour in rats, without altering thyroid function.[344]

(120)

(121)

4.20 Vasopressin and Oxytocin Analogues – Three subtypes of human arginine vasopressin receptors, hV$_{1A}$, hV$_{1B}$ and hV$_2$, were stably expressed in Chinese hamster ovary cells and characterised by [^{3}H]-AVP and a number of other non-peptide antagonist binding studies.[345] Several N-terminal V$_2$ receptor truncation mutants were shown to inhibit the function and cell surface trafficking of the coexpressed full-length V$_2$ receptor.[346] Rat versus human selectivity determinants of the V$_2$ vasopressin receptor and of its peptide ligands have been identified.[347] For example, residue 2 of species-selective peptide antagonists such as d(CH$_2$)$_5$-[D-Ile^2, Ile^4, Tyr-NH$_2$^{9}]arginine vasopressin controls their rat versus human selectivity and for species-selective agonists such as desmopressin, residues 1 and 8 modulate the binding selectivity. Pharmacological analysis of mutant receptors revealed that residues 202 and 304 fully control the species selectivity of the discriminating antagonists and a third residue (position 100) is necessary for the discriminating agonists. Essential interactions between the V_2 vasopressin renal receptor and its agonists, [Arg^8]vasopressin (AVP) and [D-Arg^8]vasopressin (DAVP), and the non-peptide antagonist OPC-31260 have also been identified.[348]

Non-peptide antagonists of oxytocin and vasopressin have been reported.[349-351] The oxytocin antagonist **122** (K_i 2.0 nM on cloned human oxytocin receptor) was absorbed in rat after an oral dose of 10 mg kg^{-1}. The pyridine N-oxide analogue (prepared as a potential metabolite) was about three-fold less potent (K_i 7.3 nM). Both the compounds had short plasma half-life ($\sim$30 min).[349] The tricyclic pyrrolobenzodiazepine derivative **123** (VPA-985) was about 200-fold more selective antagonist at the V_2 vasopressin receptor (IC$_{50}$ values 230 and 1.2 nM at V_{1a} and V_2 receptors, respectively).[350] A similar compound (**124**) was nearly equipotent at both the receptor subtypes (IC$_{50}$ values 6.3 and 4.0 nM at V_{1a} and V_2 receptors, respectively). OPC-41061 (**125**) antagonised [^{3}H]AVP binding to human V_2-receptors (K_i 0.43 nM) and V_{1a}-receptors (K_i 12.3 nM) but not to human V_{1b}-receptors.[351] In rats, OPC-41061 inhibited [^{3}H]AVP binding to V_{1a}-receptors (K_i 325 nM) and V_2-receptors (K_i 1.33 nM) showing higher receptor selectivity than with human receptors. Treatment by multiple OPC-41061 dosing for 28 days (1 and 10 mg kg, p.o.) in rats resulted in significant aquaretic effects.

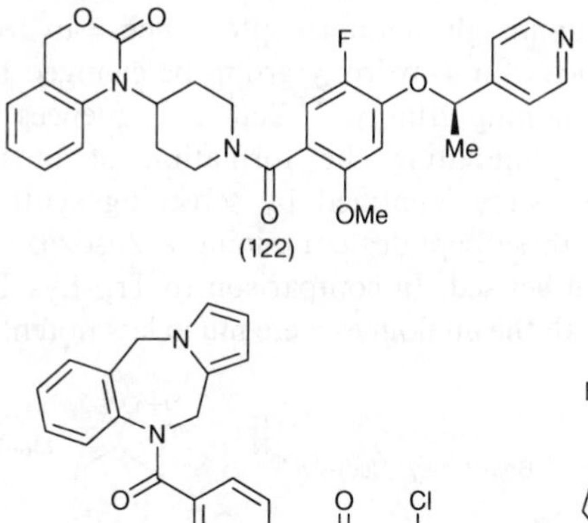

(122)

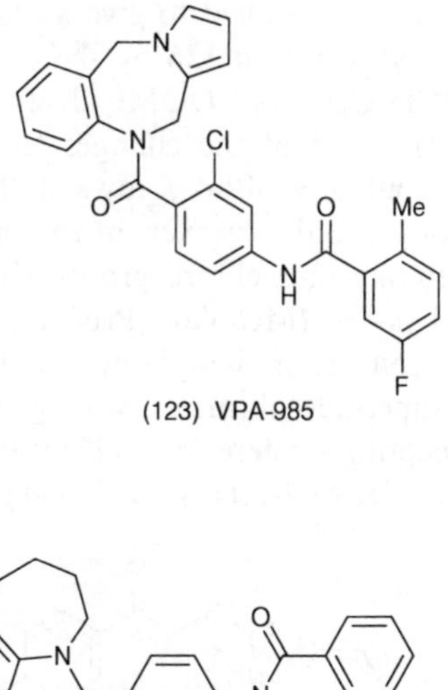

(123) VPA-985

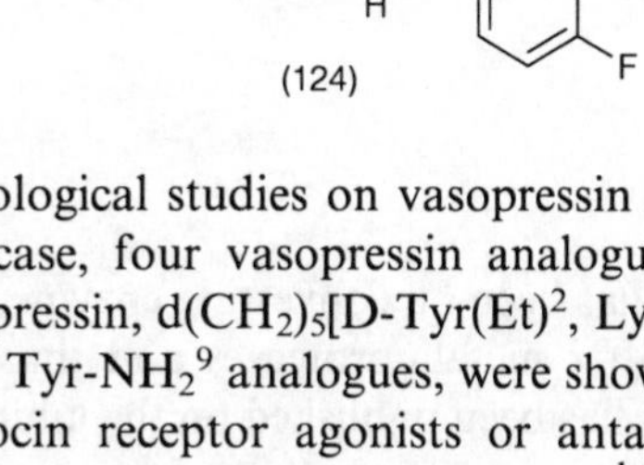

(124)

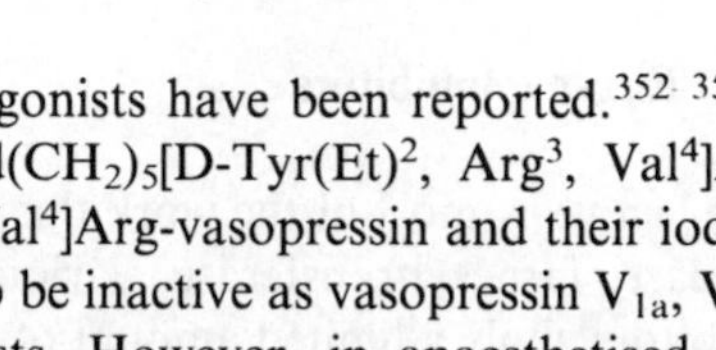

(125) OPC-41061

Biological studies on vasopressin antagonists have been reported.[352-354] In one case, four vasopressin analogues, d(CH$_2$)$_5$[D-Tyr(Et)^2, Arg^3, Val^4]Arg-vasopressin, d(CH$_2$)$_5$[D-Tyr(Et)^2, Lys^3, Val^4]Arg-vasopressin and their iodinatable Tyr-NH$_2$^{9} analogues, were shown to be inactive as vasopressin V_{1a}, V_2 or oxytocin receptor agonists or antagonists. However, in anaesthetised rats, these peptides (0.05–0.10 mg kg^{-1}) elicited a marked fall in arterial blood pressure (all four nearly equipotent). Classical V_{1a}, V_2 or oxytocin receptor antagonists did not block the vasodepressor response.[352]

4.21 Miscellaneous (Bovine Brain Peptides, Galanin, Glucagon and Immuno-modulating Peptides) – A large number of endogenous peptides were obtained by fractionating the bovine brain extracts. The peptides, formed from functional proteins (haemoglobin, myelin basic protein, cytochrome C oxidase, *etc.*) or unknown precursors, were evaluated in biological systems. A concept of 'tissue specific peptide pool' was formulated describing a system of peptidergic regulation, complementary to the conventional hormone and neuromodulatory systems.[355] Chimeric peptides linking the N-terminal galanin (GAL) fragment or its analogues to the C-terminal portion of substance P analogues or scyliorhinin I (SCY-I) analogues were synthesised. Two peptides, [cycloleucine⁴]GAL(1-13)-SP(5-11)-amide [Gly-Trp-Thr-Cle-Asn-Ser-Ala-Gly-Tyr-Leu-Leu-Gly-Pro-Gln-Gln-Phe-Phe-Gly-Leu-Met-NH₂] and GAL(1-13)-[Nle¹⁰]SCY-I(3-10)-amide [Gly-Trp-Thr-Leu-Asn-Ser-Ala-Gly-Tyr-Leu-Leu-Gly-Pro-Phe-Asp-Lys-Phe-Tyr-Gly-Leu-Nle-NH₂], inhibited the effect of galanin on the glucose-induced insulin release.[356]

Non-peptide competitive human glucagon receptor antagonist, NNC 92-1687, discovered by random screening (binding IC_{50} 20 μM, functional $K_i = 9.1$ μM) was modified to give a more potent compound **126**. Replacement of the t-butyl group in **126** by -NO₂, -OMe, -OCH₂Ph, -OCH₂C₆H₄-m-COOH and -OCH₂C₆H₄-m-COOMe groups gave less potent analogues (IC_{50} values 16-95 μM). Most of the changes to the catechol and the linker gave compounds without any affinity toward the human glucagon receptor. The 3-hydroxy group could, however, in the presence of a 4-hydroxy group be changed to a methoxy or a chloro group while retaining affinity.[357] Peptide sequences like X-Lys-Tyr-(Met/Val)-(Pro/Val)-Met stimulating the formation of inositol phosphates in lymphocyte cell lines were identified by screening synthetic hexapeptide libraries. Analogues of these peptides containing a 2-isoxazoline dipeptide isostere (*e.g.* **127**) were synthesised. In comparison to Trp-Lys-Tyr-Met-Val-Met-NH₂ (IC_{50} 0.030 μM), all the analogues were much less potent.[358]

(126) (127)

5 Enzyme Inhibitors

Like last year, most of the work this year has been on converting enzyme, HIV protease, farnesyltransferase, various matrix metalloproteases and thrombin inhibitors. Only a limited amount of work has been published on the inhibitors of renin, elastase, calpain and cathepsin D.

5.1 Aminopeptidase Inhibitors – Four new linear peptides (**128-130**), were isolated from the cyanobacterium *Microcystis aeruginosa*. Microginins 299-C (**128**) and 299-D (**129**) inhibited leucine aminopeptidase with IC_{50}s of 2.0 and 6.4 μg ml^{-1}, respectively.[359] The other two peptides (**130**, R = CH_2Cl and $CHCl_2$) did not inhibit the enzyme up to a concentration of 100 μg ml^{-1}. None of the peptides inhibited ACE, papain, trypsin, thrombin, plasmin, chymotrypsin and elastase at 100 μg ml^{-1}. Several α-mercapto-β-aminoacyl dipeptides were synthesised as selective aminopeptidase A inhibitors. Sulfonamide and carboxylate moieties known to be recognised by the S_1 subsite of the enzyme were introduced on the side chain of the α-mercapto-β-aminoacyl subunit. The dipeptides were optimised to interact with the S'_1 and S'_2 subsites by means of combinatorial chemistry.[360] The sulfonamide analogue (**131**) and the corresponding carboxylate analogue were the more potent and selective (about 150-fold less potent against aminopeptidase N) inhibitors of the series. Phthalimide derivatives like **132** were shown to inhibit aminopeptidase N (a Zn^{2+} exopeptidase, identical to the cell surface antigen CD13) (IC_{50} 0.12 μg ml^{-1}) in a selective manner (inactive against dipeptidyl peptidase (IC_{50} >100 μg ml^{-1}).[361] Compounds containing a number of other substituents in the two aromatic rings were less potent. However, some of the compounds like **133** inhibited both the enzymes (aminopeptidase N and dipeptidyl peptidase) with similar potency (IC_{50} 9.6–12.8 μg ml^{-1}).

5.2 Calpain Inhibitors – Recent advances in the development of calpain I inhibitors have been reviewed.[362] The role of calpain in skeletal-muscle protein degradation has been highlighted.[363] Analogues of a calpain inhibitor, Z-Val-Phe-H (K_i 8 nM), were prepared by replacing the P_1 and P_2 positions.[364] Many of the analogues, *e.g.* CH_3SO_2-D-Ser(Bzl)-Phe-H, CH_3SO_2-D-Ser(Bzl)-

Lys(SO$_2$Ph)-H, CH$_3$SO$_2$-D-Ser(Bzl)-Tyr(Bzl)-H, CH$_3$-CH$_2$SO$_2$-D-Ser(Bzl)-Phe-H and 2-thienyl-SO$_2$-D-Ser(Bzl)-Phe-H were either equipotent or 2-4-fold more potent than the parent dipeptide aldehyde. CH$_3$SO$_2$-D-Ser(Bzl)-Phe-CO-NHEt was >15-fold less potent. Replacement of the D-Ser(Bzl) in the P$_2$ position by Thr(Bzl), D-Cys(Bzl), D-Phgly, D-Phe, D-Trp and D-(thiophen-2-yl) did not have any significant effect on the inhibitory potency (K_i values 8–11 nM). The analogues were equipotent in inhibiting both calpain I and II and inhibited calpain I in a cell-based (Molt 4, human leukaemia T cell line) assay (K_i 0.3–0.8 µM). Many of the compounds were moderately potent inhibitors of human liver cathepsin B but none of the compounds inhibited human thrombin at a concentration of 10 µM. Replacing the Ser by D- and L-Pro led to a series of inhibitors, *e.g.* **134** and **135**, which inhibited calpain I (IC$_{50}$ 14–40 nM) but were much less potent against cathepsin B (IC$_{50}$ 14-36 µM) and α-chymotrypsin (12% inhibition 10 µM).[365] Analogues of **134** containing various substituents in the proline ring (*e.g.* *trans*-4-CN and *trans*-4-N$_3$ were about two-fold less potent against calpain I but were more selective (IC$_{50}$ cathepsin B 8700 and >10,000 nM and 6-9% inhibition of α-chymotrypsin at 10 µM). Analogues of the L-proline derivative **135** containing *trans*-4-PhSO$_2$O- and *trans*-4-PhCH$_2$O- groups in the proline ring also retained significant activity against calpain I.

(134)   (135)

5.3 Caspase Inhibitors

5.3 Caspase Inhibitors – A new member of the caspase family, caspase-14, has been identified.[366,367] Caspase-14 cDNA encodes a 257-amino acid-long protein that has significant homology to other members of the caspase family. Like other caspases, caspase-14 has a conserved active site, pentapeptide Gln-Ala-Cys-Arg-Gly. Caspase-2, a member of the ICE family, has been shown to be activated during apoptosis by another ICE member, a caspase-3 (CPP32)-like protease(s).[368] Up to 50 µM N-acetyl-Asp-Glu-Val-Asp-H, a caspase-3-preferred peptide inhibitor, inhibits caspase-2 activation and DNA fragmentation *in vivo*, but does not prevent loss of mitochondrial function, while higher concentrations of N-acetyl-Asp-Glu-Val-Asp-H (>50 µM) inhibit both.

Peptide aldehydes, Ac-Trp-Glu-His-Asp-CHO, Ac-Tyr-Val-Ala-Asp-CHO, Ac-Asp-Glu-Val-Asp-CHO, Boc-Ile-Glu-Thr-Asp-CHO and Boc-Ala-Glu-Val-Asp-CHO, were investigated against group I (caspase-1, caspase-4 and caspase-5), group II (caspase-3, caspase-7 and caspase-2) and group III (caspase-6, caspase-8, caspase-9 and caspase-10) caspases.[369] Although all these enzymes have a common requirement for Asp in the P$_1$ position, their extended specificities are different. Position P$_4$ appears to be most important determinant of specificity. Group I enzymes prefer the sequence Trp-Glu-His-

Asp, but other amino acids are also well tolerated in the P_4 position. The optimal tetrapeptide recognition motif for group II enzymes is Asp-Glu-Val-Asp, and these enzymes are highly selective, with a near absolute specificity for Asp in P_4. Group III caspases are broadly inhibited by Boc-Ile-Glu-Thr-Asp-CHO, Ac-Asp-Glu-Val-Asp-CHO and Boc-Ala-Glu-Val-Asp-CHO (K_i values 1-300 nM). Conformationally constrained inhibitors of caspase-1 (ICE) and of the human CED-3 homologue caspase-3 incorporating a P_2-P_3 dipeptide mimetic were prepared.[370] Depending on the nature of the P_4 substituent, highly selective inhibitors of both caspases were obtained. For example, **136** was more potent against caspase-1 and **137** was more potent against caspase-3.

5.4 Cathepsin Inhibitors – Inhibitors of cathepsin B, D, K and L have been reported. Cathepsin B has been implicated (in addition to matrix metallopro-teinases) in the cleavage of the cartilage proteoglycan, aggrecan, between Asn^{341} and Phe^{342}, to yield a small G1 fragment terminating in the residues Val-Asp-Ile-Pro-Glu-Asn.[371] Ester and amide derivatives of α-aza-Gly, Ala and Phe (*i.e.*, Ac-L-Phe-NHN(R)CO-X, where X = H, CH_3, or CH_2Ph, respectively) were evaluated as inhibitors of cysteine proteinases cathepsin B and papain.[372] The ester derivatives inactivated cathepsin B and papain at rates which increased dramatically with leaving group hydrophobicity and electronegativity. Amide and P_1-thioamide derivatives do not inactivate papain. Examples of cathepsin B inhibitors included Ac-Phe-NH-NH-COOCH₂CH₂Cl, Ac-Phe-NH-NH-COOCH₂CH₂Br, Ac-Phe-NH-NH-COOCH₂CCl₃, Ac-Phe-NH-NH-COOCH₂Ph and Ac-Phe-NH-NH-COOPh. These peptides also inhibited papain. The dependence of inactivation rate on leaving group electronegativity and hydrophobicity was similar for both the enzymes. None of the peptides inactivated cathepsin C, another cysteine proteinase with a rather different substrate specificity. A combinatorial peptide library approach was used to identify Arg-Ile-Trp-D-Arg-Tyr-Trp-Ala-Val, Tyr-Phe-Phe-D-Arg-Met-Phe-Gly-Val and Trp-Ile-Tyr-D-Arg-Trp-Phe-Phe-Val as moderately potent inhibitors of cathepsin B, cathepsin L and cruzipain (a cysteine protease from an intracellular protozoan *Trypanosoma cruzi*) (K_i values 0.11–0.80 μM).[373] A combination of molecular modelling studies and X-ray crystal structure (enzyme bound to an epoxysuccinyl inhibitor) approaches were used to identify to cyclohexanone-based inhibitors of cathepsin B. However, compounds like **138** (n = 1 or 2) were very weak inhibitors of the enzyme (K_i 6–7 mM).[374]

The crystal structure of a catalytically inactive form of cathepsin D (a

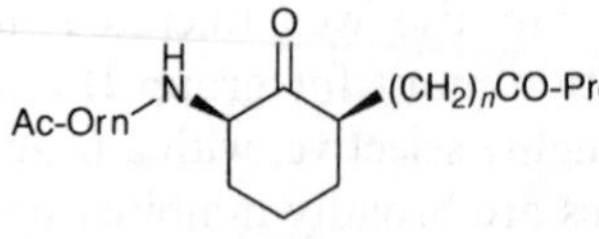

$$Ac\text{-}Orn\underset{H}{N}\text{...}(CH_2)_nCO\text{-}Pro$$

(138)

lysosomal aspartic protease involved in the turnover of cellular proteins as well as in the selective processing of MHC class II antigens, hormones and growth factors), obtained at pH 7.5, has been reported.[375] A combinatorial library approach based on a docking exercise using the X-ray crystal structure of a cathepsin D-pepstatin complex led to inhibitors of cathepsin D and another aspartyl proteinase, plasmepsin II. The library design was based on a statine template and three cyclic diamino acids as potential P_1', $P_2\text{-}P_4$ surrogates.[376] The K_i values for the statine derivative most potent against plasmepsin (139) were 490 nM and 45 µM, respectively, against plasmepsin and cathepsin D. The most potent compound against cathepsin D was 140 (K_i 100 and 1.1 µM, respectively, against plasmepsin I and cathepsin D).

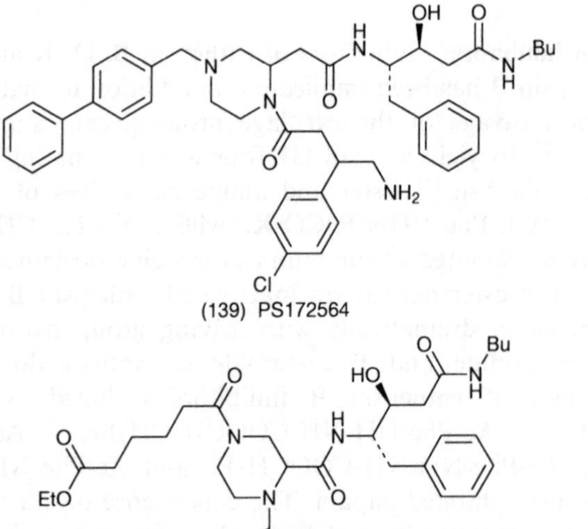

(139) PS172564

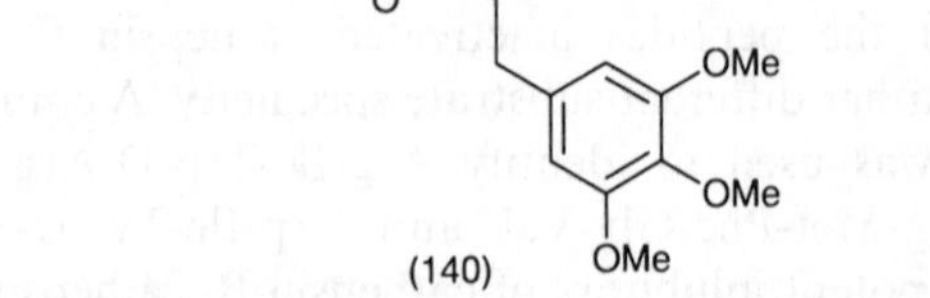

(140)

Evidence based on cathepsin-K-deficient mice has indicated a role for the enzyme in the resorption and remodelling of bone.[377] Additional evidence based on biological models suggests that the degradation of type I collagen by cathepsin K may be an important factor in bone resorption.[378] A number of inhibitors of the enzyme have been reported. The design of these compounds was based on the crystal structures of two papain-inhibitor (Z-Leu-Leu-

leucinal and Z-Leu-Leu-methoxymethyl ketone) complexes and cathepsin K-inhibitor complexes.[379-381] Analogues of **141** (R = Z-Leu) with R = Z-Ala and Z-Nva in place of Z-Leu were similar in potency to the parent compound. Many other analogues containing an *o*-benzyloxybenzoyl, *m*-benzyloxybenzoyl, *p*-benzyloxybenzoyl, and benzoyl groups in place of Z-Leu were less potent. Cathepsin K inhibitor **142** had >500-fold selectivity over human cathepsin B, L and S. The des-isobutyl analogue of **142** was much less potent.

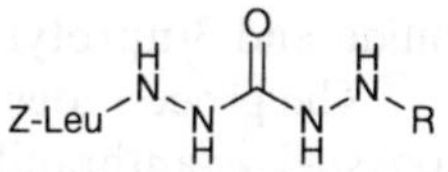

(141)

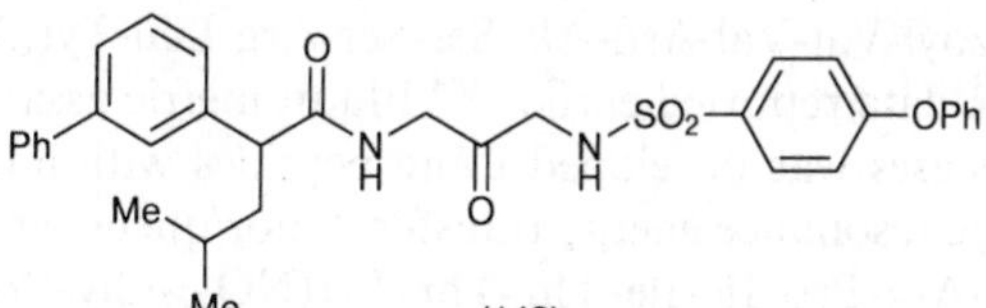

(142)

Cathepsin L (a lysosomal cysteine protease), secreted by osteoclasts, has also been suggested to participate in bone collagen degradation.[382] Another similar enzyme cathepsin L2, obtained from a brain cDNA library, was shown to be widely expression in colorectal and breast carcinomas but not in normal colon or mammary gland or in peritumoural tissues.[383] The purified enzyme cleaved Z-Phe-Arg-7-amido-4-methylcoumarin, a commonly used substrate for cysteine proteases. Cathepsin L was inhibited by peptide aldehyde derivatives like **143**. Compound **143**, a potent, selective, and reversible inhibitor of human cathepsin L with an IC_{50} of 1.9 nM, inhibited the release of Ca^{2+} and hydroxyproline from bone in *in vitro* bone culture system and also prevented bone loss in ovariectomised mice at an oral dose of 50 mg kg. Replacement of the P_3 naphthalenesulfonyl group in compound **143** (IC_{50} values against

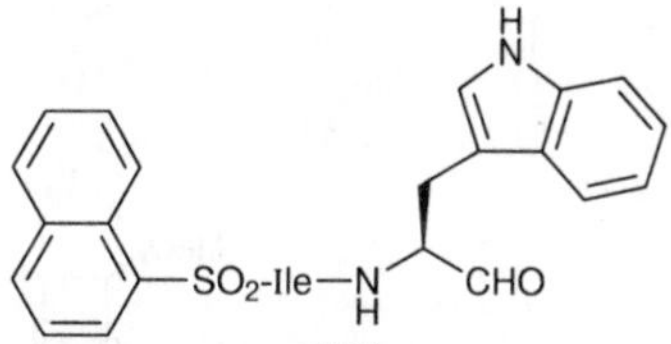

(143)

cathepsin L and cathepsin B 1.9 and 1500 nM, respectively) by Boc-, Z-, 3-Me-Ph-NH-CO-, Ph-CH$_2$NHCO-, 2-CF$_3$-PhNHCO-, 1-Nap-NHCO-, Ph-CH$_2$NHCS-, 4-Me-PhSO$_2$-, 1-Nap-SO$_2$, 2-Nap-SO$_2$ and PhCO- groups only had a small effect on cathepsin L and cathepsin B inhibitory activities.[382] Only the iPrNHCO-analogue was much less potent. Two of the analogues containing a 2-Nap-SO$_2$ and PhCO- groups at the N-terminus were much less selective. Many replacements in P_2 [Ile replaced by Leu, Ala, Val, Asp(OMe) and Phe] and P_1 (Trp-H replaced by Phe-H, Tyr-H, Leu-H, Val-H and Ala-H) positions could also be carried out without significant losses in the cathepsin L (IC_{50} values 0.97–5.3 nM) inhibitory potency. However, effects against the cathepsin B (IC_{50} values 30–3900 nM) activity resulted in several compounds like N-(1-naphthalenylsulfonyl)-Asp(OMe)-Trp-H and N-(1-naphthalenylsulfonyl)-Val-Trp-H which were only 15–30-fold more potent against cathepsin L than against cathepsin B.

5.5 Cytomegalovirus and Rhinovirus 3C Protease Inhibitors – Fluorogenic peptide substrates for human cytomegalovirus protease [*o*-aminobenzoyl-Tbg-Tbg-Asn(NMe$_2$)-Ala-Ser-Ser-Arg-Leu-Tyr(3-NO$_2$)Arg-OH and Ac-Tbg-Tbg-Asn(NMe$_2$)-Ala-AMC] were developed starting from a substrate [*o*-aminobenzoyl-Val-Val-Asn-Ala-Ser-Ser-Arg-Leu-Tyr(3-NO$_2$)-Arg-OH] and some SAR data reported earlier.[384] Fluorometric assay for the human rhinovirus 2A proteases was developed using peptides with anthranilide and 3-nitrotyrosine as the resonance energy transfer donor/quencher pair.[385] The process identified Thr-Arg-Pro-Ile-Ile-Thr-Thr-Tyr(NO$_2$)-Gly-Pro-Ser-Asp-Lys(anthranilide)-Tyr as a better substrate for HRV2 2A and Gly-Arg-Thr-Thr-Leu-Ser-Thr-Tyr(NO$_2$)-Gly-Pro-Pro-Arg-Lys(anthranilide)-Tyr as a substrate for both HRV2 2A and HRV14 2A.

Mode of binding studies for cytomegalovirus protease to inhibitors like **144** [R = Me, R$_2$ = Me$_2$, X = Me; R = H, R$_2$ = Me$_2$, X = Me; R = Me, R$_2$ = -(CH$_2$)$_4$-, X = CONHCH$_2$-C$_6$H$_4$(*p*-I)] using crystal structure and NMR techniques have been reported.[386,387] The inhibitors bound in an extended conformation. In addition, large conformational differences relative to the structure of the free enzyme were observed. Non-peptide inhibitors of cytomegalovirus protease (*e.g.* **145, 146**) were reported.[388,389] Compounds like **146** also inhibited porcine pancreatic elastase, human leukocytic elastase (IC$_{50}$ values <0.2 μM) and chymotrypsin (IC$_{50}$ 2.2 μM).

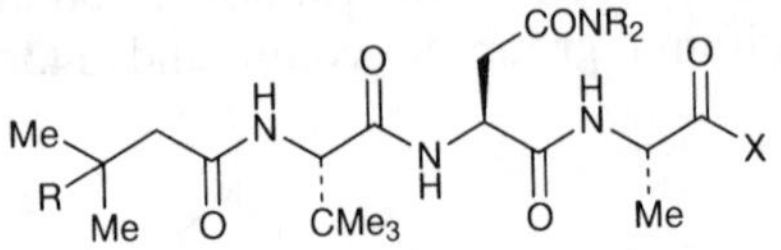

(144)

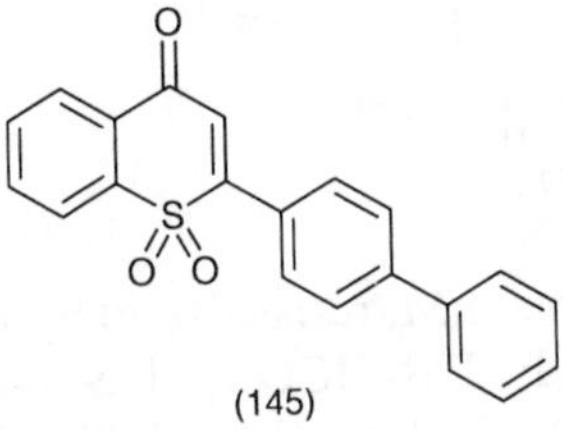

(145)

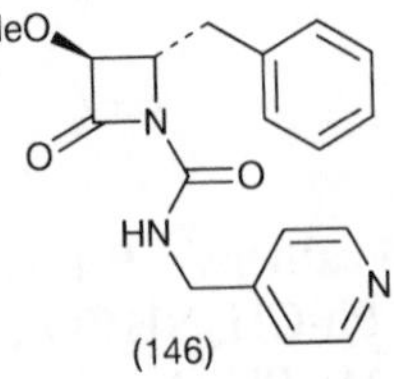

(146)

Peptide-based inhibitors of human rhinovirus 3C protease were reported.[390–392] Glutaminal derivatives like Z-Leu-Phe-Gln-H (**147**, R = -CH$_2$-CONH$_2$) (IC$_{50}$ 3.6 μM) were further modified [*e.g.* R = -NH-CHO, -NH-COMe, -NH-COEt, -NH-COCHMe$_2$, -NH-COPh, -NH-COOCMe$_3$, -NH-CONMe$_2$, -CH$_2$-CONMe$_2$, -CH$_2$-SOMe] to generate more potent inhibitors of purified HRV-14 3CP (*K*$_i$s 0.005 to 0.64 μM). The structure-based design approach using a substrate Thr-Leu-Phe-Gln-Gly-Pro (enzymic cleavage between Gln and Gly) gave irreversible inhibitors like **148**. Inhibitor **148** displayed rapid irreversible inhibition of HRV-14 3CP and potent antiviral activity against HRV-14 in cell culture ($EC_{50} = 0.056$ μM). Peptide-based inhibitors of the hepatitis C virus (serine protease) have been reported.[393-396] SAR studies on a hexapeptide Asp-Asp-Ile-Val-Pro-Cys corresponding to the N-terminal cleavage product of an hepatitis C virus dodecapeptide substrate derived from the NS5A/5B cleavage site revealed that side chains of the P_4, P_3 and P_1 residues in Ac-Asp-Asp-Ile-Val-Pro-Cys (IC_{50} 28 μM) contribute the most to binding and that the introduction of a D-amino acid at the P_5 position improves potency considerably. Replacement of the C-terminal Cys by Ala, Ile, Leu, Nleu and Nva residues resulted in less potent compounds. Conversion of the C-terminal carboxyl group in Ac-Asp-Asp-Ile-Val-Pro-Nle into $COCF_3$ did not lead to any improvement in potency (IC_{50} 160 μM). However, compounds with a C-terminal aldehyde, $-CF_2CF_3$ and -CONHBn groups were more potent (IC_{50} values 10, 79 and 2 μM, respectively). Ac-Asp-D-Asp-Ile-Val-Pro-D,L-Nle-NHBn was the most potent compound of the series (IC_{50} 0.64 μM) but it also inhibited other serine proteases like human leukocyte and porcine pancreatic elastase (IC_{50}s 0.10 and <0.06 μM, respectively). Examples of other inhibitors include peptides like Ac-Asp-D-Glu-Leu-Ile-Cha-Cys, Ac-Asp-D-Gla-Leu-Ile-Cha-Cys and Glu-Asp-Val-Val-Abu-Cys.

5.6 Converting Enzyme [Angiotensin (ACE), Neutral Endopeptidase (NEP), Endothelin and Interleukin-1β (ICE)] Inhibitors – *5.6.1 Angiotensin Converting Enzyme and Neutral Endopeptidase Inhibitors* – Several dipeptides (Ser-Tyr, Gly-Tyr, Phe-Tyr, Asn-Tyr, Ser-Phe, Gly-Phe, and Asn-Phe, IC_{50} values 4-280 μM) were identified as ACE inhibitors from a concentrate of an aqueous extract of garlic (*Allium sativum*). In comparison to captopril which lowered blood pressure in spontaneously hypertensive rats at a dose of 10 mg kg^{-1}, (po) over 1–4 hours, the peptides lowered blood pressure at a dose of 200 mg kg^{-1} (po).[397] Another ACE inhibitor [acein-1, Tyr-Leu-Tyr-Glu-Ile-Ala-Arg (corresponding to 138-144 of human serum albumin)] was isolated from the tryptic hydrolysate of human plasma.[398] The synthetic heptapeptide, hexapeptide (Tyr-Leu-Tyr-Glu-Ile-Ala) and octapeptide (Tyr-Leu-Tyr-Glu-Ile-Ala-Arg-Arg) showed dose-dependent inhibitions of ACE. New analogues of captopril and indrapril have been reported.[399,400] The captopril analogues like **149** were somewhat more potent (IC_{50} 5–8 nM) than captopril (IC_{50} 13 nM) as inhibitors of ACE and inactive against neutral endopeptidase and ECE

(149)   (150)

between 1 and 10 µM. The indrapril analogue **150** did not show any activity
against ACE, MMP-1, MMP-2 and MMP-3.

In vivo results on the dual inhibitors of ACE/NEP and aminopeptidase N/
NEP were reported.[401-403] Compound **151** (K_i 4.5 and 1.7 nM, respectively,
against ACE and NEP) inhibited NEP in rats (0.3–300 mg kg^{-1}), but inhibited
ACE only at the highest dose. The thiol dipeptide **152** (CGS30440) (IC$_{50}$ 19
and 2 nM, respectively, against ACE and NEP) at a dose of 10 mg kg^{-1} p.o.
inhibited lung tissue ACE activity in rats by 98% and 61% at 1 and 24 h and
inhibited the angiotensin-1 pressor response by 75–90% for more than 6 h.
Renal tissue NEP activity was reduced by 80% at 1 h and 73% at 24 h. In rats
supplemented with exogenous ANP, **152** (1 mg kg p.o.) elevated the concentration of circulating ANP (133%) for 4 h and increased the excretion of urine,
sodium and cyclic GMP. The aminopeptidase N/NEP inhibitors, NH$_2$-
CH(R^1)-P(O)(OH)-CH$_2$-CH(R^2)-CO-NH-CH(R^3)-COOH [R^1 = -CH$_2$Ph, -Ph
or -Me; R^2 = CH$_2$Ph(p-Ph); R^3 = -Me] [K_i values 1.4-2.9 and 5.2-15 nM against
neutral endopeptidase and aminopeptidase N, respectively], were active in
alleviating acute and inflammatory nociceptive stimuli in mice.

5.6.2 Endothelin Converting Enzyme Inhibitors – Purification and characterisation studies on a soluble human endothelin-converting enzyme-1 and an
endothelin-converting enzyme specific for big endothelin-3 have been published.[404,405] Effects of big ET-1 analogues were examined on ECE-1 activity
using solubilised membranes prepared from human ECE-1-expressed CHO-
K1 cells.[406] Among the big ET-1 analogues tested, big ET-1(18-34), [Phe^{21}]big
ET-1(18-34) and [Ala^{31}]big ET-1(18-34) inhibited ECE-1. A kinetic analysis
revealed [Phe^{21}]big ET-1(18-34) to be a competitive inhibitor (K_i = 20.6 µM)
and [Ala^{31}]big ET-1(18-34) to be a non-competitive inhibitor (K_i = 35.6 µM).
Many other analogues, big ET-1(17-26), [Ala^{20}]big ET-1(18-34), [Ala^{21}]big ET-
1(18-34), [Ala^{22}]big ET-1(18-34), [Phe^{22}]big ET-1(18-34), [Gln^{27}, Thr^{28},
Ala^{29}]big ET-1(18-34), [Phe^{21}]big ET-2(18-34), [D-Trp^{21}, D-Val^{22}]big ET-1(18-
34) and [D-Val^{22}]big ET-1(16-34), were inactive.

An ECE inhibitor and endothelin antagonist, B-90063 (**153**), isolated from
the culture supernatant of the marine bacterium *Blastobacter* sp., inhibited
human and rat ECEs (IC$_{50}$ 1–3 µM) and neutral endopeptidase and type-I and
-IV collagenases (IC$_{50}$s 66, 11 and 10 µM, respectively).[407] Three other
analogues containing a -CH$_2$OH, -CH=NNHCONH$_2$ or -CH(OH)CH$_2$-

COCH$_3$ group in place of the aldehyde group were about 5-fold less potent inhibitors of hECE. B-90063 also inhibited the binding of ET-1 to rat ET$_A$ and bovine ET$_B$ receptors (IC$_{50}$ values 43.7 and 27.2 μM, respectively). Synthetic details of naturally occurring ECE inhibitors **154** and **155** [R = -(CH$_2$)$_5$-CH(OH)(CH$_3$)] (both isolated from fermentation broths of *Saccharothrix* sp.) have been reported.[408,409] Many analogues of **154** containing replacements for the CH$_3$-CH(OH)-(CH$_2$)$_5$- group were equipotent to **154**. The thiazolyl group could only be replaced by an imidazolyl group.

A non-peptide inhibitor of ECE (**156**) (X = CCl$_3$) was obtained by a library screening approach.[410] A series of arylacetylene-containing ECE-1 inhibitors (**157**) have been prepared.[411] Analogues of **157** (70-fold selectivity for ECE-1 over NEP) containing Met, Val, Phe, D-Nal(2) and biphenylalanine residue in place of Leu (although still more potent against ECE) were less selective. Compound **158** was one of the more potent and selective inhibitor of ECE (IC$_{50}$ rhECE 33 nM, NEP 6500 nM, NEP/ECE 197).

5.7 Elastase Inhibitors – Hydrazinopeptide analogues [CONH links replaced by a CONHNH] of a human leukocyte elastase substrate, Z-Ala-Ala-Pro-Val-Ala-Ala-NH^{i}Pr, were prepared by replacing Ala^5, Val^4, or Pro^3 residues, respectively, by the corresponding α-L-hydrazino acid.[412] On incubation with HLE, hydrazinopeptide Z-Ala-Ala-Pro-Val-NHAla-Ala-NH^{i}Pr proved to be a substrate and was cleaved between Val^4 and NHAla^5. In contrast, Z-Ala-Ala-Pro-NHVal-Ala-Ala-NH^{i}Pr and Z-Ala-Ala-NHPro-Val-Ala-Ala-NH^{i}Pr

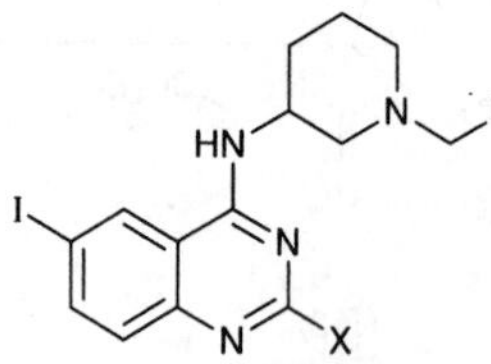

(156)

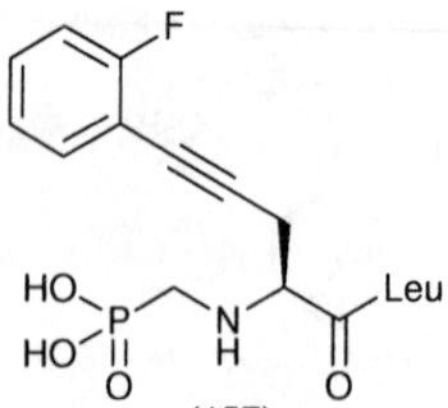

(157)

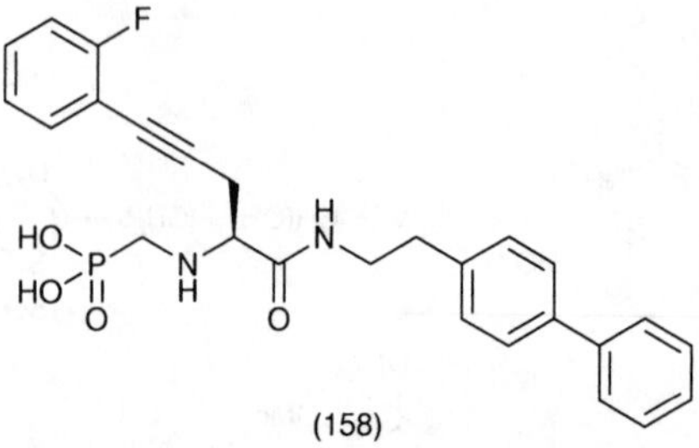

(158)

proved to bind to human leukocyte elastase without being cleaved, featuring properties consistent with reversible competitive inhibition.

A series of pentafluoroethyl ketone-based, orally active peptidic inhibitors of human neutrophil elastase were prepared starting from **159** (K_i 20 nM, 74% inhibition of lung haemorrhage at a dose of 25 mg kg^{-1} in hamster). The compounds like **160** containing 5- or 6-membered lactams as Val-Pro replacements were much less potent (K_i 14–1100 μM). The P_2 proline residue was replaced by azetidine, homoproline, thiazolidine, isoquinoline and 4-substituted proline residues.[413] The azetidinone and homoproline containing compounds were slightly less potent than compound **159** in the *in vitro* (K_i values 34 and 150 nM, respectively) and *in vivo* assays. The homoproline analogue of **159** showed a longer duration of action in the lung haemorrhage model (4 hours 50 mg kg^{-1}). Two series of β-peptidyl trifluoromethyl alcohols Z-Val-NHCH(Y)-CH(OH)-CF$_3$ [Y = -CH$_2$CH$_2$Ph or -CH(CH$_3$)$_2$] were evaluated as inhibitors of human leukocyte elastase and HIV-1 protease.[414] The two

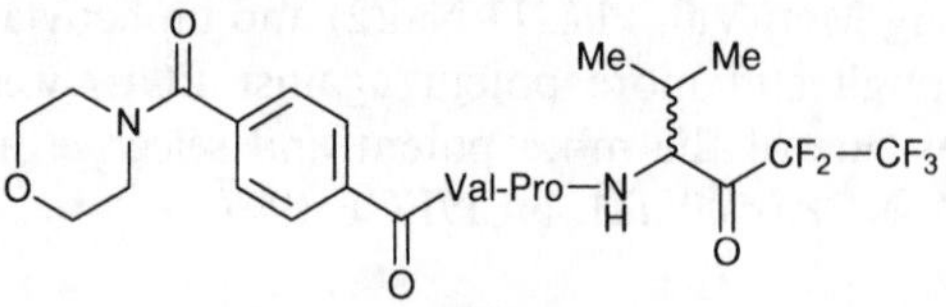

(159)

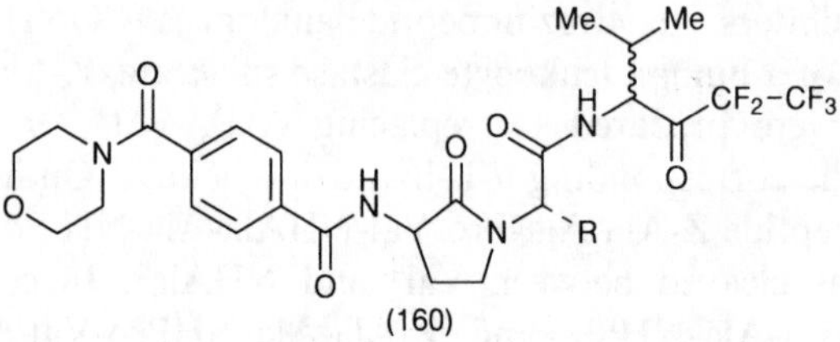

(160)

isomers of Z-Val-NH-(R,S)CH(CHMe$_2$)-COCF$_3$ were the most potent inhibitors against human leukocyte elastase (K_i values 2.37 and 8.3 µM, respectively, for the S and R isomers). Both the compounds were 10-100-fold less potent against porcine pancreatic elastase and HIV-1 protease. The most potent compounds against HIV-1 protease were Z-Val-NH-(R,S)CH(CH$_2$-CH$_2$Ph)-COCF$_3$ (K_i values 15 µM for both the S and R isomers). Non-peptidic inhibitors of elastase like **161** and **162** (R = Et or n-Pr) have been reported.[415,416]

(161)   (162)

5.8 Farnesyltransferase Inhibitors – X-Ray crystal structures of farnesyl protein transferase complexed with Ac-Cys-Val-Ile-selenoMet and α-hydroxy-farnesylphosphonic acid are reported.[417,418] Reduced pseudopeptides related to the C-terminal tetrapeptide of the Ras protein were synthesised as inhibitors of Ras protein farnesyltransferase.[419] Compound **163** was one of the more potent inhibitors (IC$_{50}$ 0.123 nM). The methyl ester of **163** selectively inhibited anchorage-independent growth of Rat1 cells transformed by v-*ras* at 2.5–5 µM. N-terminal modifications led to non-thiol analogues like **164** [IC$_{50}$ 5.4 nM].[420] Another series of N-substituted inhibitors were based on one of the squalene synthase inhibitors reported earlier.[421,422] One of the more potent analogues of this series was **165** which inhibited the enzyme in whole cells as well as in the *in vitro* system and also inhibited colony formation of activated H-*ras*-transformed NIH3T3 cells when grown in soft agar (IC$_{50}$ 27.5 µM). Furthermore, **165** suppressed tumour growth in nude mice transplanted with activated H-ras-transformed NIH3T3 cells at a dose of 40–80 mg kg^{-1}. Non-thiol-containing inhibitors of farnesyltransferase (*e.g.* **166**, **167**) are described.[423,424] In addition to 0.4 nM *in vitro* potency, **166** displayed 350 nM potency in whole cells. Many other analogues containing D-Met, MeMet, methionine sulfone, glutamine, O-methylhomoserine, norleucine and S-methyl-cysteine were less potent. The phenol-derived Cys-Val-Phe-Met analogues like **167** were much less potent. As in previous years, many inhibitors like **168** have been reported.[425-429] In the nude mouse, **168** was orally active in several human tumour xenograft models including tumours of colon, lung, pancreas, prostate, and urinary bladder origin.

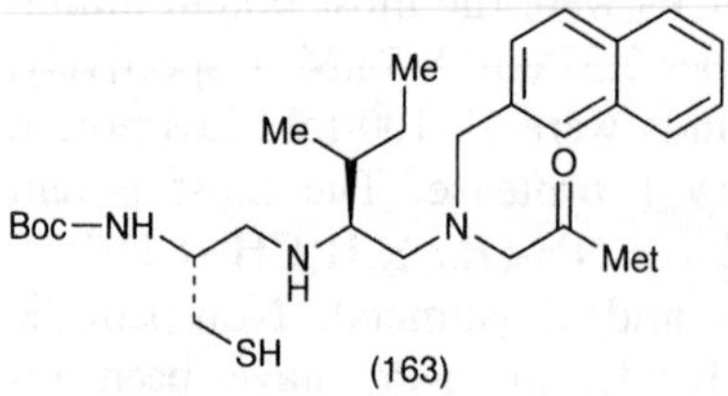

(163)

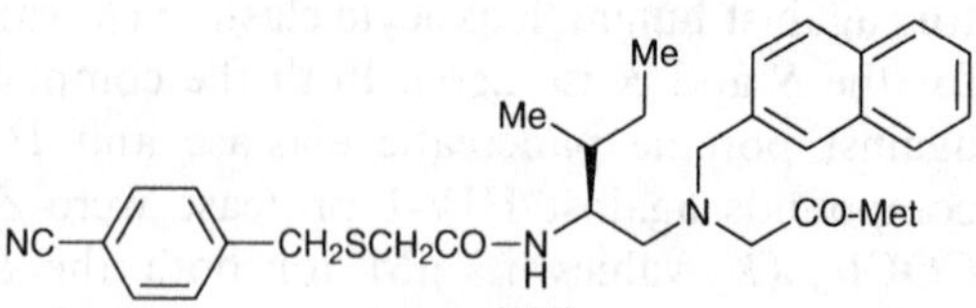

(164)

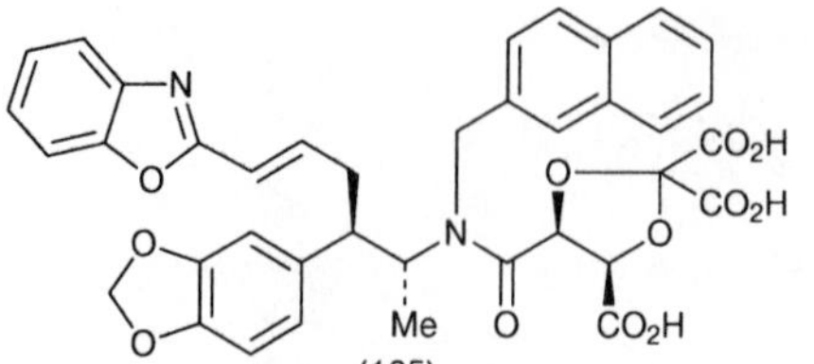

(165)

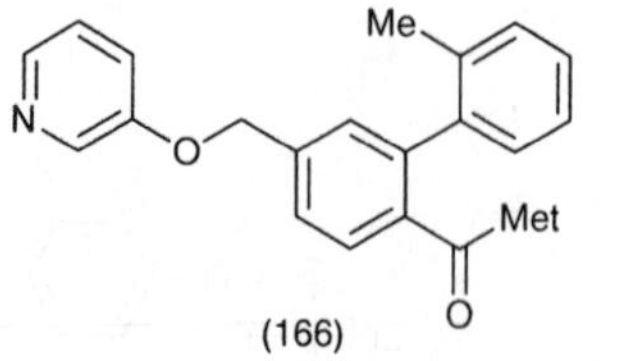

(166)

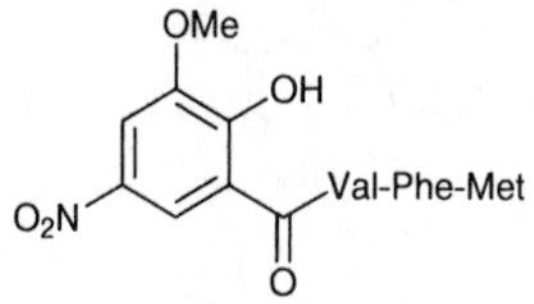

(167)

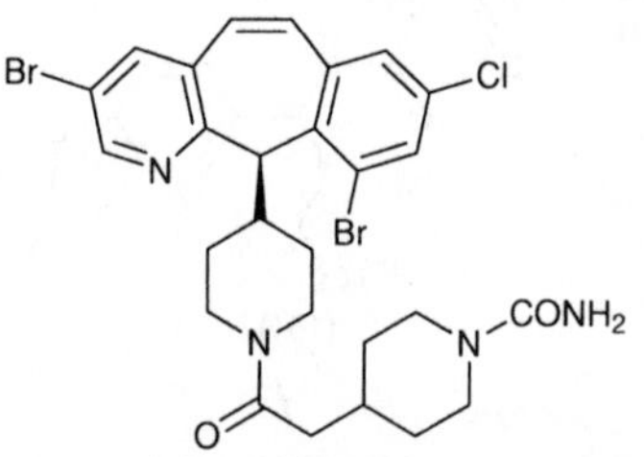

(168) SCH66336

5.9 HIV Protease Inhibitors – Various aspects of HIV research, *e.g.* resistance to the existing protease inhibitors, design, discovery and development of cyclic urea and dihydropyrone sulfonamides class of HIV protease inhibitors, have been reviewed.[430-432] Since the discovery that HIV infection is initiated by interaction of the virion envelope glycoprotein (gp120/41) with at least two cellular receptors (the CD4 molecule and a 7-transmembrane domain G-protein coupled chemokine receptor), significant effort has been directed towards finding agents which can block this interaction and, thus, inhibit HIV infection.[433-439] This work has also been reviewed recently.[440] Although CCR5 and CCR4 are believed to be the primary receptors for entry of HIV-1, additional chemokine receptors, including one encoded by cytomegalovirus have been shown by *in vitro* assays to serve as co-receptors for HIV and SIV.[441-443]

A number of publications on the peptidic and non-peptidic inhibitors of the enzyme have appeared. The dimeric nature of the HIV-1 protease was used to design inhibitors that target the dimeric interface of the enzyme.[444] Several analogues of the type HO-Phe-Asn-Leu-Thr-Ser-NH-CO-CH$_2$-CH$_2$-(CH$_2$)$_n$ -CH$_2$-CH$_2$-CO-Asn-Pro-Gln-Ile-Thr-Leu-Trp-OH (n = 8, 10, 12), HO-Phe-Asn-Leu-Thr-Ser-NH-CO-(CH$_2$)$_n$-CH=CH-(CH$_2$)$_n$-CO-Asn-Pro-Gln-Ile-Thr-Leu-Trp-OH (n = 5, 6, 7), HO-Phe-Asn-Leu-Thr-Ser-NH-CO-(CH$_2$)$_n$-CH= CH-(CH$_2$)$_n$-CO-Asn-Pro-Gln-Ile-Thr-Leu-Trp-OH (n = 5, 6, 7) and HO-Phe-

Asn-Leu-Thr-Ser-NH-CO-$(CH_2)_n$-C≡C-$(CH_2)_n$-CO-Asn-Pro-Gln-Ile-Thr-Leu-Trp-OH (n = 5, 6, 7) were synthesised. All the compounds were weak inhibitors of HIV-1 protease (IC_{50} values 2–51 µM).

Peptide bond replacements in sequences like -Val-Leu(or Phe)ψPhe-Val- and -Val-Val-Leu(or Phe)ψPhe-Val-Val- were attempted.[445] Many of the compounds like Boc-Val-Val-Leuψ$(CH_2$-$CH_2)$Phe-Val-Val-NH_2, Val-Val-Leuψ$(CH_2$-$CH_2)$Phe-Val-Val-NH_2, Boc-Pheψ$(CH_2$-NH)Phe-Ile-Phe-OMe, Boc-Pheψ$(CONNH_2$-NH)Gly-Ile-Phe-OMe, Boc-Pheψ$(CONNH_2$-NH)Gly-Phe-Ile-Phe-OMe, Boc-Pheψ$(CHOHCH_2$-NH)Phe-Ile-Phe-OMe, Z-Pheψ$(CHOHCH_2$-NH)Phe-Ile-Phe-OMe, Z-Pheψ$(CHOHCH_2$-NH)Gly-Ile-Phe-OMe and Boc-Pheψ$(CONOH)$Gly-Ile-Phe-OMe were weak inhibitors of the enzyme (10-100% inhibition at 10 µM). Some of the more potent compounds included Boc-Pheψ$(CHOHCH_2NH)$Phe-Glu(NHOH)-Phe-OMe and Boc-Pheψ$(CHOHCH_2NH)$Phe-Asp(NHOH)-Phe-OMe (IC_{50} 1.3 and 8.7 µM, respectively) [Glu(NHOH) and Asp(NHOH) refer to side chain OH derivatives of Gln and Asn]. Various cyclic peptides like **169** (n = 1 or 2) and **170** (n = 2 or 3) were either inactive or poor inhibitors of the enzyme. Backbone cyclic peptides were obtained by screening peptide libraries for the ability to inhibit nuclear import of NLS-BSA in digitonin-permeabilised HeLa and Colo-205 cultured cells. This led to the discovery of a backbone cyclic peptide **171** [IC_{50} 35 nM] which reduced HIV-1 production by 75% in infected non-dividing cultured human T-cells and was relatively resistant to tryptic digestion.[446]

Boc-Lys-β-hPhe-Phe-Lys-NH_2
　|　　　　　　　　|
CO—$(CH_2)_n$—CO
(169)

Boc-Lys-Phe-ψ(CHOH—CH_2NH)Phe-Lys-NH_2
　|　　　　　　　　　　　　　|
CO————$(CH_2)_n$————CO
(170)

CH_2—CH_2—CH_2—CH_2—CO-Val—NH—$(CH_2)_6$
　|　　　　　　　　　　　　　　　　　|
CO-Lys-Lys-Lys————————N—CH_2—CO-Lys-Leu-NH_2
(171)

Various peptidomimetic and non-peptidic inhibitors of HIV protease are being developed to overcome the problem of viral resistance. This is being attempted by modifications of the existing inhibitors like ritonavir and amprenavir. Computational studies using HIV-1 protease mutants (Met[46]Ile, Leu[63]Pro, Val[82]Thr, Ile[84]Val, Met[46]Ile/Leu[63]Pro, Val[82]Thr/Ile[84]Val and Met[46]Ile/Leu[63]Pro/Val[82]Thr/Ile[84]Val) and known inhibitors of the enzyme (ABT-538 and VX-478) are being used to design inhibitors with better binding affinity towards both mutant and wild type protease.[447] Examples of the new inhibitors reported this year include compounds **172–178**.[448-456] In the enzyme inhibition assay **172** was nearly equipotent to amprenavir but compounds like **173** were about 20-fold less potent.[448,449] Compound **174** (ABT-378) inhibited wild-type and mutant HIV protease (K_i 1.3-3.6 pM), blocked the replication of laboratory and clinical strains of HIV type 1 (EC_{50} 0.006–0.017 µM), and maintained high potency against mutant HIV selected by ritonavir *in vivo* (EC_{50} ≤ 0.06 µM).[450] Similarly, compound **175** also retained activity against mutated viral strains.[452] Analogues of **177** with X = Z, Z-Gly, Z-Ala, Z-Leu or

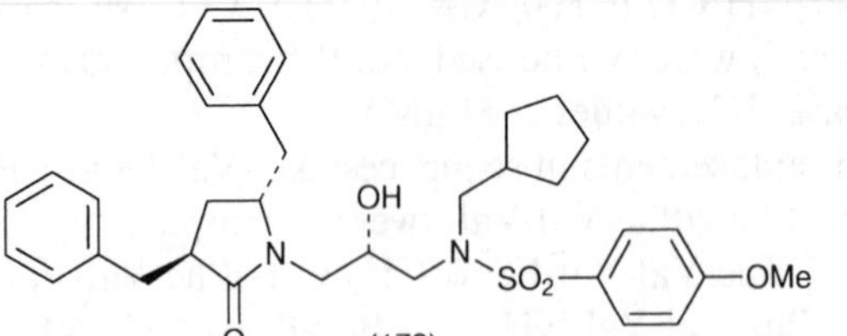

(172)

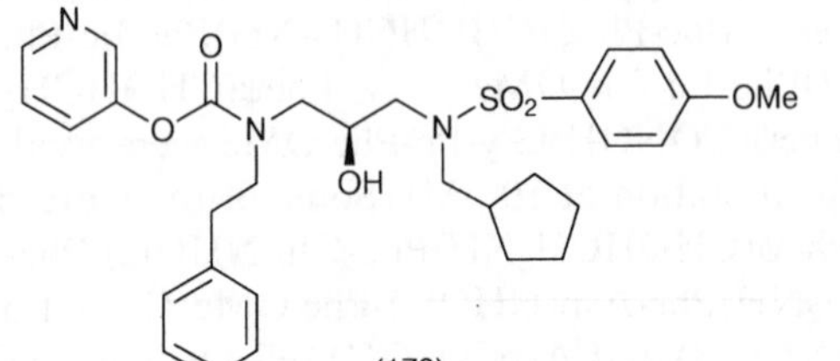

(173)

Z-Phe inhibited HIV protease with similar potency (K_i values 1.1–2.6 nM). However, only one of the compounds (X = Z-Ala) showed significant activity against feline immunodeficiency virus protease (K_i values 8.3–41 nM against various mutants).[456] The corresponding Z-Gly and Z-Leu analogues were about 5-fold less potent and the Z-Phe analogue was a poor inhibitor of the enzyme (K_i values 3700–7000 nM). Non-peptide inhibitors reported this year include a number of cyclic urea and cyanoguanidines derivatives like **178** [R = n-butyl, 3-methylbutyl, cyclobutylmethyl, cyclopentylmethyl, cyclohexylmethyl, 3-hydroxybenzyl and 3-(hydroxymethyl)benzyl].[457–461] Some of the cyanoguanidine derivatives were more potent than the corresponding urea derivatives and also demonstrated antiviral activity in tissue culture assays.

5.10 Matrix Metalloproteinase Inhibitors – Accumulating evidence has highlighted the role of matrix metalloproteinases (a multigene family of zinc-dependent enzymes involved in the degradation of extracellular matrix components) and tissue inhibitors of these enzymes in cancer, invasion and metastasis. Most matrix metalloproteinases are produced as latent zymogens in which catalytic activity is tightly regulated. Regulation involves both the control of zymogen activation and the inhibition of the active enzyme by the tissue inhibitors of metalloproteinases. Thus the activity of these enzymes can be blocked at two different levels. A synthetic furin inhibitor has been shown to inhibit matrix metalloproteinase 2 maturation and HT1080 invasiveness.[462] Crystal structure studies on the complex formed by the membrane type 1-matrix metalloproteinase with the tissue inhibitor of metalloproteinases-2 have been reported.[463] Levels of certain matrix metalloproteinases such as stromelysin-3 and gelatinase are elevated in tumour-associated stroma compared to non-involved tissue. Experiments in models of breast cancer have shown that matrix metalloproteinase inhibitors can significantly reduce the growth rate of both primary and secondary tumours, and can block the

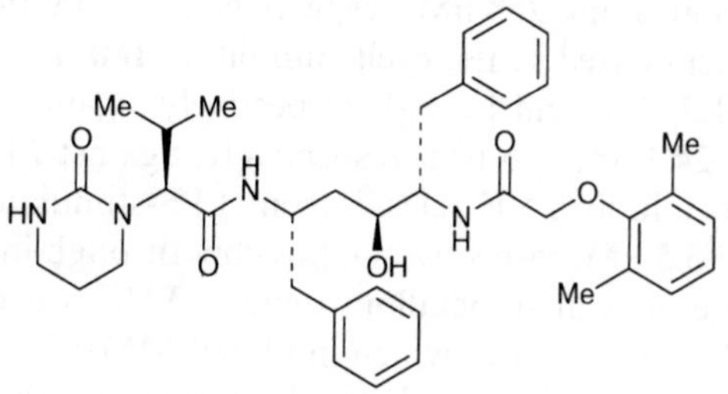

(174)

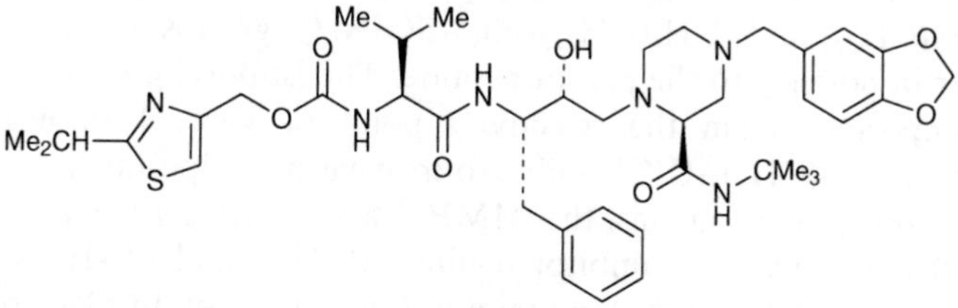

(175) A-160621

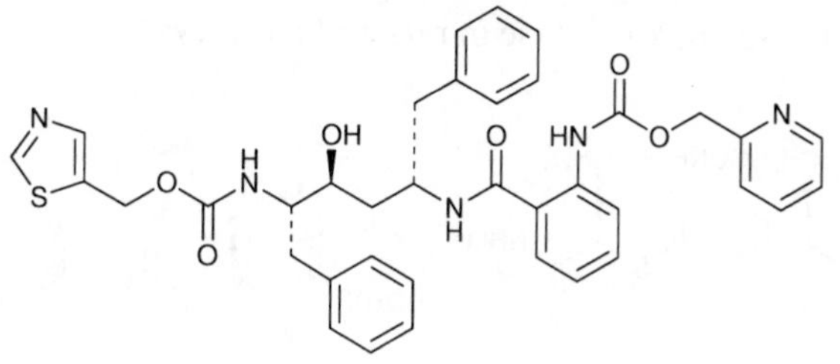

(176)

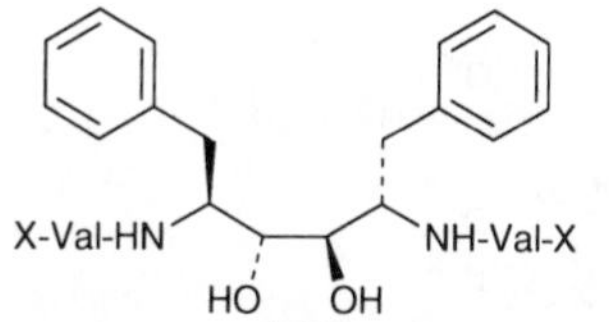

(177)

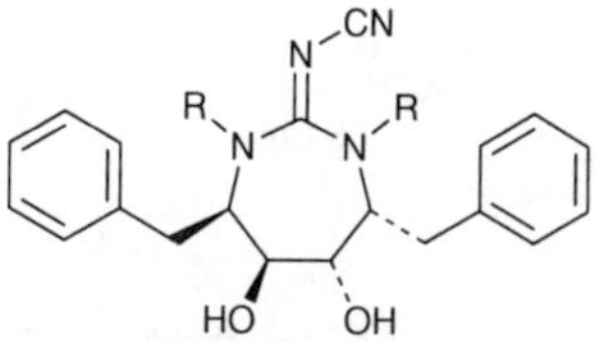

(178)

process of metastasis. Some of these aspects have been reviewed.[464-467] Several matrix metalloproteinase inhibitors including marimastat have now started clinical trials in patients with advanced malignancy.[468]

Most of the matrix metalloprotease inhibitors published this year are hydroxamate derivatives. Many of these are conformationally restrained cyclic peptides or succinate-derived hydroxamic acids incorporating a macrocyclic ring.[469-472] In the cyclic dipeptide series of inhibitors (*e.g.* **179**), the 14-membered (**179**) and 16-membered compounds were more potent.[469] Cyclic peptide **179** was a more selective inhibitor of MMP-8 (K_is 2500, 8100, 13500, 17 and 6600 nM, respectively, for MMP-1, -2, -3, -8 and -9). An analogue of **179** containing a biphenyl substituent (in place of the Ph) in the P_1' position (to fit in the larger S_1' pocket) resulted in a slightly less selective inhibitor (K_i

values 1448, 249, 2400, 9 and 288 nM, respectively, for MMP-1, -2, -3, -8 and -9). In the succinic acid based series, cyclic inhibitors **180** (n = 3; Leu side chain in place of Ile) (K_i 1.2, 32.7 and 1.8 nM, respectively, against MMP-1, -3 and -9) and **181** (K_i 2.8, 24.1 and 2.6 nM, respectively, against MMP-1, -3 and -9) were also found to inhibit TNF release from LPS-stimulated human whole blood (IC_{50} 1.2 and 6.5 µM, respectively), possibly through inhibition of TNF-α converting enzyme or related metalloproteinase TNF processing enzyme.[470] In compound **180** (n = 3, 4, 5 or 6) which inhibited MMP-1, -2, -3 and -7 (IC_{50} 1.8-12 nM), the C-terminal methyl amide was not critical for activity. Compounds (n = 4) containing an -NH_2-pyridyl, -NH-CH_2-CH_2-S-CH_3, -NH-CH_2-CH_2-N(CH_3)$_2$ and -NH-(CH_2)$_2$-Ph(4-SO_2NH_2) groups at the C-terminus were similar in potency to the parent peptide. The isobutyl group appeared to be more important from the selectivity point of view. For, example its replacement with -(CH_2)$_4$-O-CH_2-Ph group gave a compound (n = 4) which was much more potent against the MMP-2 and -3 (IC_{50}s 0.3 and 1.8 nM, respectively) and a weaker inhibitor against MMP-1 and -7 (IC_{50}s 180–210 nM).[471] Analogue of **180** (n = 4) containing Leu side chain in place of Ile and Ph in place of -NHMe) (IC_{50}s 3.3-13 nM against MMP-1, 2, 3 and 7) displayed oral activity at 10–30 mg kg^{-1} dose in rats and monkeys.[472]

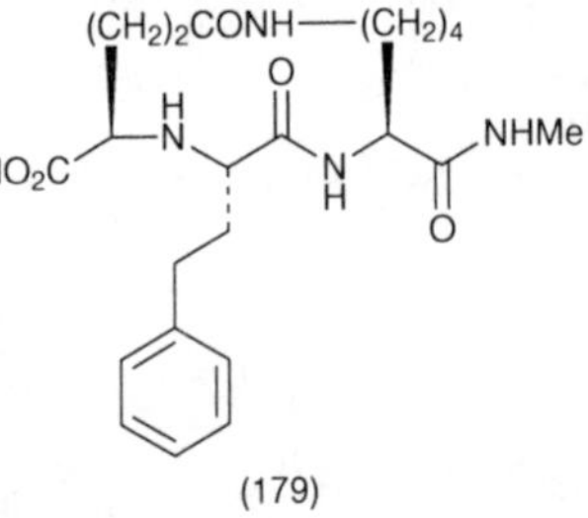

(179)

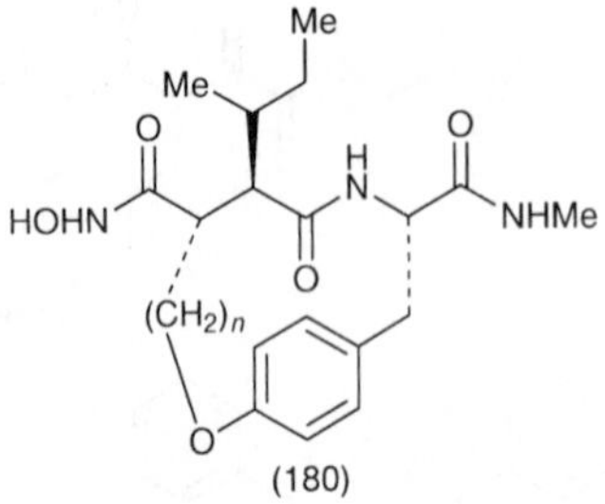

(180)

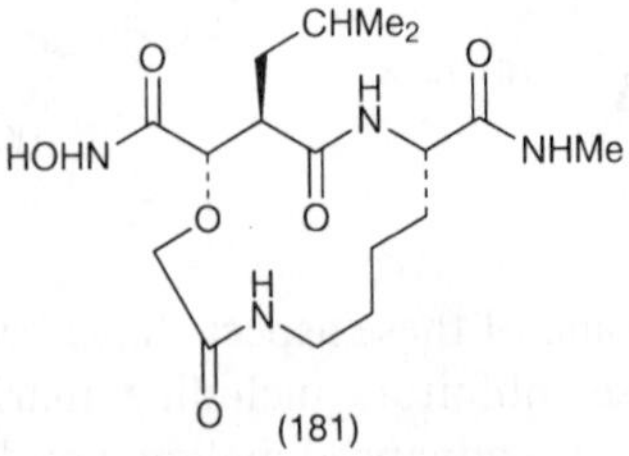

(181)

Several linear succinate- and malonate-derived hydroxamic acids have also been prepared as matrix metalloprotease inhibitors.[473-479] Examples of linear succinate derivatives include compounds like **182–184** and examples of malonate derivatives include compounds like **185–187**. Replacement of the hydroxamate group in **182** by OH, CH_3ONH, BnONH and CH_3OCO-CH_2ONH groups resulted in much less potent compounds but replacement of the isobutyl side chain by butyl, hexyl or octyl groups did not lead to any significant potency differences against collagenase, 92 kDa gelatinase and

72 kDa gelatinase inhibitory potency. N-Methylation of the Trp residue led to a big reduction (>100-fold) in enzyme inhibitory activity but its replacement by Trp(Me), Phe, 3-Bal, 1-Nal, 2-Nal, 4-Pal, 3-Qal, 8-Qal or *tert*-Leu resulted in minor differences in potency.[473] Replacement of the *n*-octyl group in **184** (IC$_{50}$s 11 and 1040 nM, MMP-1 and -3, respectively) by *i*-butyl or *n*-pentyl groups resulted in significant loss in potency. Conversion of the -CH$_2$-COOMe group to -CH$_2$-CH$_2$-O-CH$_3$ group into improve metabolic stability resulted in a compound more potent against MMP-3 (IC$_{50}$s 1890 and 250 nM for MMP-1 and -3, respectively).[475] Analogues of the aminomalonate **185** [X = Ala or Leu and R$_2$ = -NH-(*S*)CH(Ph)CH$_3$, -NH-(*R*)CH(Ph)CH$_3$ and -NH-(*R*)CH(c-C$_6$H$_{11}$)CH$_3$] inhibited MMP-9 more potently (*K*$_i$ 5–18 nM) than MMP-8 (*K*$_i$ 0.8–2.5 μM).[476] The binding mode of a MMP-8 inhibitor **186** (*K*$_i$ 0.3 μM) was determined by X-ray crystallographic analysis.[477]

Non-peptidic inhibitors of matrix metalloproteases published this year include compounds **188-191**.[480–484] The aryl sulfonamide derivatives like **189** (IC$_{50}$ 19–32 nM) containing biaryl, tetrazole, amide, and triple bond were found to be selective inhibitors of MMP-9 and -2. The compounds were inactive against MMP-1, -3, -7, ACE, ECE and NEP (IC$_{50}$ >1 μM). In addition, more potent compounds of the series were orally active in animal models of tumour growth and metastasis.[482] Analogues of the diazepine hydroxamate derivative **191** [R$_1$ = -CH$_2$Ph, -C(O)Ph, C(O)Ph-4-OCF$_3$, -C(O)Ph-2-Ph, -C(O)CH$_2$NH-Boc, -C(O)CH$_2$NH$_2$, -C(O)Bu^t, -COOBu^t, -CONHPh] were active against MMP-1 (IC$_{50}$s 91–703 nM), MMP-9 (IC$_{50}$s

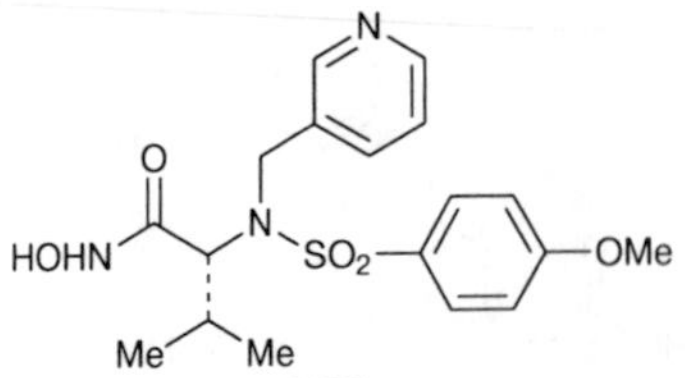

(188) 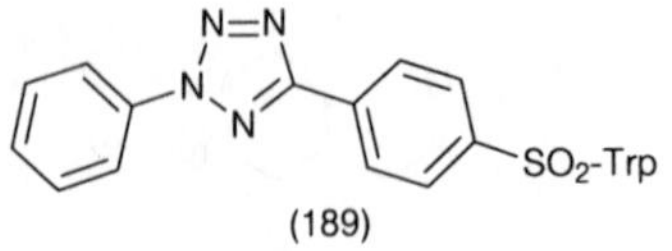   (189)

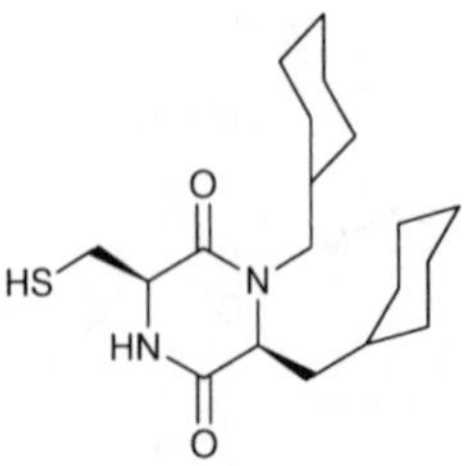

(190) 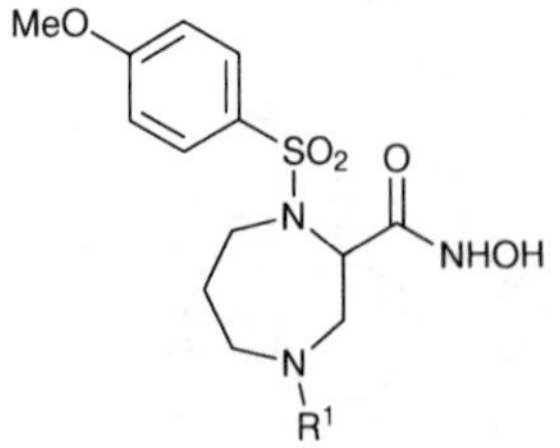   (191)

1.2–157 nM) and MMP-13 (IC$_{50}$s 1.3–65 nM).[484] In general, the compounds were less potent against MMP-1 and nearly equipotent against MMP-9 and MMP-13.

5.11 Phosphatase Inhibitors (Ser/Thr or Tyr) – Using crystallography studies on crystals of protein-tyrosine phosphatase 1B bound either to a high affinity ligand [bis-(*p*-phosphophenyl)methane] or phosphotyrosine, a new aryl phosphate binding site (in addition to the active site) was identified adjacent to the active site.[485] The subsite preferences of human PTP-1B were investigated by screening a synthetic combinatorial library (Asp-X-X-X-mTyr-Leu-Ile-Pro) containing nonhydrolysable peptidomimetics of PTP-1B substrates. Using an iterative approach, Asp-Asp-Glu-Trp-(mTyr)-Leu-Ile-Pro was found to be the most effective inhibitory sequence.[486] Sulfotyrosyl peptides corresponding to known high-affinity *in vitro* substrates were inhibitors of a non-receptor type enzyme, PTP1B, and the receptor-type enzyme, CD45. Many of the casein sulfotyrosyl peptides [Asn-Ala-Asn-Glu-Glu-Glu-(sTyr)-Ser-Ile-Gly-Ser-Ala, Asn-Glu-Glu-Glu-(sTyr)-Ser-Ile-Gly-Ser-Ala, Glu-Glu-Glu-(sTyr)-Ser-Ile-Gly-Ser-Ala, Glu-Glu-(sTyr)-Ser-Ile-Gly-Ser-Ala, Asn-Ala-Asn-Glu-Glu-Glu-(sTyr)-Ser-Ile and Asn-Ala-Asn-Glu-Glu-Glu-(sTyr)-Ser] inhibited both PTP1B (IC$_{50}$s 1-8 μM) and CD45 (IC$_{50}$s 27–60 μM).[487] Further truncation and amino acid substitution led to less potent compounds like Glu-(sTyr)-Ser-Ile-Gly-Ser-Ala, Asn-Ala-Asn-Ala-Ala-Ala-(sTyr)-Ser-Ile-Gly-Ser-Ala and Asn-Ala-Asn-Glu-Glu-Arg-(sTyr)-Ser-Ile-Gly-Ser-Ala against both the enzymes. Some of the sulfotyrosyl peptides based on insulin receptor peptides Thr-Arg-Asp-Ile-(sTyr)-Glu-Thr-Asp, Asp-Ile-(sTyr)-Glu-Thr-Asp, Asp-Ile-(sTyr)-Glu-Thr and Asp-Ile-(sTyr)-Glu, Asp-Ile-(sTyr) were nearly equipotent against both the enzymes (IC$_{50}$s 5–15 μM). Several of the Ac-Asp-Glu-(sTyr)-X [X = D-diphenylalanine, Phe(p-F), Phe(2-F), Pal, Nal, Nle, Nva] type of compounds were 2–50-fold more selective inhibitors of PTP1B. Multiple sulfotyrosyl peptides like sTyr-sTyr-sTyr, Asp-sTyr-sTyr and Glu-sTyr-sTyr also inhibited both enzymes with similar potency (IC_{50}s 6–20 µM).

A thioether-cyclised peptide (192), based on the EGF receptor autophosphorylation sequence Asp-Ala-Asp-Glu-pTyr-Leu, containing pTyr mimetic fluoro-O-malonyltyrosine, was a potent inhibitor of PTP1B (K_i 170 nM).[488] The corresponding linear peptide containing O-malonyl-Tyr residue, Ac-Asp-Ala-Asp-Glu-Tyr(O-malonyl)-Leu-NH_2, was much less potent (K_i 13 µM). Non-peptidyl inhibitors of protein tyrosine phosphatase 1B include compounds 193-195.[488-491] The IC_{50} values for α,α-difluoromethylenephosphonic acid derivatives (193, n = 3 and 4) and the glutamic acid derivative 194 were 2–6 µM.[489,490]

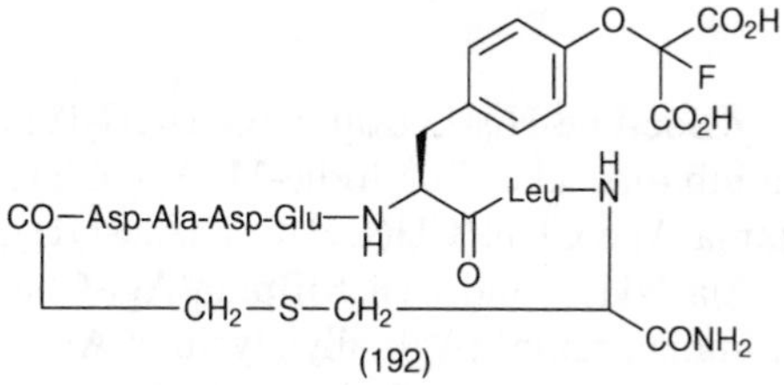

(192)

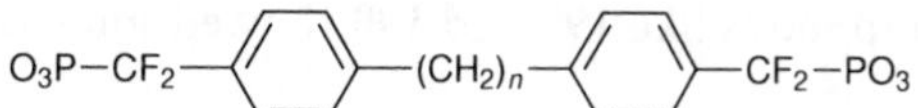

(193)

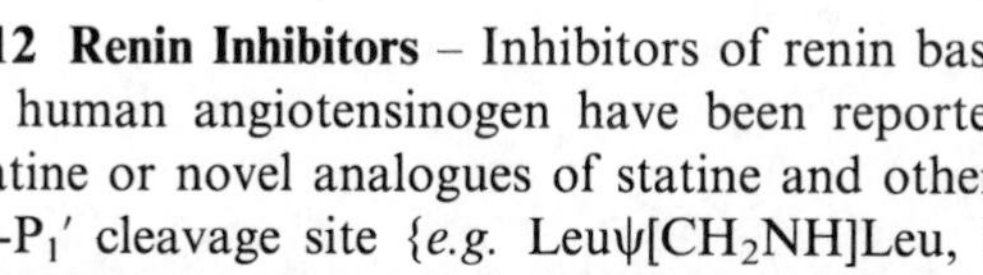

(194)

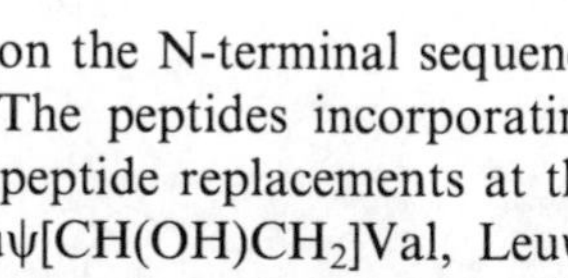

(195)

5.12 Renin Inhibitors – Inhibitors of renin based on the N-terminal sequence of human angiotensinogen have been reported. The peptides incorporating statine or novel analogues of statine and other dipeptide replacements at the P_1-P_1' cleavage site {*e.g.* Leuψ[CH_2NH]Leu, Leuψ[$CH(OH)CH_2$]Val, Leuψ-[$CH(OH)CH_2$]Leu and Leuψ[$CH(NH_2)CH_2$]Val} have been found to inhibit renin from different species.[492-494] Various compounds like 196 and 197 inhibited both human and rat renin. *In vitro*, peptide Piv-His-Pro-Phe-His-Leuψ[$CH(OH)CH_2$]Leu-Tyr-Tyr-Ser-NH_2 was a potent inhibitor of rat plasma renin (IC_{50} 0.21 nM) but was a much weaker inhibitor of human renin (IC_{50} 45 nM). Boc-Phe-His-Leuψ[$CH(OH)CH_2$]Val-Ile-His-OH was more potent against human renin (IC_{50} 2800 and 3.6 nM, rat and human renin, respectively). Peptide Boc-His-Pro-Phe-His-Leuψ[$CH(OH)CH_2$]Leu-Val-Ile-His-NH_2 was a highly effective inhibitor of both rat and human renin *in vivo*. When infused (1 mg kg^{-1} h^{-1}) into two-kidney, one-clip chronic renal hypertensive rats it lowered blood pressure and suppressed both plasma renin and angiotensin II.

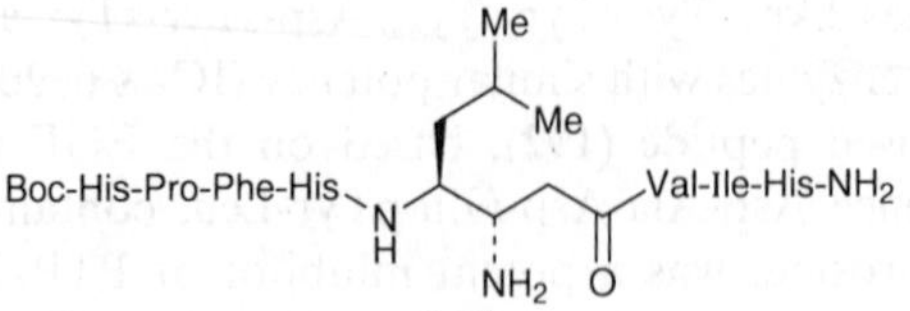

(196)

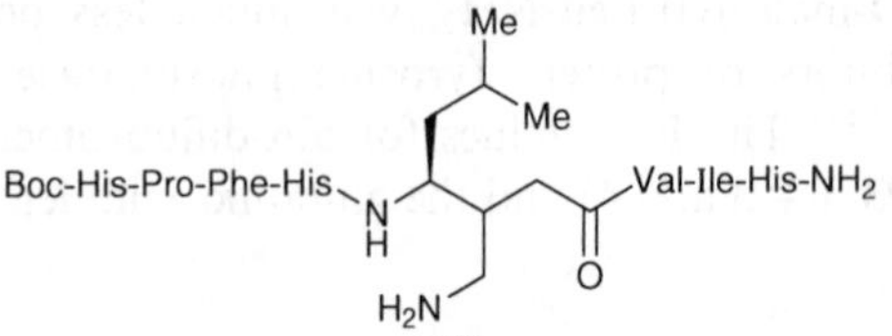

(197)

SAR studies, using Boc-Phe-His-Leuψ[CH(OH)CH$_2$]Val-Ile-His as the lead structure, led to inhibitors like Boc-hPhe-MeApe-Chaψ[CH(OH)CH$_2$]Ala-NH-CH$_2$-2-Py, Dnma-Ape-Chaψ[CH(OH)CH$_2$]Ala-NH$_2$, (R,S)Bpma-Ape-Chaψ[CH(OH)CH$_2$]Ala-NH$_2$ and (R,S)Bpma-Agl-Cha([CH(OH)CH$_2$]Val-NH$_2$ (IC$_{50}$s1-4 nM, human renin) [Agl, allylglycine; Ape, 2S-aminopentanoic acid; Bpma 2-benzyl-2-(4'-pyridylmethyl)acetyl; Dnma, di-2,2-(1'-naphthyl-methyl)acetyl]. Compounds like **198** and **199** showed some oral absorption.[494]

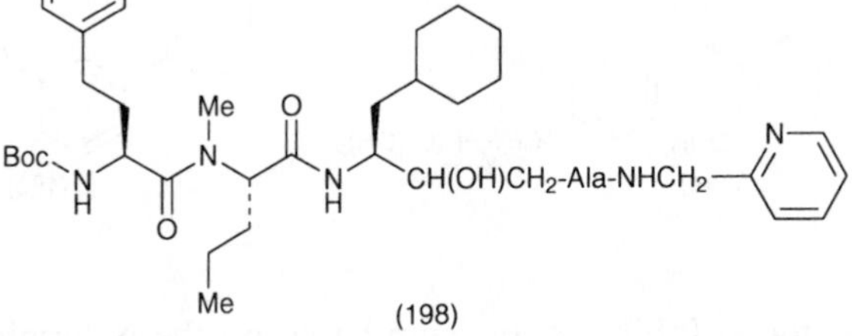

(198)

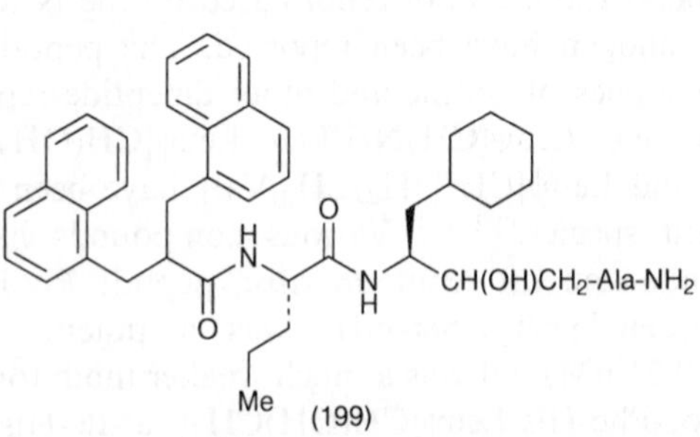

(199)

5.13 Thrombin Inhibitors (Serine Protease) – In addition to its role as a proteolytic agent in the coagulation cascade, other activities of thrombin and thrombin receptor peptide have been reported. For example, the involvement of thrombin receptor in cell invasion associated with tumour progression and normal embryonic development has been highlighted. The receptor (a member of the protease-activated receptor family) is preferentially expressed in highly metastatic human breast carcinoma cell lines and breast carcinoma biopsy specimens.[495] Introduction of thrombin receptor antisense cDNA considerably inhibited the invasion of metastatic breast carcinoma cells in culture through a reconstituted basement membrane. Thrombin cleaved recombinant human thrombopoietin between Arg^{191}-Thr^{192} and Arg^{117}-Thr^{118} residues. In a cell proliferation assay, the generation of thrombopoietin(1-191) (fragments still connected with a disulfide bridge) raised the *in vitro* activity but further cleavage at Arg^{117}-Thr^{118} reduced or destroyed the activity.[496]

New information of thrombin in platelet activation has been presented.[497,498] Thrombin cleaves its G protein-linked seven-transmembrane domain receptor, thereby releasing a 41-amino acid peptide [Met-Gly-Pro-Arg-Arg-Leu-Leu-Leu-Val-Ala-Ala-Cys-Phe-Ser-Leu-Cys-Gly-Pro-Leu-Leu-Ser-Ala-Arg-Thr-Arg-Ala-Arg-Arg-Pro-Glu-Ser-Lys-Ala-Thr-Asn-Ala-Thr-Leu-Asp-Pro-Arg] and generating a new amino terminus that acts as a tethered ligand for the receptor. The 41-amino acid peptide has been shown to activate platelets.[497] Human thrombin caused a dose-related platelet accumulation in guinea pigs. Responses of similar magnitude were induced by Ser-Phe-Leu-Leu-Arg-Asn and Ala-Phe(p-F)-Arg-Cha-HArg-Tyr-NH$_2$ (high-affinity thrombin receptor-activating peptide).[498] A peptide sequence within GPIbα (residues 269-287, Asp-Glu-Gly-Asp-Thr-Asp-Leu-Tyr-Asp-Tyr-Tyr-Pro-Glu-Glu-Asp-Thr-Glu-Gly-Asp) and some of its fragments [Asp-Glu-Gly-Asp-Thr-Asp-Leu-Tyr-Asp-Tyr-Tyr, Pro-Glu-Glu-Asp-Thr-Glu-Gly-Asp, Asp-Tyr-Tyr-Pro-Glu-Glu-Asp-Thr-Glu-Gly-Asp, Asp-Leu-Tyr-Asp-Tyr-Tyr-Pro-Glu-Glu-Asp] inhibited α-thrombin binding to platelets.[499] One of the smallest peptides Asp-Tyr-Tyr-Pro-Glu was only 2-fold less potent than the parent peptide. A few analogues of the pentapeptide [Asp-Tyr-Tyr-Pro-Glu, Asp-Ala-Ala-Pro-Glu, Ala-Tyr-Tyr-Pro-Glu and Ala-Ala-Ala-Pro-Glu] were comparable in potency to the parent pentapeptide, but a number of others without a C-terminal glutamic acid residue [Ala-Ala-Ala-Pro-Glu-NH$_2$, Ala-Ala-Ala-Pro-Gln, Ala-Ala-Ala-Pro-Asp, Ala-Tyr-Tyr-Pro-Asp, Ala-Tyr-Tyr-Pro-Gly and Asp-Ala-Ala-Pro-Gly] were inactive (IC$_{50}$ >400 μM).

As in previous years, a number of publications on thrombin inhibitors have appeared. Progress in the field has been reviewed.[500] New analogues based on the D-Phe-Pro-Arg-H containing modifications in the P$_1$ to P$_3$ positions have been reported.[501-503] Examples of such compounds include **200–202**. Analogues of D-diphenylAla-Pro derivative **200** containing various different substituted benzylamide groups [2,5-Me-4-CH$_2$NH$_2$, 2,3,5,6-Me-4-CH$_2$NH$_2$, 2,5-Me, 4-NMe$_2$, 4-SO$_2$NH$_2$, 2,3-Cl, 2,3-Me, 2,3-OMe, 3,5-Cl, 2,6-OMe, 3,5-OMe, 2-Me, 3-Me, 3-OMe, 3-Cl, 3-Br, 3-OH and 2,5-Cl] indicated that 2,5-lipophilic substituents were optimal for benzyl P$_1$ moieties. Both the oxyacetic amide **(201)** (K_i 0.74 nM) and 9-hydroxy-9-fluorenecarboxy (L-372,460, **202**, $K_i = 1.5$ nM) derivatives showed oral bioavailability (20–80%) in different animal species. Compound **202** was a weak inhibitor of trypsin (K_i 860 nM) and did

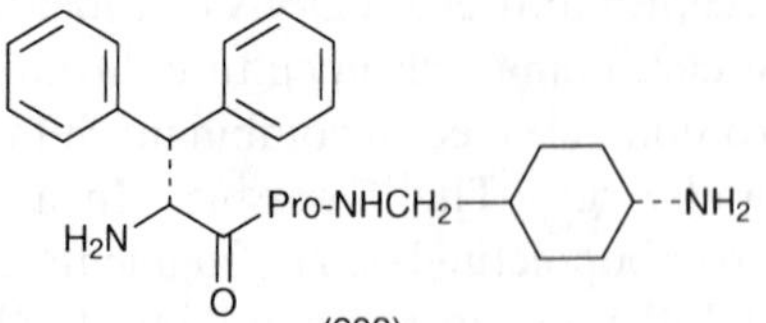

(200)

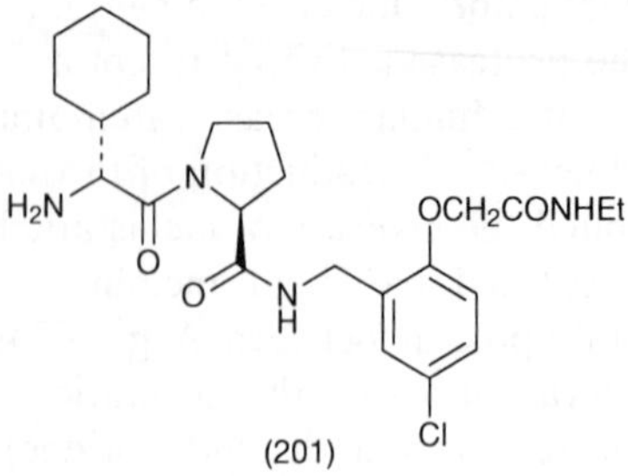

(201)

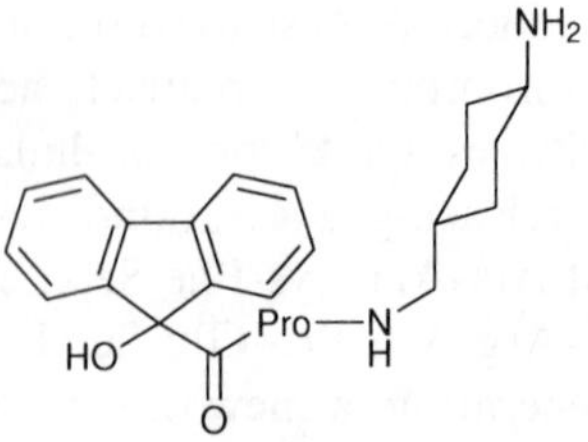

(202) L-372,460

not inhibit several other serine proteases (*e.g.* plasmin, tPA, activated protein C, plasma kallikrein and chymotrypsin, K_i (20 μM).

Thrombin inhibitors containing a guanidinium group in the P_1 position and conformationally restricting (*e.g.* lactam) residues in the P_2-P_4 positions are reported.[504-509] Compounds 203–205 were more potent inhibitors of thrombin than other serine proteases. Analogues of 204 (K_i 0.6 nM) in which the phenylpropionyl group (P_3 position) was replaced by other hydrophobic groups [*e.g.* Ph-CH$_2$-CH$_2$-CH$_2$-CO-, naphthyl-CH$_2$-CH$_2$-CO-, 4-Me-Ph-CH$_2$-CH$_2$-CO-, 3,4-dichloro-Ph-CH$_2$-CH$_2$-CO-, 2-F-Ph-CH$_2$-CH$_2$-CO- and 2-CF$_3$-Ph-CH$_2$-CH$_2$-CO-] showed comparable potency to 204. Several of the com-

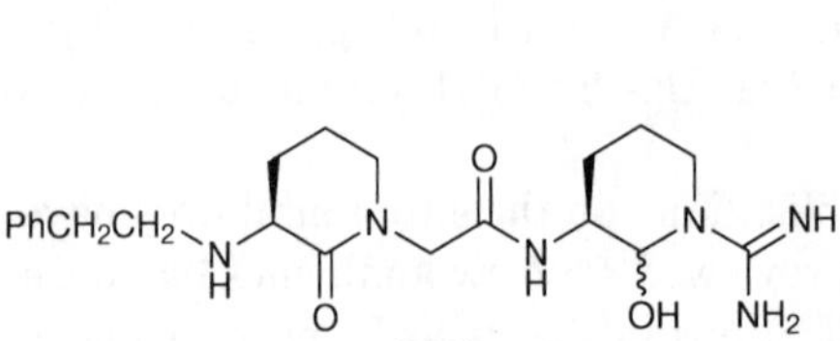

(203)

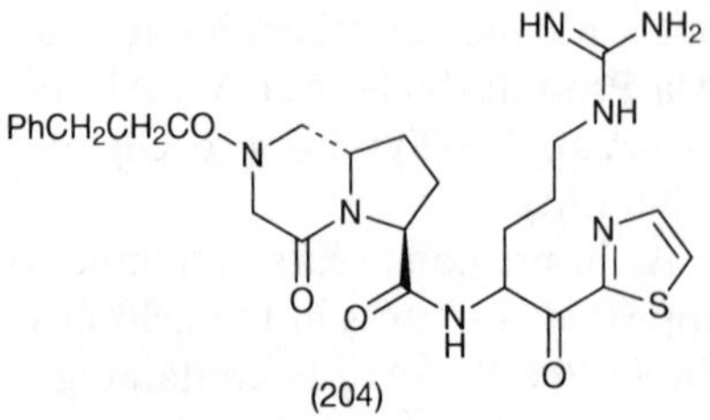

(204)

pounds like 205 (IC$_{50}$ against thrombin and trypsin 1 and 11940 nM, respectively) showed much greater selectivity but none of the compounds showed any significant oral activity when administered to rats at a dose of 30 mg kg. Examples of other inhibitors containing conformationally restricting moieties in the P_3-P_2 regions include compounds 206–208.[510-512] Compound 207 containing a β-alanylguanidine as a P_1 replacement was a weak inhibitor of thrombin and showed only a 10-fold selectivity (IC$_{50}$s 0.24 and 3.4 μM

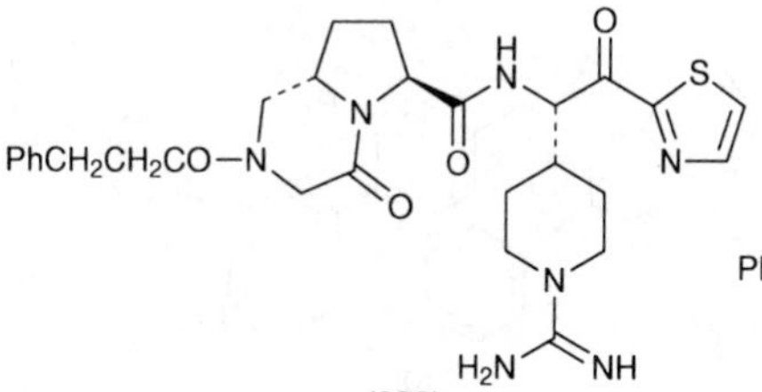

(205)

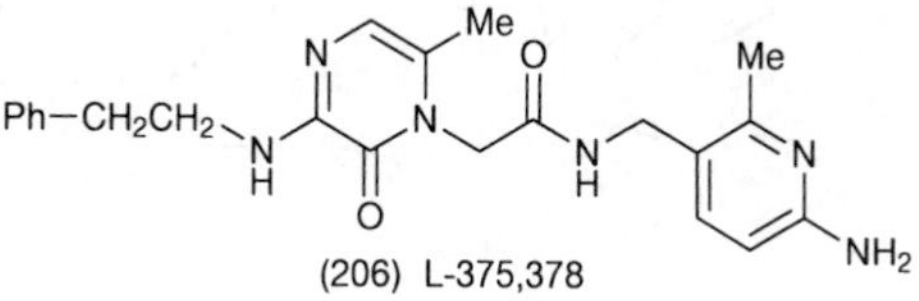

(206) L-375,378

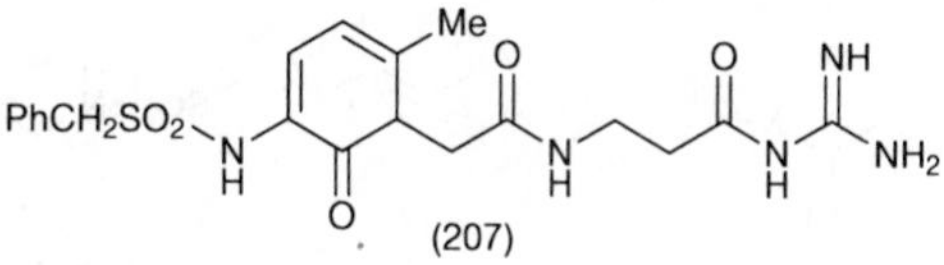

(207)

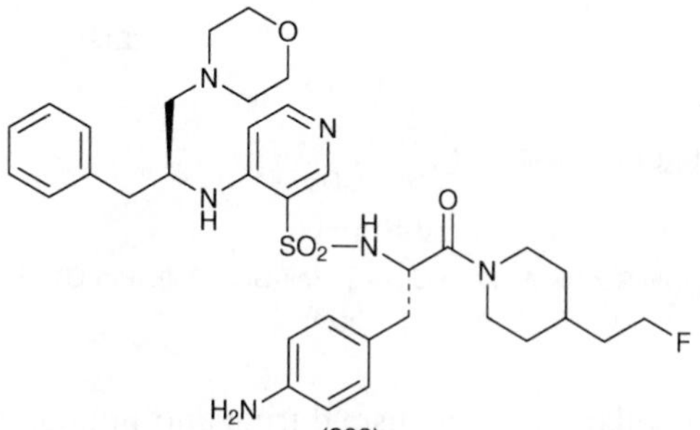

(208)

against thrombin and trypsin, respectively). Compound **208** (thrombin K_i 22 nM) showed improved pharmacokinetics in the rat (61% oral bioavailability, elimination half-life 59 min).

Non-peptidic inhibitors of thrombin have been reported.[513-516] Various examples from different classes include compounds **209-212**. The benzodiazepinone derivatives like **209** were weak inhibitors of thrombin. Compounds like **211** were highly unstable in plasma. Compound **212** containing a Phe(p-CH$_2$NH$_2$) residue in the P_1 position was one of the more potent and selective inhibitors of thrombin (K_i values against thrombin and trypsin 6.6 and 14,200 nM, respectively) and showed good oral bioavailability in rats ($\sim$70%) but low oral bioavailability in dogs (10-15%). Bifunctional peptide boronate inhibitors of thrombin like **213** and an analogue containing an isothiouronium [-(CH$_2$)$_3$-S-C(NH)NH$_2$] group in place of the bromopropyl group have been reported.[517]

5.14 Miscellaneous [Dipeptidyl-peptidase, Protein Tyrosine Kinase, Serine Proteases Like Trypsin and Tryptase] Inhibitors – Publications on dipeptidyl-peptidase IV (a peptidase activity similar to the T-cell-activation antigen CD26) (serine proteases), including a purification strategy for dipeptidyl-peptidase IV-β using the CD26α-) cell line C8166, have appeared.[518] The proteolytic processing of chemokines by CD26/DPP IV has been suggested to

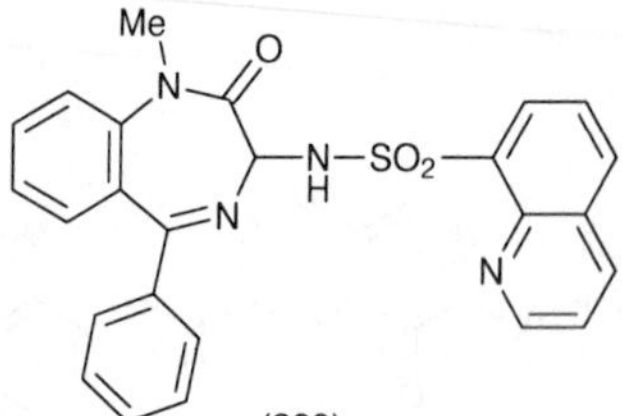

(209)

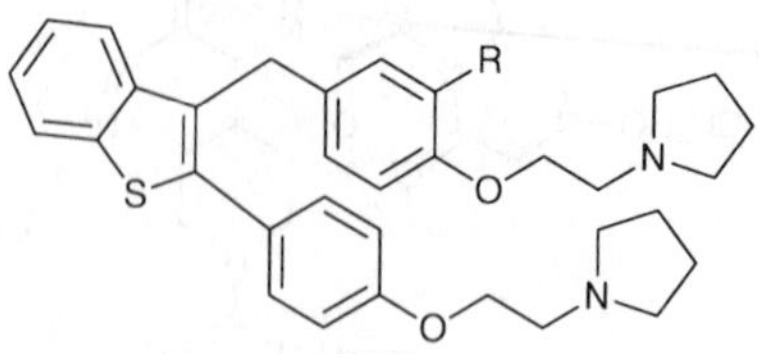

(210)

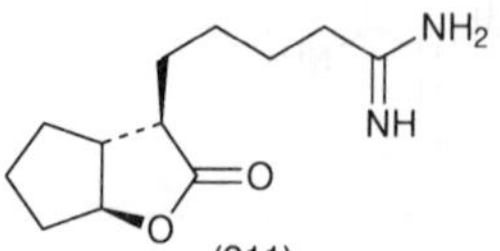

(211)

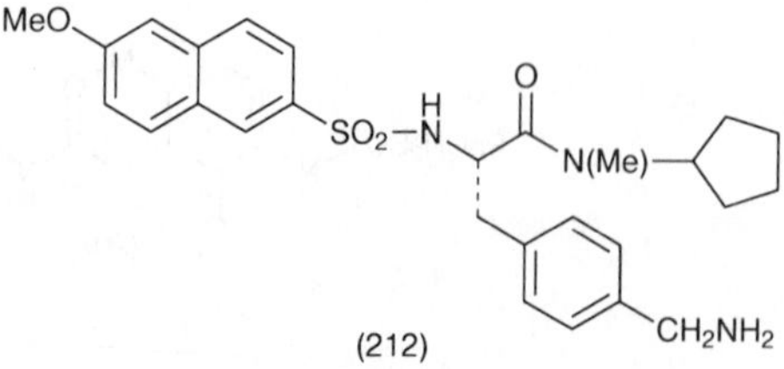

(212)

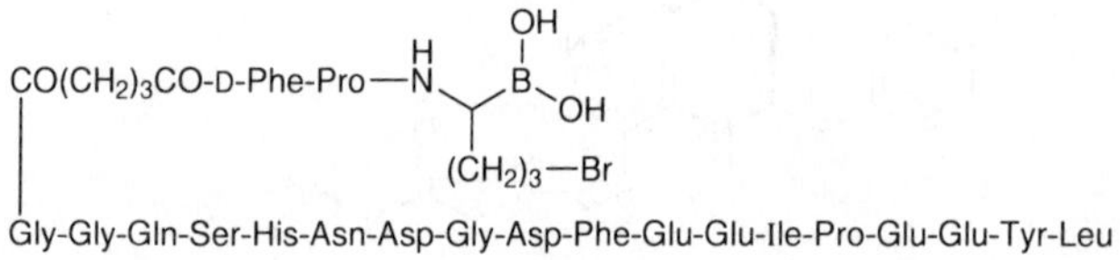

CO(CH₂)₃CO-D-Phe-Pro—N(H)—B(OH)(OH) ...

Gly-Gly-Gln-Ser-His-Asn-Asp-Gly-Asp-Phe-Glu-Glu-Ile-Pro-Glu-Glu-Tyr-Leu

(213)

be an important regulatory mechanism during anti-inflammatory and antiviral responses.[519] Derivatives of Ala-Pro like **214** and **215** have been reported as inhibitors of dipeptidyl peptidase IV.[520]

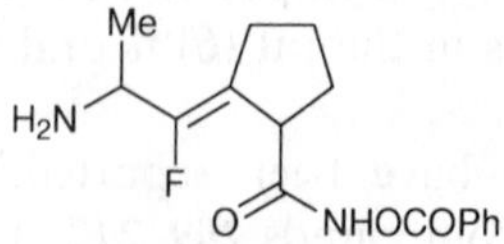

(214)

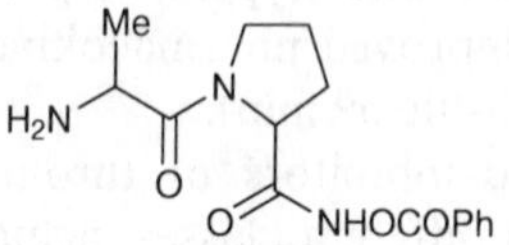

(215)

Peptide based substrates and inhibitors of protein tyrosine kinases have been reported.[521-524] An investigation on a series of cyclic peptides based on the amino acid sequence surrounding the autophosphorylation site of pp60(c-src) led to peptides like c(Asp-Asn-Gln-Tyr-Ala-Ala-Arg-Gln-D-Phe-Pro) and c(Asp-Asn-Gln-Tyr-Ala-Phe-Phe-Gln-D-Phe-Pro). The change of residues 6 and 7 resulted in a 42-fold increase in affinity.[521] A linear version of this peptide was 13-fold less potent an inhibitor than the cyclic peptide. Based on the heptapeptide (Glu-Asp-Asn-Glu-Tyr-Thr-Ala) of the autophosphorylation site of pp60(c-src), linear and cyclic tetradecapeptides [Glu-Asp-Asn-Glu-Tyr-Thr-Ala-Glu-Asp-Asn-Glu-Tyr-Thr-Ala, c(Glu-Asp-Asn-Glu-Tyr-Thr-Ala-Glu-Asp-Asn-Glu-Tyr-Thr-Ala), Glu-Asp-Asn-Glu-Phe-Thr-Ala-Glu-Asp-Asn-Glu-Tyr-Thr-Ala, Glu-Asp-Asn-Glu-Tyr-Thr-Ala-Glu-Asp-Asn-Glu-Phe-Thr-Ala, c(Glu-Asp-Asn-Glu-Phe-Thr-Ala-Glu-Asp-Asn-Glu-Tyr-Thr-Ala)] were synthesised and tested for their ability to behave as phosphoacceptor substrates for Lyn, a tyrosine kinase purified from spleen and belonging to the Src family.[522] The data showed that all the peptides could be phosphorylated by Lyn.

Based on a substrate of $p60^{c\text{-src}}$ protein tyrosine kinase, Tyr-Ile-Tyr-Gly-Ser-Phe-Lys-NH_2 (obtained by combinatorial methods), conformationally constrained substrate based inhibitors of the enzyme were synthesised.[523] Replacement of the Tyr^3 by Nal(2) resulted in an inhibitor of the enzyme (IC_{50} 66 μM). Many linear and cyclic peptide analogues of the substrate containing a Tyr or Nal(2) residue, [Tyr-c(Pen-Tyr-Gly-Ser-Phe-Cys)-Lys-Lys-NH_2, Tyr-Pen-Nal(2)-Gly-Ser-Phe-Cys-Lys-Lys-NH_2, Tyr-c(Pen-Nal(2)-Gly-Ser-Phe-Cys)-Lys-Lys-NH_2, c(Cys-Ile-Tyr-Gly-Ser-Phe-Cys)-Lys-Lys-NH_2, Tyr-c(D-Pen-Tyr-Gly-Ser-Phe-Cys)-Lys-Lys-NH_2, Tyr-c(D-Pen-Tyr-Gly-Ser-Phe-D-Cys)-Lys-Lys-NH_2, Tyr-c(D-Pen-Nal(2)-Gly-Ser-Phe-Cys)-Lys-Arg-NH_2 and Tyr-c(D-Pen-Nal(2)-Gly-Ser-Phe-Cys)-Lys-Arg-NH_2, showed 20–30-fold improvement in the enzyme inhibitory activity. Introduction of four stereoisomers of β-MeNal(2) in place of the Nal(2) in Tyr-c(D-Pen-Nal(2)-Gly-Ser-Phe-Cys)-Lys-Arg-NH_2 did not show any significant differences in the enzyme inhibitory activity. The IC_{50} values for the (2R,3R)-, (2R,3S)-, (2S,3S)-and (2S,3R)-β-Me-Nal(2) analogues were in the range of 1–2.9 μM. Two other analogues, Tyr-c(D-Pen-Tyr(3-I)-Gly-Ser-Phe-Cys)-Lys-Arg-NH_2 and Tyr-c(D-Pen-Tyr(3,5-di-I)-Gly-Ser-Phe-Cys)-Lys-Arg-NH_2, were somewhat more potent (IC_{50} values 0.13 and 0.54 μM, respectively). The four β-MeNal(2) containing peptides did not show significant activity against other Src family protein tyrosine kinases (Lck, Lyn) but the two iodinated tyrosine containing analogues were active against Lck (IC_{50} 11 μM).

SAR studies on a specific inhibitor of calmodulin-dependent protein kinase II (Lys-Lys-Ala-Leu-Arg-Arg-Gln-Glu-Ala-Val-Asp-Ala-Leu, IC_{50} 32 nM) were reported.[524] Replacement of Leu^4, Arg^6, Gln^7, Ala^9 and Val^{10} by other amino acid residues produced a marked increase in the IC_{50} value (270–30,000 nM). Although replacement of Ala^3, Glu^8, Ala^{12}, and Leu^{13} by other residues produced no significant increase in the IC_{50}, the substitution of Lys for Ala^3 decreased the IC_{50}. An analogue (Lys-Lys-Lys-Leu-Arg-Arg-Gln-Glu-Ala-Phe-Asp-Ala-Tyr), in which Ala^3 and Val^{10} were replaced with Lys and Phe, respectively, showed an IC_{50} value as low as 4 nM.

Inhibitors of serine proteases like trypsin and tryptase have been reported.[525] An inhibitor of trypsin **216** was shown to use a new mode of high-affinity binding in which a Zn^{2+} ion is tetrahedrally coordinated between two chelating nitrogens of the inhibitor and two active site (His^{57} and Ser^{195}) residues.[526] A series of benzothiazolone 1,1-dioxide derivatives like **217** (IC_{50} 64 nM) that rely predominantly upon interactions with S′ subsites of the enzyme for specificity and potency have been reported as inhibitors of human mast cell tryptase.[527] Compound **217** was a 40-fold weaker inhibitor of elastase, 100-fold weaker against trypsin, and showed no inhibition against thrombin. In the delayed-type hypersensitivity mouse inflammation model, a 5% solution of a similar compound (**218**) reduced oedema by 69% compared to control animals.

(216)   (217)

(218)

6 Phage Library Leads

Random peptide libraries are displayed on filamentous bacteriophage as fusions of either the minor coat protein, pIII, or the major coat protein, pVIII. A method of isolating the peptides displayed on a phage clone by transferring it to the N-terminus of the maltose-binding protein of *Escherichia coli* encoded by *mal*E is described.[528] Peptide-maltose-binding protein fusions are also easily affinity purified on amylose columns. A filamentous phage-based selection and functional screening method was devised using somatostatin as a model for identifying ligands specific for G protein-coupled receptors.[529] Peptide displaying phage bound to a polyclonal anti-somatostatin serum, and to several somatostatin receptor subtypes expressed on transfected CHO-K1 cells. A general method for the stabilisation of proteins that links the protease resistance of stabilised variants of a protein with the infectivity of a filamentous phage is reported.[530] In this method, a repertoire of variants of the protein to be stabilised is inserted between two domains of the gene-3-protein (minor coat protein) and the resulting pool of phage is subjected to an *in vitro* proteolysis step. The phage that host's the most stable variants should escape proteolysis most frequently and thus remain infectious, leading to a simple selection procedure.

6.1 Erythropoietin and Thrombopoietin Mimetics – Cloning, characterisation, pharmacological properties, clinical implications and clinical trial results on thrombopoietin are reviewed.[531] SAR studies on an erythropoietin mimetic peptide [Gly-Gly-Thr-Tyr-Ser-c(Cys-His-Phe-Gly-Pro-Leu-Thr-Trp-Val-Cys)-Lys-Pro-Gln-Gly-Gly] identified by a phage library approach were reported.[532,533] Alanine scanning results indicated that only the replacement of Leu[11] by Ala[11] gave a compound with similar potency to the parent peptide. Replacements of Tyr[4] or Trp[13] by Ala resulted in the least potent analogues (100–500-fold loss in potency). Thus, the two hydrophobic amino acids, Tyr[4] and Trp[13], appear essential for mimetic action. Except the Phe(p-F)[4] analogue which retained most of the activity, other compounds containing ring substituted phenylalanine derivatives [Phe(p-NO$_2$), Phe(p-NH$_2$), Phe(p-I) and

Tyr(3,5-dibromo)] and D-Tyr were much less potent. The two glycine residues from the N- and C-terminal ends of EMP1 could be deleted without significant reduction in potency, but any further truncation led to much less potent compounds. For example, Tyr-Ser-c(Cys-His-Phe-Gly-Pro-Leu-Thr-Trp-Val-Cys) was >500-fold less potent than the parent peptide.

Dimerisation of the erythropoietin receptor, in the presence of either the natural ligand or synthetic erythropoietin-mimetic peptides, is the principal extracellular event that leads to receptor activation. The crystal structure of the extracellular domain of erythropoietin receptor bound to an inactive (antagonist) [Gly-Gly-Thr-Tyr(3,5-dibromo)-Ser-c(Cys-His-Phe-Gly-Pro-Leu-Thr-Trp-Val-Cys)-Lys-Pro-Gln-Gly-Gly] peptide indicated that dimerisation still occurs, but the orientation between receptor molecules is altered relative to active (agonist) peptide complexes.[533] Erythropoietin dimers were biologically active *in vivo* as shown by their ability to increase the hematocrits of mice when injected subcutaneously. In addition, the dimers exhibited >26-fold higher activity *in vivo* than did the monomers.[534] Non-peptide ligands for the erythropoietin receptor [**219** (agonist) and **220** (antagonist)] have been reported.[535] In the UT-7/TPO cell proliferation assay, **219**, but not **220**, stimulated the proliferation of UT-7/TPO cells. When UT-7/TPO cells were cultured with **220** at 10–100 μM in the presence of 1 ng ml^{-1} human thrombopoietin, the proliferation of the cells was dose-dependently inhibited by **220**.

(219)   (220)

6.2 Receptor Ligands – Using a random pentadecamer peptide library and monoclonal antibodies AD117m and H11, glycosphingolipid mimicking peptides [Arg-Asn-Val-Pro-Pro-Thr-Phe-Asn-Asp-Val-Tyr-Trp-Ile-Ala-Phe and Val-Pro-Pro-Cys-Phe-Thr-Leu-Met-Tyr-Cys-Ala-Gly-Val-Val-Arg] were identified.[536] Binding activity of the synthetic peptides showed that 9-mer peptides [*e.g.* Val-Pro-Pro-Tyr-Phe-Thr-Leu-Met-Tyr and Val-Pro-Pro-Ser-Phe-Thr-Leu-Met-Tyr] were enough to mimic the epitope carbohydrate structure. A 15-mer library has also been used to identify TNF-α antagonists. The synthetic peptide Asp-Phe-Leu-Pro-His-Tyr-Lys-Asn-Thr-Ser-Leu-Gly-His-Arg-Pro was shown to bind to the recombinant human TNF-α using surface plasmon resonance technology and to inhibit both recombinant mouse and human TNF-α-induced cytotoxicity *in vitro* in a dose-dependent fashion.[537] A peptide [Val-Ser-Phe-Arg-Gly-Ser-Ser-Ser-Ala-Cys-Met-Leu-Val-Ser-Cys] isolated from a random 15-residue phage peptide library was a structural mimic to the P^1, P^4-diadenosine 5'-tetraphosphate binding site on its receptor.[538]

To design peptide antagonists with conformational similarity to a CTLA4-binding domain, an anti-T lymphocyte costimulatory molecule monoclonal antibody, CTLA4Mab (UC10-4F10-11), was used to select the specific binding phage clones. Two peptide motifs [Gly-Phe-Val-Cys-Ser-Gly-Ile-Phe-Ala-Val-Gly-Val-Gly-Arg-Cys, and Ala-Pro-Gly-Val-Arg-Leu-Gly-Cys-Ala-Val-Leu-Gly-Arg-Tyr-Cys] recognised by the antibody were isolated.[539] These exhibited immunopotentiating activity. Troponin C antagonists were obtained from a phage-displayed random peptide 12-mers library. The troponin C-selected peptides yield a consensus binding sequence of (Val/Leu)(Asp/Glu)X-Leu-Lys-X-X-Leu-X-X-Leu-Ala. Biotinylated peptides corresponding to library-derived sequences and similar sequences from various isoforms of troponin I were synthesised and shown to bind troponin C specifically.[540] Measurement of equilibrium dissociation constants of the peptides by surface plasmon resonance yielded dissociation constants for troponin C as low as 0.43 μM for Ser-Arg-Leu-Asp-Tyr-Leu-Lys-Ser-Ser-Leu-Leu-His-Leu-Gly-Ser-Arg and 0.86 μM for Ser-Arg-Lys-Leu-Gln-Leu-Lys-Thr-Leu-Leu-Leu-Gln-Ile-Ala-Lys-Gln; in contrast, dissociation constants for calmodulin were greater than 6 μM for all peptides studied.

6.3 Enzyme Substrates and Inhibitors – A cyclic nonapeptide, Cys-Cys-Phe-Ser-Trp-Arg-Cys-Arg-Cys, isolated from a filamentous bacteriophage-based library completely inhibited the enzymatic activity of α-chymotrypsin.[541] The binding of the phage expressing this peptide to α-chymotrypsin was inhibited by the synthetic peptide. Plasminogen activator inhibitor 1 binding peptides were obtained from a library.[542] Examples of such peptides included Leu-Leu-Leu-Trp-Pro-Leu-Ser-Glu-Lys-Pro-Val-Val-Leu-Pro-Glu, Leu-Leu-Gly-Trp-Met-Gly-Leu-Val-Glu-His-Asp-Trp-Leu-Pro-Leu and Ser-Leu-Trp-Ser-Leu-Val-Leu-Gly-Pro-Glu-Ser-Ile-Leu-Gly-Pro. One of the synthetic peptides [Leu-Leu-Leu-Trp-Pro-Leu-Ser-Glu-Lys-Pro-Val-Val-Leu-Pro-Glu] inhibited the binding of peptide carrying phages to immobilised plasminogen activator inhibitor 1 in a dose dependent manner. However, none of these phage-bound peptides prevented the interaction between plasminogen activator inhibitor 1 and its target serine protease urokinase (u-PA). To isolate peptides that block the interaction between plasminogen activator inhibitor 1 and u-PA, phages bound to immobilised PAI-1 were eluted by incubation with u-PA. This resulted in elution of a phage type consisting of 49 amino acid residues [Pro-Val-Ser-Gln-Phe-Val-Phe-Leu-Cys-Gly-His-Gln-Pro-Cys-Phe-Thr-Ser-Glu-His-Ala-His-Asp-Val-Pro-Asp-Pro-Ala-Pro-Pro-His-His-Pro-Leu-Glu-Leu-Ile-Thr-Gly-Arg-Gln-Ala-Thr-Pro-Ile-Ser-Val-Gly-Met-Ser]. u-PA inhibited the binding of this peptide to PAI-1 in a dose dependent manner.

6.4 Antibody Related Phage Peptides – Monoclonal antibodies that are entirely human are being produced from phage display and transgenic mice. A review covering the use of both technologies has appeared.[543] The use of catalytic antibody technology, based on the enormous diversity of the immune system, is described as a method for the generation of novel protein catalysts that can facilitate different chemical transformations.[544] CCR5-specific phage antibodies were obtained from a large phage display human library. The selected population of phage antibodies were subsequently used to guide molecules for a second phase of selection that was carried out in the absence of macrophage inflammatory protein-1α. This generated a panel of CCR-5-binding antibodies, of which around 20% inhibited macrophage inflammatory protein-1α binding to CD4+. The antibodies generated by this step-back selection procedure also inhibited macrophage inflammatory protein-1α-mediated calcium signalling.[545]

A large library of phage-displayed human single-chain Fv antibodies (scFv) containing 6.7×10^9 members was generated and fourteen different protein antigens were used to affinity select antibodies from the library.[546] Measurements of antibody-antigen interactions revealed several affinities below 1 nM, comparable to affinities observed during the secondary murine immune response. In particular, four different scFv recognising the ErbB2 protein had affinities ranging from 220 pM to 4 nM. Antibodies derived from the library proved to be useful reagents for immunoassays. For example, antibodies generated to the *Chlamydia trachomatis* elementary bodies stained *Chlamydia*-infected cells, but not uninfected cells.

7   Protein-Protein Interaction Inhibitors

Signal transduction cascades involve multiple enzymes and are orchestrated by selective protein-protein interactions that are essential for the progression of intracellular signalling events. Modulators of these protein-protein interactions have been used to dissect the role of individual components of each signalling cascade. Methods that have been developed for the identification of peptides that inhibit the interaction between signalling proteins and hence selectively modulate their functions have been reviewed.[547] The approaches include selection of peptide modulators using random or biased peptide libraries, and rational approaches like identification of conserved sequences in homologous domains of otherwise unrelated proteins, identification of evolutionarily conserved sequences and identification of sequences involved in intramolecular interactions. NMR-based experimental approaches for the design of functional mimetics of protein-protein interactions have also been reported.[548]

7.1   SH2 and SH3 Domain Ligands – Phosphorylated tyrosine residues of growth factor receptors that associate with intracellular proteins containing src-homology 2 (SH2) domains are integral components in several signal transduction pathways related to proliferative diseases such as cancer, atherosclerosis, and restenosis. A phosphorylated pentapeptide, pTyr-Val-Pro-Met-Leu, derived from the primary sequence of platelet-derived growth factor-β (PDGF-β) receptor blocks the association of the C-terminal SH2 domain of the p85 subunit of phosphatidylinositol 3-kinase to PDGF-β receptor (IC_{50} 0.44 μM). Further SAR studies resulted in smaller peptidomimetics with enhanced affinity. The extracellular administration of either CF_2Pmp-Val-Pro-Met-Leu or Ac-CF_2Pmp-Val-Pro-Met-NH_2 in a whole cell assay resulted in inhibition of the PDGF-stimulated association from the C-terminal SH2 domain of the p85 subunit of PI 3-kinase to the PDGF-β receptor in a dose-dependent manner.[549] In comparison to the phosphopeptide **221** which exhibited 98% inhibition of binding of the p85 N-terminal SH2 domain to a CD19 phosphopeptide at 20 μM, the pyridone phosphonate-based ligand **222** was less potent (50% inhibition at 50 μM).[550]

(221)   (222)

Dipeptide analogues as ligands for the pp60[c-src] SH2 domain are described.[551] The C-terminally modified analogues of Ac-Tyr(PO_3H_2)-Glu-N(n-C_5H_{11})$_2$ containing -N(CH_3)-(CH_2)$_6$-OH, -N(CH_3)-(CH_2)$_7$-OH and -N(CH_3)-(CH_2)$_8$-OH were somewhat more potent than the -N(n-C_5H_{11})$_2$ compound. All the other C-terminal amides containing aliphatic and aromatic substituents were significantly less potent. Replacement of the N-terminal acetyl group also led to mostly less potent compounds. Only (4-NO_2)Ph-CO- and (4-MeO)Ph-CO- analogues were equipotent to the parent peptide. Replacement of the glutamic acid residue by Met resulted in 2-fold reduction in binding affinity. The crystal structure of the free human p56[lck] SH2 domain and in complex with an inhibitor containing p-(carboxymethyl)Phe residue (cmF) as a phosphotyrosine replacement has been reported.[552] The binding mode of the acetyl-cmPhe-Glu-Glu-Ile (Kd 42 μM) was very similar to that of the pTyr-Glu-Glu-Ile (Kd 0.09 μM).

Peptidomimetic ligands of the GRB2-SH2 domain include compounds **223** and **224**. Compound **223** was one of the more potent compounds of the type 3-amino-Z-Tyr(PO_3H_2)-X_{+1}-Asn-NH_2 phosphopeptides (IC_{50} 1 nM).[553] The corresponding analogue containing a 1-amino-1-cyclopentanecarboxylic acid

(223)   (224)

in place of the 1-amino-1-cyclohexanecarboxylic acid was about 20-fold less potent. Other analogues containing replacements for the 1-amino-1-cyclohexanecarboxylic acid residue (*e.g.* Ile, Aib and 1-amino-1-cyclopropanecarboxylic acid) were 60-200-fold less potent. The phosphopeptide **223** shows a 240- and 1550-fold preferential binding to Grb2-SH2 over p85 N-terminal SH2 and Lck SH2, respectively, and at least 1250-fold selectivity to Grb2-SH2 over Shp2 SH2 and SHC PTB. A (3-aminomethyl-phenyl)-urea scaffold was used to mimic the X_{+1}-Asn part of the minimal phosphopeptide sequence, Ac-pTyr-X_{+1}-Asn-NH_2. The resulting compounds were equipotent to their peptide counterparts for the Grb2-SH2 domain.[554] The IC_{50} values were 48.6, 6.19 and 6.5 μM, respectively, for **224** [R = H, -CHMe$_2$,-CH$_2$CH$_2$CONH$_2$]

Bicyclic and tricyclic lactams were incorporated in linear peptides like Ac-Pro-Arg-Pro-Leu-Pro-Val-Ala-Pro-Gly (IC_{50} 46 μM) to generate SH3 domain binding ligands. The nonapeptides with tricyclic isomers in a predicted extended conformation (**225a,b**; 3R,6R/3R,6S) bound the SH3 domain with an IC_{50} of 46 and 75 μM, respectively. Other spirolactam-containing nonapeptides likely to adopt β-turn conformations bound with lower affinity.[555]

Ac-Pro-Arg-Pro-Leu— [structure with N positions labeled 3 and 6, S, O] —Arg-Pro-Gly

(225)

8 Advances in Formulation/Delivery Technology

Various aspects of peptide and protein delivery technology have been reviewed.[556-558] The peptide delivery issues to the cell nucleus and across the blood-brain barrier have also been reviewed and strategies like coupling the non-transportable peptide to the blood brain barrier transport vector have been highlighted.[559-561] Peptides can gain access to the brain either from the lumen of cerebral microvessels into the brain interstitium or cerebrospinal fluid by active processes (*e.g.* insulin and leptin) or through regions in the brain where the blood-brain barrier does not exist, the so-called circumventricular organs.[561] High affinity receptors in brain capillaries have been identified for peptides like insulin and leptin that could bind these blood-borne hormones and induce their endocytosis and subsequent carriage across the blood-brain barrier. To enable a Leu-enkephalin analogue dalargin (Tyr-D-Ala-Gly-Phe-Leu-Arg) and kyotorphin (Tyr-Arg) to be transported cross the blood-brain barrier the peptides were adsorbed onto the surface of poly(butylcyanoacrylate) nanoparticles and the nanoparticles were coated with polysorbate 80. In the case of dalargin, the peptide-absorbed particles (dextran 70,000 stabilised) with or without the polysorbate coating were capable of inducing analgesia.[562] Kyotrophin-loaded nanoparticles induced central analgesic effects only when its adsorption was realised with dextran 70,000-stabilised nanoparticles. A primary endothelial cell culture model of the blood-brain barrier was used to study the transport mechanisms for opiate peptides. The analysis indicated that the cyclic peptide D-Phe-Cys-Tyr-D-Trp-Arg-Thr-Pen-Thr-NH$_2$ crosses the blood-brain barrier *via* diffusion, [D-Pen^{2,5}]-enkephalin uses a combination of diffusion and a saturable transport mechanism, and biphalin ([Tyr-D-Ala-Gly-Phe-NH]$_2$) uses diffusion and the large neutral amino acid carrier.[563]

Along with other reasons, the low oral bioavailability of peptides may be due to low oral absorption from the gastrointestinal tract and first-pass excretion by the liver. The role of first pass metabolism of peptides was investigated in a liver perfusion study in rats using metkephamid (a stable analogue of Met-enkephalin) and TRH as model peptides.[564] The results indicated that approximately 30–35% of metkephamid absorbed from the intestine undergoes first-pass metabolism before entering the systemic circulation *in vivo*. In contrast, the fraction of TRH metabolised in the liver was less than 10%. Biliary excretion of peptides has also been associated with low bioavailability of peptides. Some of these compounds are transported to be taken up by hepatocytes *via* active transport systems that also recognise bile acids and/or organic anions or cations. In addition, a few of them are reported to be excreted into bile *via* the primary active transport systems located at the bile canalicular membrane. The biliary excretion mechanisms of three endothelin antagonists have been studied.[565] In addition to the above factors, active drug efflux mediated by P-glycoprotein may also adversely affect oral bioavailability.[566,567] Studies with cyclosporin, HIV protease inhibitor saquinavir and a cysteine protease inhibitor (226) are used to illustrate the significance of cytochrome P450 3A and P-glycoprotein to the oral delivery of peptides. To overcome the problem of enzymic degradation associated with peptides, a strategy of incorporating peptides in a polymer-enzyme inhibitor conjugate has been considered. For example, a polymer-pepstatin complex may prevent the degradation of peptides by pepsin. Synthetic routes to pepstatin analogues [227, R$_1$ = *iso*-butyl, *sec*-butyl, *n*-propyl; R$_2$ = *n*-butyl, *n*-hexyl, *n*-octyl] bearing an appropriate spacer group for linkage to suitable bioadhesive polymers were developed.[568]

(226)   (227)

The antidiuretic effects of two formulations of desmopressin (20 μg) were demonstrated after intranasal administration to healthy, male overhydrated volunteers.[569] Microspheres (resin material with SO$_3$ residues) of precapillary size were used to bind reversibly fibroblast growth factor for delivery *via* coronary catheter. Fibroblast growth factor was released from these microspheres (over a period of one week in *in vitro* studies) and taken up by endothelial cells, which proliferated following translocation of fibroblast growth factor to the nucleus.[570] Degradable poly(ethylene glycol) hydrogels with potential utility as delivery carriers for bioactive drugs have been developed. Proteins can be covalently linked to the hydrogel network. Upon hydrolysis, these poly(ethylene glycol) hydrogels degrade into low molecular weight derivatives, which can be easily cleared by the body, and release the protein.[571,572]

References

1. A. S. Dutta, *Specialist Periodical Reports, Amino Acids, Peptides and Proteins*, 1999, **30**, 163-284.
2. I-A. Ahn, S. W. Kim and S. Ro, *Mol. Diversity*, 1998, **4**, 23-24.
3. C. Frochot, R. Vanderesse, A. Driou, G. Linden, M. Marraud, and M. T. Cung, *Lett. Peptide Science*, 1997, **4**, 219-225.
4. F. Andre, M. Marraud, T. Tsouloufis, S. J. Tzartos and G. Boussard, *J. Peptide Sci.*, 1997, **3**, 429-441.
5. R. Vanderesse, V. Grand, D. Limal, A. Vicherat, M. Murraud, C. Didierjean and A. Aubry, *J. Am. Chem. Soc.*, 1998, **120**, 9444-9451.
6. A. Bac, K. Rivoal, M. T. Cung, G. Boussard, M. Marraud, B. Soudan, D. Tetaert and P. Degand, *Lett. Peptide Sci.*, 1997, **4**, 251-258.
7. C. T. Bui, F. A. Rasoul, F. Ercole, Y. Pham and N. J. Maeji, *Tetrahedron Lett.*, 1998, **39**, 9279-9282.
8. M. Pasto, A. Moyano, M. A. Pericas and A. Riera, *Tetrahedron Lett.*, 1998, **39**, 1233-1236.
9. I. Ojima, H. Wang, T. Wang and E. W. Ng, *Tetrahedron Lett.*, 1998, **39**, 923-926.
10. C. Anne, M-C. Fournie-Zaluski, B. P. Roques and F. Cornille, *Tetrahedron Lett.*, 1998, **39**, 8973-8974.
11. J. F. Reichwein and R. M. J. Liskamp, *Tetrahedron Lett.*, 1998, **39**, 1243-1246.
12. P. Wipf, T. C. Henninger and S. J. Geib, *J. Org. Chem.*, 1998, **63**, 6088-6089.
13. C. Bolm, J. D. Kahmann and G. Moll, *Tetrahedron Lett.*, 1997, **38**, 1169-1172.
14. C. Gennari, C. Longari, S. Ressel, B. Salom, U. Piarulli, S. Ceccarelli and A. Mielgo, *Eur. J. Org. Chem.*, 1998, 2437-2449.
15. D. Artis and M. Lipton, *J. Am. Chem. Soc.*, 1998, **120**, 12200-12206.
16. D. C. Horwell, M. J. McKiernan and S. Osborne, *Tetrahedron Lett.*, 1998, **39**, 8729-8732.
17. Q. Wang, N. A. Sasaki and P. Potier, *Tetrahedron*, 1998, **54**, 15759-15780.
18. M. Falorni, G. Giacomelli and E. Spanu, *Tetrahedron Lett.*, 1998, **39**, 9241-9244.
19. F. Polyak and W. Lubell, *J. Org. Chem.*, 1998, **63**, 5937-5949.
20. F. Gosselin and W. D. Lubell, *J. Org. Chem.*, 1998, **63**, 7463-7471.
21. B. E. Fink, P. R. Kym and J. A. Katzenellenbogen, *J. Am. Chem. Soc.*, 1998, **120**, 4334-4344.
22. J. E. Semple, *Tetrahedron Lett.*, 1998, **39**, 6645-6648.

23.   I. S. Weitz, M. Pellegrini, M. Royo, D. F. Mierke, and M. Chorev, *Lett. Peptide Sci.*, 1998, **5**, 83-86.
24.   D. E. Hibbs, M. B. Hursthouse, I. G. Jones, W. Jones, K. M. Abdul Malik and M. North, *J. Org. Chem.*, 1998, **63**, 1496-1504.
25.   Y. Feng, Z. Wang, S. Jin and K. Burgess, *J. Am. Chem. Soc.*, 1998, **120**, 10768-10769.
26.   C. Bisang, L. Jiang, E. Freund, F. Emery, C. Bauch, H. Matile, G. Pluschke and J. A. Robinson, *J. Am. Chem. Soc.*, 1998, **120**, 7439-7449.
27.   H. E. Blackwell and R. H. Grubbs, *Angew. Chem. Int. Ed.*, 1998, **37**, 3281-3284.
28.   D. Besser, G. Greiner and S. Reissmann, *Lett. Peptide Sci.*, 1998, **5**, 299-303.
29.   L. Yu, Y. Lai, J. V. Wade and S. M. Coutts, *Tetrahedron Lett.*, 1998, **39**, 6633-6636.
30.   K. D. Roberts, J. N. Lambert, N. J. Ede and A. M. Bray, *Tetrahedron Lett.*, 1998, **39**, 8357-8360.
31.   V. Cavallaro, P. Thompson and M. Hearn, *J. Peptide Sci.*, 1998, **4**, 335-343.
32.   M. Miyashita, T. Nakamori, T. Murai, H. Miyagawa, M. Akamatsu and T. Ueno, *Biosci. Biotechnol. Biochem.*, 1998, **62**, 1799-1801.
33.   J. J. Wen and C. M. Crews, *Tetrahedron Lett.*, 1998, **39**, 779-782.
34.   J. D. White, J. Hong and L. A. Robarge, *Tetrahedron Lett.*, 1998, **39**, 8779-8782.
35.   J. Poncet, L. Hortala, M. Busquet, F. Gueritte-Voegelein, S. Thoret, A. Pierre, G. Atassi and P. Jouin, *Bioorg. Med. Chem. Lett.*, 1998, **8**, 2855-2858.
36.   R. K. Pettit, G. R. Pettit and K. C. Hazen, *Antimicrob. Agents Chemother.*, 1998, **42**, 2961-2965.
37.   A. Rudi, M. Aknin, E. M. Gaydou and Y. Kashman, *Tetrahedron*, 1999, **54**, 13203-13210.
38.   K. Umezawa, K. Nakazawa, T. Uemura, Y. Ikeda, S. Kondo, H. Naganawa, N. Kinoshita, H. Hashizume, M. Hamada, T. Takeuchi and S. Ohba, *Tetrahedron Lett.*, 1998, **39**, 1389-1392.
39.   Y. Koiso, N. Morisaki, Y. Yamashita, Y. Mitsui, R. Shiral, Y. Hashimoto and S. Iwasaki, *J. Antibiot.*, 1998, **51**, 418-422.
40.   N. Morisaki, Y. Mitsui, Y. Yamashita, Y. Koiso, R. Shiral, Y. Hashimoto and S. Iwasaki, *J. Antibiot.*, 1998, **51**, 423-427.
41.   B-S. Yun, I-J. Ryoo, I-K. Lee and I-D. Yoo, *Tetrahedron Lett.*, 1998, **39**, 993-996.
42.   A. Randazzo, F. D. Piaz, S. Orru, C. Debitus, C. Roussakis, P. Pucci and L. Gomez-Paloma, *Eur. J. Org. Chem.*, 1998, 2659-2665.
43.   T. J. Gaymes, M. Cebrat, I. Z. Siemion and J. E. Kay, *FEBS Lett.*, 1997, **418**, 224-227.
44.   P. J. Milne, A. L. Hunt, K. Rostoll, J. J. Van Der Walt and C. J. M. Graz, *J. Pharm. Pharmacol.*, 1998, **50**, 1331-1337.
45.   J. V. Lopez-Liuchi, *Eur. J. Endocrinol.*, 1998, **139**, 481-483.
46.   R. Deghenghi, *J. Endocrinol Invest.*, 1998, **21**, 787-793.
47.   F. A. C. Le Noble, F. R. M. Stassen, W. J. G. Hacking and H. A. J. Struijker Boudier, *J. Hypertension*, 1998, **16**, 1563-1572.
48.   H. G. Augustin, *Trends Pharmacol. Sci.*, 1998, **19**, 216-222.
49.   T. N. C. Wells, C. A. Power and A. E. I. Proudfoot, *Trends Pharmacol. Sci.*, 1998, **19**, 376-380.
50.   T. H. Ji, M. Grossmann and I. Ji, *J. Biol. Chem.*, 1998, **273**, 17299-17302.
51.   U. Gether and B. K. Kobilka, *J. Biol. Chem.*, 1998, **273**, 17979-17982.
52.   S. Wilson, D. J. Bergsma, J. K. Chambers, A. I. Muir, K. G. M. Fantom, C.

Ellis, P. R. Murdock, N. C. Herrity and J. M. Stadel, *Br. J. Pharmacol.*, 1998, **125**, 1387-1392.

53. A. Chollet and G. Turcatti, *Lett. Peptide Sci.*, 1998, **5**, 79-82.

54. J. Hardy, K Duff, K. W. Hardy, J. Perez-Tur and M. Hutton, *Nat. Neurosci.*, 1998, **1**, 355-358.

55. R. Kisilevsky and P. E. Fraser, *Crit. Rev. Biochem. Mol. Biol.*, 1997, **32**, 361-404.

56. W. Song and D. Lahiri, *Biochem. Mol. Biol.*, 1998, **46**, 755-764.

57. H. Meziane, J-C. Dodart, C. Mathis, S. Little, J. Clemens, S. M. Paul and A. Ungerer, *Proc. Natl. Acad. Sci. USA*, 1998, **95**, 12683-12688.

58. C. Russo, G. Angelini, D. Dapino, A. Piccini, G. Piombo, G. Schettini, S. Chen, J. K. Teller, D. Zaccheo, P. Gambetti and M. Tabaton, *Proc. Natl. Acad. Sci. USA*, 1998, **95**, 15598-15602.

59. N. N. Dewji and S. J. Singer, *Proc. Natl. Acad. Sci. USA*, 1998, **95**, 15055-15060.

60. H. Steiner, A. Capell, B. Pesold, M. Citron, P. M. Kloetzel, D. J. Selkoe, H. Romig, K. Mendla and C. Haass, *J. Biol. Chem.*, 1998, **273**, 32322-32331.

61. W. Xia, J. Zhang, B. L. Ostaszewski, W. T. Kimberly, P. Seubert, E. H. Koo, J. Shen and D. J. Selkoe, *Biochemistry*, 1998, **37**, 16465-16471.

62. E. A. Mackay, A. Ehrhard, M. Moniatte, C. Guenet, C. Tardif, C. Tarnus, O. Sorokine, B. Heintzelmann, C. Nay, J-M. Remy, J. Higaky, A. VanDorsselaer, J. Wagner C. Danzin and P. Mamont, *Eur. J. Biochem.*, 1997, **244**, 414-425.

63. W. Q. Qiu, D. M. Walsh, Z. Ye, K. Vekrellis, J. Zhang, M. B. Podlisny, M. R. Rosner, A. Safavi, L. B. Hersh and D. J. Selkoe, *J. Biol. Chem.*, 1998, **273**, 32730-32738.

64. T. L. S. Benzinger, D. M. Gregory, T. S. Burkoth, H. Miller-Auer, D. G. Lynn, R. E. Botto and S. C. Meredith, *Proc. Natl. Acad. Sci. USA*, 1998, **95**, 13407-13412.

65. K. J. Marcinowski, H. Shao, E. L. Clancy and M. G. Zagorski, *J. Am. Chem. Soc.*, 1998, **120**, 11082-11091.

66. N. Yamada, K. Ariga, M. Naito, K. Matsubara and E. Koyama, *J. Am. Chem Soc.*, 1998, **120**, 12192-12199.

67. M. P. Lambert, A. K. Barlow, B. A. Chromy, C. Edwards, R. Freed, M. Liosatos, T. E. Morgan, I. Rozovsky, B. Trommer, K. L. Viola, P. Wals, C. Zhang, C. E. Finch, G. A. Krafft and W. L. Klein, *Proc. Natl. Acad. Sci. USA*, 1998, **95**, 6448-6453.

68. Y. Hirakura, Y. Satoh, N. Hirashima, T. Suzuki, B. L. Kagan and Y. Kirino, *Biochem. Mol. Biol. Int.*, 1998, **46**, 787-794.

69. O. M. A. El-Agnaf, G. B. Irvine, G. Fitzpatrick, W. K. Glass and D. J. S. Guthrie, *Biochem. J.*, 1998, **336**, 419-427.

70. C. Sato, E. M. Sigurdsson, L. Morelli, R. A. Kumar, E. M. Castano and B. Frangione, *Nat. Med.*, 1998, **4**, 822-826.

71. D. Giulian, L. J. Haverkamp, J. Yu, W. Karshin, D. Tom, J. Li, A. Kazanskaia, J. Kirkpatrick and A. E. Roher, *J. Biol. Chem.*, 1998, **273**, 29719-29726.

72. R. E. H. Hancock and R. Lehrer, *Trends Biotechol.*, 1998, **16**, 82-88.

73. J. F. Barrett and J. A. Hoch, *Antimicrob. Agents Chemother.*, 1998, **42**, 1529-1536.

74. P. Hwang and H. Vogel, *Biochem. Cell Biol.*, 1998, **76**, 235-246.

75. A. Giacometti, O. Cirioni, G. Greganti, M. Quarta and G. Scalise, *Antimicrob. Agents Chemother.*, 1998, **42**, 3320-3324.

76. D. Winder, W. H. Gunzburg, V. Erfle and B. Salmons, *Biochem. Biophys. Res. Commun.*, 1998, **244**, 608-612.

77. Y. Chan, and R. Gallo, *J. Biol. Chem.*, 1998, **273**, 28978-28985.
78. T. Rozek, R. J. Waugh, S. T. Steinborner, J. H. Bowie, M. J. Tyler and J. C. Wallace, *J. Peptide Sci.*, 1998, **4**, 111-115.
79. L. Yan and M. E. Adams, *J. Biol. Chem.*, 1998, **273**, 2059-2066.
80. I. Y. Park, C. B. Park, M. S. Kim and S. C. Kim, *FEBS Lett.*, 1998, **437**, 258-262.
81. Y. Goumon, K. Lugardon, B. Kieffer, J-F. Lefevre, A. V. Dorsselaer, D. Aunis and M. H. Metz-Boutigue, *J. Biol. Chem.*, 1998, **273**, 29847-29856.
82. S. Charpentier, M. Amiche, J. Mester, V. Vouille, J-P. Le Caer, P. Nichols and A. Delfour, *J. Biol. Chem.*, 1998, **273**, 14690-14697.
83. Y. Minami, S. Higuchi, F. Yagi and K. Tadera, *Biosci. Biotechnol. Biochem.*, 1998, **62**, 2076-2078.
84. R. Bals, X. Wang, M. Zasloff and J. M. Wilson, *Proc. Natl. Acad. Sci. USA*, 1998, **95**, 9541-9546.
85. V. Krishnakumari and R. Nagaraj, *J. Peptide Res.*, 1997, **50**, 88-93.
86. S. Y. Shin, J. H. Kang, M.K. Lee, S. Y. Kim, Y. Kim and K-S. Halm, *Biochem. Mol. Biol. Int.*, 1998, **44**, 1119-1126.
87. S. Y. Shin, M. K. Lee, K. L. Kim and K-S. Hahm, *J. Peptide Res.*, 1997, **50**, 279-285.
88. P. C. Fuchs, A. L. Barry and S. D. Brown, *Antimicrob. Agents Chemother.*, 1998, **42**, 1213-1216.
89. K. H. Lee, S. Y. Hong and J. E. Oh, *FEBS Lett.*, 1998, **439**, 41-45.
90. K. H. Lee, S. Y. Hong, J. E. Oh, M. Y. Kwon, J. H. Yoon, J. H. Lee, B. L. Lee and H. M. Moon, *Biochem. J.*, 1998, **334**, 99-105.
91. C. B. Park, H. S. Kim and S. C. Kim, *Biochem. Biophys. Res. Commun.*, 1998, **244**, 253-257.
92. K. Nihei, H. Itoh, K. Hashimoto, K. Miyairi and T. Okuno, *Biosci. Biotechnol. Biochem.*, 1998, **62**, 858-863.
93. S. G. Toske, P. R. Jensen, C. A. Kauffman and W. Fenical, *Tetrahedron*, 1998, **54**, 13459-13466.
94. D. P. Clark, J. Carroll, S. Naylor and P. Crews, *J. Org. Chem.*, 1998, **63**, 8757-8764.
95. S. Y. Hong, J. E. Oh, M. Y. Kwon, M. J. Choi, J. H. Lee, B. L. Lee, H. M. Moon and K. H. Lee, *Antimicrob. Agents Chemother.*, 1998, **42**, 2534-2541.
96. P. J. Coote, C. D. Holyoak, D. Bracey, D. P. Ferdinando and J. A. Pearce, *Antimicrob. Agents Chemother.*, 1998, **42**, 2160-2170.
97. M. Shimizu, Y. Shigeri, Y. Tatsu, S. Yoshikawa and N. Yumoto, *Antimicrob. Agents Chemother.*, 1998, **42**, 2745-2746.
98. Y. L. Zhang, H. R. Marepalli, H. Lu, J. M. Becker and F. Naider, *Biochemistry*, 1998, **37**, 12465-12476.
99. B. Devadas, S. K. Freeman, C. A. McWherter, N. S. Kishore, J. K. Lodge, E. Jackson-Machelski, J. I. Gordon and J. A. Sikorski, *J. Med. Chem.*, 1998, **41**, 996-1000.
100. M. R. Palchaudhuri, S. Wille, G. Mevenkamp, J. Spiess, B. Fuchs and F. M. Dautzenberg, *Eur. J. Biochem.*, 1998, **258**, 78-84.
101. C. W. Liaw, D. E. Grigoriadis, M. T. Lorang, E. B. De Souza and R. A. Maki, *Mol. Endocrinology*, 1997, **11**, 2048-2053.
102. N. Ohta, T. Mochizuki, M. Hoshino, L. Jun, H. Kobayashi and N. Yanaihara, *J. Peptide Res.*, 1997, **50**, 178-183.
103. J. Rivier, S. L. Lahrichi, J. Gulyas, J. Erchegyi, S. C. Koerber, A. G. Craig, A. Corrigan, C. Rivier and W. Vale, *J. Med. Chem.*, 1998, **41**, 2614-2620.

104. S. C. Koerber, J. Gulyas, S. L. Lahrichi, A. Corrigan, A. G. Craig, C. Rivier, W. Vale and J. Rivier, *J. Med. Chem.*, 1998, **41**, 5002-5011.

105. J. Rivier, J. Gulyas, A. Corrigan, V. Martinez, A. G. Craig, Y. Tache, W. Vale and C. Rivier, *J. Med. Chem.*, 1998, **41**, 5012-5019.

106. A. Ruhmann, I. Bonk, C. R. Lin, M. G. Rosenfeld and J. Spiess, *Proc. Natl. Acad. Sci. USA*, 1998, **95**, 15264-15269.

107. L. K. Malendowicz, P. Rebuffat, G. G. Nussdorfer and K. W. Nowak, *J. Steroid Biochem. Mol. Biol.*, 1998, **67**, 149-152.

108. D. J. Wustrow, T. Capiris, R. Rubin, J. A. Knobelsdorf, H. Akunne, M. D. Davis, R. MacKenzie, T. A. Pugsley, K. T. Zoski, T. G. Heffner and L. D. Wise, *Bioorg. Med. Chem. Lett.*, 1998, **8**, 2067-2070.

109. H. Matsubara, *Circulation Res.*, 1998, **83**, 1182-1191.

110. R. Bouley, J. Perodin, H. Plante, L. Rihakova, S. G. Bernier, L. Maletinska, G. Guillemette and E. Escher, *Eur. J. Pharmacol.*, 1998, **343**, 323-331.

111. J. W. Ellingboe, M. D. Collini, D. Quagliato, J. Chen, M. Antane, J. Schmid, D. Hartuppe, V. White, C. H. Park, T. Tanikella and J. F. Bagli, *J. Med. Chem.*, 1998, **41**, 4251-4260.

112. I. Anderson and G. Drew, *Br. J. Pharmacol.*, 1998, **125**, 1236-1246.

113. Y. Takata, S. Tajima, *et al.*, *J. Cardiovasc. Pharmacol.*, 1998, **32**, 834-844.

114. M. Angiolini, L. Belvisi, *et al.*, *Bioorg. Med. Chem.*, 1998, **6**, 2013-2027.

115. T. W. Moody and R. T. Jensen, *Drugs Future*, 1998, **23**, 1305-1315.

116. V. Ashwood, V. Brownhill, M. Higginbottom, D. C. Horwell, J. Hughes, R. A. Lewthwaite, A. T. McKnight, R. D. Pinnock, M. C. Pritchard, N. Suman-Chauhan, C. Webb and S. C. Williams, *Bioorg. Med. Chem. Lett.*, 1998, **8**, 2589-2594.

117. L. L. Hampton, E. E. Ladenheim, M. Akeson, J. M. Way, H. C. Weber, V. E. Sutliff, R. T. Jensen, L. J. Wine, H. Arnheiter and J. F. Battey, *Proc. Natl. Acad. Sci. USA*, 1998, **95**, 3188-3192.

118. R. R. Ryan, H. C. Weber, W. Hou, E. Sainz, S. A. Mantey, J. F. Battey, D. C. Coy and R. T. Jensen, *J. Biol. Chem.*, 1998, **273**, 13613-13624.

119. T. K. Pradhan, T. Katsuno, J. E. Taylor, S. H. Kim, R. R. Ryan, S. A. Mantey, P. J. Donohue, H. C. Weber, E. Sainz, J. F. Battey, D. C. Coy and R. T. Jensen, *Eur. J. Pharmacol.*, 1998, **343**, 275-287.

120. C. Damge and A. Hajri, *Eur J Pharmacol.*, 1998, **347**, 77-86.

121. O. Nyeki, A. Rill, I. Schon, A. Orosz, J. Schrett, L. Bartha and J. Nagy, *J. Peptide Sci.*, 1998, **4**, 486-495.

122. Z. Li, S. M. Secor, V. A. Lance, M. A. Masini, M. Vallarino and J. M. Conlon, *Gen. Comp. Endocrinol*, 1998, **112**, 108-114.

123. P. Rajasekariah, R. S. Warlow, M. E. Campbell, N. Ozsarac, P. L. Dao, M. K. Swanton and R. S. Walls, *Int. J. Biochem. Cell Biol.*, 1998, **30**, 353-367.

124. M. Filizola, O. Llorens, M. Carteni-Farina and J. J. Perez, *Bioorg. Med. Chem.*, 1998, **6**, 1491-1500.

125. J. Howl and M. Wheatley, *Lett. Peptide Sci.*, 1998, **5**, 37-41.

126. G. Greiner, U. Dornberger, I. Paegelow, B. A. Scholkens, C. Liebmann and S. Reissmann, *J. Peptide Sci.*, 1998, **4**, 92-100.

127. M. Lange, A. S. Cuthbertson, R. Towart and P. M. Fischer, *J. Peptide Sci.*, 1998, **4**, 289-293.

128. T. S. F. Saleh, R. M. J. Vianna, T. B. Creczynski-Pasa, S. Chakravarty, B. J. Mavunkel, D. J. Kyle and J. B. Calixto, *Eur. J. Pharmacol.*, 1998, **363**, 179-187.

129. Y. Abe, H. Kayakiri, S. Satoh, T. Inoue, Y. Sawada, K. Imai, N. Inamura, M.

Asano, C. Hatori, A. Katayama, T. Oku and H. Tanaka, *J. Med. Chem.*, 1998, **41**, 564-578.

130. Y. Abe, H. Kayakiri, S. Satoh, T. Inoue, Y. Sawada, N. Inamura, M. Asano, C. Hatori, H. Sawai, T. Oku and H. Tanaka, *J. Med. Chem.*, 1998, **41**, 4053-4061.

131. Y. Abe, H. Kayakiri, S. Satoh, T. Inoue, Y. Sawada, N. Inamura, M. Asano, I. Aramori, C. Hatori, H. Sawai, T. Oku and H. Tanaka, *J. Med. Chem.*, 1998, **41**, 4062-4079.

132. Y. Abe, H. Kayakiri, S. Satoh, T. Inoue, Y. Sawada, N. Inamura, M. Asano, I. Aramori, C. Hatori, H. Sawai, T. Oku and H. Tanaka, *J. Med. Chem.*, 1998, **41**, 4587-4598.

133. T. Griesbacher, R. Amann, W. Sametz, S. Diethart and H. Juan, *Br. J. Pharmacol.*, 1998, **124**, 1328-1334.

134. D. Pruneau, J-M. Luccarini, C. Fouchet, E. Defrene, R-M. Franck, B. Loillier, H. Duclos, C. Robert, B. Cremers, P. Belichard and J-L. Paquet, *Br. J. Pharmacol.*, 1998, **125**, 365-372.

135. M. L. Schubert, *Curr. Opin. Gastroenterol.*, 1998, **14**, 425-432.

136. M. Bläker, Y. Ren, M. C. Gordon, J. E. Hsu, M. Beinborn and A. S. Kopin, *Mol. Pharmacol.*, 1998, **54**, 857-863.

137. M. Beinborn, S. M. Quinn and A. S. Kopin, *J. Biol. Chem.*, 1998, **273**, 14146-14151.

138. J-C. Califano, L. Goullieux, M. Amblard, J-A. Fehrentz, N. Bernad, G. Bergé, J. Castel and J. Martinez, *Lett. Peptide Sci.*, 1997, **4**, 235-239.

139. R. D. Simmons, F. C. Kaiser, M. E. Pierson and J. R. Rosamond, *Pharmacol. Biochem. Behaviour*, 1998, **59**, 439-444.

140. M. Amblard, M. Rodriguez, M-F. Lignon, M-C. Galas, N. Bernad, A. Aumelas and J. Martinez, *Eur. J. Med. Chem.*, 1998, **33**, 171-180.

141. M-E. Million, I. Lena, S. Da Nascimento, F. Noble, V. Dauge, C. Garbay, B. P. Roques, *Lett. Peptide Sci.*, 1997, **4**, 407-410.

142. B. Pirotte, P. de Tullio, T. Podona, O. Diouf, D. Dewalque, P. Neven, B. Masereel, D-H. Caignard P, P. Renard and J. Delarge, *Eur. J. Pharm. Sci.*, 1998, **7**, 29-40.

143. J. K. Padia, M. Field, J. Hinton, K.Meecham, J. Pablo, R. Pinnock, B. D. Roth, L. Singh, N. Suman-Chauhan, B. K. Trivedi and L. Webdale, *J. Med. Chem.*, 1998, **41**, 1042-1049.

144. N. S. Sheerin and S. H. Sacks, *Curr. Opin. Nephrology Hypertension*, 1998, **7**, 305-310.

145. Z. Chen, X. Zhang, N. C. Gonnella, T. C. Pellas, W. C. Boyar and F. Ni, *J. Biol. Chem.*, 1998, **273**, 10411-10419.

146. M. Takahashi, S. Moriguchi, T. Minami, H. Suganuma, A. Shiota, Y. Takenaka, F. Tani, R. Sasaki and M. Yoshikawa, *Lett. Peptide Sci.*, 1998, **5**, 29-35.

147. T. Masaki, *Cardiovascular Res.*, 1998, **39**, 530-533.

148. C. F. van der Walle and D. J. Barlow, *Curr. Med. Chem.*, 1998, **5**, 321-335.

149. A. Bagnato and K. Catt, *Trends Endocrinol.*, 1998, **9**, 378-383.

150. K. Arvidsson, T. Nemoto, Y. Mitsui, S. Ohashi and H. Nakanishi, *Eur. J. Biochem.*, 1998, **257**, 380-388.

151. C. F. Van der Walle, S. Bansal and D. J. Barlow, *J. Pharm. Pharmacol.*, 1998, **50**, 837-844.

152. J. Sakaki, T. Murata, Y. Yuumoto, I. Nakamura, T. Frueh, T. Pitterna, G. Iwasaki, K. Oda, T, Yamamura and K. Hayakawa, *Bioorg. Med. Chem. Lett.*, 1998, **8**, 2241-2246.

153. G. Liu, K. J. Henry, Jr., B. G. Szczepankiewicz, M. Winn, N. S. Kozmina, S. A. Boyd, J. Wasicak, T. W. von Geldern, J. R. Wu-Wong, W. J. Chiou, D. B. Dixon, B. Nguyen, K. C. Marsh and T. J. Opgenorth, *J. Med. Chem.*, 1998, **41**, 3261-3275.

154. P. C. Astles, C. Brealey, T. J. Brown, V. Facchini, C. Handscombe, N. V. Harris, C. McCarthy, I. M. McLay, B. Porter, A. G. Roach, C. Sargent, C. Smith and R. J. A. Walsh, *J. Med. Chem.*, 1998, **41**, 2732-2744.

155. P. C. Astles, T. J. Brown, F. Halley, C. M. Handscombe, N. V. Harris, C. McCarthy, I. M. McLay, P. Lockey, T. Majid, B. Porter, A. G. Roach, C. Smith and R. Walsh, *J. Med. Chem.*, 1998, **41**, 2745-2753.

156. N. Cho, Y. Nara, M. Harada, T. Sugo, Y. Masuda, A. Abe, K. Kusumoto, Y. Itoh, T. Ojtaki, T. Watanabe and S. Furuya, *Chem. Pharm. Bull.*, 1998, **46**, 1724-1737.

157. K. A. Berryman, J. J. Edmunds, A. M. Bunker, S. Haleen, J. Bryant, K. M. Welch and A. M. Doherty, *Bioorg. Med. Chem.*, 1998, **6**, 1447-1456.

158. N. Murugesan, Z. Gu, P. D. Stein, S. Bisaha, S. Spergel, R. Girotra, V. G. Lee, J. Lloyd, R. N. Misra, J. Schmidt, A. Mathur, L. Stratton, Y. F. Kelly, E. Bird, T. Waldron, E. C-K. Liu, R. Zhang, H. Lee, R. Serafino, B. Abboa-Offei, P. Mathers, M. Giancarli, A. A. Seymour, M. L. Webb, S. Moreland, J. C. Barrish and J. T. Hunt, *J. Med. Chem.*, 1998, **41**, 5198-5218.

159. T. Hoshino, R. Yamauchi, K. Kikkawa, H. Yabana and S. Murata, *J. Pharmacol. Exp. Therap.*, 1998, **286**, 643-649.

160. E. H. Ohlstein, P. Nambi, D. W. P. Hay, M. Gellai, D. P. Brooks, J. Luengo, J-N. Xiang and J. D. Elliott, *J. Pharmacol. Exp. Therap.*, 1998, **286**, 650-656.

161. R. P. Nargund, A. A. Patchett, M. A. Bach, M. G. Murphy and R. G. Smith, *J. Med. Chem.*, 1998, **41**, 3103-3127.

162. K. Raun, B. S. Hansen, N. L. Johansen, H. Thogersen, K. Madsen, M. Ankersen and P. H. Andersen, *Eur. J. Endocrinol.*, 1998, **139**, 552-561.

163. A. Torsello, M. Luoni, F. Schweiger, R. Grilli, M. Guidi, E. Bresciani, R.Deghenghi, E. E. Muller and V. Locatelli, *Eur. J. Pharmacol.*, 1998, **360**, 123-129.

164. M. Maghnie, V. Spica-Russotto, M. Cappa, M. Autelli, C. Tinelli, P. Civolani, R. Deghenghi, F. Severi and S. Loche, *J. Clin. Endocrinol. Metab.*, 1998, **83**, 3886-3889.

165. E. A. Nijland, C. J. Strasburger, C. Popp-Snijders, P. S. van der Wal and E. A. van der Veen, *Eur. J. Endocrinol.*, 1998, **139**, 395-401.

166. R. J. Devita, R. Bochis, A. J. Frontier, A. Kotliar, M. H. Fisher, W. R. Schoen, M. J. Wyvratt, K. Cheng, W. W-S. Chan, B. Butler, T. M. Jacks, G. J. Hickey, K. D. Schleim, K. Leung, Z. Chen, S-H. L. Chiu, W. P. Feeney, P. K. Cunningham and R. G. Smith, *J. Med. Chem.*, 1998, **41**, 1716-1728.

167. L. Yang, G. Moriello, A. A. Patchett, K. Leung, T. Jacks, K. Cheng, K. D. Schleim, W. Feeney, W. W-S. Chan, S-H. L. Chiu and R. G. Smith, *J. Med. Chem.*, 1998, **41**, 2439-2441.

168. L. A. Cervini, C. J. Donaldson, S. C. Koerber, W. W. Vale and J. E. Rivier, *J. Med. Chem.*, 1998, **41**, 717-727.

169. N. Lamharzi, A. V. Schally, M. Koppan and K. Groot, *Proc. Natl. Acad. Sci. USA*, 1998, **95**, 8864-8868.

170. M. Madan, S. D. Berkowitz and J. E. Tcheng, *Circulation*, 1998, **98**, 2629-2635.

171. P. A. J. Henricks and F. P. Nijkamp, *Eur. J. Pharmacol.*, 1998, **344**, 1-13.

172. N. Kleiman, *Coronary Artery Dis.*, 1998, **9**, 603-616.

173. D. L. Boger, J. Goldberg, W. Jiang, W. Chai, P. Ducray, J. K. Lee, R. S. Ozer and C-M. Andersson, *Bioorg. Med. Chem*, 1998, **6**, 1347-1378.

174. *Drugs Future*, 1998, **23**, 585-590.

175. M. Merlos, A. Graul and J. Castaner, *Drugs Future*, 1998, **23**, 1297-1302.

176. M. Merlos, P. A. Leeson and J. Castaner, *Drugs Future*, 1998, **23**, 1190-1198.

177. G. Chow, S. Subburaju and M. Kini, *Arch. Biochem. Biophys.*, 1998, **354**, 232-238.

178. K. Shimokawa, L-G. Jia, J. D. Shannon and J.W. Fox, *Arch. Biochem. Biophys.*, 1998, **354**, 239-246.

179. A. Müller, F. Schumann, M. Koksch and N. Sewald, *Lett. Peptide Sci.*, 1997, **4**, 275-281.

180. M. A. Sow, A. Molla, F. Lamaty and R. Lazaro, *Lett. Peptide Sci.*, 1997, **4**, 455-461.

181. J. M. Derrick, R. G. Loudon, and T. K. Gartner, *Thrombosis Res.*, 1998, **89**, 31-40.

182. T. Boxus, R. Touillaux, G. Dive and J. Marchand-Brynaert, *Bioorg. Med. Chem.*, 1998, **6**, 1577-1595.

183. S. A. Mousa, J. M. Bozarth, W. Lorelli, M. S. Forsythe, M. J. M. C. Thoolen, A. M. Slee, T. M. Reilly and P. A. Friedman, *J. Pharmacol. Exp. Therap.*, 1998, **286**, 1277-1284.

184. S. A. Mousa, J. M. Bozarth, *et al.*, *Thromb. Res.*, 1998, **89**, 217-225.

185. S. A. Mousa, M. Forsythe, J. Bozarth, A. Youssef, J. Wityak, R. Olson and T. Sielecki, *J. Cardiovasc. Pharmacol.*, 1998, **32**, 736-744.

186. C-B. Xue, J. Roderick, S. Mousa, R. E. Olson and W. F. DeGrado, *Bioorg. Med. Chem. Lett.*, 1998, **8**, 3499-3504.

187. S. A. Mousa, R. E. Olson, J. M. Bozarth, W. Lorelli, M. S. Forsythe, A. Racanelli, S. Gibbs, K. Schlingman, T. Bozarth, R. Kapil, J. Wityak, T. M. Sielecki, R. R. Wexler, M. J. Thoolen, A. Slee, T. M. Reilly, P. S. Anderson and P. A. Friedman, *J. Cardiovasc. Pharmacol.*, 1998, **32**, 169-176.

188. K. Okumaura, T. Shimazaki, Y. Aoki, H. Yamashita, E. Tanaka, S. Banba, K. Yazawa, K. Kibayashi and H. Banno, *J. Med. Chem.*, 1998, **41**, 4036-4052.

189. Y. Hayashi, J. Katada, T. Harada, A. Tachiki, K. Ijima, Y. Takiguchi, M. Muramatsu, H. Miyazaki, T. Asari, T. Okazaki, Y. Sato, E. Yasuda, M. Yano, I. Uno and I. Ojima, *J. Med. Chem.*, 1998, **41**, 2345-2360.

190. H. Sugihara, H. Fukushi, T. Miyawaki, Y. Imai, Z. Terashita, M. Kawamura, Y. Fujisawa and S. Kita, *J. Med. Chem.*, 1998, **41**, 489-502.

191. H. U. Stilz, W. Guba, B. Jablonka, M. Just, O. Klingler, W. König, V. Wehner and G. Zoller, *Lett. Peptide Sc.*, 1998, **5**, 215-221.

192. P. N. Confalone, F. Jin and S. A. Mousa, *Bioorg. Med. Chem. Lett.*, 1998, **9**, 55-58.

193. A. Wong, S. M. Hwang, K. Johanson, J. Samanen, D. Bennett, S. W. Landvatter, W. Chen, J. R. Heys, F. E. Ali, T. W. Ku, W. Bondinell, A. J. Nichols, D. A. Powers and J. M. Stadel, *J. Pharmacol. Exp. Therap.*, 1998, **285**, 228-235.

194. K. C. Nicolaou, J. I. Trujillo, B. Jandeleit, K. Chibale, M. Rosenfeld, B. Diefenbach, D. A. Cheresh and S. L. Goodman, *Bioorg. Med. Chem.*, 1998, **6**, 1185-1208.

195. S. I. Klein, B. F. Molino, M. Czekaj, C. J. Gardner, C. Chu, K. Brown, R. D. Sabatino, J. S. Bostwick, C. Kasiewski, R. Bentley, V. Windisch, M. Perrone, C. T. Dunwiddie and R. J. Leadley, *J. Med. Chem.*, 1998, **41**, 2492-2502.

196. D. F. Kong, R. M. Califf, D. P. Miller, D. J. Moliterno, H. D. White, R. A.

Harrington, J. E. Tcheng, A. M. Lincoff, V. Hasselblad and E. J. Topol, *Circulation*, 1998, **98**, 2829-2835.

197. P. Hoffmann, A. Bernat, P. Savi and J. M. Herbert, *J. Pharmacol. Exp. Therap.*, 1998, **286**, 670-675.

198. N. P. Murphy, D. Pratico and D. J. Fitzgerald, *J. Pharmacol. Exp. Therap.*, 1998, **286**, 945-951.

199. R. A. Bednar, S. L. Gaul, T. G. Hamill, M. S. Egbertson, J. A. Shafer, G. D. Hartman, R. J. Gould and B. Bednar, *J. Pharmacol. Exp. Therap.*, 1998, **285**, 1317-1326.

200. C. Y. Cho, R. C. Youngquist, S. J. Paikoff, M. H. Beresini, A. R. Hebert, L. T. Berleau, C. W. Liu, D. E. Wemmer, T. Keough and P. G. Schultz, *J. Am. Chem. Soc.*, 1998, **120**, 7706-7718.

201. P. E. Thompson, D. L. Steer, M-I. Aguilar and M. T. W. Hearn, *Bioorg. Med. Chem. Lett.*, 1998, **8**, 2699-2704.

202. H. Minoux, N. Moitessier, Y. Chapleur and B. Maigret, *Lett. Peptide Sci.*, 1997, **4**, 463-466.

203. N. Moitessier, H. Minoux, B. Maigret, F. Chretien and Y. Chapleur, *Lett. Peptide Sci.*, 1998, **5**, 75-78.

204. G. Gasparini, P. Brooks, *et al.*, *Clin. Cancer Res.*, 1998, **4**, 2625-2634.

205. C. Fuegg, A. Yilmaz, G. Bieler, J. Bamat, P. Chaubert and F. J. Lejeune, *Nat. Med.*, 1998, **4**, 408-414.

206. B. Zheng and D. R. Clemmons, *Proc. Natl. Acad. Sci. USA*, 1998, **95**, 11217-11222.

207. K. E. Stockbauer, L. Magoun, M. Liu, E. H. Burns, Jr., S. Gubba, S. Renish, X. Pan, S. C. Bodary, E. Baker, J. Coburn, J. M. Leong and J. M. Musser, *Proc. Natl. Acad. Sci. USA*, 1999, **96**, 242-247.

208. R. M. Keenan, W. H. Miller, M. A. Lago, F. E. Ali, W. E. Bondinell, J. F. Callahan, R. P. Calvo, R. D. Cousins, S-M. Hwang, D. R. Jakas, T. W. Ku, C. Kwon, T. T. Nguyen, V. A. Reader, D. J. Rieman, S. T. Ross, D. T. Takata, I. N. Uzinskas, C. C. K. Yuan and B. R. Smith, *Bioorg. Med. Chem. Lett.*, 1998, **8**, 3165-3170.

209. R. M. Keenan, M. A. Lago, W. H. Miller, F. E. Ali, R. D. Cousins, L. B. Hall, S-M. Hwang, D. R. Jakas, C. Kwon, C. Louden, T. T. Nguyen, E. H. Ohlstein, D. J. Rieman, S. T. Ross, J. M. Samanen, B. R. Smith, J. Stadel, D. T. Takata, L. Vickery, C. C. K. Yuan and T. L. Yue, *Bioorg. Med. Chem. Lett.*, 1998, **8**, 3171-3176.

210. R. E. Gerszten, Y-C. Lim, H. T. Ding, K. Snapp, G. Kansas, D. A. Dichek, C. Cabanas, F. Sanchez-Madrid, M. A. Gimbrone, Jr., A. Rosenzweig and F. W. Luscinskas, *Circulation Res.*, 1998, **82**, 871-878.

211. B. Engelhardt, M. Laschinger, M. Schulz, U. Samulowitz, D. Vestweber and G. Hoch, *J. Clin. Invest.*, 1998, **102**, 2096-2105.

212. C. Quan, N. J. Skelton, K. Clark, D. Y. Jackson, M. E. Renz, H. H. Chiu, S. M. Keating, M. H. Beresini, S. Fong and D. R. Artis, *Biopolymers*, 1998, **47**, 265-275.

213. A. J. Souers, A. A. Virgilio, S. S. Schurer and J. A. Ellman, *Bioorg. Med. Chem. Lett.* 1998, **8**, 2297-2302.

214. K. O. Yee, M. M. Rooney, C. M. Giachelli, S. T. Lord and S. M. Schwartz, *Circulation Res.*, 1998, **83**, 241-251.

215. A. P. Mould, L. Burrows and M. J. Humphries, *J. Biol. Chem.*, 1998, **273**, 25664-25672.

216. C. Rogers, E. R. Edelman and D. I. Simon, *Proc. Natl. Acad. Sci. USA*, 1998, **95**, 10134-10139.
217. J. M. Casasnovas, T. Stehle, J-H. Liu, J-H. Wang and T. A. Springer, *Proc. Natl. Acad. Sci. USA*, 1998, **95**, 4134-4139.
218. J. Bella, P. R. Kolatkar, C. W. Marlor, J. M. Greve and M. G. Rossmann, *Proc. Natl. Acad. Sci. USA*, 1998, **95**, 4140-4145.
219. Y. Feng, D. Chung, L. Garrard, M. McEnroe, D. Lim, J. Scardina, K. McFadden, A. Guzzetta, A. Lam, J. Abraham, D. Liu and G. Endelmann, *J. Biol. Chem.*, 1998, **273**, 5625-5630.
220. B. Kutscher, M. Bernd, T. Beckers, E. E. Polymeropoulos and J. Engel, *Angew. Chem. Int. Ed, Engl.*, 1997, **36**, 2149-2161.
221. A. V. Schally and M. Comaru-Schally, *Adv. Drug Delivery Rev.*, 1997, **28**, 157-169.
222. G. Emons, O. Ortmann *et al.*, *Trends in Endocrinol.*, 1997, **8**, 355-362.
223. F. Boutignon, H. Touchet, S. David, P. Wüthrich, R. Deghenghi, H. Ong, M. Dubuc, M. Cesana and T. Maggi, *Lett. Peptide Sci.*, 1997, **4**, 423-427.
224. A. Graul, X. Rabasseda and J. Castaner, *Drugs Future*, 1998, **23**, 1057-1061.
225. A. M. Comaru-Schally, W. Brannan, A. V. Schally, M. Colcolough and M. Monga, *J. Clin. Endocrinol.*, 1998, **83**, 3826-3831.
226. A. Chen, D. Yahalom, N. Ben-Aroya, E. Kaganovsky, E. Okon and Y. Koch, *FEBS Lett.*, 1998, **435**, 199-203.
227. M. Keramida, J. M. Matsoukas, G. Agelis, D. Panagiotopoulos, J. Cladas, D. Pati, G. J. Moore, and H. R. Habibi, *Lett. Peptide Sci.*, 1998, **5**, 305-315.
228. S. Rahimipour, L. Weiner, P. B. Shrestha-Dawadi, S. Bittner, Y. Koch and M. Fridkin, *Lett. Peptide Sci.*, 1998, **5**, 421-427.
229. N. Cho, M. Harada, T. Imaeda, T. Imada, H. Matsumoto, Y. Hayase, S. Sasaki, S. Furuya, N. Suzuki, S. Okubo, K. Ogi, S. Endo, H. Onda and M. Fujino, *J. Med. Chem.*, 1998, **41**, 4190-4195.
230. R. A. H. Adan and W. H. Gispen, *Peptides*, 1997, **18**, 1279-1287.
231. S. A. Jordan and I. J. Jackson, *BioEssays*, 1998, **20**, 603-606.
232. D. H. G. Versteeg, P. Van Vanbergen, R. A. H. Adan and D. J. De Wildt, *Eur. J. Pharmacol.*, 1998, **360**, 1-14.
233. R. D. Rosenfeld, L. Zeni, A. A. Welcher, L. O. Narhi, C. Hale, J. Marasco, J. Delaney, T. Gleason, J. S. Philo, V. Katta, J. Hui, J. Baumgartner, M. Graham, K. L. Stark and W. Karbon, *Biochemistry*, 1998, **37**, 16041-16052.
234. L. L. Kiefer, J. M. Veal, K. G. Mountjoy and W. O. Wilkison, *Biochemistry*, 1998, **37**, 991-997.
235. J. M. Quillan, W. Sadee, E. T. Wei, C. Jimenez, L. Ji and J-K. Chang, *FEBS Lett.*, 1998, **428**, 59-62.
236. M. Rossi, M. S. Kim, D. G. A. Morgan, C. J. Small, C. M. B. Edwards, D. Sunter, S. Abusnana, A. P. Goldstone, S. H. Russell, S. A. Stanley, D. H. Smith, K. Yagaloff, M. A. Ghatei and S. R. Bloom, *Endocrinology*, 1998, **139**, 4428-4431.
237. A. Kask, L. Rago, F. Mutulis, R. Pahkla, J. E. S. Wikberg and H. B. Schioth, *Biochem. Biophys. Res. Commun.*, 1998, **245**, 90-93.
238. A. V. Vergoni, A. Bertolini, F. Mutulis, J. E. S. Wikberg and H. B. Schioth, *Eur. J. Pharmacol.*, 1998, **362**, 95-101.
239. A. Kask, F. Mutulis, R. Muceniece, R. Pahlka, I. Mutule, J. E. S. Wikberg, L. Rago and H. B. Schioth, *Endocrinology*, 1998, **139**, 5006-5014.

240. W. M. M. Schaaper, R. A. H. Adan, T. A. Posthuma, J. Oosterom, W.-H. Gispen and R. H. Meloen, *Lett. Peptide Sci.*, 1998, **5**, 205-208.

241. H. B. Schioth, F. Mutulis, R. Muceniece, P. Prusis and J. E. S. Wikberg, *Br. J. Pharmacol.*, 1998, **124**, 75-82.

242. V. J. Hruby, G. Han and M. E. Hadley, *Lett. Peptide Sci.*, 1998, **5**, 117-120.

243. M. F. Giblin, N. Wang, T. J. Hoffmann, S. S. Jurisson and T. P. Quinn, *Proc. Natl. Acad. Sci. USA*, 1998, **95**, 12814-12818.

244. P. J. Fairchild, *J. Peptide Sci.*, 1998, **4**, 182-194.

245. B. Hemmer, M. Vergelli, C. Pinilla, R. Houghten and R. Martin, *Immunology Today*, 1998, **19**, 163-168.

246. B. Reizis, M. Eisenstein, F. Mor and I. R. Cohen, *Immunology Today*, 1998, **19**, 212-216.

247. C. Perreault, D. C. Roy and C. Fortin, *Immunology Today*, 1998, **19**, 69-74.

248. M. Milik, D. Sauer, A. P. Brunmark, L. Yuan, A. Vitiello, M. R. Jackson, P. A. Peterson, J. Skolnick and C. A. Glass, *Nat. Biotechnol.*, 1998, **16**, 753-756.

249. B. Manoury, E. W. Hewitt, N. Morrice, P. M. Dando, A. J. B and C. Watts, *Nature*, 1998, **396**, 695-699.

250. A. Bianco, C. Zabel, P. Walden and G. Jung, *J. Peptide Sci.*, 1998, **4**, 471-478.

251. A. Bianco, C. Brock, C. Zabel, T. Walk, P. Walden and G. Jung, *J. Biol. Chem.*, 1998, **273**, 28759-28765.

252. C. I. Pearson, A. M. Gautam, I. C. Rulifson, R. S. Liblau and H. O. McDevitt, *Proc. Natl. Acad. Sci. USA*, 1999, **96**, 197-202.

253. L. Gauthier, K. J. Smith, J. Pyrdol, A. Kalandadze, J. L. Strominger, D. C. Wiley and K. W. Wucherpfennig, *Proc. Natl. Acad. Sci. USA*, 1998, **95**, 11828-11833.

254. E. C. Andersson, B. E. Hansen, H. Jacobsen, L. S. Madsen, C. B. Andersen, J. Engberg, J. B. Rothbard, G. S. McDevitt, V. Malmstrom, R. Holmdahl, A. Svejgaard and L. Fugger, *Proc. Natl. Acad. Sci. USA*, 1998, **95**, 7574-7579.

255. A. Geluk, V. Taneja, K. E. van Meijgaarden, E. Zanelli, C. Abou-Zeid, J. E. R. Thole, R. R. P. de Vries, C. S. David and T. H. M. Ottenhoff, *Proc. Natl. Acad. Sci. USA*, 1998, **95**, 10797-10802.

256. A. B. Smith, III, A. B. Benowitz, M. C. Guzman, P. A. Sprengeler and R. Hirschmann, *J. Am. Chem. Soc.*, 1998, **120**, 12704-12705.

257. A. Kushi, H. Sasai, H. Koizumi, N. Takeda, M. Yokoyama and M. Nakamura, *Proc. Natl. Acad. Sci. USA*, 1998, **95**, 15659-15664.

258. D. J. Marsh, G. Hollopeter, K. E. Kafer and R. D. Palmiter, *Nat. Med.*, 1998, **4**, 718-721.

259. T. Pedrazzini, J. Seydoux, P. Kunstner, J-F. Aubert, E. Grouzmann, F. Beermann and H.R. Brunner, *Nat. Med.*, 1998, **4**, 722-726.

260. A. Ishihara, T. Tanaka, A. Kanatani, T. Fukami, M. Ihara and T. Fukuroda, *Am. J. Physiol.*, 1998, **43**, R1500-R1504.

261. A. Kask, L. Rago and J. Harro, *Br. J. Pharmacol.*, 1998, **124**, 1507-1515.

262. H. A. Wieland, W. Engel, W. Eberlein, K. Rudolf and H. N. Doods, *Br. J. Pharmacol.*, 1998, **125**, 549-555.

263. Y, M. Tang-Christensen, P. Kristensen, C. E. Stidsen, C. L. Brand and P. J. Larsen, *J. Endocrinol.*, 1998, **159**, 307-312.

264. H. Zarrinmayeh, A. M. Nunes, P. L. Ornstein, D. M. Zimmerman, B. M. Arnold, D. A. Schober, S. L. Gackenheimer, R. F. Bruns, P. A. Hipskind, T. C. Britton, B. E. Cantrell and D. R. Gehlert, *J. Med. Chem.*, 1998, **41**, 2709-2719.

265. Y. Shigeri, M. Ishikawa, Y. Ishihara and M. Fujimoto, *Life Sci.*, 1998, **63**, PL151-PL160.
266. S. D. Bryant, S. Salvadori, P. S. Cooper and L. H. Lazarus, *Trends Pharmacol. Sci.*, 1998, **19**, 42-46.
267. T. Darland, M. M. Heinricher and D. K. Grandy, *Trends Neurological Sci.*, 1998, **21**, 215-221.
268. M. Roumy and J-M Zajac, *Eur. J. Pharmacol.*, 1998, **345**, 1-11.
269. C. Wechselberger, C. Severini, G. Kreil and L. Negri, *FEBS Lett.*, 1998, **429**, 41-43.
270. E. Masiukiewicz and B. Rzeszotarska, *Chem. Pharm. Bull.*, 1998, **46**, 1672-1675.
271. J. Malicka, M. Groth, C. Czaplewski, R. Kasprzykowska, A. Liwo, L. Lankiewicz and W. Wiczk, *Lett. Peptide Sci.*, 1998, **5**, 445-447.
272. J. R. Deschamps, C. George and J. L. Flippen-Anderson, *Lett. Peptide Sci.*, 1998, **5**, 337-340.
273. J. Olczak, K. Kaczmarek, I. Maszczynska, M. Lisowski, D. Stropova, V. J. Hruby, H. I. Yamamura, A. W. Lipkowski and J. Zabrocki, *Lett. Peptide Sci.*, 1998, **5**, 437-440.
274. M. Horikawa, Y. Shigeri, N. Yumoto, S. Yoshikawa, T. Nakajima and Y. Ohfune, *Bioorg. Med. Chem. Lett.*, 1998, **8**, 2027-2032.
275. I. Bobrova, N. Abissova, N. Mishlakova, G. Rozentals and G. Chipens, *Eur. J. Med. Chem.*, 1998, **33**, 255-266.
276. M. Cudic, J. Horvat, M. Elofsson, K-E. Bergquist, J. Kihlberg and S. Horvat, *J. Chem. Soc. Perkin I*, 1998, 1789-1795.
277. K. Shreder, L. Zhang, T. Dang, T.L. Yaksh, H. Umino, R. DeHaven, J. Daubert and M. Goodman, *J. Med. Chem.*, 1998, **41**, 2631-2635.
278. H. I. Mosberg, J. C. Ho and K. Sobczyk-Kojiro, *Bioorg. Med. Chem. Lett.*, 1998, **8**, 2681-2684.
279. R. C. Haaseth, T. Zalewska, P. Davis, H. I. Yamamura, F. Porreca and V. J. Hruby, *J. Peptide Res.*, 1997, **50**, 171-177.
280. D. Winkler, N. Sewald, K. Burger, N. N. Chung and P. W. Schiller, *J. Peptide Sci.*, 1998, **4**, 496-501.
281. S. Liao, J. Alfaro-Lopez, M. D. Shenderovich, K. Hosohata, J. Lin, X. Li, D. Stropova, P. Davis, K. A. Jernigan, F. Porreca, H. I. Yamamura and V. J. Hruby, *J. Med. Chem.*, 1998, **41**, 4767-4776.
282. K. Shreder, L. Zhang and M. Goodman, *Tetrahedron Lett.*, 1998, **39**, 221-224.
283. Y. Okada, M. Tsukatani, H. Taguchi, T. Yokoi, S. D. Bryant and L. H. Lazarus, *Chem. Pharm. Bull.*, 1998, **46**, 1374-1382.
284. C. Wang, I. J. McFadyen, J. R. Traynor and H. I. Mosberg, *Bioorg. Med. Chem. Lett.*, 1998, **8**, 2685-2688.
285. A. Olma, A. Misicka, D. Tourwé and A. W. Lipkowski, *Lett. Peptide Sci.*, 1998, **5**, 383-385.
286. A. Misicka, S. Cavagnero, R. Horvath, P. Davis, F. Porreca, H. I. Yamamura and V. J. Hruby, *J. Peptide Res.*, 1997, **50**, 48-54.
287. S. Zhang, Y. Tong, M. Tian, R. N. Dehaven, L. Cortesburgos, E. Mansson, F. Simonin, B. Kieffer and L. Yu, *J. Pharmacol. Exp. Therap.*, 1998, **286**, 136-141.
288. T. Hiranuma, K. Kitamura, T. Taniguchi, M. Kanai, Y. Arai, K. Iwao and T. Oka, *J. Pharmacol. Exp. Therap.*, 1998, **286**, 863-869.
289. B. L. Podlogar, M. G. Paterlini, D. M. Ferguson, G. C. Leo, D. A. Demeter, F. K. Brown and A. B. Reitz, *FEBS Lett.*, 1998, **439**, 13-20.
290. I. E. Goldberg, G. C. Rossi, S. R. Letchworth, J. P. Mathis, J. Ryan-Moro, L.

Leventhal, W. Su, D. Emmel, E. A. Bolan and G. W. Pasternak, *J. Pharmacol. Exp. Therap.*, 1998, **286**, 1007-1013.

291. K. Shivanandaiah, V. V. Suresh Babu and S. C. Shankaramma, *Indian J. Chem.*, 1998, **37B**, 760-767.

292. H. H. Szeto, J. F. Clapp, D. M. Disiderio, P. W. Schiller, O. O. Grigoriants, Y. Soong, D. Wu, N. Olariu, J-L. Tseng and R. Becklin, *J. Pharmacol. Exp. Therap.*, 1998, **284**, 61-65.

293. N. V. Sumbatyan, K. Gröger, O. N. Chichenkov and G. A. Korshunova, *Lett. Peptide Sci.*, 1997, **4**, 477-480.

294. A. Calderan, P. Ruzza, B. Ancona, L. Cima, P. Giusti, and G. Borin, *Lett. Peptide Sci.*, 1998, **5**, 71-73.

295. E. Okuda-Ashitaka, T. Minami, S. Tachibana, Y. Yoshihara, Y. Nishiuchi, T. Kimura and S. Ito, *Nature*, 1998, **392**, 286-289.

296. K. Carpenter and A. Dickenson, *Br. J. Pharmacol.*, 1998, **125**, 949-951.

297. J. E. Grisel, D. E. Farrier, S. G. Wilson and J. S. Mogil, *Eur. J. Pharmacol.*, 1998, **357**, R1-R3.

298. D. Tourwe, E. Mannekens, T. Nguyen, T. Diem, P. Verheyden, H. Jaspers, G. Toth, A. Peter, I. Kertesz, G. Torok, N. N. Chung and P. W. Schiller, *J. Med. Chem.*, 1998, **41**, 5167-5176.

299. P. W. Schiller, R. Schmidt, G. Weltrowska, I. Berezowska, T. M.-D. Nguyen, S. Dupuis, N. N. Chung, C. Lemieux, B. C. Wilkes, and K. A. Carpenter, *Lett. Peptide Sci.*, 1998, **5**, 209-214.

300. C. T. Dooley, P. Ny, J. M. Bidlack and R. A. Houghten, *J. Biol. Chem.*, 1998, **273**, 18848-18856.

301. C. H. J. Van Eijck, D. J. Kwekkeboom and E. P. Krenning, *Q. J. Nuclear Med.*, 1998, **42**, 18-25.

302. Z. Tulassay, *Scand. J. Gastroenterol.*, 1998, **33**, 115-121.

303. E. T. Janson, M. Stridsberg, A. Gobl, J-E. Westlin and K. Oberg, *Cancer Res.*, 1998, **58**, 2375-2378.

304. S. Schulz, S. Schulz, J. Schmitt, D. Wiborny, H. Schmidt, S. Olbricht, W. Weise, A. Roessner, C. Gramsch and V. Hollt, *Clin. Cancer Res.*, 1998, **4**, 2047-2052.

305. E. Mato, X. Matias-Guiu, A. Chico, S. M. Webb, R. Cabezas, L. Berna and A. De Leiva, *J. Clin. Endocrinol. Metab.*, 1998, **83**, 2417-2420.

306. F. Alderton, T-P. D. Fan, M. Schindler and P. P. A. Humphrey, *Br. J. Pharmacol.*, 1998, **125**, 1630-1633.

307. M. Raderer, T. Pangerl, M. Leimer, J. Valencak, A. Kurtaran, G. Hamilton, W. Scheithauer and I. Virgolini, *J. Nat. Canc. Inst.*, 1998, **90**, 1666-1668.

308. T. Sakane and J. Suzuki, *Clin. Exp. Rheumatol.*, 1998, **16**, 745-749.

309. S. Sakamoto, T. Sakamaki, T. Kanda, Y. Ito, H. Sumino, H. Masuda, Y. Ohyama, Z. Ono, M. Kurabayashi, I. Kobayashi and R. Nagai, *Res. Commun. Mol. Pathol. Pharmacol.*, 1998, **101**, 25-34.

310. A. Nagy, A. V. Schally, G. Halmos, P. Armatis, R-Z. Cai, V. Csernus, M. Kovacs, M. Koppan, K. Szepeshazi and Z. Kahan, *Proc. Natl. Acad. Sci. USA*, 1998, **95**, 1794-1799.

311. M. Koppan, A. Nagy, A. V. Schally, J. M. Arencibia, A. Plonowski and G. Halmos, *Cancer Res.*, 1998, **58**, 4132-4137.

312. C. Gilon, M. Huenges, B. Matha, G. Gellerman, V. Hornik, M. Afargan, O. Amitay, O. Ziv, E. Feller, A. Gomliel, D. Shohat, M. Wanger, O. Arad and H. Kessler, *J. Med. Chem.*, 1998, **41**, 919-929.

313. V. Brecx, P. Verheyden and D. Tourwe, *Lett. Peptide Sci.*, 1998, **5**, 67-70.

314. T-A. Tran, R-H. Mattern, M. Afargan, O. Amitay, O. Ziv, B. A. Morgan, T. E. Taylor, D. Hoyer and M. Goodman, *J. Med. Chem.*, 1998, **41**, 2679-2685.
315. R-H. Mattern, T-A. Tran and M. Goodman, *J. Med. Chem.*, 1998, **41**, 2686-2692.
316. T. J. Gillespie, A. Erenberg, S. Kim, J. Dong, J. E. Taylor, V. Hau and T. P. Davis, *J. Pharmacol. Exp. Therap.*, 1998, **285**, 95-104.
317. L. Yang, L. Guo, A. Pasternak, R. Mosley, S. Rohrer, E. Birzin, F. Foor, K. Cheng, J. Schaeffer and A. A. Patchett, *J. Med. Chem.*, 1998, **41**, 2175-2179.
318. L. Yang, S. C. Berk, S. P. Rohrer, R. T. Mosley, L. Guo, D. J. Underwood, B. H. Arison, E. T. Birzin, E. C. Hayes, S. W. Mitra, R. M. Parmar, K. Cheng, T-J. Wu, B. S. Butler, F. Foor, A. Pasternak, Y. Pan, M. Silva, R. M. Freidinger, R. G. Smith, K. Chapman, J. M. Schaeffer and A. A. Patchett, *Proc. Natl. Acad. Sci. USA*, 1998, **95**, 10836-10841.
319. S. P. Rohrer, E. T. Birzin, R. T. Mosley, S. C. Berk, S. M. Hutchins, D-M. Shen, Y. Xiong, E. C. Hayes, R. M. Parmar, F. Foor, S. W. Mitra, S. J. Degrado, M. Shu, J. M. Klopp, S-J. Cai, A. Blake, W. W. S. Chan, A. Pasternak, L. Yang, A. A. Patchett, R. G. Smith, K. T. Chapman and J. M. Schaeffer, *Sci.*, 1998, **282**, 737-740.
320. M. Ankersen, M. Crider, S. Liu, B. Ho, H. S. Andersen and C. Ctidsen, *J. Am. Chem. Soc.*, 1998, **120**, 1368-1373.
321. S. Liu, C. Tang, B. Ho, M. Ankersen, C. E. Stidsen and A. M. Crider, *J. Med. Chem.*, 1998, **41**, 4693-4705.
322. J. J. Scicinski, M. D. Barker, P. J. Murray and E. M. Jarvie, *Bioorg. Med. Chem. Lett.*, 1998, **8**, 3609-3614.
323. W. R. Baumbach, T. A. Carrick, M. H. Pousch, B. Bingham, D. Carmignac, I. C. A. F. Robinson, R. Houghten, C. M. Eppler, L. A. Price and J. R. Zysk, *Mol. Pharmacol.*, 1998, **54**, 864-873.
324. S. J. Hocart, R. Jain, W. A. Murphy, J. F. Taylor, B. Morgan and D. H. Coy, *J. Med. Chem.*, 1998, **41**, 1146-1154.
325. W. J. Rossowski, B-L. Cheng, N-Y. Jiang and D. H. Coy, *Br. J. Pharmacol.*, 1998, **125**, 1081-1087.
326. B. Holst, S. Zoffmann, C. E. Elling, S. A. Hjorth and T. W. Schwartz, *Mol. Pharmacol.*, 1998, **53**, 166-175.
327. O. Nyeki, A. Rill, I. Schon, A. Orosz, J. Schrett, L. Bartha and J. Nagy, *J. Peptide Sci.*, 1998, **4**, 486-495.
328. M. J. Seckl and E. Rozengurt, *Lett. Peptide Sci.*, 1998, **5**, 199-204.
329. J. M. Elliott, M. A. Cascieri, G. Chicchi, S. Davies, F. J. Kelleher, M. Kurtz, T. Ladduwahetty, R. T. Lewis, A. M. MacLeod, K. J. Merchant, S. Sadowski and G. I. Stevenson, *Bioorg. Med. Chem. Lett.*, 1998, **8**, 1845-1850.
330. J. M. Elliott, H. Broughton, *Bioorg. Med. Chem. Lett.*, 1998, **8**, 1851-1856.
331. H. Qi, S. K. Shah, M. A. Cascieri, S. J. Sadowski and M. MacCoss, *Bioorg. Med. Chem. Lett.*, 1998, **8**, 2259-2262.
332. C. S. J. Walpole, M. C. S. Brown, I. F. James, E. A. Campbell, P. McIntyre, R. Docherty, S. Ko, L. Hedley, S. Ewan, K.-H. Buchheit and L. A. Urban, *Br. J. Pharmacol.*, 1998, **124**, 83-92.
333. C. Walpole, S. Y. Ko, M. Brown, D. Beattie, E. Campbell, F. Dickenson, S. Ewan, G. A. Hughes, M. Lemaire, J. Lerpiniere, S. Patel and L. Urban, *J. Med. Chem.*, 1998, **41**, 3159-3173.
334. R. Cirillo, M. Astolfi, B. Conte, G. Lopez, M. Parlani, R. Terracciano, C. I. Fincham and S. Manzini, *Eur. J. Pharmacol.*, 1998, **341**, 201-209.

335. J. J. Hale, S. G. Mills, M. MacCoss, P. E. Finke, M. A. Cascieri, S. Sadowski, E. Ber, G. G. Chicchi, M. Kurtz, J. Metzger, G. Eiermann, N. N. Tsou, F. D. Tattersall, N. M. J. Rupniak, A. R. Williams, W. Rycroft, R. Hargreaves and D. E. MacIntyre, *J. Med. Chem.*, 1998, **41**, 4607-4614.

336. G. I. Stevenson, I. Huscroft, A. M. MacLeod, C. J. Swain, M. A. Cascieri, G. G. Chicchi, M. I. Graham, T. Harrison, F. J. Kelleher, M. Kurtz, T. Ladduwahetty, K. J. Merchant, J. M. Metzger, D. E. MacIntyre, S. Sadowski, B. Sohal and A. P. Owens, *J. Med. Chem.*, 1998, **41**, 4623-4635.

337. Y. Ikeura, Y. Ishichi, T. Tanaka, A. Fujishima, M. Murabayashi, M. Kawada, T. Ishimaru, I. Kamo, T. Doi and H. Natsugari, *J. Med. Chem.*, 1998, **41**, 4232-4239.

338. R. Hosoki, M. Yanagisawa, Y. Onishi, K. Yoshioka and M. Otsuka, *Eur. J. Pharmacol.*, 1998, **341**, 235-241.

339. R. T. Jacobs, A. B. Shenvi, R. C. Mauger, T. G. Ulatowski, D. Aharony and C. K. Buckner, *Bioorg. Med. Chem. Lett.*, 1998, **8**, 1935-1940.

340. H. Itadani, T. Nakamura, J. Itoh, I. Iwaasa, A. Kanatani, J. Borkowski, M. Ihara and M. Ohta, *Biochem. Biophys. Res. Commun.*, 1998, **250**, 68-71.

341. J. Cao, D. O'Donnell, H. Vu, K. Payza, C. Pou, C. Godbout, A. Jakob, M. Pelletier, P. Lembo, S. Ahmad, and P. Walker, *J. Biol. Chem.*, 1998, **273**, 32281-32287.

342 A. Cremades, R. Penafiel, V. Rausell, J. D. Rio-Garcia and D. G. Smyth, *Eur. J. Pharmacol.*, 1998, **358**, 63-67.

343. W. Chu, J. H. Perlman, M. C. Gershengorn and K. D. Moeller, *Bioorg. Med. Chem. Lett.*, 1998, **8**, 3093-3096.

344. L. Prokai, X. Ouyang, K. Prokai-Tatrai, J. W. Simpkins and N. Bodor, *Eur. J. Med. Chem.*, 1998, **33**, 879-886.

345. A. Tahara, M. Saito, T. Sugimoto, Y. Tomura, K. Wada, T. Kusayama, J. Tsukada, N. Ishii, T. Yatsu, W. Uchida and A. Tanaka, *Br. J. Pharmacol.*, 1998, **125**, 1463-1470.

346. X. Zhu and J. Wess, *Biochemistry*, 1998, **37**, 15773-15784.

347. N. Cotte, M-N. Balestre, S. Phalipou, M. Hibert, M. Manning, C. Barberis and B. Mouillac, *J. Biol. Chem.*, 1998, **273**, 29462-29468.

348. C. Czaplewski, R. Kazmierkiewicz and J. Ciarkowski, *Lett. Peptide Sci.*, 1998, **5**, 333-335.

349. M. S. Kuo, M. G. Bock, R. M. Freidinger, M. T. Guidotti, E. V. Lis, J. M. Pawluczyk, D. S. Perlow, D. J. Pettibone, A. G. Quigley, D. R. Reiss, P. D. Williams and C. J. Woyden, *Bioorg. Med. Chem. Lett.*, 1998, **8**, 3081-3086.

350. J. D. Albright, M. F. Reich, E. G. D. Santos, J. P. Dusza, F-W. Sum, A. M. Venkatesan, J. Coupet, P. S. Chan, X. Ru, H. Mazandarani and T. Bailey, *J. Med. Chem.*, 1998, **41**, 2442-2444.

351. Y. Yamamura, S. Nakamura, S. Itoh, T. Hirano, T. Onogawa, T. Yamashita, Y. Yamada, K. Tsujimae, M. Aoyama, K. Kotosai, H. Ogawa, H. Yamashita, K. Kondo, M. Tominaga, G. Tsujimoto and T. Mori, *J. Pharmacol. Exp. Therap.*, 1998, **287**, 860-867.

352 W. Y. Chan, N. C. Wo, S. Stoev, L. L. Cheng and M. Manning, *Br. J. Pharmacol.*, 1998, **125**, 803-811.

353. R. G. Evans, G. Bergstrom and A. J. Lawrence, *J. Cardiovasc. Pharmacol.*, 1998, **32**, 571-581.

354. A. Heinemann, G. Horina, R. E. Stauber, C. Pertl, P. Holzer and B. A. Peskar, *Br. J. Pharmacol.*, 1998, **125**, 1120-1127.

274   *Amino Acids, Peptides and Proteins*

355. A. A. Karelin, M. M. Philippova, E. V. Karelina, B. N. Strizhkov, G. A. Grishina, I. V. Nazimov and V. T. Ivanov, *J. Peptide Sci.*, 1998, **4**, 211-225.

356. P. Rekowski, A. Borowiec, J. Druzynska and E. Kusiak, *Lett. Peptide Sci.*, 1998, **5**, 417-420.

357. P. Madsen, L. B. Knudsen, F. C. Wiberg and R. D. Carr, *J. Med. Chem.*, 1998, **41**, 5150-5157.

358. J. Y. Lee, Y. J. Chung, Y-S. Bae, S. H. Ryu and B. H. Kim, *J. Chem. Soc. Perkin I*, 1998, 359-365.

359. K. Ishida, H. Matsuda and M. Murakami, *Tetrahedron*, 1998, **54**, 13475-13484.

360. C. David, L. Bischoff, H. Meudal, C. Llorens-Cortes, B. P. Roques and M-C. Fournie-Zaluski, *Lett. Peptide Sci.*, 1997, **4**, 411-414.

361. H. Miyachi, M. Kato, F. Kato and Y. Hashimoto, *J. Med. Chem.*, 1998, **41**, 263-265.

362 S. Chatterjee, *Drugs Future*, 1998, **23**, 1217-1225.

363. J. Huang and N. E. Forsberg, *Proc. Natl. Acad. Sci. USA*, 1998, **95**, 12100-12105.

364. S. Chatterjee, Z-Q. Gu, D. Dunn, M. Tao, K. Josef, R. Tripathy, R. Bihovsky, S. E. Senadhi, T. M. O'Kane, B. A. McKenna, S. Mallya, M. A. Ator, D. Bozyczko-Coyne, R. Siman and J. P. Mallamo, *J. Med. Chem.*, 1998, **41**, 2663-2666.

365. R. Tripathy, Z-Q. Gu, D. Dunn, S. E. Senadhi, M. A. Ator and S. Chatterjee, *Bioorg. Med. Chem. Lett.*, 1998, **8**, 2647-2652.

366. S. Hu, S. J. Snipas, C. Vincenz, G. Salvesen and V. M. Dixit, *J. Biol. Chem.*, 1998, **273**, 29648-29653.

367. M. Ahmad, S. M. Srinivasula, R. Hegde, R. Mukattash, T. Fernandes-Alnemri and E. S. Alnemri, *Cancer Res.*, 1998, **58**, 5201-5205.

368. H. Li, L. Bergeron, V. Cryns, M.S. Pasternack, H. Zhu, L. Shi, A. Greenberg and J. Yuan, *J. Biol. Chem.*, 1997, **272**, 21010-21017.

369. M. Garcia-Calvo, E. P. Peterson, B. Leiting, R. Ruel, D. W. Nicholson and N. A. Thornberry, *J. Biol. Chem.*, 1998, **273**, 32608-32613.

370. D. S. Karanewsky, X. Bai, S. D. Linton, J. F. Krebs, J. Wu, B. Pham and K. J. Tomaselli, *Bioorg. Med. Chem. Lett.*, 1998, **8**, 2757-2762.

371. J. S. Mort, M-C. Magny and E. R. Lee, *Biochem. J.*, 1998, **335**, 491-494.

372 R. Xing and R. Hanzlik, *J. Med. Chem.*, 1998, **41**, 1344-1351.

373. M. Meldal, I. Svendsen, L. Juliano, M. A. Juliano, E. D. Nery and J. Scharfstein, *J. Peptide Sci.*, 1998, **4**, 83-91.

374. J. L. Conroy, P. Abato, M. Ghosh, M. I. Austermuhle, M. R. Kiefer and C. T. Seto, *Tetrahedron Lett.*, 1998, **39**, 8253-8256.

375. A. Y. Lee, S. V. Gulnik and J. W. Erickson, *Nature Struct. Biol.*, 1998, **5**, 866-871.

376. C. D. Carroll, T. O. Johnson, S. Tao, G. Lauri, M. Orlowski, I. Y. Gluzman, D. E. Goldberg and R. E. Dolle, *Bioorg. Med. Chem. Lett.*, 1998, **8**, 3203-3206.

377. P. Saftig, E. Hunziker, O. Wehmeyer, S. Jones, A. Boyde, W. Rommerskirch, J. D. Moritz, P. Schu and K. V. Figura, *Proc. Natl. Acad. Sci. USA*, 1998, **95**, 13453-13458.

378. P. Garnero, O. Borel, I. Byrjalsen, M. Ferreras, F. H. Drake, M. S. McQueney, N. T. Foged, P. D. Delmas and J-M. Delaisse, *J. Biol. Chem.*, 1998, **273**, 32347-32352.

379. J. M. LaLonde, B. Zhao, W. W. Smith, C. A. Janson, R. L. DesJarlais, T. A. Tomaszek, T. J. Carr, S. K. Thompson, H. J. Oh, D. S. Yamashita, D. F. Veber and S. S. Abel-Meguid, *J. Med. Chem.*, 1998, **41**, 4567-4576.

380. S. K. Thompson, W. W. Smith, B. Zhao, S. M. Halbert, T. A. Tomaszek, D. G. Tew, M. A. Levy, C. A. Janson, K. J. D'Alessio, M. S. McQueney, J. Kurdyla, C. S. Jones, R. L. DesJarlais, S. S. Abel-Meguid and D. F. Veber, *J. Med. Chem.*, 1998, **41**, 3923-3927.

381. R. L. DesJarlais, D. S. Yamashita, H-J. Oh, I. N. Uzinskas, K. F. Erhard, A. C. Allen, R. C. Haltiwanger, B. Zhao, W. W. Smith, S. S. Abdel-Meguid, K. D'Alessio, C. A. Janson, M. S. McQueney, T. A. Tomaszek, M. A. Levy and D. F. Veber, *J. Am. Chem. Soc.*, 1998, **120**, 9114-9115.

382 T. Yasuma, S. Oi, N. Choh, T. Nomura, N. Furuyama, A. Nishimura, Y. Fujisawa and T. Sohda, *J. Med. Chem.*, 1998, **41**, 4301-4308.

383. I. Santamaria, G. Velasco, M. Cazorla, A. Fueyo, E. Campo and C. Lopez-Otin, *Cancer Res.*, 1998, **58**, 1624-1630.

384. P. R. Bonneau, C. Plouffe, A. Pelletier, D. Wernic and M-A. Poupart, *Analyt. Biochem.*, 1998, **255**, 59-65.

385. Q. M. Wang, R. B. Johnson, W. Sommergruber and T. A. Shepherd, *Arch. Biochem. Biophys.*, 1998, **356**, 12-18.

386. L. Tong, C. Qian, M-J. Massariol, R. Deziel, C. Yoakim and L. Lagace, *Nature Struct. Biol.*, 1998, **5**, 819-826.

387. S. R. LaPlante, D. R. Cameron, N. Aubry, P. R. Bonneau, R. Deziel, C. Grand-Maitre, W. W. Ogilvie and S. H. Kawai, *Angew. Chem. Int. Ed.*, 1998, **37**, 2729-2732.

388. D. Dhanak, R. M. Keenan, G. Burton, A. Kaura, M. G. Darcy, D. H. Shah, L. H. Ridgers, A. Breen, P. Lavery, D. G. Tew and A. West, *Bioorg. Med. Chem. Lett.*, 1998, **8**, 3677-3682.

389. C. Yoakim, W. W. Ogilvie, D. R. Cameron, C. Chabot, I. Guse, B. Hache, J. Naud, J. A. O'Meara, R. Plante and R. Deziel, *J. Med. Chem.*, 1998, **41**, 2882-2891.

390. S. E. Webber, K. Okano, T. L. Little, S. H. Reich, Y. Xin, S. A. Fuhrman, D. A. Matthews, R. A. Love, T. F. Hendrickson, A. K. Patick, J. W. Meador, III, R. A. Ferre, E. L. Brown, C. E. Ford, S. L. Binford and S. T. Worland, *J. Med. Chem.*, 1998, **41**, 2786-2805.

391. P. S. Dragovich, S. E. Webber, R. E. Babine, S. A. Fuhrman, A. K. Patick, D. A. Matthews, C. A. Lee, S. H. Reich, T. J. Prins, J. T. Marakovits, E. S. Littlefield, R. Zhau, J. Tikhe, C. E. Ford, M. B. Wallace, J. W. Meador, III, R. A. Ferre, E. L. Brown, S. L. Binford, J. E. V. Harr, D. M. DeLisle and S. T. Worland, *J. Med. Chem.*, 1998, **41**, 2806-2818.

392 P. S. Dragovich, S. E. Webber, R. E. Babine, S. A. Fuhrman, A. K. Patick, D. A. Matthews, S. H. Reich, J. T. Marakovits, T. J. Prins, R. Zhau, J. Tikhe, E. S. Littlefield, T. M. Bleckman, M. B. Wallace, T. L. Little, C. E. Ford, J. W. Meador, III, R. A. Ferre, E. L. Brown, S. L. Binford, D. M. DeLisle and S. T. Worland, *J. Med. Chem.*, 1998, **41**, 2819-2834.

393. M. Llinas-Brunet, M. Bailey, *Bioorg. Med. Chem. Lett.*, 1998, **8**, 1713-1718.

394. M. Llinas-Brunet, M. Bailey, R. Deziel, G. Fasal, V. Gorys, S. Goulet, T. Halmos, R. Mourice, M. Poirier, M-A. Poupart, J. Rancourt, D. Thibeault, D. Wernic and D. Lamarre, *Bioorg. Med. Chem. Lett.*, 1998, **8**, 2719-2724.

395. C. Steinkuhler, G. Biasiol, M. Brunetti, A. Urbani, U. Coch, R. Cortese, A. Pessi and R. De Francesco, *Biochemistry*, 1998, **37**, 8899-8905.

396. P. Ingallinella, S. Altamura, E. Bianchi, M. Taliani, R. Ingenito, R. Cortese, R. De Francesco, C. Steinkuhler and A. Pessi, *Biochemistry*, 1998, **37**, 8906-8914.

397. K. Suetsuna, *J. Nutr. Biochem.*, 1998, **9**, 415-419.

398. K. Nakagomi, A. Fujimura, H. Ebisu, T. Sakai, Y. Sadakane, N. Fujii and T. Tanimura, *FEBS Lett.*, 1998, **438**, 255-257.

399. S. Hanessian, U. Reinhold, M. Saulnier and S. Claridge, *Bioorg. Med. Chem. Lett.*, 1998, **8**, 2123-2128.

400. G. Jones, W. Jones and M. North, *Tetrahedron*, 1999, **55**, 279-290.

401. F. Anastasopoulos, R. Leung, A. Kladis, G. M. James, T. A. Briscoe, T. P. Gorski and D. J. Campbell, *J. Pharmacol. Exp. Therap.*, 1998, **284**, 799-805.

402   R. E. Chatelain, R. D. Ghai, A. J. Trapani, L. M. Odorico, B. N. Dardik, S. D. Lombaert, R. W. Lappe and C. A. Fink, *J. Pharmacol. Exp. Therap.*, 1998, **284**, 974-982.

403. H. Chen, F. Noble, P. Coric, M-C. Fournie-Zaluski and B. P. Roques, *Proc. Natl. Acad. Sci. USA*, 1998, **95**, 12028-12033.

404. K. Ahn, S. B. Herman and D. C. Fahnoe, *Arch. Biochem. Biophys.*, 1998, **359**, 258-268.

405. H. Hasegawa, K. Hiki, T. Sawamura, T. Aoyama, Y. Okamoto, S. Miwa, S. Shimohama, J. Kimura and T. Masaki, *FEBS Lett.*, 1998, **428**, 304-308.

406   W. Liu, R. Takayanagi, T. Ito, K. Oba and H. Nawata, *FEBS Lett.*, 1997, **420**, 103-106.

407. S. Takaishi, N. Tuchiya, A. Sato, T. Negishi, Y. Takamatsu, Y. Matsushita, T. Watanabe, Y. Iijima, H. Haruyama, T. Kinoshita, M. Tanaka and K. Kodama, *J. Antibiot.*, 1998, **51**, 805-815.

408. M. A. Massa, W. C. Patt, K. Ahn, A. M. Sisneros, S. B. Herman and A. Doherty, *Bioorg. Med. Chem. Lett.*, 1998, **8**, 2117-2122.

409. S-T. Huang and D. M. Gordon, *Tetrahedron Lett.*, 1998, **39**, 9335-9338.

410   K. Ahn, A. M. Sisneros, S. B. Herman, S. M. Pan, D. Hupe, C. Lee, S. Nikam, X-M. Cheng, A. M. Doherty, R. L. Schroeder, S. J. Haleen, S. Kaw, N. Emoto and M. Yanagisawa, *Biochem. Biophys. Res. Commun.*, 1998, **243**, 184-190.

411. E. M. Wallace, J. A. Moliterni, M. A. Moskal, A. D. Neubert, N. Marcopulos, L. B. Stamford, A. J. Trapani, P. Savage, M. Chou and A. Y. Jeng, *J. Med. Chem.*, 1998, **41**, 1513-1523.

412   L. Guy, J. Vidal, A. Collet, A. Amour and M-R. Ravaux, *J. Med. Chem.*, 1998, **41**, 4833-4843.

413. R. J. Cregge, S. L. Durham, R. A. Farr, S. L. Gallion, C. M. Hare, R. V. Hoffman, M. J. Janusz, H-O. Kim, J. R. Koehl, S. Mehdi, W. A. Metz, N. P. Peet, J. T. Pelton, H. A. Schreuder, S. Sunder and C. Tardif, *J. Med. Chem.*, 1998, **41**, 2461-2480.

414. A. Armour, M. Reboud-Ravaux, E. D. Rosny, A. Abouabdellah, J-P. Begue, D. Bonnet-Delpon and M. Le Gall, *J. Pharm. Pharmacol.*, 1998, **50**, 593-600.

415. S. J. F. Macdonald, D. J. Belton, D. M. Buckley, J. E. Spooner, M. S. Anson, L. A. Harrison, K. Mills, R. J. Upton, M. D. Dowle, R. A. Smith, C. R. Molloy and C. Risley, *J. Med. Chem.*, 1998, **41**, 3119-3922.

416   M. Gutschow and U. Neumann, *J. Med. Chem.*, 1998, **41**, 1729-1740.

417. S. B. Long, P. J. Casey and L. S. Beese, *Biochemistry*, 1998, **37**, 9612-9618.

418. C. L. Strickland, W. T. Windsor, R. Syto, L. Wang, R. Bond, Z. Wu, J. Schwartz, H. V. Le, L. S. Beese and P. C. Weber, *Biochemistry*, 1998, **37**, 16601-16611.

419. S. J. Desolms, E. A. Giuliani, S. L. Graham, K. S. Koblan, N. E. Kohl, S. D. Mosser, A. I. Oliff, D. L. Pompliano, E. Rands, T. H. Scholz, C. M. Wiscount, J.B. Gibbs and R. L. Smith, *J. Med. Chem.*, 1998, **41**, 2651-2656.

420   M. J. Breslin, S. J. DeSolms, E. A. Giuliani, G. E. Stokker, S. L. Graham, D. L.

Pompliano, S. D. Mosser, K. A. Hamilton and J. H. Hutchinson, *Bioorg. Med. Chem. Lett.*, 1998, **8**, 3311-3316.

421. T. Aoyama, T. Satoh, M. Yonemoto, J. Shibata, K. Nonoshita, S. Arai, K. Kawakami, Y. Iwasawa, H. Sano, K. Tanaka, Y. Monden, T. Kodera, H. Arakawa, I. Suzuki-Takahashi, T. Kamei, and K. Tomimoto, *J. Med. Chem.*, 1998, **41**, 143-147.

422. M. Yonemoto, T. Satoh, H. Arakawa, I. Suzuki-Takahashi, Y. Monden, T. Kodera, K. Tanaka, T. Aoyama, Y. Iwasawa, T. Kamei, S. Nishimura and K. Tomimoto, *Mol. Pharmacol.*, 1998, **54**, 1-7.

423. D. A. Augeri, S. J. O'Connor, D. Janowick, B. Szczepankiewicz, G. Sullivan, J. Larsen, D. Kalvin, J. Cohen, E. Devine, H. Zhang, S. Cherian, B. Saeed, S-C. Ng and S. Rosenberg, *J. Med. Chem.*, 1998, **41**, 4288-4300.

424. G. Caliendo, F. Fiorino, P. Grieco, E. Perissutti, A. Ramunno, V. Santagada, S. Albrizio, D. Califano, A. Giuliano and G. Santelli, *Eur. J. Med. Chem.*, 1998, **33**, 725-732.

425. R. Wolin, M. Connolly, J. Kelly, J. Weinstein, S. Rosenblum, A. Afonso, L. James, P. Kirschmeier and W. R. Bishop, *Bioorg. Med. Chem. Lett.*, 1998, **8**, 2521-2526.

426 A. K. Mallams, R. R. Rossman, R. J. Doll, V. M. Girijavallabhan, A. K. Ganguly, J. Petrin, L. Wang, R. Patton, W. R. Bishop, D. M. Carr, P. Kirschmeier, J. J. Catino, M. S. Bryant, K-J. Chen, W. A. Korfmacher, C. Nardo, S. Wang, A. A. Nomeir, C-C. Lin, Z. Li, J. Chen, S. Lee, J. Dell, P. Lipari, M. Malkowski, B. Yaremko, I. King and M. Liu, *J. Med. Chem.*, 1998, **41**, 877-893.

427. F. G. Njoroge, B. Vibulbhan, P. Pinto, W. R. Bishop, M. S. Bryant, A. A. Nomeir, C-C. Lin, M. Liu, R. J. Doll, V. Girijavallabhan and A. K. Ganguly, *J. Med. Chem.*, 1998, **41**, 1561-1567.

428. F. G. Njoroge, A. C. Taveras, J. Kelly, S. Remiszewski, A. K. Mallams, R. Wolin, A. Afonso, A. B. Cooper, D. F. Rane, Y-T. Liu, J. Wong, B. Vibulbhan, P. Pinto, J. Deskus, C. S. Alvarez, J. del Rosario, M. Connolly, J. Wang, J. Desai, R. R. Rossman, W. R. Bishop, R. Patton, L. Wang, P. Kirschmeier, M. S. Bryant, A. A. Nomeir, C-C. Lin, M. Liu, A. T. McPhail, R. J. Doll, V. M. Girijavallabhan and A. K. Ganguly, *J. Med. Chem.*, 1998, **41**, 4890-4902.

429. M. Liu, M. S. Bryant, J. Chen, S. Lee, B. Yaremko, P. Lipari, M. Malkowski, E. Ferrai, L. Nielsen, N. Prioli, J. Dell, D. Sinha, J. Syed, W. A. Korfmacher, A. A. Nomeir, C-C. Lin, L. Wang, A. C. Taveras, R. J. Doll, F. G. Njoroge, A. K. Mallams, S. Remiszewski, J. J. Catino, V. M. Girijavallabhan, P. Kirschmeier and W. R. Bishop, *Cancer Res.*, 1998, **58**, 4947-4956.

430 D. Boden and M. Markowitz, *Antimicrob. Agents Chemother.*, 1998, **42**, 2775-2783.

431. G. V. De Lucca and P. Y. S. Lam, *Drugs Future*, 1998, **23**, 987-994.

432. P. A. Aristoff, *Drugs Future*, 1998, **23**, 995-999.

433. N. Madani, S. L. Kozak, M. P. Kavanaugh and D. Kabat, *Proc. Natl. Acad. Sci. USA*, 1998, **95**, 8005-8010.

434. G. A. Donzella, D. Schols, S. W. Lin, J. A. Este, K. A. Nagashima, P. J. Maddon, G. P. Allaway, T. P. Sakmar, G. Henson, E. De Clercq and J. P. Moore, *Nat. Med.*, 1998, **4**, 72-77.

435. J. M. Kilby, S. Hopkins, T. M. Venetta, B. DiMassimo, G. A. Cloud, J. Y. Lee, L. Alldredge, E. Hunter, D. Lambert, D. Bolognesi, T. Matthews, M. R.

Johnson, M. A. Nowak, G. M. Shaw and M. S. Saag, *Nat. Med.*, 1998, **4**, 1302-1307.

436   E. Drakopoulou, J. Vizzavona and C. Vita, *Lett. Peptide Sci.*, 1998, **5**, 241-245.

437.   M. G. B. Drew, S. Gorsuch, J. Mann and S. Yoshida, *J. Chem. Soc. Perkin 1*, 1998, 1627-1636.

438.   D. Gizachew, D. B. Moffett, S. C. Busse, W. M. Westler, E. A. Dratz and M. Teintze, *Biochemistry*, 1998, **37**, 10616-10625.

439.   M. Pritsker, P. Jones, R. Blumenthal and Y. Shai, *Proc. Natl. Acad. Sci. USA*, 1998, **95**, 7287-7292.

440   J. C. Cairns and M. P. D'Souza, *Nat. Med.*, 1998, **4**, 563-568.

441.   L. G. Kostrikis, Y. Huang, J. P. Moore, S. M. Wolinsky, L. Zhang, Y. Guo, L. Deutsch, J. Phair, A. U. Neumann and D. D. Ho, *Nat. Med.*, 1998, **4**, 350-353.

442.   T. M. Ross and B. R. Cullen, *Proc. Natl. Acad. Sci. USA*, 1998, **95**, 7682-7686.

443.   W. Gong, O. M. Z. Howard, J. A. Turpin, M. C. Grimm, H. Ueda, P. W. Gray, C. J. Raport, J. J. Openheim and J. M. Wang, *J. Biol. Chem.*, 1998, **273**, 4289-4292.

444.   L. G. Ulysse and J. Chmielewski, *Bioorg. Med. Chem. Lett.*, 1998, **8**, 3281-3286.

445.   P. Garraouste, M. Pawlowski, T. Tonnaire, S. Sicsic, P. Dumy, E. De Rosny, M. Reboud-Ravaux, P. Fulcrand and J. Martinez, *Eur. J. Med. Chem.*, 1998, **33**, 423-436.

446   A. Friedler, N. Zakai, O. Karni, Y. C. Broder, L. Baraz, M. Kotler, A. Loyter and C. Gilon, *Biochemistry*, 1998, **37**, 5616-5622.

447.   A. C. Nair, S. Miertus, A. Tossi and D. Romeo, *Biochem. Biophys. Res. Commun.*, 1998, **243**, 545-551.

448.   F. G. Salituro, C. T. Baker, J. J. Court, D. D. Deininger, E. E. Kim, B. Li, P. M. Novak, B. G. Rao, S. Pazhanisamy, M. D. Porter, W. C. Schairer and R. D. Tung, *Bioorg. Med. Chem. Lett.*, 1998, **8**, 3637-3642.

449.   C. T. Baker, F. G. Salituro, J. J. Court, D. D. Deininger, E. E. Kim, B. Li, P. M. Novak, B. G. Rao, S. Pazhanisamy, W. C. Schairer and R. D. Tung, *Bioorg. Med. Chem. Lett.*, 1998, **8**, 3631-3636.

450   H. L. Sham, D. J. Kempf, A. Molla, K. C. Marsh, G. N. Kumar, C-M. Chen, W. Kati, K. Stewart, R. Lal, A. Hsu, D. Betebenner, M. Korneyeva, S. Vasavanonda, E. McDonald, A. Saldivar, N. Wideburg, X. Chen, P. Niu, C. Park, V. Jayanti, B. Grabowski, G. R. Granneman, E. Sun, A. J. Japour, J. M. Leonard, J. J. Plattner and D. W. Norbeck, *Antimicrob. Agents Chemother.*, 1998, **42**, 3218-3224.

451.   D. J. Kempf, H. L. Sham, K. C. Marsh, C. A. Flentge, D. Betebenner, B. E. Green, E. McDonald, S. Vasavanonda, A. Saldivar, N. E. Widwburg, W. M. Kati, L. Ruiz, C. Zhao, L. M. Fino, J. Patterson, A. Molla, J. J. Plattner and D. W. Norbeck, *J. Med. Chem.*, 1998, **41**, 602-617.

452.   X. Chen, D. J. Kempf, H. L. Sham, B. E. Green, A. Molla, M. Korneyeva, S. Vasavanonda, N. E. Wideburg, A. Saldivar, K. C. Marsh, E. McDonald and D. W. Norbeck, *Bioorg. Med. Chem. Lett.*, 1998, **8**, 3531-3536.

453.   R. S. Randad, L. Lubkowska, M. A. Eissenstat, S. V. Gulnik, B. Yu, T. N. Bhat, D. J. Clanton, T. House, S. E. Stinson and J. W. Erickson, *Bioorg. Med. Chem. Lett.*, 1998, **8**, 3537-3542.

454.   S. F. Martin, G. O. Dorsey, T. Gane, M. C. Hillier, H. Kessler, M. Baur, B. Matha, J. W. Erickson, T. N. Bhat, S. Munshi, S. V. Gulnik and I. A. Topol, *J. Med. Chem.*, 1998, **41**, 1581-1597.

455. S. Ro, S-G. Baek, B. Lee, C. Park, N. Choy, C. S. Lee, Y. C. Son, H. Choi, J. S. Koh, H. Yoon, S.C. Kim and J. H. Ok, *Bioorg. Med. Chem. Lett.*, 1998, **8**, 2423-2426.

456 T. Lee, G. S. Laco, B. E. Torbett, H. S. Fox, D. L. Lerner, J. H. Elder and C-H. Wong, *Proc. Natl. Acad. Sci. USA*, 1998, **95**, 939-944.

457. P. J. Ala, E. E. Huston, R. M. Klabe, P. K. Jadhav, P. Y. S. Lam and C-H. Chang, *Biochemistry*, 1998, **37**, 15042-15049.

458. G. De Lucca, U. T. Kim, J. Liang, B. Cordova, R. M. Klabe, S. Garber, L. T. Bacheler, G. N. Lam, M. R. Wright, K. A. Logue, S. Erickson-Viitanen, S. S. Ko and G. L. Trainor, *J. Med. Chem.*, 1998, **41**, 2411-2423.

459. P. K. Jadhav, F. J. Woerner, P. Y. S. Lam, C. N. Hodge, C. J. Eyermann, H-W. Man, W. F. Daneker, L. T. Bacheler, M. M. Rayner, J. L. Meek, S. Erickson-Viitanen, D. A. Jackson, J. C. Kalabrese, M. Schadt and C-H. Chang, *J. Med. Chem.*, 1998, **41**, 1446-1455.

460 W. Han, J. C. Pelletier and C. N. Hodge, *Bioorg. Med. Chem. Lett.*, 1998, **8**, 3615-3620.

461. R. F. Kaltenbach III, D. A. Nugiel, P. Y. S. Lam, R. M. Klabe and S. P. Seitz, *J. Med. Chem.*, 1998, **41**, 5113-5117.

462. E. Maquoi, A. Noel, F. Frankenne, H. Angliker, G. Murphy and J-M. Foidart, *FEBS Lett.*, 1998, **424**, 262-266.

463. C. Fernandez-Catalan, W. Bode, R. Huber, D. Turk, J. J. Calvete, A. Lchte, H. Tschesche and K. Maskos, *EMBO J.*, 1998, **17**, 5238-5248.

464. C. Benaud, R. B. Dickson and E. W. Thompson, *Breast Cancer Res. Treatment*, 1998, **50**, 97-116.

465. P. Brown, *Breast Cancer Res. Treatment*, 1998, **52**, 125-136.

466 M. Toi, S. Ishigaki, *Breast Cancer*, 1998, **52**, 113-124.

467. M. Polette and P. Birembaut, *Int. J. Biochem. Cell Biol.*, 1998, **30**, 1195-1202.

468. A. W. Miller, P. D. Brown, J. Moore, W. A. Galloway, A. G. Cornish, T. J. Lenehan and K. P. Lynch, *Br. J. Clin. Pharmacol.*, 1998, **45**, 21-26.

469. R. J. Cherney, L. Wang, D. T. Meyer, C-B. Xue, Z. R. Wasserman, K. D. Hardman, P. K. Welch, M. B. Covington, R. A. Copeland, E. C. Arner, W. F. DeGrado and C. P. Decicco, *J. Med. Chem.*, 1998, **41**, 1749-1751.

470 C-B. Xue, X. He, J. Roderick, W. F. DeGrado, R. J. Cherney, K. D. Hardman, D. J. Nelson, R. A. Copeland, B. D. Jaffee and C. P. Decicco, *J. Med. Chem.*, 1998, **41**, 1745-1748.

471. D. H. Steinman, M. L. Curtin, R. B. Garland, S. K. Davidsen, H. R. Heyman, J. H. Holms, D. H. Albert, T. J. Magoc, I. B. Nagy, P. A. Marcotte, J. Li, D. W. Morgan, C. Hutchins and J. B. Summers, *Bioorg. Med. Chem. Lett.*, 1998, **8**, 2087-2092.

472. G. S. Sheppard, A. S. Florjancic, J. R. Giesler, L. Xu, Y. Guo, S. K. Davidsen, P. A. Marcotte, I. Elmore, D. H. Albert, T. J. Magoc, J. J. Bouska, C. L. Goodfellow, D. W. Morgan and J. B. Summers, *Bioorg. Med. Chem. Lett.*, 1998, **8**, 3251-3256.

473. D. E. Levy, F. Lapierre, W. Liang, W. Ye, C. W. Lange, X. Li, D. Grobelny, M. Casabonne, D. Tyrrell, K. Holme, A. Nadzan and R. E. Galardy, *J. Med. Chem.*, 1998, **41**, 199-223.

474. M. Yamamoto, H. Tsujishita, N. Hori, Y. Ohishi, S. Inoue, S. Ikeda and Y. Okada, *J. Med. Chem.*, 1998, **41**, 1209-1217.

475. M. G. Natchus, M. Cheng, C. T. Wahl, S. Pikul, N. G. Almstead, R. S. Bradley, Y. O. Taiwo, G. E. Mieling, C. M. Dunaway, C. E. Snider, J. M. McIver, B. L.

Barnett, S. J. McPhail, M. B. Anastasio and B. De, *Bioorg. Med. Chem. Lett.*, 1998, **8**, 2077-2080.

476   D. Krumme, H. Wenzel and H. Tschesche, *FEBS Lett.*, 1998, **436**, 209-212.

477.   E. G. von Roedern, H. Brandstetter, R. A. Engh, W. Bode, F. Grams, and L. Moroder, *J. Med. Chem.*, 1998, **41**, 3041-3047.

478.   E.G. von Roedern, F. Grams, H. Brandstetter and L. Moroder, *J. Med. Chem.*, 1998, **41**, 339-345.

479.   S. Patel, L. Saroglou, C. D. Floyd, A. Miller and M. Whittaker, *Tetrahedron Lett.*, 1998, **39**, 8333-8334.

480   Y-C. Li, X. Zhang, R. Melton, V. Ganu and N. C. Gonnella, *Biochemistry*, 1998, **37**, 14048-14056.

481.   J. I. Levin, J. F. Dijoseph, L. M. Killar, A. Sung, T. Walter, M. A. Sharr, C. E. Roth, J. S. Skotnicki and D. Albright, *Bioorg. Med. Chem. Lett.*, 1998, **8**, 1163-1168.

482.   Y. Tamura, F. Watanabe, T. Nakatani, K. Yasui, M. Fuji, T. Komurasaki, H. Tsuzuki, R. Maekawa, T. Yoshioka, K. Kawada, K. Sugita and M. Ohtani, *J. Med. Chem.*, 1998, **41**, 640-649.

483.   A. K. Szardenings, D. Harris, S. Lam, L. Shi, D. Tien, Y. Wang, D. V. Patel, M. Navre and D. A. Campbell, *J. Med. Chem.*, 1998, **41**, 2194-2200.

484.   J. I. Levin, J. F. DiJoseph, L. M. Killar, A. Sung, T. Walter, M. A. Sharr, C. E. Roth, J. S. Skotnicki and J. D. Albright, *Bioorg. Med. Chem. Lett.*, 1998, **8**, 2657-2662.

485.   Y. A. Puius, Y. Zhao, M. Sullivan, D. S. Lawrence, S. C. Almo and Z-Y. Zhang, *Proc. Natl. Acad. Sci. USA*, 1997, **94**, 13420-13425.

486   M. C. Pellegrini, H. Liang, S. Mandiyan, K. Wang, A. Yuryev, I. Vlattas, T. Sytwu, Y-C. Li and L. P. Wennogle, *Biochemistry*, 1998, **37**, 15598-15606.

487.   S. Desmarais, Z. Jia and C. Ramachandran, *Arch. Biochem. Biophys.*, 1998, **354**, 225-231.

488.   P.P. Roller, L. Wu, Z-Y. Zhang and T. R. Burke, Jr., *Bioorg. Med. Chem. Lett.*, 1998, **8**, 2149-2150.

489.   S. D. Taylor, C. C. Kotoris, A. N. Dinaut, Q. Wang, C. Ramachandran and Z. Huang, *Bioorg. Med. Chem.*, 1998, **6**, 1457-1468.

490   Z-J. Yao, B. Ye, X-W. Wu, S. Wang, L. Wu, Z-Y. Zhang and T. R. Burke, Jr., *Bioorg. Med. Chem.*, 1998, **6**, 1799-1810.

491.   C. C. Kotoris, M-J. Chen and S. D. Taylor, *Bioorg. Med. Chem. Lett.*, 1998, **8**, 3275-3280.

492.   D. M. Jones, J. Sueiras-Diaz, M. Szelke, B. J. Leckie, S. R. Beattie, J. Morton, S. Neidle and R. Kuroda, *J. Peptide Res.*, 1997, **50**, 109-121.

493.   J. Sueiras-Diaz, D. M. Jones, M. Szelke, B. J. Leckie, S. R. Beattie, C. Beattie and J. J. Morton, *J. Peptide Res.*, 1997, **50**, 239-247.

494.   J. Sueiras-Diaz, D. M. Jones, M. Szelke, J. Deinum, L. Svensson, C. Westerlund and M. Sohtell, *J. Peptide Res.*, 1997, **50**, 248-261.

495.   S. Even-Ram, B. Uziely, P. Cohen, S. Grisaru-Granovsky, M. Maoz, Y. Ginzburg, R. Reich, I. Vlodavsky and R. Bar-Shavit, *Nat. Med.*, 1998, **4**, 909-914.

496   T. Kato, A. Oda, Y. Inagaki, H. Ohashi, A. Matsumoto, P. Ozaki, Y. Miyakawa, H. Watarai, K. Fuju, A. Kokubo, T. Kadoya, Y. Ikeda and H. Miyazaki, *Proc. Natl. Acad. Sci. USA*, 1997, **94**, 4669-4674.

497.   M. I. Furman, L. Liu, S. E. Benoit, R. C. Becker, M. R. Barnard and A. D. Michelson, *Proc. Natl. Acad. Sci. USA*, 1998, **95**, 3082-3087.

498. P. J. S. Chiu, G. G. Tetzloff, C. Foster, M. Chintala and E. J. Sybertz, *Eur. J. Pharmacol.*, 1997, **321**, 129-135.

499. H. Krutzsch, S. B. Williams, L. P. McKeown and H. R. Gralnick, *Thrombosis Res.*, 1997, **88**, 333-335.

500 P. E. J. Sanderson and A. M. Naylor-Olsen, *Curr. Med. Chem.*, 1998, **5**, 289-304.

501. S. F. Brady, K. J. Stauffer, W. C. Lumma, G. M. Smith, H. G. Ramjit, S. D. Lewis, B. J. Lucas, S. J. Gardell, E. A. Lyle, S. D. Appleby, J. J. Cook, M. A. Holahan, M. T. Stranieri, J. J. Lynch, Jr., J. H. Lin, I-W. Chen, K. Vastag, A. M. Naylor-Olsen and J. P. Vacca, *J. Med. Chem.*, 1998, **41**, 401-406.

502. W. C. Lumma, Jr., K. M. Witherup, T. J. Tucker, S. F. Brady, J. T. Sisko, A. M. Naylor-Olsen, S. D. Lewis, B. J. Lucas and J. P. Vacca, *J. Med. Chem.*, 1998, **41**, 1011-1013.

503. T. J. Tucker, S. F. Brady, W. C. Lumma, S. D. Lewis, S. J. Gardell, A. M. Naylor-Olsen, Y. Yan, J. T. Sisco, K. J. Stauffer, B. J. Lucas, J. J. Lynch, J. J. Cook, M. T. Stranieri, M. A. Holahan, E. A. Lyle, E. P. Baskin, I-W. Chen, K. B. Dancheck, J. A. Krueger, C. M. Cooper and J. P. Vacca, *J. Med. Chem.*, 1998, **41**, 3210-3219.

504. R. Krishnan, E. Zhang, K. Hakansson, R. K. Arni, A. Tulinsky, M. S. L. Lim-Wilby, O. E. Levy, J. E. Semple and T. K. Brunck, *Biochemistry*. 1998, **37**, 12094-12103.

505. J. E. Semple, *Bioorg. Med. Chem. Lett.*, 1998, **8**, 2501-2506.

506. T. D. Owens and J. E. Semple, *Bioorg. Med. Chem. Lett.*, 1998, **8**, 3683-3688.

507. J. E. Semple, D. C. Rowley, T. D. Owens, N. K. Minami, T. H. Uong and T. K. Brunck, *Bioorg. Med. Chem. Lett.*, 1998, **8**, 3525-3530.

508. Y. St-Denis, C. E. Augelli-Szafran, B. Bachand, K. A. Berryman, J. DiMaio, A. M. Doherty, J. J. Edmonds, L. Leblond, S. Levesque, L. S. Narasimhan, J. R. Penvose-Yi, J. R. Rubin, M. Tarazi, P. D. Winocour and M. A. Siddiqui, *Bioorg. Med. Chem. Lett.*, 1998, **8**, 3193-3198.

509. J. S. Plummer, K. A. Berryman, C. Cai, W. L. Cody, J. DiMaio, A. M. Doherty, J. J. Edmonds, J. X. He, D. R. Holland, S. Levesque, D. R. Kent, L. S. Narasimhan, J. R. Rubin, S. T. Rapundalo, M. A. Siddiqui, A. J. Susser, Y. St-Denis and P. D. Winocour, *Bioorg. Med. Chem. Lett.*, 1998, **8**, 3409-3414.

510. P. E. J. Sanderson, T. A. Lyle, K. J. Cutrona, D. L. Dyer, B. D. Dorsey, C. M. McDonough, A. M. Naylor-Olsen, I-W. Chen, Z. Chen, J. J. Cook, C. M. Cooper, S. J. Gardell, T. R. Hare, J. A. Krueger, S. D. Lewis, J. H. Lin, B. J. Lucas, Jr., E. A. Lyle, J. J. Lynch, Jr., M. T. Stranieri, K. Vastag, Y. Yan J. A. Shafer and J. P. Vacca, *J. Med. Chem.*, 1998, **41**, 4466-4474.

511. A. E. P. Adang, H. Lucas, A. P. A. DeMan, R. A. Engh and P. D. J. Grootenhuis, *Bioorg. Med. Chem. Lett.*, 1998, **8**, 3603-3608.

512. J. Ambler, E. Baker, L. Brown, P. Butler, D. Farr, K. Dunnet, D. Le Grand, D. Janus, D. Jones, K. Menear, M. Mercer, G. Smith, M. Talbot and M. Tweed, *Bioorg. Med. Chem. Lett.*, 1998, **8**, 3583-3588.

513. D. Dumas, G. Leclerc, J. J. Baldwin, S. D. Lewis, M. Murcko and A. M. Naylor-Olsen, *Eur. J. Med. Chem.*, 1998, **33**, 471-488.

514. D. J. Sall, S. L. Briggs, N. Y. Chirgadze, D. K. Clawson, D. S. Gifford-Moore, V. J. Klimkowski, J. R. McCowan, G. F. Smith and J. H. Wikel, *Bioorg. Med. Chem. Lett.*, 1998, **8**, 2527-2532.

515. H. Finch, N. A. Pegg, J. McLaren, A. Lowdon, R. Bolton, S. J. Coote, U. Dyer, J. G. Montana, M. R. Owen, M. Dowle, D. Buckley, B. C. Ross, C. Campbell, C.

Dix, C. Mooney, C-M. Tang and C. Patel, *Bioorg. Med. Chem. Lett.*, 1998, **8**, 2955-2960.

516.   K. Lee, W-H. Jung, C. W. Park, C. Y. Hong, I. C. Kim, S. Kim, Y. S. Oh, O. H. Kwon, S-H. Lee, H. D. Park, S. W. Kim, Y. H. Lee and Y. J. Yoo, *Bioorg. Med. Chem. Lett.*, 1998, **8**, 2563-2568.

517.   E. Skordalakes, S. Elgendy, C. A. Goodwin, D. Green, M. F. Scully, V. J. Kakkar, J-M. Freyssinet, G. Dodson and J. J. Deadman, *Biochemistry*, 1998, **37**, 14420-14427.

518.   J. Blanco, C. Nguyen, C. Callebaut, E. Jacotot, B. Krust, J-P. Mazaleyrat, M. Wakselman and A. G. Hovanessian, *Eur. J. Biochem.*, 1998, **256**, 369-378.

519.   P. Proost, I. De Meester, D. Schols, S. Struyf, A-M. Lambeir, A. Wuyts, G. Opdenakker, E. De Clercq, S. Scharpe and J. V. Damme, *J. Biol. Chem.*, 1998, **273**, 7222-7227.

520.   J. Lin, P. J. Toscano and J. T. Welch, *Proc. Natl. Acad. Sci. USA*, 1998, **95**, 14020-14024.

521.   J. S. McMurray, R. J. A. Budde, S. Ke, N. U. Obeyesekere, W. Wang, L. Ramdas and C.A. Lewis, *Arch. Biochem. Biophys.*, 1998, **355**, 124-130.

522.   P. Ruzza, A. Donella-Deana, A. Calderan, G. Zanotti, L. Cesaro, L. A. Pinna and G. Borin, *J. Peptide Sci.*, 1998, **4**, 33-45.

523.   J. Alfaro-Lopez, W. Yuan, B. C. Phan, J. Kamath, Q. Lou, K. S. Lam and V. J. Hruby, *J. Med. Chem.*, 1998, **41**, 2252-2260.

524.   A. Ishida, Y. Shigeri, Y. Tatsu, K. Uegaki, I. Kameshita, S. Okuno, T. Kitani, N. Yumoto and H. Fujisawa, *FEBS Lett.*, 1998, **427**, 115-118.

525.   J. Whisstock, R. Skinner and A. M. Lesk, *Trends Biochem. Sci.*, 1998, **23**, 63-67.

526.   B. A. Katz, J. M. Clark, J. S. Finer-Moore, T. E. Jenkins, C. R. Johnson, M. J. Ross, C. Luong, W. R. Moore and R. M. Stroud, *Nature*, 1998, **391**, 608-612.

527.   K. D. Combrink, H. B. Gulgeze, N. A. Meanwell, B. C. Pearce, P. Zulan, G. S. Bisacchi, D. G. M. Roberts, P. Stanley and S. M. Seiler, *J. Med. Chem.*, 1998, **41**, 4854-4860.

528.   M. B. Zwick, L. L. C. Bonnycastle, K. A. Noren, S. Venturini, E. Leong, C. F. Barbas, III, C. J. Noren and J. K. Scott, *Anal. Biochem.*, 1998, **264**, 87-97.

529.   M. Rousch, J. T. Lutgerink, J. Coote, A. De Bruine, J-W. Arends and H. R. Hoogenboom, *Br. J. Pharmacol.*, 1998, **125**, 5-16.

530.   V. Sieber, A. Pluckthun and F. X. Schmid, *Nature Biotechnol.*, 1998, **16**, 955-960.

531.   A. J. J. Wood, *New Eng. J. Med.*, 1998, 746-754.

532.   D. L. Johnson, F. X. Farrell, F. P. Barbone, F. J. McMahon, J. Tullai, K. Hoey, O. Livnah, N. C. Wrighton, S. A. Middleton, D. A. Loughney, E. A. Stura, W. J. Dower, L. S. Mulcahy, I. A. Wilson and L. K. Jolliffe, *Biochemistry*, 1998, **37**, 3699-3710.

533.   O. Livnah, D. L. Johnson, E. A. Stura, F. X. Farrell, F. P. Barbone, Y. You, K. D. Liu, M. A. Goldsmith, W. He, C. D. Krause, S. Pestka, L. K. Jolliffe and I. A. Wilson, *Nature Structr. Biol.*, 1998, **5**, 993-1003.

534.   A. J. Sytkowski, E. D. Lunn, K. L. Davis, L. Feldman and S. Siekman, *Proc. Natl. Acad. Sci. USA*, 1998, **95**, 1184-1188.

535.   T. Kimura, H. Kaburaki, T. Tsujino, Y. Ikeda, H. Kato and Y. Watanabe, *FEBS Lett.*, 1998, **428**, 250-254.

536.   T. Taki, D. Ishikawa, H. Hamasaki and S. Handa, *FEBS Lett.*, 1997, **418**, 219-223.

537.   C. L. Chirinos-Rojas, M. W. Steward, and C. D. Partidos, *J. Immunol.*, 1998, **161**, 5621-5626.

538. R. H. Hilderman, G. Liu and J. K. Zimmerman, *Eur. J. Biochem.*, 1998, **258**, 396-401.

539. T. Fukumoto, N. Torigoe, S. Kawabata, M. Murakami, T. Uede, T. Nishi, Y. Ito and K. Sugimura, *Nature Biotechnol.*, 1998, **16**, 267-270.

540. H. H. Pierce, F. Schachat, P. W. Brandt, C. R. Lombardo and B. K. Kay, *J. Biol. Chem.*, 1998, **273**, 23448-23453.

541. M. Krook, C. Lindbladh, J.A. Eriksen and K. Mosbach, *Mol. Diversity*, 1998, **3**, 149-159.

542. H. Gardsvoll, A-J. van Zonneveld, A. Holm, E. Eldering, M. van Meijer, K. Dano and H. Pannekoek, *FEBS Lett.*, 1998, **431**, 170-174.

543. T. S. Vaughan, J. K. Osbourn and P. R. Tempest, *Nature Biotechnol.*, 1998, **16**, 535-539.

544. I. Fujii, S. Fukuyama, Y. Iwabuchi and R. Tanimura, *Nature Biotechnol.*, 1998, **16**, 463-467.

545. J. K. Osbourn, J. C. Earnshaw, K. S. Johnson, M. Permentier, V. Timmermans and J. McCafferty, *Nature Biotechnol.*, 1998, **16**, 778-781.

546. M. D. Sheets, P. Amersdorfer, R. Finnern, P. Sargent, E. Lindqvist, R. Schier, G. Hemingsen, C. Wong, J, C. Gerhart and J. D. Marks, *Proc. Natl. Acad. Sci. USA*, 1998, **95**, 6157-6162.

547. M. C. Souroujon and D. Mochly-Rosen, *Nat. Biotechnol.*, 1998, **16**, 919-924.

548. J. Song and F. Ni, *Biochem. Cell Biol.*, 1998, **76**, 177-188.

549. S. R. Eaton, W. L. Cody, A. M. Doherty, D. R. Holland, R. L. Panek, G. H. Lu, T. K. Dahring and D. R. Rose, *J. Med. Chem.*, 1998, **41**, 4329-4342.

550. J-N. Fu and A. L. Castelhano, *Bioorg. Med. Chem. Lett.*, 1998, **8**, 2813-2816.

551. G. J. Pacofsky, K. Lackey, K. J. Alligood, J. Berman, P. S. Charifson, R. M. Crosby, G. F. Dorsey, Jr., P. L. Feldman, T. M. Gilmer, C. W. Hummel, S. S. Jordon, C. Mohr, L. M. Shewchuk, D. D. Sternbach and M. Rodriguez, *J. Med. Chem.*, 1998, **41**, 1894-1908.

552. L. Tong, T. C. Warren, S. Lucas, J. Schembri-King, R. Betageri, J. R. Proudfoot and S. Jakes, *J. Biol. Chem.*, 1998, **273**, 20238-20242.

553. C. Garcia-Echeverria, P. Furet, B. Gay, H. Fretz, J. Rahuel, J. Schoepfer and G. Caravatti, *J. Med. Chem.*, 1998, **41**, 1741-1744.

554. J. Schoepfer, B. Gay, G. Caravatti, C. Garcia-Echeverria, H. Fretz, J. Rahuel and P. Furet, *Bioorg. Med. Chem. Lett.*, 1998, **8**, 2865-2870.

555. D. J. Witter, S. J. Famiglietti, J. C. Cambier and A. L. Castelhano, *Bioorg. Med. Chem. Lett.*, 1998, **8**, 3137-3142.

556. S. Hansen and J. Wakonen, *Eur. J. Pharm. Sci.*, 1998, **6**, 337-341.

557. A. Fasano, *Trends Biotechnol.*, 1998, **16**, 152-157.

558. S. D. Putney and P. A. Burke, *Nature Biotechnol.*, 1998, **16**, 153-157.

559. C. W. Pouton, *Adv. Drug Delivery. Rev.*, 1998, **34**, 51-64.

560. W. M. Pardridge, *J. Neurochemi.*, 1998, **70**, 1781-1792.

561. M. J. McKinley and B. J. Oldfield, *Trends Endocrinol. Metab.*, 1998, **9**, 349-354.

562. U. Schroeder, P. Sommerfeld, S. Ulrich and B. A. Sabel, *J. Pharm. Sci.*, 1998, **87**, 1305-1307.

563. R. D. Egleton, T. J. Abbruscato, S. A. Thomas and T. P. Davis, *J. Pharm. Sci.*, 1998, **87**, 1433-1439.

564. Y. Taki, T. Sakane, T. Nadai, H. Sezaki, G. L. Amidon, P. Langguth and S. Yamashita, *J. Pharm. Pharmacol.*, 1998, **50**, 1013-1018.

565. S. Akhteruzzaman, Y. Kato, A. Hisaka and Y. Sugiyama, *J. Pharmacol. Exp. Therap.*, 1999, **288**, 575-581.

566. V. J. Wacher, J. A. Silverman, Y. Zhang and L. Z. Benet, *J. Pharm. Sci.*, 1998, **87**, 1322-1330.
567. A. E. Kim, J. M. Dintaman, D. S. Waddell and J. A. Silverman, *J. Pharmacol. Exp. Therap.*, 1998, **286**, 1439-1445.
568. M. Kratzel, R. Hiessbock and A. Bernkop-Schnurch, *J. Med. Chem.*, 1998, **41**, 2339-2344.
569. N. Eller, C. J. Kollenz and G. Hitzenberger, *Int. J. Clin. Pharmacol. Therap.*, 1998, **36**, 139-145.
570. M. Arras, H. Mollnau, R. Strasser, R. Wenz, W. D. Ito, J. Schaper and W. Schaper, *Nature Biotechnol.*, 1998, **16**, 159-162.
571. X. Zhao and J. M. Harris, *J. Pharm. Sci.*, 1998, **87**, 1450-1458.
572. M. J. Roberts and J. M. Harris, *J. Pharm. Sci.*, 1998, **87**, 1440-1445.

4
Cyclic, Modified and Conjugated Peptides

BY J. S. DAVIES

1 Introduction

Scientists world-wide endeavouring in this field of activity have maintained their productivity for 1998 to the same extent as in recent years. A very diverse set of original papers have been scanned, directed mainly by abstracts listed in C. A. Selects[1] on Amino Acids, Peptides and Proteins (up to Issue 12, June 1999). Most of the major Journals's index pages were also scanned to enhance the comprehensive coverage. At hand also during the reviewing process were the Symposia Proceedings from the 1st International Peptide Symposium at Kyoto[2] and the 25th European Peptide Symposium at Budapest,[3] but the contents have not been reported on here, in line with past precedent.

Fragmented information about progress towards the total synthesis of the vancomycin glycopeptide antibiotics have graced the pages of this Chapter over many volumes. The year 1998 saw two groups, D. A. Evans and co-workers at Harvard University, and K. C. Nicolaou's group at Scripps Institute, La Jolla achieve their ultimate goal of total syntheses for the aglycon of vancomycin. The details of each synthesis are recorded under the sub-section on glycopeptide antibiotics, but an impression of the impact made by achieving this impressive goal can be gleaned from two mini-reviews[4,5] of the skilful work carried out at the geographical extremities of the U.S.A.

In the diverse structures discussed in this Chapter over many years, slippage in the standards of abbreviations and nomenclature used in the literature can often be experienced. A compilation[6] of abbreviations and symbols is therefore welcomed.

2 Cyclic Peptides

2.1 General Considerations – Higher plants have been revealed as a rich source of cyclic peptides, and have justified a 24-reference review[7] on conformational aspects. Structure-activity relationships for synthetic macrocyclic peptidomimetics have also been reviewed[8]. Ever since their first discovery in the 1960s. the piperazic acids have appeared in a range of compounds covered by this Chapter (see K. J. Hale *et al.*, *Tetrahedron*, 1996 **52**, 1047 for relevant

Amino Acids, Peptides and Proteins, Volume 31
© The Royal Society of Chemistry, 2000

bibliography). Their synthesis and unique conformational properties have now been reviewed[9].

Extended native chemical ligation[10] has provided a novel method to synthesise cyclic peptides with amide backbones without the necessity of a Cys residue as exemplified in Scheme 1. A one-pot reaction[11] without active esters

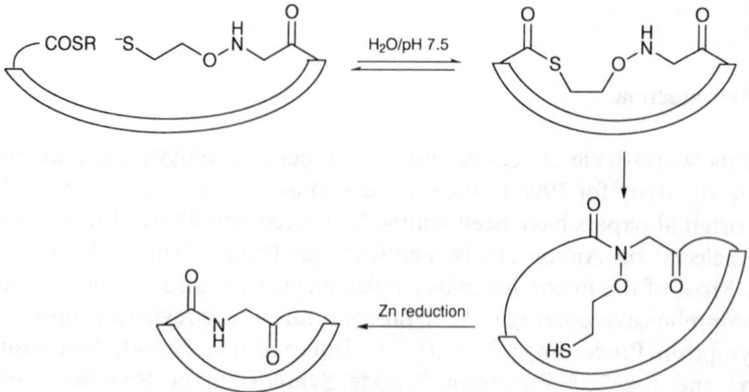

Scheme 1

and a need for protecting groups should be a welcome development in cyclic peptide synthesis. Scheme 2 outlines the synthesis of a cyclotetrapeptide from 2

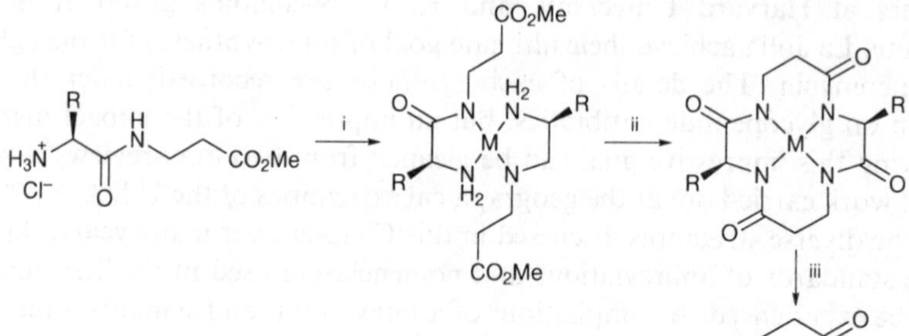

Reagents: i, $Na_2[PdCl_4]NiCl_2\cdot6H_2O$; ii, [PPN]Cl; iii, HCl/MeOH or $CuCl_2\cdot2H_2O$, 6NaOMe/MeOH

Scheme 2

equivalents of dipeptide, but is probably slightly specialised to be universally accepted. Asymmetric ring opening of polymer-bound *meso*-epoxides with trimethylsilyl azide ($TMSN_3$) has provided[12] stereochemical templates (1) for the synthesis of cyclic RGD pharmacophores in combinatorial libraries. Quite significant stereochemical specificities towards $\alpha\beta_3$ integrins were obtained in some peptidomimetics. Cyclomonomer-v-cyclohigher homologue formation has been assessed[13] *via* the insertion of Pro and Pro-derived mimetics into

(1) and its *S,R* analogue

Table 1

		Peptide 1			Peptide 2		
		Linear	Cyclic	Oligomeric	Linear	Cyclic	Oligomeric
(A)	BOP/DIEA	0	42	56	4	24	67
(B)	BOP/HOBt/DIEA	2	51	46	16	30	54
(C)	HBTU/HOBt/DIEA	0	35	56	13	32	54
(D)	DIC/HOBt	33	61	6	87	7	5

thymopentin analogues, Arg-Lys(Ac)-Ala-Val-Tyr, Val-Arg-Lys(Ac)-Ala-Val-Tyr and Ala$_5$, Ala$_6$ and Ala$_7$. Both L and D in the C-terminal positions were assessed, using HAPyU for cyclisation and contrary to general belief the Pro residues inserted into all penta- and hexa-peptides hindered cyclic monomer formation, but Ala$_7$ readily formed cyclic monomers. To overcome some side-reactions identified in the generation of cyclic libraries, a more detailed[14] study of solid phase side-chain to side-chain cyclisations involving Lys→Glu and Glu→Lys side-chains has been carried out. Table 1 summarises the results with four coupling conditions on 4-methyl benzhydryl resin of peptide Fmoc-Lys-Phe-D-Ala-Pro-Glu-Gly (Peptide 1) and its analogue (Peptide 2) with positions of Lys and Glu reversed. It is concluded that DIC (diisopropylcarbo-diimide) over extended reaction times would be the best conditions, but Peptide 1 cyclises easier than Peptide 2. A novel backbone amide lin-ker[15](BAL) has been used to synthesise a number of C-terminal modified peptides as well as cyclic peptides typified by cyclo(Arg-D-Phe-Pro-Glu-Asp-Asn-Tyr-Glu-Ala-Ala) where macrocyclisation was carried out by PyAOP/HOAt.

Crude mixtures of difficult tetrapeptide cyclisations have been analysed[16] by electrospray ionisation mass spectrometry (ESI-MS), concentrating on the tetrapeptide H-Leu-Pro-Leu-Pro-OH in order to assess the cyclomonomer-v-cyclodimer formation ratios. MALDI-MS Techniques have also been instru-mental[17] in determining the cyclisation efficiency on TentaGel resin, even on a single bead of resin. The studies showed that the 8-mer gave the highest yield with rings > eleven amino acids or fewer than six residues giving more dimers and linear peptide adducts.

The original Ellman's reagent, 5,5'-dithiobis(2-nitrobenzoic acid) bound through two sites to a suitable solid support (PEG-PS, modified Sephadex™, or controlled-pore-glass) gives a convenient fast cyclisation[18] of disulfide links over a wide range of pH (2.7 to 6.6). The reaction is summarised in Scheme 3. Facile synthesis of cyclic di-, tri-, tetra- and pentasulfide containing peptides

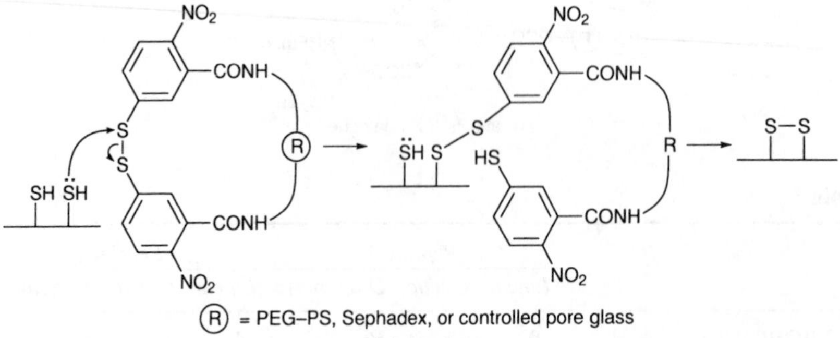

(R) = PEG–PS, Sephadex, or controlled pore glass

Scheme 3

have been carried out[19] using bis-(tetrabutylammonium) hexasulfide (BTH), starting from Cys-containing peptides on-resin.

2.2 Dioxopiperazines (Cyclic Dipeptides) – Twelve new polychlorinated dioxopiperazines, typified by structures (2)-(4) have been isolated[20] from the sponge *Dysidea chlorea*. Two potent cytotoxic epidithiapiperazinediones ambewelamides A(5) and B(6) have been isolated[21] for the first time from a lichen, *Usnea* sp. and can be considered to be new members of highly modified

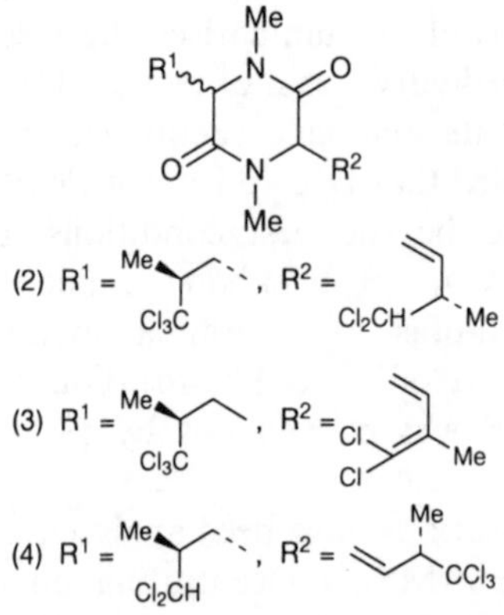

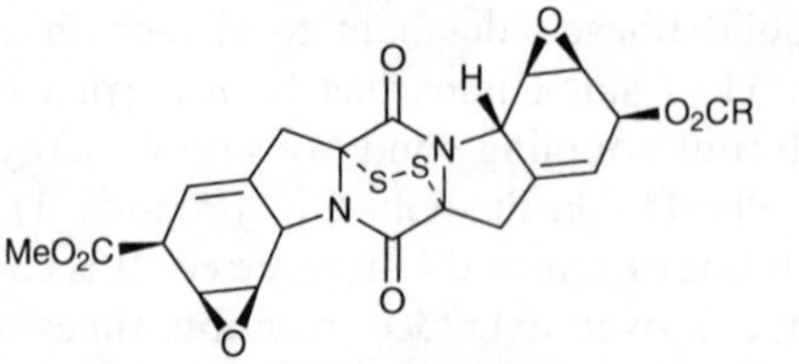

(2) $R^1 =$ (Me, Cl_3C), $R^2 =$ (Cl_2CH, Me)

(3) $R^1 =$ (Me, Cl_3C), $R^2 =$ (Cl, Cl, Me)

(4) $R^1 =$ (Me, Cl_2CH), $R^2 =$ (Me, CCl_3)

(5) R = $(CH_2)_2Me$
(6) R = $(CH_2)_4Me$

phenylalanine dioxopiperazines. Investigations[22] have revealed that structures assigned to D-residues in dioxopiperazines isolated from the sponge *Calyx* CF *podatypa* arise from non-enzymic epimerisation of the corresponding L-Pro analogues. The authors feel that assigning absolute stereochemistry by simple comparison of sign and magnitude of specific optical rotation should be actively discouraged.

High yielding, solution phase dioxopiperazine libraries have been generated[23] in a one-pot reaction (involving the 3 steps of Scheme 4) based on the Ugi condensation reaction. Best yields were with disubstituted Gly derivatives, but no dioxopiperazine formed from Pro. Reactions[24] of C-terminal Trp

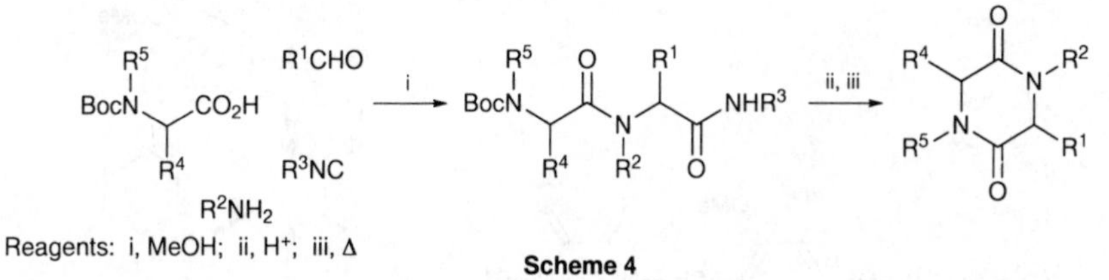

Reagents: i, MeOH; ii, H⁺; iii, Δ

Scheme 4

dipeptides with singlet oxygen at pH 4.7, followed by work up with Me_2S give rise to dioxopiperazines having structures exemplified by (7). The formation of the dioxopiperazine ring is a last stage which occurs spontaneously in aq. acid solution. Aprotic dipolar protophobic solvents, and catalysis by alkylammonium carboxylate salts reduces[25] the t_α for formation of dioxopiperazines from H-Ala-Pro-NH_2·CF_3COOH from 20 days in methanol to 36 mins in acetonitrile. Synthetic methods have been developed to extend the C(5)-C(5a) exomethylene unit in bicyclomycins (8) without loss of biological activity. The methyl analogue ($R = Me$) proved to be the most active analogue. When the OH group at the 6-position was replaced[27] by weaker H-bonding groups the biological activity was lowered, suggesting a specific role for the OH in recognising the transcription termination factor rho. Spirodioxopiperazines such as (9) are the result[28] of cyclisation of dipeptide amino-esters attached at the spiro atom.

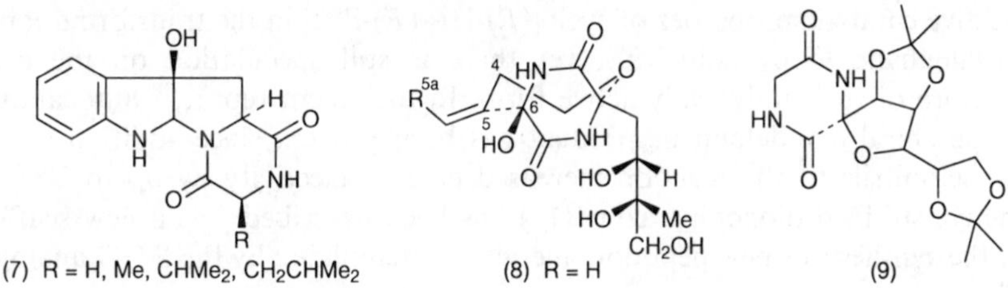

(7) R = H, Me, CHMe₂, CH₂CHMe₂   (8) R = H   (9)

The cyclisation of H-Ala-Pro-NH_2 to cyclo(Ala-Pro) has been studied[29] as a model for cleavage of an N-terminus peptide bond with spontaneous formation of a dioxopiperazine. Cyclisation is a multistep process, a *trans-cis* isomerisation of the amide bond followed by NH_2 attack on the amide carbonyl. Buffers only affect the deprotonation stage. Dioxopiperazine formation has been encountered[30] as a side reaction when Fmoc is removed during the synthesis of backbone cyclised peptides containing N-carboxy-methyl amino acids. Similarly cyclic dipeptides formed as side-reactions can be conveniently adopted[31] as useful scaffolds for combinatorial chemistry using the 'Backbone Amide Linker' which attaches the growing peptide chain through a backbone N instead of a carboxyl group. Typical dioxo-piperazines are represented by structure (10). The dioxopiperazine chiral auxiliary intermediate (11), capable of multigram synthesis of homochiral phenylalanine,[32] can be processed to give an efficient deracemisation of ($\pm$)-

(10)   (11)

phenylalanine. It has been reported[33] that exo/endo selectivities in Diels-Alder reactions are enhanced in cyclic didehydropeptides compared to their acyclic counterparts.

The energetics of HNCO elimination from cyclo(Pro-Gly) radical cation has been studied[34] by electron ionisation tandem mass spectrometry and theoretical calculations, and a stereochemical analysis[35] on dioxopiperazines of enalapril and lisinopril has observed a tendency for the side-chains to bend over the piperazinedione ring. Cyclo(DL-Pro-Gly) has been used[36] as a test compound to analyse error propagation from NMR-derived internuclear distances, while cyclo (Ala-Ala) and cyclo(D-Ala-L-Ala) have been shown[37] to form stable host-guest complexes with an amide macrocycle.

There have been several reports since the discovery in 1981 of the enantio-selective catalytic properties of cyclo[(R)-His-(R)-Phe] in the transformation of benzaldehyde to cyanohydrins, yet there is still speculation on the exact structure of the catalytically active form. In the recent report,[38] autocatalysis by the chiral mandelonitrile product has been proven, since addition of (S)-mandelonitrile to the reaction increased enantioselectivity by up to 20% ee. The constrained dioxopiperazine (12) has been described[39] as a new scaffold for the synthesis of new peptidomimetics, as exemplified by the RGD analogue (13).

(12) R = OH, R^1 = Me
(13) R = H, R^1 = NH−CH$_2$−CONH−

2.3 Cyclotripeptides – Homodetic cyclic tripeptides still remain a rarity, but as published titles of papers have broadened the description to include extended permutations of macrocycles containing three amino acid residues, the work has been included here. Thus the Palauan sponge *Microciona eurypa* has yielded[40] the cyclic isodityrosine tripeptide eurypamide A (14) and related analogues. The dihydroxyarginine residue in (14) has not been previously found. In total number of residues the novel methylamine-bridged enkephalin (15) would be more at home in the next sub-section, but it only has three residues in the macrocycle. Enkephalin analogue (15) was found[41] to have affinities of 1.6, 2.1 and 340 nM respectively for isolated μ, δ, and κ opioid receptors, which makes it a promising lead compound for further investigations.

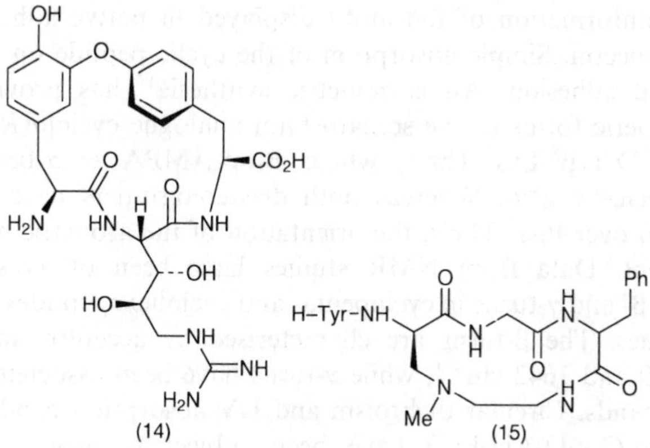

(14)

(15)

2.4 Cyclotetrapeptides – The cyclic core of hibispeptins A(16) and B(17) have been isolated[42] from the root bark of *Hibiscus syriacus* and includes a unique unit, 2-amino-3-(2-hydroxy-5-aminoacetyl)pentanoic acid. Cyclic β^3-tetrapeptides have been modelled and constructed[43] to form nanotube mimics of transmembrane channels. Cyclo(β^3-HTrp)$_4$, cyclo(-β^3-HTrp-β-Hleu)$_2$, and cyclo-(β^3-Hleu)$_4$ based on the general structure (18) were synthesised from

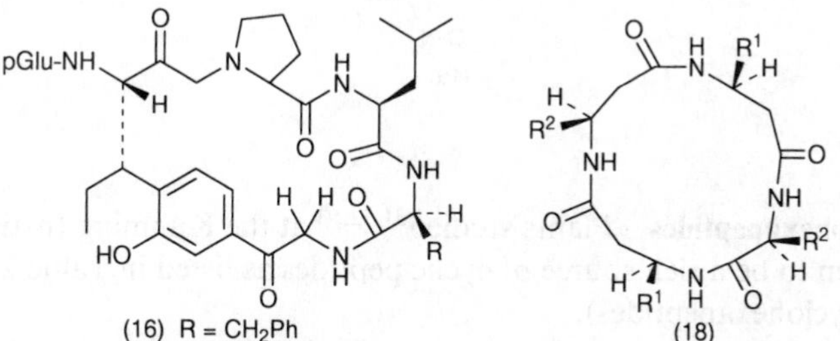

(16) R = CH$_2$Ph
(17) R = CH$_2$CHMe$_2$

(18)

linear precursors off-resin, cyclisation being carried out using HATU. Both the Trp-containing analogues showed remarkable ion-transport activities but the other analogue was inactive. A range of data from various techniques has enabled modelling studies[44] to be carried out on dansyl (DNS) derivatives of the cyclic enkephalin analogues cyclo(D-Dab-Gly-Trp-D-Leu) and cyclo(D-

Dab-Gly-Trp-D-Leu). The results showed a β-turn at Gly^2-Trp^3 in both cases.

2.5   Cyclopentapeptides – In cyclo(Phe-Phe-Aib-Leu-Pro), the Aib residue can be considered to be equivalent to a D-residue. X-Ray crystallography[45] has found a γ-turn at the Aib position, and a *cis* peptide bond between Leu^4-Pro^5. In solution, NMR studies were able to pick out the two conformational *cis-trans* isomers around the Leu^4-Pro^5 bond. Cyclo[-D-Val-Arg-Gly-Asp-Glu(ε-Ahx-Tyr-Cys-NH$_2$)] grafted][46] on to bovine serum albumin (BSA) can mimic the conformation of the motif displayed in native adhesion proteins such as fibronectin. Simple adsorption of the cyclic peptide on the BSA also enhanced cell adhesion. An asymmetric synthesis[47] has provided the two diastereoisomeric forms of the somatostatin analogue cyclo[α(R and S)Bn-o-AMPA-Phe^7-D-Trp^8-Lys^9-Thr^{10}] where Bn-o-AMPA is α-benzyl-*o*-amino-methylphenylacetic acid. Whereas both diastereoisomers have a stable βII′ conformation over Phe^7-Thr^{10}, the orientation of the aromatic bridge units is quite different. Data from NMR studies have been of assistance in the analysis[48] of β- and γ-turns in cyclopenta- and cyclohexapeptides by FTIR and CD techniques. The β-turns are characterised by acceptor amide I bands between 1629 and 1642 cm^{-1}, while γ-turns have been associated with 1615-1625 cm^{-1} bands. Circular dichroism and UV absorption band shapes from cyclo(Gly-Pro-Gly-D-Ala-Pro) have been subject to assessment[49] by the polarisability tensor theory.

A range of cyclic peptides, from penta- up to deca-, have been synthesised[50] on-resin while attached to the resin *via* carbonate, or carbamate linkers, depicted by (19).

(19)

2.6   Cyclohexapeptides – Plants studied[51,52,53] at the Kunming Institute China have proven to be a rich source of cyclic peptides as listed in Table 2 (not all of them are cyclohexapeptides).

The 4(R),11(R) analogue of cyclocinamide A (20) has been synthesised[54] to ascertain the exact stereochemistry at C(4) and C(11). Ring closure to form the core ring was at the C-carboxyl of the bromotryptophan residue using pentafluorophenyldiphenyl phosphinate. The final product's spectra were different from the natural form leading to the suggestion that the configuration at C(4) and C(11) is S. Derivatives of natural β-amino acids have been utilised[55] using trichloroethyl ester groups for C-terminal protection and

Table 2

Cyclic peptide	Name	Source
cyclo(Pro-Gly-Leu-Val-Ile-Tyr)	Glabrin A	*Annona glabra*[51]
cyclo(Pro-Gly-OxMet-Val-Ala-Val-Tyr-Gly-Thr)	Glabrin B	" "
cyclo(Trp-Ala-Gly-Val-Ala)	Vaccarin A	*Vaccana segetalis*[52]
cyclo(Pro-Gly-Leu-Ser-Phe-Ala-Phe)	Vaccarin B	" "
cyclo(Pro-Val-Trp-Ala-Gly-Val)	Vaccarin D	" "
cyclo(Ala-Tyr-Asn-Phe-Gly-Leu)	Dianthin A	*Dianthus superbus*[53]
cyclo(Ile-Phe-Phe-Pro-Gly-Pro)	Dianthin B	" "

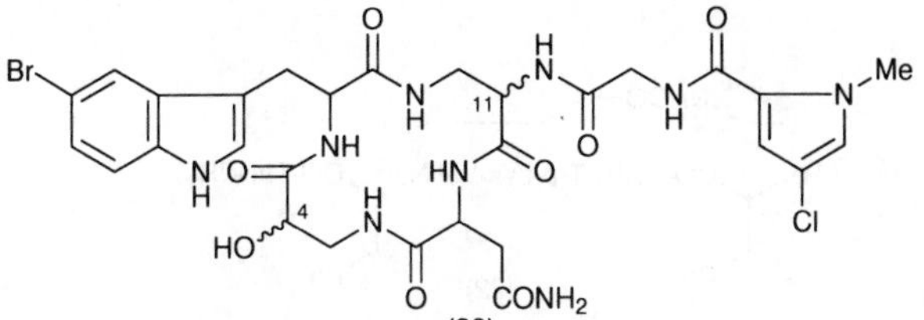

(20)

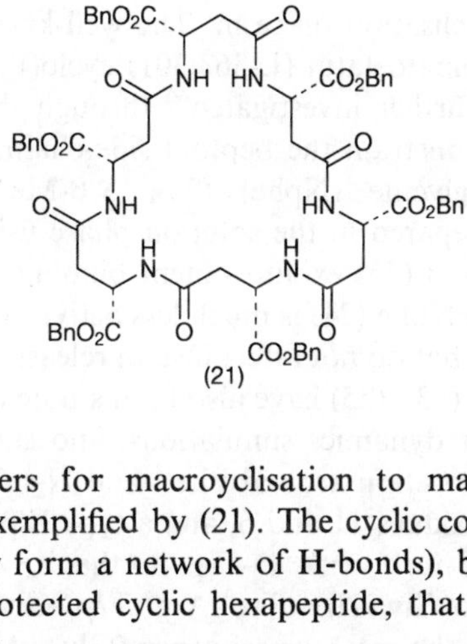

(21)

pentafluorophenyl esters for macroyclisation to make cyclic trimers and hexamers, the latter exemplified by (21). The cyclic compounds proved to be very insoluble (as they form a network of H-bonds), but NMR analysis does suggest, for the de-protected cyclic hexapeptide, that there is a left handed helix turn present.

Cyclo(Pro-Xxx-Lys-Arg-Gly-Asp), with Xxx = Ala, Ser, Leu or Tyr has been used[56] as a model to scrutinise the production of libraries created on-resin, but with a variety of side-chain linkers on to the resin. The cyclohexyl-oxycarbonyl group (Choc), which is removed with anhydrous HF, has been used[57] to create bicyclic epitope peptides of glycoprotein D of HSV 1 as summarised in Scheme 5. Bridging *via* the side-chains of Asp[5] and Lys[10] using PyBOP reagent has also been carried out[58] on αMSH(1-13) and α-MSH(4-10). A backbone cyclic somatostatin analogue (22) has been synthesised[59] which shows high selectivity for the SSTR5 receptor, and is highly stable against enzymic degradation. Its major conformation is defined by a type II′β-turn at D-Trp-Lys and a *cis* amide bond at Val-(Phe). PyBOP at room temperature

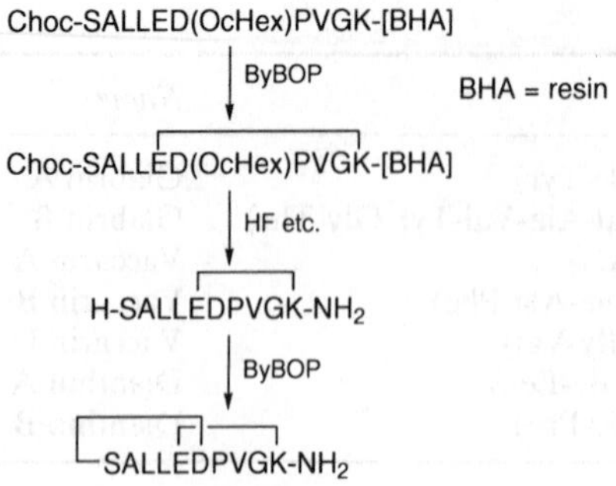

Choc-SALLED(OcHex)PVGK-[BHA]

ByBOP   BHA = resin

Choc-SALLED(OcHex)PVGK-[BHA]

HF etc.

H-SALLEDPVGK-NH$_2$

ByBOP

SALLEDPVGK-NH$_2$

Scheme 5

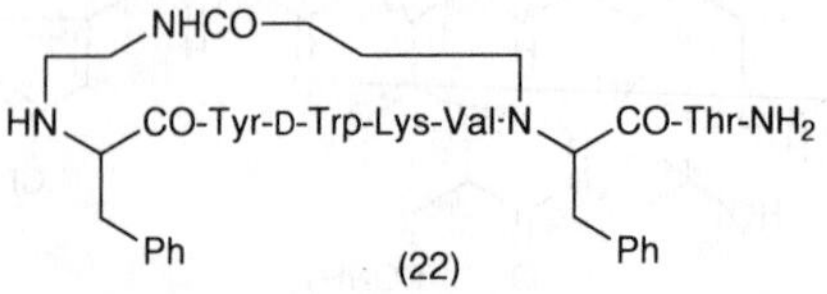

(22)

brought about macrocyclisation on-resin. The well-known Merck cyclic hex-apeptide analogue of somatostatin (L-363,301) cyclo(Pro^6-Phe^7-D-Trp^8-Lys^9-Thr^{10}-Phe^{11}) has been further investigated[60] through the utilisation of β-Me chiral substitutions to constrain the peptoid side-chain. Molecules with Pro^6 replaced by N-phenylglycine (Nphe) (23), *S*-β-MeNphe (24) and (*R*)-β-MeNphe (25) were prepared in the solution phase using DPPA for cyclisa-tion. Compounds (23) and (25) exhibit potent binding activities to the hsst 2 receptor, but the *S*-β-MeNphe (24) is much less active. The compounds inhibit growth hormone release but do not affect insulin release.

The conformations of (23)–(25) have also been studied using NMR, distance geometry and molecular dynamics simulations, and show two sets of NMR signals due to *cis-trans* rotation around Phe^{11} and N-substituted glycine, with a type VI β-turn over residues 11 and 6, and a type II'β-turn over residues 11 and 6, and a type II'β-turn with D-Trp in the i + 1 position. Solution conformations[62] of cyclo(Gly^1-His^2-Phe^3-Arg^4-Trp^5-Gly^6) and its D-Phe ana-logue, corresponding to the message sequence (Gly-α-MSH 5-10) of α-MSH, have been obtained by NMR techniques in DMSO at 300 K. They reveal, for the L-analogue, a dynamic equilibrium of conformers characterised by a γ-bend at Gly^6, two γ-bends at Gly^6,Phe^3, and a conformer with a single β-turn and a γ-bend. Two β-turns over residues 2-5 and 5,6,1 and 2 dominate the equilibrium in the D-Phe analogue. A cyclopeptide Ac-cyclo(Cys-His-Leu-Asp-Cys)-Ile-Trp-OH designed to mimic the C-terminal hexapeptide of en-dothelin has also undergone[63] detailed NMR analysis, and was found to adopt a conformation closely matching the crystal structure of endothelin 1 and BQ 123. Yet this analogue seems to have lost all pharmacological activity at the endothelin B receptor. Another series of cyclic peptides bearing the disulfide link designed[64] as agonists of the five known somatostatin receptors can also be converted into potent antagonists, by incorporating an L^5,D^6 antagonist motif as in H-Nal-cyclo(D-Cys-Pal-D-Trp-Lys-Val-Cys]-Nal-NH$_2$ which has an affinity value of 75 nM for hsst$_2$.

Results of *ab initio*, semi-empirical and empirical calculations on cyclo(Asp-Trp-Phe-Dap-Leu-Met) have been compared[65] with its X-ray structure. The AMBER calculations come out well in support of the two β-turns over Leu-Met and Trp-Phe. Cyclohexapeptides, cyclo[Lys-X^2-Lys-Phe(*p*NO$_2$)-Lys-X^6] where $X^2 = $ Arg, $X^6 = $ Ser(β-Glc) or Phe(pNO$_2$) have been prepared[66] and immobilised on epoxysilane-functionalised glass support in order to study their molecular recognition with amino acids. L-Glutamine turned out to be the best recognised amino acid by all three receptors.

2.7 Cycloheptapeptides – One new cycloheptapeptide, annomuricatin B, cyclo(Pro-Asn-Ala-Trp-Leu-Gly-Thr), has been isolated[67] from the seeds of *Annona muricata*, while the marine sponge (*Axinella* cf *carteri*)'s axinostatin 4, cyclo(Pro-Leu-Thr-Pro-Leu-Trp-Val), has been synthesised[68] to prove the configuration of the Trp, Val and Thr units. The BOP reagent was found to be the best coupling agent, but the synthetic material lacked cytotoxicity, which now follows the recent hunches that the cytotoxicity of the natural material might be due to highly cytotoxic trace impurities, possibly synergistic with axinostatin 4. When cyclic peptides spanning the major antigenic determinant of HIV glycoprotein 41 (gp41) synthesised[69] bearing a disulfide link as in (26) are replaced by ones cyclised *via* lactam side-bridges as in (27), the latter show decreased antigenic activity. The lactam bridge was constructed *via* BOP/HOBt couplings.

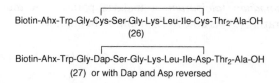

Biotin-Ahx-Trp-Gly-Cys-Ser-Gly-Lys-Leu-Ile-Cys-Thr$_2$-Ala-OH

(26)

Biotin-Ahx-Trp-Gly-Dap-Ser-Gly-Lys-Leu-Ile-Asp-Thr$_2$-Ala-OH

(27) or with Dap and Asp reversed

2.8 Cyclooctapeptides – The latex of *J. Chevalieri Beille*, a small shrub in Senegal, used to treat infected wounds, has been a rich source of cyclic peptides. The latest series found[70] are chevalierin A, B and C, B being the methionyl oxidised form of A, cyclo(Gly-Ile-Pro-Ile-Leu-Ala-Ile-Met), while C is a cyclic nonapeptide, cyclo(Tyr-Thr-Ile-Phe-Asp-Ile-Phe-Gla-Ala). Synthetic strategy to A, B and C utilised Fmoc technology for the linear precursors on chlorotrityl resin, followed by cyclisation of the precursor peptides at a C-terminal Gly residue using HBTU under high dilution. The sclerotoid ascostromata *Eupenicillium shearii* has yielded[71] shearamide A, cyclo(Pro-Thr-Val-Pro-Phe-Gly-Trp-Ile) which displayed moderate activity in assays against *H. zea* larvae. Two new antiproliferative cyclopeptides, axinellins A and B have been isolated[72] from the marine sponge *Axinella carteri*. NMR, FAB MS and Marfey's method for configurational assay have confirmed cyclo(Phe-Pro-

Asn-Pro-Phe-Thr-Ile) as axinellin A and cyclo(Phe-Pro-Thr-Leu-Val-Pro-Trp-Pro) as axinellin B. Apart from the cyclolinopeptides A and B, linseed mill cake has now been shown[73] to contain another immunosuppressive cyclolinopeptide X, cyclo(Pro-Pro-Phe-Phe-Ile-Leu-Leu-X) where X is N-methyl-4-aminoproline. The latter substitutes for two residues of cyclolinopeptide A and provides a non-planar *cis* amide bond. The roots of *Silene szechuensis*, a folk herb of Yunnan, are a source[74] of at least three cyclo-octapeptides, silenin A, cyclo(Pro-Leu-Ser-Phe-Pro-Tyr-Leu-Val), B, cyclo(Phe-Leu-Ala-Pro-Leu-Pro-Phe-Pro), and C, cyclo(Tyr-Ala-Phe-Pro-Gly-Phe-Tyr-Pro).

During the screening[75] of a soluble cyclic octapeptide library, cyclo(Ala-His-Pro-Gln-Phe-Pro-Ala-Glu)-Lys-NH$_2$ was identified as a tight binding ligand (IC$_{50}$ = 128 nM) to streptavadin. Hymenamide G has been reported[76] to be a cyclooctapeptide, but unavailability of the original source precludes reporting of its exact structure.

New syndiotactic cyclic octapeptides cyclo(D-Phe-Asp-D-Phe-Asn-D-Phe-Asp-D-Phe-Asn) and cyclo(D-MeAla-Asp-D-MeAla-Asn-D-MeAla-Asp-D-MeAla-Asn) have been prepared[77] on-resin with the C-terminal Asp residue protected as an allyl ester attached to the resin *via* its side-chain carboxyl. Cyclisation with HOAt/HATU secured a result in the D-Phe analogue, but the N-methylated analogue cyclised best with DIC/HOAt.

An N-acetyloctapeptide (28) containing the hexapeptide fragment of Proline-Rich Protein (PRP) has been cyclised[78] between the side-chain of Glu^1 and Lys^8 using azide methodology. Immunotropic activity of the cyclic form was similar to a linear nonapeptide. A template bound cyclic peptide (29) aims[79] at stabilising a β-I-turn around the tetrapeptide Asn-Pro-Asn-Ala motif believed to be central to the immunodominancy of circumsporozite surface protein of *Plasmodium falciparum*, a malaria parasite. A multiple-antigen peptide incorporating (29) is able to elicit antibodies in mice.

Ac-Glu-Tyr-Val-Pro-Leu-Phe-Pro-Lys-NH$_2$
(28)

(29)

2.9 Cyclononapeptides and Cyclodecapeptides

– While examples of cyclononapeptides have been discussed in Table 2 (Ref. 51) and subsection 2.8 (Ref. 70) of this Chapter, only one report[80] was retrieved on cyclodecapeptides. This reported the synthesis of analogues of antamanide in which one of the Phe residues in positions 6 or 9 were substituted by Gly residues. [Gly^9]-Antamanide was cytotoxic while [Gly^6]-antamanide was inactive. Both analogues were inactive *in vivo* against phalloidin.

2.10 Higher Cyclic Peptides – Following up on some recent breakthroughs in synthetic techniques for coupling highly hindered residues on-resin, [MeSer^1]-cyclosporin (30, R^1 = H R^2 = Me), [MeThr^1]cyclosporin (30, R^1 = Me, R^2-Me) and [3-OHMeLeu^1] cyclosporin (30, R^1 = CHMe$_2$, R^2 = Me) have been synthesised[81] using DIPCDI/HOAt coupling conditions. Site selective methylation at the 11-position also proved successful. An X-ray determination[82] of demethyl-ated cyclosporin A (cyclosporin E, 30, R′= -CH(Me)CH$_2$CH=CH-Me, R^2 = H) has been reported, and a new cyclosporin, [Ala^2,Val^{11}] cyclosporin (30, R′= CH(Me)CH$_2$CH=CH-Me, R^2 = H, and Ala in position 2) has been isolated[83] from submerged culture of *Mycelium sterilae* MS2929. [Ø-^{3}H-Bmt]-Cyclosporin A (30, R^1 = CH(Me)CH$_2$CH = CH-CH$_2$^{3}H, R^2 = Me) has been prepared[84] from protected bromo-precursors in 97% radiochemical purity and a specific activity of 4.3 Ci mmol^{-1}. When FAB MS and NMR methodologies were applied[85] to the oxidation products of reacting cyclosporin A with hydrogen peroxide, the results indicated that the parent compound was modified to give unit (31) at position 1.

(30)   (31)

Gramicidin S (32, X = H) still commands attention, and work[86] on the parent antibiotic, as well as its di-D-cyclohexyl alanyl analogues, has revealed that the reluctance of urethane derivatives (32, X = ROCO) of the Orn side-chains to N-methylate with MeI/Ag$_2$O may be due to the H-bonds detected between the δ-NH of Orn and the CO groups of the D-Phe residues as represented in structure 32. Derivatives of gramicidin S in which one or both of the two Orn residues have been subject to picolinoylation have been prepared.[87] [Orn-(PyCO)$^{2,2'}$]-Gramicidin S formed a 1:1 complex with Cu^{2+} and Zn^{2+} in MeOH, while for [Orn(Boc)^2, Orn(PyCO)$^{2'}$]-gramicidin S stepwise formation of 1:1 and 2:1 Cu$^+$ complexes was observed.

A template molecule designed as part of the template-assembled synthetic protein (TASP) concept has been constructed[88] by selective oxime formation between bicyclic peptide (33) and a helical 18-residue peptide, OHCCO-Phe-Ser-Arg-Ser-Asp-Glu-Leu-Thr-Arg-His-Ile-Arg-Ile-His-Thr-Gly-Lys(CO-CH$_2$ONHFmoc)-Gly-NH$_2$, followed by oxidation of Ser hydroxyl. The template preserves the essential feature of the native zinc finger motif. The efficiency of phosphorylation of tyrosine residues located at positions 5 and 12

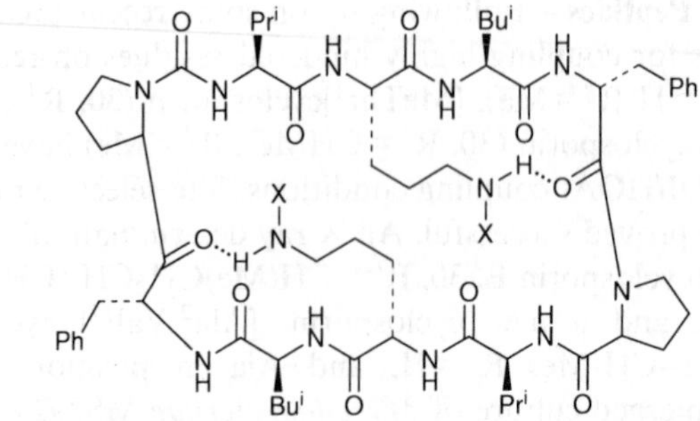

(32)

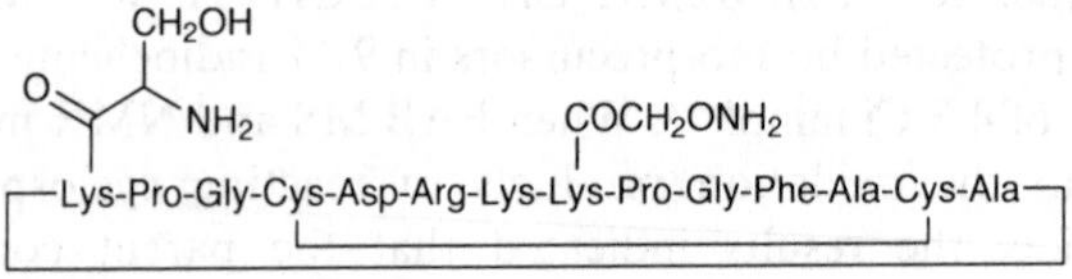

(33)

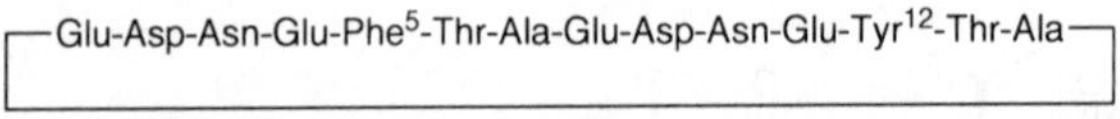

(34)

by the lyn oncogene has been studied[89] using the cyclic peptide (34) and the analogue with Phe[5] and Tyr[12] interchanged. These analogues were phosphonylated two-fold better than their linear analogues.

2.11 Peptides Containing Thiazole/Oxazole Rings – A great deal of activity has been reported in this area, both in the identification of natural examples and in the total synthesis of previously known members. The marine environment has proved a rich source. Thus from the *Theonella* sponge keranamides K and L have been identified[90] as (35) and (36) respectively. The antibiotic borophycin, isolated from cyanobacterium, *Nostoc spongiaeforme* var. *tenue*, has been co-extracted[91] with four modified cyclo peptides, tenuecyclamides A and B (37, differing only in their stereochemistry), and C and D (38) that differ in the oxidation state of the methionyl residue. Comoramides A (39) and (40), together with mayotamides A (41) and B (42) have been identified[92] in extracts from the marine ascidian *Didemnum molle*. All four components exhibit mild cytotoxicity.

Four new structures, patellamide G (43) and ulithiacyclamides E-G (44–46), have been isolated[93] from the ascidian *Lissoclinum patella*, while the marine sponge *Mycale* sp. is the source[94] of glutathione adducts, mycalolides A (47) and (B) (48).

A total synthesis of micrococcin P₁ (49) has been achieved[95] from 4 separate fragments. Fragment A was first coupled to C and then added to fragments B and D synthesised independently. A full report[96] has now appeared of the total synthesis of 14,15-anhydropristinamycin II_B (50), a member of the virginia-mycin family of antibiotics, while a re-synthesis[97] of cyclodidemnamide with D-Val instead of the L-Val originally used in 1996 gave identical physical data

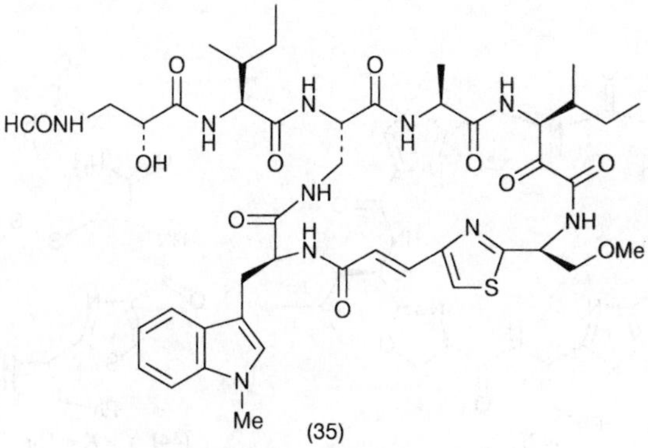

(35)

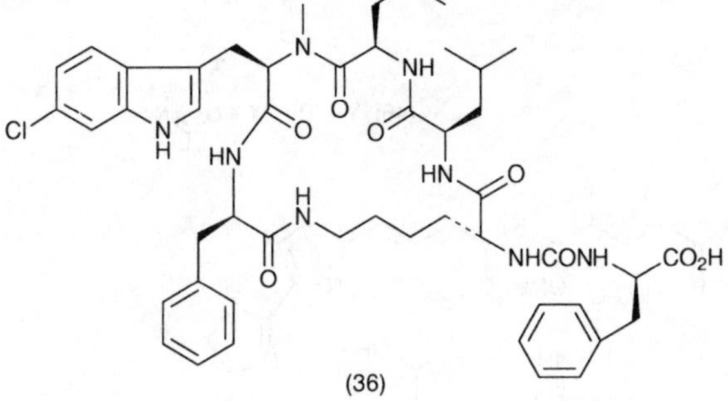

(36)

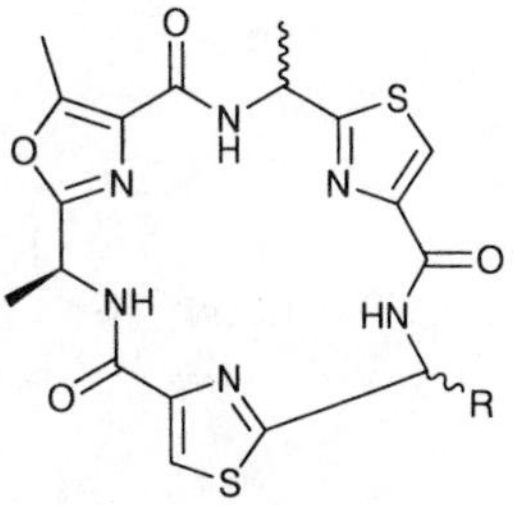

(37) R = Me (D or L)
(38) R = (CH$_2$)$_2$SMe,
or R = (CH$_2$)$_2$-S(O)Me

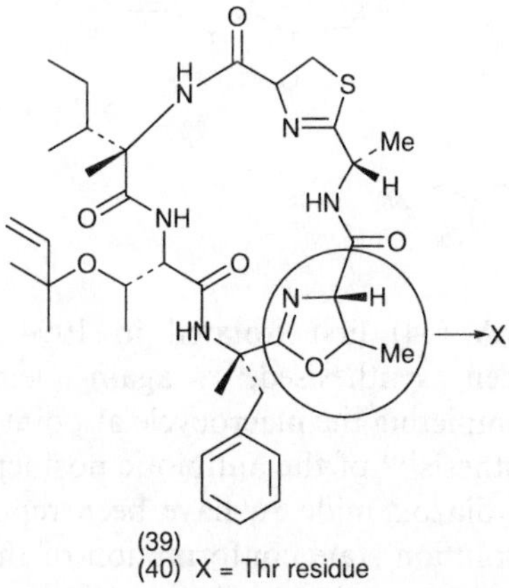

(39)
(40) X = Thr residue

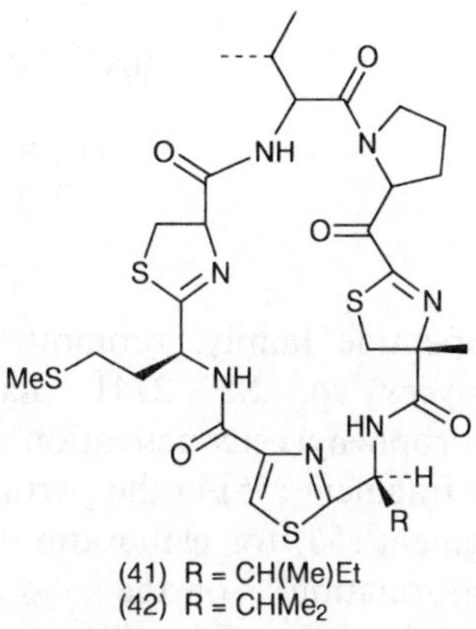

(41) R = CH(Me)Et
(42) R = CHMe$_2$

to the natural form. Total synthesis and confirmation of stereochemistries have been achieved[98] for raocyclamide A (51) and B (52), inter-related *via* Scheme 6. Nostocyclamide (53), the main antibacterial from the genus *Nostoc*, has been synthesised[99] from its oxazole/thiazole components, which were condensed to form a linear precursor, cyclised to (53) *via* the pentafluorophenyl ester derivative at the carboxyl group of the oxazole residue. Another member of

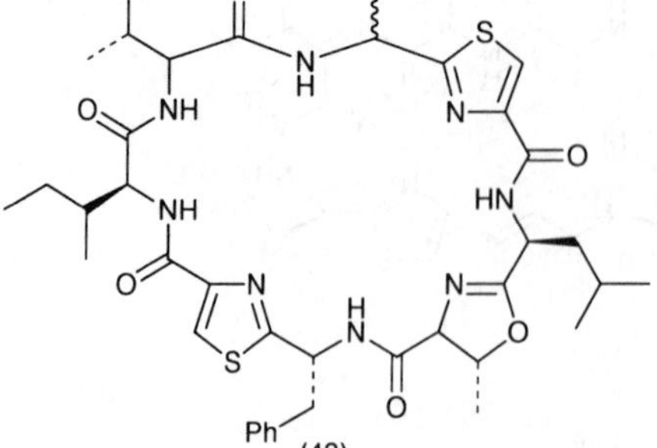

(43)

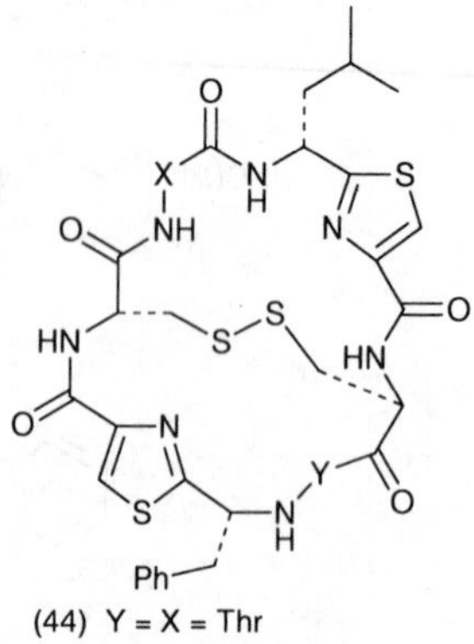

(44) Y = X = Thr

(45) Y = [structure], X = Thr

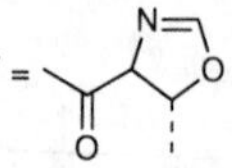

(46) Y = Thr, X = [structure]

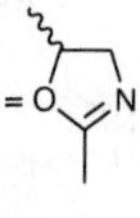

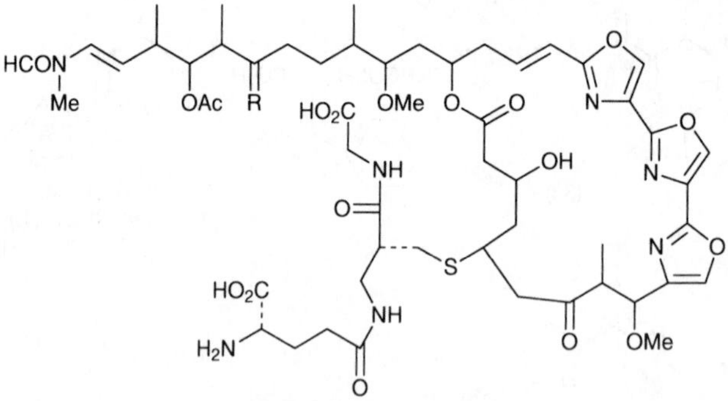

(47) R = O
(48) R = CHO[structure]

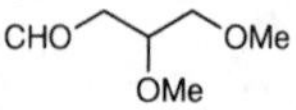

the thiopeptide family, promothiocin A (54) first isolated in 1994 from *Streptomyces* sp. SF 2741, has been synthesised,[100] again using a pentafluorophenyl ester activation for completing the macrocycle at point (X).

A key fragment (55) in the partial synthesis[101] of the antibiotic nosiheptide, and fragment (56) for elaboration[102] to diazonamide A, have been reported. Novel information[103] on the solid and solution state conformation of the 21-membered lissoclinamide (57) reveals a dominant combination of a type II β-turn at the prolyloxazoline moiety and a β-loop at the thiazoline-D-Phe-thiazoline sequence. Replacement of a single oxazoline ring with a thiazoline heterocycle increased cytotoxicity. Cyclo[Ile-Ser-(Gly)Thz-Ile-Thr-Gly(Thz)], related to patellamide A, has been synthesised,[104] and reacted with Cu(II), to form mono- and dinuclear complexes. The purple dicopper(II) derivative was observed from EPS spectra and force-field simulations to have the saddle-shaped structure similar to the known solution structure of patellamide A.

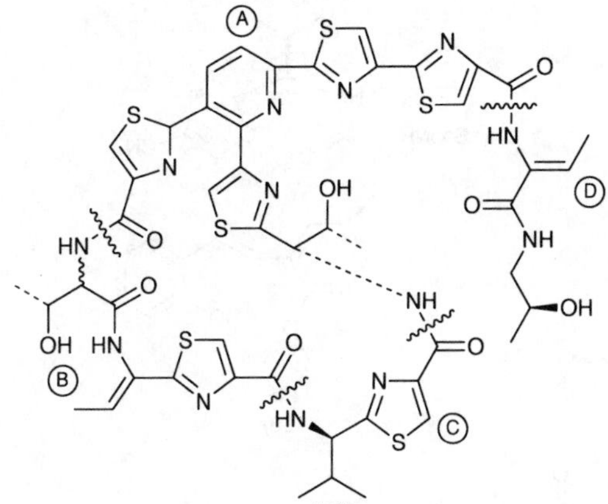

(49)

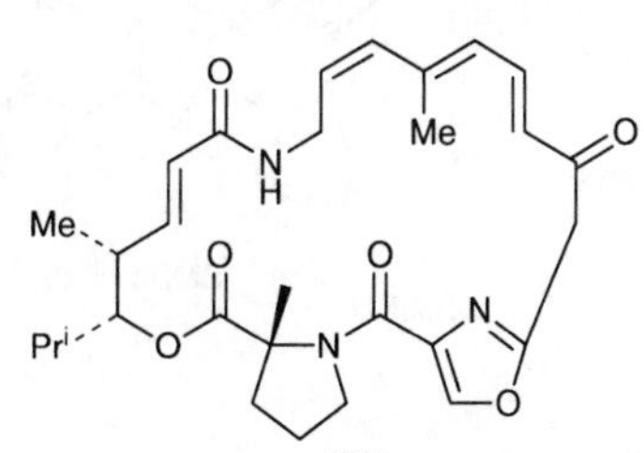

(50)

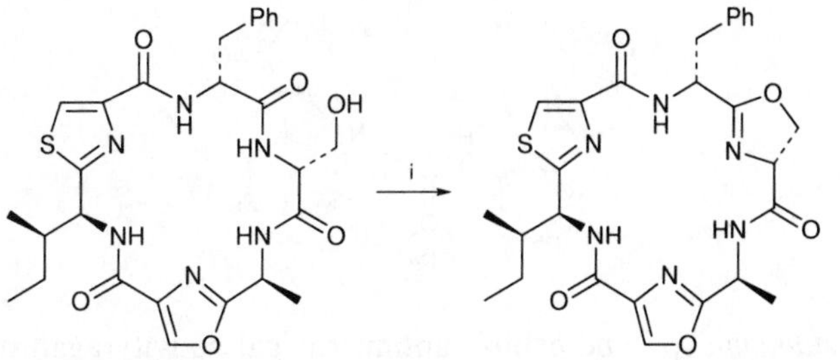

(51)

(52)

Reagent: i, Burgess reagent at 66 °C

Scheme 6

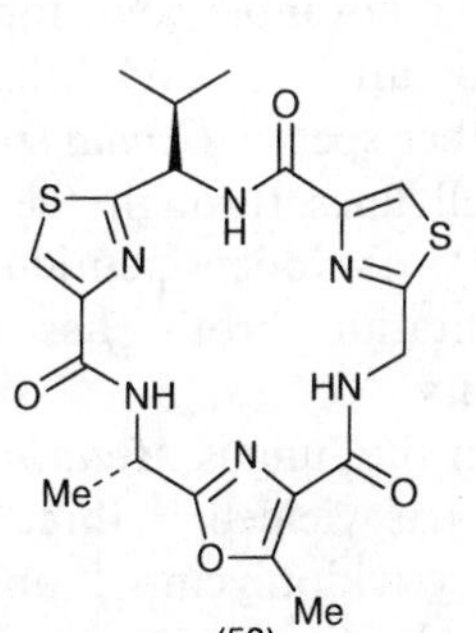

(53)

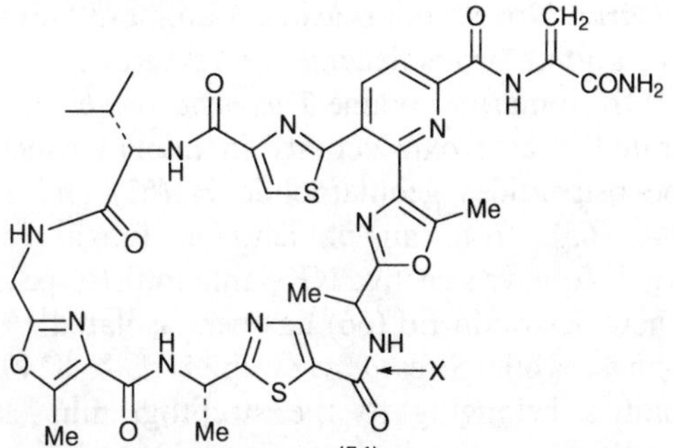

(54)

2.12 Cyclodepsipeptides – There does not seem to be a limit to the diverse structures trawled from the literature especially those originating from the marine environment. Biologically active cyclodepsipeptides have been the subject of a review.[105] Extracts from the New Zealand deep water sponge *Lamellomorpha strongylata* have already yielded cytotoxic compounds such as calyculins A, B, E and F, calyculinanides A and B and swinholide. Its main cytotoxic fraction has now been characterised[106] as theonellapeptolide IIIe (58). Exumolides A (59) and (B) (60) have been obtained[107] from the marine

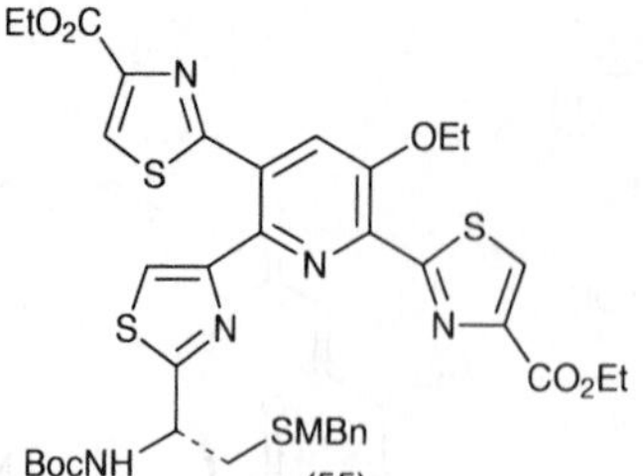

(55)

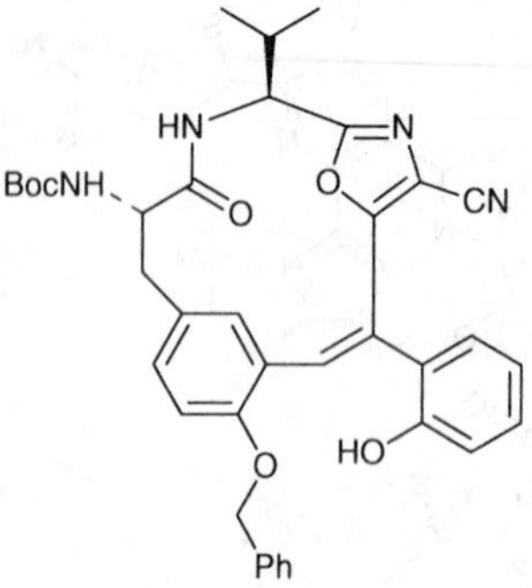

(56)

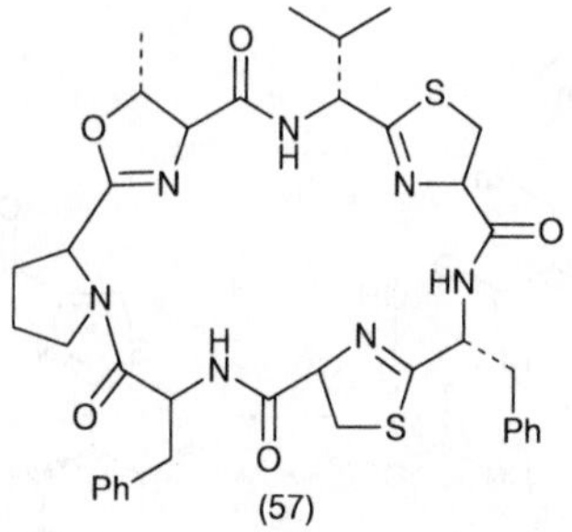

(57)

fungus *Scytalidium* sp., and exhibit antimicroalgal activity against unicellular chlorophyte *Dunaliella* sp. A variant of an alkyne amino acid found in onchidin has been identified[108] in dolastatin 17 (61) from the sea hare *Dolabella auricularia*. Novel amino acids, 4-amino-3,5-dihydroxyhexanoic acid, formyl-leucine and chloro-isoleucine are features in the structure[109] of cyclolithistide A (62) from marine sponge *Theonella swinhoei*. Another species *Geodia* sp. has generated[110] cytotoxic activity in human cancer cell lines through the new cyclodepsipeptides geodiamolide H (63) and I (64). Cyclodepsipeptide Sch 217048 (65), from an unidentified fungal fermentation broth, has been shown[111] to have selective NK_2 antagonist-type activity.

A new destruxin Ed (66) has been isolated[112] from the fungus *Metarhizium anisopliae*, while *Streptomyces* sp. HIL Y-8240155 has yielded[113] three new antibiotics belonging to the streptogramin class, grividomycins I and II diastereoisomers bearing structure (67), and III (68). Also from a *Streptomyces* sp. culture broth, a potent inducer of apoptosis in pancreatic carcinoma cells, named polyoxypeptin (69), has been characterised.[114] Topostatin (70), a new topoisomerase inhibitor, has been isolated[115] from *Thermomonospora alba* strain 1520, and a lipopeptide biosurfactant from *Bacillus* subtilis OKB 105 is a mixture[116] of closely related octamers having the general structure (71). Feeding experiments[117] utilising ^{13}C-labels support the hypothesis that salina-nide A, a potent anti-inflammatory agent, is probably derived from the hexapeptide H-Thr-D-Ile-Hpg-MePhe-D-aThr-Ser-OH involving a non-ribo-somal peptide synthetase pathway.

Syntheses of cyclodepsipeptides abound this year. Thus, after proving that

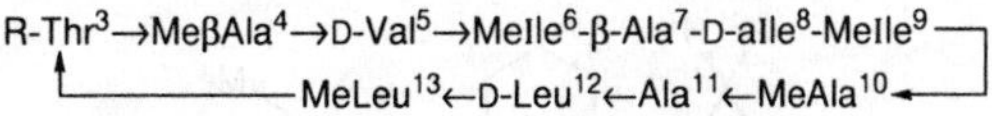

R-Thr³→MeβAla⁴→D-Val⁵→MeIle⁶-β-Ala⁷-D-aIle⁸-MeIle⁹ ─┐

└── MeLeu¹³←D-Leu¹²←Ala¹¹←MeAla¹⁰ ←┘

58 R = MeOAc–Val¹–D-MeLeu²

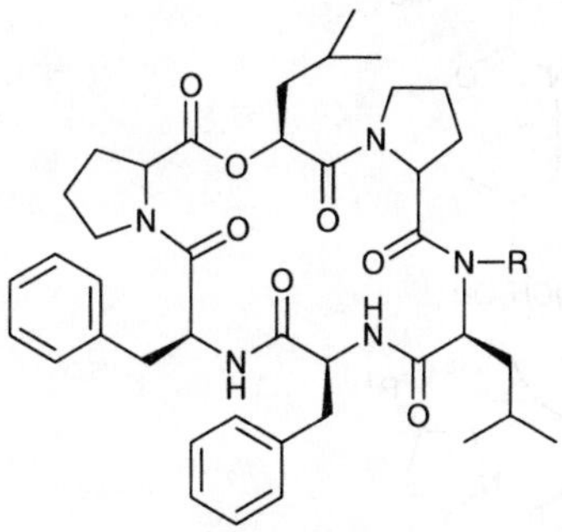

(59) R = Me
(60) R = H

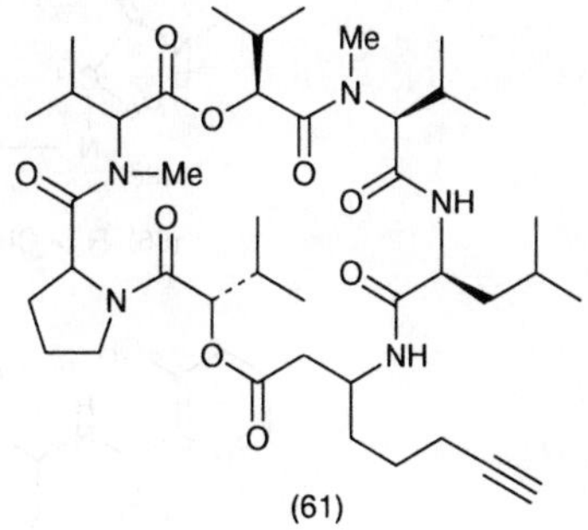

(61)

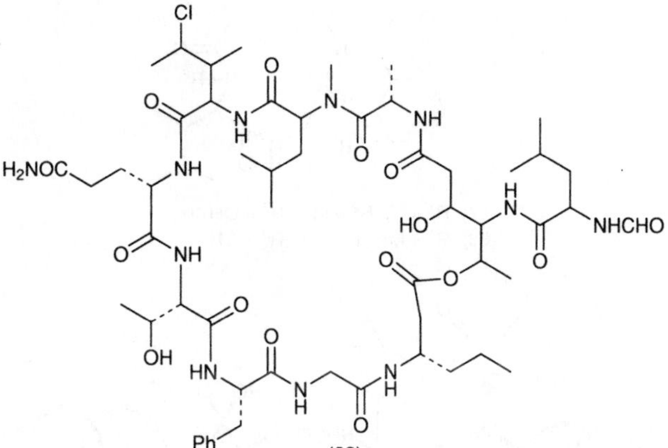

(62)

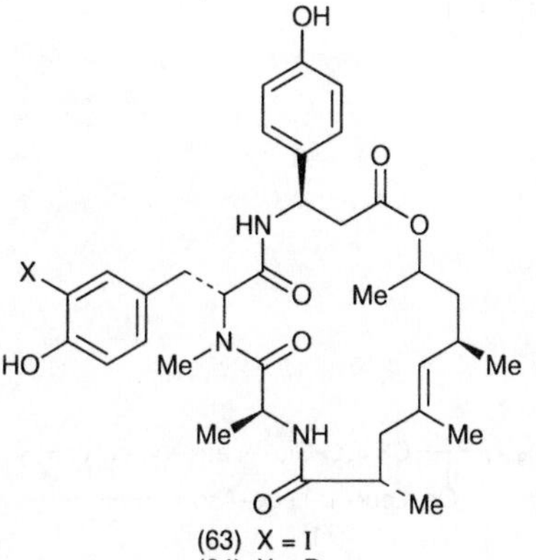

(63) X = I
(64) X = Br

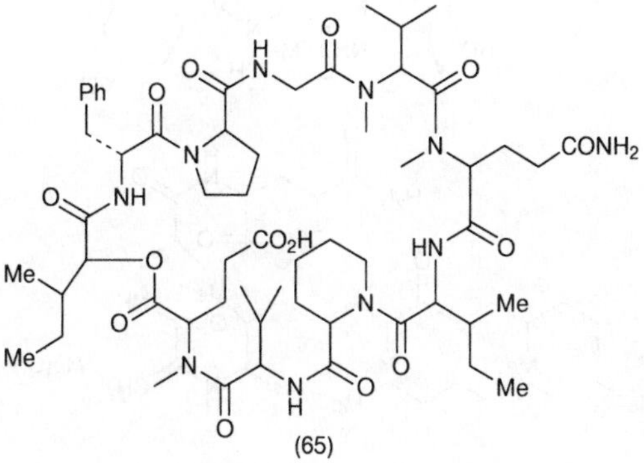

(65)

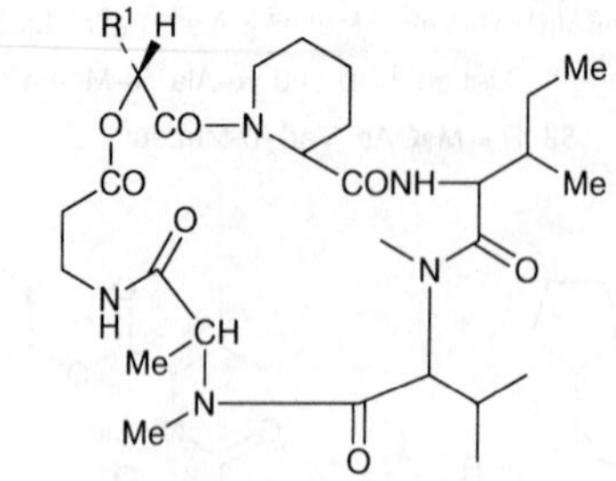

(66) R^1 = CH$_2$CH(OH)CH$_2$OH

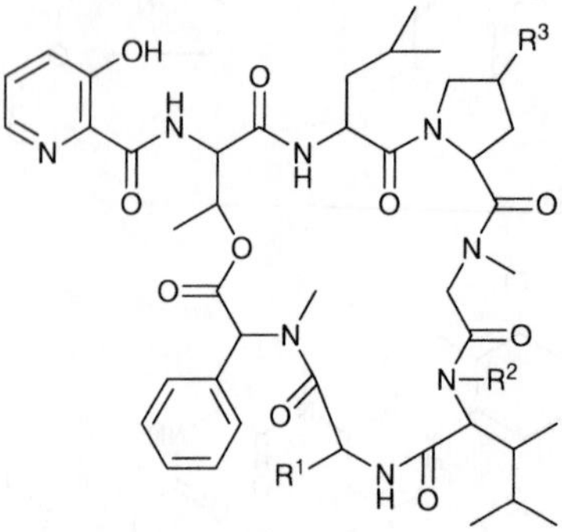

(67) R^1 = H, R^2 = Me, R^3 = OH
(68) R^1 = Me, R^2 = H, R^3 = OH

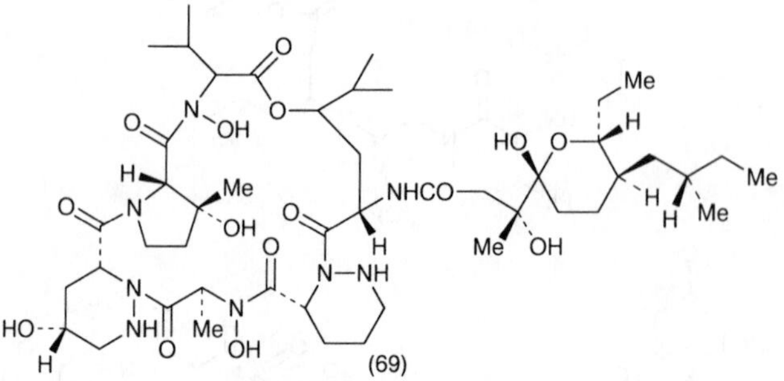

(69)

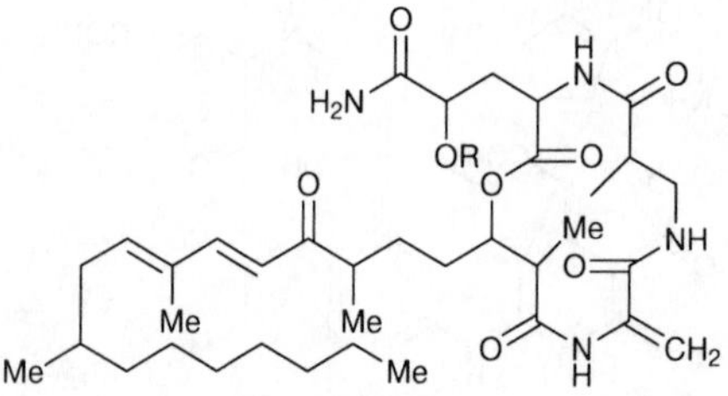

(70)

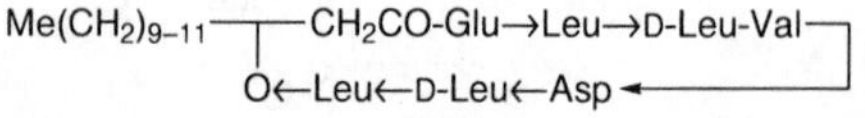

Me(CH$_2$)$_{9-11}$ ─┬─ CH$_2$CO-Glu→Leu→D-Leu-Val ─┐
                    O←Leu←D-Leu←Asp ◄───────────────┘

(71)

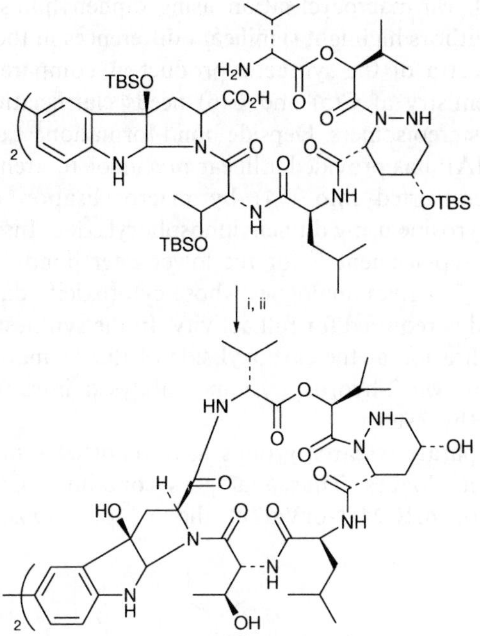

Reagents: i, HOAt, HATU; ii, Pr$^i_2$NEt

Scheme 7

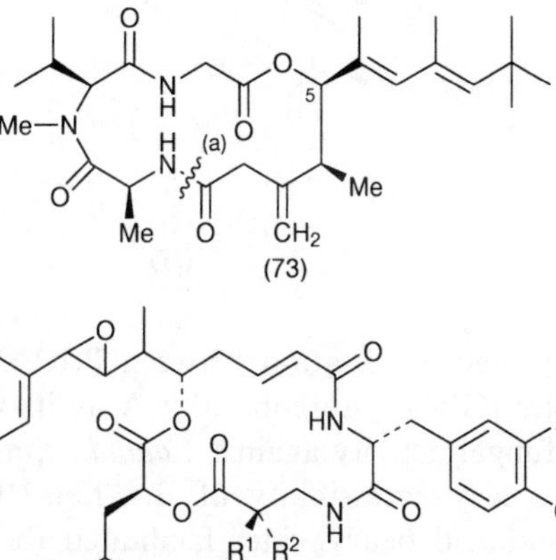

(73)

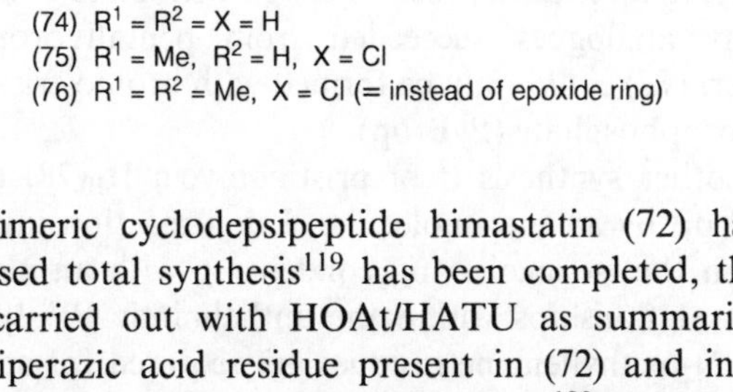

(74) R^1 = R^2 = X = H
(75) R^1 = Me, R^2 = H, X = Cl
(76) R^1 = R^2 = Me, X = Cl (= instead of epoxide ring)

the Trp-matrix in the dimeric cyclodepsipeptide himastatin (72) has the D-configuration,[118] a revised total synthesis[119] has been completed, the key macrocyclisations being carried out with HOAt/HATU as summarised in Scheme 7. The hydroxypiperazic acid residue present in (72) and in other cyclodepsipeptides such as monamycin has been synthesised[120] enantiospecifically from D-mannitol. A total synthesis[121] of antillatoxin (73) has been successful, *via* macrocyclisation using diphenylphosphorazidate at point (a). But the authors highlight significant differences in the optical rotations and the NMR spectra of the synthetic product as compared to the natural version. Stereochemistry at C(5) and C(6) needs clarificiation from subsequent syntheses of stereoisomers. Depside bond formation[122] using diisopropylcarbodiimide/DMAP has provided a linear precursor to arenastatin A(74), which then can be converted into (74) by macrocylisation at the amino group of methoxytyrosine using diphenylphosphorylazide. Inserting an isosteric methylene ether replacement[123] for the lower ester bond in the structure of cryptophycin A (75) gives analogues whose cytotoxicity data imply that the original ester bond is required for full activity. In the synthesis[124] of cryptophycin (76), macrocyclisation at the carboxyl side of the aromatic amino acid residue was carried out *via* 2-hydroxypyridine catalysed intramolecular aminolysis of a trichloroethyl ester.

Two separate research groups have reported syntheses for aureobasidin A (77) and analogues. Thus analogues containing Glu (and its esters), δHO-NorVal, or δHOMeNorVal at the 6,7 or 8 positions in (77) have been

(77)

prepared[125] by macrocyclisation between Pro^5 and AA^6. Amongst the lipophilic analogues [Glu^6] – aureobasidin A as its C$_6$ alcohol ester showed the strongest antifungal activity against *Candida* spp. In the other synthesis[126] of analogues, the easy accessibility of Boc-Leu-HOMeVal-(2R)-hydroxy-(3R)-methylpentanoic acid benzyl ester facilitated the syntheses. Coupling of the N-methylated residues proceeded well with HATU while the macrocyclisation of many analogues succeeded from pentafluorophenyl ester precursors. However, (77) could only be formed with bromo-tris-pyrrolidinophosphonium hexafluorophosphate (PyBrop).

In another synthesis[127] of pristinamycin II$_B$(78) the macrocyclic ring was 'zipped' up by an intramolecular Takai reaction involving an iodomethylene group on the isoxazole ring condensing with an aldehyde group. Structure-activity relationships within the anthelmintic PF 1022A (79) have shown[128] that all N-methylleucine residues are required for activity, and increasing the N-alkyl groups gave less activity. Final cyclisation steps used in the analogues were mediated *via* BOPCl. An ethylene bridge has also been intro-

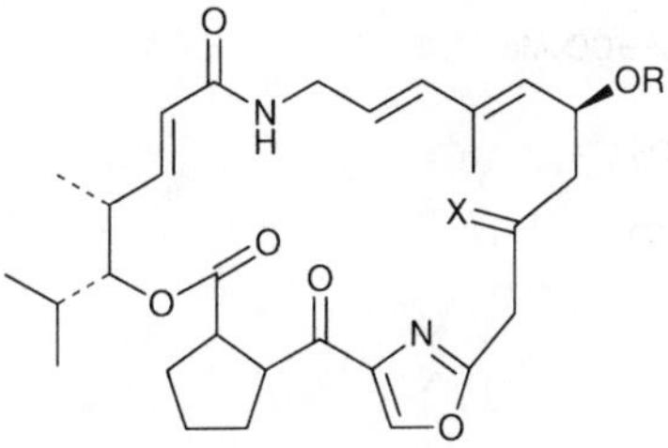

(78) R = DPTBS, X = H, OH

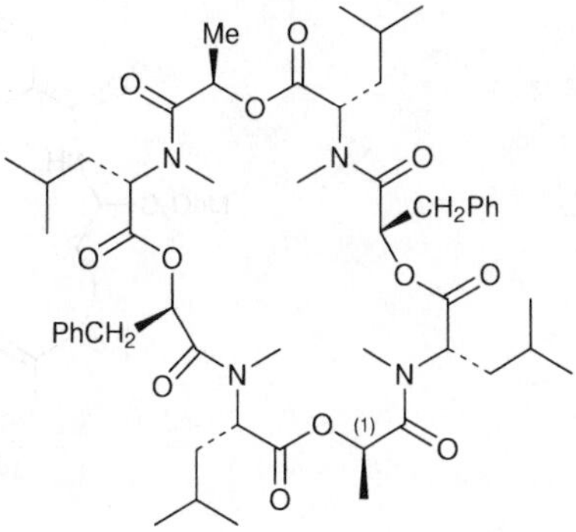

(79)

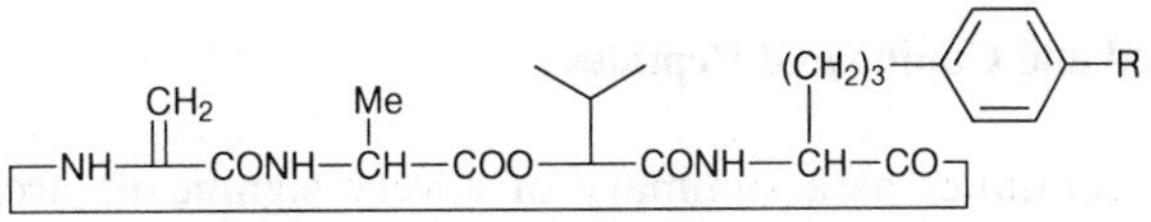

(80) R = OMe
(81) R = H
(82) R = OH

duced[129] from position (1) in (79) to its nearest N-alkyl amide to give an ε-lactam ring and a more rigid conformation. Cyclisation via an oxime resin was carried out as a trial for the rapid preparation of a variety of analogues for structure-activity studies. Fmoc-solid phase technology has been used[130] to prepare linear precursors for the synthesis of AM Toxins I (80) and II (81) and an analogue (82). A 2,5-diaminopropanoic acid residue was used as the N-terminal precursor to Δ-Ala, and the cyclisation was carried out using HATU.

The search for new armoury to counteract severe Gram positive infections in hospitals has renewed interest in pristinamycin I_A analogues. (1-Des-3-hydroxypicolinyl) Pristinamycin I_A has been shown[131] to undergo two unexpected transformations, including reductive amination using NaBH₄CN. A systematic examination[132] of sandramycin analogues containing intercalation chromophores (1-isoquinolyl and 3-isoquinolyl) has enabled the binding constant to be established with calf thymus DNA. Attempts to solve the crystal structure of uncomplexed actinomycins D and Z₃ have been successful[133] through the use of solvates with methanol/ethyl acetate (D) and with water and benzene (Z₃). The breakthrough in the final elucidation was use of low temperature, an intense X-ray beam and a new ab initio method of solving structures. X-Ray studies[134] have also been carried out on a variety of 18-membered cyclodepsipeptide rings such as (83) or its dipyridyl analogue. Single crystal structures showed similar relatively flat-ring structures, which showed self-assembly by stacking one over the other if pyridine was present in the molecule.

(83)

3   Modified and Conjugated Peptides

This section continues as a summary of a very significant area of research, mainly involving additional non-peptidic elements. As this has also been the traditional sub-section for discussion of the glycopeptide antibiotics, it deserves the Chapter's 'Oscar Award' this year as a result of the reports of the great accomplishments of two research groups recording the total synthesis of vancomycin aglycone.

3.1 Phosphopeptides – There is almost an equal distribution of papers reported this year between exponents of 'global phosphorylation' approaches *vis à vis* supporters of the 'building block' approach. The formation of H-phosphonate by-products, *e.g.* Thr(PO$_2$H$_2$) instead of Thr(PO$_3$H$_2$) during global phosphorylation with both dibenzyl- and di-t-butyl N,N-phosphoramidite has been investigated[135] and rectified by the use of aq. iodine for the oxidation step to make the phosphorodiester. Phospho-, H-phosphono- and methylphosphono-homoserine analogues have been formed from H-Gly-Gly-Hse-Ala-OH by on-resin phosphorylation and phosphorylation using phosphoamidites. By-products observed, apart from unmodified peptides, were H-phosphonopeptides during phosphorylation, and the phosphorylated peptide during H-phosphonopeptide formation. Attempts[137] have been made to phosphorylate LY303366, the semi-synthetic analogue of echinocardin B, in order to increase its solubility in water for i.v. administration. Phosphorylation of its phenolic hydroxyl group was examined using tetrabenzyl pyrophosphate with a number of bases, with 3M LiOH in DMF at −30 °C proving to be the best conditions. It has also been possible[138] to prepare L-phosphotyrosine mimics, from pre-synthesised pentapeptide sequences to study pp60$^{c\text{-src}}$ inhibition. 4-Formylphenylalanine incorporated into a sequence can be transformed, using (t-BuO)$_2$P(H)=O/CsF followed by trifluoroacetic acid, into the -Ph-CH(OH)-P(OH)$_2$=O phosphotyrosine side-chain mimetic.

The 'building-block' approach depends on the efficient availability of suitably-protected phosphorylated units. Thus a convenient synthesis[139] of Fmoc-O-monobenzylphosphonotyrosine, by monodebenzylation of the dibenzyl derivative using sodium iodide, is welcomed. For the synthesis of thiophosphotyrosine containing peptides, the Boc protected building blocks Boc-(O-dimethylthiophosphono)-Tyr-OH and Boc-(O-dicyanoethylthiophosphono)-Tyr-OH have been assessed.[140] Three Fmoc-phosphotyrosine derivatives have been studied[141] in the form of the N,N'-dialkyldiamides, Fmoc-Tyr(P(NRR')₂O)-OH, where R,R' are n-propyl, isopropyl or isobutyl. The isobutyl derivative turned out to be best in terms of its ease of preparation and excellence in cleavage properties. Phospho-L-azatyrosine can be substituted[142] into peptides *via* the Fmoc-derivative (84).

(84)

A systematic study[143] of the fragmentation pattern of phosphopeptides in an electrospray ion trap mass spectrometer has been presented. In some cases, *e.g.* phosphotyrosine and phosphothreonine peptides, ions showing no loss of phosphate could be detected, while in other cases H_3PO_4 and/or H_3PO_3 are lost. Phosphoserine containing peptides tend to fragment by relatively simple patterns, predominantly loss of H_3PO_4.

3.2 Glycopeptide Antibiotics – As indicated in the introduction to this Chapter, the highlight of the year, no doubt, was the total syntheses of the aglycone unit (85) of vancomycin by the groups of Evans[144] and Nicolaou[145] who published their syntheses side by side in *Angewandte Chemie*. Although taking us into the year 1999, Nicolaou's group have also reported the total synthesis[146] of vancomycin (86) through the chemical glycosidation of the vancomycin aglycone. Many of the man years to complete this great challenge have been reported on in this Chapter over many years, so space does not allow us to comprehensively cover the total synthesis in this coverage. However, from the last steps of the two approaches summarised in Scheme 8 can be seen the different strategies used by the two groups in achieving their goals.

Reconstruction of vancomycin (86) from a pseudoaglycon bearing only a monosaccharide unit has been achieved[147] by glycosylation of an acyl/allylated derivative using the sulfoxide (87). A naturally occurring vancomycin analogue [LY26426(A82846B)] where the amino sugar stereochemistry differs and a third sugar residue is attached, has been subject[148] to intensive SAR studies *via* reductive alkylation to give structural diversity in the side-chain. Potency of the new analogues against staphylococci and streptococci was as good or better than vancomycin. Lipases from *Pseudomonas* sp have been used[149] to give good yields of vancomycin acids without degradation. Teicoplanin, from the ristocetin family, has also been modified[150] by enzymatic deacylation, followed by reductive alkylation, but no gain in activity was achieved. A series of 7d-aminomethylated derivatives of teicoplanin aglycone has contained

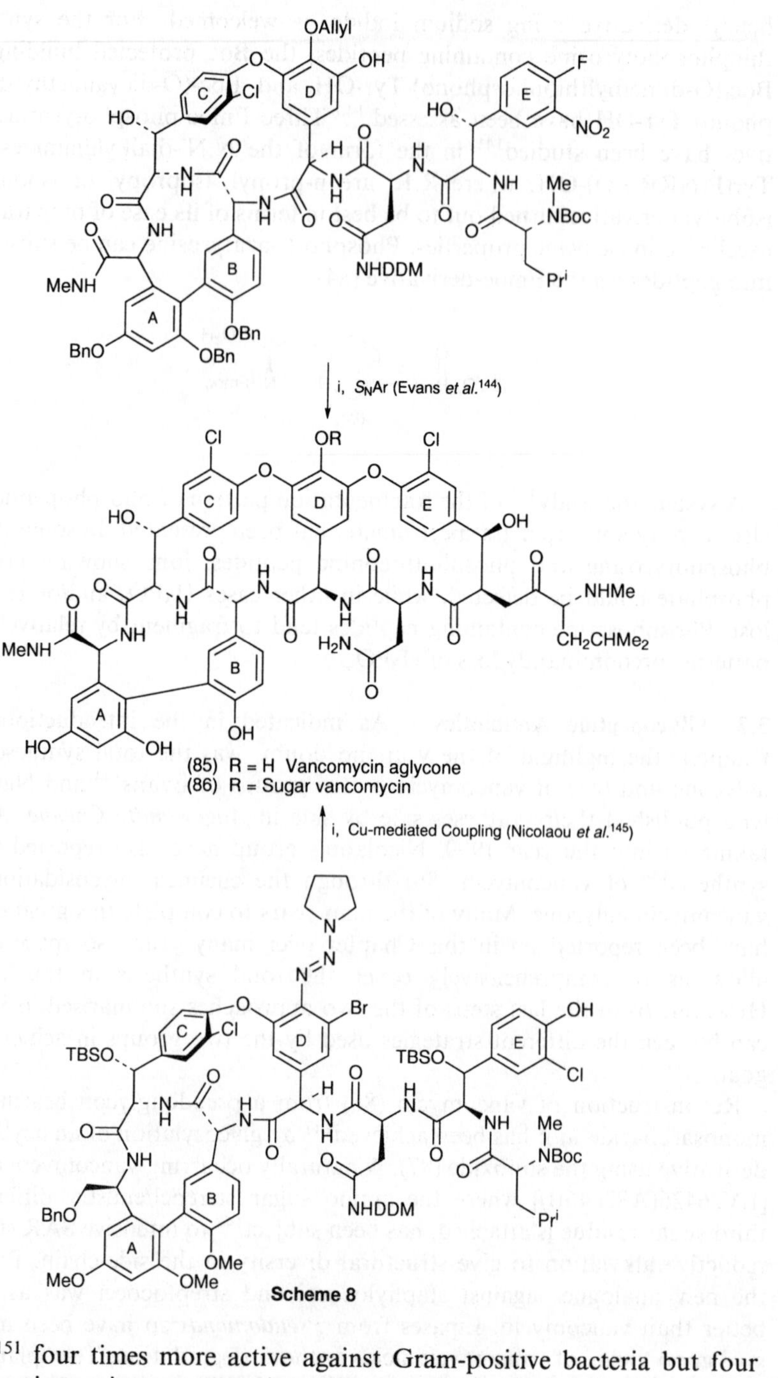

(85) R = H Vancomycin aglycone
(86) R = Sugar vancomycin

Scheme 8

examples[151] four times more active against Gram-positive bacteria but four times less active against vancomycin-resistant enterococci.

In a series of reductively alklyated derivatives of glycopeptide antibiotic A-40926, derivative (88) had excellent activity[152] against all susceptible Gram-

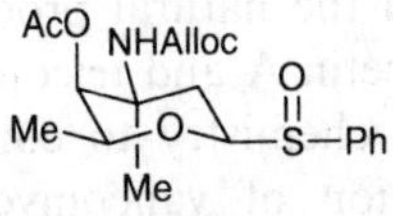

(87)

positive bacteria tested. An efficient acid-catalysed conversion[153] of complestatin (89) into chloropeptin (90) has been explained and used to assign the
complete stereochemistry of the former. Chemical modification of glycopeptides, with special emphasis on erythromycin, has been the subject of a
review.[154]

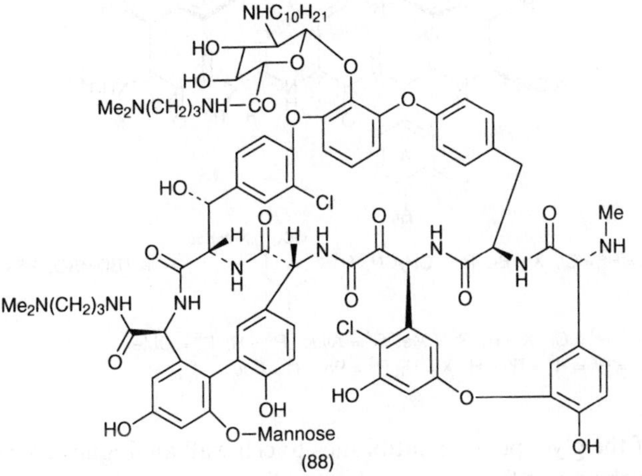

(88)

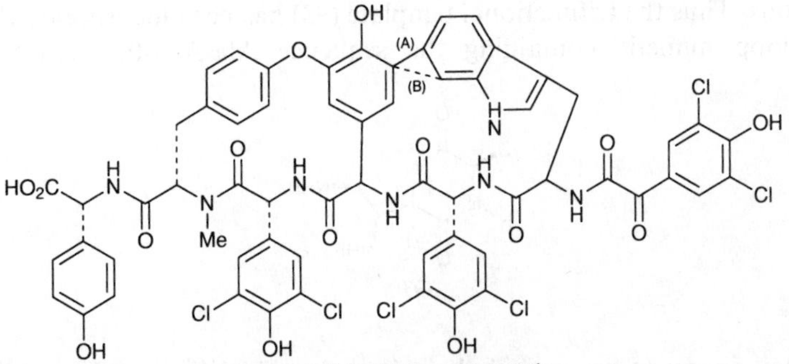

(89) C–C link at (A) – complestatin
(90) C–C link at (B) – chloropeptin

Although total syntheses have been achieved, fundamental studies to piece
together vancomycin-type skeletons continue. Thus the preparation of an
appropriately protected, fully functionalised vancomycin CDE ring (91) has
been reported[155] and involved two aromatic nucleophilic substitution reactions
for sequential CD and DE ring macrocyclisations. An efficient chloro-deami-
nation procedure has been developed[156] to produce (92a) with the correct

atropostereochemistry found in the natural product. The synthesis of (92b), the BCDF ring system of ristocetin A and teicoplanin, has involved an S_NAr reaction using areneruthenium chemistry to construct the DE bicyclic aryl ether system.[157] The degradation of vancomycin to a series of modified aglucovancomycin derivatives has provided[158] molecules for studying the thermal atropisomerism. In all cases selective isomerism of the DE ring system atropisomers was observed under conditions where CD and AB stereochemistries were unaffected.

(91) $R^1 = R^2 = Cl$, $X = Br$, $R^3 = CH_2CN$, $R^4 = $, $R^5 = TBDMSO$, $R^6 = OMe$

(92a) $R^1 = R^2 = Cl$, $X = H$, $R^3 = Me$, $R^4 = Alloc$, $R^5 = H$, $R^6 = OMe$
(92b) $R^1 = R^2 = R^5 = R^6 = H$, $X = H$, $R^3 = Ph$, $R^4 = Boc$

Binding of the glycopeptide antibiotics to cell wall analogues continues to be of interest in the quest for better understanding, and for the design of stronger associations. Thus the trifunctional template (93) has been incorporated[159] in a protein loop mimetic containing the sequence Ala-Asn-Pro-Asn-Ala-Ala.

(93)

Head to tail dimers of vancomycin have been described[160] to exploit additional cooperative interactions when binding to bacterial cell wall precursors at a surface. Non-covalent complexes between vancomycin and cell-wall precursor analogues Ac-D-Ala-D-Ala-OH and Ac-L-Ala-L-Ala($^2H_3$)-OH have been probed[161] using electrospray mass spectrometry. Strong interaction was proven for the D-D-analogue, but none for the L-L form. However, the conditions of the electrospray MS experiments need to be carefully controlled to give conditions free from non-specific aggregations. Tighter binding trivalent ligands derived from D-Ala-D-Ala-OH have been designed[162] for high affinity associations with vancomycin, and nineteen binding constants of an all D-tetrapeptide library to vancomycin have been measured[163] by on-line affinity capillary electrophoresis-electrospray MS (ACE-MS). NMR Studies[164] have revealed that the sugar residues in vancomycin lead to significant conformational changes in the antibiotic aglycone. They affect the orientation of the aromatic rings, with respect to the backbone and influence the alignment of the amide protons which are important for dimerisation and cell-wall binding.

3.3 Glycopeptides – A general review[165] concentrating on both N- and O-glycopeptides has appeared, while another review[166] concentrates more on the synthesis of glycosidic linkages. An unusual glycopeptide linkage has been discovered[167] within the bicyclic structure (94) of theopalaumide, from the sponge *Theonella swinhoei*. Theopalaumide and an isomer inhibit the growth of *Candida albicans*.

(94)

3.3.1 O-Linked Glycopeptides – Koenigs-Knorr type glycosylations[168] of Fmoc-Ser-OPfp and Fmoc-Thr-OPfp with protected β-D-Gal(1-3)-D-Gal-N₃ using Ag salts result in only α and/or β isomers in excellent yields. Using AgCO₃ at $-40\,^{\circ}$C the ratio α:β was 5:1 while with Ag₂CO₃/AgClO₄ at room temperature gave the β-isomer exclusively. Glycosylation[169] of Fmoc-Ser/Thr/ Tyr with β-D-glucose pentaacetate took place most efficiently in the presence of BF₃·Et₂O. The Fmoc-derivatives could then be converted into their tri-chloro-phenyl esters to make e.g. [Tyr(β-D-Glc)⁵] dermorphin and [Ser(β-D-Glc)⁵,Tyr⁷] dermorphin. Flα-antigen (95), a member of the tumour-associated O-linked mucin glycopeptides, has been synthesised[170] *via* two alternative routes, one based on linking between Ser/Thr and GalNAc, then coupling to a

(95) R = H, Me

lactosamine unit. The other route involved assembly from perbenzylated lactal prior to construction of the glycopeptide linkage.

β-D-Galactopyranosyl and α-D-glucopyranosyl-β-D-galactopyranosyl moieties carrying silyl, isopropylidene and 4-methoxybenzyl protecting groups have been attached[171] to 5-hydroxy-L-norvaline to form successful Fmoc building block derivatives for the solid phase synthesis of fragments from type II collagen. Fmoc-Glycopeptide esters with unprotected side-chains have been prepared[172] using the Rink amide resin, and have provided an insight into the different positions the GlcNAc moiety can take for the glycopeptides to function as substrates for subtilisin-catalysed glycopeptide condensations. Glycopeptides that mimic the action of oligosaccharides have been rapidly identified[173] by a combinatorial library set up from three different glycosyl amino building blocks, Fmoc-Thr[α-Ac$_4$Man(1→3)α-2-*O*-Bz-4,6-Ac$_2$Man]-OPfp, Fmoc-Thr(α-Ac$_4$Man)-OPfp and Fmoc-Asn(β-Ac$_3$GlcNAc)-OPfp and PEGA solid support. The most active glycopeptides that display specificity for the lectin in hemagluttination assays were, T(α-D-Man)ALKPTHV, LHGGFT(α-D-Man)HV, T(α-D-Man)EHKGSKV, GT(α-D-Man)FPGLAV, and T(α-D-Man)LFKGFHV. A small library of fucosyl threonine peptides has been set up[174] as sialyl Lewis X mimetics. In the E-selectin assay, derivative (96) proved most active. On-resin condensation[175] over 11 days using an excess of (97) with previously synthesised (98) has yielded the lipoglycopeptide (99) which contains the hemophilic recognition region of mouse epthelial cadherin. Coupling was carried out using PfPyU, as in Scheme 9. Glycosylated derivatives of the cell-adhesion/platelet aggregation motif -Gly-Arg-Gly-Asp-Ser have been synthesised[176] from Fmoc-building blocks but biological activity was generally reduced by adding sugar residues. However, glycoconjugates of arginine-vasopressin were found[177] to have greater *in vivo* antidiuretic activity than the parent peptide, although it is believed that their action can be explained *via* their ready hydrolysis to the arginine-vasopressin.

NMR, CD spectrophotometry, fluorescence anisotropy and molecular dynamic calculations have all been used[178] to study the conformations of (100) and (101), models of the repeating C-terminal domain of RNA polymerase. Data indicate that glycosylation, as in (101), leads to the formation of a non-random turn-like structure, not present in (100). A similar structure inducing effect has been noticed[179] in glycosylated threonine-containing oligopeptides of

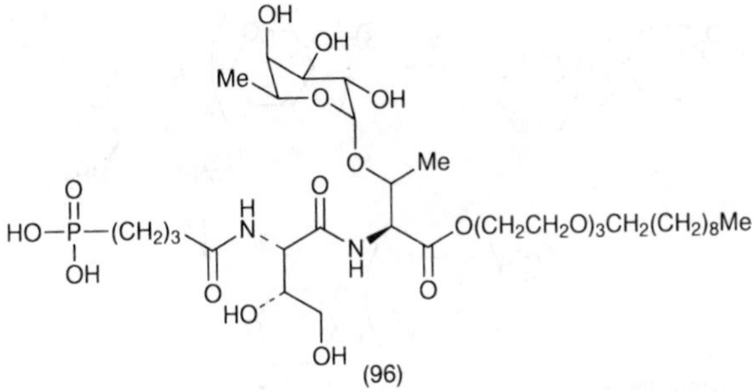

(96)

(Cet)$_2$Glu←Succ-Ser(Bu^t)-His(Trt)-Ala-Val-OH
(97)

+

H-Ser(Bu^t)-Ser(α-Ac$_3$GalNAc)-Asn(Trt)-Gly-Glu(OBu^t)-Ala-Val-Glu(OBu^t)-Hycrom
(98)

| i, ii

(Cet)$_2$Glu←Succ-Ser-His-Ala-Val-Ser-Ser(α-Ac$_3$GalNAc)-Asn-Gly-Glu-Ala-Val-Glu-OH
(99)

Reagents: i, PfPyU; ii, TFA followed by Pd(PPh$_3$)$_4$

Scheme 9

increasing chain length. Peptide models chosen for the study were the homo oligomers, Z-(Thr)$_n$NHMe (n = 1–4), Z-[Gal(Ac$_4$)β]Thr$_n$-NHMe (n = 1–5) and Z-[Galβ]Thr$_n$-NHMe (n = 1–5).

Ac-Ser-Tyr-Ser-Pro-Thr(R)-Ser-Pro-Ser-Tyr-Ser-NH$_2$

(100) R = H
(101) R = β-O-GlcNac

3.3.2 N-Glycopeptides

3.3.2 N-Glycopeptides – A short review[180] of new methodology for synthesising glycosyl asparagine derivatives, their adaptation to solid phase protocol, and their tranglycosylation with endo-b-N-acetylglucosaminidase has appeared. Eel calcitonin having 11 sugar residues attached was synthesised *via* the techniques reviewed. A thioester method was used by the same authors[181] to form [Asn(GlcNAc)^3]-eel calitonin. An alternative protocol[182] to the usual building block assembly on a polymer has been developed. It involves the glycol assembly method, whereby a polymer mounted N-acetylglucosamine unit can be attached *via* the side-chain aspartyl carboxyl group using conventional peptide coupling methods such as IIDQ as symbolised in Scheme 10. A key stage in the synthesis was the iodosulfonamidation of the polymer-bound glucal.

An EEDQ-mediated coupling[183] of 2-acetamido-2,3-di-O-acetyl-6-(2α-t-Boc-amino)benzamido-2,6-dideoxy-β-D-glucopyranosylanine with N-Bz-Asp-Tyr(NO$_2$)-OMe has yielded a fluorescence-quenched glycopeptide having anthranilamide/3-nitrotyrosine as the donor/acceptor pair. An N-glucoaspar-

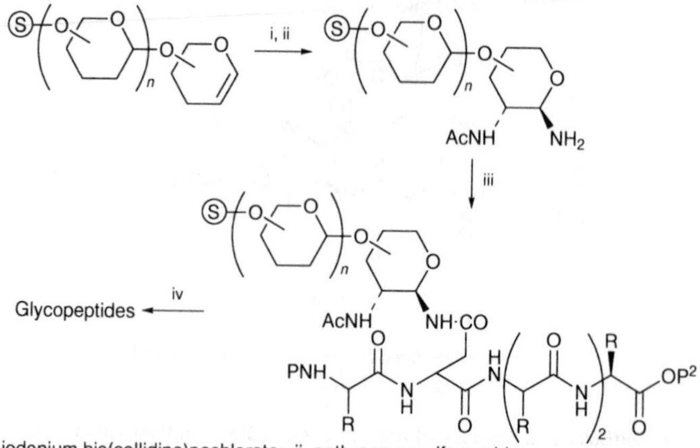

Reagents: i, iodonium bis(collidine)pechlorate; ii, anthracene sulfonamide;
iii, Asp-containing-protected peptide; iv, cleavage from resin

Scheme 10

agine analogue[184] prepared from 2-amino-1,5-anhydro-2-deoxy-glucitol hydrochloride and Z-Asp(OH)-OBn has been used in the synthesis of a mimic of the
V3-loop structure of the principal neutralising determinant of HIV-1 *viz.* the
sugar modified asparagine-containing Ac-SXNTRKSIHIGPGRAF-NH$_2$.
Natural sources of N-linked oligosaccharides have been obtained[185] by mild
hydrazinolysis of natural glycoproteins, and after processing can be linked to
side-chain activated Asp derivative, Fmoc-Asp(ODhbt)-OBu^t. Subsequent
acetylation of carbohydrate OH groups and cleavage of the t-butyl ester
yielded building blocks incorporated into a multiple column peptide synthesis
protocol. A typical result is exemplified by structure (102). The protected
building block (103), bearing the acid labile TBDMS group, has been more
successful[186] for the solid phase synthesis of a glycopeptide corresponding to
residue 447-460 of protein S than the acetylated derivative. The trifluoromethyl
asparagine analogue (104) has been synthesised and fully characterised,[187] as a
potential enzymically-stable binding motif in the synthesis of drugs. Fmoc-N-
Acetylglucosaminyl asparagine has been incorporated[188] into the solid phase
protocol to make N-acetylglucosaminyl Peptide T (Ala-Ser-Thr-Thr-Thr-
Asn(GlcNAc)-Tyr-Thr). This was then followed by further transglycosylation
by a microbial endoglycosidase to yield glycosylated Peptide T, which blocks
infection of human T cells by human immunodeficiency virus. Reports,
concentrating more on the syntheses of carbohydrate moieties of glycopeptides
include a new effective synthesis[189] of [β-D-GlcNAc(1→4)norMurNAc-L-
Abu-D-IsoGln] and [β-D-GlcNstearoyl(1→4)norMurNAc-Abu-D-IsoGln]
and the lipophilic analogues[190] (105) and (106) of N-acetylnormuramyl-L-2-
aminobutanoyl-D-isoglutamine.

A comparison[191] has been made, using FTIR and NMR experiments,
between N-glycosylated cyclic peptides cyclo[Gly-Pro-Xxx(GlcNAc)-Gly-
δ-Ava] and their non-glycosylated parent compounds cyclo(Gly-Pro-Xxx-Gly-
δ-Ava) where Xxx was Asn or Gly and δ-Ava = NH-(CH$_2$)$_4$-CO. Type II

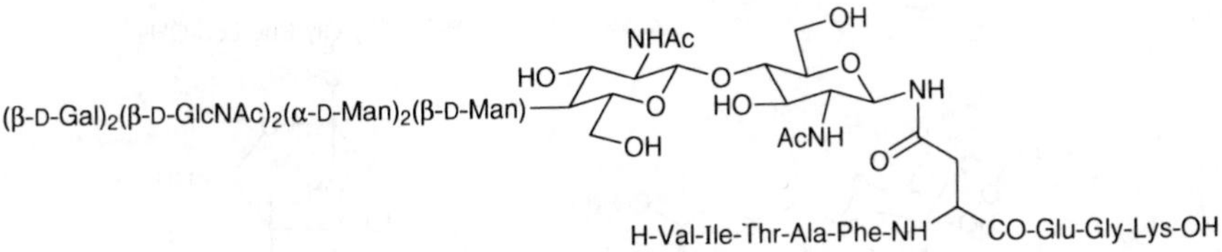

NHAc
OH
HO
O
HO
NH
(β-D-Gal)$_2$(β-D-GlcNAc)$_2$(α-D-Man)$_2$(β-D-Man)
O
OH
AcNH
O

H-Val-Ile-Thr-Ala-Phe-NH CO-Glu-Gly-Lys-OH
(102)

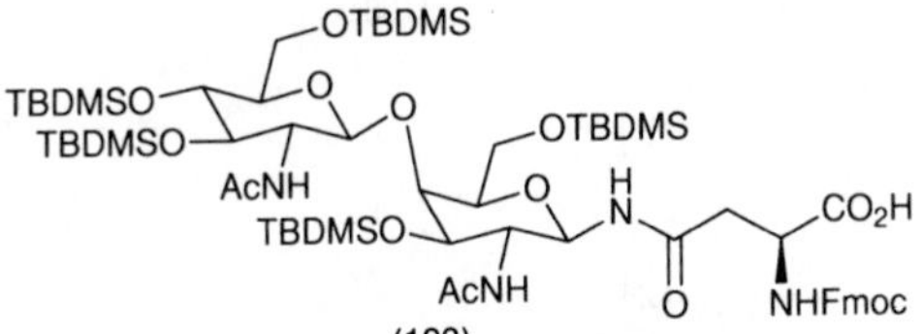

OTBDMS
TBDMSO
TBDMSO
O
O
OTBDMS
AcNH
O
H
N
CO$_2$H
TBDMSO
AcNH
O
NHFmoc
(103)

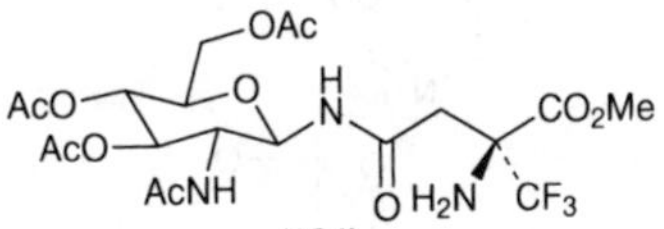

OAc
AcO
O
H
N
CO$_2$Me
AcO
AcNH
O
H$_2$N CF$_3$
(104)

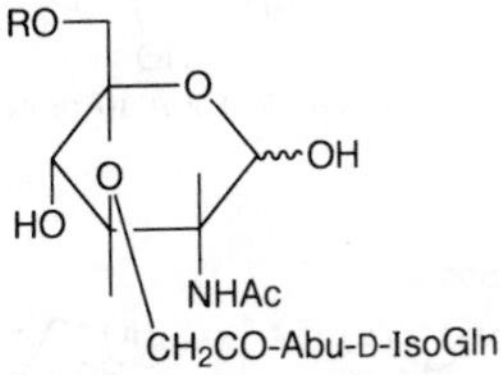

RO
O
O
OH
HO
NHAc
CH$_2$CO-Abu-D-IsoGln
(105) R = Me(CH$_2$)$_{16}$CO
(106) R = [Me(CH$_2$)$_{13}$]$_2$CHCO

β-turns were found in both families encompassing the Pro-Xxx and Pro-Xxx(GlcNAc) residues respectively, and no great differences in conformations were found.

3.3.3 C-Glycosides and Other Neoglycoconjugates – Increased activity in this area justifies a new sub-section this year. Replacement of the oxygen atom in O-glycosidic peptides by a methylene group has been reported before, but direct replacement of the asparagine N-link by methylene has been reported for the first time this year. An example[192] of this replacement is the C-glycoside (107) prepared *via* a tandem Horner-Emmons-Wadsworth olefination and a Michael addition between an aspartyl β-ketophosphonate and a 4,6-O-benzylidene GlcNAc sugar. Leu-Enkephalin analogues, such as (108) or its analogue with carbohydrate N-alkylated onto Gly^3, have been synthesised[193] using N-glycated glycine as building blocks. As N-alkylated derivatives they give rise to *cis-trans* isomers which can be estimated by NMR techniques. The α-carbon of Gly has been the link position used[194] to provide protected building blocks such as (109). A solid phase synthesis[195] of an analogue (110) of mouse Hb (67–76) with a metabolically stable C-glycoside link has been carried out using an azido acid as building block. A stereoselective synthesis[196] of the N-alkylated glycine derivative (111) provides a building block with a C-glycosidic linkage isosteric to a Ser O-glycoside. Methylene isosteres of α- and β-galactopyranosyl-L-serine have been produced[197] from a BF$_3$·Et$_2$O catalysed coupling of tetrabenzyl α-D-galactopyranosyl trichloroacetimidate with a silyl enol ether carrying an oxazolidine ring. The C-analogue (112) of 2-acetylamino-2-

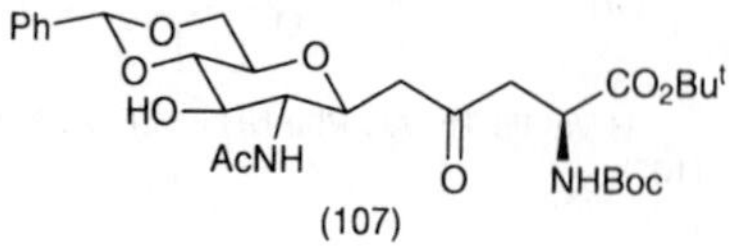

H-Tyr-Gly-Gly-Phe-Leu-OH

(107)

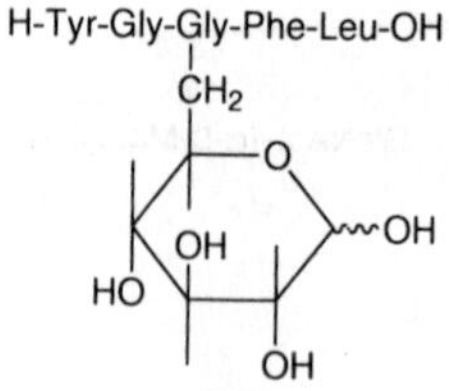

(108)

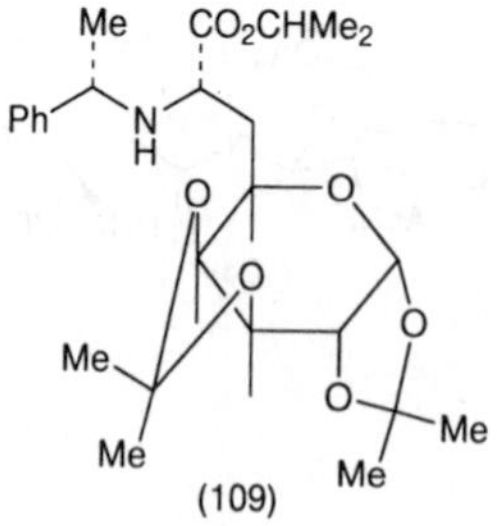

(109)

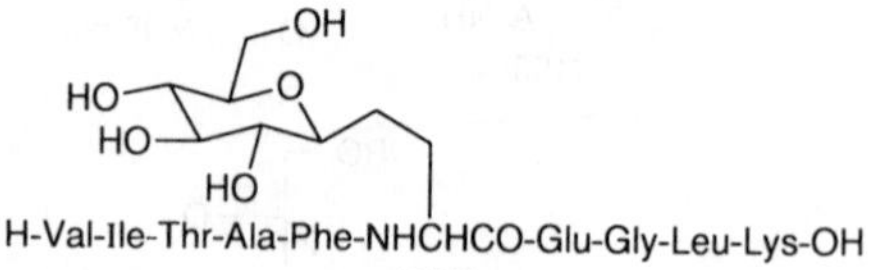

H-Val-Ile-Thr-Ala-Phe-NHCHCO-Glu-Gly-Leu-Lys-OH

(110)

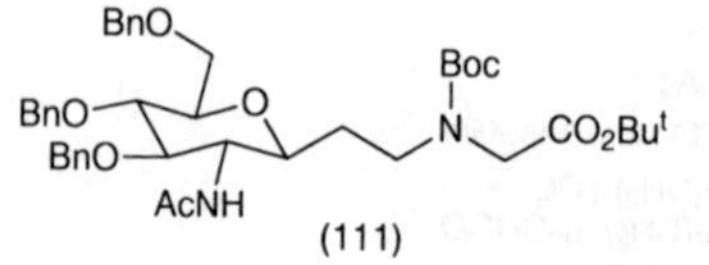

(111)

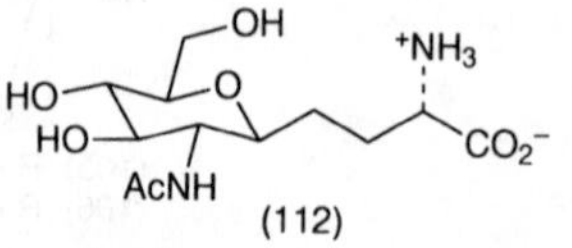

(112)

deoxy-β-D-glucopyranosyl L- and D-serine has been obtained from glucosamine.[198]

A mimic[199] of natural asparagine glycosylation has come in the form of the asymmetric disulfide building block (113), while a sulfur link has been incorporated[200] into a site selective glycosylating agent (114) which reacts with side-chain lysine amino groups. Conjugation of small carbohydrates to peptides can be carried out using maleimido-thiol chemistry as described[201] for the maltose derivatised peptide (115). Neoglycoconjugates such as (116) have been

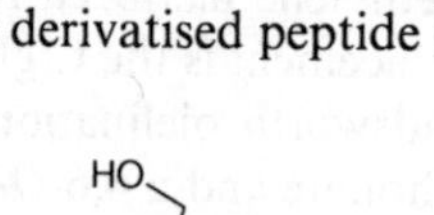

(113)

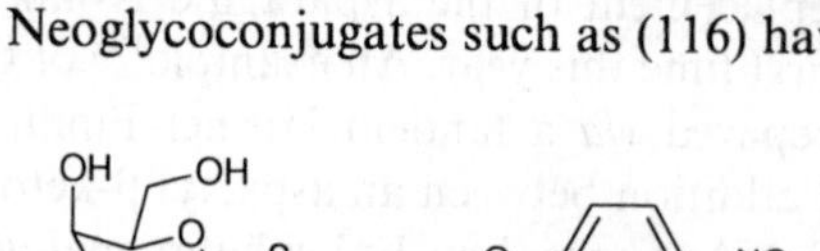

(114)

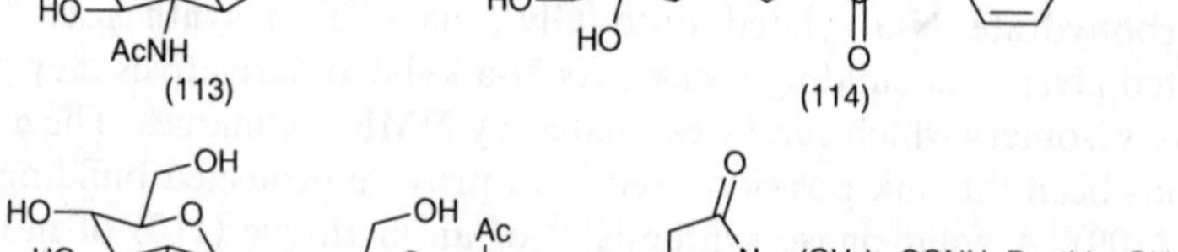

N—(CH$_2$)$_5$·CONH–Peptide–OH

(115)

CO-(Leu)$_3$-Ile-Leu-Gly-Val-(Leu)$_2$-Gly-Val-(Leu)$_2$-Thr-(Leu)$_8$-OH

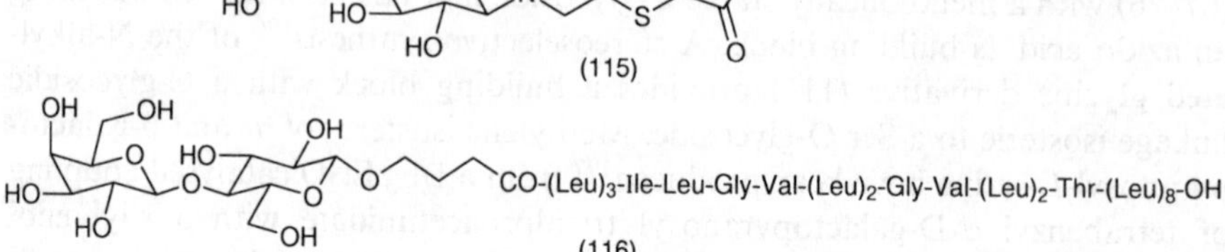

(116)

synthesised[202] on solid support, and contain a dimerisation motif of glyco-phorin A transmembrane domain.

3.4 Lipopeptides – Nostofungicidine (117), a novel lipopeptide from the blue-green alga *Nostoc commune*, was isolated[203] from a methanolic extract and characterised using extensive NMR techniques. Butanol extracts of the marine cyanobacterium *Lyngbya majuscula* have been found[204] to contain two new lipopeptides, carmabin A (118) and B (119), both characterised again using NMR and MS techniques. Solid state fermentation of vegetative mycelia of strain YL-03706F has yielded[205] antibiotic YM-170320 (120).

Hydrazone chemical ligation has been instrumental[206] in linking together

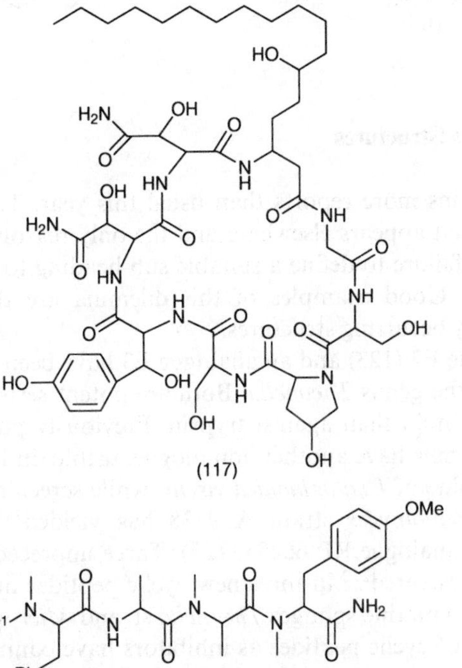

(117)

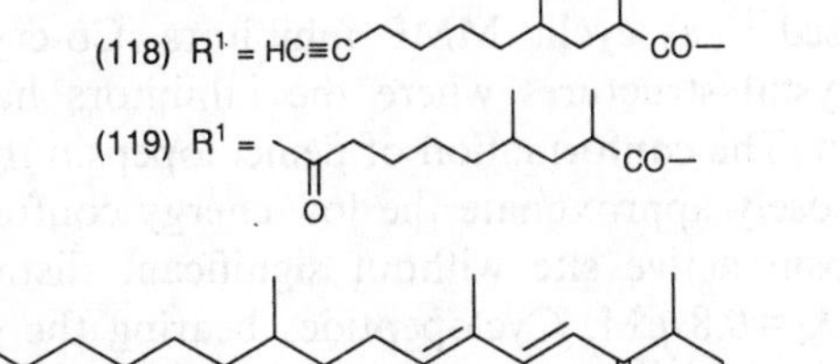

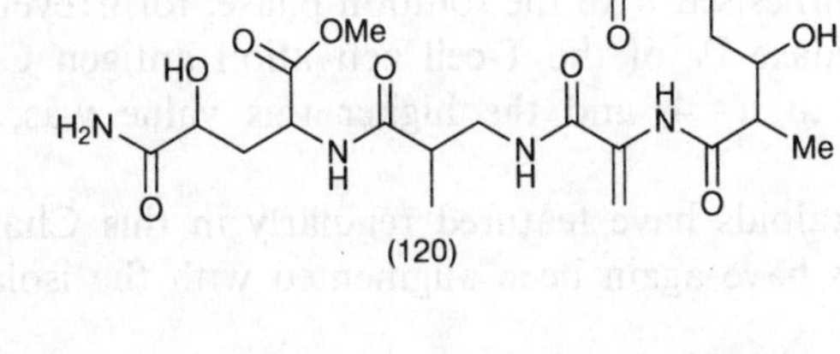

(118) R¹ = HC≡C

(119) R¹ =

(120)

hydrazine lysine side-chains with lipopeptide aldehydes while allylic esters and their mild removal by Pd^0 has ensured[207] the preparation of the N-terminus of a human $G_{\alpha o}$ protein as exemplified by (121). Basic and acid-labile lipopeptides such as (122) (C-terminus of human-N-Ras protein) can be readily synthesised[208] without hydrolysis and β-elimination if the p-acetoxybenzyloxycarbonyl (AcOZ) urethane protecting group is used. This can be removed by enzymes, *e.g.* acetyl esterases from oranges.

CO-Gly-Cys-Thr-Leu-Ser-Ala-OH
       |
       CO

(121)

OC
|
H-Gly-Cys-Met-Gly-Leu-Pro-Cys-OMe

(122)

4   Miscellaneous Structures

This section contains more reports than usual this year. The work is no less important than what appears elsewhere and the only reason it appears in this sub-section is our failure to define a suitable sub-heading to accommodate the diverse structures. Good examples of this dilemma are the following four reports of naturally occurring structures.

Cyclotheonamide E2 (123) and an analogue E3 have been isolated[209] from a marine sponge of the genus *Theonella*. Both are potent serine proteases, more active against thrombin than against trypsin. Previously published structures of usitiloxins A–D now have another homologue, usitiloxin F (124), isolated[210] as a minor metabolite of *Ustilaginoidea virens*, while screening for neurotensin antagonists in *Streptomyces* strain A 9738 has yielded[211] the constrained neurotensin (8-13) analogue RP 66453 (125). Three unprecedented amino acid units have been discovered[212] in three new cyclic peptides microsclerodermins C–E (126–128) from marine sponges *Theonella* sp and *Microscleroderma* sp.

The application of cyclic peptides as inhibitors have conjured up some new type of macrocycles. Thus the cyclic inhibitors SC-903 (129) and SE-205 (130) have been synthesised[213] as cyclic MMP inhibitors. Co-crystallisation with MMP-3 showed crystal structures where the inhibitors had bound in the predicted orientation. The conformation of penicillopepsin inhibitor (131) has been shown[214] to closely approximate the low-energy conformation required for the penicillopepsin active site without significant distortion. Hence its binding affinity of $K_i = 0.8$ μM. Cyclopeptides bearing the general structure (132) have been synthesised[215] in the solution phase, for irreversible-inhibition of dipeptidyl peptidase IV of the T-cell activation antigen CD26. Ring size varied from n = 2 to n = 4, and the higher this value was, the better the inhibition.

Cyclopeptide alkaloids have featured regularly in this Chapter. Naturally occurring examples have again been augmented with the isolation[216] of two

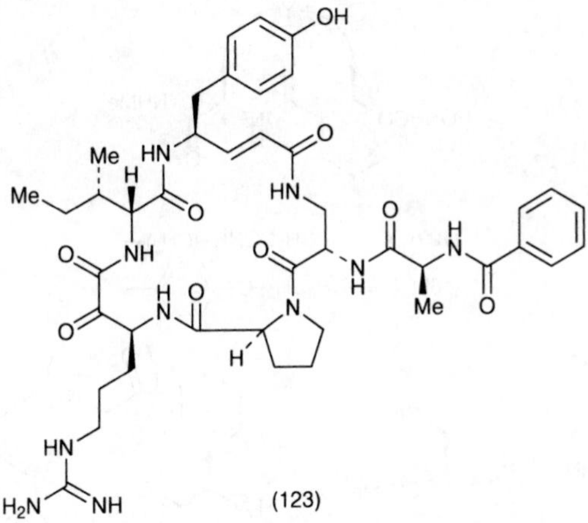

(123)

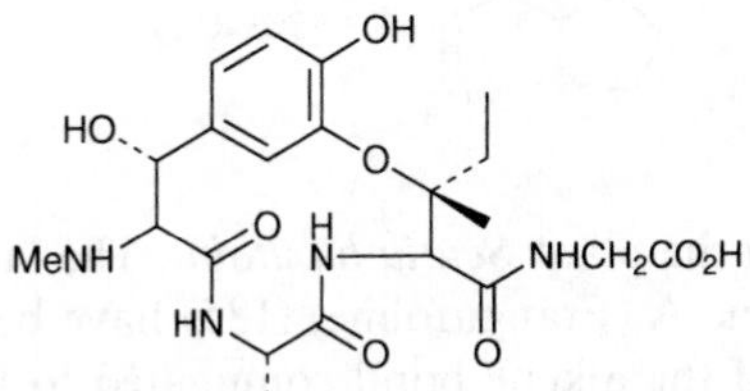

(124)

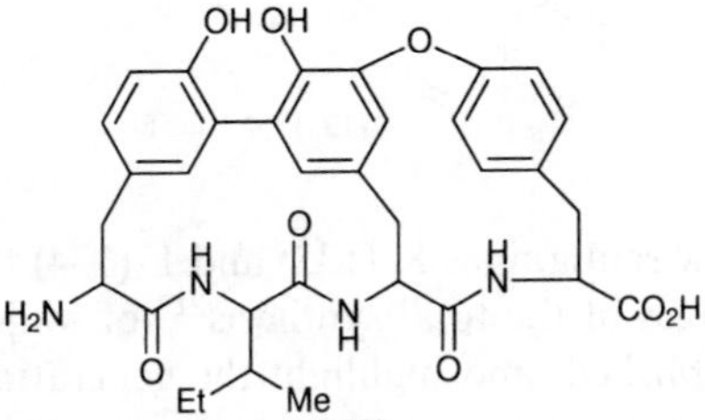

(125)

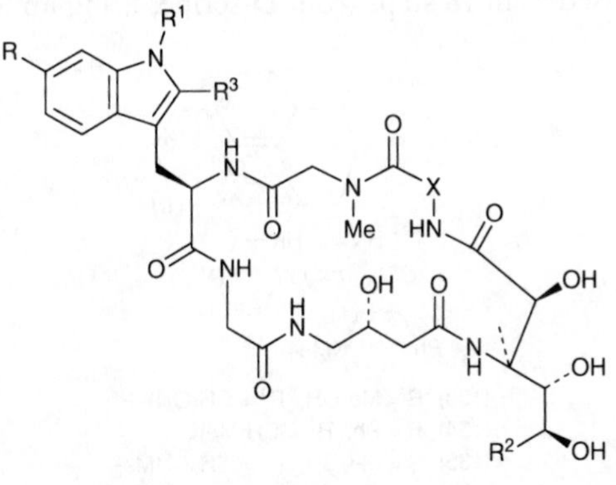

(126) R = Cl, R^1 = CONH$_2$, R^2 = PhCH=CH–CH$_2$, R^3 = H, X = –CH$_2$

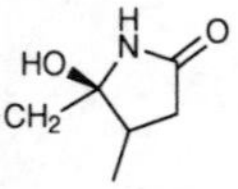

(127) R = Cl, R^1 = R^3 = H, R^2 = PhCH=CH–CH$_2$, X = –CH$_2$

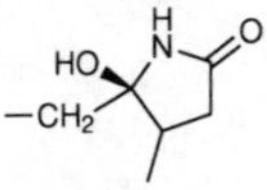

(128) R = R^1 = H, R^2 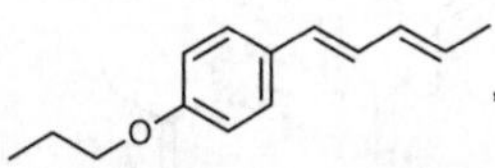 = , R^3 = CO$_2$H, X =

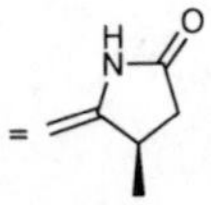

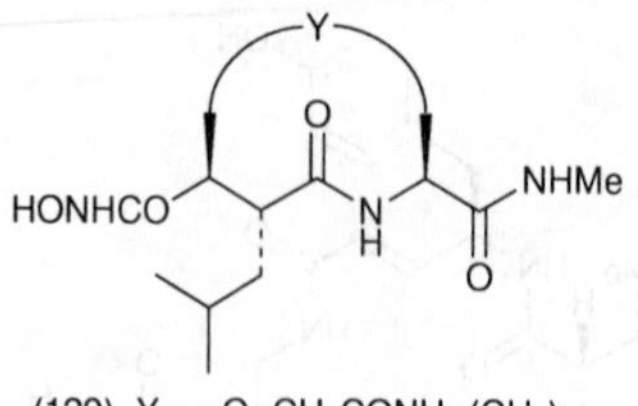

(129) Y = –O–CH$_2$CONH–(CH$_2$)$_3$–

(130) Y = –CH$_2$CH$_2$–O–

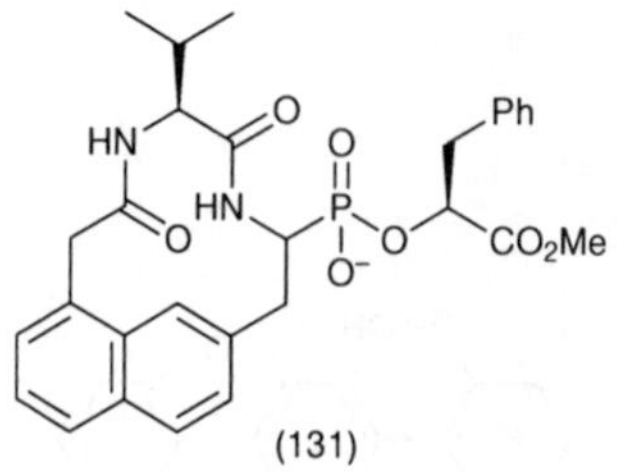

(131)

2CF$_3$CO$_2^-$

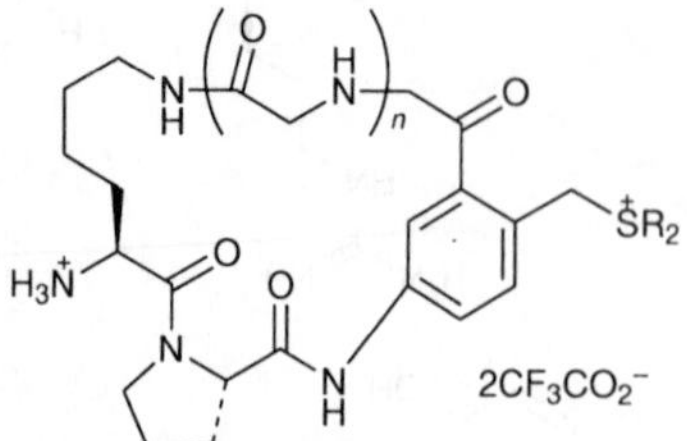

(132)

new scutianines K (133) and L (134) from the plant *Scutia buxifolia*. The later stages of the total synthesis[217] of sanjoinine A (Frangufoline) (135) have been published, and highlight the generation of the alkene bond conjugated to the aromatic ring. Another analogue, sanjoinine G1(136) and its C-11 epimer have also been synthesised[218] in 18 steps from D-serine, using an S$_N$Ar reaction with

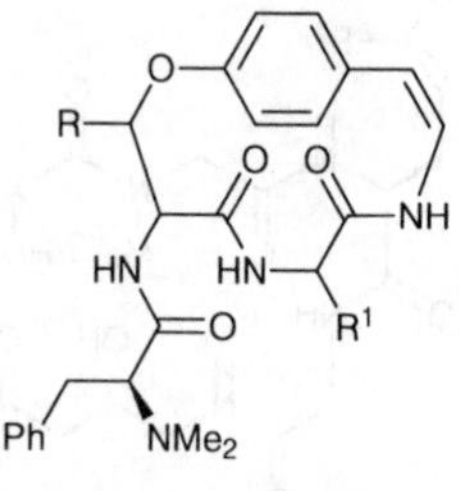

(133) R = Me$_2$CH, R^1 = CH(OH)–Ph
(134) R = Ph, R^1 = CH(Me)Et
(135) R = Me$_2$CH, R^1 = CH$_2$CHMe$_2$

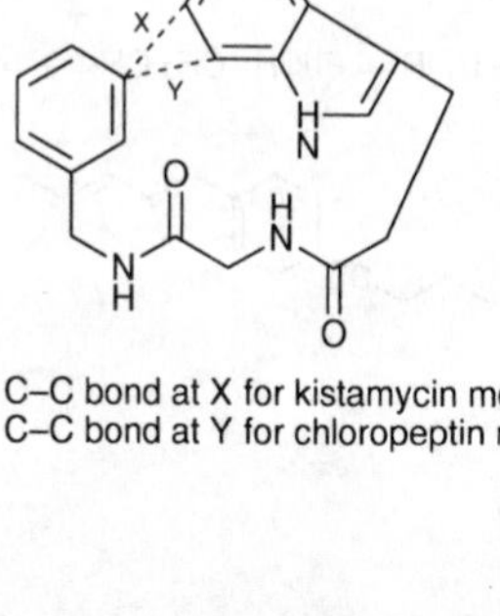

(136) C–C bond at X for kistamycin model
(137) C–C bond at Y for chloropeptin model

4-fluorobenzonitrile, and macrocyclisation using pentafluorophenyl ester activation. The key steps in this general approach to 14-membered cyclopeptide alkaloids have also been featured in another report.[219] The so-called 'eastern' sub structures of kistamycin and chloropeptin I, II have been synthesised[220] *via* the model compounds (136) and (137). Critical steps in the syntheses were the linking up of the phenyl ring to the indole phenyl residue *via* brominated aromatic precursors mediated by $Ni(Ph_3P)_2Cl_2/Zn/Ph_3P/DMF$.

Macrocyclisation *via* the S_NAr methodology to make aryl-aryl bonds, has been extended to include aryl-alkyl ether bond formation as illustrated[221] by the multistep formation of (138) *via* Scheme 11. Similar philosophy has been used in the synthesis[222] of peptidomimetic β-turns on a Rink resin support as summarised in Scheme 12, which could be extended to the generation of β-turn libraries.

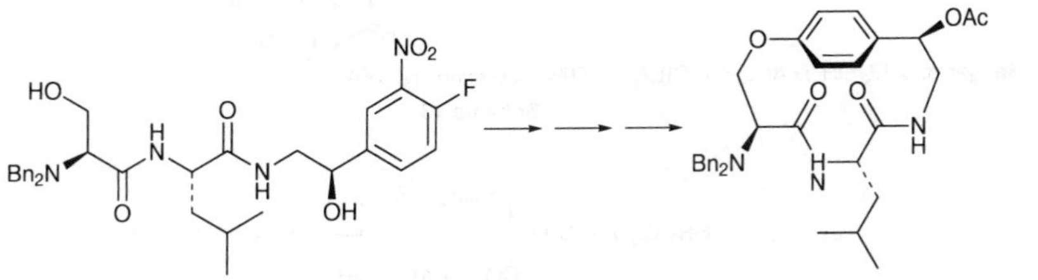

Scheme 11

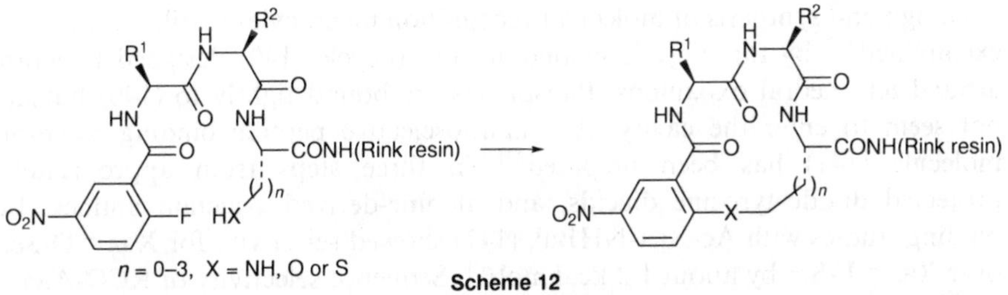

(138)

$n = 0-3$, $X = NH$, O or S

Scheme 12

Amongst the many advantages of having an authoritative review[223] written on lantibiotics and microcins are the structural listings of these compounds into categories (A), (B), (C), (D) and (E), so that the reader can appreciate the wealth of structures generated *via* the post-translational modification of polypeptides. A straightforward *in situ* conversion of serine into bromoalanine using $P(Ph)_3/CBr_4$ has contributed to an efficient solid phase synthesis[224] of known lanthionine-containing cyclic peptides as summarised in Scheme 13. In a similar scheme,[225] intramolecular substitution of the chloro-group of β-chloro-alanine with the thiol group of cysteine has been accomplished in good yields. The β-chloroalanine in this case was prepared on-resin by converting a TBDMS-protected OH group of inhomoserine into a chloro group with triphenylphosphine dichloride. A cyclic thio-ether peptide (139) which binds anti-cardiolipin antibodies has been synthesised[226] on a HMPB/MBHA resin.

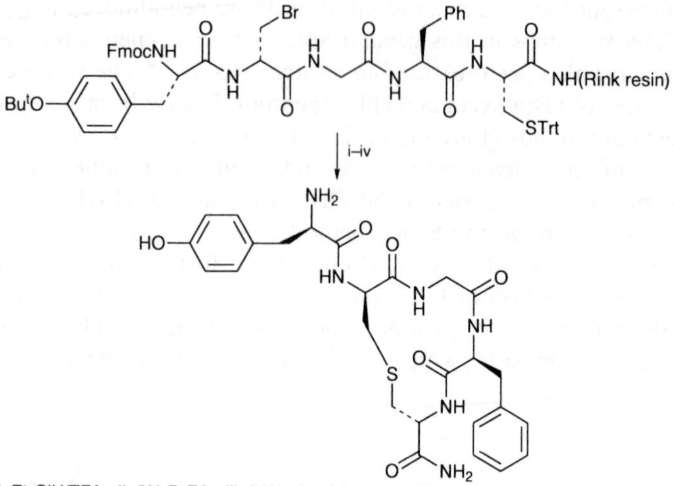

Reagents: i, Et₃SiH/TFA; ii, 5% DIEA; iii, 20% piperidine; iv, TFA

Scheme 13

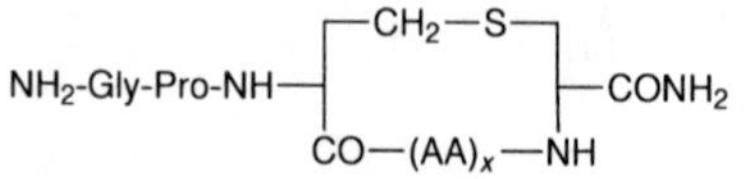

NH₂-Gly-Pro-NH—[CH₂—S]—CONH₂
                  CO—(AA)ₓ—NH

AA = Ile-(Leu)₂-Ala-αMePro-Asp-Arg-

(139)

Design and synthesis of molecular recognition macrocyles is still buoyant, as exemplified[227] by the bicyclic guanidium macrocycle (140) designed to wrap around tetrahedral oxoanions. Phosphates are bound tightly to (140) but do not seem to enter the cavity. An enantioselective peptide binding receptor molecule (141) has been prepared[228] in three steps from appropriately protected diiodo-tyrosine diacids, and alanine-derived γ-lactam amines. In binding studies with Ac-Xaa-NHBuᵗ, (141) showed selectivity for Xaa = D-Ser over Xaa = L-Ser by about 1.2 kcal mol⁻¹. Sequence selectivity of RCO-Aaa³-Aaa²-Aaa¹ with the binding receptor (142) has also been investigated,[229] and showed sequence specificity including a special affinity for R = cyclopropyl or Me₂N in the acylated tripeptides.

Full details[230] of the protocols necessary to synthesise the tricyclic designer somatostatin antagonist molecule (143) have now been reported. Five-dimensional orthogonal amino group protection had to be utilised. After last year's reported synthesis, the crystal structures[231] and self-assembling properties of three adamantane (Adm) containing cyclodepsipeptides have been analysed. While the 18-membered cyclo (Adm-Ser)₂ showed a simple structure with no internal H-bonds, the higher structures cyclo(Adm-Ser-Val)₂ and cyclo(Adm-Ser-Ser)₂ were more interesting. The Val analogue had a dimeric structure with water-filled channels, while all Ser analogues revealed a unique antiparallel β-sheet structure. Synthesis, crystal structures, conformational studies and molecular recognition studies have been reported[232] for the aromatic bridged

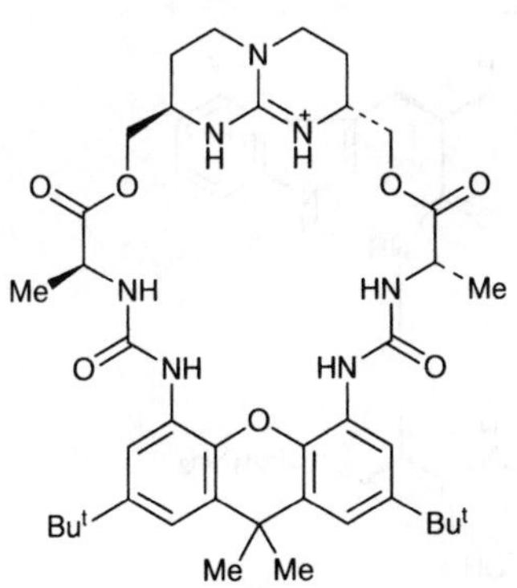
(140)

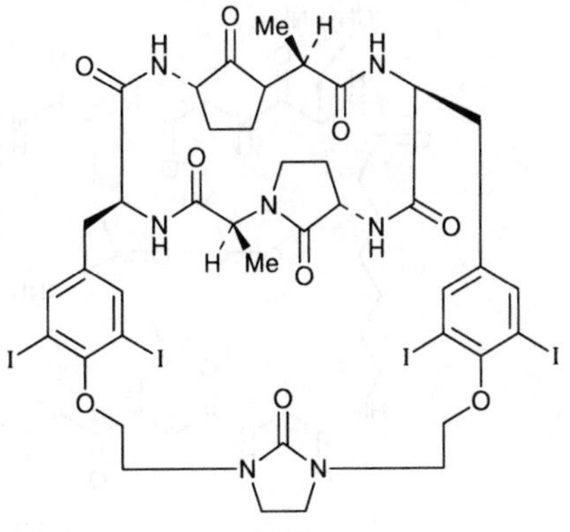
(141)

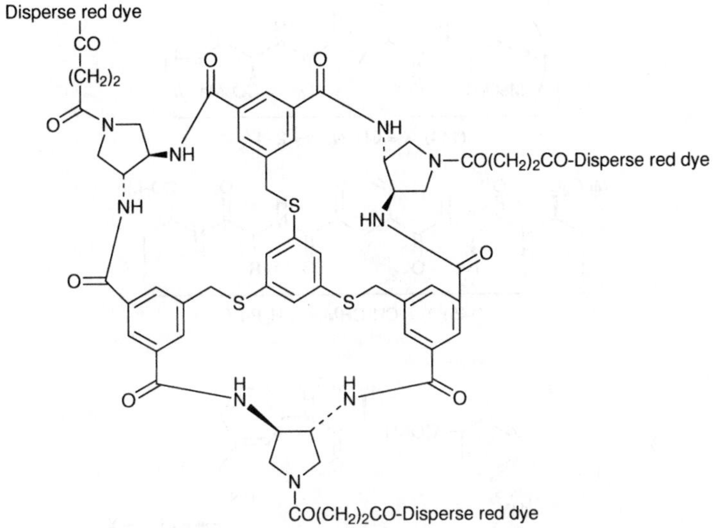
Disperse red dye
N—CO(CH2)2CO-Disperse red dye
CO(CH2)2CO-Disperse red dye
(142)

cystine peptides (144) and (145). Compounds based on (144) and (145) show a β-turn-like conformation in solution. In host-guest complexation studies, molecule (144) X = N, n = 2 showed the maximum affinity of $K_{assoc} = 3.69 \times 10^2\ M^{-1}$ with glutaric acid, $HOOC(CH_2)_3COOH$. The amino-benzoyl-lysine cyclotetramer (146) has been included[233] as a template-assembled synthetic 4α-helix peptide bundle, to ascertain its capacity to enhance α-helicity in an amphiphilic tridecapeptide. Helicity is enhanced by 64-75% when the peptide is connected at points X to the template.

In the search for materials which can enantioselectively bind small molecules, thus providing the possibility of chiral separations, a 60 member library of resolving resins has been constructed[234] from the template (147). And finally, to end the Chapter this year two sets of structures that really fit into this only as 'wild-card' invitees. Cyclic oligocarbamate libraries such as (148)

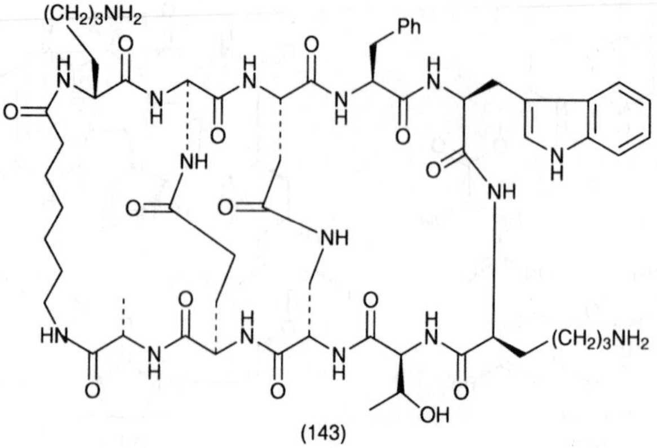

(143)

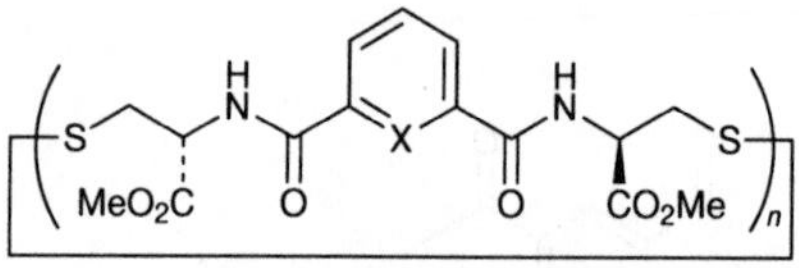

(144) X = CH, N; *n* = 2–4

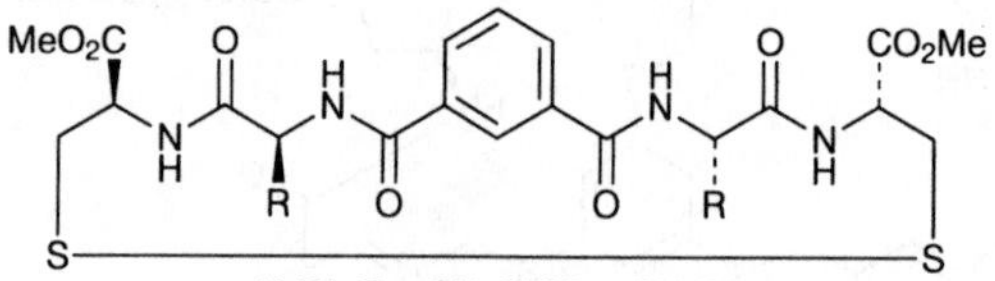

(145) R = CH₂CHMe₂, CH₂Ph

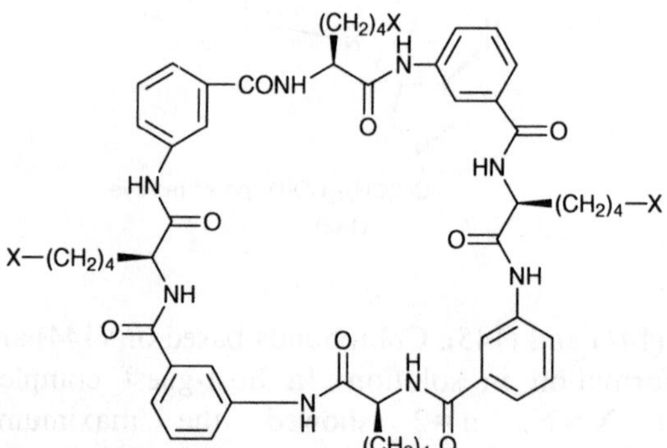

(146)

have been set up[235] using 'one bead, one peptide' protocols, and several classes of ligands for integrin GPIIb/IIIa binding were discovered. Two cyclic ligands had activities within a factor of 3 of the snake venom protein kistrin. With a view also to adapt for combinatorial library work, the cyclic oligocarbamate series (149) n = 1–4 have also been synthesised[236] using automated solid phase techniques, using carbonates of β-amino alcohols with Fmoc protocols.

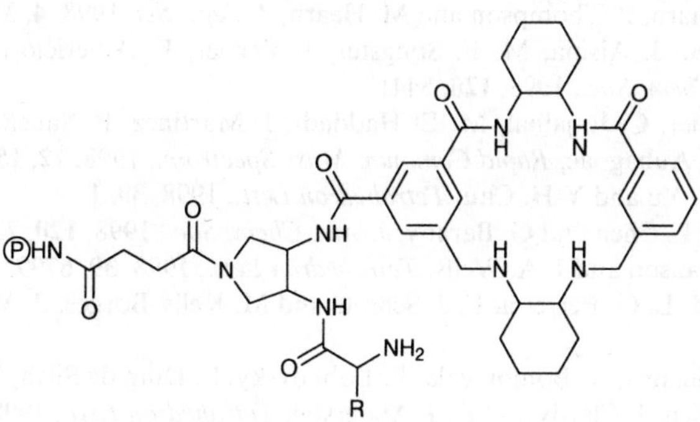

(147)

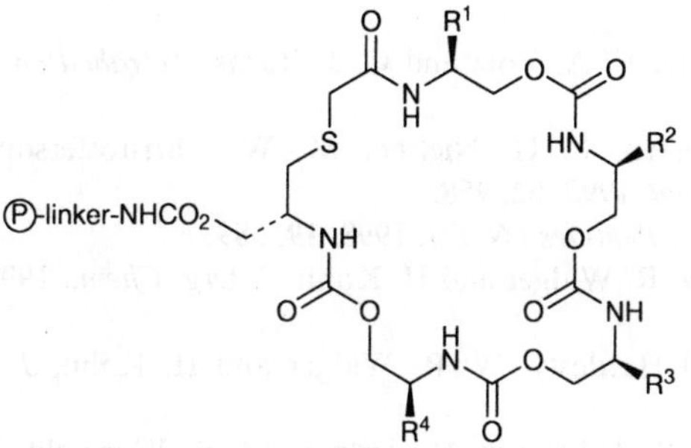

(148)

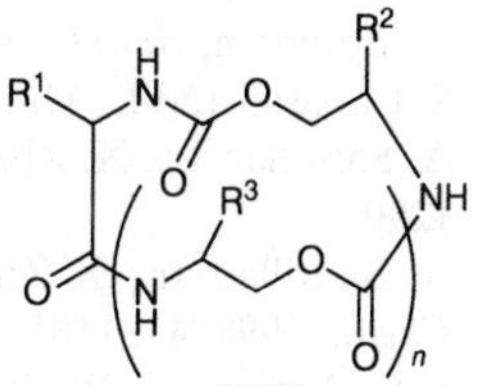

(149)

References

1. CA Selects on Amino Acids, Peptides and Proteins, published by the American Chemical Society and Chemical Abstracts Service, Columbus, Ohio.
2. 'Peptide Science – Present and Future' – Proceedings of the 1st International Peptide Symposium, Kyoto, Japan, ed. Y Shimonishi, Kluwer Academic, 1999, 842 pp.
3. 'Peptides 1998', Proceedings of the 25th European Peptide Symposium at Budapest, eds. S. Bajusz and F. Hudecz, Akadémiai Kiadó Budapest.1999, 902 pp.
4. A. J. Zhang and K Burgess, *Angew. Chem. Int. Ed.*, 1999, **38**, 634.
5. K. Rueck-Braun, *Nachr. Chem. Teck. Lab.*, 1998, **46**, 1182.
6. J. Hlavacek, *Chem. Listy*, 1998, **92**, 486.
7. H. Morita, *Int.Congr. Ser.*,1998, **1157**, 467.
8. R. P. McGeary and D. F. Fairlie, *Curr. Opin. Drug Discovery Dev.*, 1998, **1**, 208.
9. M. A. Ciufolini and N. Xi, *Chem. Soc. Rev.*, 1998, **27**, 437.
10. Y. Shao, W. Lu and S. B. H. Kent, *Tetrahedron Lett.*, 1998, **39**, 3911.
11. K. Haas, W. Ponikwar, H. Noth and W. Beck, *Angew. Chem. Int. Ed.*, 1998, **37**, 1086.
12. D. A. Annis, O. Helluin and E. N. Jacobsen, *Angew. Chem. Int. Ed.*,1998, **37**, 1907.
13. J. Klose, A. Ehrlich and M. Bienert, *Lett. Pept. Sci.*, 1998, **5**, 129.

14. V. Cavallaro, P. Thompson and M. Hearn, *J. Pept. Sci.* 1998, **4**, 335.
15. K. Jensen, J. Alsina, M. F. Songster, J. Vagner, F. Albericio and G. Barany. *J. Am. Chem. Soc.*, 1998, **120**, 5441.
16. F. Cavelier, C. Enjalbal, M. El Haddadi, J. Martinez, P. Sanchez, J. Verduccil and J. L. Aubagnac, *Rapid Commun. Mass Spectrom.*, 1998, **12**, 1585.
17. Z. Yu, C. Yu and Y-H. Chu, *Tetrahedron Lett.*, 1998, **39**, 1.
18. I. Annis, L. Chen and G. Barany, *J. Am. Chem. Soc.*, 1998, **120**, 7226.
19. D.A. Erlanson and J .A. Wells. *Tetrahedron Lett.*, 1998, **39**, 6799.
20. X. Fu, M. L. G. Ferreira, F. J. Schnitz and M. Kelly-Borges, *J. Nat. Prod.*, 1998, **61**, 1226.
21. D. E. Williams, K. Bombuwala, E. Lobkovsky, E. Dilip de Silva, V Karunaratne, T. M. Allen, J. Clardy and R. J. Anderson, *Tetrahedron Lett.*, 1998, **39**, 9579.
22. S. D. Bull, S. G. Davies, R. G. Parkin and F. Sanchez-Sancho, *J. Chem. Soc. Perkin Trans. 1*, 1998, 2313.
23. C. Hulme, M. M. Morrissette, F. A. Volz and C. J. Burns, *Tetrahedron Lett.*, 1998, **39**, 1113.
24. U. Anthoni, C. Christophersen, P. H. Nielsen, M. W. Christofferson and D. Sorensen, *Acta Chem. Scand.* 1998, **52**, 958.
25. S. Capasso and L. Mazzarella, *Peptides (N.Y.)*, 1998, **19**, 389.
26. A. Santillan Jr., X. Zhang, W. R. Widger and H. Kohn, *J. Org. Chem.*, 1998, **63**, 1290.
27. A. Santillan Jr., X Zhang, J Hardesty, W. R. Widger and H. Kohn, *J. Med. Chem.*, 1998, **41**, 1185.
28. J. C. Estevez, J. W. Burton, R. J. Estevez, H. Ardron, M. R. Wormald, R. A. Dwek, D. Brown and G. W. J. Fleet, *Tetrahedron: Asymmetry*, 1998, **9**, 2137.
29. S. Capasso, A. Vergara and L. Mazzarella, *J. Am. Chem. Soc.*, 1998, **120**, 1990.
30. D. Besser, G. Greiner and S. Reissmann, *Lett, Pept. Sci.*,1998, **5**, 299.
31. M. del Frenso, J. Alsina, M. Royo, G. Barany and F. Albericio, *Tetrahedron Lett.*, 1998, **39**, 2639.
32. S. D. Bull, S. G. Davies, S. W. Epstein and J. V. A. Ouzman, *Tetrahedron: Asymmetry* 1998, **9**, 2795.
33. B. A. Burkett, C. L. L. Chai and D. C. R. Hockless, *Aust. J. Chem.*,1998, **51**, 993.
34. Y. Ling and C. Lifshitz, *J. Mass Spectrom.*, 1998, **33**, 25.
35. A. Demeter, T. Fodor and J. Fischer, *J. Mol. Struct.*,1998, **471**, 161.
36. Z. Dzakula, N. Juranic, M. L. DeRider, W. M. Westler, S. Macura and J. L. Markley, *J. Magn. Reson.*, 1998, **135**, 454.
37. C, Allott, H. Adams, C. A. Hunter, J. A. Thomas, P. L. Bernad Jr, and C. Rotger, *Chem. Commun.*, 1998, 2449.
38. E. F. Kogut, J. C. Thoen and M. A. Lipton, *J. Org. Chem.*, 1998, **63**, 4604.
39. J. F. Pons. J. L. Fauchere, F. Larraty, A. Molla and R. Lazaro, *Eur. J. Org. Chem.*, 1998, 853.
40. M. V. R. Reddy, M. K. Harper and D. J. Faulkner, *Tetrahedron*, 1998, **54**, 10649.
41. K. Shreder, L. Zhang, T. Dang, T. L. Yaksh, H. Umeno, R. DeHaven, J. Daubert and M. Goodman, *J. Med. Chem.*, 1998, **41**, 2631.
42. B. S. Yun, I. J. Ryoo, I. J. K. Lee and I. D. Yoo, *Tetrahedron Lett.*, 1998, **39**, 993; *Tetrahedron*, 1998, **54**, 15155.
43. T. D. Clark, L. K. Buchler and M. R. Ghadiri, *J. Am. Chem. Soc.*, 1998, **120**, 651.
44. J. Malicka, M. Groth, C. Czaplewski, R. Kasprzykowska, A. Liwo, L. Lankiewicz and W. Wiczk, *Lett. Pept. Sci.*, 1998, **5**, 445.

45. G. Zonotti, M. Saviano, T. Tancredi, F. Rossi, C. Pedone and E. Benedetti, *J. Pept. Res.*, 1998, **51**, 460.

46. D. Delforge, M. Art, J. Dewelle, M. Raes and J. Remacle, *Lett. Pept. Sci.*, 1998, **5**, 87.

47. V. Breex, P. Verheyden and D. Tourwe, *Lett. Pept. Sci.*, 1998, **5**, 67.

48. E. Vass, M. Kurz, R. K. Konat and M. Hollosi, *Spectrochim. Acta, 1998,* **54A**, 773.

49. H. Ito, *J. Chem. Phys*, 1998, **108**, 93.

50. J. Alsina, F. Rabanal, C. Chiva, E. Giralt and F. Albericio, *Tetrahedron*, 1998, **54**, 10125.

51. L. C-Ming, T. N-Hua, M. Qing, Z. H-Lan, H. X-Jiang, L. H-Ling and Z. Jun, *Phytochemistry* 1998, **47**, 1293.

52. R. Zhang, C. Zhou, N. Tan and J. Zhou, *Yunnan Zhiwu Yanjiu*, 1998, **20**, 105.

53. Y-C. Wang, N-H. Tan, J. Zhou and H-M. Wu, *Phytochemistry* 1998, **49**, 1453.

54. P. A. Gineco and M. Reilly, *Tetrahedron Lett.*, 1998, **39**, 8925.

55. J. L. Matthews, K. Gademann, B. Jaun and D. Seebach, *J. Chem. Soc., Perkin Trans. 1*, 1998, 3331.

56. P. Romanovskis and A. F. Spatola, *J. Pept. Res.*, 1998, **52**, 356.

57. G. Mezo, N. Mihala, G. Koczan and F. Hudecz, *Tetrahedron*, 1998, **54**, 6757.

58. W. M. M. Schaaper, R. A. H. Adan, T. A. Posthuma, J. Oosterom, W-H. Gispen and R. H. Meloen, *Lett. Pept. Sci.*, 1998, **5**, 205.

59. C. Gilon, M. Huenges, B. Matha, G. Gellerman, V. Hornik, M. Afargan, O. Amitay, O. Ziv, E. Feller, A Gamliel, D. Shohat, M. Wanger, O. Arad and H. Kessler, *J. Med. Chem.* 1998, **41**, 919.

60. T-A. Tran, R-H. Mattern, M. Afargan, O. Amitay, O. Ziv, B. A. Morgan, J. E. Taylor, D. Hoyer and M. Goodman, *J. Med. Chem.*, 1998, **41**, 2679.

61. R. H. Mattern, T. A. Tran and M. Goodman, *J. Med. Chem.*, 1998, **41**, 2686.

62. M. S. Prachand, M. M. Dhingra, A. Saran, E. Coutinho, J. Bodi, H. SuliVargha and K. Medzihardszky, *J. Peptide Res.*, 1998, **51**, 251.

63. C. F. Van Der Walle, S. Bansal and D. J. Barlow, *J. Pharm. Pharmacol.*, 1998, **50**, 837.

64. S. J. Hocart, R. Jain, W. A. Murphy, J. E. Taylor, B. Morgan and D. H. Coy, *J. Med. Chem.*, 1998, **41**, 1146.

65. G. Alagona, C. Ghio and S. Monti, *THEOCHEM*, 1998, **426**, 339.

66. D. Leipert, D. Nopper, M Bauser, G. Gauglitz and G. Jung, *Angew. Chem., Int. Ed.*, 1998, **37**, 3308.

67. C-M. Li, N-H. Tan, H-L. Zheng, M. Qing, X-J. Hao, Y-N. He and Z. Jun, *Phytochemistry*, 1998, **48**, 555.

68. R. B. Bates, S. Caldera and M. D. Ruane, *J. Nat. Prod.*, 1998, **61**, 405.

69. D. Limal, J.P.Briand, P. Dalbon and M. Jolivet, *J. Pept. Res.*, 1998, **52**, 121.

70. C. Baraguey, C. Auvin-Guette, A. Blond, F. Cavelier, F. Lenzenven, J. L. Pousset, and B. Bodo, *J. Chem., Perkin Trans. 1*, 1998, 3033.

71. G. N. Belofsky, J. B. Gloer, D. T. Wicklow and P. F. Dowd, *Tetrahedron Lett*, 1998, **39**, 5497.

72. A. Randazzo, F. Dal Piaz, S. Orru, C. Debitus, C. Roussakis, P. Pucci and L. Gomez-Paloma, *Eur, J. Org. Chem.* 1998, 2659.

73. B. Picur, M. Lisowksi and I. Z. Siemion, *Lett., Pept. Sci,*. 1998, **5**, 183.

74. R. Zhang, C. Zou, Y. He, N. Tan and J. Zhou, *Yunnan Zhiwu Yanjiu.* 1997, **19**, 304.

75. X. Zang, Z. G. Yu and Y.H. Chu, *Bioorg. Med. Chem. Lett.*, 1998, **8**, 2327.

76. S. L. Belagali, H. K. Kumar, B. Poojary and M. Himaja, *Chim. Acta Tuc.*, 1998, **26**, 59.

77. M. E. Polaskova, N. J. Ede and J. N. Lambert, *Aust. J. Chem.*, 1998, **51**, 535.

78. I. Wirkus-Romanowska, M. Ciurak, H. Miecznikkowska, K. Rolka, M. Janusz, S. Szymaniec, K. Krukowska, J. Lisowski and G. Kupryszewski, *Pol. J. Chem*, 1998, **72**, 2394.

79. C. Bisang, L. Jiang, E. Freund, F. Emery, C. Bauch, H. Matile, G. Pluschke and J. A. Robinson, *J. Am. Chem. Soc.*, 1998, **120**, 7439.

80. P. Amodeo, G. Saviano, G. Borin, A. Calderan, P. Ruzza and T. Tancredi, *J. Peptide Res.* 1998, **51**, 180.

81. P. Raman, S.S. Stokes, Y.M. Angell, G. Flentke and D. H. Rich, *J.Org. Chem.*, 1998, **63**, 5734.

82. M. Husak, B. Kratochil, M. Buchta, L. Cvak and A. Jegaron, *Collect. Czech. Chem. Commun.*, 1998, **63**, 115.

83. M. Buchta, A. Jegorov, L. Cvak, V. Havlicek, M. Budesinsky and P. Sedmera, *Phytochemistry*, 1998, **48**, 1195.

84. B. Cerny, A. Jegorov, J. Polivkova, P. Sedmera and V. Havlicek, *J. Labelled, Compd. Radiopharm.*, 1998, **41**, 267.

85. W.T. Liu, K. Marat, Y. Ren, R. T. Eng and P. Y. Wong, *Clin. Biochem*, 1998, **31**, 173.

86. M. Kawai, T. Yamamoto, K. Yamada, M. Yamaguchi, S. Kurobe, H. Yamamura, S. Araki, Y. Butsugan, K. Kobayashi, R. Katakai, K. Saito and T. Nakajima, *Lett. Pept. Sci.*, 1998, **5**, 5.

87. K. Yamada, H. Ozaki, N. Kanda, H, Yamamura, S. Araki and M. Kawai, *J. Chem. Soc., Perkin Trans. 1*, 1998, 3999.

88. G. Tuchscherer, C. Lehmann and M. Mathieu, *Angew. Chem. Int. Ed.*, 1998, **37**, 2990.

89. P. Ruzza, A. Donella-Deana, A. Calderan, G. Zanotti, L. Cesaro, L. A. Pinna and G. Borin, *J. Pept. Sci.*, 1998, **4**, 33.

90. H. Uemoto, Y. Yahiro, H. Shigemori, M. Tsuda, T. Takao, Y. Shimonishi and J. Kobayashi, *Tetrahedron*, 1998, **54**, 6719.

91. R. Banker and S. Carmeli, *J. Nat. Prod*, 1998, **61**, 1248.

92. A. Rudi, M. Aknin, E. M. Gaydou and Y. Kashman, *Tetrahedron*, 1998, **54**, 13203.

93. X. Fu, T. Do, F. J. Schmitz, V. Andrusevich and M. H. Engel, *J. Nat. Prod.*, 1998, **61**, 1547.

94. S. Matsunaga, Y. Nogata and N. Fusetani, *J. Nat. Prod.*, 1998, **61**, 663.

95. K. Okumara, A. Ito, D. Yoshioka, C-g. Shin, *Heterocycles*, 1998, **48**, 1319.

96. D.A. Entwhistle, S.I. Jordan, J. Montgomery and G. Pattenden, *Synthesis*, 1998, 603.

97. M.C. Norley and G. Pattenden, *Tetrahedron Lett.*, 1998, **39**, 3087.

98. D. J. Freeman and G. Pattenden, *Tetrahedron Lett.*, 1998, **39**, 3251.

99. C. J. Moody and M. C. Bagley, *J. Chem. Soc., Perkin Trans. 1*, 1998, 601.

100. C. J. Moody and M.C. Bagley, *Chem. Commun.*, 1998, 2049; *Synlett.*, 1998, 361.

101. K. Umemura, H. Noda, J. Yoshimura, A. Konn, Y. Yonezawa, C-g.Chung, *Bull. Chem. Soc. Jpn.*, 1998, **71**, 1391.

102. S. Jeong, X. Chen and P. G. Harran, *J. Org. Chem.*, 1998, **63**, 8640.

103. P. Wipf, P.C. Fritch, S. J. Geib and A. M. Sefler, *J. Am. Chem. Soc.*, 1998, **120**, 4105.

104. P. Comba, R. Cussack, D. P. Fairlie, L. R. Gahan, G. R. Hanson, U. Kazmaier and A. Ramlow, *Inorg. Chem.*, 1998, **37**, 6721.
105. J. Vinsova and E. Kasafirek, *Chem. Listy*, 1998, **92**, 197.
106. S. Li, E. J. Dumdei, J. W. Blunt, M. H. G. Munro, W. T. Robinson and L. K. Parnell, *J. Nat. Prod.*, 1998, **61**, 724.
107. K. M. Jenkins, M. K. Renner, P. R. Jensen and W. Fenical, *Tetrahedron Lett.*, 1998, **39**, 2463.
108. G. R. Pettit, J. P. Xu, F. Hogan and R. L. Cerny, *Heterocycles*, 1998, **47**, 491.
109. D. P. Clark, J. Carroll, S. Naylor and P. Crews, *J. Org. Chem.*, 1998, **63**, 8757.
110. W. F. Tinto, A. J. Lough, S. McLean, W. F. Reynolds, M. Yu and W. R. Chan, *Tetrahedron*, 1998, **54**, 4451.
111. V. R. Hegde, M. S. Puar, T. M. Chan, P. Dai, P. R. Das and M. Patel, *J. Org. Chem.*, 1998, **63**, 9584.
112. A. Jegorov, P. Sedmera, V. Havlicek and V. Matha, *Phytochemistry*, 1998, **49**, 1815.
113. T. Mukhopadhyay, C. M. M. Franco, R. G. Bhat, S. N. Sawant, B. N. Ganguli, R. H. Rupp, H. W. Fehlhaber and V. Teetz, *Tetrahedron*, 1998, **54**, 7625.
114. K. Umezawa, K. Nakazawa, T. Uemura, Y. Ikeda, S. Kondo, H. Naganawa, N. Kinoshita, H. Hashizume, M. Hamada, T. Takeuchi and S. Ohba, *Tetrahedron Lett.*, 1998, **39**, 1389.
115. K. Suzuki, S. Yahara, Y. Kido, K. Nagao, Y. Hatano and M. Uyeda, *J. Antibiot.*, 1998, **51**, 999.
116. M. Kowall, J. Vater, B. Kluge, T. Stein, P. Franke and D. Ziessow, *J. Colloid Interface Sci.*, 1998, **204**, 1.
117. B. S. Moore and D. Seng, *Tetrahedron Lett.*, 1998, **39**, 3915.
118. T. M. Kamenecka and S. J. Danishefsky, *Angew. Chem. Int. Ed.*, 1998, **37**, 2993.
119. T. M. Kamenecka, and S. J. Danishefsky, *Angew. Chem. Int. Ed.*, 1998, **37**, 2995.
120. K. J. Hale, N. Jogiya and S. Manaviazar, *Tetrahedron Lett.*, 1998, **39**, 7163.
121. F. Yokokawa and T. Shioiri, *J. Org. Chem.*, 1998, **63**, 8638.
122. J. D. White, J. Hong and L. A. Robarge, *Tetrahedron Lett.*, 1998, **39**, 8779.
123. B. H. Norman, T. Hemscheidt, R. M. Schultz and S. L. Andis, *J. Org. Chem.*, 1998, **63**, 5288.
124. A. H. Fray, *Tetrahedron: Asymmetry*, 1998, **9**, 2777.
125. T. Kurome, T. Inoue, K. Takesako, I. Kato, K. Inami and T. Shiba, *J. Antibiot.*, 1998, **51**, 359.
126. U. Schmidt, A. Schumacher, J. Mittendorf and B. Riedl, *J. Pept. Res.*, 1998, **52**, 143.
127. P. Breuilles and D. Uguen, *Tetrahedron Lett.*, 1998, **39**, 3149.
128. J. Scherkenbeck, A. Harder, A.Plant, and H. Dyker, *Biorg. Med. Chem. Lett.*, 1998, **8**, 1035.
129. F. E. Dutton and B. H. Lee, *Tetrahedron Lett.*, 1998, **39**, 5313.
130. M. Miyashita, T. Nakamori, T. Murai, H. Miyagawa, M. Akamatsu and T. Ueno, *Biosci., Biotechnol., Biochem.*, 1998, **62**, 1799.
131. J. C. Barriere, E. Bacque, G. Puchault, Y. Quenet, C. Molherat, J. Cassayre and J-M. Pais, *Tetrahedron*, 1998, **54**, 12859.
132. D. L. Boger, J-H. Chen, K.W. Saionz and Q. Jin, *Bioorg. Med. Chem.*, 1998, **6**, 85.
133. M. Schafer, G. M. Sheldrick, I. Behner and H. Lackner, *Angew. Chem., Int. Ed.*, 1998, **37**, 2381.

134. D. Ranganathan, H.V. Darshan, R. Gilardi and I. L. Karle, *J. Am. Chem. Soc.* 1998, **120**, 10793.
135. J.W. Perich, *Lett. Pept. Sci.*, 1998, **5**, 49.
136. A. Tholey, R. Pipkorn, M. Zeppezauer and J. Reed, *Lett. Pept. Sci.*, 1998, **5**, 263.
137. U. E. Udodong, W. W. Turner, B. A. Astelford, F. Brown Jr., M. T. Clayton, S. E. Dunlap, S. A. Frank, J. L. Grutsch, L. M. LaGrandeur, D. E. Verral and J. A. Werner, *Tetrahedron Lett.*, 1998, **39**, 6115.
138. J. H. Lai, T. H. Marsilje, S. Choi, S. A. Nair and D. G. Hangauer, *J. Peptide Res*, 1998, **51**, 271.
139. B. K. Handa and C. J. Hobbs, *J. Pept. Sci.*, 1998, **4**, 138.
140. E-K. Kim, H. L. Choi and E-S. Lee, *Arch. Pharmacal. Res.*, 1998, **21**, 330.
141. M. Ueki, M. Goto, J. Okumara and Y. Ishii, *Bull. Chem. Soc. Jpn.*, 1998, **71**, 1887.
142. Z. J. Yao, B. Ye, K. Miyoshi, A. Otaka and T.R. Burke, *Synlett*, 1998, 428.
143. J. P. DeGnore and J. Qin, *J. Am. Soc. Mass Spectrom.*, 1998, **9**, 1175.
144. D. A. Evans, M.R. Wood, B.W. Trotter, T.I. Richardson, J.C. Barrow and J. L. Katz, *Angew. Chem. Int. Ed.*, 1998, **37**, 2700; D. A. Evans, C. J. Dinsmore, P. S. Watson, M. R. Wood, T. I. Richardson, B. W. Trotter and J. L. Katz, *ibid.*, 1998, **37**, 2704.
145. K. C. Nicolaou, S. Natarajan, H. Li, N. F. Jain, R. Hughes, M. E. Solomon, J. M. Ramanjulu, C. N. C. Boddy and M. Takayanagi, *Angew. Chem. Int. Ed.*, 1998, **37**, 2708; K. C. Nicolaou, N. F. Jain, S. Natarajan, R. Hughes, M. E. Solomon, H. Li, J. M. Ramanjulu, M. Takayanagi, A. E. Koumbis and T. Bando, *ibid.*, 1998, **37**, 2714; K. C. Nicolaou, M. Takayanagi, N. F. Jain, S. Natarajan, A. E. Koumbis, T. Bando and J. M. Ramanjulu, *ibid.*, 1998, **37**, 2717.
146. K. C. Nicolaou, H. J. Mitchell, N. F. Jain, N. Winssinger, R. Hughes and T. Bando, *Angew. Chem. Int. Ed.*, 1999, **38**, 240.
147. M. Ge, C. Thompson and D. Kahne, *J. Am. Chem. Soc.*, 1998, **120**, 11014.
148. M. J. Rodriguez, M. J. Snyder, N. J. Zweifel, S. C. Wilkie, D. R. Stack, R. D. G. Cooper, T. I. Nicas, D. L. Mullen, T. F. Butler and R. C. Thompson, *J. Antibiot.*, 1998, **51**, 560.
149. M. Adamczyk, J. Grote and S. Rege, *Biorg. Med. Chem. Lett.*, 1998, **8**, 885.
150. N. J. Snyder, R. D. G. Cooper, B. S. Briggs, M. Zmijewski, D. L. Mullen, R. E. Kaiser and T. I. Nicas, *J. Antibiot.*, 1998, **51**, 945.
151. A. Y. Pavlov, M.N. Preobrazhenskaya, A. Malabarba, R. Ciabatti and L. Colombo, *J. Antibiot.*, 1998, **51**, 73.
152. A. Y. Pavlov, M. N. Preobrazhenskaya, A. Malabarba and R. Ciabatti, *J. Antibiot.*, 1998, **51**, 525.
153. H. Jayasuriya, G. M. Salituro, S. K. Smith, J. V. Heck, S. J. Gould, S. B. Singh, C. F. Homnick, M. K. Holloway, S. M. Pitzenberger and M. A. Patane, *Tetrahedron Lett.*, 1998, **39**, 2247.
154. Y. A. Pavlov and M. N. Preobrazhenskaya, *Bioorg. Khim.*, 1998, **24**, 644.
155. D. L. Boger, R. T. Beresis, O. Loiseleur, J. H. Wu and S. L. Castle, *Biorg. Med. Chem. Lett.*, 1998, **8**, 721.
156. C. Vergne, M. Bois-Choussy and J. P. Zhu, *Synlett*, 1998, 1159.
157. A. J. Pearson and M. V. Chelliah, *J. Org. Chem.*, 1998, **63**, 3087.
158. D. L. Boger, S. Miyazaki, O. Loiseleur, R.T. Beresis, S. L. Castle, J. H. Wu and Q. Jin, *J. Am. Chem. Soc.*, 1998, **120**, 8920.

159. A. M. A Van Wageungen, T. Staroske and D. H. Williams, *Chem. Commun.*, 1998, 1171.
160. T. Staroske and D. H. Williams, *Tetrahedron Lett.*, 1998, **39**, 4917.
161. T. Staroske, A. J. R. Heck, P. J. Derrick and D. H. Williams, *J. Mass Spectrom. Soc. Jpn.*, 1998, **46**, 69.
162. J. Rao, J. Lahiri, L. Isaacs, R.M. Weis and G. M. Whitesides, *Science* 1998, **280**, 708.
163. Y. M. Dunayevskiy, Y. V. Lyubarskaya, Y.H. Chu, P. Vouros and B. L. Karger, *J. Med. Chem.*, 1998, **41**, 1201.
164. S. G. Grdadolnik, P. Pristovsek and D. F. Mierke, *J. Med. Chem.*, 1998, **41**, 2090.
165. G.-J. Boons and R. L. Polt, *Carbohyd. Chem.*, ed. G.-J. Boons, Blackie, London, 1998, p. 223.
166. C. M. Taylor, *Tetrahedron*, 1998, **54**, 11317.
167. E. W. Schmidt, C. A. Bewley and D. J. Faulkner, *J. Org Chem.*, 1998, **63**, 1254.
168. J. Satyanarayana, T. L. Gururaja, G. A. Naganagowoda, N. Ramasubbu and M. J. Levine, *J. Pept. Res.*, 1998, **52**, 165.
169. K. M. Sivanandaiah, V. V. S. Babu and S. C. Shankaramma, *Indian J. Chem. Sect. B. Org. Chem.*, 1998, **37B**, 760.
170. X-T. Chen, D. Sames and S. Danishefsky, *J. Am. Chem. Soc.*, 1998, **120**, 7760.
171. J. Broddefalk, K. E. Bergquist and J. Kihlberg, *Tetrahedron*, 1998, **54**, 12047.
172. K. Witte, O. Seitz and C. H. Wong, *J. Am. Chem. Soc.*, 1998, **120**, 1979.
173. P. M. St.Hilaire, T. L. Lowary, M. Meldal and K. Bock, *J. Am. Chem. Soc.*, 1998, **120**, 13312.
174. T. F. J. Lampe, G. Weitz-Schmidt and C. H. Wong, *Angew. Chem. Int. Ed.*, 1998, **37**, 1707.
175. J. Habermann and H. Kunz, *Tetrahedron Lett.*, 1998, **39**, 4797; 1998, **39**, 265.
176. E. Harth-Fristchy, S. Dufour, M. Si-Tar, N. Chignard, V. Biberovic and D. Cantacuzene, *J. Pept., Res.* 1998, **52**, 51.
177. H. Susaki, K. Suzuki, M. Ikeda, H. Yamada and H. K. Watanabe, *Chem. Pharm. Bull.*, 1998, **46**, 1530.
178. E. E. Simanek, D-H. Huang, L. Pasternak, T. D. Machajewski, O. Seity, D. S. Millar, H. J. Dyson and C. H. Wong, *J. Am. Chem. Soc.*, 1998, **120**, 11567.
179. L. Biondi, F. Filira, M. Gobbo, E. Pavin and R. Rocchi, *J. Pept. Sci.*, 1998, **4**, 58.
180. T. Inazu, K. Haneda and M. Mizuno, *Yuki Gosei Kagaku Kyokai*, 1998, **56**, 210.
181. M. Mizuno, I. Muramoto, T. Kawakami, M. Seike, S. Aimoto, K. Haneda and T. Inazu, *Tetrahedron Lett.*, 1998, **39**, 55.
182. J. Y. Roberge, X. Beebe and S. J. Danishefsky, *J. Am. Chem. Soc.*, 1998, **120**, 3915.
183. V. Ferro, L. Weiler and S. G. Withers, *Carbohydr. Res.*, 1998, **306**, 531.
184. A. Schafer, G. Klich, M. Schreiber, H. Paulsen and J. Thiem, *Carbohydr. Res.*, 1998, **313**, 107.
185. E. Meinjohanns, M. Meldal, H. Paulsen, R. A. Dwek and K. Bock, *J. Chem. Soc., Perkin Trans. 1*, 1998, 549.
186. J. Holm, S. Linse and J. Kihlberg, *Tetrahedron*, 1998, **54**, 11995.
187. P. Laurent, L. Hennig, K. Burger, W. Hiller and M. Neumayer, *Synthesis*, 1998, 905.
188. K. Yamamoto, K. Fujimori, K. Haneda, M. Mizuno, T. Inazu and H. Kumagai, *Carbohydr. Res.*, 1998, **305**, 415.

189. M. Ledvine, D. Zyka, J. Jezek, T. Trnka and D. Saman, *Collect. Czech. Chem. Commun.*, 1998, **63**, 577.
190. M. Ledvina, J. Jesek, D. Saman and V. Hribalova, *Collect. Czech. Chem. Commun.*, 1998, **63**, 590.
191. E. Vass, E. Lang, J. Samu, Zs. Majer, M. Kajter-Peredy, M. Mak, L. Radics and M. Hollosi, *J. Mol. Struct.*, 1998, **440**, 59.
192. R. M. Werner, L. M. Williams and J. T. Davis, *Tetrahedron Lett.*, 1998, **39**, 9135.
193. M. Cudic, J. Horvat, M. Elofsson, K. E. Bergquist, J. Kihlberg and S. Horvat, *J.Chem. Soc., Perkin Trans. 1*,1998, 1789.
194. P. Coutrot, C. Grison and F. Coutrot, *Synlett*, 1998, 393.
195. U. Tedebark, M. Meldal, L. Panza and K. Bock, *Tetrahedron Lett.*, 1998, **39**, 1815.
196. M. A. Dechantsreiter, F. Burkhart and H. Kessler, *Tetrahedron Lett.*, 1998, **39**, 253.
197. A. Dondoni, A. Marra and A. Massi, *Chem. Commun.*, 1998, 1741.
198. T. Fuchss and R.R. Schmidt, *Synthesis*, 1998, 753.
199. W. M. Macindoe, A. H. van Oijen and G-J. Boons, *Chem. Commun.*, 1998, 847.
200. L. Anderson, G. Stenhagen and L. Baltzer, *J. Org. Chem.*, 1998, **63**, 1366.
201. P. R. Hansen, C. E. Olsen and A. Holm, *Bioconj. Chem.*, 1998, **9**, 126.
202. S. Ando, J-I. Aikawa, Y. Nakahara and T. Ogawa, *J. Carbohydr. Chem.*, 1998, **17**, 633.
203. S. Kajiyama, H. Kanzaki, K. Kawazu and A. Kobayashi, *Tetrahedron Lett.*, 1998, **39**, 3737.
204. G. J. Hooper, J. Orjala, R. C. Schatzman and W. H. Gerwick, *J. Nat. Prod.*, 1998, **61**, 529.
205. T. Sugawara, A. Tanaka, K. Tanaka, K. Nagai, K. Suzuki and T. Suzuki, *J. Antibiot.*, 1998, **51**, 435.
206. O. Melnyk, M. Bossus, D. David, C. Rommens and H. Gras-Masse, *J. Pept. Res.*, 1998, **52**, 180.
207. T. Schmittberger, A. Cotte, and H. Waldmann, *Chem. Commun.*, 1998, 937.
208. E. Nägele, M. Schelhaas, N. Kuder and H. Waldmann, *J. Am. Chem. Soc.*, 1998, **120**, 6889.
209. Y. Nakao, N. Oku, S. Matsungaga and N. Fusetani, *J. Nat. Prod.*, 1998, **61**, 667.
210. Y. Koiso, N. Morisaki, Y. Yamashita, Y. Mitsui, R. Shirai, Y. Hashimoto and S. Iwasaki, *J. Antibiot.*, 1998, **51**, 418.
211. G. Helynck, C. Dubertret, D. Frechet and J. Leboul, *J. Antibiot.*, 1998, **51**, 512.
212. E. W. Schmidt and D. J. Faulkner, *Tetrahedron*, 1998, **54**, 3043.
213. C-B. Xue, X. He, J. Roderick, W. F. Degrado, R. J. Cherney, K. D. Hardman, D. J. Nelson, R. A. Copeland, B. D. Jaffee and C. P. Decicco, *J. Med. Chem.*, 1998, **41**, 1745.
214. J. H. Meyer and P. A. Bartlett, *J. Am. Chem. Soc.*, 1998, **120**, 4600.
215. C. Nguyen, J. Blanco, J. P. Mazaleyrat, B. Krust, C. Callebaut, E. Jacotot, A. G. Hovanessian and M. Wakselman, *J. Med. Chem.*, 1998, **41**, 2100.
216. A. F. Morel, E. C. S. Machado, J. J. Moreira, A. S. Menezes, M. A. Mostardeiro, N. Zanatta and L. A. Wessjohann, *Phytochemistry*, 1998, **47**, 125.
217. D. Xiao, S. P. East and M. M. Joullié, *Tetrahedron Lett.*, 1998, **39**, 9631.
218. S. P. East, F. Shao, L. Williams and M. M. Joullié, *Tetrahedron*, 1998, **54**, 13371.
219. S. P. East and M. M. Joullié, *Tetrahedron Lett.*, 1998, **39**, 7211.
220. A-C. Carbonnelle, E. G. Zamora, R. Beugelmans and G. Roussi, *Tetrahedron Lett.*, 1998, **39**, 4471.

221. T. Laib and J. P. Zhu, *Tetrahedron Lett.*, 1998, **39**, 283.

222. Y. Feng, Z. Wang, S. Jin and K. Burgess, *J. Am. Chem. Soc.*, 1998, **120**, 10768.

223. D. Kaiser, R. W. Jack and G. Jung, *Pure Appl. Chem.*, 1998, **70**, 97.

224. J. P. Mayer, J. Zhang, S. Groeger, C-F. Liu and M. A. Jarosinski, *J. Peptide Res.*, 1998, **51**, 432.

225. L. Yu, Y. Lai, J. V. Wade and S. M. Coutts, *Tetrahedron Lett.*, 1998, **39**, 6633.

226. D. S. Jones, C. A. Gamino, M. E. Randow, E. J. Victoria, L. Yu and S. M. Coutts, *Tetrahedron Lett.*, 1998, **39**, 6107.

227. V. Alcazar, M. Segura, P. Prados and J. de Mendoza, *Tetrahedron Lett.*, 1998, **39**, 1033.

228. S. S. Yoon, *Bull. Korean Chem. Soc.*, 1998, **19**, 254.

229. S. S. Yoon, *Bull. Korean Chem. Soc.*, 1998, **19**, 252.

230. R. Hirschmann, W. Q. Yao, B. Arison, L. Maechler, A. Rosegay, P. A. Sprengeler and A. B. Smith, *Tetrahedron*, 1998, **54**, 7179.

231. I. L. Karle, D. Ranganathan and V. Haridas, *J. Am. Chem. Soc.*, 1998, **120**, 6903.

232. D. Ranganathan, V. Haridas and I. L. Karle, *J. Am. Chem. Soc.*, 1998, **120**, 2695.

233. A. K. Wong, M. P. Jacobsen, D. J. Winzer and D. P. Fairlie, *J. Am. Chem. Soc.*, 1998, **120**, 3836.

234. M. D. Weingarten, K. Sekanina and W. Clark Still, *J. Am. Chem. Soc.*, 1998, **120**, 9112.

235. C. Y. Cho, R. S. Youngquist, S. J. Paikoff, M. H. Beresini, A. R. Hebert, L. T. Berleau, C. W. Liu, D. E. Wemmer, T. Keough and P. G. Schultz, *J. Am. Chem. Soc.*, 1998, **120**, 7706.

236. R. Warrass, K-H. Wiesmuller and G. Jung, *Tetrahedron Lett.*, 1998, **39**, 2715.

5
Metal Complexes of Amino Acids and Peptides

BY E. FARKAS AND I. SÓVÁGÓ

1 Introduction

This review covers the papers published on the metal complexes of amino acids, peptides and related ligands in 1997 and 1998. The major source of the papers reviewed here was CA Selects on Amino Acids, Peptides and Proteins,[1] but title pages of the most common Journals in the fields of inorganic, bioinorganic and coordination chemistry were also checked. The number of references cited in this chapter is 346, which, as compared to that of the previous review in Volume 29, reflects more than a 30% increase in the scientific activity of this area in the last two years.

The results published on the metal complexes of amino acids and peptides are treated in separate sections, although there are several publications dealing with the coordination chemistry of both peptides and corresponding amino acids. The papers, as in previous Volumes, can be classified into three major categories: (a) synthesis and structural studies; (b) solution equilibria; (c) reactivity and kinetics.

The interest in the metal complexes of amino acids and other biologically related ligands arose some decades ago and the most important findings are now available in various reviews. The most recent reviews published in 1997 and 1998 cover some specific aspects of the coordination chemistry of amino acids and peptides. It has been widely accepted that the non-coordinating side chains of amino acids or peptides do not have significant effects on the complex formation processes and metal binding ability of the ligands. There are, however, some specific amino acid sequences which result in the enhanced stability of the metal complexes, and the copper(II) complexes of peptides containing aromatic side chains, disulfide bridges or proline residues have been reviewed.[2] As a consequence of the clinical applications of cisplatin and related platinum containing drugs, the investigation of platinum(II) and palladium(II) complexes of amino acids and derivatives is a subject of continuous interest and an excellent review has been published in this field recently.[3] The increasing biological applications of the radioactive isotopes of rhenium and technetium[4] and the possible applications of lanthanides[5] have prompted the

Amino Acids, Peptides and Proteins, Volume 31
© The Royal Society of Chemistry, 2000

publication of reviews in these areas. The birth of 'Bioorganometallic Chemistry' is one of the major developments of chemistry in the last few decades and the most important findings in the bioorganometallic chemistry of amino acids and peptides have been reviewed recently.[6] The amide functions of peptide molecules have a versatile coordination chemistry, the structural aspects of which have also been reviewed.[7]

2   Amino Acid Complexes

2.1   Synthesis and Structural Studies – In spite of the enormous number of studies carried out over the years on the synthesis of amino acid complexes formed with some 3d metal ions, especially copper(II), numerous papers on such systems are still being published.[8-36] In some cases reinvestigations of previously studied complexes were made with the help of better technical and/or computational facilities,[8,9] whilst in other cases mixed metal complexes were synthesized,[10,11] and bis-chelated Cu(II) complexes of amino acids with hydrophobic residues were characterized by infrared spectroscopy.[12] The crystal structure of [Cu(L-Glu)(H$_2$O)]·H$_2$O was re-examined with greater precision. The results reveal that the copper(II) atom is coordinated in a square-planar geometry with the amino nitrogen and the carboxylate oxygen of the L-Glu, the oxygen of a water molecule, and the side-chain carboxylate oxygen of a neighbouring L-Glu in the equatorial positions. Weak coordination of two additional Glu oxygens to both axial positions complete a distorted octahedron. The [Cu(L-Glu)(H$_2$O)] units, linked by coordination of the side-chain carboxylate group, form an infinite left-handed single helix. If D-Glu is used, a right-handed helix results.[8] Re-examination of the crystal structure of Cu(II) complexes with β-Ala showed the formation of two different centrosymmetric carboxylato-bridged dimeric units of the chloride containing compound, [Cu$_2$(β-Ala)$_4$Cl$_2$]Cl$_2$·H$_2$O (**1**). The coordination geometry around each copper(II) center is square pyramidal, with four oxygen atoms in the basal plane and the chloride ion at the apical positions. The amino groups are in the protonated, non-coordinated form. The coordination polyhedra of the two dimers are slightly different, namely, the Cu(1)–Cu(1)^i distance is somewhat shorter and the Cu(1)–Cl bond length is longer than the corresponding distances for Cu(2)–Cu(2)^{ii} and Cu(2)–Cl.[9]

The mixed-metal complex of glutamate [CuMg(L-Glu)$_2$(H$_2$O)$_3$]·2H$_2$O was prepared for the first time and characterized by X-ray diffraction (**2**). The complex has a one-dimensional polymeric structure, in which Cu(II) and Mg(II) ions are bridged by glutamate. Each Cu atom exhibits distorted square pyramidal five-coordination, with two Glu ions acting as bidentate ligands through the amino N and the α-carboxylate O atoms in *trans* positions in the equatorial plane, and with a water molecule in the apical position. Each Mg atom is coordinated in a slightly distorted octahedral geometry by two γ-carboxylate O atoms from two glutamate ions and by four water molecules.[10] The heteronuclear complex of copper(II) and yttrium(III), [CuY(A-

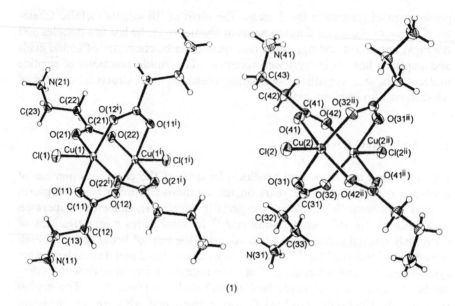

(1)

(2)

(3)

la)$_4$(H$_2$O)$_5$](ClO$_4$)·3H$_2$O **(3)**, was obtained from aqueous solution at pH 4. In this binuclear complex copper and yttrium ions are connected by four bridging carboxylate groups from four Ala ligands, the amino N atoms of which are not coordinated. The coordination number of yttrium is eight and that of copper is five.[11]

(4)

Metal ion–amino acid–2,2′-bipyridine (bpy) or 1,10-phenanthroline (phen) complexes have traditionally been studied as simple biological models. Two dissimilar copper ions were identified in the crystal lattice of the ternary complex (L-Asp)(phen)copper(II) hydrate **(4)**. These copper ions are in chains running parallel to the *b* crystal axis. The metal ions in the (A) chains are connected by the Asp ligand, and those in the (B) chains by a chemical path that involves a carboxylate bridge and a hydrogen bond. Both chains are held together by a complex network of hydrogen bonds and by hydrophobic interactions between aromatic phen molecules.[13] Studies on the structural dependence of aromatic ring stacking in copper(II)–bpy/phen–L-XPhe derivatives (L-XPhe, where X = H, NO$_2$, OH, NH$_2$ or halogens) resulted in the synthesis of four new compounds, [Cu(L-NH$_2$Phe)(bpy)]NO$_3$·H$_2$O, [Cu(L-Tyr)(phen)]ClO$_4$·2.5H$_2$O, [Cu(L-Phe)(phen)]Cl·3H$_2$O and [Cu(L-Phe)(bpy)]-ClO$_4$·H$_2$O. X-ray diffraction results show that all of these complexes have a similar distorted square-pyramidal structure around the central Cu(II) ion. However, the first two complexes have a structure involving aromatic ring stacking, the last one has a structure without it and the third has both types of structures in the unit cell. These differences are well demonstrated by the structures of [Cu(L-NH$_2$Phe)(bpy)]NO$_3$·H$_2$O **(5)** and [Cu(L-Phe)(bpy)]-ClO$_4$·H$_2$O **(6)**. For the above systems solution studies were also performed.[14] Intramolecular aromatic ring stacking also exists in ternary copper(II) complexes formed with phen/bpy and 4,5-dihydroxyphenylalanine (dopa) or 2,4,5-trihydroxyphenylalanine (topa). According to the X-ray results the stacking

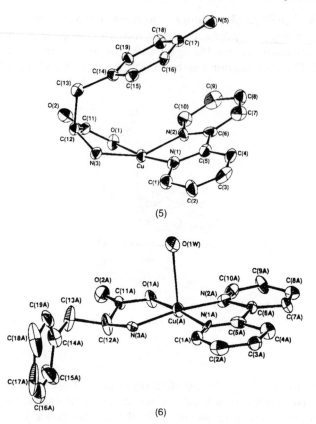

(5)

(6)

with the topa ring having three hydroxyl groups is somewhat stronger than that with the dopa.[15] Solution spectroscopic studies on ternary copper(II) complexes formed with 3,5-diiodo-L-tyrosine(L-I$_2$Tyr) or 3-iodo-L-tyrosine (L-ITyr) and diamines (phen, dpy, 2-(aminomethyl)pyridine, histamine or ethylenediamine) and X-ray analysis on [Cu(bpy)(L-ITyr)(H$_2$O)]·NO$_3$·-CH$_3$OH·H$_2$O and [Cu(histamine)(L-I$_2$TyrO$^-$)(H$_2$O]$_2$·2H$_2$O showed that the intramolecular aromatic ring stacking interactions were unambiguously favoured by the iodo groups. Moreover, the iodine atoms were also found to affect hydrogen bond formation of the hydroxyl group.[16]

Approximately regular trigonal bipyramidal geometry, which is novel for copper(II) complexes containing a discrete amino acid chelate, was reported for the copper(II) ternary complexes formed with L-Val or α-aminoisobutyrate and N,N,N′,N″,N″-pentamethyldiethylene-triamine.[18] In a study of copper(II)–L-Met, L-Trp–nucleotide and nickel(II)–L-Cys–nucleotide (5′AMP, 5′GMP or 5′IMP) ternary systems, four new complexes were synthesized and characterized by elemental analysis and IR spectroscopy. The L-Met and L-Trp coordinate to the copper(II) ion by their carboxylate and amino groups. An interesting dimeric complex containing the two copper atoms bridged by the nucleotide was obtained from a mixture of Cu–5′AMP and Trp. In the complex Ni–5′IMP–L-Cys bidentate coordination of the amino acid by its carboxylate and thiol groups occurs.[19] Unusual tridentate coordination of Gly was found in a polymeric copper(II) complex $[Cu(Gly)(NCS)(H_2O)]_n$. The glycinato anion chelates a copper atom *via* its N and O atoms and the second O atom of the carboxylate group links another Cu atom at a long distance of 290.8(5) pm (7).[20]

(7)

Hydrolysis and co-condensation of Boc-His(Boc)-CONH-$(CH_2)_3$Si$(OEt)_3$ with tetraethoxysilane, *via* a sol-gel procedure, results in a material bearing His molecules covalently bonded on an inorganic support (silica). The copper(II) complex of this material was prepared. This new biomimetic complex, in which the metal ion is coordinated by two non-inequivalent His imidazoles, exhibits high specific surface area and shows significant catalytic activity for DTBC (3,5-di-tert-butylcatechol) oxidation in the presence of dioxygen.[21]

Modification of amino acids can increase selectivity towards various metal ions or improve their recognition by living organisms. For this reason, significant attention is dedicated to metal complexes of different derivatives of amino acids. Two derivatives of L-Val methyl ester (Scheme 1) and the complexes Ni(etv)$_2$, Cu(mtv)$_2$·2H$_2$O and Pt(Hmtv)Cl$_2$ containing amino-N and sulfur donor atoms were prepared. The antifungal activities of the labile, octahedral nickel(II) and copper(II) complexes were higher than that of the square planar, much more inert platinum(II) complex.[22] With the aim of synthesizing new anti-hypertensive compounds five long-chain (C$_8$ to C$_{12}$) alkyl ether derivatives of S-(–)-Tyr and five long-chain (C$_8$ to C$_{12}$) alkyl ester derivatives of S-(+)-Glu and their copper(II) complexes were synthesized. Spectroscopic evidence suggests the *trans*-configuration of the bidentate (N,O coordinated) ligands in these square-planar complexes.[23]

Reaction of Gly with bis(imidazol-2-yl)nitromethane resulted in a new tetradentate ligand (carboxymethylamino)bis(imidazole-2-yl)methane

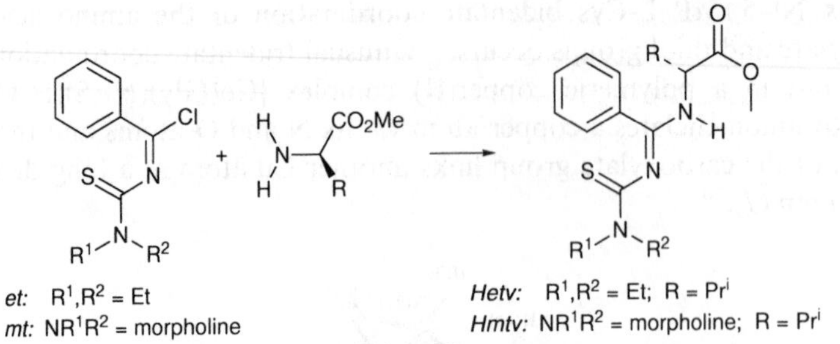
Cl
H
CO2Me
N
+
H
R
S
N
N
R1
R2
R
O
O
N
H
S
N
N
R1
R2
et: R1,R2 = Et
mt: NR1R2 = morpholine
Hetv: R1,R2 = Et; R = Pri
Hmtv: NR1R2 = morpholine; R = Pri

Scheme 1

(Hglyim) **(8)**. To synthesize model complexes which are able to mimic the active sites of hemocyanin and tyrosinase, several dinuclear copper(II) complexes of this ligand were prepared and one of them, $[Cu_2(glyim)_2Cl_2]\cdot 6H_2O$, was characterized by crystal structure analysis. In this dimeric complex the two copper ions are linked by two ligands with each copper in a square pyramidal N_3OCl environment and the chloride ion occupying the apical position. Two of the coordinating nitrogens are from the imidazole groups of one ligand, while an amine nitrogen and a carboxyl oxygen are donated by the other ligand. This 2+2 ligand-sharing arrangement results in a rather short distance between the two copper(II) ions.[24]

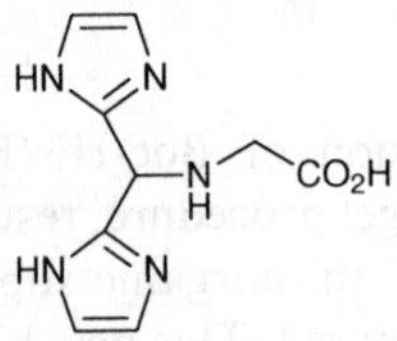
HN
N
N
H
CO2H
HN
N

(8) Hglyim

Complexes with tridentate Schiff base dianions are known key intermediates in the metabolic reaction of amino acids. For this reason complexes of amino acid derived Schiff bases were studied in numerous cases.[25-31] In the (imidazole)(N-salicylidene-β-Ala)–copper(II) complex $[Cu_2(sal\text{-}\beta\text{-}Ala)_2(Im)_2]$[25] the Schiff base dianions bridge the Cu(II) ions. A mononuclear complex was, however, isolated from the 1-methylimidazole-[N-salicylidene-(*R,S*)-Val]-copper(II) system.[26] Dinuclear Cu(II) complexes were also isolated from the N-salicylideneglycinate-urea/thiourea-copper(II) and N-salicylidenemethylalaninate-urea-copper(II) systems. The X-ray structures of the latter complexes show that the phenoxy-O ligands bridge the connecting monomeric units in all three cases. Replacement of urea by thiourea, however, leads to a change of the molecular structure. The ligands are in *trans* position in the former case **(9a)** and are in *cis* position in the latter case **(9b)**.[27] The NSC^- anion was used to generate a dinuclear copper(II) complex in the N-salicylidene-(*R,S*)-α-alaninate-isocyanate-copper(II) system. The crystal structure of this complex consists of centrosymmetric binuclear $[Cu_2$(N-salicylidene-(*R,S*)-

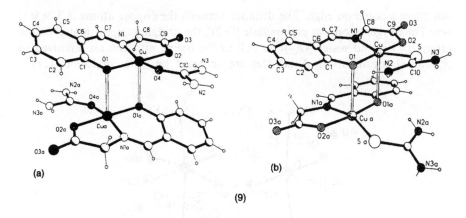

(9)

α-Ala)$_2$(μ-NCO)$_2$]$^{2-}$ complex anions connected with the K$^+$ ions by electrostatic forces **(10)**.[28] A nonenzymatic copper(II)-mediated transamination reaction of L-amino acids (Ala, Leu, Lys, Glu or Gln) *via* Schiff base formation with pyridoxal 5-phosphate results in dimeric end products. The final complex was isolated and determined by X-ray crystal structure analysis.[30] The nickel(II) complex of the Schiff base derived from alanine was used in the synthesis of L- and D-β-3,4-dichlorophenyl-α-methylalanine.[31]

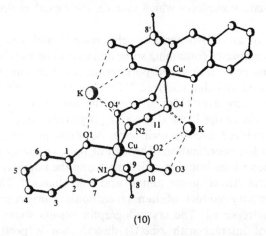

(10)

The two isomeric complexes bis[N-(carbamoylethyl)-β-alaninato-O^1,N,O^2]-copper(II) hydrate and bis[N-(carbamoylethyl)-α-alaninato-O^1,N,O^2]copper(II) hydrate crystallize in different space groups, in C2 and C2/c, respectively.[32] Template-directed Mannich condensation of Cu(Phe)$_2$ with formaldehyde and nitroethane led to a nitro-substituted precursor complex which upon reduction gave the crystallographically characterized ((2S)-5-amino-2-benzyl-6-hydroxy-5-methyl-3-azahexanoate)copper(II).[33]

Oxime analogues of amino acids are efficient ligands for Cu(II) and Ni(II) ions.[34-37] X-ray evidence shows that the dimer formed between copper(II) and 2-cyano-2-(hydroxyimino)acetic acid is composed of two octahedra, which do not share a common edge. The distance between the copper atoms is 379.5(1) pm. The oxygen atoms of carboxylate [O(2)], the bridging deprotonated oxime group [O(1')] and water [O(4)] as well as the oxime nitrogen are involved in equatorial coordination. Axial sites are occupied by water molecules **(11)**.[34]

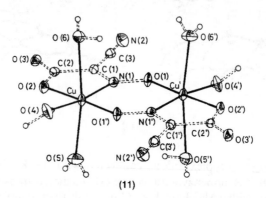

(11)

Strong hydrogen bonding stabilises the *cis* configuration of two ligands in the planar complex NiH.$_1$L isolated from the nickel(II)–2-hydroxyimino-propanamide system.[35] Results of another study involving the nickel(II) and copper(II) complexes of 2-cyano-2-(hydroxyimino)acetamide reveal the first examples of oximato complexes which contain the ligand molecules in *trans* positions.[36]

The syntheses of zinc(II)–aspartate, –glutamate,[38] and zinc(II)–molybde-num(VI) peroxo complexes containing some amino acid ligands have also been reported.[39] A number of amino acid porphyrins and their zinc(II) complexes were also synthesized and characterized.[40] Five-coordinated zinc(II) is present in the monomeric complexes **(12a)** in which the amino-N is coordinated to zinc(II). Amino acid-bridged chiral zinc(II) porphyrin dimers **(12b)**, which show significant induced circular dichroism, however, do not involve direct amino acid–metal ion coordination. The sign of the CD curves is determined by the configuration of the bridging amino acids and the intensity of CD bands is affected by steric effects of the amino acid side chains.[40] The first X-ray structure of a ternary adduct of human carbonic anhydrase II with the activator Phe was reported. The crystallographic results show that the Phe molecule does not interact with zinc(II) directly but is positioned in the hydrophobic region of the active site cavity.[41] The reaction of a Zn(II)-Cys complex and inorganic sulfide results in cysteine-capped ZnS nanocrystalline semiconductors which were prepared and characterized.[42] Manganese(II) complexes of amino acids Gly, L-Leu, L-Val, L-Gln, L-Thr, L-Trp, L-Phe, L-Met, L-Asp and L-Glu, [Mn{RCH(NH$_2$)COO}$_2$]·xH$_2$O were synthesized. All of these were found to show catalase-type activity.[43]

Complexes of other 3d block metal ions such as iron(II)/iron(III), cobal-t(III), different oxidation states of vanadium and chromium were also synthe-sized.[44-63] Simple amino acids are not really good chelators of these metal ions, and there may be strong competition between complex formation and hydro-

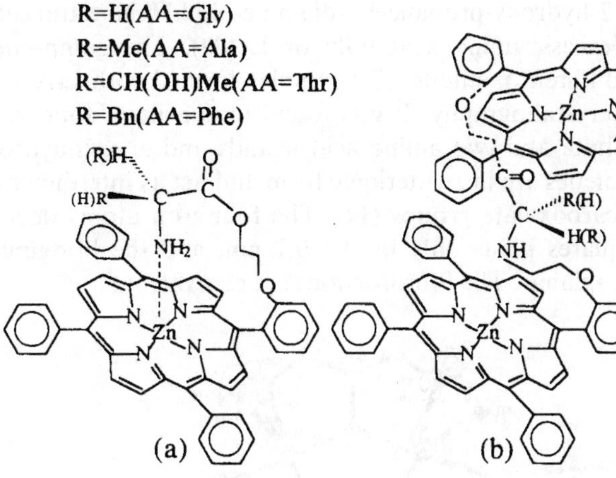

R=H(AA=Gly)
R=Me(AA=Ala)
R=CH(OH)Me(AA=Thr)
R=Bn(AA=Phe)

(a) o-AA-C₂-(TPP)Zn

(b) o,o-C₂-AA-C₂-(TPP)₂Zn₂

(12)

lytic processes but the possible importance of these interactions in the different adducts prompted the study of these systems. Moreover some amino acid derivatives and analogues can be more effective ligands of the above mentioned metals than the parent molecules. Amino acid methyl esters (L-LeuOMe and L-ValOMe) were found to coordinate to ferric porphyrins in the axial positions (13). The complexes were isolated in the solid state and were found to be quite stable in solution. They are the first examples of low-spin amino acid ester-ligated iron porphyrin complexes.[44]

Oxo- and carboxylate-bridged tri-, tetra- and higher nuclear iron clusters have received special attention as models for the iron storage protein ferritin. Tetranuclear iron(III) complexes were prepared by reaction of $Fe(NO_3)_3$, 0.5

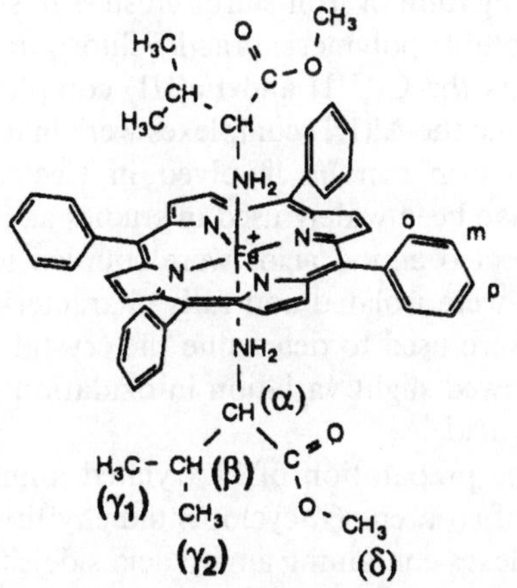

(13)

equivalents of 2-hydroxy-propane-1,3-diamine-N,N,N′,N′-tetraacetic acid (H_5(dhpta)) and excess amino acid (Gly or L-Ala). The compounds were characterized by different methods. The complex formed with L-Ala was also studied by X-ray crystallography. It was found to consist of four iron atoms bridged by two dhpta and two amino acid ligands and an oxohydroxo unit. The two L-Ala moieties are in zwitterionic form and act as interdimer bridging ligands with the carboxylate groups (14). The four iron atoms deviate from their best least-squares plane only by 0.3–0.7 pm, and the bridging oxygen atoms are nearly coplanar. The four iron ions are equivalent.[45]

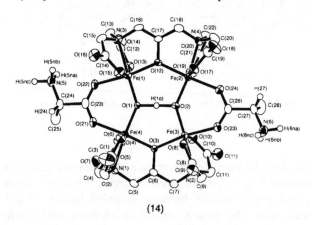

(14)

Some 3:1 site-differentiated [Fe_4S_4] clusters were treated with equimolar amounts of amino acid methyl ester (methyl ester of L-Cys, L-Tyr, L-His or D,L-Ser) in dimethyl sulfoxide solution and the amino acid substituted site-differentiated clusters were obtained. When 0.5 mol equivalents of L-Cys methyl ester were employed cysteinato-bridged dimeric clusters were formed (Scheme 2). The ligand binding of iron in some of the prepared materials is thought to mimic the ligation of iron-sulfur clusters in some metalloproteins.[46] Among the water insoluble polymeric uracil/5-fluorouracil-His-Al(III)/Cr(III)/Fe(III) ternary adducts the Cr(III) and Fe(III) complexes showed significant antitumor activity, while the Al(III) complexes were inactive.[47,48]

As is well known, iron can be involved in electron transfer processes. Ferrocene moieties have been widely used in studies as redox probe molecules. A series of ester-protected amino acids were coupled to ferrocene carboxylic acid and all products were isolated and fully characterized (Scheme 3). X-ray diffraction methods were used to determine the crystal structures. In solution CV measurements showed slight variation in oxidation potential, with respect to the attached amino acid.[49]

A new route for the preparation of N-arylated amino acids, *via* synthesis and decomplexation of (η-arene)(η-cyclopentadienyl)iron(II) hexafluorophosphate sandwich complexes containing amino acid side chains, was reported.[50]

Cobalt(III) mesoporphyrin IX and cobalt(III) protoporphyrin IX–amino acid ternary complexes[51] as well as facial and meridional isomers of *cis*-(ethyl-

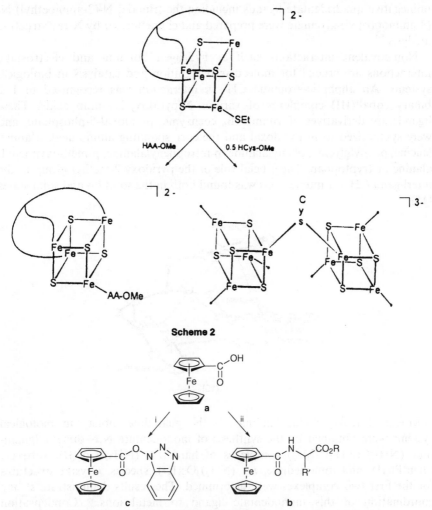

Scheme 2

Reagents: i, CH$_2$Cl$_2$, DCC, HOBt, 3 h; ii, CH$_2$Cl$_2$, DCC, HOBt, AA ester (free base), 2 days

Scheme 3

enediamine-N,N'-di-3-propionato)cobalt(II) complexes with aliphatic α-S-amino acids were prepared.[52] Cobalt (III) complexes of N,N-bis(carboxymethyl) derivatives of L-Phe, L-Glu or L-Leu with the amino acids Gly, L-Leu, L-Phe or L-Glu were studied by electronic absorption spectroscopy, CV and X-ray crystal structure determination. An octahedral coordination mode was found in all cases. The tertiary amine-N, the three carboxylate-O atoms of the tetradentate ligand and the amino-N and carboxylate-O of the amino acid were found to be coordinated to the cobalt(III) ion, but differences in the side chains cause differences in the outer coordination sphere. This is reflected in a relationship between the character of the side-chain and the rate of the electron transfer reaction.[53] Dinitrocobalt(III) complexes with different uninegative quadridentate ligands including the tripodal N-(2-aminoethyl)-N-(3-aminopropyl)-glycinate were prepared and characterized by X-ray structure analysis.[54]

Non-covalent interactions such as hydrogen bonding and electrostatic interactions are crucial for molecular recognition and catalysis in biological systems. An aliphatic-aromatic CH$\cdots\pi$ interaction was recognized in 1:2 binary cobalt(III) complexes of various N-pyridoxy-L-amino acids. These ligands are derivatives of vitamin B_6 coenzyme, pyridoxal-5-phosphate, and were synthesized from pyridoxal and the corresponding amino acids alanine, leucine, phenylglycine, phenylalanine, p-nitrophenylalanine, p-methoxyphenyl-alanine or tryptophan. The crucial role of the pyridoxy 2-methyl group in the interligand CH$\cdots\pi$ interactions was found both in the solution and solid states **(15)**.[55]

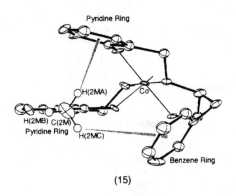

(15)

Models for Arg–metal binding *via* the guanidine moiety in biological systems were obtained by the synthesis of monodentate N,N-dimethylguani-dine (NH=C(NH$_2$)NMe$_2$) complexes of hard, *i.e.* (NH$_3$)$_5$Co(III), soft, *i.e.* (dien)Pt(II), and intermediate, *i.e.* (NH$_3$)$_5$Os(III), species. Crystal structures for the first two complexes were determined. The results demonstrate strong coordination of this monodentate ligand to metal ions.[56] Condensation reactions of cobalt(III)–amino acidato complexes with formaldehyde result in the formation of cobalt(III) complexes containing new hexadentate ligands.[57]

Amongst the early and middle 3d transition metal ion complexes synthe-sized, those of vanadium are the most popular.[58-62] [VO(His)$_4$]SO$_4$·2H$_2$O was the first isolated oxovanadium(IV) complex of an essential amino acid.[58] Oxovanadium alkoxides incorporating chelation by propane-1,3-diol or gly-cerol and salicylaldimines of α-amino acids Gly, L-Ala, L-Val or L-Phe as second ligands were prepared for the first time. X-ray results of the two complexes formed in the presence of N-salicylidene-Gly showed that both propane-1,3-diol and glycerol coordinate as bidentate ligands to the metal ion and an undissociated alcoholic group of glycerol lies *trans* to the oxo oxygen atom as is demonstrated in **(16)**.[59]

[VO(N-salicylidene-L-Trp)(H$_2$O)] and [VO(N-salicylidene-L-Trp)(quinolin-8-ol)·2H$_2$O] were isolated from aqueous solutions containing oxovana-

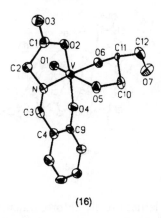

(16)

dium(IV), L-Trp and salicylaldehyde and also containing quinolin-8-ol, respectively. From water–pyridine solution, however, a different complex, $[Hpy^+]_4$ $[C_{14}H_{13}N_2^+]_2[V_{10}O_{28}^{6-}]$, was prepared. The formation of the decavanadate complex clearly shows the unexpected, unusual reactivity of these systems.[60] A new series of ternary vanadium(V) hydroxylamido complexes with Gly, Ser, Gly-Gly and Im were prepared and characterized both in solution and solid state. The vanadium(V) centers are seven-coordinate pentagonal bipyramidal geometry containing two bidentate hydroxylamido ligands, one oxo ligand and the organic ligand in all complexes. The hydroxylamido groups are coordinated side-on with the hydroxylamido nitrogen *cis* to Gly, Ser or Gly-Gly but *trans* to Im in the equatorial plane. The carboxylate or peptide oxygen atoms of Gly, Ser or Gly-Gly are coordinated *trans* to the oxo group. The molecular structure of the complex $[VO(NH_2O)_2(Gly)]\cdot H_2O$ is shown as an example **(17)**.[61]

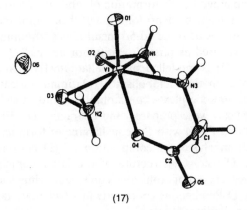

(17)

Salicylaldiminates of L-Ala, L-Val and L-Phe and their complexes incorporating the binuclear core were prepared by electroreduction of $V_2O_3^{4+}$. These complexes contain a formal $[V^V, V^{IV}]$ pair. The X-ray structure of the complex containing the salicylaldimine of L-Ala $Et_4N[V_2O_3(L\text{-alsal})_2]\cdot MeCN$ **(18)**

(18)

confirms that the rigid lattices really enforce valence localization in the solid state. Fluid solutions, however, promote valence delocalization on the EPR time scale.[62]

Some interesting organometallic complexes containing zero-valent chromium and tungsten were prepared by reactions of amino acids and derivatives (Gly, sarcosine, dimethylglycine, L-*tert*-Leu, 2-phenylglycine) with chromium and tungsten carbonyls. The complexes were characterized both in the solution and solid states. The geometry of the complex anion is, in each case, distorted octahedral consisting of four carbonyls and a typical five-membered amino acid type chelate but the orientation of the Gly chelates in the tungsten and chromium complexes are different. In the chromium complex, the glycinate ring packs so that it is *cis* to the glycinate rings in adjacent molecules (19a) but in the case of tungsten the ring packs *trans* to other rings (19b). Strong intermolecular hydrogen bonds hold the complexes together in infinite helical chains. A greatly enhanced dissociation rate of CO in the complexes of glycinate or sarcosinate relative to that of N,N-dimethylglycinate was found. This result was explained by the formation of labile transient amido complexes with glycine and sarcosine by deprotonation of their amine moieties.[63]

Numerous complexes of some 4d-5d metal ions with amino acids and their derivatives were prepared as possible antitumor agents.[63-87] In many studies, the inert platinum(II) is modelled by a palladium(II) analogue. Due to the involvement of Met in the metabolism of platinum anticancer drugs, some platinum or palladium complexes containing Met and other sulfur-containing amino acid complexes were prepared and characterized. For example, compounds of Pd(II) and Pt(II) with D-penicillamine of formula $[M_2(D\text{-}Pen)_2Cl_2]$ and the tripalladium cluster of composition $Pd_3(D\text{-}Pen)_3\cdot(KCl)_{7.8}\cdot(H_2O)_{19/8}$ were synthesized The crystal structure of the latter compound shows four chemically but not crystallographically equivalent trinuclear clusters in the unit cell (20). The D-Pen ligands coordinate in a tridentate fashion and each S atom bridges two palladium atoms.[64]

Ternary complexes formed by the reaction of equimolar amounts of L-Cys derivatives (Figure 1) and $[Pt(en)(H_2O)_2]^{2+}$ or $[Pd(en)(H_2O)_2]^{2+}$ have been studied. The thioether-S and pyridyl-N donor atoms coordinate while the amino-N and carboxylate-O atoms remain uncoordinated and may be at-

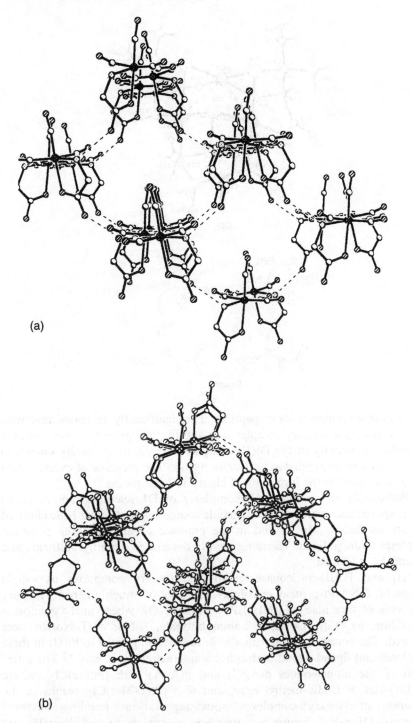

(a)

(b)

(19)

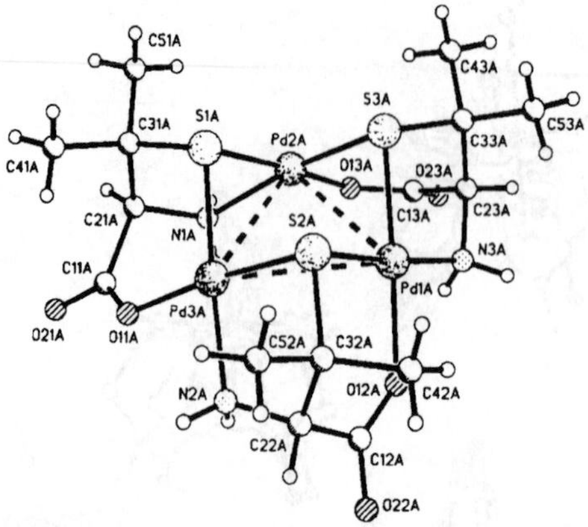

(20)

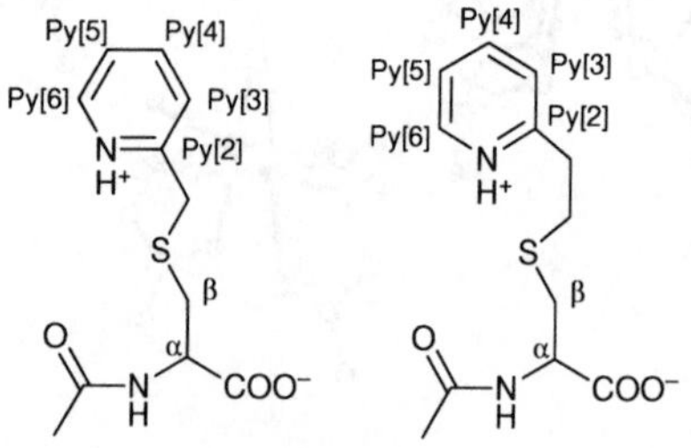

Figure 1

tached to other amino acids or peptides. The significantly decreased reactivity observed for these ternary complexes in substitution processes may improve the biological activity of the Pd(II) complexes which are generally known to react too fast to be applied as antitumor agents. The presence of excess L-Cys derivatives results in the formation of binary 1:2 complexes.[65]

Palladium(II) and platinum(II) complexes of DL-selenomethionine (sem) were prepared and characterized. Dihalide complexes containing N,Se chelated sem are formed in MeCN and in the presence of dmf, but the platinum complexes undergo ligand rearrangement in deuterated dimethyl sulfoxide and $[Pt(sem)_2]^{2+}$ is formed.[66]

Pt(II) and Pd(II)-nucleobase–amino acid ternary complexes as simple models of DNA-Pt-protein crosslinks are also of much interest. Ternary complexes of formulae cis-$[(NH_3)_2Pt(nucl)(aa)]NO_3$, where nucl = guanosine or cytidine, aa = Gly, L-Ala, L-2-aminobutyrate, L-Nor or L-Norleu were prepared. The amino acids coordinate via their -NH_2 groups to Pt(II) in these complexes and ligand-ligand hydrophobic interactions also exist.[67] The interaction of the dinucleotides d(ApG) and d(ApA) with [Pd(aa)Cl$_2$], where aa = L/D-His or L-His methyl ester, and with [Pt(L-Met)Cl$_2$] results in the formation of ternary complexes possessing various bonding modes.[68] N-Benzoyl-DL-α-Val dianion was found to coordinate to palladium(II) and platinum(II) ions through a deprotonated amide nitrogen and a carboxylate oxygen in various ternary complexes. This bonding mode was supported by the X-ray structure of the palladium(II) complex formed by the reaction of Pd(bpy)Cl$_2$ with N-benzoyl-DL-α-Val in aqueous solution at pH ~ 9.[69] It is known that bis(phosphonates) show a high affinity towards bone and other calcified tissues. For this reason a new platinum phosphonato complex (21) with geminal bis(phosphonomethylene) groups as carriers to the bone matrix was synthesized. The binding properties of the platinum complex towards DNA were highly influenced by the presence of Ca^{2+} ion.[70]

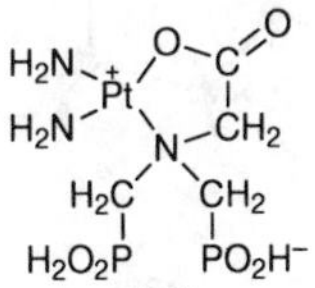

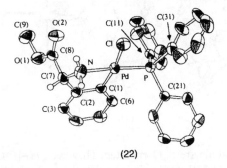

(21)

The importance of cyclometallated compounds in organic synthesis initiated the preparation of optically active cyclopalladated complexes (22) of α-mono-substituted primary benzylamines such as 2-phenylglycine methyl ester.[71] The

(22)

chelate assisted oxidative addition of L-Cys methyl or ethyl ester to Pt(PPh$_3$)$_4$ results in the synthesis of the complex [Pt(H)(SCH$_2$CH(COOR)NH$_2$) (PPh$_3$)], with displacement of three moles of triphenylphosphine, when an excess of the Cys ester is used. With Cys, an excess is not necessary. If, however, the bulkier derivatives HSCH$_2$CH$_2$NMe$_2$ and HSCMe$_2$CH(COOMe)NH$_2$ are used reaction does not occur even with excess ligand.[72] Oxidative addition of methyl N-benzoyl-2-bromoglycinate to bis-(dibenzylideneacetone)palladium, in the presence of bpy, and to (Ph$_3$P)$_2$Pt(η^2-C$_2$H$_4$) gives the α-metallated Gly esters. The structure of the bpy containing ternary complex is shown in the representative example (23).[73]

Compared to the numerous complexes of platinum(II) with amino acids as ligands, there are much fewer examples of platinum(IV) complexes. The reaction of Na$_2$PtCl$_6$ with Gly in water yields *cis*-[PtCl$_2$(Gly)$_2$]. The reaction of [Na(18-cr-6)]$_2$[PtCl$_6$]·3H$_2$O with Gly and D-(+)-Ala in water gives [Na(18-cr-6)(H$_2$O)$_2$][PtCl$_4$(Gly)]·(18-cr-6) (24) and [Na(18-cr-6)][PtCl$_4$(Ala)], respectively

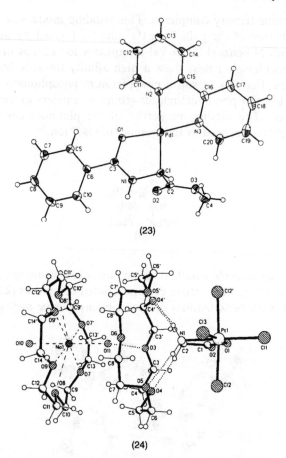

(23)

(24)

in which the amino acids are N,O-chelated. However, *cis*-[PtCl₄(GlyH)₂]· 2(18-cr-6) and *cis*-[PtCl₄(AlaH)]₂·(18-cr-6) in which the amino acids act as N-donor ligands form when the reaction mixture is irradiated with light from a halogen lamp. It is interesting that an additional crown ether molecule is linked *via* strong hydrogen bonds to the Gly containing adducts, but not to the analogous Ala complexes which may be a consequence of the steric effect of the methyl group of Ala.[74]

L-Azetidine-2-carboxylic acid is a homologue of L-proline, from which it differs by the loss of one methylene group from the ring. The structure of the complex formed between L-azetidine-2-carboxylate and the toxic 4d cadmium(II) ion consists of a one-dimensional polymer chain bridged by chloride ions and a carboxyl group. Four chloride ions coordinate to a cadmium ion and form a square planar complex. The planes extend in the direction of the *b*-axis like an infinite folding screen (25).[75]

Selected half-sandwich complexes of molybdenum[76] and tungsten[77] containing amino acids were prepared. X-ray diffraction results show that the geometry around the molybdenum in [(η⁵-C₅Me₅)(ON)(I)MoNH₂C(H)-

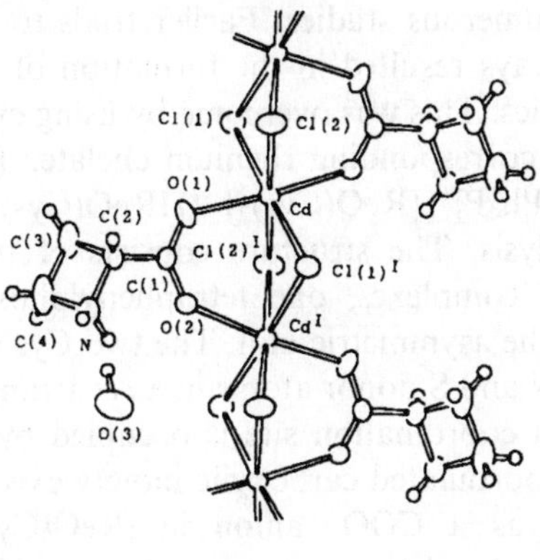

(25)

[(tBu)CO$_2$] is tetragonal pyramidal, while it is pseudo-octahedral in [W(η-C$_5$Me$_5$)Cl$_3$(NH$_2$CPh$_2$CO$_2$)] and [W(η-C$_5$Me$_5$)Cl$_3$\{NH$_2$CH(CH$_2$C$_6$H$_4$-OH-p)CO$_2$\}], the structure of which is shown in (26). Typical amino acid type

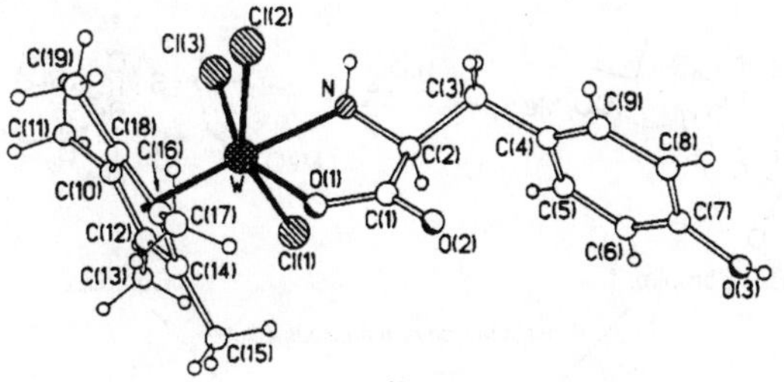

(26)

N,O-coordination was proved in all the above adducts. In contrast, amino acids (Gly, Ala, Pro, Val, Leu, Ser, Asn, Gln and Glu) coordinate in MoO(O$_2$)$_2$(α-amino acid)(H$_2$O) type complexes as zwitterions *via* one of the carboxylate oxygens. This is demonstrated by the crystal structure of MoO(O$_2$)$_2$(Ala)(H$_2$O) (27).[78]

Clinical applications of compounds of the metastable ^{99m}Tc and $^{186/188}$Re

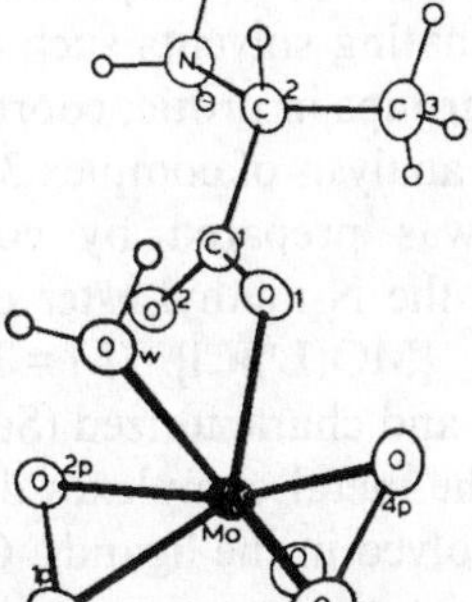

(27)

isotopes initiated numerous studies. Earlier trials to obtain technetium complexes with Cys always resulted in the formation of a product contaminated with polymeric species. This was overcome by using cystine as the precursor of Cys to prepare the corresponding rhenium chelate. The isolated tetraphenylphosphonium salt $[Ph_4P]^+[\{ReO(Cys)_2\}^-\{HReO(Cys)_2\}]\cdot 4H_2O$ was characterized by X-ray analysis. The structure consists of two distorted octahedral oxorhenium(V)-Cys complexes, one tetraphenylphosphonium ion and four water molecules in the asymmetric unit. The two Cys moieties are coordinated to ReO^{3+} through N and S donor atoms in a *cis* arrangement in the equatorial plane, and the sixth coordination site is occupied by one of the carboxylate oxygens. The non-coordinated carboxylic moiety exists in the COOH form in $[HReO(Cys)_2]$, but as a COO^- anion in $[ReO(Cys)_2]^-$.[79] D-Penicillamine methyl ester was also used to prepare oxorhenium(V) complexes, and it was confirmed (Scheme 4) that the strong tendency to form neutral oxorhenium(V)

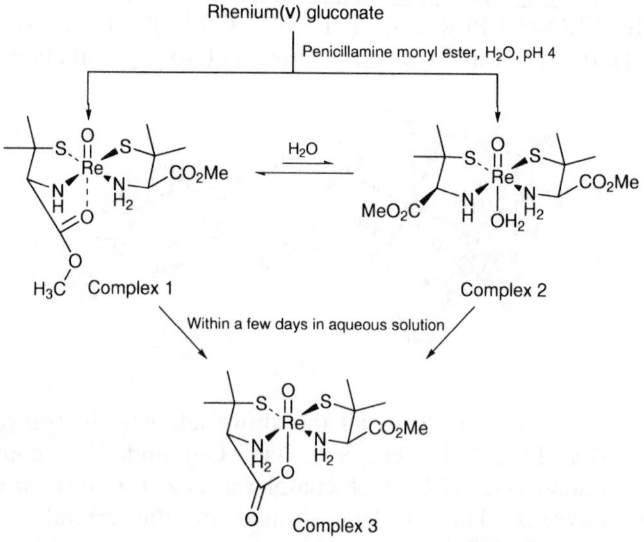

Scheme 4

complexes with this ligand resulted in the deprotonation of one of the amine groups in aprotic, non-coordinating solvents such as $CHCl_3$ (complex 1) and hydrolysis of one of the ester groups in protic, coordinating solvents like water occurred (complex 3).[80] X-ray analysis of complex 3 was performed.

A new class of ligands was prepared by coupling reactions between N-protected amino acids and the N-methyl ester of dithiocarbazic acid. The oxo- and nitrido-complexes $[MO(L^n)Cl]$ (M = Tc or Re, n = 1–5) and $[TcN(L^n)(PPh_3)]$ were isolated and characterized (Scheme 5). It was found that the formation and yields of the metal complexes depended on the protecting groups and R substituents involved in the ligands. One of the oxo- and one of the nitrido-complexes of technetium were studied by X-ray diffraction. Square-pyramidal geometry was found in both complexes, the ligands occu-

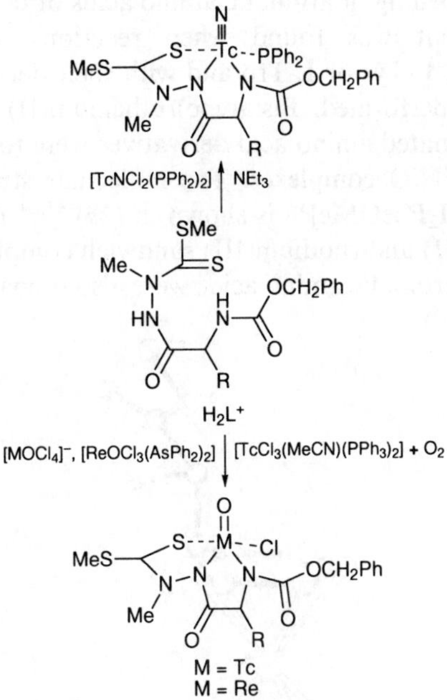

Scheme 5

pying three of the four basal positions leaving the fourth position free for the coordination of a monodentate coligand.[81]

α-Amino acids are oxidatively metabolized to α-keto acids presumably through α-imino acids as intermediates. On the other hand α-imino acids can be stabilized by chelation to a metal ion. An α-imino acidato ruthenium(II) complex, *i.e.* N-benzyl-1,2-didehydroglycinato bis(phen)ruthenium(II), was obtained by controlled-potential anodic oxidation of the corresponding α-amino-acidato ruthenium(II) complex. Its crystal structure **(28)** shows a

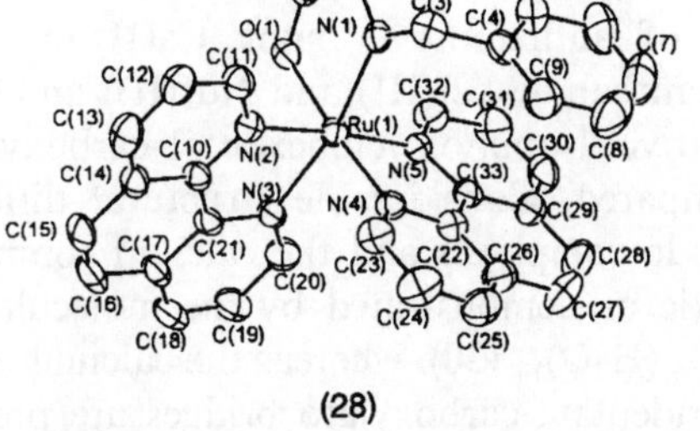

(28)

shorter Ru-N(1) bond length and a smaller N(1)-Ru(1)-O(1) bond angle than those in the α-amino-acidato complexes, and stacking interaction between the α-imino-acidato moiety and one of the phen rings occurs.[82]

Organometallic labelling of aromatic amino acids or derivatives with an (η^6-arene)Ru(II) fragment was found when reactions of [Ru(acetone)$_3$(η^6-cymene)]$^{2+}$ with Phe, L-Tyr or L-Trp and with their derivatives in CH_2Cl_2 or in CF_3COOH were performed. Bis(arene)ruthenium(II) sandwich complexes containing η^6-coordinated amino acid derivatives were formed rather than the alternative κN or $\kappa^2 N,O$ complexes. The molecular structure of the cation [Ru(η^6-cymene)(η^6-H$_2$PheOMe]$^{3+}$ is shown in (29).[83,84] η^5-Pentamethylcyclo-pentadienyliridium(III) and rhodium(III) sandwich complexes containing derivatives of the above aromatic amino acids were also prepared.[85]

(29)

Dimeric rhodium(II) complexes with α-amino acids are very rare. Reaction of [Rh$_2$(CO$_3$)$_4$]$^{4+}$ with protonated (S)-Pro or (S)-Leu, however, resulted in the formation of dimeric complexes in which the amino acid molecules are coordinated as bridging ligands *via* their carboxylato groups. The complexes were prepared and characterized by elemental analysis and different spectroscopic methods.[86] Interesting ruthenium donor/acceptor complexes with amino acid (Gly, Phe, Ileu) bridges were prepared for modelling of biological electron transfer.[87]

Due to the ability of lanthanides to mimic Ca(II) in diverse situations the X-ray structures of lanthanide (Er(III) and Nd(III)) and Ca(II) complexes of an amino acid derivative, 1-amino-cyclohexane-1-carboxylic acid (Acc^6), were determined and compared. Considerable structural differences were found between the lanthanide complexes and the calcium complex. The lanthanide complexes are dimeric as demonstrated by the molecular structure of [Er$_2$-(Acc^6)$_4$(H$_2$O)$_8$](ClO$_4$)$_6$·(H$_2$O)$_{11}$ (30), whereas the calcium complex is polymeric (31). Both bi- and tridentate carboxylato bridges are present in the calcium complex. In contrast, only bidentate carboxylato groups are present in the erbium complex and bi- and unidentate carboxylato groups in the neodymium complex. The coordination number is eight in the lanthanide complexes but six and seven for the three different metal environments in the calcium complex.[88]

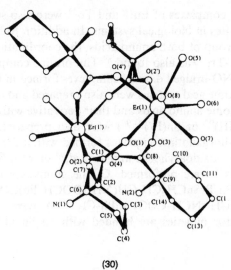

(30)

(31)

L-Phe and L-Trp complexes of Eu^{3+} and Tb^{3+} were also studied as probes of calcium binding sites in biological systems. In addition to the interaction with the carboxylate group of both amino acids, the coordination of the side chain imino group of L-Trp was also found.[89] Thioproline complexes of La^{3+}, Pr^{3+} and Nd^{3+},[90] and NO-bridged dimeric complexes formed in the Eu^{3+} – DL- and L-alaninehydroxamic acid systems were also prepared and characterized.[91]

Complexes of some amino acids and their derivative with p block metal ions (Al(III),[48,49] Ge(III)[92] or Sn(IV)[93-95]) were also prepared. For example, the synthesis of new diorganotin complexes formed with a derivative of α-amino-butyric acid,[93] N-2-oxidoarylidene amino acid,[94] or with Schiff bases derived from amino acids[95] were performed. Of the complexes formed, the crystal structures of $[Ph_2BrSnCH_2CH_2CH(NHCOOCH_2Ph)COOCH_3]$ **(32)** and $[Ph_2Sn(2-OC_{10}H_6CH=NCH_2COO)]SnPh_2Cl_2$ **(33)** were determined.[95] The donor and acceptor moieties are bonded with an Sn–O bond in the latter complex.

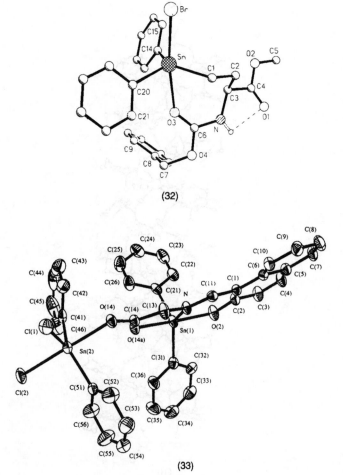

(32)

(33)

2.2 Solution Studies – In addition to pH-potentiometry, different spectro-scopic methods, *e.g.* UV-visible, ESR, NMR and in some cases electrochemical methods or calorimetry were used to determine the stoichiometry, stability and bonding modes of species formed in solution. In many cases both solution and solid state studies were carried out. Owing to the great importance of such studies an enormous amount of data has accumulated over the years especially for simple binary systems. For the period of coverage equilibrium results for metal complexes of interesting derivatives of amino acids, and for ternary systems, have been published. Molecular recognition studies have also been prominent in recent years.

Because of the tendency of vanadium(III) to undergo oxidation and hydro-lysis, little is known about vanadium(III) amino acid complexes. CD and potentiometric measurements for the vanadium(III)-His system showed the predominant formation of an interesting dinuclear species, V_2OL_4 (L = His) in the pH range of 6-8.5. Each V(III) is tridentately coordinated by a L-His, bidentately coordinated by another with the two monomeric units being bridged by an oxoligand.[96] Complex formation between Al(III) and various bidentate and potentially tridentate amino acids and their derivatives in aqueous solution, in agreement with expectation, showed a weak interaction through the carboxylate function in most cases. For Asp, however, which contains two carboxylates and a central amino donor, strong evidence for tridentate coordination was suggested.[97] Unusual coordination modes of Asn and Gln derivatives to palladium(II) was found. This involves deprotonation and coordination of the amide nitrogen of the side-chain at high pH (pK_a value is ca. 10) in the palladium-bpy-N-sulfonyl-Asn/Gln complex.[98]

Potentiometric measurements were used to study the formation and the $HClO_4$ induced dissociation of Cys from $[Mo_2O_2(\mu\text{-}S)_2(Cys)_2]^{2-}$ and $[Mo_2O_2(\mu\text{-}O)_2(Cys)_2]^{2-}$. More extensive studies on the complex containing the di-α-sulfido core indicated formation of two protonated species, which contain one or both of the Cys carboxylates in protonated form. The final product is $[Mo_2O_2(\mu\text{-}S)_2(H_2O)_6]^{2+}$.[99] Equilibrium studies on complexation between Cys and cadmium(II) supported the formation of stable chelates in the species formed.[100] Determination of the formation constants of the binary complexes of the thioamino acid derivatives, Cys, glutathione, penicillamine with Cu(I) is severely hampered by precipitation. Fortunately, ternary complexes containing these ligands and copper(I) have been investigated and this allows a reassess-ment of the binary systems over a wide pH range. These low-molecular mass complexes with Cu(I) and redox equilibria were also incorporated into computer simulations of copper speciation in blood plasma.[101]

Organotin(IV) compounds have attracted considerable attention, *e.g.* because of their biological activity and industrial importance.[102-104] Stoichio-metries and stability constants were determined for the complexes formed between dibutyltin(IV) and different amino acids and peptides,[102] between trimethyltin(IV) and N-substituted Gly in aqueous solution[103] and also between triphenyltin(IV) and various amino acids in 75% dioxane-water.[104] The results show chelate formation in dibutyltin(IV) containing systems and monodentate coordination of amino acids most likely *via* the amino group in triorganotin(IV) containing systems. At lower pH, however, the zwitterionic forms of the amino acids coordinate through the carboxyl oxygen atom to the trimethyltin(IV) ion.[103] Complex formation between lanthanide and actinide ions and selected amino acids was investigated in several cases.[105-108] The formation of outer sphere complexes was only found in the Eu(III)–Ala system, but inner sphere coordination was previously detected in the Eu(III)–Gly system.[106] Formation of a heteronuclear species was proposed in the Zn(II)–Ag(I)–L-His equilibrium system.[109]

Solution studies on nickel(II)–diamino acids (2,3-diaminopropionic acid, 2,4-diaminobutyric acid) and on nickel(II)–His had already been carried out in many cases. These complexes were reduced on a mercury electrode and their electrochemical properties were reported in a recent publication.[110] Whilst most equilibrium studies are made in aqueous solution, studies of solvent effects on stabilities of the complexes are very rare. Investigation of cadmium(II)–Gly/Phe systems in ethanol/water solution produced some new data in this field.[111]

Taurine is a naturally occurring analogue of β-Ala containing a $-SO_3^-$ group instead of $-COO^-$. This change results in a much lower complex forming ability of taurine relative to that of amino acids. Monodentate coordination of taurine was found in its Cd(II) and Ag(I) complexes.[112] Substitution of the planar, weakly basic, mononegative carboxylate function in amino acids by the more bulky, tetrahedral, dinegative and more basic phosphonate group changes the metal binding abilities. Cu(II), Ni(II) and Zn(II) complexes of the phosphonate analogues were studied in most cases[113-115] although studies on platinum(II) and palladium(II) complexes were also carried out.[116] Monophosphonate generally forms complexes in which the metal binding mode is bidentate using the $\{NH_2, PO_3^{2-}\}$ donor set.[113] In platinum(II) and palladium(II) containing systems, however, formation of complexes with monodentate coordination *via* the amino N atom was also observed.[116] Moreover, in the presence of zinc(II), the monophosphonate analogue of Asp, Met or Ala forms cyclic phosphonoamidates (**34**) in neutral and slightly alkaline solution.[113,114]

(34)

When it is sterically permitted, tridentate coordination of aminophosphonates results in the formation of very stable complexes.[117,118] The herbicide activity of N-(phosphonomethyl)Gly $((HO)_2\text{-PO-CH}_2\text{-H}_2\text{N}^+\text{-CH}_2\text{-COOH}$, glyphosate) is well known. This ligand forms quite stable copper(II) complexes in which its tridentate coordination can be assumed.[119] In order to assess the Al(III)-binding abilities of phosphonylated proteins and peptides, the interactions of Al(III) with O-phosphoserine and O-phosphotyrosine were studied by pH-metric and ^{31}P NMR measurements. O-Phosphoserine was found to bind Al(III) in a monodentate manner *via* the phosphate moiety, and in a tridentate manner by the simultaneous coordination of all donor groups. The Tyr derivative, however, binds Al(III) only at the phosphate function.[120]

An investigation of the interaction between natural sarcosine derivatives, creatine, creatine phosphate and creatinine showed that creatine phosphate is the only creatine-based ligand which is able to coordinate with vanadyl(IV) in aqueous solution. In organic solvents, however, the interaction of creatine and creatinine was also demonstrated.[121] Willardiine is an analogue of Phe containing the uracil residue. Potentiometric and spectroscopic studies showed that the primary metal binding site of the ligand is the α-amino-carboxylate chelating set. The uracil moiety, however, can coordinate to the metal ion in basic solution giving rise to intermolecular bridging (35).[122]

(35)

Aminohydroxamic acids are very effective chelating derivatives of amino acids. They coordinate 'soft' metal ions *via* their amino-N and hydroxamate-N atoms and 'hard' metal ions *via* their hydroxamate oxygens. The formation of mono(hydroxamato)trioxomolybdenum(VI) and bis(hydroxamato)dioxo-molybdenum(VI) was found in molybdenum(VI)-β-alaninehydroxamic acid and in molybdenum(VI)-glutamic acid-γ-hydroxamic acid systems. Tridentate coordination of aspartic acid-β-hydroxamic acid *via* the hydroxamate and carboxylate oxygens, however, resulted in the formation of a dinuclear complex, $[Mo_2O_5(LH)_2]^{2-}$ in addition to the formation of $[MoO_3(LH)]$.[123] A new dihydroxamate based siderophore analogue, 1,4,8,11-tetraazacyclotetra-decane-12,14-dioxo-4,8-bis(N-methyl-acetohydroxamic acid) was synthesized. The stability and electrochemical properties of its iron(III) complex were investigated in aqueous solution. The results were also supported by molecular mechanics calculations[124]

As was formerly reported, insertion of an oximic function into a simple amino acid results in an effective chelating agent for Cu^{2+} and Ni^{2+}.[33-36] If, however, both oximic and hydroxamic functions are inserted into a simple amino acid a really powerful chelating agent for Cu(II) and Ni(II) is created.[125]

Numerous solution studies were made for various ternary systems during the period of coverage.[126-144] In many cases the stoichiometries and formation constants were determined, concentration distribution curves were calculated, and the relative stabilities of the ternary complexes compared to those of the corresponding binary systems were evaluated.[126-136] The main conclusions for selected systems are the following: the stability is as expected from statistical considerations – in ternary complexes formed in M(II)-dipicolinic acid-Gly systems,[126] in complexes of Cu(II), Zn(II), Cd(II) with Gly, DL-Ala, DL-Leu and derivatives of 2-cyanomethyl benzimidazole or barbituric acid,[127] or in mixed ligand complexes formed between different metal ions, cefadroxil antibiotic and amino acids.[128] On the other hand, the formation of ternary complexes is favoured in copper(II)-(S)-leucinehydroxamic acid-Val, Pro, Phe, Trp systems.[129] The influence of additional coordination sites such as side chain -OH, -CONH$_2$, and -COOH on the formation of ternary complexes was also discussed.[131] If more than four donor atoms of the two coordinating ligands are in chelatable positions, five-coordinated ternary copper(II) complexes are often formed. This results in increased stability of the ternary complexes and characteristic redshifts in the corresponding absorption spectra. Measurable axial coordination was found, *e.g.* in copper(II)-dien-alanine-hydroxamic acid,[137] in copper(II)-polyamine-diaminocarboxylate ternary complexes,[138] and in ternary systems containing a macrocycle as one of the ligands.[139,140] Metal ion mediated ligand-ligand interactions (electrostatic, stacking) increase the stability of ternary complexes in several systems, *e.g.* in copper(II)-picolinic acid-amino acid ternary complexes,[141] and in copper(II)-3,5-diiido-L-tyrosinate-amino acidate (L-Arg, ω-protonated L-Lys, L-Asp, L-Glu or L-Ala) complexes.[142] This latter finding implies that the 3'- and 5'-iodine atoms of thyroxine may be essential for effectively interacting with the target binding site through various weak interactions. Intramolecular hydrogen bonding between the OH oxygen on the olefinic alcohol and the N-H proton of coordinated sarcosine results in the stabilization of one rotamer of the η^2-coordinated olefin in *cis*[(N-olefin)-Pt(2-methyl-3-buten-2-ol)(sarcosine)Cl] complex (**36**).[143] NMR methods were used to study the effect of olefin structure (R in CH$_2$=CHR) on the strength and stereoselectivity of binding of η^2-coordinated prochiral olefins in platinum(II)–olefin–amino acid complexes.[144]

Molecular/chiral recognition in many cases occurs *via* the formation of ternary complexes, *e.g.* molecular recognition between a copper(II) complex of β-cyclodextrin and aromatic amino acids,[145] chiral recognition between the copper(II) complex of 6-deoxy-6-N-(2-methylaminopyridine)-β-cyclodextrin

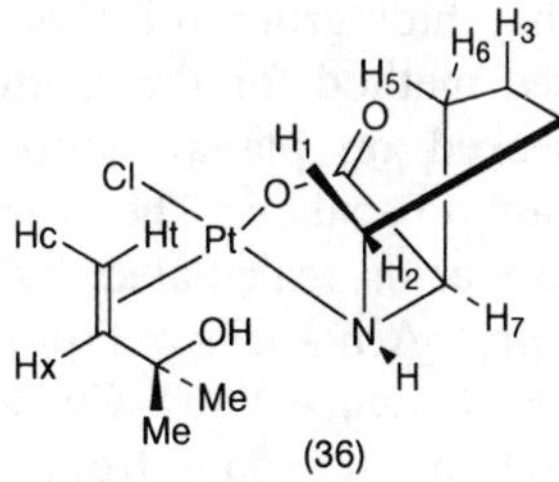

(36)

and L/D-Trp,[146] and between copper(II) complexes with L/D-Cys derivatives of β-cyclodextrin and L/D-TrpO⁻.[147] In the latter case, high performance resolution of racemate mixtures of L/D-TrpO⁻ was achieved only when the complex of the D-Cys derivative was used as the eluent. Synthetic achiral gadolinium(III) porphyrin complexes extracted chiral amino acids from an aqueous solution in the form of 1 : 1 adducts.[148] Excellent chiral recognition of the unprotected amino acids with lanthanide(III) tris(β-diketonate) complexes under neutral conditions was found. The formation of ternary complexes was supported by different methods.[149] Complexation of α-amino acid esters, β-amino ethers, and amines with a series of zinc(II) 1,19-bilindione derivatives (37) was studied. Interesting helical chirality in the bilindione framework was

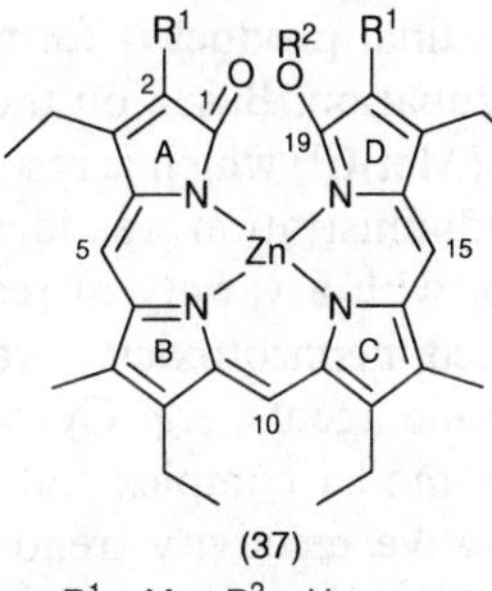

(37)

a, R¹ = Me, R² = H
b, R¹ = CH₂CH₂OH, R² = H
c, R¹ = CH₂CH₂OAc, R² = H
d, R¹ = Me, R² = Me
e, R¹ = CH₂CH₂OH, R² = Me
f, R¹ = CH₂CH₂OAc, R² = Me
g, R¹ = CH₂CH₂CH₃, R² = Me
h, R¹ = Me, R² = CHMe₂

triggered by the binding of chiral co-ligands.[150] EPR results show that the binding of bis(L/D-Lys)copper(II) and bis(L/D-Arg)copper(II) on highly oriented DNA fibers depends on the conformation of DNA and also on the type and chirality of the amino acids. The results demonstrate that DNA recognizes the difference in the chirality of even a small amino acid molecule.[151]

Ab initio calculations on the [Gly-Cu]⁺, [Ser-Cu]⁺, and [Cys-Cu]⁺ complexes showed that the preferred binding site of Cu⁺ involves chelation between the carbonyl oxygen and the amino nitrogen. Additional chelation with the alcohol group of Ser or the thiol group of Cys leads to larger binding energies.[152] A newly proposed method for the estimation of conformational energy was successfully checked on planar copper(II) bis- and aquabis-complexes with N-alkylated amino acids.[153] The order of lithium ion affinities for the 20 common α-amino acids, calculated by a kinetic method, was established as: Arg > His > Gln > Asn > Lys > Trp > Glu > Asp > Tyr > Met > Phe > Thr > Pro > Ser > Ile > Leu > Val > Cys > Ala > Gly.[154] The only ligand field calculations based on the data from the absorption and low temperature sharp-line excitation spectra were made of *trans*(imidazole)-bis(L-His)chromium(III) nitrate. The results indicate that the imidazole nitrogen is a weak σ-donor and its π-interaction can be interpreted by back-donation to the electron withdrawing empty molecular orbitals.[155]

2.3 Kinetics and Mechanism – Kinetic studies on complex formation between metal ions and amino acids or their derivatives were performed in several cases. The biological relevance initiated studies of the reactivity between different platinum compounds and thiol containing amino acid derivatives. The reaction pathways determined for the formation of Pt(II)-L-Met adducts in aqueous and in saline solutions of cisplatin/its monoaquo species and Met indicates that the monoaquo species is much more reactive toward Met than cisplatin itself. Moreover, following the formation of highly unstable and very reactive adducts, the Pt(Met)$_2$ final product is formed in the reaction mixture only after several hours of incubation. Based on the results, the Met-platinum adduct (particularly Pt(NH$_3$)$_2$(Met)Cl) which is responsible for the nephrotoxicity observed after cisplatin administration was identified.[156] The reactivity of Pt$_2$(bpy)$_2$(μ-N-acetyl-L-cys-S)$_2$ with a variety of rescue agents (including Cys and Gly) for cisplatin induced nephrotoxicity was determined in aqueous solution. It was found that some agents, *e.g.* Cys, efficiently cleaved the Pt-S cysteinato bond in the above model complex but no reaction was observed with Gly within 3 h. The relative reactivity trend and toxicity of the tested rescue agents were evaluated.[157] Kinetic analysis of the reaction between *cis*-diamminedichloroplatinum(II) and L-Cys at different pH and concentrations of reactants resulted in the model shown in Scheme 6. Rate constants for the formation of different species were determined. The effects of pH and ratio of reactants on the products predominantly formed in the reactions were also evaluated.[158]

A reaction mechanism consisting of five parallel paths is reported for chelate formation between Pd(II) and S-carboxymethyl-L-cysteine.[159] Structural and reactivity studies were made to establish reactivity features for *trans*-dichloroplatinum(II) compounds, *trans*-[PtCl$_2$(NH$_3$)$_2$] and *trans*-[PtCl$_2$(NH$_3$)quinoline] in their Met and Met-guanine containing adducts.[160]

Spectroscopic and kinetic results gave evidence for complex formation between the intermediate trivalent platinum complexes PtCl$_6^{3-}$ and PtCl$_5^{2-}$, generated by photoreduction of the PtCl$_6^{2-}$ complex and α- or β-Phe. The coordination of amino acids occurred through the COO$^-$ group.[161] An associative activation mode for substitution was proposed as a result of a

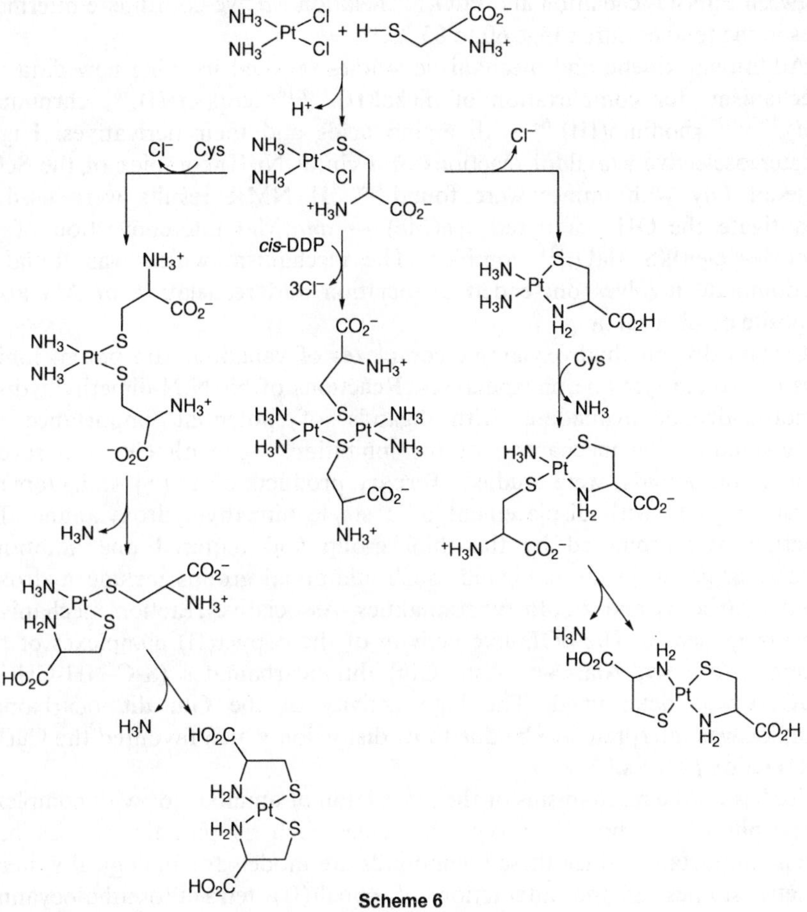

Scheme 6

kinetic study of the interaction of L-Cys with diaquaethylenediamineplatinum(II) perchlorate in aqueous solution.[162] The *cis*-(R,S)-[Pd(egta)]$^{2-}$ (egta^{4-} = glycine, N,N'-(1,2-ethanediylbis(oxy-2,1-ethanediyl)bis[N-carboxymethyl]) ion has the same *cis* stereochemistry as other bis(amino acid) complexes of palladium(II) (**38**). However, the pendant non-coordinated carboxylates generate internal associative reaction pathways for a switch

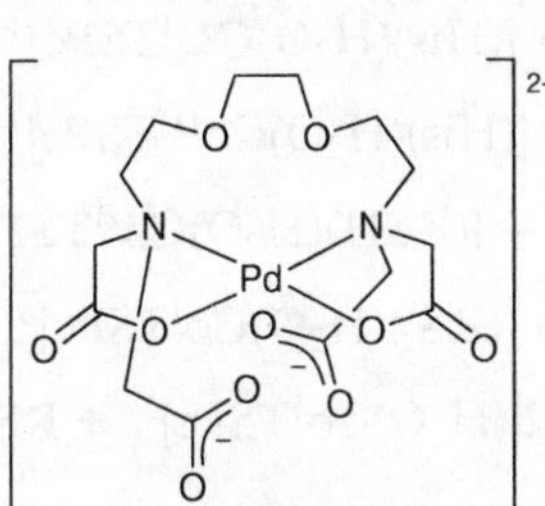

(38)

between PdN_2O_2 chelation and PdO_3N chelation *via* five-coordinate intermediates in the temperature range 60 to 85 °C.[163]

Additional kinetic and mechanistic studies resulted in some new data and mechanisms for complexation of nickel(II),[164,165] copper(II),[165] chromium-(III),[166,167] rhodium(III)[168] with amino acids and their derivatives. Highly diastereoselective aza-aldol reactions of a chiral Ni(II)-complex of the Schiff base of Gly with imines were found.[169] ^{1}H NMR results were used to investigate the OH^- catalysed *syn*(Me) $\leftrightarrow$ *anti*(Me) interconversion of the $[Co(Mecyclen)](S-AlaO)]^{2+}$ complex. The mechanism which was found to predominate involves one-ended dissociation and rechelation of Ala at an opposite octahedral face.[170]

Certain dimethylhydroxylamine complexes of vanadium are potent inhibitors of protein tyrosine phosphatases. Reactions of bis(N,N-dimethylhydroxamido)hydroxooxovanadate with ligands of potential importance for understanding the mechanism of the inhibition, *e.g.* with Cys and several related compounds, were studied. Ternary products of 1:1:1 stoichiometry formed rapidly with displacement of a single dimethylhydroxylamine. The reaction was promoted by the thiol group and required one additional functional group in the co-ligand. Such additional groups include hydroxyl, amino, carboxyl and amido functionalities. Associative reaction mechanisms were proposed.[171] The SOD-like activity of the copper(II) complexes of five amino acid (Gly, Ala, Ser, Asn, Glu) dithiocarbamates ($S_2C-NH-CH(R)-COOH$) was determined. The high activity of the Glu dithiocarbamate complex was interpreted to be due to its distorsion which favoured the Cu(II)/Cu(I) redox process.[172]

Studies on the mechanisms of the interaction of amino acids with complexes of porphyrins or porphyrin-type molecules such as phthalocyanines have special importance, since these compounds are models for biological systems. Kinetic studies of the interaction of cobalt(II) tetrasulfophthalocyanine, $[Co^{II}TSPc]^{4-}$ with His and Cys showed first order kinetics in both cases. However, while the coordination of His (most probably axial coordination) to the $[Co^{II}TSPc]^{4-}$ facilitated the air oxidation of the complex to the cobalt(III) species (1), (2), coordination of Cys resulted in electron transfer from Cys to the central cobalt(II) in $[(RSH)(H_2O)Co^{II}TSPc]^{4-}$ and, following the loss of the oxidized $RS\cdot$ species from the axial position, cystine and $[Co^{I}TSPc]^{5-}$ species were formed, (3) – (5).

$$[(H_2O)_2Co^{II}TSPc]^{4-} + His \rightarrow [(His)(H_2O)Co^{II}TSPc]^{4-} + H_2O \qquad (1)$$

$$[(His)(H_2O)Co^{II}TSPc]^{4-} \xrightarrow{O_2} [(His)(H_2O)Co^{III}TSPc]^{3-} \qquad (2)$$

$$[(H_2O)_2Co^{II}TSPc]^{4-} + RSH \rightarrow [(RSH)(H_2O)Co^{II}TSPc]^{4-} + H_2O \qquad (3)$$

$$[(RSH)(H_2O)Co^{II}TSPc]^{4-} \rightarrow [(RS^{\cdot})(H_2O)Co^{I}TSPc]^{5-} + H^+ \qquad (4)$$

$$2[(RS\cdot)(H_2O)Co^{I}TSPc]^{5-} \rightarrow 2[(H_2O)Co^{I}TSPc]^{5-} + RSSR \qquad (5)$$

Electron transfer between $[Co^{II}TSPc]^{4-}$ and His was excluded.[173]

The kinetics and mechanism of electron transfer reactions between chiral Co^{III} and optically active Fe^{II} complexes were studied by CD spectroscopy. The ligands used all had the same basic structure of N,N'-[(pyridine-2,6-diyl)bis(methylene)]bis[amino acid]. The substituents were systematically varied to obtain evidence for the intimate mechanism of the electron transfer. An inner-sphere mechanism with a transition state containing the carboxylate group on the inert Co^{III} complex as the bridge was proposed.[174] Evidence was found for the formation of a tyrosine radical by light-induced electron transfer reactions in a Ru(II)-polypyridine complex with a covalently attached tyrosyl moiety. In the presence of an external electron acceptor the excited state produces Ru(III) and a reduced acceptor in the first step which is followed by an intramolecular electron transfer from the tyrosyl moiety to Ru(III). The model system was created to mimic electron transfer reactions in photosystem II.[175] The photoreduction of iron(III) to iron(II) porphyrin *via* one-electron transfer from an axial ligand has been the subject of several recent investigations. Irradiation of acidic anaerobic aqueous solutions of iron(III)tetrakis(2-N-methylpyridyl)porphyrin, in the presence of mono- or di-basic amino acids or their N-acylated derivatives, generates the iron(II) porphyrin and the corresponding acyloxy radical; subsequent decarboxylation of the latter gives the corresponding ammonioalkyl or amidoalkyl radical. The rate of the photoreduction depends on several factors, *e.g.* on the separation between and carboxylic groups in the amino acid.[176]

A facile and versatile method for the chemical dehydrogenation of α-amino acids chelated to ruthenium(II) using cerium(IV) nitrate as oxidizing reagent was reported.[177] The reaction mechanism for the formation of (α-imino-acidato)bis(phen)ruthenium(II) was published.[177] Because of the similar behaviour of macromolecules and enzymes, micellar catalysis is nowadays of special interest. The oxidation of Leu by vanadium(V) in the presence of micelle[178] and the kinetics of the interaction of Cd(II)-His complex with ninhydrin in the absence of and in the presence of cationic and anionic micelles were studied. Anionic micelles catalysed the latter reaction, whilst the cationic micelles inhibited it.[179]

The kinetics of oxidation of amino acids by high oxidation state metal ions, *e.g.* α-amino-4-imidazolepropionic acid or L-Asp, L-Phe, L-Ser and L-Met by Ce(IV),[180,181] L-Asp, L-His and L-Glu by Mn(III),[182,183] L-Asp, L-Glu, L-Asn, L-Arg, L-Leu by MoO$_4^-$,[184-190] L-Glu by Tl(III),[191] Cys and penicillamine by Fe(III),[192] Ser, Thr by Cr$_2$O$_7^{2-}$,[193-195] is discussed in many papers. Metal ion catalysed oxidation of amino acids by O$_2$ or by other oxidizing agents is also discussed.[196,197] Model compounds for antitumor-active platinum(IV) prodrugs, *trans*-[Pt(CN)$_4$X$_2$]$^{2-}$ (X = Cl or Br) can be reduced to [Pt(CN)$_4$]$^{2-}$ by L-Met. The oxidation product of Met is methionine S-oxide. The reduction by Met is considerably slower than reduction by thiol-containing biomolecules, *e.g.* Cys, penicillamine or glutathione in neutral solution, but the rates are comparable under acidic conditions. In the proposed mechanism the thioether groups of Met interact with coordinated halide, mediating the electron transfer to the platinum(IV) centre in the transition state.[198] The central metal ion was oxidized to chromium(VI) in chromium-(III)–DL-Val[199] and in chromium(III)–L-Arg[200] complexes by periodate. It was proposed that electron transfer proceeded through an inner-sphere mechanism *via* coordination of IO_4 to chromium(III).

Leucine *p*-nitrophenyl ester undergoes slow spontaneous first-order hydrolysis at pH 6.2. Chiral cyclopalladated complexes were found to catalyse the enantioselective hydrolysis of the above ester and the catalytic effect was strongly dependent on the absolute configuration of the stereogenic centers in both the Leu ester and the complex. Molecular modelling suggests that the stereoselectivity results from the hydrophobic/stacking interaction between the leaving 4-nitrophenolate and the pyridine moiety of the complex **(39)**. Less efficient enantioselectivity was observed for the hydrolysis of Met ester.[201] The catalytic activity of lanthanide ions on the hydrolysis of alkyl esters and amides of α-amino acids was also studied.[202]

(39)

3 Peptide Complexes

3.1 Synthesis and Structural Studies on Peptide Complexes – Peptide molecules are versatile and selective ligands for metal binding and the study of their complex formation reactions has received increasing attention in the last two years. It has already been widely accepted that stable complex formation with peptides requires the simultaneous deprotonation and coordination of terminal amino and neighbouring amide groups. The coordination geometry and structural parameters of the various peptide complexes are, however, significantly influenced by the presence and location of side-chain residues. As a consequence, it is not possible to outline the general features of the coordination chemistry of all peptide molecules, but the most recent and most important findings can be classified as functions of the nature of the central metal ion and the amino acid sequences of the peptide molecules.

It has already been reported that platinum(II) is one of the most effective metal ions to promote amide deprotonation and coordination, but the slow formation kinetics and the existence of various isomers made it difficult to characterize the platinum(II)–peptide interactions. The complexes H[Pt(digly)Cl], K[Pt(digly)]Cl (**40**), H[Pt(digly)Cl$_2$] (**41**) and K[Pt(digly)Cl$_2$] have been synthesized recently and their structures elucidated by X-ray crystallography[203] (digly stands for the dianion of Gly-Gly).

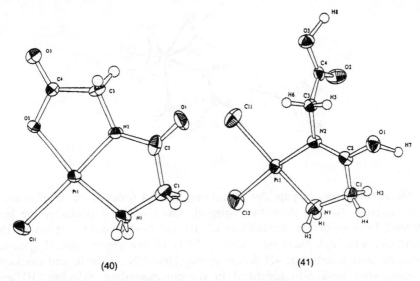

(40)   (41)

The binding mode of Gly-Gly in the species (**40**) is described by the usual coordination of terminal amino and deprotonated amide nitrogen and carboxylate oxygen donor atoms. Species (**41**) was obtained in strongly acidic media and is characterized by bidentate coordination of 2N donor atoms, while the carboxylate and carbonyl-O donor atoms are protonated. The latter binding mode of amide group, namely deprotonated amide-N and non-coordinated, protonated carbonyl-O, is rather unique in metallopeptide chemistry. Antitumor activity and toxicity of the complexes were also tested and compared with those of cisplatin.[203]

Vanadium is known to be an important metal for the function of a number of enzymatic reactions and it was reported to show insulin-mimetic properties. These observations stimulated a great deal of interest in the complex formation of oxovanadium(IV), vanadate(V) and peroxovanadate(V) with peptides.[60,204,205] The systems VO^{2+}–L (where L = Gly-Gly, Gly-Gly-Gly, Ala-Gly, Gly-Ala, Ala-Ala, Gly-Gly-Ala and Ala-Gly-Gly) have been studied by the combination of various spectroscopic techniques.[204] In the case of dipeptides the existence of various isomers was detected with the involvement of amide nitrogens in metal binding in alkaline solutions. For tripeptides there was no evidence of amide deprotonation and coordination, oligomeric complexes being the major species. Hydroxamidovanadate derivatives of the dipeptides Gly-Gly and Gly-Tyr and the tripeptide Gly-Gly-Gly were studied in solution and in the solid state[205] and were used to mimic the properties of peroxovanadates. The structure of the species [V(O)(ONH$_2$)$_2$(Gly-Gly)] **(42)** was determined by X-ray diffraction methods and bidentate (NH$_2$,CO) chelation of the peptides was shown in all cases. The coordination geometry of vanadium is described as a pentagonal bipyramid with amino coordination in the plane of the two hydroxylamine ligands and carbonyl oxygen are in the apical position.

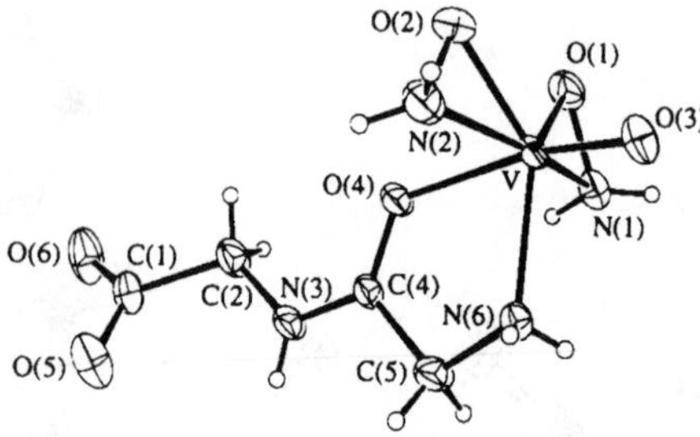

(42)

Metal complexes of peptides containing histidyl residues attracted continuous interest. Among them the tripeptide Gly-Gly-His is probably the best studied, because of the outstanding Cu(II) and Ni(II) binding ability of the molecule. The high thermodynamic stability of the species [MLH$_{-2}$] arises from the coordination of 4N donor atoms (NH$_2$, 2N$^-$, N(im)), and the same binding sites were also identified in the corresponding palladium(II) and gold(III) complexes.[206] The crystal structure of the species [Au(Gly-Gly-HisH$_{-2}$)]Cl·H$_2$O **(43)** was determined by X-ray crystallography and almost the same structure was reported for the corresponding palladium(II) complex.

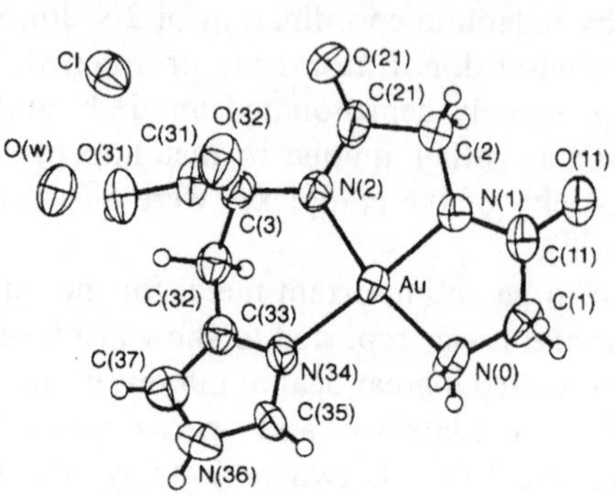

(43)

It is important to note that in contrast with common tripeptides the carboxylate residue is non-bonded in these species and can be protonated in slightly acidic media (pH < 3). Both complexes undergo further deprotonation in alkaline solutions: in the case of [PdLH$_{-2}$] the deprotonation is attributed to the N1(H) group of N3-coordinated imidazole (pK ~ 11.30), while two deprotonation reactions were reported to occur for the gold(III) species with $pK_1 = 8.63$ (N1(H) of imidazole) and $pK_2 = 11.50$ (coordinated terminal amino group).[206] Multinuclear NMR studies (^{1}H, ^{13}C and [195]Pt) were used to identify the products and binding sites of His-Ala and His-Gly-Ala in the reaction with various palladium(II) and platinum(II) containing species.[207] The results show that all palladium(II) complexes ([PdCl$_4$]$^{2-}$, [Pd(dien)]$^{2+}$ and [Pd(en)]$^{2+}$) can easily substitute all protonation sites of histidyl residues (terminal NH$_2$ and N1 and N3 of imidazole) in very acidic media (pH < 1). The formation of dimeric species *via* (N1,N3)-imidazole bridging and the existence of (NH$_2$,N3)-chelates were also detected. In the case of platinum(II) containing systems the terminal carboxylate oxygen atoms were identified as the primary ligating groups of peptides followed by the formation of the thermodynamically preferred Pt–N bonded species.

The reaction of zinc(II) ions with the terminally protected tetrapeptide Boc-Glu-Thr-Ile-His-OMe as a mimic of the active site of proteases was studied by NMR and fast atom bombardment mass spectrometry.[208] The results revealed that the imidazole nitrogen donor atom of the histidyl residue is the preferred zinc binding site of the molecule, but the formation of a bis complex *via* the coordination of imidazolyl and glutamyl amide nitrogen donor atoms was also proposed. The complexation of zinc(II) with His-His-Gly-Gly and His-Gly-His-Gly was studied by ^{1}H NMR spectroscopy over a wide pH range.[209] The formation of 6-membered chelates (NH$_2$, N3(im)) was detected in slightly acidic media (pH < 6) in both cases. The complex formation processes of the two tetrapeptides were, however, different in neutral and slightly alkaline media. In the case of His-His-Gly-Gly the first amide group (from His-His) underwent deprotonation and coordination around pH 7 and the coordination of the first His residue was replaced by the second one in a 3N species (NH$_2$,N$^-$,N3). The fourth coordination site of zinc(II) was probably occupied by the imidazole-N of another Zn(II)-His-His-Gly-Gly complex. Deprotonation and coordination of the amide groups were not observed in the zinc(II) complexes of the other tetrapeptide, His-Gly-His-Gly, the chelate from the N-terminus being intact even in alkaline solutions. The protonation sites and various transition metal complexes of a linear octapeptide containing alternate His and Gly residues (HGHGHGHG) were studied by CD spectroscopy.[210] It was found that the addition of three protons to the free peptide (to the amino and 2 imidazole nitrogen atoms) strongly stabilizes a folded structure characterized by hydrogen bonding. The addition of metal ions caused similar changes and the folding ability of the various metal ions correlated well with their imidazole binding strength: Cu(II) $\cong$ Ni(II) > Cd(II) > Zn(II).

Another large group of oligopeptides studied recently contained both histidyl and cysteinyl residues and they were used to mimic the binding properties of various metalloenzymes and zinc finger proteins. Nickel(II) complexes of various tripeptides of the type XCH (where X = Gly or Lys, C = Cys as free thiol, S-protected or disulfide and H = His with free carboxylate or carboxamide at the C-termini) have been prepared and structurally characterized.[211] The carboxamide terminal tripeptide ligands with S-protected thiols behave similarly to the well-known Ni(II)-Gly-Gly-His-NH$_2$ system, forming air-stable, diamagnetic, square planar nickel(II) complexes with 4N-coordination (Scheme 7a). The analogous carboxylate terminal peptides have the same binding sites, but undergo spontaneous oxidative decarboxylation in the presence of nickel(II) (Scheme 7b). All complexes containing free thiol groups reacted with O$_2$ to yield disulfide-bridged dimeric complexes (Scheme 7c). The reaction of nickel(II) with the completely unprotected forms of tripeptides in the presence of air resulted in disulfide bond formation, but without decarboxylation.

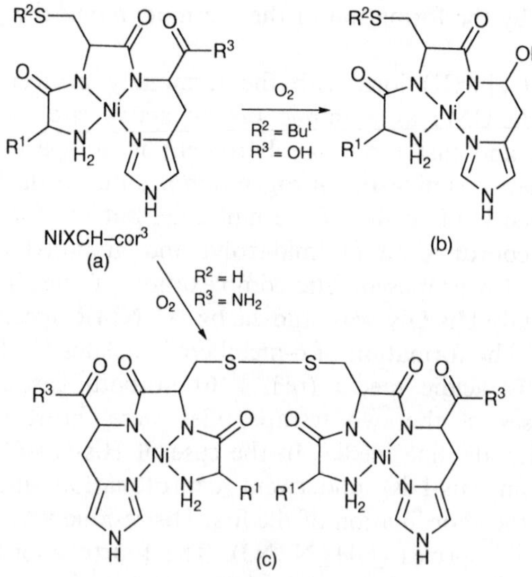

Scheme 7

The N-protected pentadecapeptides having Cys$_2$His$_2$ or Cys$_2$Asp$_2$ metal binding motifs have been synthesized and their interaction with cadmium(II), cobalt(II) and zinc(II) studied by spectroscopic techniques.[212] The formation of 1:1 complexes was detected in all cases and the relative affinity of metal ions to bind peptides followed the order: Cd(II) $\sim$ Zn(II) > Co(II). The coordination geometry of cobalt(II) complexes was described as tetrahedral in the peptides containing histidyl residues. Surprisingly, the peptides containing Asp residues did not show spectral changes upon the addition of cobalt(II), supporting the view that this peptide does not bind cobalt(II) tetrahedrally and that the Cys residue is not a metal binding site. Mercury and copper complexes of a series of tetrapeptides containing cysteinyl and histidyl residues have been synthesized and characterized by elemental analysis, conductivity measurements and various spectroscopic techniques.[213,214] The results were interpreted in terms of the formation of macrochelates, in which the cysteinyl sulfur and histidyl imidazole nitrogen donor atoms are the major metal binding sites. The high affinity of copper(II) and nickel(II) for peptides containing two cysteine and/or histidine residues has been shown by electrospray ionization mass spectrometry and metal chelate affinity techniques, too.[215]

Peptides containing cysteinyl residues as the possible metal binding sites in addition to the peptide backbone are still the focus of interest.[216-220] The bis(cysteinyl) sequences, Cys-X_n-Cys, are especially important among these, because they occur naturally in various zinc containing enzymes, zinc fingers and also in metallothioneins. Zinc(II) complexes of the terminally protected forms of the peptides Cys-Cys, Cys-Gly-Cys, Cys-Phe-Cys and Cys-Gly-Ile-Cys have been prepared and their structures elucidated by 2D NMR spectroscopy.[217] Sulfur atoms of Cys residues were identified as the exclusive metal binding sites and the coordination of zinc(II) ions resulted in the folding of peptides. The similarity between the model complexes and their natural equivalents suggests that zinc does not simply stabilize the structure, but is also an essential component for the folding of the corresponding peptide segments.[218] *In vivo* anti-influenza virus activity of a 19-amino acid peptide (a zinc finger peptide) has been monitored and it was concluded that zinc finger peptides may provide a new class of antiviral agents effective against influenza virus.[219] The mercury complexes of stoichiometries [(Boc-Cys-Pro-Leu-Cys-OMe)(S-*tert*-C_4H_9)Hg]$^-$ and [(Boc-Cys-Pro-Leu-Cys-OMe)Hg] have been prepared and the coordination geometries of the metal ions were identified as trigonal planar and linear, respectively.[220] On the basis of molecular dynamics calculations it was concluded that the trigonal planar metal ion regulates hydrogen bonding modes of peptides in the same manner as tetrahedral, but different from linear, metal ions.

Alkali and alkaline earth metal ions are generally considered as very weak Lewis acids in the interactions with amino acids or peptides. On the other hand, these metal ions are also required in specific stoichiometries for cellular processes to promote enzyme activity and protein function. For example, sodium ion must interact with the appropriate peptides or proteins in order to carry out its regulatory or structural functions. These interactions can be quite complex,[221-224] because peptides and proteins are polyfunctional ligands. The sodium ion affinities of several cyclic and linear dipeptides and of selected derivatives have been determined in the gas-phase[221] and the following order of peptide affinities was obtained (in kJ mol^{-1}): *cyclo*-Gly-Gly (143) < *cyclo*-Ala-Gly (149) < *cyclo*-Ala-Ala (151) < AcGly (172) < Gly-Gly (177) < Ala-Gly (178) < Gly-Ala (179) < Ala-Ala (180) < Gly-Gly-OEt (181) < Gly-Gly-NH_2 (183). The differences in the Na$^+$-ion affinities were explained by the different binding sites, namely the monodentate coordination of carbonyl oxygen donors was suggested for cyclic dipeptides, while linear molecules can coordinate sodium ions in a multidentate arrangement involving at least two carbonyl and possibly the amino groups. Both theoretical calculations and mass spectrometric measurements indicated that sodiated oligoglycines do not form salt bridge structures in the gas-phase.[222] The higher cross sections of the protonated forms of oligoglycines relative to those of sodiated species were interpreted in terms of more compact and spherical arrangements of the protonated ligands.

The affinity of alkaline earth metal ions for peptides is significantly higher than that of sodium ion, which makes the preparation of complexes in the solid phase much easier. The first calcium(II) complex of a linear synthetic peptide, Boc-Gly-Nleu-Nleu, has been recently prepared and its structure determined by X-ray diffraction techniques.[223] The formation of a trimeric complex (44), $Ca_3(Boc-Gly-Nleu-NleuO)_6$, has been identified, and the

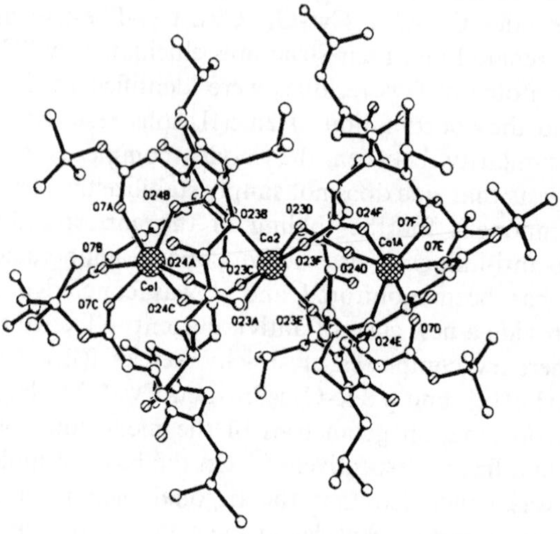

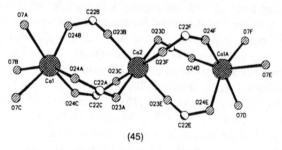

(44)

calcium binding site has been described by a bischelate complex with the arrangement of oxygen donor atoms corresponding to a pseudo-octahedral geometry (45). It is important to note that complex (44) is rather hydrophobic

(45)

on the surface, where the methyl groups of the Boc-termini and the isobutyl side chains are located, while the charged carboxylate oxygen atoms are positioned in the central core of the globular complex. Calcium(II) complexes of derivatives of tetraproline have been studied by various experimental techniques and it was found that the peptides show very different conformational versatilities and metal binding abilities in solution.[224]

Several oligopeptides and their analogues are directly used in clinical therapy or are considered as the active sites of peptide hormones and metalloenzymes. Melanostatin (Pro-Leu-GlyNH$_2$) is a tripeptide neuro-hormone that is responsible for modulation of dopamine receptors. The structure of the copper(II) complex of melanostatin was studied in DMSO solution by NMR relaxation rate analysis and EPR spectral simulation.[225] In agreement with expectations only 1 : 1 complexes are formed at all metal ion to ligand ratios and in the major species the central metal ion is coordinated by 4N donor atoms (NH$_2$ + 3N). The EPR parameters of the species, however, suggested a strong tetrahedral distortion of the usual square planar arrangement of 4N complexes. Gonadotropin releasing hormone (GnRH = Glp-His-Trp-Ser-Tyr-Gly-Leu-Arg-Pro-Gly-NH$_2$) is a decapeptide and plays an important role in the mediation of neuroendocrine control of reproductive processes. It has also been proposed that several metal ions including copper(II) significantly influence the activity of the hormone. The potency of the copper(II), nickel(II) and zinc(II) complexes of the peptide molecule in binding with GnRH receptors was studied under 'in vivo' conditions.[226] It was found that Cu(II)-GnRH competed a little more effectively than the free hormone, while a small, but opposite change was observed for the nickel(II) and zinc(II) complexes.

Glutathione (γ-Glu-Cys-Gly) is the most common naturally occurring tripeptide and plays key physiological roles. The interaction of zinc(II) and glutathione was studied by differential pulse polarography at pH 8.5 in borate buffer.[227] The formation of dinuclear complexes with 2 : 2 = Zn(II) : GSH stoichiometry was suggested. The proposed structure of this species contains bridging thiol groups, while the other binding sites of the four-coordinated metal ions are occupied by the glutamyl amino and glycyl carboxylate residues. Cadmium(II) complexes of glutathione were studied under similar conditions and in contrast with the zinc(II) containing systems the formation of both 1 : 2 and 2 : 2 complexes was detected.[228] The binding mode of the mononuclear bis complex is described *via* the coordination of thiol and glycyl carboxylate residues with uncoordinated and protonated amino groups (**46a**). Metal binding sites of the dinuclear complex (**46b**), however, were interpreted in the same way as described for the zinc(II) complexes.[227,228]

The Cd–S interaction is of major biological significance due to the high

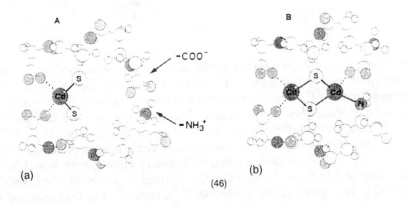

(a)   (b)

(46)

toxicity of cadmium. In all probability, glutathione plays an important role in detoxification of cadmium by chelating and transporting the metal ion. As a consequence, the influence of glutathione and phytochelatins on the formation of CdS bionanocrystallites have been studied.[229] The general formula of phytochelatins is (γ-Glu-Cys)$_n$Gly (n = 2–11) showing the structural similarity with glutathione. The results revealed that major differences exist in the glutathione- and phytochelatin-capped CdS nanocrystallites. In the case of glutathione significant size-heterogenety was observed, while much less dispersion was characteristic of phytochelatins. On the other hand, significantly less glutathione was required to complex cadmium than in the case of phytochelatins.

At present, many studies on the coordination chemistry of rhenium and technetium are directed to the production of ^{99m}Tc and [186]Re or [188]Re radiopharmaceuticals, targeted *via* bioactive molecules including various peptides.[4,230-236] The molecules that bind specifically the cancer cells are labelled with ^{99m}Tc for diagnosis, while the isotopes of rhenium are used in the radiotherapy. Oxorhenium(V) and oxotechnetium(V) complexes of a tetrapeptide analogue, dimethyl-Gly-Ser-Cys-GlyNH$_2$, were prepared and characterized by various experimental techniques.[230] The complexes exist as two isomers, the serine –CH$_2$OH group being in the *syn* or *anti* conformations with respect to the Re-oxo bond. The ratio of the isomers is 1 : 1.1 at room temperature. The molecular structure of the *syn* isomer of the rhenium complex (47) was determined by X-ray crystallography. The metal ion is

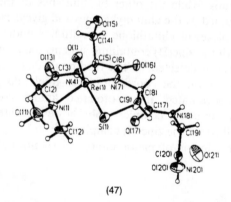

(47)

coordinated in a distorted square pyramidal geometry with two oxo moieties in the apical position and the terminal amino and two deprotonated amide nitrogen (from the Ser and Cys residues) and thiolate sulfur donor atoms being the metal binding sites of the ligand. The same species were formed with oxotechnetium(V), but the ratio of isomers is approximately 1 : 1.

Oxorhenium(V) complexes of two cysteine containing peptides, Gly-Gly-Cys and Cys-Gly, have been studied by capillary electrophoresis and X-ray absorption spectroscopy.[231] Gly-Gly-Cys formed 1 : 1 species with ReO(V) units with SN$_3$ coordination mode (48a). In contrast, the formation of bis complexes was described with Cys-Gly, in which the terminal amino and thiol groups (48b) are the major metal binding sites. The rhenium(V) derivatives of two other Cys containing ligands, Gly-Cys and Gly-Cys-Gly, have also been prepared and stoichiometries and binding modes of the ligands determined.[232]

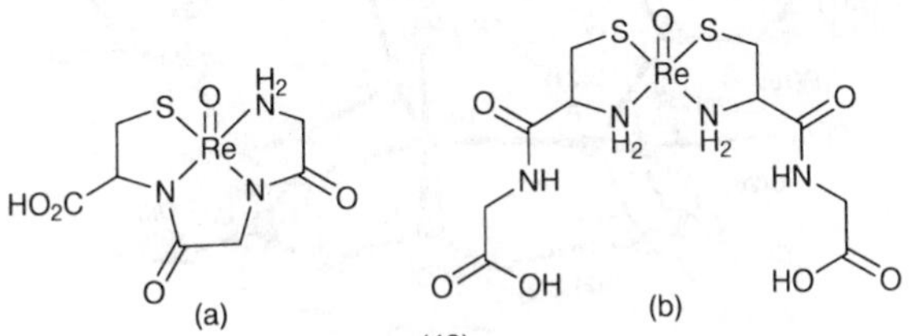

(48)

One of the most common groups of chelating agents used for binding of oxotechnetium(V) and oxorhenium(V) contains the N_2S_2 donor set. The synthesis of a new series of pseudotripeptide based tetradentate ligands with the same metal binding sites have been published and their potential applications are discussed.[233] Another promising route to improve the coordination properties and biological applicability of complexing agents is the introduction of new strong metal chelating groups into the structural frameworks of biomolecules. Derivatives of cysteine containing dipeptides with phosphine groups at the N-termini have recently been prepared and their rhenium and technetium complexes studied.[234,235] The crystal structures of two rhenium(V) complexes have been determined[235] and distorted octahedral coordination of the metal ions was shown in both cases (49,50).

A large number of papers have been published on ternary complexes of transition elements containing peptides as one (or both) of the coordinating

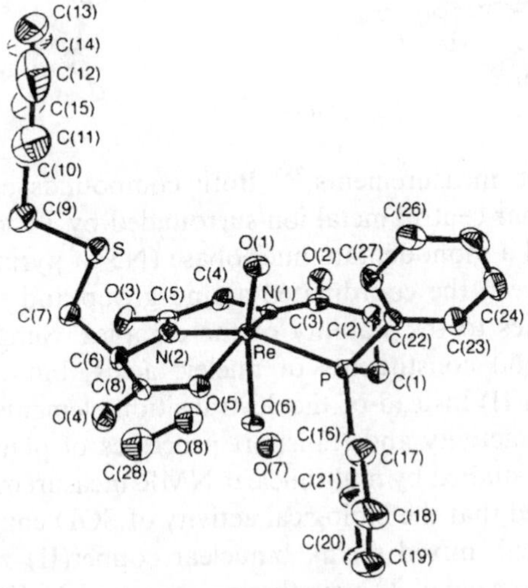

(49)

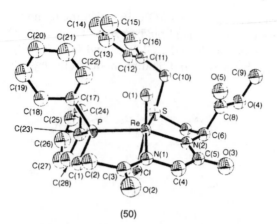

(50)

ligands. Some of these papers provide structural characterization of mixed ligand systems,[237-240] while the other group of studies is devoted to solution equilibria of ternary systems and they will be discussed in Section 3.3. Ternary systems containing dipeptides and nucleobases or nucleotides are the simplest models to get reliable information on the metal ion assisted interactions of proteins and nucleic acids. The mixed complexes, (Gly-Gly)(isocytosine)copper(II)-dihydrate (**51a**) and (Gly-Gly)(6-methylisocytosine)copper(II)-monohydrate (**51b**) have been prepared and characterized by X-ray diffraction and

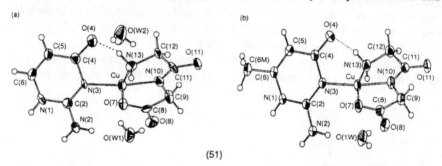

(51)

various spectroscopic measurements.[237] Both compounds contain a slightly distorted square planar central metal ion surrounded by a tridentate dipeptide (NH$_2$,N$^-$,COO$^-$) and a monodentate nucleobase (N3 of pyrimidine) ligand. A hydrogen bond between the coordinated amino group and the oxo group of isocytosine contributes to the stability of the complex. Similar ternary complexes of peptides and constituents of nucleic acids, but containing palladium(II) or platinum(II) instead of the 3d transition elements, can be used to mimic the biological activity and transport processes of platinum containing drugs and these were studied by multinuclear NMR measurements.[238,239]

It is widely accepted that the biological activity of SOD enzymes is linked to an imidazolate-bridged, mixed metal, binuclear copper(II)–zinc(II) centre at the active site of the enzyme. The synthetic compound Na[(Gly-Gly)Cu–Im–Zn(II)(Gly-Gly)] has been prepared and characterized recently.[240] The results suggest that the imidazolate-bridged complex is stable in the pH range 7 to 10, but breaks down on the copper(II) side in slightly acidic media.

The metal ion coordination of deprotonated amide groups of peptide molecules is probably the most crucial factor in the coordination chemistry of peptides. As a consequence, model compounds having the amide functions and other ligating groups in chelatable positions are structurally related to peptides and have received increasing attention recently. Captopril, enalaprilat and lisinopril (Figure 2) are angiotensin-converting enzyme inhibitors and their copper and iron complexes were studied by UV-spectroscopy and electrospray mass spectrometry.[241] The formation of 1:1 complexes as the major species was detected in all cases, but iron was not as effective as copper for the formation of metal complexes.

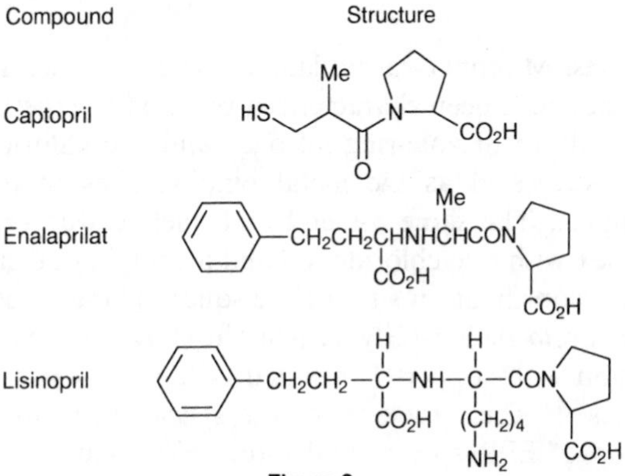

Figure 2

The binding of lisinopril to zinc(II), copper(II) and nickel(II) has been studied by a potentiometric method in solution and the crystal structure of a dimeric copper(II) complex (**52**) was also determined.[242] The protonated complex ([CuHL]$^+$) was detected as the major species in neutral or weakly acidic solution containing a tridentate lisinopril molecule coordinated *via* the carboxylate-O, secondary amine-N and carbonyl-O donor atoms, while the amino group of the lysyl residue is protonated. The same binding mode of the ligand is described in the dimeric species [Cu$_2$(HL)$_2$(H$_2$O)$_2$]Cl$_2$ obtained in the solid state (**52**) containing distorted square pyramidal metal ions bridged *via* the prolyl carboxylate of another ligand. The fifth coordination site of the square pyramid is occupied by a water molecule.

It has already been reported that the sulfur donor captopril reacts to form polymeric species with gold(I) and the interaction of this polymeric complex with selenocyanate, selenourea and thiourea was studied by NMR spectroscopy.[243] Fast redox reactions of the intermediates were reported in the reaction with selenium containing drugs, while thiourea gave a stable ternary complex [thiourea-Au(I)-captopril]. Famotidine is an effective stress-ulceration inhibitor and it offers a set of potential coordination sites for metal ions. It has been shown that famotidine is an excellent chelating agent for palladium(II)

O(9)
Cl(1) — O(8)
O(10) O(7)

N(3)   C(13a) O(7a)
N(3a) O(8a) O(9a)
C(12) C(7) C(12a) Cl(1a)
C(13) C(6)   O(10a)
C(8)   O(5) C(11a) O(3a)
C(10) N(2) C(9) O(2a) C(3a)
C(14) C(2) C(1)   C(10a)
C(15) C(4) C(5)   C(4a) C(14a)
C(16)   O(1) O(5a) O(4) N(1a) C(15a) C(21a)
C(21) N(1) O(1a)   C(2a)   C(20a)
20) C(18) O(3) O(2) Cu(1) N(2a) C(1a) C(16a)
C(19) C(17) C(3)   O(6a) C(17a) C(19a)
O(4a) C(5a) C(8a)   C(18a)
O(6) C(9a)   C(7a)
C(6a)

(52)

and platinum(II) ions. Mononuclear palladium(II) and dinuclear platinum(II) famotidine complexes have been characterized by various experimental techniques.[244] Thioether sulfur, thiazole-ring nitrogen and two chloride ions in a *cis* arrangement were suggested as the metal binding sites in the monomeric palladium(II) complex. The same 5-membered chelate was proposed in the platinum(II) complex with one chloride ion and a bridging guanidine nitrogen occupying the other coordination sites of the square planar metal ion. A new cyclic octapeptide, *cyclo*-Ile-Ser-(Gly)Thz-IleThr-(Gly)Thz, has been synthesized and its reaction with copper(II) resulted in the formation of mono- and dinuclear complexes.[245] The structures of the species in solution were elucidated by simulation of EPR spectra and force field calculations. The formamide ligand can be considered to be the most simple model of peptide functions. The primary products formed in the reactions between formamide and copper(I) ion in the gas phase correspond mainly to formamide–Cu^+, (formamide)$_2Cu^+$, formamide–$CuNH_3^+$ and HNCO–$CuHCu^+$ complexes. Theoretical calculations support the findings that attachment of Cu^+-ion takes place preferentially at the carbonyl group, while interactions at the C–N bonds are energetically disfavoured.[246]

An interesting feature of the coordination chemistry of oligopeptides is that they can stabilize the unusually high oxidation states of various transition metal ions. Similar capacity of various oxamides and related ligands has been reported for the complexation reactions with copper(III).[247] The stabilization of the trivalent oxidation state of copper is attributed to the increasing number of deprotonated amide functions in the coordination sphere of the metal ions. A perfect correlation between Cu^{III}–Cu^{II} redox potentials and the visible absorption maxima of the copper(II) complexes has been found and explained by the relative gain in the crystal field stabilization energy during oxidation reactions.

Numerous strapped and capped metalloporphyrins have already been prepared for the investigation of ligand binding and catalytic activity of heme proteins and related ligands. A novel, water-soluble peptide-porphyrin assembly has been prepared, in which a conformationally stable peptide is strapped over the face of a modified zinc(II)-porphyrin system (53). The sequence of the peptide is designed such that the four histidine side chains can bind two metal ions on then same side of the helix (Pep1).[248]

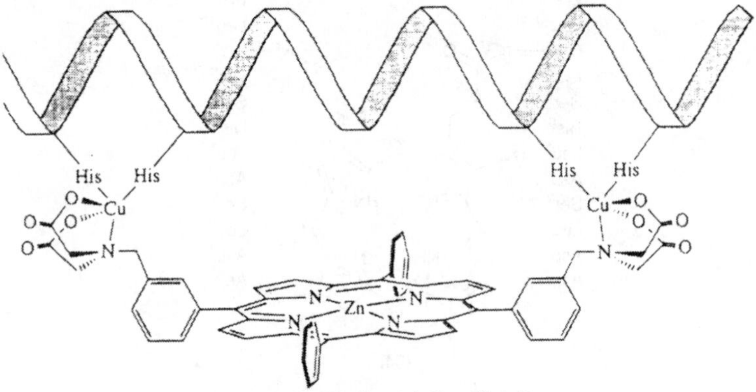

Assembly I: [Zn(Por1)]-Cu$_2$-(Pep1)

Pep1 = Ac-YEAHAKAHAAAAAAHAEAHAKK-CONH$_2$

His-X$_3$-His   His-X$_3$-His

Pep2 = Ac-YEAHAKAHAAAAAAAAEAAAKK-CONH$_2$

His-X$_3$-His

(53)

To prove the existence of the closed strap assembly the analogous compound with only two His residues (Pep2) has also been prepared. Another group of synthetic studies on heme–peptide complexes was devoted to new hemoprotein models designed to investigate the effect of protein structure and composition on the activity of heme centers.[249,250] Mimochrome I and II (54) are representative examples of these model compounds and the results obtained on their iron(III) and cobalt(III) complexes show that the composition and folding of the peptide chains critically affect the overall structural organization of the models.

Multinuclear and paramagnetic NMR spectroscopic measurements are increasingly used experimental techniques for the elucidation of the metal binding sites of proteins, peptides and related ligands. For example, the use of ^{113}Cd NMR chemical shift proved to be very successful structural probe in the identification of tetrathiolate metalloproteins.[251] The metal binding sites of *Rhus vernificera* stellacyanin have been elucidated by paramagnetic ^{1}H NMR measurements on the nickel(II) substituted derivatives.[252] Evidence was obtained which indicates that the metal ion is coordinated to Cys, two His and the amide oxygen of Gln residues. Bleomycin and tallysomycin are closely related anticancer antibiotics activated *via* their iron(II) complexes. Zinc(II) can be used as a structural probe for the identification of the iron binding sites of the molecules. 2D NMR studies and molecular mechanics/molecular

NH$_2$ NH$_2$
Leu^{14} Leu^{14}
Thr^{13} Thr^{13}
Ile^{12} Ile^{12}
Lys^{11} Lys^{11}
Leu^{10} Leu^{10}
Lys^9—N$^\varepsilon$ N$^\varepsilon$—Lys^9
Lys^8 Lys^8
Ser^7 Ser^7
His^6 His^6
Leu^5 Leu^5
Asp^4 Asp^4
Ser^3 Ser^3
Leu^2 Leu^2
Asp^1 Asp^1
Ac   Ac

(54)

dynamics calculations on the zinc(II) complexes of the antibiotics revealed that metal ion coordination takes place *via* 5N-donors and the chirality of the free and complexed ligands has been assessed.[253]

3.2 Kinetics and Reactivity – Metal-ion-assisted Transformations of Peptide Molecules – Structural characterization and solution equilibria of peptide complexes have been the subject of numerous studies in the literature for a long time, but only a few papers were available on the kinetic behaviour of complex formation reactions between metal ions and peptides. On the other hand, the investigation of the metal ion promoted transformations of peptides including the oxidation, decomposition, formation or hydrolytic cleavage of peptides received increased attention in the last two years. The sulfur containing amino acids are generally at the center of these studies, but the significance of derivatives of histidine or other amino acids is also increasing.

S-Nitrosothiols of thionitrites (RSNO) have been the focus of much attention in very recent times, because they are believed to play an important role in the functions associated with nitric oxide in the body. It has been known that metal ions promote decomposition of these molecules and kinetic studies have been performed for the decomposition of S-nitrosothiols by mercury(II) and silver(I) salts.[254] The ligands involved S-nitroso derivatives of amino acids, N-acetyl amino acids and peptides (*e.g.* S-nitrosocysteine, S-nitroso-N-acetyl-penicillamine, S-nitroso-glutathione, *etc.*). It was found that the mercury(II) catalyzed decomposition is first order in both reactants and the products are nitrous acid and the corresponding thiol–Hg^{2+} complex. Silver(I) reacts similarly, but in a much slower process and the reaction is characterized by a second order kinetic term. There was very little variation in the rate constants with RSNO structure, except the case of S-nitroso-N-acetyl-penicillamine, which underwent the fastest reaction and this was explained by the highest thermodynamic stability of the corresponding Hg(II) complexes.

The kinetics of complex formation between $[Pd(H_2O)_4]^{2+}$ and glutathione was investigated in the presence of sodium dodecyl sulfate (SDS), an effective surfactant. Acceleration and retardation of complex formation in the presence of SDS was observed in strongly acidic (up to pH 3.5) and in slightly acidic (3.5 < pH < 5) pH ranges, respectively. This was explained by electrostatic interactions and the results suggest that kinetic measurements can be used for the precise detection of surfactant concentrations at which micelle formation occurs.[255]

It is generally accepted that DNA coordination at guanine N7 is of key importance in the mechanism of action of platinum containing anticancer drugs. On the other hand, platinum(II) compounds are known to exhibit relatively high kinetic affinity for sulfur containing ligands and concentration dependent nephrotoxicity of cisplatin has been ascribed to its reaction with Cys or Met residues of metalloproteins. As a consequence, the interactions of platinum(II) anticancer drugs and their palladium(II) analogues with sulfur containing ligands and the ternary systems containing both nitrogen and sulfur donors attracted much attention in the last two years. The intramolecular migration of $[Pt(dien)]^{2+}$ from the thioether sulfur to imidazole nitrogen donor atoms of His-Met has been proved by HPLC and NMR studies.[256] The concentration of the kinetically favoured S-bound complex reached a maximum in 2-3 hours and then slowly isomerised to the thermodynamically favoured imidazole(N1)-bonded species with a half-life of about 40 hours (at 313 K and pH 6.5). Taking into account that the *in vivo* half-life of platinum(II) drugs is several days the reactivity of histidyl residues should be considered under biological conditions. Similar conclusions were reached in another study in which the selective displacement of the S-bound methionine adduct of $[Pt(dien)]^{2+}$ was promoted by histidine containing ligands and the formation of Pt–N3(imidazole) adducts was observed.[257] The studies performed on the interaction of $[Pt(en)]^{2+}$ with di- and tripeptides containing histidyl and methionyl residues led to the same conclusions.[258,259] $[Pt(en)(H_2O)_2]^{2+}$ has two coordination sites easily available for binding of side-chain donor groups, thus the speciation of the systems is more complicated. Both HPLC and NMR studies indicate, however, the kinetic preference of platinum(II) for thioether binding, which results in the formation of a macrochelate if Met residues are not N-terminal. The structure (55) of the 12-

(55)

membered macrochelate formed in the reaction of $[Pt(en)]^{2+}$ with *cyclo*-His-Met involves the coordination of imidazole(N1) and thioether donor atoms. The unusually large chelate ring exhibits a remarkable long-term kinetic stability even in alkaline solutions, although the coordination of nitrogen donors is thermodynamically more favoured under these conditions.

The reaction of the carboplatin analogue [Pt(en)(Me-Mal)] with methionine containing peptides was also studied by HPLC and NMR techniques.[260] The presence of the species shown in Scheme 8 was detected in the reaction with N-acetylmethionine and similar species were suggested for dipeptides of methionine. The ring-opened adducts (see **a** on Scheme 8) were surprisingly stable ($t_{1/2} \sim 8.5$ hours at 310 K) and it was explained by intramolecular hydrogen bonding. For peptide complexes the importance of ring-opened adducts is characteristic for internal or C-terminal Met residues, whereas the reaction with N-terminal methionyl residues may lead to stable (S,N)-chelated species. The latter complexes can undergo further activation with amine release due to the high *trans* influence of coordinated sulfur donor atoms.[260]

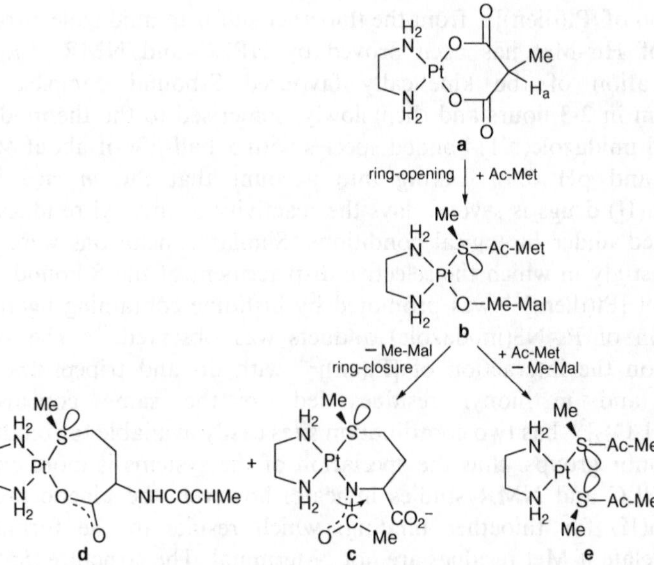

Scheme 8

The interactions of thiols with platinum complexes are often considered to have a negative effect on antitumor activity of the complexes and to be responsible for drug inactivation. The reactions of *cis*-[PtCl$_2$(NH$_3$)(2-pic)] and *cis*-[PtCl$_2$(NH$_3$)(3-pic)] with 5'-GMP and glutathione was studied to compare the effects of nitrogen and thiolate coordination.[261] The rate for binding of glutathione is greatly slowed for the first ligand due to steric requirements, but is still faster than that of 5'-GMP. On the other hand, bis(5'-GMP) adducts of platinum(II) complexes were able to form in the presence of glutathione at neutral pH in a competitive reaction. The cleavage of the Pt–S bond in the complex [Pt(terpy)(glutathione)] was induced by the addition of copper(II) ions.[262] This suggests that the active platinum(II) complexes may be regenerated under biological conditions and the effect of copper(II) was interpreted by the coordination of thiol and neighbouring amide nitrogen donors. The reactivity of the platinum(II) complex [Pt(terpy)Cl]$^+$ towards various sulfur containing ligands was also studied.[263] The formation of stable adducts was described for thiol containing ligands (cysteine and thioglycolic acid) and the kinetic parameters of the reaction have been determined. However, in the case of the thioether ligand methionine the platinum(II) complex [Pt(terpy)Cl]$^+$ proved to be unreactive.[263]

Another interesting feature of palladium(II) complexes of peptides is that the metal ion is able to promote selective cleavage of peptide bonds. The half-life for the hydrolysis of the unactivated amide bond in neutral aqueous solution is around 500 years. Recent studies revealed that palladium(II) can significantly promote cleavage of the amide bond involving the carboxylate moieties of methionyl or histidyl residues. The reactions of palladium(II) complexes, [Pd(en)Cl$_2$] (N,N-chelate) and [Pd(Met)Cl$_2$] (S,N-chelate), with histidine containing di- and tripeptides have been studied by NMR spectroscopy.[264] It was found that efficient hydrolysis of X-His-Y peptides occurs at the His-Y amide bonds if X is an internal or N-protected amino acid. If X is unprotected the stable tridentate coordination of the peptide results in the release of the en ligand and inhibits hydrolytic processes. Likewise the hydrolysis will not occur if histidine is the C-terminal amino acid. If His is the N-terminal amino acid in the sequence His-X-Y then the His-X bond will hydrolyse at low pH values. Regioselective cleavage of a nonadecapeptide, representing the C-terminal segment of the protein myohemerythrin, by *cis*-[Pd(en)(H$_2$O)$_2$]$^{2+}$ has been detected by NMR spectroscopy. It was found that both His(5) and His(9) amino acid residues of the oligopeptide can bind [Pd(en)(H$_2$O)]$^{2+}$ units, but only the complex formed at the His(5) residue promotes the cleavage of Val(3)-Pro(4) amide bond. This regioselectivity was attributed to the electrostatic repulsion between [Pd(en)(H$_2$O)]$^{2+}$ entities and protonated lysyl residues.[265]

A dinuclear thiolate bridged palladium(II) complex was reported to accelerate catalytic hydrolysis of methionine containing peptides at the amide bond involving the carboxylic group of Met.[266] The reaction proceeds very rapidly, a half-life t$_{1/2}$ (7 min was obtained for Ac-Met-Ala at 50 °C. Similar tendencies were observed in the case of other methionyl containing peptides, but the formation of a Pd–S(thioamide) bond was reported to inhibit the hydrolysis.[267] The palladium(II) complexes of a series of bidentate N-donor ligands were used to investigate the hydrolytic cleavage of Met containing peptides to monitor the effect of various palladium(II) species.[268] The results revealed that at least one aqua ligand *cis* to the substrate is required for the observation of hydrolytic reactions of peptides and the more labile ligands are the better promoters.[269] The effect of palladium(II) complexes of dipeptides on the rate and selectivity of hydrolytic cleavage of other peptides has also been studied.[270] The dipeptides containing *N*-terminal methionyl residues (Met-X) and their N-protected forms (Ac-Met-Y) were applied as substrates. It was found that Lys, Glu, His or Asp in the dipeptide blocked the hydrolysis, while some enhancement of hydrolytic rate was observed with Ser (as X) compared to Gly or Ala. On the other hand, the more bulky leaving group (Y in Ac-Met-Y) resulted in decreased rate constants.

Various palladium(II) complexes were used to study the hydrolytic processes of myoglobin.[271] The results revealed that binding of palladium(II) aqua complexes to a side-chain residue seems to be a necessary, but not sufficient, condition for cleavage of a proximate peptide bond. Of the 153 amide bonds in myoglobin only 13 are cleaved by $[Pd(dtco-OH)(H_2O)_2]^{2+}$ (dtco-OH = dithia-cyclooctan-3-ol) and the sites of cleavage were clustered in the vicinity of Met(55) and (131), His(64), (81) and (93) and Arg (31) and (139).

In addition to the studies devoted to the investigation of metal ion promoted hydrolytic cleavage of peptides, metal ion assisted oxidative reactions represent another important category in the coordination chemistry of peptides. Copper(III) complexes of the tetrapeptides Gly_2-His-Gly and Aib_2-His-Gly (Aib = α-aminoisobutyric acid) have recently been characterized and the terminal amino, two deprotonated amide nitrogens and imidazole(N3) were suggested as the metal binding sites.[272] Both peptide complexes undergo amine deprotonation with pK values of 8.79 and 8.81, respectively. The low pK value for amine deprotonation was explained by the effect of the size of the chelate rings and by the coordinated imidazole residue. Similar metal ion promoted amine deprotonation was described in the gold(III) complexes of Gly-Gly-His as discussed in Section 3.1.[206] Earlier studies on the copper(III) complexes of tripeptides revealed that peptides undergo a rapid oxidative decarboxylation of the C-terminal His residue. Decomposition of the copper(III) complexes of Gly_2-His-Gly and Aib_2-His-Gly has also been observed resulting in the formation of tetrapeptides containing internal α-hydroxyhistidyl residue as an intermediate and α,β-dehydrohistidyl moiety as the final product after oxidation and dehydration, respectively. It has been shown in another study that copper(II) is able to catalyse oxidation of His to 2-oxo-His residues in peptides in the presence of ascorbate and molecular oxygen.[273] The ternary species formed in the cobalt(II)–L–O_2 system (where L = histamine and its pseudopeptide deivatives, glycylhistamine and sarcosylhistamine) were studied by potentiometric and spectroscopic methods.[274] Three different protonation states of the oxygenated species were observed in all systems between pH 7 and 11.2. In the case of histamine the uptake of O_2 was attributed to the $[CoL_2]$ binary complex, while $[CoLH_{-1}]$ and $[CoL_2H_{-1}]$ were also suggested as active species in the case of the dipeptides. A comparison of stability constants revealed that 4N-coordinated complexes have a greater affinity for oxygen uptake than the 3N-species. The stabilizing effect of deprotonated amide nitrogen on the oxygenated complexes of peptides has also been demonstrated. The role of coordinated amide nitrogen in oxygen uptake was also studied in the cobalt(II)-diglycyl-ethylenediamine system.[275] The $[CoL_2]$ binary complex has been described as the active form of reversible oxygen binding, but metal ion promoted deprotonation of the amide bond was not observed.

Another interesting feature of the cobalt(II) complexes of peptides is that the species formed upon irradiation with light in the system containing cobalt(II) ions and the tripeptide with C-terminal His residues are capable of DNA strand scission without exogenous chemical activating agents other than the ambient dioxygen.[276] Similar studies on DNA cleavage have been performed using the nickel(II) complexes of tripeptides of histidine.[277] The results indicate that the metallopeptides degrade through two pathways resulting from an initial C4′–H deoxyribose damage. The catalytic activity of the nickel(II) complex of the tripeptide, Lys-Gly-His, has been demonstrated in the guanine-specific modification of both single- and double-stranded oligonucleotides *via* the autooxidation of sulfite.[278] Copper(II) induced single-strand scission of DNA has been studied in the presence of the bis(2-pyridylmethyl) derivatives of diglycine and triglycine.[279] Crystal structures of the copper(II) complexes of the peptide analogues have been determined and the results indicate the same coordination geometry for both complexes containing tetragonal pyramidal copper(II) ions. A synthetic nuclease **(56)** has been

(56)

designed by the synthesis of a metal-peptide conjugate.[280] Complexes of rhodium(III) bind in the major groove of DNA and upon photoactivation cleave DNA at the rhodium intercalation site. A 16-residue peptide tethered to the intercalator was designed to mimic the active sites of metal containing hydrolases and two histidine residues were placed in positions 7 and 11 to create a zinc(II) coordination site. DNA cleavage was found to depend on the presence of rhodium(III) complex, the tethered peptide and zinc(II). It has been shown in another study that ATCUN-motif can also be used to design proteins with the ability for selective cleavage of DNA.[281]

The oxidation of thiols is generally catalysed by various transition metal ions. The kinetics of the oxidation of glutathione (and L-cysteine and D-penicillamine) by the [pentaamminechromatocobalt(III)] complex ion has been studied by spectrophotometric measurements.[282] The reactions occur in two stages: (a) the formation of [pentaamminehydroxocobalt(III)] and free chromate in a rapid stage and (b) the oxidation of thiol by free chromate in a slower stage. The formation of Cr(V) and Cr(IV) species was suggested, but the oxidation state of cobalt(III) was not changed during the reaction. Reversible oxidation of poly(L-cysteine) immobilized on controlled pore glass (PLC-CPG) was demonstrated using H_2O_2, aerated buffer and copper(II) as oxidizing agents.[283] It was shown that both hydrogen-peroxide and copper(II) oxidized PLC-CPG, but oxidation with copper(II) eliminated fewer cadmium(II) binding sites than H_2O_2. The oxidation with copper(II) was explained by the formation of disulfides and copper(I) complexes.

Cobalt(III) Schiff-base complexes are known to inhibit the enzymatic activity of human α-thrombin. It has been shown recently that the attachment of an active site directed peptide to the cobalt chelate leads to selective irreversible inhibition of thrombin.[284] An important conclusion of the study was that the strategy of linking enzyme-directed groups of metal chelates should be applicable to other systems of therapeutic or biological interest.

It has been accepted for a long time that low molecular weight transition metal complexes (especially copper(II)) can catalyse the dismutation of superoxide anion to oxygen and peroxide, as well as react rapidly with hydrogen peroxide to give molecular oxygen and water. Recent studies on the copper(II) complexes with ligands of biological relevance have shown that indeed these species behave as SOD analogues. Among the ligands that coordinate copper(II) cyclopeptides have attracted particular interest owing to their constrained geometry. Copper(II) complexes of two cyclopeptides, *cyclo*(Gly-His)$_4$ and *cyclo*(Gly-His-Gly)$_2$, have been characterized thermodynamically and structurally and SOD activity of the various species has been determined.[285] Solution equilibria of the systems are rather complicated, but the complexes [Cu{*cyclo*(Gly-His)$_4$}]$^{2+}$, [Cu{*cyclo*(Gly-His)$_4$H$_{-2}$}] and [Cu{*cyclo*(Gly-His-Gly)$_2$H$_{-2}$}] were identified as the major species. The coordination of four nitrogen donor atoms was shown in all cases, which corresponds to the binding of four imidazole residues in the first species, and two imidazoles and two deprotonated amide nitrogens in the deprotonated complexes. The imidazole coordinated complex showed higher redox potential and better catalytic activity than the other two species.

Metal ions and various metal complexes can influence not only cleavage, but formation of peptide bonds. A theoretical study on the free energies of peptide bond hydrolysis and formation revealed that formation of a dipeptide from two amino acids is about eight times more difficult than the subsequent condensation of an amino acid to a dipeptide or a longer chain. Moreover, condensation of an amino acid to a peptide is five times more difficult than joining two smaller peptides.[286] As a consequence, metal ions can play an important catalytic role in the formation of peptide chains under biological conditions. The catalytic effect of magnesium(II) ions has been demonstrated in a study, when the condensation of diglycine resulted in the enhanced formation of tetraglycine and hexaglycine in the presence of the metal ion.[287] Salt induced peptide formation reactions[288] and metal ion-assisted peptide cyclization[289] represent other interesting areas of peptide synthesis. A facile metal-mediated synthesis of 12-, 14-, 16- and 18-membered cyclic tetrapeptides has recently been described.[290] Square planar complexes of nickel(II), palladium(II) and copper(II) were found to be the active intermediates in the reaction. In addition to the catalytic role of classical metal complexes in peptide synthesis, the half-sandwich organometallic derivatives of ruthenium, rhodium and iridium also possessed catalytic activity in the formation of oligopeptides from α-amino acid esters.[291] Another interesting application of organometallic compounds in metallopeptide chemistry is their use for the labelling of peptides at the amino terminus.[292]

Fluorescent indicators are increasingly used for the quantitative analysis of metal ion concentration of various biological samples. The central problem in the production of new sensors lies in selectivity. The selective metal binding ability of specific peptide sequences can be used for the design of metal ion chemosensors. A family of peptides modeled for zinc finger domains has recently been designed and characterized.[293,294] The chemosensors for divalent zinc comprise a synthetic peptide template and a covalently attached fluorescent indicator. To achieve highly selective binding of Cu(II) or Ni(II) the amino terminus of serum albumin can be exploited (ATCUN motif: Amino Terminal Cu, Ni motif = X-X-His). A family of penta- and hexapeptides based upon the ATCUN motif has been prepared, in which the terminal amino acid residues are substituted with a fluorophore attachment. Fluorescence emission spectra of the peptides indicated tight metal binding with 1:1 stoichiometry for all metal ions, but while copper(II) quenched the fluorescence of the resulting peptides the coordination of nickel(II) enhanced the fluorescent resonance energy transfer.[295,296] A ferrocenyl group has been shown to mediate electron transfer between electrodes and redox sites of proteins. The synthesis of a nonnatural amino acid, L-ferrocenylalanine (ferAla), makes it possible to prepare site-specific redox-active peptides and proteins.[297]

Selective transport of metal ions across a membrane is known to play an essential role in many biological processes. Simple dipeptides form stable complexes with several transition metal ions and are promising candidates for selective transport of divalent metal ions. N-Monoalkylated and -dialkylated dipeptides of the amino acids Gly, Phe and Leu were synthesized and investigated as the carriers for the transport of copper(II), zinc(II) and nickel(II) from an aqueous buffer solution pH 5.6 to 0.1 M HCl *via* a bulk chloroform membrane. The high selectivity for copper(II) was explained by the neutral charge of the copper complex and it was possible to modulate transport efficiency by the N-alkyl groups.[298]

Many organisms contain proteins which regulate the size and shape of inorganic crystals in their skeletal elements. Thus it should be possible to construct a polypeptide that could bind to a specific surface of a growing mineral crystal and alter its shape. The synthesis of a potential calcite binding α-helical peptide (CBP1) containing a special array of aspartyl residues has been reported and its effect on calcite growth was monitored. The results suggest that the helical form of the peptide recognizes specific crystal surface characteristics, whereas the unfolded form acts nonspecifically as a polyanion.[299]

Rhodium(III) complexes are known as effective catalysts in hydrogenation reactions. High diastereoselectivity in the rhodium catalysed hydrogenation of dehydrodipeptides has been achieved in protic solvents and it was explained by electrostatic interactions between the ligands of catalyst and the dehydrodipeptides.[300] The formation of ion-pair complexes between the cationic metalloporphyrins of palladium(II) and zinc(II) and anionic pentapeptides has been detected and their role in electron transfer discussed.[301] Nickel(II) peroxide is a useful and strong oxidant of various organic substrates and cleaves selectively the carbon–nitrogen bond of glycyl residues in dipeptides to give the corresponding amides. The reaction provides further insight into the catalytic activity of certain enzymes and may have potential for the synthesis of α,β-dehydrodipeptides.[302]

3.3 Solution Equilibria – Stability Constants of Metal-ion–Peptide Complexes – Complex formation processes of oligopeptides with non-coordinating side chains are generally characterized by the metal binding of terminal amino and amide nitrogen donor atoms. This results in the formation of 1N, 2N, 3N and 4N-complexes and the stability constants of these species are only slightly influenced by the presence of the various hydrophobic side chains. However, it has been shown in a recent study that in some cases even the non-coordinating amino acid side chains can enormously increase the stability of copper(II)–peptide complexes.[303] A combined pH-metric and spectroscopic study on the copper(II) complexes of Asn-Ser-Phe-Arg-Tyr-NH$_2$ revealed a very high enhancement in the stability of the 4N-complex. The stability constants of the copper(II) complexes of pentapeptide analogues obtained by systematic Ala substitution revealed that the increase in stability comes from the specific amino acid sequence of the pentapeptide and was explained by the formation of a loop with the involvement of the C- and N-termini and by the intramolecular interaction between the Arg and Tyr residues. Equilibrium studies on the dibutyltin(IV) complexes of peptides with non-coordinating side chains have also been performed and the usual tridentate coordination of the dipeptides was established.[304] Most solution studies on the transition metal complexes of peptides were, however, devoted to the investigation of the effect of side-chain donor groups on complex formation processes. The metal ion coordination of these residues generally results in significant selectivity in the metal binding ability of peptides and this aspect of the coordination chemistry of peptides has been reviewed for ligands containing imidazole, carboxylate or various sulfur donor atoms in the side chains.[305]

Aspartame, a synthetic sweetener, is the methyl ester of a dipeptide (L-Asp-L-PheOMe). Its coordination chemistry, on the one hand, can be described as that of a common dipeptide, and on the other hand as a derivative of β-alanine due to the extra carboxylate of aspartyl residue. Copper(II) complexes of aspartame and its degradation products (Asp and Phe) have been studied by potentiometric measurements and stability constants of the various species reported.[306] Earlier studies on the copper(II) complexes of dipeptides of tyrosine revealed that the phenolate-O of the Tyr residue can act as a bridging ligand. The effect of Tyr residues on the complex formation processes of hexa- and heptapeptides has recently been studied.[307] The amino acid sequences of these peptides are Arg-Tyr-Leu-Gly-Tyr-Leu and Arg-Tyr-Leu-Gly-Tyr-Leu-Gln, which correspond to the segments (90-95 and 90-96) of α-casein and contain two phenolate oxygen donor atoms in addition to the peptide backbone. The results obtained for the copper(II) complexes revealed that both peptides are effective chelating agents and that the complex formation is characterized by the coordination of terminal amino and sequential amide nitrogens by increasing pH. The development of a weak charge transfer band around 390 nm in the absorption spectra at pH 8 was attributed to a $Cu(II) \leftarrow O^-$ (phenolate) bond in a dimeric complex, but this is a minor species supporting the view that the biological importance of a Tyr residue of the protein could be of minor importance. The heptapeptide segments of bovine and human β-casomorphin – Tyr-Pro-Phe-Val-Glu-Pro-Ile and Tyr-Pro-Phe-Pro-Gly-Pro-Ile, respectively – contain N-terminal Tyr residues as additional donor sites for metal binding. The coordination chemistry of these peptides is, however, governed by the presence of Pro residues, which work as a break-point for metal binding. The presence of Pro as the second amino acid residue resulted in the formation of rather stable dimeric complexes *via* (NH_2,CO)-coordination and phenolate bridging. The existence of amide bonded species was detected at high pH values, but only in the case of heptapeptides with two Pro residues.[308]

The side-chain OH groups of seryl or threonyl residues generally have only a minor effect on the thermodynamic stability of transition metal complexes of peptides. Recent studies on the copper(II) complexes of tri- and tetrapeptides containing α-hydroxymethylated amino acids revealed, however, that the presence of OH-groups results in significant enhancement of the stability constants of complexes.[309-311] In the case of the copper(II) complexes of tripeptides containing α-hydroxymethyl-L-serine the same species were formed as with Ala-Ala-Ala, but the increase in stability was up to 3.7 orders of magnitude compared to that of trialanine. In basic solutions dissociation and metal binding of the alcoholate group was also suggested.[309] Copper(II) complexes of tetrapeptides containing α-hydroxymethyl-arginine or -ornithine residues have been studied by potentiometric and spectroscopic techniques and the results were compared to those of common tetrapeptides. In this case, significant enhancement of stability was reported for the complexes containing 2N or 3N donor atoms.[310] In the case of α-hydroxymethyl-ornithine, deprotonation and coordination of the alcoholic group was detected in copper(II) complexes in basic solutions.[311]

As already discussed in Section 3.1 of this Chapter most of the studies on the metal complexes of peptides were devoted to the role of sulfur and/or imidazole-N donor atoms. The results obtained for the copper(II) and nickel(II) complexes of the thioamide analogues of tetraalanine revealed that insertion of a sulfur atom into the amide group significantly increases the metal binding ability of the molecules.[312] A comparison of the stability constants of the parent tetrapeptide complexes and those of the systematically substituted thioamide analogues showed that the most effective ligands are obtained when thiocarbonyl is inserted into the first, second or third peptide bond, whereas the thiocarbonyl moiety in the fourth position has much less influence on the binding modes and thermodynamic stabilities of metal complexes. Copper(II) and nickel(II) complexes of tripeptides containing methionine in all possible locations (Met-Gly-Gly, Gly-Met-Gly, Gly-Gly-Met, Met-Met-Ala, Met-Gly-Met and Met-Met-Met) were studied by potentiometric and spectroscopic measurements.[313] The metal ion speciation of the various systems are very similar to those of common tripeptides (*e.g.* triglycine), but the thioether group of methionine also makes some contribution to the metal binding. In the case of copper(II) complexes coordination of the sulfur atom is reflected in the development of a Cu(II)←S charge transfer band. The development of this band is a function of the number and location of thioether residues and is observed in the species $[CuL]^+$, $[CuLH_{-1}]$ and $[CuLH_{-2}]^-$ in the case of peptides containing N-terminal, internal and C-terminal methionyl residues, respectively. Nickel(II) forms stable, square planar, diamagnetic species with all tripeptides, $[NiLH_{-2}]^-$, and a weak interaction of thioether residue was observed only in the octahedral species $[NiL]^+$ of peptides with N-terminal methionyl residues.[313]

The thiol group of cysteine is generally the primary binding site for soft or borderline metal ions and its coordination significantly influences the conformation of peptide molecules.[314-316] Zinc(II) complexes of eight tripeptides and one tetrapeptide with cysteine and/or histidine at both termini have been investigated by potentiometric measurements in solution and prepared in the solid state.[314] The tripeptides were used in the fully protected forms, making cysteine thiolate and histidine imidazole the only donor functions. All peptides formed 1:1 zinc(II) complexes in solution, but in the case of bis(cysteinyl) derivatives 2:2 complexes were also formed. Metal binding of the peptides was described *via* the formation of large chelate rings (12- to 17-membered) and the complexes of bis(cysteinyl) ligands had the highest thermodynamic stability. Iron(II) complexes of cysteine containing peptides were studied in aqueous micellar solution and the spectroscopic properties of the complexes were used to mimic the active center of reduced rubredoxin.[315]

Copper(II) complexes of the hexapeptide Lys-Leu-Ala-His-Phe-Gly, containing an internal His residue, were studied by potentiometric and spectroscopic techniques.[317] The terminal amino group and the imidazole nitrogen in a large chelate ring were identified as the primary binding sites in slightly acidic media. Deprotonation and coordination of the amide nitrogens were detected with increasing pH, replacing the imidazole-N coordination at high pH values. Metal ion coordination of the ε-amino group of the lysyl residue was not observed, but the formation of binuclear complexes with well-separated copper(II) centers was detected by EPR spectroscopy. The formation of a similar macrochelate *via* the coordination of terminal amino and side-chain imidazole residues has been observed in the palladium(II) complexes of ProGlyAlaHis.[318] Palladium(II), however, has a much higher affinity for peptides than copper(II). This is reflected in the very high stability constants of palladium(II) complexes and the very low pH range of complex formation (pH < 1). Increasing the pH of the palladium(II)-tetrapeptide solution resulted in the deprotonation and coordination of amide nitrogens.

Several derivatives of amino acids and peptides containing more than one imidazolyl residues have been prepared to mimic the binding sites of metalloenzymes. Copper(II) and zinc(II) complexes of the ligand N,N'-di-L-histidyl-ethane-1,2-diamine (**57**) and the parent compound histamine were characterized by pH-metric and spectroscopic studies.[319] The formation of dimeric species, $[M_2L_2]^{4+}$ (**57a**) was detected with both metal ions having bis(histamine)-like coordination at physiological pH. The species $[MLH_{-2}]$ (**57b**) was also formed with both metal ions, but in the case of zinc(II) it was

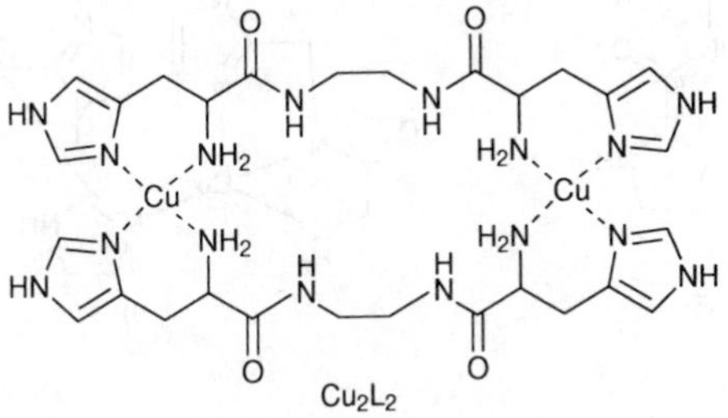

Cu₂L₂
(a)

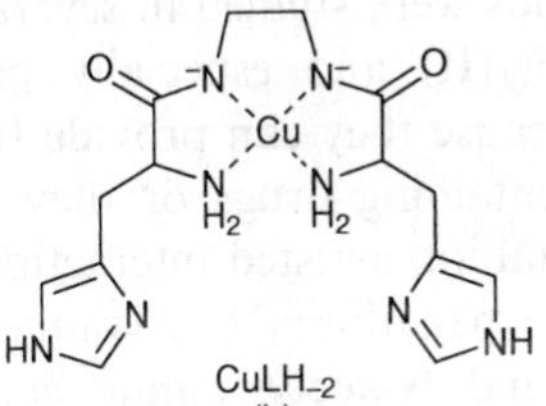

CuLH₋₂
(b)

(57)

identified as a mixed hydroxo complex, while in the case of copper(II) the extra base-consuming process was assigned to deprotonation and coordination of amide groups. The existence of tetrameric species was also suggested in the intermediate pH range.

The ligands containing two imidazole rings linked by aliphatic carbon chains can form especially stable complexes with 3d transition elements. Metal complexes of GlyBIMA [glycinamido-bis(imidazol-2-yl)methane (**58a**)] and ProDIPA [prolinamido-bis(pyridin-2-yl)methane, (**58b**)] were studied by potentiometric and spectroscopic techniques.[320] The chelating side chains (BIMA and DIPA) were reported as the exclusive metal binding sites in acidic media, but the metal binding ability of GlyBIMA was much higher than that of ProDIPA. The presence of the terminal amino group, however, resulted in the ambidentate behaviour and tridentate coordination of the ligands with increasing pH. In the case of the copper(II)-GlyBIMA system it resulted in the formation of an imidazole bridged dimeric species (**59**), while only monomeric complexes were obtained with ProDIPA. Deprotonation and coordination of the amide nitrogen was reported to occur in the nickel(II), cobalt(II) and zinc(II) complexes.

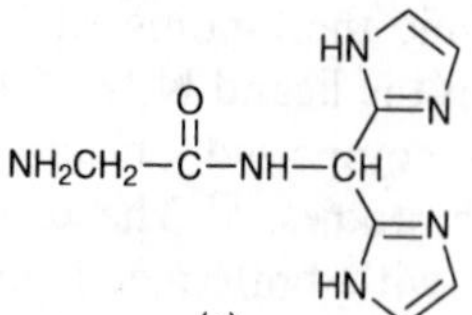

(a)

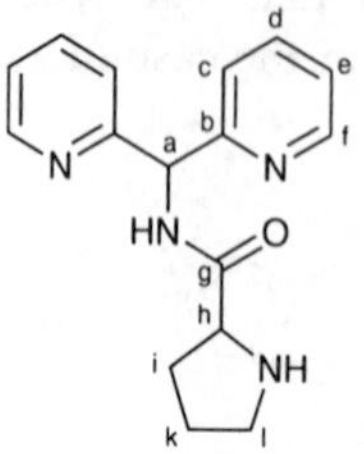

(b)
(58)

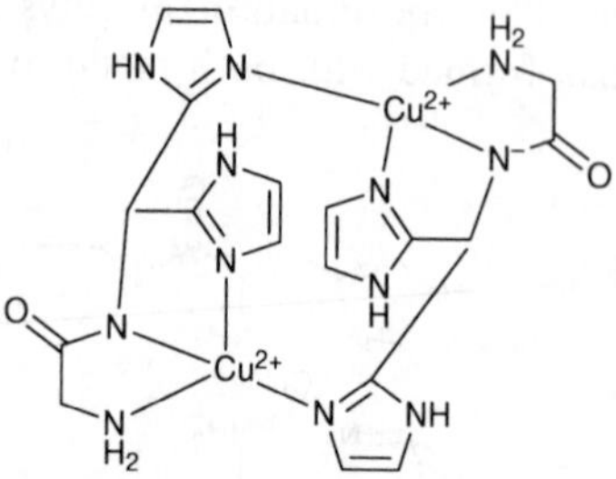

(59)

Ternary complexes of transition metal ions with peptides and various other biologically important ligands were studied in several laboratories. The mixed ligand complexes of copper(II) and especially palladium(II) received the greatest attention,[321-328] because they can provide further information on the mechanism of platinum containing drugs or they can be considered as the simplest models for the metal ion assisted interactions of proteins and nucleic acids. Ternary complexes of palladium(II)–triamine systems (dien and terpy) with N-alkyl nucleobases and N-acetyl amino acids have been studied by potentiometric and NMR measurements.[322] In the case of AcHis both mono- and dinuclear species were formed *via* the metal ion coordination of imidazole nitrogen donors. The existence of linkage isomers was shown for the mono-nuclear species, which correspond to the metal binding of N1 or N3 of imidazole. The ratio of the two isomers depends on steric factors influenced both by the triamine and the imidazole containing ligands. In the case of AcLys the ε-amino group of lysyl residue can be considered as a metal binding site. The two triamine complexes of palladium(II), however, have very different affinities towards the coordination of aliphatic amino nitrogen and only hydroxo complex formation was detected in the [Pd(terpy)]$^{2+}$–AcHis system. A similar study has been performed on ternary complexes of nucleobases and derivatives of amino acids, but the tridentate triamines were replaced by the dipeptide Gly-Met.[323] A single crystal X-ray structure of the palladium(II) complex of Gly-Met revealed the tridentate coordination of the ligand *via* the terminal amino, deprotonated amide nitrogen and thioether sulfur donor atoms. The formation of the same ternary species were identified in the reaction with AcHis, but the stability constants in the Pd(II)–Gly-Met–AcHis system were lower than those of the ternary systems with dien or terpy. The stability constants for the ternary palladium(II) complexes of the bidentate diamine (1,3-diaminopropane) and amino acids or peptides have also been reported.[324] Binary and ternary palladium(II) complexes of Gly-His-Lys have been studied by a variety of NMR techniques.[325] The complex [Pd(Gly-His-Lys)H $_1$Cl] was prepared in the reaction of $K_2[PdCl_4]$ with the tripeptide and 3N donor atoms (amino-, amide- and imidazole-N) were shown to be the metal binding sites. The fourth coordination site is occupied by chloride ion, but it can easily be substituted by various nitrogen donors of nucleotides. Similar studies have been performed on the copper(II) complexes of diastereomeric pairs of dipeptides (Ala-Ala, Ala-Trp and Leu-Phe) and oligonucleotides. The formation of ternary complexes was ascertained and their effect on the intramolecular ligand-ligand interactions has been discussed.[326]

Stability constants of the ternary complexes of copper(II) with Gly-DL-Leu and amino acids and their esters were determined by potentiometric measurements. The kinetics of the base hydrolysis of amino acid esters were also monitored and metal ion promoted hydrolysis in the ternary systems was found.[327] Equilibrium data for the ternary systems of copper(II) with oxidized glutathione and alkyl substituted derivatives of phen were determined by potentiometric titrations and binding modes of the complexes were characterized by spectroscopic methods. The results suggest that the equatorial binding sites of copper(II) are occupied by 2N donor atoms from phen and an (N,O) donor set from Glu residue of oxidized glutathione. The formation of dimeric species and the Cu–S coordination were excluded.[328]

Equilibrium studies on the interaction of dipeptides with dinuclear copper(II) complexes have been performed to mimic the catalytic activity of proteolytic enzymes.[329] The macrocycle 1,4,7,13,16,19-hexaaza-10,22-dioxa-cyclotetracosane (OBISDIEN) is a bis-chelating ligand and can form dinuclear complexes with copper(II) *via* the coordination of three nitrogen donor atoms. The fourth coordination sites of the metal ions can bind the dipeptide molecules *via* the (NH$_2$,CO) or (NH$_2$,N$^-$) binding sites (**60**). It has been shown that binding of Gly-Gly or Ala-Ala in the dinuclear complex is more effective than that of Gly-Leu and the possible biological implications of the results have been discussed. The metal complexes of a new binuclear chelating agent (tetraaza macrocycle, cyclophane) have also been studied and different coordination modes of copper(II) and zinc(II) was shown.[330]

A series of Fe^{III}(coproporphyrin)–peptide complexes has been studied by spectrophotometric measurements and binding constants of the various peptides to iron(III) have been calculated.[331] A strong correlation between the side-chain hydrophobicity of the peptides and the free energy of ligation or the reduction potentials of the heme–peptide complexes was obtained.

The investigation of the solution equilibria of the systems containing vanadium in different oxidation states and various peptide molecules received increasing attention.[332-334] The speciation and the kinetic properties of the complexes formed in the reaction of vanadate(V) and Pro-Ala and Ala-Gly have been characterized by potentiometry and ^{51}V, ^{13}C and ^{1}H NMR spectroscopy.[332] Both peptides were found to bind vanadium as tridentate ligands, *via*

(60)

(NH$_2$,N$^-$,COO$^-$)-binding, but the formation of complexes with Ala-Gly was much faster than those of Pro-Ala. Oxovanadium(IV) complexes of the dipeptides Gly-Asp and Asp-Gly and related ligands were studied by potentiometric and various spectroscopic techniques.[333] In the weakly acidic pH range only the carboxylate, carbonyl and amino groups were suggested as the possible metal binding sites, but the formation of bis complexes was influenced very much by the location of the Asp residue in the molecules. The coordination of amide nitrogen was suggested to occur above pH 6–7. However, in the case of the VO(IV)–salicyl-Gly system the metal ion induced deprotonation and coordination of amide groups was suggested to occur around pH 4, which is very close to that observed for the corresponding copper(II) complexes. This means that the phenolate group in a chelating position is an effective anchor for both oxovanadium(IV) and copper(II) to promote amide deprotonation and coordination.[334]

The modifications of the N- or C-termini and the amide groups of peptide molecules led to the synthesis of some new organic compounds, which are promising chelating agents. Transition metal complexes of the phosphonodipeptides and amino acids, Gly(P), Gly-Gly(P) and Met-Ala(P) have been studied by potentiometric and NMR measurements.[116,335] The complex formation processes of derivatives of glycine with platinum(II) and palladium(II) were very similar to those of the carboxylic analogues.[116] Similar results were obtained for the metal complexes of (*S,S*) and (*S,R*) diastereomers of Met-Ala(P). Hence, only the amino-N, carbonyl-O and phosphonic-O donor atoms were involved in metal binding in the nickel(II), cobalt(II) and zinc(II) complexes. In the case of copper(II), metal ion promoted amide coordination also has been suggested, but the binding of the thioether residue was ruled out with all 3d transition elements. On the other hand, the thioether residue was considered to be the primary metal binding site of Met-Ala(P) in the palladium(II) and platinum(II) complexes. The interaction of calcium(II) ions with L-glutamyl-L-serine phosphate was studied in another report and it was demonstrated that the ligand is able to chelate calcium(II) and to enhance its intestinal absorption under *in vivo* conditions.[336]

The amide derivatives of simple amino acids and their N-substituted derivatives were found to be very efficient ligands for binding of copper(II), nickel(II) and palladium(II).[337,338] The remarkable stability of the palladium(II) and nickel(II) complexes of amides was attributed to the cooperative deprotonation of two amide groups and formation of diamagnetic, square planar complexes. Other important derivatives of amides and related ligands are the ketoamide and oxime analogues of peptides. The investigation of the copper(II) complexes of pyruvyl-L-methionine revealed that ketoamide derivatives of amino acids are effective chelating agents.[339] The complex $[CuL_2H_{-2}]$ was detected as the major species around physiological pH and its binding mode shows that coordination of the amide nitrogen can occur in the absence of an extra anchoring group (**61**). The increase of pH results in further

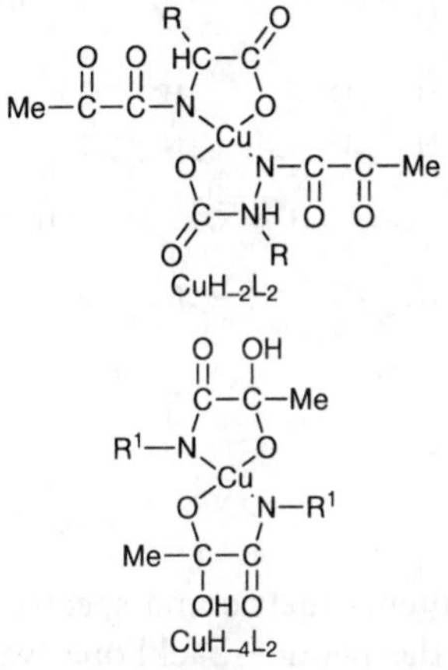

$CuH_{-2}L_2$

$CuH_{-4}L_2$

(61)

deprotonation and in the formation of $[CuL_2H_{-4}]$, in which the hydrated pyruvate residue is deprotonated. Several oxime analogues of amino acids and related ligands have recently been synthesized and their transition metal complexes were studied in solution and in the solid state. Some of these results[34-37] have already been discussed in Section 2, while other studies are cited here[340,341] and all data support the view that oxime analogues of natural amino acids are more efficient ligands than the parent molecules.

Desferrioxamine B (DFA) is a well-known siderophore analogue containing amide and hydroxamate functional groups and used as an effective drug in the chelation therapy of iron and aluminium. Its copper(II), nickel(II), zinc(II) and molybdenum(VI) complexes have been studied in aqueous solution.[342] In agreement with earlier findings only the hydroxamate moieties took part in metal binding. In the case of nickel(II) and zinc(II) three hydroxamate groups were suggested to occupy the octahedral coordination sphere of the metal ions, while two hydroxamate residues were proposed to coordinate to copper(II) and dioxomolybdenum(VI).

Another important development of metallopeptide chemistry is described in the increasing number of publications on the interactions of metal ions with large peptide molecules. Peptide nucleic acids (PNA) are synthetic analogues of DNA in which the natural phosphate-deoxyribose backbone is replaced by a pseudo-peptide chain.[343] The interaction of copper(II) with a tetrapeptide with two thymine (dThy, **62**) and an octapeptide with four thymine (tThy, **62**)

(62)

residues was studied by potentiometric and spectroscopic measurements. The nitrogen donor atoms of the peptide backbone were identified as the major metal binding sites, but the presence of nucleobases resulted in the enhancement of metal binding ability of the ligands.

A disulfide bridged two-stranded α-helical peptide has been designed, which undergoes a folding transition upon binding of lanthanide ions to two specific binding sites engineered in the molecule. The relationship between structural properties and metal binding behaviour is discussed.[344] Albumin, the most abundant protein in mammalian blood serum, has very high metal binding ability and is connected to the X-Y-His N-terminal amino acid sequence. The second specific metal binding site of human, bovine and porcine albumins was studied by CD and EPR spectroscopic measurements.[345] The second copper(II) site was described by a tetragonal (2N,4O)-coordination geometry and binding of copper(II) was stronger than nickel(II) or cadmium(II), but weaker than zinc(II) binding to porcine albumin. Binding of the metal ions at the second site, however, did not effect the copper(II) binding ability at the first coordination site of albumin. Casein phosphopeptides may function as carriers for different minerals especially calcium. Calcium binding constants of these peptides and their dephosphorylated analogues have been determined using capillary electrophoresis techniques.[346] The maximum calcium binding ability was achieved at pH 7 and phosphorylated peptides exhibited higher binding constants than their dephosphorylated analogues.

References

1. Chemical Abstracts Selects on Amino Acids, Peptides and Proteins, published by the American Chemical Society and Chemical Abstract Service, Columbus, Ohio.
2. W. Bal, M. Dyba and H. Kozlowski, *Acta Biochim. Pol.*, 1997, **44**, 467.
3. T.G. Appleton, *Coord. Chem. Rev.*, 1997, **166**, 313.
4. S. Liu, D.S. Edwards and J.A. Barrett, *Bioconjugate Chem.*, 1997, **8**, 621.
5. H. Tsukube and S. Shinoda, *Bol. Soc. Chil. Quim.*, 1997, **42**, 237.
6. K. Severin, R. Bergs and W. Beck, *Angew. Chem. Int. Ed.* 1998, **37**, 1634.
7. O. Clement, B.M. Rapko and B.P. Hay, *Coord. Chem. Rev.*, 1998, **170**, 203.
8. M. Mizutani, N. Maejima, K. Jitsukawa, H. Masuda and H. Einaga, *Inorg. Chim. Acta*, 1998, **283**, 105.
9. J. Jezierska, T. Glowiak, A. Ozarowski, Y.V. Yablokov and Z. Rzaczynska, *Inorg. Chim. Acta*, 1998, **275-276**, 28.
10. T. Lu, X. Li, Z. Mao, W. Qiu, L. Ji and K. Yu, *Polyhedron*, 1998, **17**, 75.
11. F. Gao, R.Y. Wang, T.Z. Jin, G.X. Xu, Z.Y. Zhou and X.G. Zhou, *Polyhedron*, 1997, **16**, 1357.
12. A. Cuevas, I. Viera, M.H. Torre, E. Kremer, S.B. Etcheverry and E.J. Baran, *Acta Farm. Bonaerense*, 1998, **17**, 213.
13. C.D. Brondino, R. Calvo, A.M. Atria, E. Spodine, O.R. Nascimento and O. Pena, *Inorg. Chem.*, 1997, **36**, 3183.
14. T. Sugimori, H. Masuda, N. Ohata, K. Koiwai, A. Odani and O. Yamauchi, *Inorg. Chem.*, 1997, **36**, 576.
15. S. Suzuki, K. Yamaguchi, N. Nakamura, Y. Tagawa, H. Kuma and T. Kawamoto, *Inorg. Chim. Acta*, 1998, **283**, 260.
16. F. Zhang, T. Yajima, H. Masuda, A. Odani and O. Yamauchi, *Inorg. Chem.*, 1997, **36**, 5777.
17. C.M. Fan, L.J. Bai, L.H. Wei, W.L. Yang and D.W. Guo, *Transition Met. Chem.*, 1997, **22**, 109.
18. T. Murakami and S. Kita, *Inorg. Chim. Acta*, 1998, **274**, 247.
19. B. Onoa and V. Moreno, *Transition Met. Chem.*, 1998, **23**, 485.
20. M.A.S. Goher, L.A. Al-Shatti and F.A. Mautner, *Polyhedron*, 1997, **16**, 889.
21. M. Louloudi, Y. Deligiannakis and N. Hadjiliadis, *Inorg. Chem.*, 1998, **37**, 6847.
22. J.J. Criado, E.R. Fernández, E. Garcia, M.R. Hermosa and E. Monte, *J. Inorg. Biochem.*, 1998, **69**, 113.
23. M.L.A. Silva and E.C.A. Felício, *Bull. Soc. Chim. Fr.*, 1997, **134**, 645.
24. C.J. Campbell, W.L. Driessen, J. Reedijk, W. Smeets and A.L. Spek, *J. Chem. Soc. Dalton Trans.*, 1998, 2703.

25. G. Plesch, V. Kettmann, J. Sivy, O. Svajlenová and C. Friebel, *Polyhedron*, 1998, **17**, 539.

26. G. Plesch, C. Friebel, S.A. Warda, J. Sivy and O. Svajlenová, *Transition Met. Chem.*, 1997, **22**, 433.

27. S.A. Warda, P. Dahlke, S. Wocadlo, W. Massa and C. Friebel, *Inorg. Chim. Acta*, 1998, **268**, 117.

28. C. Friebel, G. Plesch, V. Kettmann, J.K. Smogrovic and O. Svajlenová, *Inorg. Chim. Acta*, 1997, **254**, 273.

29. Z.H. Chohan, M. Praveen and A. Ghaffar, *Met.-Based Drugs*, 1997, **4**, 267.

30. T. Ishida, Y. In, C. Hayashi, R. Manabe and A. Wakahara, *Bull. Chem. Soc. Jpn.*, 1997, **70**, 2375.

31. A.S. Sagiyan, S.Z. Sagyan, G.L. Grigoryan, T.F. Savel'eva, Yu.N. Belokon and S.K. Grigoryan, *Khim. Zh. Arm.*, 1997, **50**, 149.

32. W. Chen and M.C. Lim, *Acta Crystallogr., Sect. C: Cryst. Struct. Commun.*, 1997, **C53**, 539.

33. N.D. Villanueva, J.R. Bocarsly and J.W. Ziller, *Inorg. Chem.*, 1997, **36**, 4585.

34. T.Y. Sliva, A. Dobosz, L. Jerzykiewicz, A. Karaczyn, A.M. Moreeuw, J.Swiatek-Kozlowska, T. Glowiak and H. Kozlowski, *J. Chem. Soc. Dalton Trans.*, 1998, 1863.

35. T.Y. Sliva, T. Kowalik-Jankowska, V.M. Amirkhanov, T. Glowiak, C.O. Onindo, I.O. Fritskii and H. Kozlowski, *J. Inorg. Biochem.*, 1997, **65**, 287.

36. T.Y. Sliva, A.M. Duda, T. Glowiak, I.O. Fritsky, V.M. Amirkhanov, A.A. Mokhir and H. Kozlowski, *J. Chem. Soc. Dalton Trans.*, 1997, 273.

37. B. Cervera, J.L. Sanz, M.J. Ibánez, G. Vila, F. LLoret, M. Julve, R. Ruiz, X. Ottenwaelder, A. Aukauloo, S. Poussereau, Y. Journaux and M.C. Munoz, *J. Chem. Soc. Dalton Trans.*, 1998, 781.

38. Y. Zhang, J. Bai, M. Lu and A. Lu, *Huaxue Shijie*, 1997, **38**, 82.

39. M.S. Sastry and S.S. Gupta, *Proc.- Indian Acad. Sci., Chem. Sci.*, 1997, **109**, 173.

40. H. Liu, J. Huang, X. Tian, X. Jiao, G.T. Luo and L. Ji, *Inorg. Chim. Acta*, 1998, **272**, 295.

41. F. Briganti, V. Iaconi, S. Mangani, P. Orioli, A. Scozzafava, G. Vernaglione and C.T. Supuran, *Inorg. Chim. Acta*, 1998, **275-276**, 295.

42. W. Bae and R.K. Mehra, *J. Inorg. Biochem.*, 1998, **70**, 125.

43. M. Devereux, M. Jackman, M. McCann and M. Casey, *Polyhedron*, 1998, **17**, 153.

44. C. Morice, P.L. Maux and G. Simonneaux, *Inorg. Chem.*, 1998, **37**, 6100.

45. T. Tanase, T. Inagaki, Y. Yamada, M. Kato, E. Ota, M. Yamazaki, M. Sato, W. Mori, K. Yamaguchi, M. Mikuriya, M. Takahashi, M. Takeda, I. Kinoshita and S. Yano, *J. Chem. Soc. Dalton Trans.*, 1998, 713.

46. J.E. Barclay, M.I. Diaz, D.J. Evans, G. Garcia, M.D. Santana and M.C. Torralba, *Inorg. Chim. Acta*, 1997, **258**, 211.

47. K.K. Narang, V.P. Singh and D. Bhattacharya, *Polyhedron*, 1997, **16**, 2491.

48. K.K. Narang, V.P. Singh and D. Bhattacharya, *Transition Met. Chem.*, 1997, **22**, 333.

49. H.B. Kraatz, J. Lusztyk and G.D. Enright, *Inorg. Chem.*, 1997, **36**, 2400.

50. R.M.G. Roberts and E. Johnsen, *J. Organomet. Chem.*, 1997, **544**, 197.

51. O.P. Goel, S.J. Johnson and L.D. Wise, *PCT Int. Appl. WO 97 05,152*, 1997.

52. S.R. Grguric, S.R. Trifunovic and T.J. Sabo, *J. Serb. Chem. Soc.*, 1998, **63**, 669.

53. H. Kumita, K. Jitsukawa, H. Masuda and H. Einaga, *Inorg. Chim. Acta*, 1998, **283**, 160.

54. Y. Kitamura, N. Azuma, K. Minamoto, S. Murakami and Y. Tanabe, *Polyhedron*, 1997, **16**, 3757.

55. K. Jitsukawa, K. Iwai, H. Masuda, H. Ogoshi and H. Einaga, *J. Chem. Soc. Dalton Trans.*, 1997, 3691.

56. D.P. Fairlie, W.G. Jackson, B.W. Skelton, H. Wen, A.H. White, W.A. Wickramasinghe, T.C. Woon and H. Taube, *Inorg. Chem.*, 1997, **36**, 1020.

57. P.J. Brothers, G.R. Clark, H.R. Palmer and D.C. Ware, *Inorg. Chem.*, 1997, **36**, 5470.

58. P.A.M. Williams and E.J. Baran, *Transition Met. Chem.*, 1997, **22**, 589.

59. S. Mondal, S.P. Rath, K.K. Rajak and A. Chakravorty, *Inorg. Chem.*, 1998, **37**, 1713.

60. J.C. Pessoa, M.T. Duarte, R.D. Gillard, C. Madeira, P.M. Matias and I. Tomaz, *J. Chem. Soc. Dalton Trans.*, 1998, 4015.

61. A.D. Keramidas, S.M. Miller, O.P. Anderson and D.C. Crans, *J. Am. Chem. Soc.*, 1997, **119**, 8901.

62. S. Mondal, P. Ghosh and A. Chakravorty, *Inorg. Chem.*, 1997, **36**, 59.

63. D.J. Darensbourg, J.D. Draper and J.H. Reibenspies, *Inorg. Chem.*, 1997, **36**, 3648.

64. G. Cervantes, V. Moreno, E. Molins and M. Quirós, *Polyhedron*, 1998, **17**, 3343.

65. T. Rau, R. Alsfasser, A. Zahl and R. Eldik, *Inorg. Chem.*, 1998, **37**, 4223.

66. G. Faraglia, D. Fregona and S. Sitran, *Transition Met. Chem.*, 1997, **22**, 492.

67. E. Katsarou, C. Charalambopoulos and N. Hadjiliadis, *Met.-Based Drugs*, 1997, **4**, 57.

68. M. Vicens, A. Caubet and V. Moreno, *Met.-Based Drugs*, 1997, **4**, 43.

69. Y.Q. Gong, Y.F. Cheng, J.M. Gu and X.R. Hu, *Polyhedron*, 1997, **16**, 3743.

70. T.J. Einhäuser, M. Galanski, E. Vogel and B.K. Keppler, *Inorg. Chim. Acta*, 1997, **257**, 265.

71. Y. Fuchita, K. Yoshinaga, Y. Ikeda and J.Kinoshita-Kawashima, *J. Chem. Soc. Dalton Trans.*, 1997, 2495.

72. J. Real, A. Polo and J. Duran, *Inorg. Chem. Commun.*, 1998, **1**, 457.

73. B. Kayser, C. Missling, J. Knizek, H. Nöth and W. Beck, *Eur. J. Inorg. Chem.*, 1998, 375.

74. D. Steinborn, H. Junicke and F.W. Heinemann, *Inorg. Chim. Acta*, 1997, **256**, 87.

75. T. Kobayashi, Y. Inomata, T. Takeuchi and F.S. Howell, *Inorg. Chim. Acta*, 1997, **258**, 109.

76. M. Maurus, B. Aechter, W. Hoffmüller, K. Polborn and W. Beck, *Z. Anorg. Allg. Chem.*, 1997, **623**, 299.

77. C. Redshaw, V.C. Gibson, W. Clegg, A.J. Edwards and B. Miles, *J. Chem. Soc. Dalton Trans.*, 1997, 3343.

78. C. Djordjevic, N. Vuletic, B.A. Jacobs, M. Lee-Renslo and E. Sinn, *Inorg. Chem.*, 1997, **36**, 1798.

79. M. Chatterjee, B. Achari, S. Das, R. Banerjee, C. Chakrabarti, J.K. Dattagupta and S. Banerjee, *Inorg. Chem.*, 1998, **37**, 5424.

80. S. Kirsch, B. Noll, H. Spies, P. Leibnitz, D. Scheller, T. Krueger and B. Johannsen, *J. Chem. Soc. Dalton Trans.*, 1998, 455.

81. M. Cattabriga, A. Marchi, L. Marvelli, R. Rossi, G. Vertuani, R. Pecoraro, A. Scatturin, V. Bertolasi and V. Ferretti, *J. Chem. Soc. Dalton Trans.*, 1998, 1453.

82. T. Mori, M. Yamaguchi, M. Sato and T. Yamagishi, *Inorg. Chim. Acta*, 1998, **267**, 329.

83.   J.M. Wolff and W.S. Sheldrick, *J. Organomet. Chem.*, 1997, **531**, 141.
84.   J.M. Wolff and W.S. Sheldrick, *Chem. Ber. Rec.*, 1997, **130**, 981.
85.   D.A. Herebian, C.S. Schmidt, W.S. Sheldrick and C. van Wüllen, *Eur. J. Inorg. Chem.*, 1998, 1991.
86.   M. Koralewicz, F.P. Pruchnik, A. Szymaszek, K.Wajda-Hermanowicz and K.Wona-Grzegorek, *Transition Met. Chem.*, 1998, **23**, 523.
87.   T.C. Tsai and I.J. Chang, *J. Am. Chem. Soc.*, 1998, **120**, 227.
88.   K. Aparna, S.S. Krishnamurthy, M. Nethaji and P. Balaram, *Polyhedron*, 1997, **16**, 507.
89.   C.R. Carubelli, A.M.G. Massabni and S.R. De A. Leite, *J. Braz. Chem. Soc.*, 1997, **8**, 597.
90.   R. Li and S. Zhong, *Huaxue Shijie*, 1997, **38**, 515.
91.   E. Galdecka, Z. Galdecki, P. Gawryszewska and J. Legendziewicz, *New J. Chem.*, 1998, **22**, 941.
92.   Z.B. Zhang, L.S. Li, R.Y. Chen, Q.M. Wang and Q. Zeng, *Chin. Chem. Lett.*, 1997, **8**, 9.
93.   K. Dölling, A. Krug, H. Hartung and H. Weichmann, *Z. Naturforschung*, 1997, **52b**, 9.
94.   L.E. Khoo, Y. Xu, N.K. Goh, L.S. Chia and L.L. Koh, *Polyhedron*, 1997, **16**, 573.
95.   M. Nath and R. Yadav, *Bull. Chem. Soc. Jpn.*, 1997, **70**, 1331.
96.   K. Bukietynska, Z. Karwecka and H. Podsiadly, *Polyhedron*, 1997, **16**, 2613.
97.   T. Kiss, I. Sóvágó, I. Tóth, A. Lakatos, R. Bertani, A. Tapparo, G. Bombi and R.B. Martin, *J. Chem. Soc. Dalton Trans.*, 1997, 1967.
98.   G. Battistuzzi, M. Borsari, L. Menabue, M. Saladini and M. Sola, *Inorg. Chim. Acta*, 1998, **273**, 397.
99.   A. Mederos, D.M. Saysell, J. Sanchiz and A.G. Sykes, *J. Chem. Soc. Dalton Trans.*, 1998, 2723.
100.   E. Bottari and M.R. Festa, *Talanta*, 1997, **44**, 1705.
101.   L.C.Tran-Ho, P.M. May and G.T. Hefter, *J. Inorg. Biochem.*, 1997, **68**, 225.
102.   M.M. Shoukry and M.M.H. Mohamed, *Main Group Met. Chem.*, 1997, **20**, 281.
103.   A. Takahashi, T. Natsume, N. Koshino, S. Funahashi, Y. Inada and H.D. Takagi, *Can. J. Chem.*, 1997, **75**, 1084.
104.   M.M. Shoukry and M.M.H. Mohamed, *J. Coord. Chem.*, 1998, **43**, 217.
105.   S. Jerico, C.R. Carubelli, A.M.G. Massabni E.B. Stucchi, S.R.A. Leite and O. Malta, *J. Braz. Chem. Soc.*, 1998, **9**, 487.
106.   H.B. Silber and Y. Nguyen, *J. Alloys Compd.*, 1998, **275-277**, 811.
107.   B.B. Tewari, *J. Indian Chem. Soc.*, 1998, **75**, 256.
108.   J.S. Chen, G. Lu, D.Y. Wei, M. Yao and L.F. Shen, *Huaxue Xuebao*, 1998, **56**, 892.
109.   F. Fang, X. Zhang and X. Guo, *Huaxue Tongbao*, 1997, 40.
110.   J. Urbanska and H. Kozlowski, *J. Coord. Chem.*, 1997, **42**, 197.
111.   E. Gao, Y. Sun and Z. Shi, *Huaxue Yanjiu Yu Yingyong*, 1997, **9**, 582.
112.   E. Bottari and M.R. Festa, *Talanta*, 1998, **46**, 91.
113.   B. Kurzak, E. Matczak-Jon and M. Hoffmann, *J. Coord. Chem.*, 1998, **43**, 243.
114.   E. Matczak-Jon, B. Kurzak, W. Sawka-Dobrowolska, B. Lejczak and P. Kafarski, *J. Chem. Soc. Dalton Trans.*, 1998, 161.
115.   P. Buglyó, T. Kiss, M. Dyba, M. Jezowska-Bojczuk, H. Kozlowski and S. Bouhsina, *Polyhedron*, 1997, **16**, 3447.

116. L. Bláha, I. Lukeš, J. Rohovec and P. Hermann, *J. Chem. Soc. Dalton Trans.*, 1997, 2621.

117. B. Kurzak, A. Kamecka, K. Kurzak, J. Jezierska and P. Kafarski, *Polyhedron*, 1998, **17**, 4403.

118. Boduszek, M. Dyba, M. Jezowska-Bojczuk, T. Kiss and H. Kozlowski, *J. Chem. Soc. Dalton Trans.*, 1997, 973.

119. P.G. Daniele, C. De Stefano, E. Prenesti and S. Sammartano, *Talanta*, 1997, **45**, 425.

120. E. Kiss, A. Lakatos, I. Bányai and T. Kiss, *J. Inorg. Biochem.*, 1998, **69**, 145.

121. S.B. Etcheverry and P.A.M. Williams, *J. Inorg. Biochem.*, 1998, **70**, 113.

122. E. Chruscinska, E. Garribba, G. Micera, A. Panzanelli and M. Biagioli, *Talanta*, 1998, **47**, 343.

123. E. Farkas, K. Megyeri, L. Somsák and L. Kovács, *J. Inorg. Biochem.*, 1998, **70**, 41.

124. M.A. Santos, M. Gaspar, M.L.S.S. Gonçalves, M.T. Amorim, *Inorg. Chim. Acta*, 1998, **278**, 51

125. A. Dobosz, I.O. Fritsky, A. Karaczyn, H. Kozlowski, T. Yu. Sliva and J. Swiatek-Kozlowska, *J. Chem. Soc. Dalton Trans.*, 1998, 1089.

126. M.M. Khalil, S.A. Mohamed and A.M. Radalla, *Talanta*, 1997, **44**, 1365.

127. A.H. Amrallah, N.A. Abdalla and E.Y. El-Haty, *Talanta*, 1998, **46**, 491.

128. M.M. Shoukry, W.M. Hosny, A.A. Razik and R.A. Mohamed, *Talanta*, 1997, **44**, 2109.

129. F. Dallavalle, G. Folesani, E. Leporati and R. Borromei, *J. Coord. Chem.*, 1998, **44**, 225.

130. G.N. Mukherjee and T. Ghosh, *J. Indian Chem. Soc.*, 1997, **74**, 8.

131. M. Palanichamy and M. Anbu, *Proc. Indian Acad. Sci., Chem. Sci.*, 1997, **109**, 105.

132. I.T. Ahmed, A.A.A. Boraei and O.M. El-Roudi, *J. Chem. Eng. Data*, 1998, **43**, 459.

133. M. Padmavathi and S. Satyanarayana, *Indian J. Chem., Sect. A: Inorg., Bioinorg., Phys., Theor. Anal. Chem.*, 1997, **36A**, 1001.

134. Y.S. Zhang, Z.M. Wang, H.K. Lin, S.R. Zhu, H.W. Sun, D.G. Wang and R.T. Chen, *Gaodeng Xuexiao Huaxue Xuebao*, 1998, **19**, 1992.

135. S.A. Zaidi and F. Khan, *J. Electrochem. Soc. India*, 1998, **47**, 137.

136. A. Ilyas and F. Khan, *J. Electrochem. Soc. India*, 1997, **46**, 173.

137. B. Kurzak, D. Kroczewska and J. Jezierska, *Polyhedron*, 1998, **17**, 1831.

138. L. Lomozik, L. Bolewski and R. Dworczak, *J. Coord. Chem.*, 1997, **41**, 261.

139. H.K. Lin, S.R. Zhu, Z.F. Zhou, X.C. Su, Z.X. Gu and Y.T. Chen, *Polyhedron*, 1998, **17**, 2363.

140. M.B. Inoue, L. Machi, M. Inoue and Q. Fernando, *Inorg. Chim. Acta*, 1997, **261**, 59.

141. A. Mohan, K. Radha and M.S. Mohan, *Asian J. Chem.*, 1998, **10**, 50.

142. F. Zhang, T. Yajima, A. Odani and O. Yamauchi, *Inorg. Chim. Acta*, 1998, **278**, 136.

143. K.F. Morris, L.E. Erickson, B.V. Panajotova, D.W. Jiang and F. Ding, *Inorg. Chem.*, 1997, **36**, 601.

144. L.E. Erickson, P. Hayes, J.J. Hooper, K.F. Morris, S.A. Newbrough, M. Van Os and P. Slangan, *Inorg. Chem.*, 1997, **36**, 284.

145. R. Jin and X. He, *Fenxi Kexue Xuebao*, 1997, **13**, 177.

146. R.P. Bonomo, V. Cucinotta, F. D'Alessandro, G. Impellizzeri, G. Maccarrone,

E. Rizzarelli, G. Vecchio, L. Carima, R. Corradini, G. Sartor and R. Marchelli, *Chirality*, 1997, **9**, 341.
147. T. Campagna, G. Grasso, E. Rizzarelli and G. Vecchio, *Inorg. Chim. Acta*, 1998, **275-276**, 395.
148. H. Tamiaki, N. Matsumoto and H. Tsukube, *Tetrahedron Lett.*, 1997, **38**, 4239.
149. H. Tsukube, S. Shinoda, J. Uenishi, T. Kanatani, H. Itoh, M. Shiode, T. Iwachido and O. Yonemitsu, *Inorg. Chem.*, 1998, **37**, 1585.
150. T. Mizutani, S. Yagi, A. Honmaru, S. Murakami, M. Furusyo, T. Takagishi and H. Ogoshi, *J. Org. Chem.*, 1998, **63**, 8769.
151. M. Chikira, M. Inoue, R. Nagane, W. Harada and H. Shindo, *J. Inorg. Biochem.*, 1997, **66**, 131.
152. S. Hoyau and G. Ohanessian, *J. Am. Chem. Soc.*, 1997, **119**, 2016.
153. N. Raos, *Croatica Chemica Acta*, 1997, **70**, 913.
154. U.N. Andersen and G. Bojesen, *J. Chem. Soc., Perkin Trans. 2*, 1997, 323.
155. K. W. Lee, K.I. Eom and S.J. Park, *Inorg. Chim. Acta*, 1997, **254**, 131.
156. O. Heudi, A. Cailleux and P. Allain, *Inorg. Biochem.*, 1998, **71**, 61.
157. K.A. Mitchell and C.M. Jensen, *Inorg. Chim. Acta*, 1997, **265**, 103.
158. R.N. Bose, S.K. Ghosh and S. Moghaddas, *J. Inorg. Biochem.*, 1997, **65**, 199.
159. V.M. Vasic, M.S. Tosic, T. Jovanovic, L.S. Vujisic and J.M. Nedeljkovic, *Polyhedron*, 1998, **17**, 399.
160. U. Bierbach and N. Farrell, *J. Biol. Inorg. Chem.*, 1998, **3**, 570.
161. V.F. Plyusnin, V.P. Grivin, L.F. Krylova, L.D. Dikanskaja, Y.V. Ivanov and H. Lemmetyinen, *J. Photochem. Photobiol., A*, 1997, **104**, 45.
162. S. Ghosh, G.S. De and A.K. Ghosh, *Indian J. Chem., Sect. A: Inorg., Bio-inorg., Phys., Theor. Anal. Chem.*, 1997, **36A**, 863.
163. F.T. Lin, R.A. Kortes and R.E. Shepherd, *Transition Met. Chem.*, 1997, **22**, 243.
164. A. Thomas, E. Wolcan, M.R. Féliz and A.L. Capparelli, *Transition Met. Chem.*, 1997, **22**, 541.
165. H.C. Malhotra and A. Kumar, *J. Indian Chem. Soc.*, 1997, **74**, 220.
166. K.U. Din and G.J. Khan, *J. Indian Chem. Soc.*, 1997, **74**, 393.
167. D. Subrahmanyam and V.A. Ramam, *J. Indian Counc. Chem.*, 1997, **13**, 40.
168. A.K. Ghosh, P.S. Sengupta and G.S. De, *Indian J. Chem., Sect. A: Inorg., Bio-inorg., Phys., Theor. Anal. Chem.*, 1997, **36A**, 611.
169. V.A. Soloshonok, D.V. Avilov, V.P.Kukhar, L.V. Meervelt and N. Mischenko, *Tetrahedron Lett.*, 1997, **38**, 4671.
170. D.A. Buckingham, C.R. Clark and A.J. Rogers, *Inorg. Chem.*, 1997, **36**, 3791.
171. F. Nxumalo and A.S. Tracey, *J. Biol. Inorg. Chem.*, 1998, **3**, 527.
172. R. Cao, N. Traviseo, A. Fragoso, R. Villalonga, A. Diaz, M.E. Martinez, J. Alpizar and D.X. West, *J. Inorg. Biochem.*, 1997, **66**, 213.
173. M. Sekota and T. Nyokong, *Polyhedron*, 1997, **16**, 3279.
174. K. Bernauer, D.Hugi-Cleary, H.J. Hilgers, H.A. Khalek, N. Brügger and C. Kressl, *Inorg. Chim. Acta*, 1998, **275-276**, 1.
175. A. Magnuson, H. Berglund, P. Korall, L. Hammarström, B. Akermark, S. Styring and L. Sun, *J. Am. Chem. Soc.*, 1997, **119**, 10720.
176. B.C. Gilbert, J.R.L. Smith, A.F. Parsons and P.K. Setchell, *J. Chem. Soc. Perkin Trans. 2*, 1997, 1065.
177. T. Mori, T. Yamamoto, M. Yamaguchi and T. Yamagishi, *Bull. Chem. Soc. Jpn.*, 1997, **70**, 1069.
178. N. Bhargava and A. Pandey, *Oxid. Commun.*, 1998, **21**, 263.

179. Md.Z.A. Rafiquee, R.A. Shah, Kabir-Ud-Din and Z. Khan, *Int. J. Chem. Kinet.*, 1997, **29**, 131.

180. N. Mishra and K.C. Nand, *Oxid. Commun.*, 1998, **21**, 98.

181. S. Bayulken and A.S. Sarac, *Z. Phys. Chem.*, 1998, **205**, 181.

182. K.S. Rangappa, S. Chandraju and N.M.M. Gowda, *Synth. React. Inorg. Met-Org. Chem.*, 1998, **28**, 275.

183. K.S. Rangappa, S. Chandraju and N.M.M. Gowda, *Int. J. Chem. Kinet.*, 1998, **30**, 7.

184. B.R. Sahu, R. Chourey, S. Pandey and V.R. Shastry, *Oxid. Commun.*, 1998, **21**, 574.

185. R.B. Chougale, R.G. Panari and S.T. Nandibewoor, *Oxid. Commun.*, 1998, **21**, 565.

186. R.G. Panari, R.B. Chougale and S.T. Nandibewoor, *Oxid. Commun.*, 1998, **21**, 503.

187. B.R. Sahu, V.R. Chourey, R. Vijay, S. Pandey and V.R. Shastry, *Asian J. Chem.*, 1998, **10**, 320.

188. A. Arrizabalaga, F.J. Andres-Ordax, M.Y. Fernandez-Aranguiz and R. Peche, *Int. J. Chem. Kinet.*, 1997, **29**, 181.

189. V.R. Chourey, S. Pande, L.V. Shastry and V.R. Shastry, *Asian J. Chem.*, 1997, **9**, 435.

190. V.R. Chourey and V.R. Shastry, *Res. J. Chem. Environ.*, 1997, **1**, 49.

191. S.K. Asthana, S.K. Mishra and K.C. Nand, *Oxid. Commun.*, 1997, **20**, 132.

192. M.J. Sisley and R.B. Jordan, *Adv. Chem. Ser.*, 1997, **253**, 267.

193. E. Karim and M.K. Mahanti, *Oxid. Commun.*, 1998, **21**, 559.

194. Z. Khan, Md.Z.A. Rafiquee and Kabir-ud-Din, *Transition Met. Chem.*, 1997, **22**, 350.

195. V. Sharma, P.K. Sharma and K.K. Banerji, *J. Indian Chem. Soc.*, 1997, **74**, 607.

196. M.S. Ramachandran, R. Santhanalakshmi and D. Easwaramoorthy, *Oxid. Commun.*, 1998, **21**, 230.

197. L. Pecci, G. Montefoschi, G. Musci and D. Cavallini, *Amino Acids*, 1997, **13**, 355.

198. T. Shi, J. Berglund and L.I. Elding, *J. Chem. Soc. Dalton Trans.*, 1997, 2073.

199. A.A. Abdel-Khalek, El-Said M. Sayyah and H.A. Ewais, *Transition Met. Chem.*, 1997, **22**, 375.

200. A.A. Abdel-Khalek, El-Said M. Sayyah and H.A. Ewais, *Transition Met. Chem.*, 1997, **22**, 557.

201. A.D. Ryabov, G.M. Kazankov, S.A. Kurzeev, P.V. Samuleev and V.A. Polyakov, *Inorg. Chim. Acta*, 1998, **280**, 57.

202. T. Takarada, R. Takahashi, M. Yashiro and M. Komiyama, *J. Phys. Org. Chem.*, 1998, **11**, 41.

203. N. Nagao, T. Kobayashi, T. Takayama, Y. Koike, Y. Ono, T. Watanabe, T. Mikami, M. Suzuki, T. Matumoto and M. Watabe, *Inorg. Chem.*, 1997, **36**, 4195.

204. J. Costa Pessoa, S.M. Luz and R.D. Gillard, *J. Chem. Soc., Dalton Trans.*, 1997, 569.

205. P.C. Paul, S.J. Angus-Dunne, R.J., Batchelor, F.W.B. Einstein and A.S. Tracey, *Can. J. Chem.*, 1997, **75**, 183.

206. S.L. Best, T.K. Chattopadhyay, M.I. Djuran R.A. Palmer, P.J. Sadler, I. Sóvágó and K. Várnagy, *J. Chem. Soc., Dalton Trans.*, 1997, 2587.

207. P. Tsiveriotis, N. Hadjiliadis and G. Stavropoulos, *Inorg. Chim. Acta*, 1997, **261**, 83.

208. T.Yamamura, H. Nakamura, S. Nakajima, T. Sasaki, M. Ushiyama, M. Ueki and H. Hirota, *Inorg. Chim. Acta*, 1998, **283**, 243.

209. J. Ueda, A. Hanaki, N. Yoshida and T. Nakajima, *Chem. Pharm. Bull.*, 1997, **45**, 1108.

210. R.P. Bonomo, L. Casella, L. De Gioia, H. Molinari, G. Impellizzeri, T. Jordan, G. Pappalardo, R. Purello and E. Rizzarelli , *J. Chem. Soc., Dalton Trans.*, 1997, 2387.

211. S.A. Ross and C.J. Burrows, *Inorg. Chem.*, 1998, **37**, 5358.

212. N. Romero-Isart, N. Duran, M. Capdevila, P. Gonzalez-Duarte, S. Maspoch and J.L. Torres, *Inorg. Chim. Acta*, 1998, **278**, 10.

213. V. Magafa, G. Stavropoulos, P. Tsiveriotis and N. Hadjiliadis, *Inorg. Chim. Acta*, 1998, **272**, 7.

214. V. Magafa, S.P. Perlepes and G. Stavropoulos, *Transition Met. Chem.*, 1998, **23**, 105.

215. J. Volz, F. U. Bosch, M. Wunderlin, M. Schumacher, K. Melchers, K. Bensch, W. Steinhilber, K.P. Schäfer, G. Tóth, B. Penke and M. Przybylski, *J. Chromatogr. A*, 1998, **800**, 29.

216. L. Guo, R. Vogler and F. Bitguel, *Neimenggu Daxues Xueabo, Ziran Kexueban*, 1998, **29**, 672.

217. A. Meisner, W. Haehnel and H. Vahrenkamp, *Chem. Eur. J.*, 1997, **3**, 261.

218. M. Junker, K.K. Rodgers and J.E. Coleman, *Inorg. Chim. Acta*, 1998, **275-6**, 481.

219. A.K. Judd, A. Sanchez, D.J. Bucher, J.H. Huffman, K. Bailey and R.W. Sidwell, *Antimicrob. Agents Chemother.*, 1997, **41**, 687.

220. T. Yamamure, T. Watanabe, A. Kikuchi, T. Yamane, M. Ushiyama and H. Hirota, *Inorg. Chem.*, 1997, **36**, 4849.

221. B.A. Cerda, S. Hoyau, G. Ohanessian and C. Wesdemiotis, *J. Am. Chem. Soc.*, 1998, **120**, 2437.

222. T. Wyttenbach, J.E. Bushnell and M.T. Bowers, *J. Am. Chem. Soc.*, 1998, **120**, 5098.

223. E.A. Jefferson, P. Gantzel, E. Benedetti and M. Goodman, *J. Am. Chem. Soc.*, 1997, **119**, 3187.

224. P. De Santis, A. Palleschi, A. Scipioni, M. Camalli, R. Spagna and G. Zanotti, *Biopolymers*, 1998, **45**, 257

225. R. Basosi, E. Gaggeli, N. Gaggeli, R. Pogni and G. Valensin, *Inorg. Chim. Acta*, 1998, **275-6**, 274.

226. Kochman, A. Gajewska, H. Kochman, H. Kozlowski, E. Masiukiewicz and K. B. Rzeszotarska, *J. Inorg. Biochem.*, 1997, **65**, 277.

227. M.S. Diaz-Cruz, J. Mendieta, A. Monjonell, R. Tauler and M. Esteban, *J. Inorg. Biochem.*, 1998, **70**, 91.

228. M.S. Diaz-Cruz, J. Mendieta, R. Tauler and M. Esteban, *J. Inorg. Biochem.*, 1997, **66**, 29.

229. W. Bae and R.K. Mehra, *J. Inorg. Biochem.*, 1998, **69**, 33.

230. E. Wong, T. Fauconnier, S. Bennett, J. Valliant, T. Nguyen, F. Lau , L.F.L. Lu, A. Pollak, R.A. Bell and J.R. Thornback, *Inorg. Chem.*, 1997, **36**, 5799.

231. R. Jankowsky, S. Kirsch, T. Reich, H. Spies and B. Johannsen, *J. Inorg. Biochem.*, 1998, **70**, 99.

232. S. Kirsch, R. Jankowsky, H. Spies and B. Johannsen, *Forschungszent. Rossendorf, [Ber.] FZR* 1997, (FZR–200), 79.

233. M. Scheunemann and B. Johannsen, *Tetrahedron Lett.*, 1997, **38**, 1371.

234. U. Mazzi, *Transition Met. Chem.*, 1997, **22**, 430.

235. M. Santimaria, U. Mazzi, S. Gatto, A. Dolmella, G. Bandoli and M. Nicolini, *J. Chem. Soc., Dalton Trans.*, 1997, 1765.

236. M. Rayopadhye, T.D. Harris, K. Yu, D. Glowacka, P.R. Damphousse, J.A. Barrett, S.J. Heminway, D.S. Edwards and T.R. Carroll, *Bioorg. Med. Chem. Lett.*, 1997, **7**, 955.

237. A. Garcia-Raso, J.J. Fiol, B. Adrover, V. Moreno, E. Molins and I. Mata, *J. Chem. Soc., Dalton Trans.*, 1998, 1031.

238. E. Katsarou, A. Troganis and N. Hadjiliadis, *Inorg. Chim. Acta*, 1997, **256**, 21.

239. E. Katsarou, A. Kolstad, N. Hadjiliadis and E. Sletten, *J. Inorg. Biochem.*, 1998, **70**, 265.

240. R.N. Patel and K.B. Pandeya, *J. Inorg. Biochem.*, 1998, **72**, 109.

241. M.T. Fernandez, M.M. Silva, L. Mira, M.H. Florencio, A. Gill and K.R. Jennings, *J. Inorg. Biochem.*, 1998, **71**, 93.

242. E.B. Gonzalez, E. Farkas, A.A. Soudi, T. Tan, A.I. Yanowsky and K.B. Nolan, *J. Chem. Soc., Dalton Trans.*, 1997, 2377.

243. M.N. Akhtar, A.A. Isab and A.R. Al-Arfaj, *J. Inorg. Biochem.*, 1997, **66**, 197.

244. G.B. Onoa and V. Moreno, *J. Inorg. Biochem.*, 1998, **72**, 141.

245. P. Comba, R. Cusack, D.P. Fairlie, L.R. Gahan, G.R. Hanson, U. Kazmaier and A. Ramlow, *Inorg. Chem.*, 1998, **37**, 6721.

246. A. Luna, B. Amekraz, J. Tortajada, J.P. Morizur, M. Alcami, O. Mó and M. Yánez, *J. Am. Chem. Soc.*, 1998, **120**, 5411.

247. R. Ruiz, C. Surville-Barland, A. Aukauloo, E. Anxolabehere-Mallart, Y. Journaux, J. Cano and M.C. Munoz, *J. Chem. Soc., Dalton Trans.*, 1997, 745.

248. T.B. Karpishin, T.A. Vanelli and K.J. Glover, *J. Am. Chem. Soc.*, 1997, **119**, 9063.

249. R. Fattoruso, C.De Pasquale, G. Morelli and C. Pedone, *Inorg. Chim. Acta*, 1998, **278**, 76.

250. A. Lombardi, F. Nastri M. Sanseverino, O. Maglio, C. Pedone and V. Pavone, *Inorg. Chim. Acta*, 1998, **275-276**, 301.

251. B.J. Goodfellow, M.J. Lima, C. Ascenso, M. Kennedy, R. Sikkink, F. Rusnak, I. Moura and J.J.G. Moura, *Inorg. Chim. Acta*, 1998, **273**, 279.

252. C.O. Fernández, A.I. Sannazzaro, L.E. Díaz and A.J. Vila, *Inorg. Chim. Acta*, 1998, **273**, 367.

253. A.M. Calafat, H. Won and L.G. Marzili, *J. Am. Chem. Soc.*, 1997, **119**, 3656.

254. H.R. Swift and D.L.H. Williams, *J. Chem. Soc., Perkin Trans.*, 1997, 1933.

255. M.S. Tosic, V.M. Vasic, J.M. Nedeljkovic and L.A. Ilic, *Polyhedron*, 1997, **16**, 1157.

256. C.D.W. Fröhling and W.S. Sheldrick, *Chem. Commun.*, 1997, 1737.

257. S. U. Milinkovic and M.I. Djuran, *Gaz. Chim. Ital.*, 1997, **127**, 69.

258. C.D.W. Fröhling and W.S. Sheldrick, *J. Chem. Soc., Dalton Trans.*, 1997, 4411.

259. A.F.M. Siebert and W.S. Sheldrick, , *J. Chem. Soc., Dalton Trans.*, 1997, 385.

260. Z. Guo, T.W. Hambley, P. Murdoch, P.J. Sadler and U. Frey, *J. Chem. Soc., Dalton Trans.*, 1997, 469.

261. Y. Chen, Z. Guo, J.A. Parkinson and P.J. Sadler, *J. Chem. Soc., Dalton Trans.*, 1998, 3577.

262. C-C. Cheng and C-H. Pai, *J. Inorg. Biochem.*, 1998, **71**, 109.

263. Z.D. Bugarcic, B.J. Djordjevic and M.I. Djuran, *J. Serb. Chem. Soc.*, 1997, **62**, 1031.

264. S.U. Milinkovic, T.N. Parac, M.I. Djuran and N.M. Kostic, *J. Chem. Soc., Dalton Trans.*, 1997, 2771.

265. T.N. Parac and N.M. Kostic, *Inorg. Chem.*, 1998, **37**, 2141.
266. G.B. Karet and N.M. Kostic, *Inorg. Chem.*, 1998, **37**, 1021.
267. T.N. Parac and N.M. Kostic, *J. Serb. Chem. Soc.*, 1997, **62**, 847.
268. X. Chen, X. Luo, Y. Song, S. Zhou and L. Zhu, *Polyhedron*, 1998, **17**, 2271.
269. X. Chen, L. Zhu, X. You and N.M. Kostic, *J. Biol. Inorg. Chem.*, 1998, **3**, 1.
270. X. Luo, X. Chen, Y. Song and L. Zhu, *Chemistry Lett.*, 1998, 1079.
271. L. Zhu, R. Bakhtiar and N.M. Kostic, *J. Biol. Inorg. Chem.*, 1998, **3**, 383.
272. M.R. McDonald, F.C. Fredericks and D.W. Margerum, *Inorg. Chem.*, 1997, **36**, 3119.
273. M. Khossravi and R.T. Borchardt, *Pharm. Res.*, 1998, **15**, 1096.
274. T. Gajda, B. Henry and J.J. Delpuech, *Inorg. Chem.*, 1997, **36**, 1850.
275. A. Kufelnicki, M. Swiatek, A. Wogt and J. Skarzewski, *J. Coord. Chem.*, 1998, **43**, 21.
276. D.C. Ananias and E.C. Long, *Inorg. Chem.*, 1997, **36**, 2469.
277. Q. Liang, D.C. Ananias and E.C. Long, *J. Am. Chem. Soc.*, 1998, **120**, 248.
278. J.G. Muller, R.P. Hickerson, R.J. Perez and C.J. Burrows, *J. Am. Chem. Soc.*, 1997, **119**, 1501.
279. T. Kobayashi, T. Okuno, T. Suzuki, M. Kunita, S. Ohba and Y. Nishida, *Polyhedron*, 1998, **17**, 1553.
280. M.P. Fitzsimons and J.K. Barton, *J. Am. Chem. Soc.*, 1997, **119**, 3379.
281. B. Sarkar, Design of Proteins with ATCUN Motif which Specifically Cleave DNA, in Cytotoxic, Mutagenic and Carcinogenic Potential of Heavy Metals Related to Human Environment, ed. N.D. Hadjiliadis, *NATO ASI Series*, Vol. 26, p. 477, Kluwer Academic, 1997.
282. D.A. Dixon, T.P. Dasgupta and N.P. Sadler, *Inorg. React. Mech.*, 1998, **1**, 41.
283. M. Howard, H.A. Jurbergs and J.A. Holcombe, *Anal. Chem.*, 1998, **70**, 1604.
284. T. Takeuchi, A. Böttcher, C.M. Quezada, M.I. Simon, T.J. Meade and H.B. Gray, *J. Am. Chem. Soc.*, 1998, **120**, 8555.
285. R.P. Bonomo, G. Impellizzeri, G. Pappalardo, R. Purello, E. Rizzarelli and G. Tabbi, *J. Chem. Soc., Dalton Trans.*, 1998, 3851.
286. R.B. Martin, *Biopolymers*, 1998, **45**, 351.
287. Y. Yamagata and K. Inomata, *Origins Life Evol. Biosphere*, 1997, **27**, 339.
288. Y. Suwannachot and B.M. Rode, *Origins Life Evol. Biosphere*, 1998, **28**, 79.
289. L. Zhang and J.P. Tam, *Tetrahedron Lett.*, 1997, **38**, 4375.
290. K. Haas, W. Ponikwar, H. Nöth and W. Beck, *Angew. Chem. Int. Ed.*, 1998, **37**, 1086.
291. W. Hoffmüller, M. Maurus, K. Severin and W. Beck, *Eur. J. Inorg. Chem.*, 1998, 729.
292. W. Bauer, M. Prem, K. Polborn, K. Sünkel, W. Steglich and W. Beck, *Eur. J. Inorg. Chem.*, 1998, 485.
293. G.K. Walkup and B. Imperiali, *J. Am. Chem. Soc.*, 1997, **119**, 3443.
294. G.K. Walkup and B. Imperiali, *J. Org. Chem.*, 1998, **63**, 6727.
295. A. Torrado, G.K. Walkup and B. Imperiali, *J. Am. Chem. Soc.*, 1998, **120**, 609.
296. D.A. Pearce, G.K. Walkup and B. Imperiali, *Bioorg. Med. Chem. Lett.*, 1998, **8**, 1963.
297. M. Kira, T. Matsubara, H. Shinohara and M. Sisido, *Chem. Lett.*, 1997, 89.
298. M.C. Cleij, P. Scrimin, P. Tecilla and U. Tonellato, *J. Org. Chem.*, 1997, **62**, 5592.
299. D.B. DeOliviera and R.A. Laursen, *J. Am. Chem. Soc.*, 1997, **119**, 10627.
300. I. Yamada, K. Fukui, Y. Aoki, S. Ikeda, M. Yamaguchi and T. Yamagishi, *J. Organometallic Chem.*, 1997, **539**, 115.

301. M. Aoudia and M.A.J. Rodgers, *J. Am. Chem. Soc.*, 1997, **119**. 12859.
302. C.J. Easton, S.K. Eichinger and M.J. Pitt, *Tetrahedron*, 1997, **53**, 5609.
303. W. Bal, M. Dyba, F. Kasprzykowski, H. Kozlowski, R. Latajka, L. Lankiewicz, Z. Mackiewicz and L.D. Pettit, *Inorg. Chim. Acta*, 1998, **283**, 1.
304. M.M. Shoukry and M.M.H. Mohamed, *Main Group Met. Chem.*, 1997, **20**, 281.
305. I. Sóvágó and K. Várnagy, Metal Binding Selectivity of Oligopeptides, in Cytotoxic, Mutagenic and Carcinogenic Potential of Heavy Metals Related to Human Environment, ed. N.D. Hadjiliadis, *NATO ASI Series*, Vol. 26, p. 537, Kluwer Academic, 1997.
306. S. Kholeif and G. Anderegg, *Inorg. Chim. Acta*, 1997, **257**, 225.
307. E. Chruscinska, M. Dyba, G. Micera, W. Ambroziak, J. Olczak, J. Zabrocki and H. Kozlowski, *J. Inorg. Biochem.*, 1997, **66**, 19.
308. E. Chruscinska, J. Olczak, J. Zabrocki, M. Dyba, G. Micera, D. Sanna and H. Kozlowski, *J. Inorg. Biochem.*, 1998, **69**, 91.
309. T.Kowalik-Jankowska, M. Stasiak, M.T. Leplawy and H. Kozlowski, *J. Inorg. Biochem.*, 1997, **66**, 193.
310. E. Lodyga-Chruscinska, G. Micera, D. Sanna, H. Kozlowski, K. Kaczmarek, J. Olejnik and M.T. Leplawy, *J. Inorg. Biochem.*, 1998, **72**, 187.
311. E. Chruscinska, K. Kaczmarek, J. Olejnik, M.T. Leplawy, A. Panzanelli and G. Micera, *Inorg. Chim. Acta*, 1998, **269**, 279.
312. T. Kowalik-Jankowska, M. Jasionowski, L. Lankiewicz and H. Kozlowski, *J. Inorg, Biochem.*, 1997, **66**, 45.
313. K. Várnagy, B. Bóka, I. Sóvágó, D. Sanna, P. Marras and G. Micera, *Inorg. Chim. Acta*, 1998, **275-6**, 440.
314. P. Gockel, M. Gelinsky, R. Vogler and H. Vahrenkamp, *Inorg. Chim. Acta*, 1998, **272**, 115.
315. W-Y. Sun, N. Ueyama and A. Nakamura, *Biopolymers*, 1998, **46**, 1.
316. P. Tsiveriotis and N. Hadjiliadis, Interaction of Platinum(II), Palladium(II) and Mercury(II) Salts with Histidine and/or Cysteine Containing Peptides, in Cytotoxic, Mutagenic and Carcinogenic Potential of Heavy Metals Related to Human Environment, ed. N.D. Hadjiliadis, *NATO ASI Series*, Vol. 26, p. 537, Kluwer Academic, 1997.
317. B. Gyurcsik, I. Vosekalna and E. Larsen, *Acta Chem. Scand.*, 1997, **51**, 49.
318. P. Tsiveriotis, N. Hadjiliadis and I. Sóvágó, *J. Chem. Soc., Dalton Trans.*, 1997, 4267.
319. I. Török, T. Gajda, B. Gyurcsik, G.K. Tóth and A. Péter, *J. Chem. Soc., Dalton Trans.*, 1998, 1205.
320. K. Várnagy, I. Sóvágó, W. Goll, H. Süli-Vargha, G. Micera and D. Sanna, *Inorg. Chim. Acta*, 1998, **283**, 233.
321. I. Sóvágó, A. Kiss, E. Farkas and B. Lippert, Potentiometric and Spectroscopic Studies on the Ternary Complexes of Copper(II) and Palladium(II) with Peptides and Nucleobases, in Cytotoxic, Mutagenic and Carcinogenic Potential of Heavy Metals Related to Human Environment, ed. N.D. Hadjiliadis, *NATO ASI Series*, Vol. 26, p. 521, Kluwer Academic, 1997.
322. A. Kiss, E. Farkas, I. Sóvágó, B. Thormann and B. Lippert, *J. Inorg. Biochem.*, 1997, **68**, 85.
323. M. Wienken, A. Kiss, I. Sóvágó, E.C. Fusch and B. Lippert, *J. Chem. Soc., Dalton Trans.*, 1997, 563.
324. S.M. El-Medani, S.M. Shohayeb and M.M. Shoukry, *Transition Met. Chem.*, 1998, **23**, 287.

325. I. Rombeck and B. Lippert, *Inorg. Chim. Acta*, 1998, **273**, 31.
326. R.P. Bonomo, V. Cucinotta, G. Grasso, G. Maccarone and L. Mastruzzo, *J. Inorg. Biochem.*, 1998, **70**, 1.
327. M.M. Shoukry, E.M. Khiry and R.G. Khalil, *Transition Met. Chem.*, 1997, **22**, 465.
328. P. Piu, G. Sanna, A. Masia, M.A. Zoroddu and R. Seeber, *J. Chem. Soc., Dalton Trans.*, 1997, 2369.
329. M.T.B. Luiz, B. Szpoganicz, M. Rizotto, A.E. Martell and M.G. Basalotte, *Inorg. Chim. Acta*, 1997, **345**, 254.
330. M.B. Inoue, F. Medrano, M. Inoue, A. Raitsimring and Q. Fernando, *Inorg. Chem.*, 1997, **36**, 2335.
331. D.L. Huffman, M.M. Rosenblatt and K.S. Suslick, *J. Am. Chem. Soc.*, 1998, **120**, 6183.
332. M. Fritzsche, K. Elvingson, D. Rehder and L. Petterson, *Acta Chem. Scand.*, 1997, **51**, 483.
333. J. Costa Pessoa, T. Gajda, R.D. Gillard, T. Kiss, S.M. Luz, J.J.G. Moura, I. Tomaz, J.P. Telo and I. Török, *J. Chem. Soc., Dalton Trans.*, 1998, 3587.
334. T. Kiss, K. Petrohán, P. Buglyó, D. Sanna, G. Micera, J. Costa Pessoa and C. Madeira, *Inorg. Chem.*, 1998, **37**, 6389.
335. I. Lukes, L. Bláha, F. Kesner, J. Rohoves and P. Hermann, *J. Chem. Soc., Dalton Trans.*, 1997, 2629.
336. L. Galzigna, N. Domergue and A. Previero, *J. Peptide Res.*, 1998, **52**, 15.
337. F. Dallavalle, G. Folesani, E. Leporati and L.H. Abdel-Rahman, *J. Coord. Chem.*, 1997, **42**, 189.
338. R.W. Hay, N. Govan, A. Perotti and O. Carugo, *Transition Met. Chem.*, 1997, **22**, 389.
339. T. Yu. Sliva, A.M. Duda, V.M. Amirkhanov, I. O. Fritsky, T. Glowiak and H. Kozlowski, *J. Inorg. Biochem.*, 1997, **65**, 67.
340. A.M. Duda, A. Karaczyn, H. Kozlowski, I.O. Fritsky, T. Glowiak, E.V. Prisyazhnaya, T. Yu. Sliva and J. Swiatek-Kozlowska, *J. Chem. Soc., Dalton Trans.*, 1997, 3853.
341. I.O. Fritsky, H. Kozlowski, E.V. Prisyazhnaya, Z. Rzaczynska, A. Karaczyn, T. Yu. Sliva and T. Glowiak, *J. Chem. Soc., Dalton Trans.*, 1998, 3629.
342. E. Farkas, H. Csóka, G. Micera and A. Dessi, *J. Inorg. Biochem.*, 1997, **65**, 281.
343. J. Szyrwiel, P. Mlynarz, H. Kozlowski and M. Taddei, *J. Chem. Soc., Dalton Trans.*, 1998, 1263.
344. W.D. Kohn, C.M. Kay, B.D. Sykes and R.S. Hodges, *J. Am. Chem. Soc.*, 1998, **120**, 1124.
345. W. Bal, J. Christodoulou, P.J. Sadler and A. Tucker, *J. Inorg. Biochem.*, 1998, **70**, 33.
346. H. Meisel and C. Olieman, *Anal. Chim. Acta*, 1998, **372**, 291.

6
Current Trends in Protein Research

BY JENNIFER A. LITTLECHILD

1 Introduction

The article highlights some of the advances made in our understanding of protein structure and function during 1997 and 1998. The wealth of information available only allows this chapter to give a flavour of what has been reported during this period. Topics covered include the importance of protein conformation in disease, human enzymes, proteins, RNA and DNA binding proteins and metalloproteins.

2 Protein Conformation and Disease

This has been an ever increasing area of interest since it was realised that many diseases are known to arise from the conformational instability of a particular protein. A review appeared in the *Lancet* in 1997[1] that categorised an increasing number of diverse degenerative and other diseases that reflect shared molecular mechanisms of initiation and self-association with consequent tissue deposition and damage. Of particular significance are two types of prion encephalopathies such as the bovine form of bovine spongiform encephalopathy (BSE) and its human equivalent new variant Creutzfeldt-Jakob disease (nvCJD). A review of Prion diseases and the BSE crisis by Prusiner[2] covers this area which is still very much under discussion. The basic problem of protein folding, which we still do not fully understand, is fundamental to our knowledge of many human diseases many of which result in protein aggregation with the β-pleated linkage being a common mechanism. Recent synchrotron X-ray studies have confirmed the β-structured helical array parallel to the fibril axis in six different *ex vivo* amyloid fibrils and two synthetic fibrillar preparations.[3] The amyloid fibrils comprise a structural superfamily which have a common protofilament structure. It is thought that the protein must unfold to some extent before amyloid formation and this is proposed to extent to an intermediate 'molten globule' stage when studies were carried out with two natural variants of the protein lipozyme that were known to form amyloids.[4] Different activation processes can induce protein conformational change. With the prion proteins it is suggested that an activated intermediate is formed by the binding of an unidentified intermediate, protein X.[5] This protein is believed

Amino Acids, Peptides and Proteins, Volume 31
© The Royal Society of Chemistry, 2000

413

to be involved in the P$_r$Pc to P$_r$Psc transformation and provides an explanation for the protective effect of basic residues in some isoforms of prion proteins in sheep and humans. In prion diseases the transformation is an autocatalytic process carried out by a co-operative mechanism.[6] Other molecules have been reported to accelerate fibril formation in Alzheimer's disease. A specific proteoglycan, perlecan, binds to critical αβ isoforms and accelerates fibril formation and provides subsequent fibril stabilisation.[7] There is much to learn about how these ordered protein aggregates are involved with the diseased state. Evidence suggests that the formation of amyloid plaques in both prion and Alzheimer's diseases is an accompanying consequence rather than a direct cause of the disease. A summary of the understanding of Alzheimer's disease has been presented by Kisilevsky.[8] Proline-substituted pentapeptide, homologous to the amyloid β peptide of Alzheimer's disease, is presented as a means to block the formation of amyloid fibrils in a rat brain model and act as a potential therapeutic agent.[9]

Fibril formation can be initiated by relatively minor changes in a protein molecule. Lui and co-workers have found that swapping a 15 residue terminal helix between two molecules of RNase A results in a fibril that can bind Congo Red.[10] Reviews of prion biology have been published in 1998.[11]

A Prions viewer is available on the World Wide Web http://mad-cow.org/ ~tom/prion-structure-folder/viewers.html. This contains a collection of structural information on prions and associated proteins.

3   Protein Chaperonines

3.1   Small Heat Shock Protein – This protein was discovered as part of a genome sequencing and structural bioinformatics study on the archaeon *Methanococcus jannaschii*. Many groups are now considering carrying out structural studies on all of the proteins found in a particular genome in order to obtain more information on protein function. This particular protein complex displays a chaperone function by protecting the cell against high temperatures. The heat shock proteins all appear to have a 100 residue sequence motif which is homologous to the α-crystallin found in vertebrate eye lens. A complex of 24 subunits of this protein form a hollow sphere with an inner diameter of 65 Å and an outer diameter of 120 Å as shown in Figure 1. The whole particle has an octahedral symmetry. There are small windows around the particle which are thought to be large enough for small molecules or extended polypeptide chains to diffuse in and out.[12]

3.2   Protein Disulfide Oxidoreductase – This protein is involved in the rate limiting step of protein folding in the formation of disulfide bonds. The structure of this protein has given some insight into the mechanism of reduction or formation of disulfides. The structure of the protein from the hyperthermophilic archaeon *Pyrococcus furiosus* has been determined[13] and shows two thioredoxin fold motifs both with an active site. The N-terminal

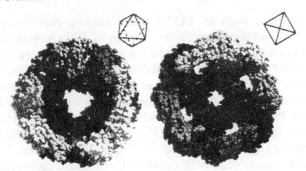

Figure 1 *Overall structure of small heat shock protein. The interior of the sphere is viewed along the three fold axis (left) and the four fold axis (right). The front one-third of each sphere is cut off to reveal the inside of the hollow sphere* (Reproduced with permission from *Nature*, 1988, **394**, 595; © 1998 Macmillan Magazines Ltd; see http://www.nature.com)

active site has a sequence CQYC and the C-terminal active site a sequence CPYC and it is suggested that the N-terminal disulfide is less stable than the C-terminal one.

3.3 Thermosome – Another protein found in archaea, which appears to be related to the chaperonine GroeEL of *E. coli*, is a class II chaperonin which does not require an additional protein such as GroES for its function. Two papers in 1997 and 1998 described the structure of the so called thermosome from *Thermoplasma acidophilum*.[14,15] The protein complex is made up of rings of alternating α and β subunits. There is a central cavity which can accommodate proteins of ~50 kDa in size. The nucleotide ATP binds at the top of the complex and its hydrolysis drives domain rotation necessary for substrate binding release.

4 Protein Catalysts

The use of enzymes in chemical synthesis is already well established. Biotransformation reactions to obtain optically pure drugs and drug intermediates are becoming an acceptable and economically viable alternative to conventional organic synthesis. Optically pure building blocks for new important drugs such as the carbocyclic ring produced from γ lactam are already being commercially exploited.

The search for suitable catalysts can either be by screening a range of microorganisms for the desired activity[16] or by redesign of known catalysts. This usually requires that the three dimensional structure of the enzyme is already known.

4.1 Enzyme Modifications – Covalent modification of cysteine residues in proteins can be used to attach a variety of cofactors, metals and EDTA based metal-binding ligands. Such an EDTA metal binding derivative has been complexed to the C-terminal regulatory domain of RNA polymerase.[17] The resulting construct was used to investigate the interaction between cyclic AMP and the cyclic AMP receptor protein (CRP) and the C-terminal domain when the complex was bound to DNA. It was shown using this technique that protein-protein interactions appear to regulate key aspects of RNA-polymerase function. Pyridoxamine-amino acid chimeras have been used to make semi-synthetic aminotransferase mimics.[18] A pyridoxamine cofactor has been incorporated into an alanine residue in the C-peptide of RNase S. The rate accelerations and enantioselectivities obtained with these constructs suggest that the structure of the noncovalent RNase S protein-peptide complex plays an important role in modulating the properties of these pyridoxamine-containing catalysts. Site-directed mutagenesis has been an important tool to examine amino acid residues important for catalysis in a wide variety of enzymes. Kuang and co-workers were able to modulate the rate, enantioselectivity and substrate specificity of semisynthetic transaminases based on lipid binding proteins using site-directed mutagenesis.[19] A series of constructs were made in which the position of the pyridoxamine attachment within the intestinal fatty acid binding protein (IFABP) cavity was varied by site-directed mutagenesis. Another study by Davies and Distefano describes a semisynthetic metalloenzyme based on a protein cavity that catalyses the enantioselective hydrolysis of ester and amide substrates.[20] A protein-1,10-phenanthroline conjugate was produced using adipocyte lipid-binding protein to produce ALBP-Phen. The ability of the conjugate to bind copper(II) was demonstrated by fluorescence spectroscopy. ALBP-Phen-Cu(II) catalyses the enantioselective hydrolysis of several unactivated amino acid esters under mild conditions (pH 6.1, 25 °C) at rates 32–280 fold above the background rate in buffered aqueous solution. ALBP-Phe-Cu(II) also promotes the hydrolysis of an aryl amide substrate at 37 °C and at a rate 1.6×10^4 fold above background. Michaelis-Menten kinetics, which are characteristic of enzymatic processes, are demonstrated for the amide hydrolysis reaction.

5   Enzyme Catalysis

Since the last review we have seen the structures of many more enzymes solved to high resolution using X-ray methods. This has led to a greater understanding of enzymatic mechanism. By using cryo-cooling of protein crystals and the use of enzyme complexed with reaction intermediates and inhibitors it is possible to trap the enzyme 'in action'.

5.1   Glutamine PRPP Amidotransferase – One example which has recently been studied is the enzyme glutamine PRPP amidotransferase. This enzyme catalyses the first step in *de novo* purine biosynthesis and is involved in the conversion of 5-phosphoribosyl-(α)1-pyrophosphate (PRPP) to 5-phosphoribosyl-(β)1-amine (PRA). The nitrogen source for this reaction is the amide group of glutamine as with many other biosynthetic enzymes. A paper reporting the crystal structure of this enzyme[21] describes the active conformer of the enzyme where a channel of 20 Å in length is observed for channelling NH_3 between the two active sites of the enzyme. This enzyme has two catalytic domains: one of these belongs to the Ntn hydrolase family and the other to the type I phosphoribosyltransferase family. The enzyme exists in an open inactive conformer without ligands and a closed active conformer. It is also subject to complex allosteric control which has now been more fully understood.[22,23] In order to understand the structural basis for the nucleotide specificity and synergism the structure of the ternary enzyme.ADP.GMP complex was determined. The enzyme regulates the purine biosynthetic pathway through feedback inhibition of the end products, AMP, GMP, ADP and GDP. The enzyme is able to make use of the structural similarity between the nucleotide inhibitors and the PRPP substrate. One of the important steps in understanding the mechanism was the synthesis of optically pure carbocyclic PRPP.[24] This analogue was used to trap the closed active conformation of the enzyme. The enzyme active site and interactions with this substrate analogue are shown in Figure 2.

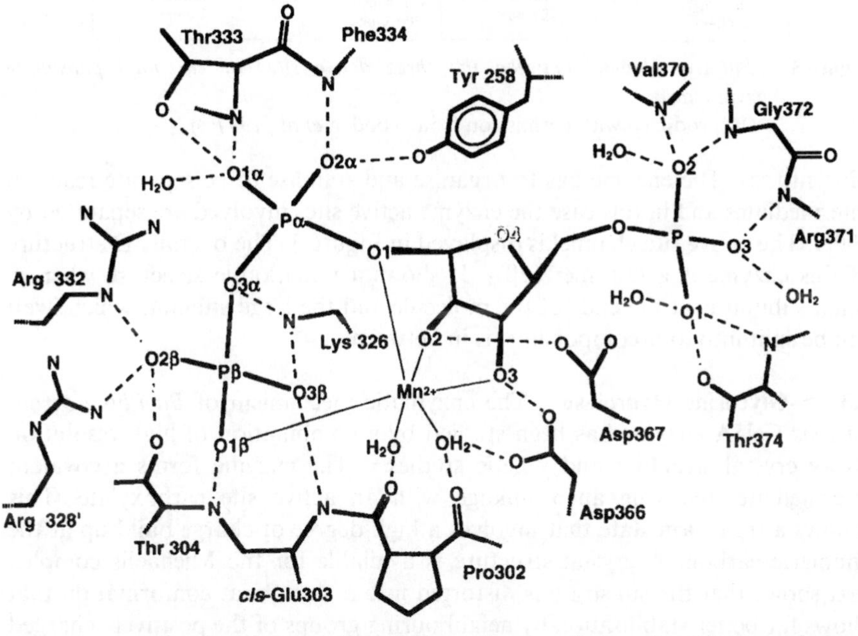

Figure 2 *A schematic diagram showing the active site of glutamine PRPP amidotransferase*
(Reproduced with permission from Krahn *et al.*, 1997[21])

5.2 Carbamoyl Phosphate Synthetase – Another interesting enzyme which also has a channel for NH_3 is carbamoyl phosphate synthetase.[25,26] This enzyme plays a critical role in the biosynthesis of arginine and pyrimidine

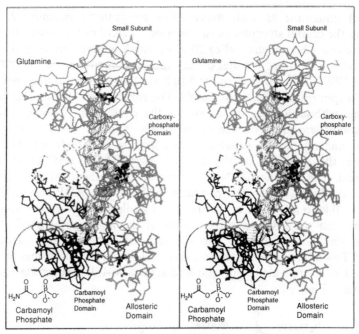

Figure 3   *Putative channel between the three active sites of carbamyl phosphate synthetase*
(Reproduced with permission from Thoden *et al.*, 1997[25])

biosynthesis. The enzyme has to organise and stabilise three separate reaction intermediates and in this case the enzyme active sites involved are separated by 100Å. The active site channel is displayed in Figure 3. The overall 3D structure of this enzyme is a tetramer $(\alpha\beta)_4$. It shows a remarkable structure where a small subunit is at the ends of the molecule and the large subunit, which itself can be split into four components, is in between.

5.3   α-Glycoside Hydrolase – The enzymatic mechanism of *Bacillus agaradhaerens* Cel5A enzyme has been studied by a combination of high resolution X-ray crystal structure and kinetic studies.[27] The enzyme forms a covalent intermediate involving an α linkage with an active site carboxylate. This follows a transition state that involves a high degree of charge build up at the anomeric carbon. A crystal structure is available for the Michaelis complex that shows that the substrate is distorted into a skew boat conformation that allows far better stabilisation by neighbouring groups of the positively charged transition state. In the crystal structure of the covalent intermediate/enzyme intermediate the pyran ring is in a chair conformation. A water molecule is positioned for nucleophilic attack above the anomeric carbon in order to liberate the final product. The reaction mechanism proposed is shown in Figure 4. In addition it was also possible to obtain crystal structures of the enzyme product and the free enzyme. Figure 4 shows the electron density and its interpretation as observed along the reaction pathway of the Michaelis

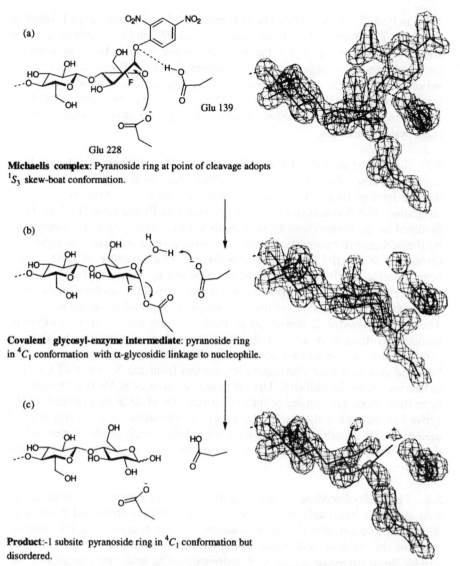

(a)

Glu 139

Glu 228

Michaelis complex: Pyranoside ring at point of cleavage adopts $^1S_3$ skew-boat conformation.

(b)

Covalent glycosyl-enzyme intermediate: pyranoside ring in $^4C_1$ conformation with α-glycosidic linkage to nucleophile.

(c)

Product:-1 subsite pyranoside ring in $^4C_1$ conformation but disordered.

Figure 4 *Snapshots along a reaction coordinate:* (a) *Michaelis complex of intact substrate,* (b) *trapped covalent glycosyl-enzyme intermediate, and* (c) *product. The electron density maps shown are maximum-likelihood-weighted $2F_o$–F_c syntheses, contoured at approximately 0.44 e/A^3*
(Reproduced with permission from Davies *et al.*, 1998[27])

complex of intact substrate, trapped covalent glycosyl-enzyme intermediate and product.

5.4 Diaminopimelate Epimerase – Diaminopimelate (DAP) epimerase is a PLP independent amino acid racemase which catalyses the epimerisation of

L,L-DAP to D,L-meso DAP. The D,L-meso DAP is a precursor of L-lysine in bacteria. The enzyme has been isolated from *Haemophilus influenzae*. The structure is interesting since it has an interdomain disulfide bond between the two homologous domains which are comprised of eight β-strands and two α-helices.[28] A catalytic role is proposed for the two cysteine residues which are involved in the disulfide bridge. An aziridine derivative of the substrate DAP is known to inactivate the enzyme which is thought to have alkylated one of these cysteine residues.

5.5   β-Ketoacyl-acyl Carrier Protein Synthase II and Yeast Thiolase I from *Escherichia coli* – These two protein structures have been reported recently and despite having only 18% sequence identity appear to have very similar structures. The β-ketoacyl-acyl carrier protein (ACP) synthase II (KAS II) is involved in the biosynthesis of fatty acids where it catalyses chain elongation by the addition of two-carbon units. The second yeast thiolase I catalyses the formation of acetyl CoA from ketoacyl-CoA. The structure of KAS II has been reported by Huang *et al.*[29] and of thiolase I by Mathieu *et al.*[30] In both structures a hydrophobic curved pocket extends from the protein surface to a cysteine residue at the floor of the cleft which is proposed to be the active site. The cysteine residue is conserved in both enzymes and in several other β-ketoacyl synthases. A conserved histidine residue is proposed to form a catalytic pair similar to that observed in the cysteine protease enzymes. The histidine acts as a base abstracting the proton from the Sγ atom of Cys 163 enhancing its nucleophilicity. The catalysed reaction of KAS II is thought to be in three steps, (I) transfer of the acyl group of acyl-ACP to a cysteine at the active site yielding a thioester, (II) carbanion formation as a consequence of malonyl-ACP decarboxylation, (III) nucleophilic attack of the carbanion of the carbonyl carbon atom of the thioester, resulting in carbon-carbon bond formation.

5.6   Phenol Hydroxylase – This enzyme can accept phenol as a substrate as well as simple hydroxyl-, amino-, halogen-, or methyl-substituted phenols as substrates and catalyse the hydroxylation of these compounds. The enzyme contains the cofactors FAD and NADPH. The structure of the enzyme from *Trichosporon cutaneum* has been determined.[31] The first two domains of the protein bind FAD and the substrate phenol and the function of the C-terminal domain is unknown. The enzyme is active as a dimer and a 'lid' can close over the active site. The substrate phenol lies deep in an active site pocket anchored by H-bonds to a Tyr and Asp residues. The ortho carbon of the substrate is in a favourable position for hydroxylation to occur. It is proposed that an open conformation is adopted before the cofactor NADPH binds and the enzyme closes when hydroxylation takes place. In the closed conformation a His residue is close to the FAD cofactor phosphate group but it is 20 Å away in the open conformation. The isoalloxamine ring has a different conformation in the open and closed forms of the enzyme as shown in Figure 5. Interestingly in this structure one subunit of the dimer is in the open conformation and the

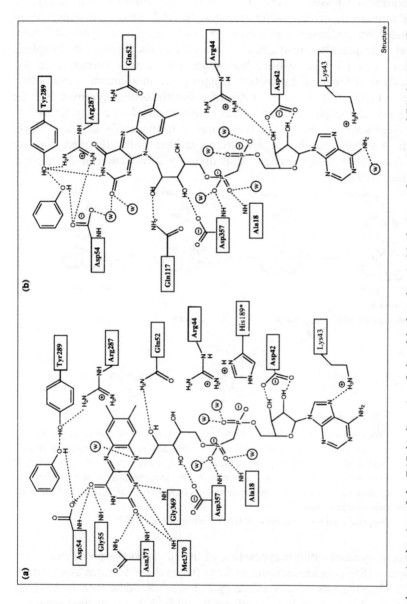

Figure 5 *Schematic drawing of the FAD-binding site of phenol hydroxylase in* (a) *the* in *conformation and* (b) *the* out *conformation. The two conformations are stabilised by hydrogen bonds, but the bonding pattern is significantly different in the two states (Reproduced with permission from Enroth et al., 1998[31])*

other subunit in closed giving some insight into the movements of the enzyme during the catalytic mechanism.

5.7 Terpernoid Cylases – Terpenoid cyclases are enzymes that catalyse a complex mechanism of cyclisation cascades. In 1997 the crystal structures of the pentalenene synthase[32] and tobacco 5-exp-aristolochene synthase[33] were described. The pentalenene synthase is involved in the farnesyl diphosphate cyclisation to a tricyclic hydrocarbon which serves as a precursor to the pentalenolactone family of antibiotics. The proposed mechanism of cyclisation is shown in Figure 6. The second enzyme 5-epi-aristolochene synthase uses the same farnesyl diphosphate which in this case is cyclised to a bicyclic hydrocarbon, 5-epi-aristolochene, which is a precursor to the antifungal phytoalexin capsidiol. A diverse number of sesquiterpenes have been identified that are derived from farnesyl diphosphate. This diversity is discussed by Sacchettini and Poulter, 1997.[34]

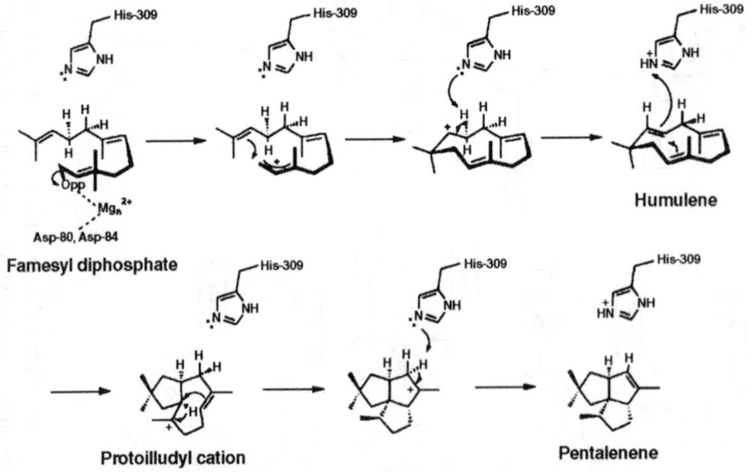

Figure 6 *Proposed mechanism for cyclisation of famesyl pyrophosphate to intermediates humulene and protoilludyl cation, with subsequent rearrangement into pentalene. The pyrophosphate leaving group is omitted for clarity; however, it may remain bound in the enzyme in the enzyme active site during the cyclisation cascade and contribute to the electrostatic stabilization of carbocation intermediates*
(Reproduced with permission from Lesburg *et al.*, 1997[32])

Triterpene cyclases catalyse cyclisation of the 30 carbon isoprene substrates squalene and (S)-2,3-oxidosqualene to form polycyclic substrates as shown in Figure 7. The structure of a squalene cyclase was also described in 1997 by Wendt *et al.*[35] The enzyme has a double α barrel fold that is distinct from the single α barrel fold of the sesquiterpene cyclases. This enzyme has to cyclise a larger 30 carbon polyisoprene substrate. An aspartic acid residue in the protein is proposed to be the general acid responsible for the protonation step as shown in Figure 8.

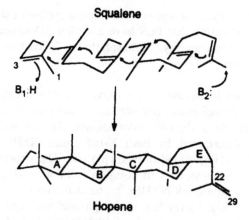

Figure 7 *The proposed reaction steps in squalene-hopene cyclases involving carbocationic intermediates. The general acid B1H protonates (H) squalene at C3, whereas the general base B2 deprotonates at C29 of the hopenyl cation, in a side reaction, the cation is hydroxylated forming hopan-22-ol*
(Reproduced with permission from Wendt *et al.*, 1997[35])

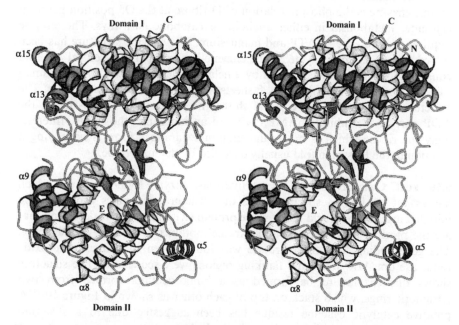

Figure 8 *Stereoview of squalene-hopene cyclase chain fold with labelled NH2- and COOH termini, inhibitor position (L) and channel entrance (E)*
(Reproduced with permission from Wendt *et al.*, 1997[35])

Plant sesquiterpene cyclases and some diterpene cyclases exhibit substantial sequence identity indicating that they have a similar 3D structure for cyclisation of either 10 carbon geranyl diphosphate or 20 carbon geranylgeranyl diphosphate substrates.[33,36]

5.8   Δ[5]-3-Ketosteroid Isomerase – This enzyme has been extensively studied from the bacterium *Pseudomonas testosteroni*. It catalyses the conversion of Δ[5] to Δ[4]-3-ketosteroid *via* a dienolic intermediate. The structure of the free enzyme has been determined by both X-ray[37] and NMR methods.[38] An aspartic acid residue is implicated in base catalysis. The structure of a mutant of the enzyme has been studied in complex with a product analogue 19-nortesterone hemisuccinate (19-NTHS).[39] The structure of the enzyme which consists of a curved, eight stranded, mixed parallel and anti-parallel β-sheet and three α-helices binds one molecule of 19 NTHS in a hydrophobic cavity. The complex structure shows that a tyrosine residue accepts a hydrogen bond from a second aspartic acid residue and donates a hydrogen bond to the 3-keto group of 19 NTHS. A conformation change is observed between the bound and unbound states of the enzyme.

5.9   Ribokinase – This is a new class of kinase enzyme. The reaction catalysed by this enzyme is the phosphorylation of D-ribose at the O5 position before it can enter metabolism in either anabolic or catabolic pathways. The enzyme requires the presence of ATP and magnesium. The enzyme structure has been solved for the *E. coli* enzyme.[40] The enzyme has two parts, a central α/β fold consisting of nine β sheets flanked by α helices which forms a classic Rossman fold and a second part formed by β sheets which interact to form a dimer. The structure is shown in Figure 9 and shows the positions of the D-ribose, the ADP part of the inhibitor AMP.PNP and an inorganic phosphate. Interestingly the ribose which is in the uncommon α furanose form is occupying a position in the Rossman fold usually occupied by the nucleotide.

5.10   YcaC Gene Product – This is a hydrolase enzyme found in *E. coli* which has as yet no known specificity. As the structural genomics programs are initiated around the world many more protein structures will be described that are the product of an 'open reading frame' obtained from DNA sequencing. The gene coding for this hydrolase was found when the dimethylsulfoxide reductase dmsABC open and flanking regions were sequenced. The structure shows this enzyme to be formed as a homo-octamer formed from two tetrameric rings, which stack on top of each other as shown in Figure 10.[41] A putative catalytic cysteine residue has been suggested due to a structure similarity to N-carbamyl sarcosineamidohydrolase.

5.11   *E. coli* Signal Peptidase/β-Lactam Type Inhibitor Complex – This protein is a membrane bound endopeptidase and the C-terminal fragment which is located in the periplasmic space has been studied with a bound β-lactam inhibitor.[42] The structure is shown in Figure 11 and shows a covalent bond

Figure 9 *Schematic representation of a ribokinase dimer*
(Reproduced with permission from Sigrell *et al.*, 1998[40])

Figure 10 *A ribbon diagram of YcaCgp viewed down the fourfold axis. The bottom four subunits are omitted for clarity. The proposed active site is located between two subunits and is represented by an asterisk*
(Reproduced from *Structure*, **6**, 1329; © 1998 with permission from Elsevier Science)

between a serine proposed to act as the acylating nucleophile during catalysis and the carbonyl carbon of the inhibitor. The four membered lactam ring is cleaved and this enzyme uses lysine as the catalytic base. The mitochondrial and bacterial enzymes use a catalytic lysine however, the equivalent enzyme found in the endoplasmic reticulum uses histidine.

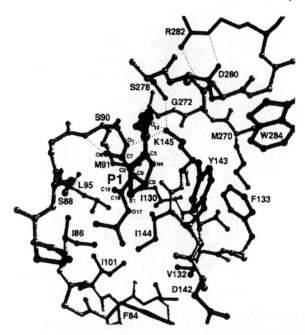

Figure 11   *A ball-and-stick representation of the active-site residues of signal peptidase.*
The β-lactam (5S,6S penem) inhibitor is covalently bound to the Oγ of Ser
90, with the carbonyl oxygen (O8) of the cleaved β-lactam (the bond between
C7 and N4 has been cleaved) sitting in the oxyanion hole formed by the main-
chain nitrogen of Ser 90 (S90). The methyl group (C16) of the inhibitor,
labelled P1, sits in the S1 substrate-binding site
(Reproduced with permission from *Nature*, 1998, **396**, 186; © 1998
Macmillan Magazines Ltd; see http://www.nature.com)

6   RNA–Protein Complexes

Our understanding of how protein complexes with RNA has increased over
the last two years due to an increased amount of structural information now
available. The structure of RNA has more variation than the regular double
helix structure of DNA. It, however, is a very important molecule which can
carry genetic information in viruses and can interact with proteins in many
important cellular macromolecular complexes. A protein motif called RNP is
often observed in many proteins that bind RNA. It is made up of 80 amino
acid residues and has two motifs RNP1 and RNP2 and folds into a compact
βαββαβα fold.

6.1   Spliceosomal Protein/RNA Complex – Small RNA protein particles are
involved in the 'splicing' of DNA where they remove regions which are not
coding for proteins. A recent study describes the 2.4 Å crystal structure of
spliceosomal U2B″–U2A′ proteins complexed with RNA.[43] This has provided
additional information from that obtained from earlier work, published by the

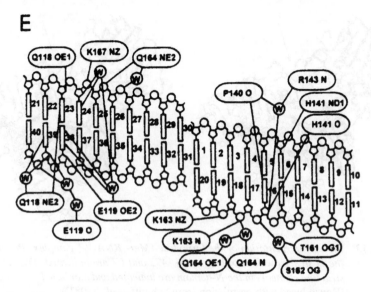

Figure 12 *Schematic diagram showing the interactions between the Xlrbpa-2 dsRBD and dsRNA*
(Reproduced with permission from Ryter and Schultz, 1998[44])

same group and it has provided an insight into the specificity of binding. The U2B″ protein is 75% homologous to a U1A protein studied earlier and contains the RNP module. The U2A′ protein contains leucine rich repeats, and the complex binds to a stem loop structure of RNA.

6.2 Double Stranded RNA Binding Protein from *Xenopus laevis* – Some proteins are able to bind to double stranded RNA. The structure of a 65 amino acid residue protein module complexed to 10 base pairs of double stranded RNA has been solved at 1.9 Å resolution. It demonstrates the molecular basis of recognition by the protein.[43] There are three distinct areas of the protein that interact with the RNA by direct or water mediated interactions. Two of these are in the minor groove of the RNA and one in the major groove. These are shown in Figure 12.

6.3 Protein Synthesis – RNA protein complexes play an important role in the protein synthesis machinery of the cell. Important information is now becoming available on many of these complexes including the multiprotein/RNA complex on which protein synthesis proceeds – the ribosome.

6.3.1 Transfer RNA/protein complexes – Transfer (t)RNA, or the 'adaptor molecule' as it is often called, plays an important role in bringing the correct amino acids to the ribosome where they are built up into protein dictated by the code on the messenger RNA molecule. In bacteria the first tRNA to be brought to the ribosome is formyl methionine tRNA. Methionyl-tRNA$_f^{Met}$

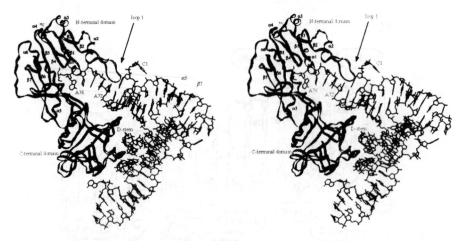

Figure 13   *Stereo representation of the formylase-fMet-tRNA$_8^{Met}$ complex. Positions of*
the D-stem as well as of bases A76, A72 and C1 are indicated. The secondary
structure elements in the N-domain are indicated and numbered
(Reproduced with permission from Schmitt *et al.*, 1998[45])

transformylase/formyl methionyl-tRNA$_f^{Met}$: a recently reported structure of
this complex reveals the details of recognition.[45] The formylated tRNA starts
the protein chain, interacts with initiation factors and binds to the 'P' site of
the ribosome. The transformylase enzyme recognises a unique feature found in
the acceptor stem of this tRNA. The structure of this complex is shown in
Figure 13 which also gives some indication of the relative size of these two
molecules.

tRNA also binds specifically to its aminoacyl tRNA synthetase enzyme and
the structures of these complexes have been studied over some years mainly by
the group of Cusack. A review describing several of these structures has been
published by Cusack.[46] In addition another complex of the tRNA^{Pro} anticodon
recognition has been described.[47]

6.3.2 Thermus Thermophilus Prolyl-tRNA Synthetase/tRNA^{Pro} Anticodon –
Amino acid tRNA synthetase enzymes fall into several different groups. The
prolyl-tRNA synthetase is a class IIa enzyme and a recent structure shows that
this synthetase, as others in its class, has a C-terminal anticodon binding
domain which binds to the anticodon stem loop of the tRNA from the side of
the major groove with three bases splayed out compared to the five seen with
class IIb aminoacyl tRNA synthetase enzymes.[47] In the case of prolyl-tRNA
synthetase only G35 and G36 are needed for recognition whereas in an
example of class IIb, lysyl-tRNA synthetase three anticodon bases C34, U35,
U36 make specific interactions.

6.3.3 tRNA Endonuclease – This enzyme has been studied from the archaeon
Methanococcus jannaschii.[48] It is involved in splicing of tRNA introns. The
structure is a 'dimer of dimers' and appears to have four catalytic sites of

which only two are thought to be active. The endonuclease specifically recognises a specific structure in the tRNA termed a bulge-helix-bulge. The enzyme is thought to be related to equivalent enzymes found in higher cells although its method of recognition of the tRNA and splicing is different.[49]

6.3.4 Ribosomal Protein S4 – This is a primary rRNA binding protein. The structure of the protein alone is described here; however, the region of rRNA to which it binds is well known. The protein structure has been determined by X-ray and NMR methods from a deletion mutant of *Bacillus stearothermophilus* S4 protein. Davies *et al.*[50] have described the X-ray structure and have found the protein to be composed of two domains. Domain 1 is a four helix bundle and domain 2 shows close homology to the ETS domain found in eukaryotic transcription factors. There is one face of the molecule that is positively charged and includes many conserved residues which suggests it is the RNA binding surface. The structure has also been solved independently by NMR methods[51] and shows a similar structure with a small interdomain rotation.

6.3.5 Structure of the Large Ribosomal Subunit – One of the most exciting RNA/protein structures is emerging – that of the ribosome. Recently it has been possible to interpret the diffraction data collected from these large macromolecular assemblies.

The first 9 Å map of a 50S large ribosomal subunit was reported.[52] The interpretation of this model was based on using phases from electron microscopy to locate the heavy atom derivatives containing 18 tungsten, 11 tungsten and tantalum clusters. The large ribosomal subunit was from a halophilic organism, *Haloarcula marismortui*, which grows in high salt conditions. Crystals and some initial interpretation of this structure were first described by the group of Yonath.[53] This is only the beginning of the structural elucidation of the ribosome particle and much will be reported on this subject in future years.

7   DNA Protein Complexes

Many DNA protein complexes have been studied since the last review in 1996. Several of these have involved the study of polymerase enzymes such as the DNA polymerase I from two thermophilic bacteria *Bacillus stearothermophulus*[54] and *Thermus aquaticus*.[55] Both structures have been determined in the presence of DNA and allow the determination of the structural features for sequence specific molecular recognition. The polymerase I enzymes are responsible for repair of DNA lesions in prokaryotic organisms.

7.1   DNA Topoisomerase – The topoisomerases remove supercoils from DNA during recombination, replication and transcription by introducing a break in one DNA strand. The structure of the human core enzyme and C-terminal fragment have been determined in complex with DNA.[56] The N-terminal fragment of the topoisomerase I with DNA had been described in 1997 by the same group[57] and together they provide some insight into the DNA binding and manipulation of eukaryotic topoisomerase enzymes. The study of Redinbo *et al.*[56] also describes docking studies with an anticancer drug, camptothecin.

7.2   HIV Reverse Transcriptase – Structural studies of HIV reverse transcriptase have been important in providing information for design of inhibitors to be used in the treatment of AIDS. However, patients undergoing therapy have become resistant to these drugs due to mutations taking place to the enzyme. A recent structure of a covalently trapped catalytic complex of HIV-reverse transcriptase has provided some understanding of this drug resistance.[58] Another important HIV-1 reverse transcriptase complex has been solved by Jaeger *et al.*, where a RNA pseudoknot inhibitor is complexed with the enzyme.[59]

7.3   DNA Binding Protein Dps – This protein from *E. coli* is an interesting oligomeric structure which appears to be related to bacterioferritins.[60] This protein protects DNA from oxidative damage. This is an interesting example of how a protein class can be used for two different functions. It is suggested that sequestration of iron into the negatively-charged cavity in a mechanism similar to that observed in ferritin is a significant factor in the protection of DNA. Hexagonal sheets of Dps dodecamers have been observed by electron microscopy when the Dps protein is complexed with DNA.

7.4   Sso7d DNA Binding Protein – Non specific DNA binding proteins found in the archaeon *Sulpholobus* are involved in the stabilisation and organisation of DNA. The structure's of Sso7d from *S. solfataricus*[61] and a related protein Sac7d from *S. acidocaldarius*[62] show the proteins to interact into the minor groove of the DNA causing a sharp kink of 61°.

8   Human Proteins

The ability to clone and overexpress many human proteins and enzymes has led to an ever increasing wealth of information on their structure and function. The human genome project will provide information on the DNA sequence of all of our genes. It is increasingly apparent that single amino acid changes can result in human disease. We are just beginning to understand the complexity of many biological processes and the subtle changes that can make us predisposed to various diseases.

8.1   Eps15, Homology Domain 2 – Small binding domains are known that can target a specific peptide sequence. The Eps15 homology (EH) domain is able to recognise proteins containing a Asn-Pro-Phe sequence. Proteins that contain the EH domain are involved in regulation of endocytosis and actin cytoskeletal organisation. Eps15 is a multidomain protein that contains not only three EH domains but other domains specific for other proteins. The structure of the domain 2 determined by NMR methods[63] consists of two EH domains connected by a short antiparallel β-sheet. The second EH domain has a tightly bound calcium ion. Multidomain proteins are not amenable to crystallisation and NMR methods are useful to determine individual or small duplicated domains as reported here. Also, valuable information can be gained concerning protein recognition and in this case changes in chemical shift are observed upon addition of peptide.

8.2 Hepatocyte Growth Factor, Nk-1 Fragment – This factor has been implicated in the invasion of tumours and binds specifically to the proto-oncogene product c-met, a tyrosine protein kinase. This appears to be produced in increased amounts in colorectal carcinomas. Part of this protein structure has been described from both NMR and X-ray methods and the results are similar. The NMR structure represents just the N-terminal domain[64] whereas the X-ray structure represents the N-domain and the first so called 'Kringle domain'.[65] The N-terminal site is the place of receptor and heparin binding. Similar domains are found in other proteins like plasminogen and macrophage stimulating protein.

8.3 Insulin-like Growth Factor Receptor (IGF-IR) – The N-terminal three domains of the insulin like growth factor have been studied by X-ray methods.[66] The IGF domain has also been studied by NMR methods.[67] The IGF-IR receptor is involved in normal growth and development as well as malignant transformation. The first three domains in the X-ray structure called L1, Cys rich and L2 adopt an extended bilobal structure which surround a central space which could accommodate a ligand. The L domains at each end consist of a single stranded right handed β-helix, capped at either ends by short α-helices and disulfide bonds. This gives an idea of how these growth factors and insulin might interact with cell surface receptors to produce their metabolic and mitogenic effects.

8.4 HERG K$^+$ Channel, N Terminal eag Domain – Channels for K$^+$ are located in cardiac muscle and the nervous system. A genetic disorder which affects these channels (QT syndrome) results in cardiac arrhythmia and sudden death. Interestingly the structure of the eag domain of HERG (human eag-related protein) is different from other K$^+$ channels but is similar to that of the bacterial light sensing protein, photoactive yellow protein.[68]

8.5 Arylsulfatase A – This family of enzymes are able to hydrolyse sulfate ester bonds in several different compounds. There are several distinct human disorders associated with a deficiency in a particular sulfatase. Sulfate deficiency is the result of a defect in an essential post-translational oxidation of the Cβ of a conserved free cysteine in the sulfatases to produce L-Cα formyl glycine. The crystal structure of human arylsulfatase A has been described.[69]

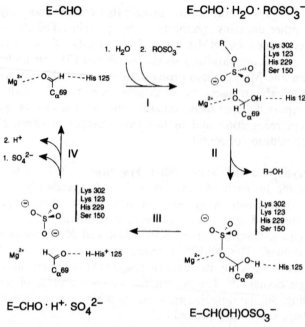

Figure 14   *Catalytic mechanism of arylsulfatAse proposed in analogy to that deduced for*
             E.coli alkaline phosphatase
             (Reproduced with permission from Lukatela *et al.*, 1998[69])

This human enzyme is a homo octamer composed of 4 dimers $(\alpha_2)_4$. The
enzyme hydrolyses the 3-sulfate group of cerebroside-3-sulfate, a major
component of the myelin sheet. The active site that contains the modified
cysteine residue lies at the bottom of a positively-charged cavity which is
located at the C-terminal end of the major β-sheet. A mechanism for the
enzyme has been proposed for sulfate ester hydrolysis, in which formylglycine
takes up a water to form the hydrate. One of the hydroxyl groups of the
aldehyde hydrate is activated by Mg^{2+} for nucleophilic attack at the sulfur of
the incoming substrate $ROSO_3^-$. After the release of the product ROH and
the formation of the covalent intermediate, elimination of HSO_4^- takes place
and the formyl-glycine aldehyde reforms. This proposed mechanism is shown
in Figure 14.

8.6   Cell Cycle Control Phosphatase Cdc25A – Cyclin dependent kinases are
regulated by phosphorylation. Their activation depends on the phosphatase
Cdc25. which specifically dephosphorylates Tyr14 and Thr14. The structure of
the catalytic domain of Cdc25A has been reported.[70] This enzyme is a
substrate for Cdk2 cyclin E. It is found that the catalytic domain has a
different fold from other phosphatases but is identical to that of rhodanese. It
does, however, have a motif Cys-(X)5-Arg which is found in other phosphatase
enzymes. Figure 15 shows the structure of the Cdc25 and some related
proteins, including rhodanese.

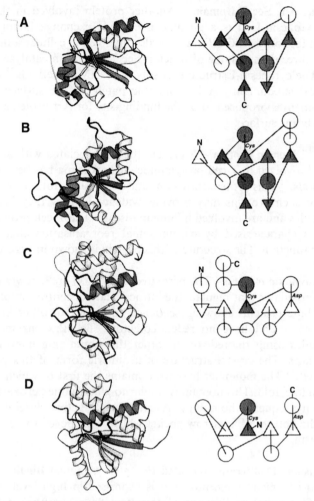

Figure 15 *Topological comparison of Cdc25 and related proteins*
(Reproduced with permission from Fauman *et al.*, 1998[70]; © Cell Press)

8.7 Complement Component C3 Fragment – This protein is involved in the complement mediated human defence system and needs to be proteolytically activated for attachment to pathogens. The proteolysis exposes an internal, reactive Cys17-Gln20 thioester.[71] The thioester undergoes nucleophilic attack by a His residue producing an activated intermediate, acyl-imidazole-Gln20. The Gln20 undergoes a second nucleophilic attack by pathogenic surface nucleophiles which results in covalent attachment of the C3 fragments. This then allows binding to B cells of the immune system. The structure of the C3d fragments is an α-α barrel formed from consecutive α helices alternating from the inside to the outside of the barrel.[72] Other related proteins have been shown to also have this unusual α barrel fold such as glucoamylase, endoglucanase and protein farnesyltransferase β-subunit.

8.8   Cytohesin-1, Sec 7 Domain – Another protein involved in the immune system is cytohesin-I. It is a guanine nucleotide exchange factor for ADP ribosylation fact or (Arf) GTPases. It is also involved in cellular adhesion. The protein has three domains one of which is Sec 7 which can catalyse nucleotide exchange. The cytohesin-I structure as solved by NMR methods[73] consists of two α helical subdomains. A hydrophobic patch on the surface of the C-terminal domain is proposed to be the binding site for Arf proteins that occur on the membrane surface.

8.9   p14^{TCL1} Human Protein – A protein that is associated with an oncogene product and involved in T-cell prolymphocytic leukaemia has been studied by X-ray methods.[74] This protein shows an unusual β-barrel fold. It is hoped that the study of such proteins may provide evidence of how they interact with other molecules and are involved in human disease. The T-cell prolymphocytic leukaemia is characterised by chromosomal reorganisation that interrupts proper gene function. The structure of this protein is shown in Figure 16.

8.10   Peroxiredoxin hORF6 – The peroxiredoxins (Prx) are a newly discovered family of peroxidases that regulate the intracellular concentration of hydrogen peroxide. These proteins show no sequence homology to other known anti-oxidant proteins and have no redox cofactors. The Prx enzymes reduced hydroperoxides using thioredoxin or other thio-containing intermediates as electron donors. The crystal structure of a dimeric form of this protein has been reported.[75] The monomer has two domains the first of which contains a 'thioredoxin like' fold. This domain is similar to glutathione peroxidase despite a low primary sequence homology. A catalytically active cysteine residue is located at the bottom of a narrow pocket which is proposed to be involved in the mechanism.

8.11   Peroxisome Proliferator-activated Receptor-γ, Ligand Binding Domain – This protein is a nuclear receptor and is expressed in high levels in adipose tissue and macrophages. Proteins of this type have a central domain that binds DNA and a C-terminal domain that is involved in ligand binding, dimerisation and transactivation functions. The structure of the ligand binding and co-activator assembly of this receptor has been determined.[76] This structure is in complex with a rosiglitazone antidiabetic ligand which is a thiazolidinedione which activates this receptor and is used for the treatment of type 2 diabetes.

8.12   Plasmin – Plasmin is a well known component of blood that dissolves clots. It is formed from a precursor protein plasminogen. Tissue-type plasminogen activator (TPA) and other activators such as streptokinase are used clinically as thrombolytic agents. The structure of the human plasmin catalytic domain in complex with streptokinase has been recently studied.[77] The whole plasminogen molecule has five so called 'kringle' domains and a catalytic domain which has been found to be similar to the trypsin protease family. The

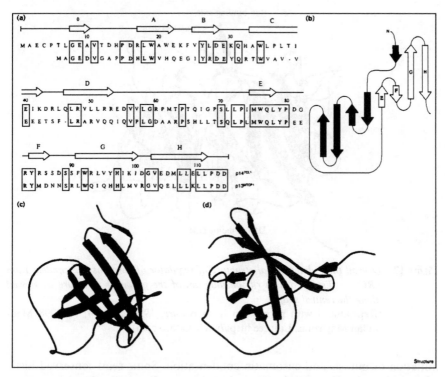

Figure 16 *The sequence and structure of p14^{TCLI}. (a) Sequence alignment and secondary structure assignment of p14^{TCLI} (top row) and p13^{MTCPI} (bottom row). Conserved residues are boxed. The first strand is not really part of the barrel and is therefore labelled O. (b) Topology diagram of the P14^{TCLI} barrel structure. The two repeated motifs strands A-D and strands E-H are shown. (c,d) Two ribbon representations of the P14^{TCLI} crystal structure. Two different orientations are shown: one looking sideways (c) onto the barrel and one looking from the top (d)*
(Reproduced with permission from Hoh *et al.*, 1998[74])

streptokinase, which is not a protease, but when bound to plasminogen can activate proteolysis to the active plasmin, has three protein domains. These domains which are mainly β-strands with a α-helix, encircle the catalytic domain of plasmin.

8.13 Human Regulator Protein of Chromosome Condensation 1 – This protein is an important factor in nucleo-cytoplasmic transport and the cell cycle. The structure of this interesting protein has been determined by Renault *et al.*,[78] and it is found to be an interesting seven-bladed β propeller structure as shown in Figure 17. The protein is known to serve as a guanine nucleotide exchange factor for a protein called Ran which is a nuclear homologue of the protein Ras. It is also known to bind DNA *via* a protein/protein complex. Mutations in the protein affect pre-messenger RNA processing and transport, mating,

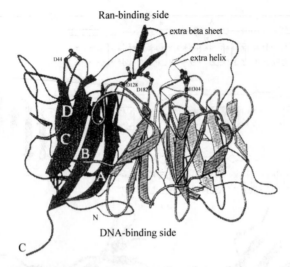

Ran-binding side

Figure 17   *Overall three-dimensional structure of regulator of chromosome condensation
(RCC1) displayed as a ribbon diagram of the propeller structure as viewed
along the central shaft*
(Reproduced with permission from *Nature*, 1998, **392**, 97; © 1998 Macmillan Magazines Ltd; see http://www.nature.com)

initiation of mitosis and chromatin condensation. Some semi-conserved histidine residues that connect the 'blades' of the propeller and residues important
for interaction with the Ran protein have been located. There is a structural
similarity to the β-subunit of heterotrimeric G proteins.

8.14 β-Tryptase – This human enzyme is released in allergic and inflammatory
disorders and has been implicated in asthma. It cleaves peptides such as
bronchodilatory neuropeptides and vasoactive intestinal peptide. It is resistant
to serine protease inhibitors but is inhibited by a small inhibitor found in the
leech. Studies of this enzyme[79] have shown in to form a tetramer with a central
oval pore. The enzyme is a target for the synthesis of inhibitors that can be
used to treat allergic and inflammatory disorders.

8.15 XRCCI, BRCT Domain – This appears to be a protein motif that is
found in many proteins involved in DNA repair, DNA recombination and the
cell cycle.[80,81] In humans it is important for promoting the association of DNA
polymerase β and DNA ligase III. The dimer interface in the C-terminal
BRCT domain has shown the importance of a conserved tryptophan residue
located at the centre of a hydrophobic pocket.[82] Mutations of this residue to
an arginine are known to result in a predisposition to cancer.

8.16 Deoxyhypusine Synthase – Hypusine is a modification that occurs to a
lysine residue in a eukaryotic initiation factor eIF5A. It appears to affect
protein synthesis and is involved in cell proliferation. The enzyme doxyhypu-

sine synthase uses spermidine and lysine in eIF5A as substrates and the cofactor NAD. The structure of the human enzyme has been described[82] and is is a tetramer with a classic Rossman fold. The reaction of the enzyme is thought to proceed by the dehydrogenation of spermidine, followed by formation of an enzyme-imine intermediate between the 4-aminobutyl group of spermidine and the ε-amino group of a lysine residue in the active site of the enzyme. The butyl amine moiety is then transferred to the specific lysine of the protein substrate. The enzyme is considered a useful target for synthesis of inhibitors for HIV-1 replication control and other diseases resulting in cell hyper-proliferation.

9   Haem Binding/Fe Binding Proteins

9.1   Nitric Oxide Synthase – Nitric oxide which has received considerable attention in its role as a key signalling molecule, is produced by nitric oxide synthase which is a large protein consisting of a haem binding domain, a calmodulin binding linker region and a NADPH cytochrome P450 reductase-like flavin domain. The structure of the haem binding domain of the endothelial enzyme has a zinc ion positioned equidistant from each haem of the dimer which is tetrahedrally coordinated to symmetrically related cysteine residues at the bottom of the dimer interface.[84] Another structure was described in the same year by Crane *et al.*, which is an inducible nitric oxide synthase has an inter subunit disulfide bond.[85] Both enzymes require H_4B as a cofactor. The endothelial enzyme has been crystallised with an inhibitor S-ethylisothiourea and shows that in this case the H_4B site is occupied by the substrate L-arginine which suggests the ability of this site to stabilise the pterin cation radical. This gives some insight into the role of H_4B in nitric oxide synthase catalytic mechanism. The hydrogen bonding network at the H_4B site is shown in Figure 18. It is proposed that the uniqueness of the H_4B-eNOS interaction and the ability to bind L-Arg at the pterin site suggests the ability of this site to stabilise a positively charged state of the pterin ring.

9.2   Cytochrome C_{544} – This cytochrome is from an organism *Nitrosomas europaea* that oxidises ammonia as its sole carbon source. The protein is unusual since it contains four haem groups. The recent structure has revealed the positions of these haem groups[86] which are similar to that observed in a hydroxyamine oxidoreductase from which cytochrome C_{554} is believed to accept two electrons. Two di-haem pairs are arranged with haem planes that are parallel in the pair and perpendicular to the haems in the other pair.

9.3   Fe-only Hydrogenase (CpI) from *Clostridium pasteurianum* – Hydrogenases described previously contain both nickel and iron and sometimes also selenium. Peters *et al.* have described the structure of an iron only hydrogenase.[87] The CpI enzyme has a [Fe-S] cluster which contains six Fe atoms

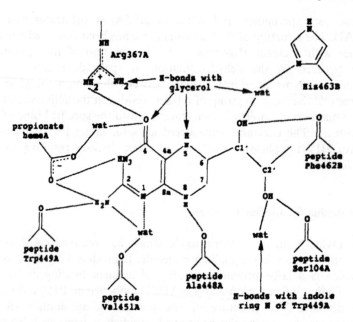

Figure 18   *Hydrogen bond networks at the H4B-binding site*
(Reproduced with permission from Raman *et al.*, 1998[84])

arranged in two groups. These are coordinated to the protein by cysteine residues. One of these forms a cysteinate thiol to bridge between the [2Fe] subcluster and the [4Fe-4S] subcluster. This is the first time that this type of arrangement has been described. A mechanism is proposed for proton reduction by CpI where an active site Fe displaces a bound terminal water ligand and generates a Fe-hybrid intermediate. A cysteine at the active site could then act as a proton donor for the formation of dihydrogen.

9.4   Fhu A – This protein forms a hollow, 22 stranded antiparallel β barrel structure with a nearly globular plug at one end.[88,89] It is involved in *E. coli* in the uptake of siderophore ferrichrome which is the way in which bacteria acquire iron from the environment. The fold resembles that found in other membrane proteins like the porins and the ferrichrome binding site is located in the upper part of the plug domain. Comparison of the apo and unliganded structure shows subtle changes in the ligand binding site which are propagated and amplified across the plug. The translocation of ferrichrome-iron from its external binding site to the inside of the cell is thought to occur along a channel opened by loop rearrangement when a complex is formed with another protein TonB, or so called energy transducing molecule, which is anchored to the cytoplasm membrane. The structure described by Ferguson *et al.* shows a lipopolysaccharide molecule associated with Fhu A consisting of two linked phosphorylated glucosamines, six fatty acid chains, two octose and two heptose residues in the inner core and three hexose molecules in the outer core.

9.5 Naphthalene 1,2-Dioxygenase – Naphthalene 1,2-dioxygenase is an enzyme involved in the oxidation of naphthalene to (+)-cis(1R,2S)-dihydroxy-1,2-dihydronapthhalene. The structure of this enzyme was described in 1998[90] and is the prototype for this family of enzymes. The structure is a hexameric 'mushroom' shape with β3 subunits forming a stem and α3 subunits forming a cap. The α subunit contains a Reiske [2Fe-2S] centre in one domain. In the Reiske centre one iron coordinates with two cysteines and the other with two histidines. The two sulfide ions bridge the two iron atoms to form a flat rhombic arrangement. A similar iron centre found in bovine heart mitochondrial cytochrome bc1 ferrodoxin can be superimposed with that found in this enzyme. In the catalytic domain a narrow gorge allows access to the non-haem ferrous ion coordinated to two histidines, a bidentate aspartic acid and water in a distorted octahedral bipyramid.

10 Copper Containing Proteins

10.1 Catechol Oxidase – This enzyme has a binuclear copper centre and is found in plants. It catalyses the oxidation of a range of o-quinones using molecular oxygen. The enzyme structure shows that the dicopper site is found at the centre of a four helix bundle.[91] Each of these coppers is coordinated by three histidine residues contributed from the four helices of the α bundle. The coppers move apart upon reduction with the separation being 2.9 Å in the oxidised structure and 4.4 Å in the reduced structure. A structure has also been obtained by the same group of the enzyme in complex with phenylthiourea where the oxidised CO-PTU complex indicates that the catechol substrate and dioxygen can bind simultaneously.

10.2 Hemocyanin – Hemocyanin is the oxygen carrying molecule of molluscs and arthropods. The oxygen bound form of this protein is blue and two Cu(I) ions are oxidised to Cu(II) with the production of peroxide. A C-terminal fragment, called Odg, has been crystallised and its structure solved.[92] This fragment is made up of two domains, one of which has 15 α helices and the other an antiparallel β sandwich of six β-sheets. The active site containing the copper is located between the two domains. This structure is from the mollusc Octopus. Each copper is liganded by three His residues and there is a bridging peroxo group between them. The protein fold of the antropodan hemocyanin is different but the mechanism for binding of the copper is the same.

11 Other Proteins

11.1 Botulinum and Tetanus Neurotoxins – The structures of these two toxins have been elucidated in 1997 and 1998. The tetanus receptor binding fragment was described by Umland *et al.*[93] and more recently the botulinum neurotoxin type A by Lacy *et al.*[94] The neurotoxin is made as a single chain which is then cleaved to form the light catalytic domain and the heavy chain. The heavy chain appears as two distinct subdomains. One of these domains, called the translocation domain, undergoes an acid induced conformational change that allows it to penetrate the endosome and form a pore for the catalytic domain to pass through. The catalytic domain is a zinc containing enzyme where the metal is co-ordinated by two histidine and two glutamic acid residues and its role is to cleave the presynaptic protein SNAP-25. This in the case of botulinum blocks the release of the neurotransmitter acetylcholine resulting in muscle paralysis.

11.2 Colicin – Colicin is a bacteriocin that can bind to and pass through the cell membrane of *E. coli* forming a pore. Structures of colicin la and colicin N have been described.[95,96] The protein is made up of several domains: the N-terminal domain specifies the translocation route, and the central domain the receptor domain and the C-terminal domain the toxicity. The pore forming domain is a bundle of ten α helices which form a hydrophobic core containing a buried helical hairpin and a channel filled with water molecules that stretches from one side of the molecule to the other. Figure 19 shows the pore forming and the receptor domain of colicin N.[96]

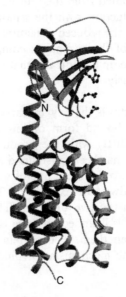

Figure 19   *Schematic representation of colicin N. The cluster of positively charged residues in the putative binding cleft is shown in ball-and-stick representation. The receptor binding domain is at the top and the pore forming domain at the bottom*
(Reproduced from *Structure*, **6**, 863; © 1998 with permission of Elsevier Science)

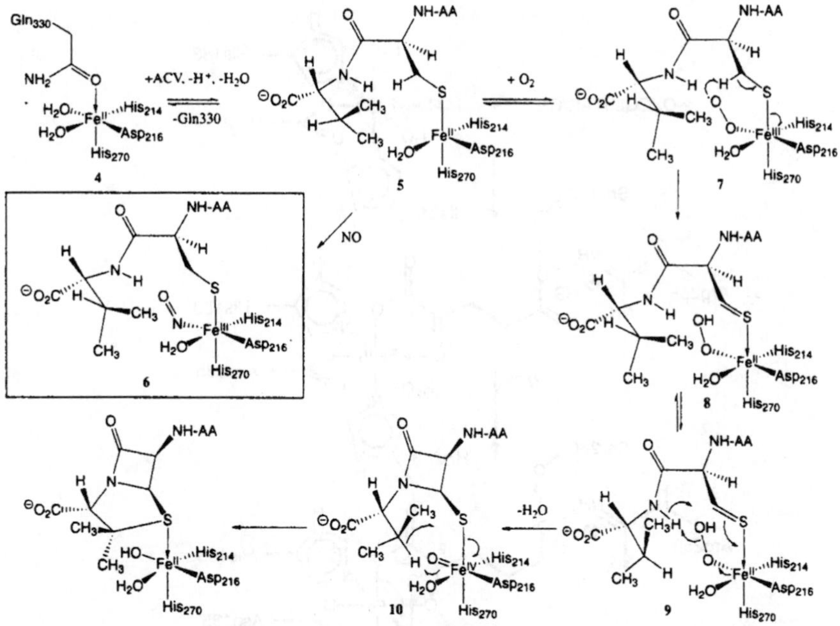

Figure 20 *Mechanism for isopenicillin N formation and the formation of the Fe:ACV:-NO:Isopenicillin N-synthase complex*
(Reproduced with permission from *Nature*, 1997, **387**, 827; © 1997 Macmillan Magazines Ltd; see http://www.nature.com)

11.3 Penicillin Biosynthetic Enzymes – The structure of isopenicillin N-synthase complexed with substrate has aided in our understanding of penicillin formation.[97] The enzyme deacetoxycephalosporin C synthase has also recently been described[98] and shows a similar motif but differs in the use of dioxygen. The mechanisms of the two enzymes are shown in Figures 20 and 21. The former uses four-electron reduction of molecular oxygen to water, whereas the latter uses only a two-electron reduction of its substrate. The two enzymes share only 19% sequence identity but they are structurally similar, having a so called 'jelly role' structure. However, there are differences as might be expected around the active sites. Study of these enzymes will help understand their mechanism and hopefully be important in the synthesis of new antibiotics.

12 Summary

Some of the main advances in our understanding of proteins have been described. We are beginning to understand the importance of a change of protein conformation that can lead to various human diseases. Our knowledge of the structure of human proteins will increase our understanding of the

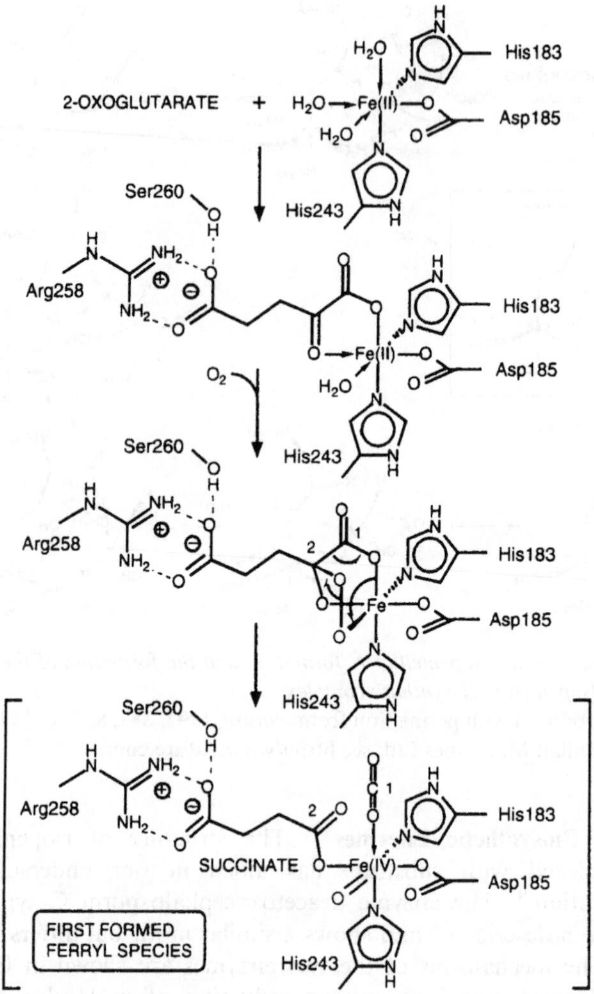

Figure 21 *Scheme to show the proposed mechanism of the ferryl intermediate in deacetoxycephalosporin C and related 2-oxoacid-dependent ferrous enzymes. Although the process is drawn as a concerted reaction in the last step, a stepwise process is also possible. The model does not exclude the presence of the penicillin substrate during ferryl formation*
(Reproduced with permission from *Nature*, 1998, **394**, 805; © 1998 Macmillan Magazines Ltd; see http://www.nature.com)

importance of molecular recognition for protein/protein, protein/DNA, protein/RNA, and protein carbohydrate interactions. The importance of enzymes in biotechnology and synthesis of fine chemicals is an area of increasing development.

References

1. R.W. Carrell and D.A. Lomas, *Lancet*, 1997, **350**, 134–138.
2. S.B. Prusiner, *Science*, 1997, **278**, 245–251.
3. M. Sunde, L.C. Serpell, M. Bartlam, P.E. Fraser, M.B. Pepys and C.C.F. Blake, *J. Mol. Biol.*, 1997, **273**, 729–739.
4. D.R. Booth, M. Sunde, V. Bellotti, C.V. Robinson, W.L. Hutchinson, P.E. Fraser, P.N. Hawkins, C.M. Dobson, S.E. Radford, C.C.F. Blake and M.B. Pepys, *Nature*, 1997, **385**, 789–793.
5. M. Kaneko, L. Zulianello, M. Scott, C.M. Cooper, A.C. Wallace, T.L. James, F.E. Cohen and S.B. Prusiner, *Proc. Natl. Acad. Sci. USA*, 1997, **94**, 10069–10074.
6. M. Laurent, FEBS *Lett.*, 1997, **407**, 1–6.
7. G.M. Castillo, C. Ngo, J. Cummings, T.N. Wight and A.D. Snow, *J. Neurochem.*, 1997, **69**, 2452–2465.
8. R. Kisilevsky, *Nat. Med.* 1998, **4**, 772–773.
9. C. Soto, E.M. Sigurdsson, L. Morelli, R. Asok Kumar, E.M. Castano and B. Frangione, *Nat. Med.*, 1998, **4**, 822–826.
10. Y. Lui, P.J. Hart, M.P. Schlunegger and D. Eisenberg, *Proc. Natl. Acad. Sci. USA*, 1998, **95**, 3437–3442.
11. S.B. Prusiner, M.R. Scott, S.J. De Armond and F.E. Cohen, *Cell*, 1998, **93**, 337–348.
12. K.K. Kim, R. Kim and S-H. Kim, *Nature*, 1998, **394**, 595–599.
13. B. Re, G. Tibbelin, D. de Pascale, M. Rossi, S. Bartolucci and R. Ladenstein, *Natl. Struct. Biol.*, 1998, **5** 602–611.
14. M. Klumpp, W. Baumeister and L. Essen, *Cell*, 1997, **91**, 263–270.
15. L. Ditzel, J. Löwe, D. Stock, K-O. Stetter, H. Huber, R. Huber and S. Steinbacher, *Cell*, 1998, **93**, 125–138.
16. A. Brabban, J.A. Littlechild and P. Wisdom, *J. Indust. Microbiol.*, 1996, **16**, 8–14.
17. K. Murakami, M. Kimura, J.T. Owens, C.F. Meares and A. Ishihoma, *Proc. Natl. Acad. Sci. USA*, 1997, **94**, 1709–1714.
18. R.S. Roy and B. Imperiali, *Protein Sci.*, 1997, **10**, 691–698.
19. H. Kuang, R.R. Davies and M.D. Distefano, *Bioorg. Med. Chem. Lett.*, 1997, **7**, 2055–2060.
20. R.R. Davies and M.D. Distefano, *J. Am. Chem. Soc.*, 1997, **119**, 11643–11652.
21. J.M. Krahn, J.H. Kim, M.R. Burns, R.J. Parry, H. Zalkin and J.L. Smith, *Biochemistry*, 1997, **36**, 11061–11068.
22. S. Chen, D.R. Tomchick, D. Wolle, P. Hu, J.L. Smith, R.L. Switzer and H. Zalkin, *Biochemistry*, 1997, **36**, 10718–10726.
23. C.R.A. Muchmore, J.M. Krahn, J.H. Kim, H. Zalkin and J.L. Smith, *Protein Science*, 1998, **7**, 39–51.
24. R.J. Parry, M.R. Burns, S. Jiralerspong and L. Alemany, *Tetrahedron*, 1997, **53**, 7077–7088.
25. J.B. Thoden, H.M. Holden, G. Wesenberg, F.M. Raushel and I. Rayment, *Biochemistry*, 1997, **36**, 6305–6316.
26. J.B. Thoden, S.G. Miran, J.C. Phillips, A.J. Howard and F.M. Raushel, *Biochemistry*, 1998, **37**, 8825–8831.
27. G.J. Davies, L. Mackenzie, A. Varrot, M. Dauter, A.M. Brzozowski, M. Schülein and S.G. Withers, *Biochemistry*, 1998, **37**, 11707–11713.

28. M. Cirilli, R. Zheng, G. Scapin. and J.S. Blanchard, *Biochemistry*, 1998, **37**, 16452–16458.

29. W. Huang, J. Jia, P. Edwards, K. Dehesh, G. Schneider and Y. Lindquist, *EMBO J.*, 1998, **17**, 1183–1191.

30. M. Mathieu, Y. Modis, J.P. Zeelen, C.K. Engel, R.A. Abagyan, A. Ahlberg, B. Rasmussen, V.S. Lamzin and R.K. Wierenga, *J. Mol. Biol.*, 1997, **273**, 714–728.

31. C. Enroth, H. Neujahr, G. Schneider and Y. Lindqvist, *Structure*, 1998, **6**, 605–617.

32. C.A. Lesburg, G. Zhai, D.E. Cane and D.W. Christianson, *Science*, 1997, **277**, 1820–1824.

33. C.M. Starks, K. Back, J. Chappell, and J.P. Noel, *Science*, 1997, **277**, 1815–1820.

34. J.C. Sacchettini and C.D. Poulter, *Science*, 1997, 277, 1788–1789.

35. K.U. Wendt, K. Poralla and G.E. Schulz, *Science*, 1997, **277**, 1811–1815.

36. K.U. Wendt and G.E. Schulz, *Structure*, 1998, **6**, 127–133.

37. H.S. Cho, G. Choi, K.Y. Choi and B.H. Oh, *Biochemistry*, 1998, **37**, 8325–8330.

38. Z.R. Wu, S. Ebrahimian, M.E. Zawrotny, L.D. Thomburg, G.C. Perez-Alvarado, P. Brothers, R.M. Pollock and M.F. Summers, *Science*, 1997, **276**, 415–418.

39. M.A. Massiah, C. Abeygunawardana, A.G. Gittis and A.S. Mildvan, *Biochemistry*, 1998, **37**, 14701–14712.

40. J.A. Sigrell, A.D. Cameron, T.A. Jones and S.L. Mowbray, *Structure*, 1998, **6**, 183–194.

41. C. Colovas, D. Cascio and T.O. Yeates, *Structure*, 1998, **6**, 1329–1337.

42. M. Paetzel, R.E. Dalbey and N.C.J. Strynadka, *Nature*, 1998, **396**, 186–190.

43. S.R. Price, P.R. Evans, and K. Nagai, *Nature*, 1998, **394**, 645–650.

44. J.M. Ryter and S.C. Schultz, *EMBO J.*, 1998, **17**, 6819–6826.

45. E. Schmitt, M. Panvert, S. Blanquet and Y. Mechulam *EMBO J.*, 1998, **17**, 6819–6826.

46. S.Cusack, A. Yaremchuk, and M. Tukalo, in *The Many Faces of RNA*, ed. Eggleston, D.S., Prescott, C.D. Pearson, N.D., London: Academic Press, 1998, 55–65.

47. S. Cusack, A. Yaremchuk, A. Kriklivig and M. Tukalo, *Structure*, 1998, **6**, 101–108.

48. H. Li, C.R. Trotta and J. Abelson, *Science*, 1998, **280**, 279–284.

49. J. Lykke-Anderson and R.A. Garrett, *EMBO. J.*, 1997, **16**, 6290–6300.

50. C. Davies, R.B. Gerstner, D.E. Draper, V. Ramakrishnan and S.W. White, *EMBO J.*, 1998, 17, 4545–4558.

51. M.A. Markus, R.B. Gerstner, D.E. Draper and D.A. Tochia, *EMBO J.*, 1998, **17**, 4559–4571.

52. N. Ban, B. Freeborn, P. Nissen, P. Penczek, R.A. Grassucci, R. Sweet, J. Frank, P.B. Moore and T.A. Steitz, *Cell*, 1998, **93**, 1105–1115.

53. A. Yonath and F. Franceschi, *Structure*, 1998, **6**, 679–684.

54. J.R. Kiefer, C. Mao, J.C. Braman and L.S. Beese, *Nature*, 1998, **391**, 304–307.

55. Y. Li, S. Korolev and G. Waksman, *EMBO J.*, 1998, **17**, 7514–7525.

56. M.R. Redinbo, L. Stewart, P. Kuhn, J.J. Champoux and W.G.J. Hol, *Science*, 1998, **279**, 1504–1513.

57. L. Stewart, M.R. Redinbo, X. Qiu, W.G.H. Hol and J.J. Champoux, *Science*, 1997, **279**.

58. H. Huang, R. Chopra, G.L. Verdine and S.C. Harrison, *Science*, 1998, **282**, 1669–1675.
59. J. Jaeger, T. Restle and T.A. Steitz, *EMBO J.*, 1998, **17**, 4535–4542.
60. R.A. Grant, D.J. Filman, S.E. Finkel, P. Kolter and J.M. Hogle, *Nat. Struct. Biol.*, 1998, **5** 294–303.
61 Y-G. Gao, S-Y. Su, H. Robinson, H. Padmanabhan, S. Lim, B.S. McCrary, S.P. Edmondson, J.W. Shriver and A.H-J. Wang, *Nat. Struct. Biol.*, 1998, **5**, 7820786.
62. H. Robinson, Y-G. Gao, B.S. McCrary, S.P. Edmonson, J.W. Shriver and A.H-J. Wang, *Nature*, 1998, **392**, 202–205.
63. T. de Beer, R.E. Carter, K.E. Lobel-Rice, A. Sarkin and M. Overdium *Science*, 1998, **281**, 1357–1360.
64. H. Zhou, M.J. Mazzulla, J.D. Kaufman, S.J. Stahl, P.T. Wingfield, J.S. Rubin, D.P. Bottaro and R.A. Byrd, *Structure*, 1998, **6**, 109–116.
65. M. Ultsch, N.A. Lokker, P.J. Godowski and A.M. de Vas, *Structure*, 1998, **6**, 1383–1393.
66. T.P.J. Garrett, N.M. McKern, M. Lou, M.J. Frenkel, J.D. Bentley, G.O. Lovrecz, T.C. Elleman, T.J. Cosgrove and C.W. Ward, *Nature*, 1998, **394**, 395–399.
67. W. Kalus, M. Zweckstetter, C. Renner, Y. Sanchez, J. Georgescu, M. Grol, D. Demuth, R. Schumacher, C. Dony, K. Lang and T.A. Holak, *EMBO J.*, 1998, **17**, 6558–6572.
68. J.H.M. Cabral, S.L. Cohen, B.T. Chait, M. Li and R. Mackinnon, *Cell*, 1998, **95**, 649–655.
69. G Lukatela, N. Krauss, K. Theis, T. Selmer, V. Gieselmann, K. von Figura and W. Saenger, *Biochemistry*, 1998, **37**, 3654–3664.
70. E.B. Fauman, J.P. Cogswell, B. Lovejoy, W.J. Rocque, W. Holmes, V.G. Montana, H. Piwnica-Worms, M.J. Rink and M.A. Saper, *Cell*, 1998, **93**, 617–625.
71. S.K. Law and A.W. Dodds, *Protein Sci.*, 1997, **6**, 263–274.
72. B. Nagar, R.G. Jones, R.J. Diefenbach, D.E. Isenman and J.M. Rini, *Science*, 1998, **280**, 1277–1281.
73. S.F. Betz, A., Schnuchel, H. Wang, E.T. Olejniczak, R.P. Meadows, B.P. Lipsky, E.A.S. Harris, D.E. Stauton and S.W. Fesik, *Proc. Natl. Acad. Sci. USA*, 1998, **95**, 7909–7914.
74. F. Hoh, Y-S. Yang, L. Guignard, A. Padilla, M-H. Stern, J-M. Lhoste and H. van Tilbeurgh, *Structure*, 1998 **6**, 147–155.
75. H-J. Choi, S.W. Kang, C-H. Yang, S.G. Rhee and S-E. Ryu, *Nat. Struct. Biol.*, 1998, **5**, 400–406.
76. R.T. Nolte, G.B. Wisely, S. Westin, J.E. Cobb, M.H. Lambert, R. Kurokawa, M.G. Rosenfeld, T.M. Willson, C.K. Glass and M.V. Milburn, *Nature*, 1998, **395**, 137–143.
77. X. Wang, X. Lin, J.A. Loy, J. Tang and X.C. Zhang, *Science*, 1998, **281**, 1662–1665.
78. L. Renault, N. Nasser, I. Vetter, J. Becker, C. Klebe, M. Roth and A. Wittinghofer, *Nature*, 1998, **392**, 97–101.
79. P.J.B. Pereira, A. Bergner, S. Macedo-Ribeiro, R. Huber, G. Matschiner, H. Fritz, C.P. Sommerhoft and W. Bode, *Nature*, 1998, **392**, 306–311.
80. P. Bork, K. Hofmann, P. Bucher, A.F. Neuwald, S.F. Altschul and E.V. Koonin, *FASEB J.*, 1997, **11**, 68–76.

81. R.A. Nash, K.W. Caldecott, D.E. Barnes and T. Lindahl, *Biochemistry*, 1997, **36**, 5207–5211.

82. X.D. Zhang, S. Moreru, P.A. Bates, P.C. Whitehead, A.I. Coffere, K. Hairbucher, R.A. Nash, M.J. Sternberg, T. Lindahl and P.S. Freemont, *EMBO J.*, 1998, **17**, 6404–6411.

83. D-I. Liao, E.C. Wolff, M.H. Park and D.R. Davies, *Structure*, 1998, **6**, 23–32.

84. C.S. Raman, H. Li, P. Martásek. V. Krail, B.S.S. Masters and T.L. Poulos, *Cell*, 1998, **95**, 939–950.

85. B.R. Crane, A.S. Arvai, D.K. Ghosh, C.Q. Wu, E.D. Getzoff, D.J. Stuehr and J.A. Tainer, *Science*, 1998, **279**, 2121–2126.

86. T.M. Iverson, D.M. Arciero, B.T. Hsu, M.S.P. Logan, A.B. Hooper and D.C. Rees, *Nat. Struct. Biol.*, 1998, **5**, 1005–1012.

87. J.W. Peters, W.N. Lanzilotta, B.J. Lemon and L.C. Seefeldt, *Science*, 1998, **282**, 1853–1858.

88. K.P. Locher, B. Rees, R. Mitscher, A. Moulinier, J.P. Rosenbusch and D Moras, *Cell*, 1998, **95**, 771–778.

89. A.D. Ferguson, E. Hofmann, J.W. Coulton, K. Diederichs and W. Wette, *Science*, 1998, **282**, 2215–2220.

90. K. Kauppi, K. Lee, E. Carredano, R.W. Parales, D.T. Gibson, H. Eklund and S. Ramaswamy, *Structure*, 1998, **6**, 571–586.

91. T. Klabunde, C. Eiken, C. Sacchettini and B. Krebs, *Nat. Struct. Biol.*, 1998, **12**, 1084–1090.

92. M.E. Cuff, K.I. Miller, K.E. van Holde and W.A. Hendrickson, *J. Mol. Biol.*, 1998, 278, 855–870.

93. T.C. Umland, L.M. Wingert, S. Swaminathan, W.F. Furey, J.J. Schmidt and M. Sax, *Nat. Struct. Biol.*, 1997, **4**, 788–792.

94. D.B. Lacy, W. Tepp, A.C. Cohen, B.R. DasGupta and R.C. Stevens, *Nat. Struct. Biol.*, 1998, **5**, 898–902.

95. M. Weiner, D. Freymann, P. Ghosh and R.M. Stroud, *Nature*, 1997, **354**, 409–410.

96. I.R. Vetter, M.W. Parker, A.D. Tucker, J.H. Lakey, F. Pattus and D. Tsemoglou, *Structure*, 1998, **6**, 863–874.

97. P.I. Roach, I.J. Clifton, C.M.H. Hensgens, N. Shibata, C.J. Schofield, J. Hajdu and J.E. Baldwin, *Nature*, 1997, **387**, 827–830.

98. K. Valegard, A.C. Terwisscha van Scheltinga, M.D. Lloyd, T, Hara, S. Ramaswamy, A. Perrakis, A. Thompson, H. Lee, J.E. Baldwin, C.J. Schofield, J. Hajdu and I. Andersson, *Nature*, 1998, **394**, 805–808.